PROCESSING AND UTILIZATION OF HIGH SULFUR COALS

COAL SCIENCE AND TECHNOLOGY

Series Editor:

Larry L. Anderson
Department of Fuels Engineering, University of Utah, Salt Lake City, UT 84112, U.S.A.

Vol. 1: Geochemistry of Coal (Bouška)
Vol. 2: Fundamentals of Coal Beneficiation and Utilization (Tsai)
Vol. 3: Coal: Typology, Chemistry, Physics and Constitution (Van Krevelen)
Vol. 4: Coal Pyrolysis (Gavalas)
Vol. 5: Free Radicals in Coals and Synthetic Fuels (Petrakis and Grandy)
Vol. 6: Coal Combustion Chemistry-Correlation Aspects (Badin)
Vol. 7: The Chemistry of Coal (Berkowitz)
Vol. 8: Natural Gas Substitutes from Coal and Oil (Qader)
Vol. 9: Processing and Utilization of High Sulfur Coals (Attia, Editor)

COAL SCIENCE AND TECHNOLOGY 9

PROCESSING AND UTILIZATION OF HIGH SULFUR COALS

Proceedings of The First International Conference on Processing and Utilization of High Sulfur Coals, October 13–17, 1985, Columbus, Ohio, U.S.A.

edited by

YOSRY A. ATTIA

The Ohio State University, Mining Engineering Division, Columbus, Ohio, U.S.A.

ELSEVIER Amsterdam – Oxford – New York – Tokyo 1985

ELSEVIER SCIENCE PUBLISHERS B.V.
1, Molenwerf
P.O. Box 211, 1000 AE Amsterdam, The Netherlands

Distributors for the United States and Canada:

ELSEVIER SCIENCE PUBLISHING COMPANY INC.
52, Vanderbilt Avenue
New York, NY 10017

Library of Congress Cataloging-in-Publication Data

International Conference on Processing and Utilization of
High Sulfur Coals (1st : 1985 : Columbus, Ohio)
Processing and utilization of high sulfur coals.

Bibliography: p.
1. Coal--Sulfur content--Congresses. 2. Coal--
Desulphurization--Congresses. 3. Coal--Environmental
aspects--Congresses. I. Attia, Yosry A., 1945-
II. Title.
TP325.I73 1985 662.6'22 85-16308
ISBN 0-444-42545-4

ISBN 0-444-42545-4 (Vol. 9)
ISBN 0-444-41970-5 (Series)

Printed in The Netherlands

THE FIRST INTERNATIONAL CONFERENCE ON
PROCESSING AND UTILIZATION OF HIGH SULFUR
COALS, OCTOBER 13-17, 1985, COLUMBUS, OHIO U.S.A.

Sponsored by

The Ohio State University, Columbus, Ohio
Mining Engineering Division, Metallurgical Engineering
Department, and the Department of Conferences and Institutes

And

Northwestern University, Evanston, Illinois
Departments of Chemical Engineering and
Civil Engineering

Southern Illinois University, Carbondale, Illinois
Department of Mining Engineering, and The
Coal Extraction and Utilization Research Center

Receptions Hosted By

Dean, College of Engineering, The Ohio State University
Columbus, Ohio

And

Sherex Chemical Company, Dublin, Ohio

THE FIRST INTERNATIONAL CONFERENCE ON PROCESSING AND UTILIZATION OF HIGH SULFUR COALS
OCTOBER 13-17, 1985, COLUMBUS, OHIO U.S.A.

ORGANIZING COMMITTEE

Yosry A. Attia, Chairman
The Ohio State University, Columbus, Ohio

D. Berger
Ohio State Dept. of Development
Columbus, Ohio

Y.P. Chugh
Southern Illinois University
Carbondale, Illinois

P.R. Dugan
Ohio State University
Columbus, Ohio

R.B. Engdahl
Battelle-Columbus Laboratories
Columbus, Ohio

R. Essenhigh
Ohio State University
Columbus, Ohio

J.A. Fitzpatrick
Northwestern University
Evanston, Illinois

D.W. Fuerstenau
University of California
Berkeley, California

J. Gozon
Ohio State University
Columbus, Ohio

H.A. Hamza
CANMET, Devon Coal Research Center
Devon, Alberta, Canada

W.A. Kneller
University of Toledo
Toledo, Ohio

C.W. Kruse
Illinois State Geological Survey
Champaign, Illinois

J.S. Laskowski
University of British Columbia
Vancouver, Canada

J. Lytle
Illinois State Geological Survey
Champaign, Illinois

R. Markuszewski
Ames Laboratory
Ames, Iowa

J.H. Oxley
Battelle-Columbus Laboratories
Columbus, Ohio

D. Pride
Ohio State University
Columbus, Ohio

D.R. Skidmore
Ohio State University
Columbus, Ohio

G.R. St.Pierre
Ohio State University
Columbus, Ohio

G. Thodos
Northwestern University
Evanston, Illinois

L. Winget
Ohio State Dept. of Development
Columbus, Ohio

FOREWORD

Problems facing the utilization of high sulfur coals at present are immense and deserve the attention of leading scientists and engineers, public policy makers and administrators, coal producers and users, and environmentalists. Much of the blame for acid rain is directed at the use of high sulfur coals in electric power generation and various industrial activities. Acid rain is emerging as a major threat to the survival of mankind. It not only gradually destroys plant and animal life in forests and water resources, but it also depletes the ozone layer in the atmosphere. Weather patterns make acid rain a global phenomenon, for acid rain clouds are not containable to the source and move across state and national borders. Of concurrent concern are the adverse economic impacts of inadequately utilized natural resources. Coal is a major energy resource for many industrialized nations. Switching to different energy sources would mean the wastage of valuable and readily available coal resources, massive unemployment and major losses to local economies. The challenge is to develop technologies to enable cost-effective use of high sulfur coals while meeting the environmental regulations to alleviate the acid rain problem. Hence the purpose of this international conference is to document, encourage, and stimulate the development of technological solutions around the world.

This book presents the proceedings of the conference where investigators reported the latest discoveries, developments toward commercial practice, and state-of-the-art reviews of scientific principles and applications of various technologies for handling, processing, and utilization of high sulfur coals. Seventy four papers were presented at the conference by representatives from industry, academia, and private and governmental research and consulting organizations from eleven countries. Mainly because of time constraints, not all of the papers presented at the conference are included in this work. Major technical areas covered in this book are: characterization of sulfur in coals, environmental problems and impacts of high sulfur content on coal utilization, selective mining and blending, coal desulfurization by physical and chemical cleaning techniques, biological coal desulfurization, handling of coal fines, sulfur capture during combustion, control of sulfur dioxide emission, and new applications for high sulfur coals. This book will be of particular interest and benefit to researchers, graduate students and all persons involved in coal production, processing and utilization; public policy making and administration related to energy conservation, economic development, and environmental protection; and investors in industry and new technology.

The response of scientists and engineers to this conference has been extremely gratifying and encouraging. I would like to thank all of the authors and speakers for their efforts in preparing high quality papers in a relatively short time. I would like to acknowledge the help of the many organizing committee members in organizing the sessions, and reviewing the manuscripts. In particular, I would like to thank Ms. Louise Larew for her enthusiastic support and coordination of the conference, and Ms. Lori Henderson for her assistance in editing the manuscripts and handling conference correspondence. My thanks also to Dr. M. Elzeky for his assistance in editing the manuscript. The support of the conference sponsors and reception hosts is gratefully acknowledged. In particular, I would like to thank Drs. J. Inloe, G.R. St.Pierre, Y.P. Chugh, J. Fitzpatrick, Dean D. Glower, and Dr. T. Wheeler. all of whom were instrumental in obtaining the financial support. I also appreciate the initial support for my idea of this conference by my colleagues of the Mining Engineering Division and Assoc. Dean R. Redmond of the College of Engineering. And last but not least, I would like to thank my wife and daughter for their patience and understanding during my work on this book.

Yosry A. Attia

June 27, 1985
Columbus, Ohio

Editor

TABLE OF CONTENTS

PART 9. BIOLOGICAL DESULFURIZATION OF COAL

PART 10. NEW APPLICATIONS FOR HIGH SULFUR COALS

PART 1

CHARACTERIZATION OF SULFUR IN COALS

IMAGE ANALYSIS OF PYRITE IN OHIO COAL: RELATION BETWEEN PYRITE GRAIN-SIZE DISTRIBUTION AND PYRITIC SULFUR REDUCTION

Richard W. Carlton

Dept. of Natural Resources, Div. of Geological Survey, Fountain Square, Columbus, Ohio, USA 43224

ABSTRACT

The size characteristics of pyrite in coal have long been known to influence the amount of pyritic sulfur reduction that will occur when a coal is washed by float-sink methods. However, manual microscopic methods of determining the size characteristics of pyrite embedded in coal are tedious and time consuming, and, as a result, practical application of this knowledge has been very limited. Modern automated image analysis offers a potentially rapid, nontiring method of determining the size distribution of pyrite associated with coal.

Data collected on 26 Ohio coal samples crushed to 14 x 28 mesh indicate a strong negative correlation between cumulative percent pyrite in certain pyrite size ranges as determined by image analysis and pyritic sulfur reduction of coal washed by float-sink methods. In some samples the actual pyritic sulfur reduction may be much higher or lower than predicted by the equation defining the sulfur reduction/size distribution correlation. These anomalous samples result from the degree and abundance of pyrite clustering in individual composite coal-pyrite grains.

INTRODUCTION

The size characteristics of pyrite in coal have long been known to influence the amount of pyritic sulfur reduction that will occur when a coal is washed by float-sink methods. One of the more important studies showing this relationship was published by McCartney and others in 1969 (ref. 1). Their results showed that a significant relationship exists between the mean grain size of pyrite determined by manual microscopic methods and the pyritic sulfur reduction of the washed coal. They also found a very strong correlation between the proportion of pyrite in coal particles containing more than 50 percent pyrite and pyritic sulfur reduction. However, manual microscopic methods of determining the size characteristics of pyrite embedded in coal are tedious and time consuming, and, as a result, practical application of this knowledge has been very limited. Modern automated image analysis offers a potentially rapid, nontiring method of determining the size characteristics of pyrite associated with coal.

Most of Ohio's coal contains significant amounts of sulfur in the form of pyrite (FeS_2). Because of this high sulfur content and the laws controlling SO_2 emissions, approximately 60 percent of the coal produced in Ohio in 1983 was washed. In 1983 the five most productive coal seams in Ohio, in order of decreasing tonnage, were the Pittsburgh (No. 8), the Meigs Creek (No. 9), the Middle Kittanning (No. 6), the Clarion (No. 4A) and the Lower Kittanning (No. 5) (ref. 2). The average pyritic sulfur content for each of these five seams is shown in Table 1. In each case the proportion of total sulfur contributed by pyrite is very large. If ways can be found to remove this pyrite economically, Ohio will be much closer to having a competitive coal-mining industry again.

TABLE 1

Most productive coal seams in Ohio in 1983 and their average sulfur content.[1]

Seam (in order of decreasing tonnage mined in 1983)	Average percent pyritic sulfur	Average percent total sulfur	Percent of total sulfur contributed by pyrite	No. of samples
Pittsburgh (No. 8)	2.67	4.56	59	69
Meigs Creek (No. 9)	1.73	3.59	48	29
Middle Kittanning (No. 6)	2.18	3.59	61	111
Clarion (No. 4A)	2.55	4.29	59	41
Lower Kittanning (No. 5)	3.38	4.79	71	99

[1]Refs. 3,4,5.

Washed coals in Ohio are commonly between 6 inches (15.24 cm) and 28 mesh (0.595 mm) in size. In this size range, gravity-separation techniques such as jigs and heavy-media vessels or cyclones generally are used to separate the noncoal material from the coal. One of the basic problems with these methods, using coal of this size, is that much of the pyrite is finer than 28 mesh and remains as inclusions in the clean coal. Assuming coal has a specific gravity of 1.35, a composite coal-pyrite grain with 7 percent by volume[1] pyrite will have a specific gravity of 1.61. Thus, even with 100 percent efficiency of gravity separation, coal with a trace to about 7 percent pyrite will remain in the clean coal fraction if separation at 1.60 specific gravity is used. Seven

[1]The area measurements of randomly distributed pyrite on a polished surface are equal to the volume percent pyrite. Unless noted otherwise, all pyrite percentages are volume percentages.

percent by volume pyrite on a weight basis is equal to 21.87 percent pyrite and nearly 11 percent pyritic sulfur.

Generally one way to remove more pyrite is to crush the coal to finer sizes. However, beyond a certain point--which differs from one coal to the next--finer grinding will not yield substantially greater amounts of pyrite. For example, coal crushed to a given size and containing coarse pyrite may have many composite grains with less than 7 percent pyrite. These grains will float in 1.60 specific gravity liquids (Fig. 1). However, if the single grain is broken into three smaller pieces, the proportion of pyrite to coal in some fragments will probably increase. If the pyrite makes up very much more than 7 percent it will sink, thus increasing the washability of the coal (Fig. 1).

Conversely, if the pyrite is very fine in comparison to the coal grain size and is evenly disseminated throughout the coal particle, breaking it into smaller pieces offers no advantage to coal washing. Most particles will probably remain near 7 percent pyrite (Fig. 1).

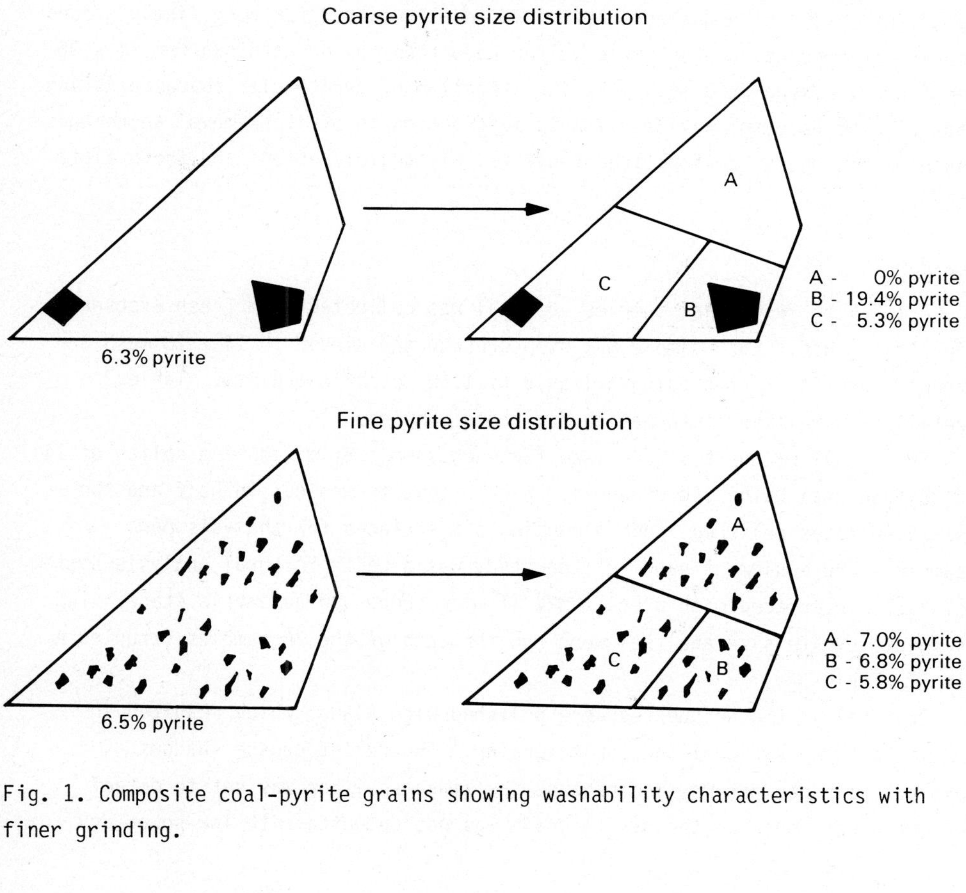

Fig. 1. Composite coal-pyrite grains showing washability characteristics with finer grinding.

This relationship is the basic reason why pyrite size distribution can be used to predict pyritic sulfur reduction. For coal crushed to a given size, the pyrite content of the washed product will either remain the same or decrease when compared with the pyrite content of the unwashed coal. The extent of this decrease in pyritic sulfur is largely dependent on the size distribution of the pyrite.

Many of the newer techniques suggested for pyrite removal prior to combustion require finely ground coal. If medium- to high-sulfur coals in Ohio are to be used effectively, fine cleaning will be necessary. Data gained through microscopic analyses will be very important.

PURPOSE

In this investigation, the relationship is examined between pyrite size distribution as determined by automated image analysis and the float-sink characteristics of coal ground to 14 x 28 mesh. This size range was used because microscopic study of coarser coal using our equipment lacks the necessary precision, and because coal ground to less than 28 mesh is generally treated separately in wash plants or discarded. However, the microscopic study of pyrite size characteristics is ideally suited for very finely ground coal. If prediction of pyritic sulfur reduction can be attained for 14 x 28 mesh gravity-separated coal, the investigation of pyrite size characteristics may also be able to provide valuable information in pyrite removal techniques such as biological destruction of pyrite, oil agglomeration, and froth flotation.

METHODS

A total of 26 channel samples of coal was collected from fresh exposures in strip mines. Each sample was step crushed and sieved to 14 x 28 mesh and then float-sink tested using a liquid of 1.60 specific gravity. Table 2 details the relevant information for these samples.

Three coal briquettes were made for each sample by mixing 28g splits of 14 x 28 mesh coal with 7.5g of epoxy. Each briquette was cut in half and the sawed surfaces polished, thus producing six surfaces for analysis per sample. The equipment used for the study was a Leitz Textural Analysis System (T.A.S.) interfaced with a Leitz MPV II microscope and automatic stage. Two separate pyrite size analyses were run for each of the 26 samples (studies A and B, Table 3).

Originally, the briquettes were polished with Al_2O_3, which resulted in relief between the coal and pyrite grains. The relief caused shadows at pyrite grain perimeters, and because the image analyzer was set to detect bright areas, much of the pyrite grain was not detected with low-power dry

TABLE 2
Sample location, coal thickness, and pyritic-sulfur reduction data.

OGS file no.	Coal	County	Township	Coal thickness (inches)	14 x 28 mesh coal: Weight % pyritic sulfur, Unwashed	14 x 28 mesh coal: Weight % pyritic sulfur, Float	14 x 28 mesh coal: % pyritic sulfur reduction using 1.60 specific gravity liquid
1351	L. Kittanning	Holmes	Walnut Creek	36.00	2.97	1.53	48.5
1352	M. Kittanning	Holmes	Berlin	26.00	7.23	1.42	80.4
1353	M. Kittanning	Coshocton	Mill Creek	28.00	2.86	0.58	79.7
1354	L. Kittanning	Coshocton	Mill Creek	36.00	5.81	2.50	57.0
1355	L. Kittanning	Coshocton	Lafayette	23.25	1.88	0.96	48.9
1356	M. Kittanning	Coshocton	Lafayette	44.50	3.03	0.96	68.3
1358	M. Kittanning	Muskingum	Washington	20.75	4.71	1.92	59.2
1359	U. Freeport	Muskingum	Wayne	47.25	3.58	1.30	63.7
1360	M. Kittanning	Perry	Pike	55.50	2.51	1.63	35.1
1361	M. Kittanning	Perry	Bearfield	41.00	1.82	1.06	41.8
1379	M. Kittanning	Vinton	Swann	49.50	3.28	0.61	81.4
1380	L. Kittanning	Vinton	Swann	26.75	5.94	1.28	78.5
1382	L. Kittanning	Vinton	Clinton	53.00	3.40	0.96	71.8
1383	M. Kittanning	Jackson	Madison	12.25	2.10	0.56	73.3
1384	L. Kittanning	Jackson	Madison	27.00	1.60	0.37	76.9
1385	L. Kittanning	Jackson	Bloomfield	34.00	2.36	1.10	53.4
1386	L. Kittanning	Stark	Pike	38.50	3.24	0.46	85.8
1387	L. Kittanning	Stark	Osnaburg	42.50	5.79	2.69	53.5
1388	M. Kittanning	Stark	Osnaburg	41.00	1.58	0.64	59.5
1389	M. Kittanning	Stark	Sugar Creek	28.00	2.82	0.66	76.6
1390	M. Kittanning	Stark	Paris	25.50	1.54	0.51	66.9
1391	M. Kittanning	Columbiana	Columbiana, West	54.50	1.98	0.42	78.8
1392	L. Kittanning	Stark	Paris	24.75	6.94	1.03	85.2
1393	M. Kittanning	Columbiana	Franklin	21.25	3.65	0.18	95.1
1395	M. Kittanning	Columbiana	Center	36.00	5.54	0.64	88.4
1396	M. Kittanning	Columbiana	Hanover	37.50	3.43	0.76	77.8

objectives. High-power oil-immersion objectives considerably reduced the relief effect and were used in the first study (study A). Later it was found that diamond polishing almost totally eliminated the relief and allowed low-power objectives to be used (study B).

TABLE 3

Operating conditions for studies A and B using a Leitz image analyzer, MPV II petrographic microscope, and polished coal briquettes.

	Study A	Study B
Microscope objective	32x oil	4x dry
Area of single field in square micrometers	5,491.4	92,426.2
Number of fields measured	7,200	980
Number of pyrite grains measured	882 to 16,514	517 to 11,587
Smallest detectable pyrite	~ 1 μm[1]	~ 9 μm[2]
Maximum grain size measured	72 μm[3]	300μm[3]
Totally automatic analysis	no	yes
Polishing compound	Al_2O_3 slurry	diamond slurry
Number of surfaces measured per sample	6	5

[1]Smaller pyrite disregarded.
[2]1 to 9 μm pyrite accounted for by determining the difference between total pyrite percent found using the 32x objective and the percent found using the 4x objective.
[3]Grains larger than this were placed in an oversize category.

In each size analysis, pyrite grain areas were converted to equivalent circular diameters (ECD). An ECD is the diameter of a circle with the same area as the pyrite grain being measured. The computer is programmed to place the grain area into a size category according to its ECD, resulting in the microscopic equivalent of a sieve size analysis (Fig. 2).

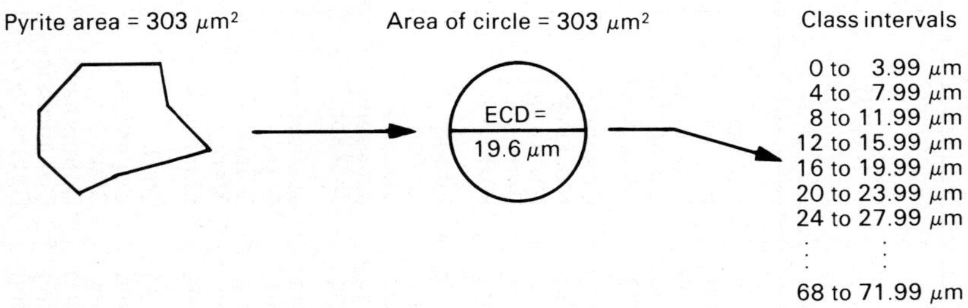

Fig. 2. Microscopic size analysis. All areas are summed in each class; at the end of the analysis the total pyrite area for each class interval is known.

RESULTS

Linear regression analyses were performed between pyritic sulfur reduction percentages obtained from float-sink tests and cumulative pyrite percentages determined by image analysis. Sulfur reduction values were considered to be the dependent variable (y) and the pyrite percentages the independent variable (x).

In study A the strongest correlation between sulfur reduction and cumulative percent pyrite was at 24 micrometers ECD. In study B the correlation coefficient was strongest at 66 micrometers ECD (Table 4). The statistical linear relationships at 24 and 66 micrometers ECD are shown in Fig. 3. The standard error of the estimate (Sy/x) indicates the scatter in the y direction of the observed data points about the regression line in a manner analogous to the standard deviation.

TABLE 4
Correlation coefficients between pyritic sulfur reduction (y) and cumulative pyrite size distribution percents (x) at selected ECD. Each r value represents a correlation between 26 samples.

Study A		Study B	
<2 μm ECD	r = -0.86	<15 μm ECD	r = -0.69
<12 μm ECD	r = -0.93	<39 μm ECD	r = -0.88
<24 μm ECD	r = -0.95	<66 μm ECD	r = -0.90
<36 μm ECD	r = -0.93	<105 μm ECD	r = -0.87
<60 μm ECD	r = -0.89	<129 μm ECD	r = -0.88
<72 μm ECD	r = -0.76	<147 μm ECD	r = -0.85

If the sample regression line is accepted as a good indicator of the true regression line, then twice the standard error of the estimate can be used to approximate the prediction interval at the 95 percent confidence level (ref. 6). Given a value of x (pyrite percent at 24 micrometers ECD), the pyritic sulfur reduction (y) can be predicted to within about ± 10 percent in study A and to about ± 13 percent in study B.

The mean pyrite diameter of each sample was computed using Folk's graphic mean (M_z) from the cumulative curve data for study B (ref. 7).

$$M_z = (\emptyset\, 16 + \emptyset\, 50 + \emptyset 84)/3 \qquad (1)$$

Sixteen of the 26 samples in study B required minor extrapolations of the data to reach the Ø 84 percentile or Ø 16 percentile (Fig. 4). One sample, number 1390, in study B required significant extrapolation of the cumulative curve to reach the Ø 16 percentile and was not used. In study A nearly all samples needed extensive extrapolations and the data were not used.

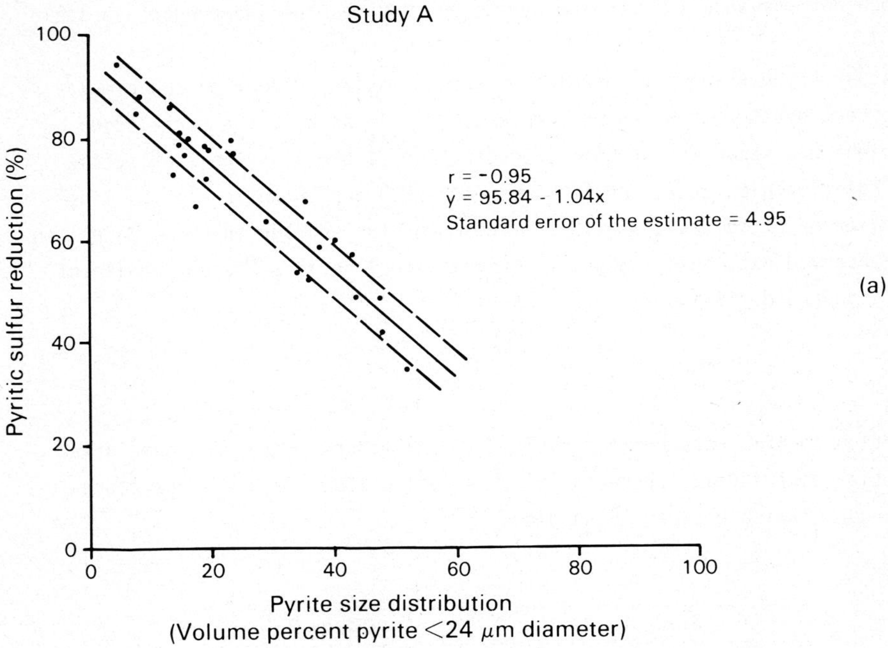

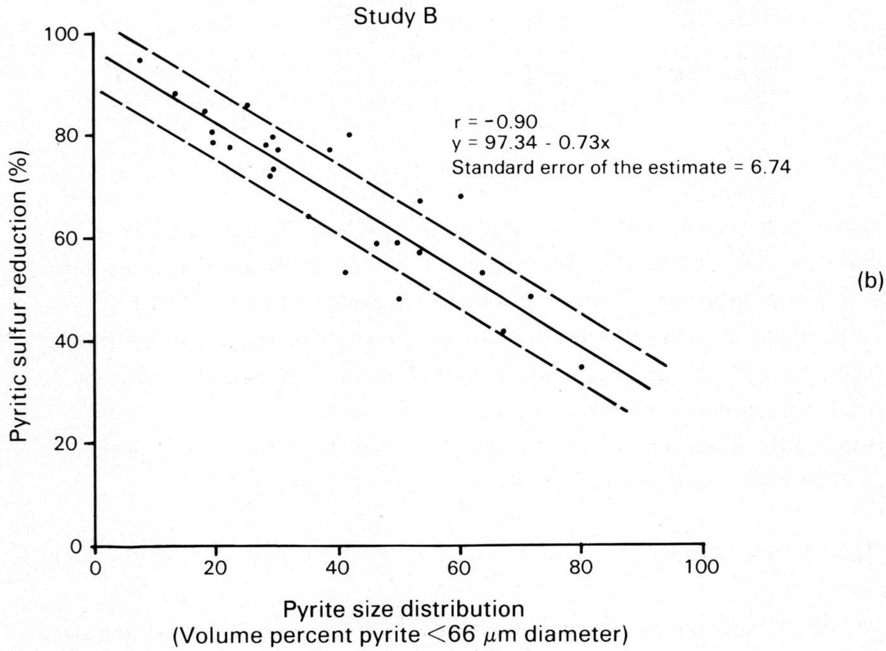

Fig. 3. Regression of (a) pyritic sulfur reduction percent on volume percent pyrite at 24 micrometers, and (b) pyritic sulfur reduction percent on volume percent pyrite at 66 micrometers.

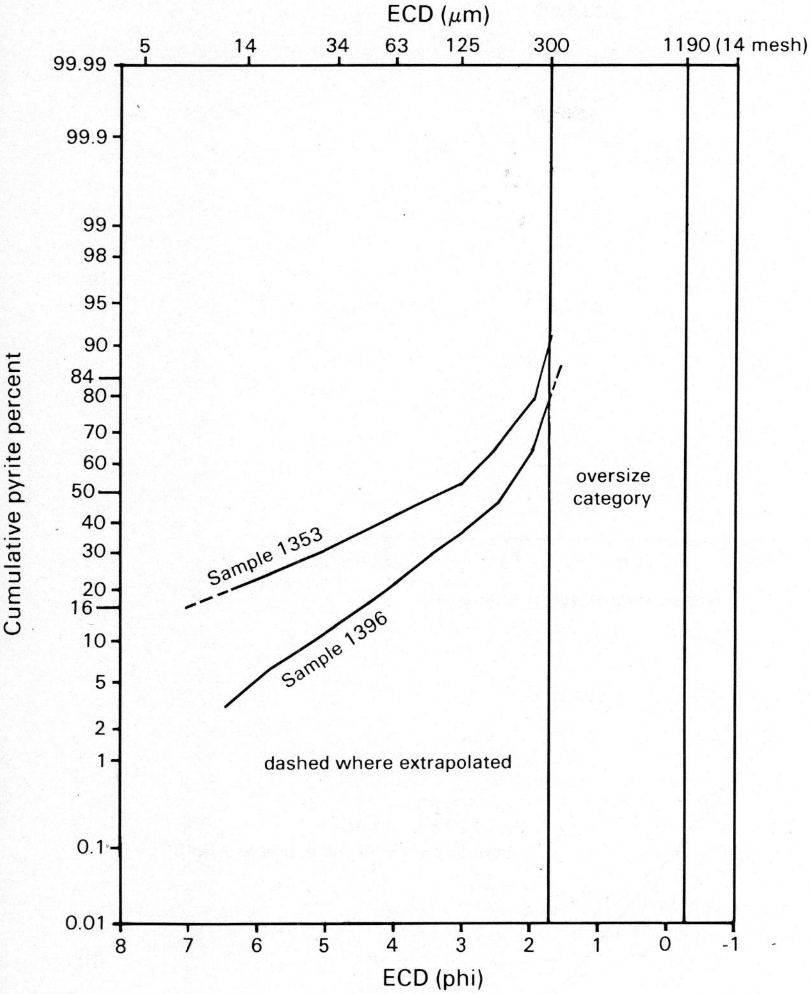

Fig. 4. Typical cumulative curves of pyrite size distribution data from study B. Phi values = - $\log_2 D$ where D = pyrite grain diameter in micrometers.

The relationship between cumulative percent pyrite and mean grain size is curvilinear. In this case it is helpful to convert grain diameters to phi values and plot on arithmetic-base paper. When the grain diameters in micrometers are plotted on logarithmic-base paper or in phi values on arithmetic-base paper, the relationship is linear and a linear regression analysis can be performed on the data (fig. 5).

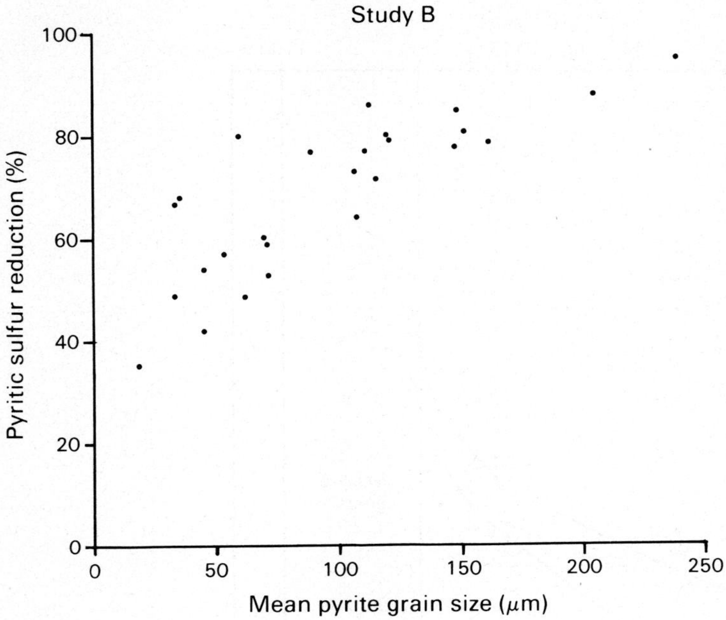

(a)

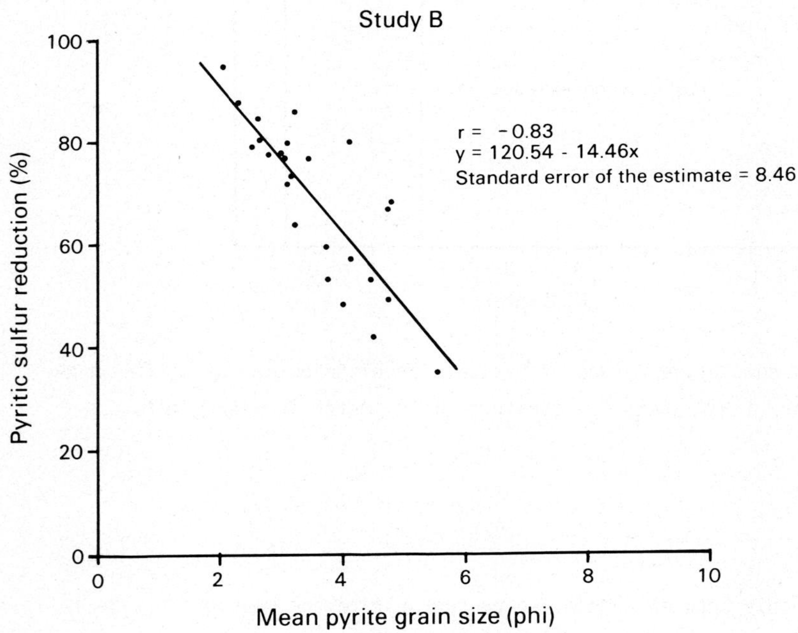

(b)

Fig. 5. The relationship of (a) the pyritic sulfur reduction percent to the mean pyrite grain size in micrometers, and (b) the regression of pyritic sulfur reduction percent on mean pyrite grain diameter in phi units.

DISCUSSION

Precision

The distribution of pyrite in coal is highly variable. Samples sufficiently representative of a coal bed at a specific site must be large, yet microscopic analysis necessitates crushing and splitting so as to obtain a very small sample which will be representative of the larger sample. Most microscopic work on coal is performed on samples crushed to 14 or 20 mesh. These sizes are not sufficiently fine to distribute pyrite evenly over the polished surface of the briquette. In addition, pyrite in Ohio coal generally contributes much less than 5 percent by volume to the coal sample. For example, 3 percent by weight pyritic sulfur, which is typical of Ohio coal, equals 5.61 weight percent pyrite but only 1.57 volume percent pyrite (assuming coal has a specific gravity of 1.35). Thus a large area must be scanned by the microscope to insure that sufficient pyrite grains are measured for acceptable reproducibility of the measurements. As in thin-section analysis, precision depends on the number of grains measured. However, errors in size analysis cannot be assessed in precisely the same way because the distribution statistics required are a function of many size classes and the proportion in each class is dependent on those in adjacent classes. According to Adams (ref. 8), the error in estimation of any percentile is proportional to the standard deviation of the grain-size distribution. Hence coal samples with poorly sorted pyrite (large standard deviation) should require a larger number of counts than well-sorted samples in order to estimate a given statistic such as the volume percent pyrite less than 24 micrometers ECD. A test was performed on sample 1383 to determine the precision of the pyrite size-distribution data. Six complete analyses were performed on this sample with 7,200 frames being measured for each analysis. The number of pyrite grains measured ranged from 1,852 to 2,142. The mean and standard deviation of the cumulative pyrite percents were calculated for each cumulative class size. The six cumulative pyrite percents for the less-than-24-micrometer size range for sample 1383 are 16.5, 13.2, 16.3, 23.0, 12.7, and 14.4. The mean and standard deviation for these values are 16.0 and $\pm$ 3.8 percent, respectively. Although over 1,800 pyrite grains were measured in each analysis, the precision could be improved by counting a great many more grains. If the precision of the microscopic analysis is increased, better predictability of the pyritic sulfur reduction should be attainable.

Microscopic grain sizes

Pyrite grain-size distributions determined by automated image analysis are not true distributions of the actual grain sizes. Spheres cut by random section will possess a diameter ranging from near zero to the maximum diameter of the particle. Krumbein and Pettijohn (ref. 9) as well as others have shown that the apparent arithmetic-mean diameter of spheres in thin section is equal to $\pi/4$ times the true arithmetic-mean diameter. Conversion equations have been proposed by many investigators in an effort to characterize this relationship and extend it to naturally occurring nonspherical grains of unequal size, but none have met with general acceptance (ref. 6). An added complication in using the automated method described in this study involves the measurement of partial grain areas. Invariably some pyrite grains fall on the border of the measurement field. In these cases the grain area measured can be slightly to greatly reduced from its total cross-sectional area. Even so, there are empirical linear relationships between size distributions determined by automated image analysis and sieve/settling sizes under certain conditions (ref. 10 and Carlton, work in progress).

The reason the strongest correlations occur at markedly different ECD's in the two studies is presumably due to the different operating conditions used to measure the pyrite. In study A a single frame is about 70 x 78 micrometers in size, whereas in study B a single frame is 289 x 320 micrometers in size, or nearly 17 times larger. The latter is much more likely to measure more "entire" grains than the smaller field of view. Therefore it is important to realize that the regression equations and ECD's cited in studies A and B are correct only when the operating conditions in effect when the relationship was established are used.

Clustering

There are certain pyrite size-distribution characteristics which can be readily identified by casual observation through the microscope but which have not yet been quantified. For example, in Fig. 6, the data point identified by a halo was found by image analysis to have 42 percent pyrite less than 66 micrometers ECD. Using the regression line as the predictor, this sample should have about 66 percent pyritic sulfur reduction when washed. Actual washing, however, indicated a reduction of 80 percent. In cases where actual reduction is much higher than predicted, many of the pyrite grains are clustered, so that the volume percent pyrite of the composite grain is much greater than 7 percent--the approximate pyrite volume percent needed to sink a composite grain in a liquid of 1.60 specific gravity (Fig. 7). The clustered small pyrite grains behave as a single large grain when washed. This clustering causes increased pyritic sulfur reductions in some samples, even though only very small individual pyrite grains are detected by the image analyzer. In other instances, actual pyritic sulfur reduction is much lower than predicted by the pyrite size-distribution data (data points far below the regres-

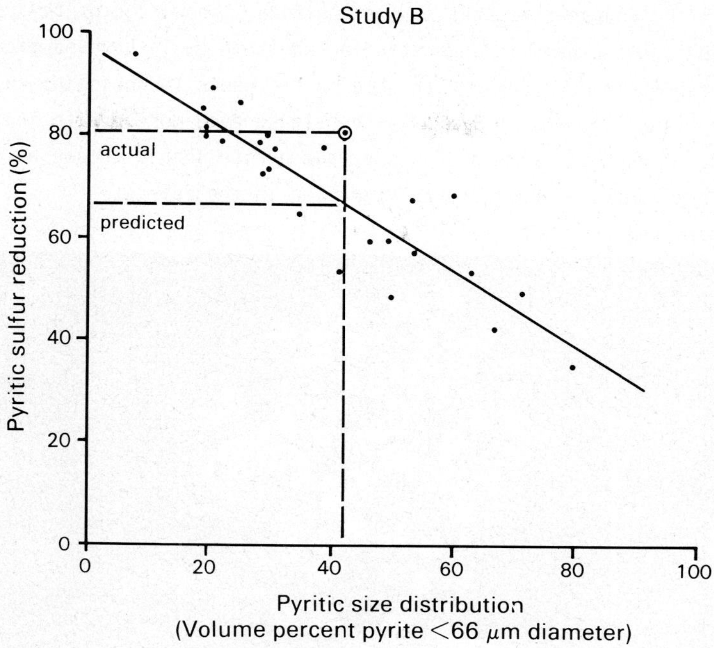

Fig. 6. Errors in prediction of pyritic sulfur reduction percent. Data point with halo has a predicted pyritic sulfur reduction percent of 66, but float-sink tests indicate an 80 percent reduction of pyrite for this sample.

sion line, Fig. 6). Samples in which this occurs generally have pyrite grains scattered throughout the coal but rarely clustered in single coal particles in amounts greater than 5 or 6 percent (Fig. 8).

CONCLUSIONS

Most of Ohio's major coal seams contain abundant pyrite. This pyrite generally accounts for more than 50 percent of the total sulfur content of the coal and in some cases is very fine grained and cannot readily be removed from coarsely ground coal. One way of releasing more pyrite from the coal matrix is to crush the coal to finer sizes. However, some coals will release more pyrite than others because of differences in their pyrite size distributions. This preliminary study has shown that the pyrite size distribution in 14 x 28 mesh coal strongly influences the pyritic sulfur reduction in float-sink tests and that automated image analysis can predict the pyritic sulfur reduction to within ± 10 percent in these coals. Better prediction is probably possible if substantially more than 2,000 pyrite grains are counted and the clustering factor is taken into account.

Even with the coarse coal that is cleaned today, knowledge of the pyrite size distribution can help in understanding why some coals have particularly high or low sulfur reductions. With some of the newer technologies which require very finely ground coal and which release even more pyrite from the coal matrix, microscopic size analysis of the pyrite should become an even more important aspect of coal characterization in the future.

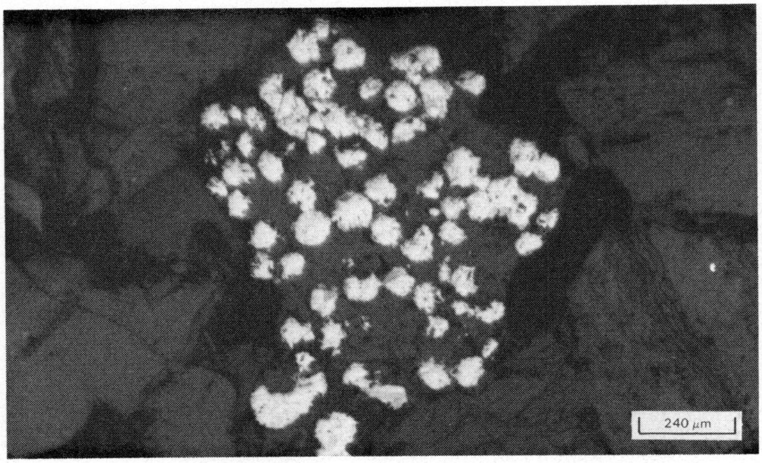

Fig. 7. Clustering of pyrite in excess of 7 percent. Pyrite occupies about 43 percent of the coal grain.

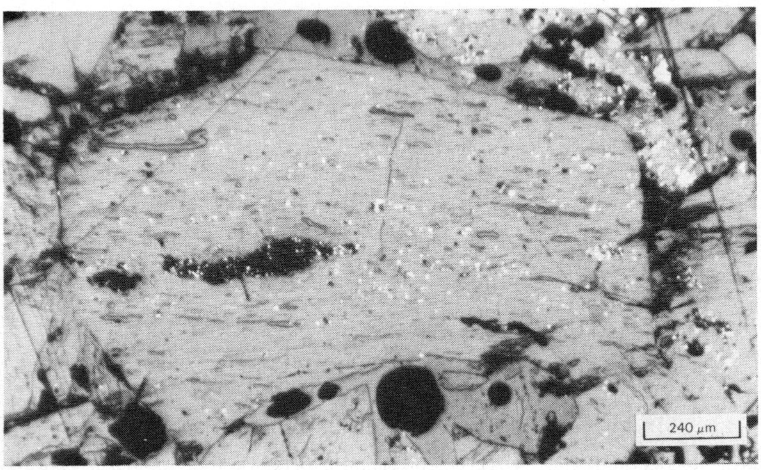

Fig. 8. Very fine, sparsely distributed pyrite occupying about 4.5 percent of the large central grain.

REFERENCES

1 J.T. McCartney, H.J. O'Donnell, and S. Ergun, Pyrite size distribution and coal-pyrite particle association in steam coals, U.S. Bureau of Mines, Rept. of Investigations 7231, 1969, 18 p.

2 M.R. Sneeringer, 1983 Report on Ohio mineral industries with directories of reporting coal and industrial mineral operators, Ohio Geol. Survey, 1984, 83 pp.

3 G. Botoman and D.A. Stith, Analysis of Ohio coals, Ohio Geol. Survey Information Circ. 47, 1978, 148 pp.

4 G. Botoman and D.A. Stith, Analysis of Ohio coals, 1977-1978, Ohio Geol. Survey Information Circ. 50, 1981, 54 pp.

5 D.A. Stith, G. Botoman, and R. Carlton, Coal washability study, final report, Ohio Geol. Survey Open-File Rept. 82-2, 1982, 68 pp.

6 J.A. Harrell and K.A. Eriksson, Empirical conversion equations for thin-section and sieve derived size distribution parameters, Jour. Sed. Petrology, 49 (1979) 273-280.

7 R.L. Folk, Petrology of sedimentary rocks, Hemphill's, Austin, Texas, 1965, 159 pp.

8 J. Adams, Sieve size statistics from grain measurements, Jour. Geology, 85 (1977) 209-227.

9 W.C. Krumbein and F.J. Pettijohn, Manual of sedimentary petrography, Appleton-Century-Crofts, Inc., New York, 1938, 549 p.

10 W. Petruk, Correlation between grain sizes in polished section with sieving data and investigation of mineral liberation measurements from polished sections, Minerals Research Program, Canada Centre for Mineral and Energy Technology, Processing Contribution No. 98, 1978, c272-c277.

PYRITE AND ASH LIBERATION FROM SELECTED COALS

M. T. ZITTERBART, D. G. NICHOLS, and D. E. LOWENHAUPT

Conoco Inc., Coal Research Division, 4000 Brownsville Rd., Library, Pennsylvania 15129

ABSTRACT

To determine the effect of crushing on pyrite and ash liberation, float/sink tests were run at 1.4 and 2.0 specific gravities on 8, 28, 50, and 100 mesh fractions of Pittsburgh, Ohio No. 6A and No. 8, and Illinois No. 6 coal samples. These samples were also examined microscopically to determine the approximate size of the pyrite particles, and the amount of pyrite liberated from the coal.

Statistical analyses of the data have shown that the percent pyrite liberation was inversely correlated to the mean grind size for all the coals studied. Ohio and Pittsburgh seam coal data showed that the percent pyrite removed by float/sink tests was directly correlated with the degree of pyrite liberation as determined microscopically. Ohio and Illinois coal samples have shown that some 75% of the pyrite was liberated upon finer grinding of the coal. Percent ash removals increased for the Pittsburgh and Illinois coals when ground from 8 to 28 mesh, but little improvement in ash removal was attained upon further grinding to 100 mesh. The trade-off between the increased costs due to grinding and the handling of fines versus the benefit of sulfur/ash removal is an important consideration in the possible use of this data.

INTRODUCTION

The work presented here is based on a fundamental experimental investigation to determine the effect of crushing on the potential for reduction of pyrite and ash from selected high sulfur coals. Float/sink tests were carried out on coal samples which had been consistently ground to achieve a controlled size distribution. A petrographic microscope was used to determine the extent to which pyrite particles were liberated due to the controlled grinding.

Some literature data are also presented to check the potential of size reduction as a means for pyrite or ash liberation from coal. Overall, however, a trade-off exists between the increased costs due to grinding versus the benefit of sulfur and ash removal from the coal.

BACKGROUND

Microscopic analysis of coal pyrite

Automated microscope analysis, which measures by reflectance the diameters of particles exposed on a polished surface, has been reported to be the most effective method of determining pyrite size distribution (ref. 1). Microscopic analyses resulted in a mean pyrite particle size of 60 microns and 250 microns, respectively, for Ohio No. 8 Pittsburgh seam coal and a northern West Virginia Pittsburgh seam coal, both of which had been ground to -8 mesh. Upon further grinding of the Pittsburgh seam coal (WV) to -10 mesh, the mean pyrite size was reduced to 100 microns.

McCartney et al. (ref. 2) correlated the size of pyrite particles for coals and the amount of pyrite liberated from the coal with the reduction in pyrite accomplished by float/sink tests.

Out of sixty-one raw coals analyzed, the microscopic analysis of eleven -14 mesh Pittsburgh seam coal samples from West Virginia showed an average pyrite size of 96 microns with 36 to 83% of the pyrite liberated from the coal. (Note that the 96 micron size here agrees closely with the 100 micron size cited above.) Cleaning these samples at 1.60 specific gravity resulted in an average pyrite reduction of 55%. Five -14 mesh coal samples from the Ohio No. 8 Pittsburgh seam contained pyrite having 55 micron average size with 40 to 70% of the pyrite microscopically seen to be free from the coal. (Note that the 55 micron size here agrees closely with the 60 microns cited above.) Approximately 50% pyrite reductions were achieved for the Ohio coals by cleaning at 1.60. Microscopic analyses of six -14 mesh Illinois No. 6 coal samples determined an average pyrite particle size of 44 microns, with 21 to 71% of the pyrite liberated from the coal. Float/sink tests at 1.60 resulted in an average of 41% of the pyrite removed.

The data for all sixty-one coals were plotted to show a linear relationship between pyrite removal and mean pyrite particle size having correlation coefficient of 0.89. It was expressed that the larger the size of the pyrite particles, the more likely that pyrite will dominate in some coal particles and will cause them to sink in the separating liquid.

Automated image analysis

Automated image analysis (AIA) combined with scanning electron microscopy (SEM) has been used at Ames Laboratory to characterize pyrite and mineral matter in raw and cleaned Pittsburgh seam and Illinois No. 6 coal samples (ref. 3). AIA of a raw Pittsburgh seam coal revealed that 50% of the mineral matter was pyrite, with some 70% of the pyrite having particle size of 21 microns or larger. A 78% pyrite reduction resulted from cleaning the Pittsburgh seam coal at 1.3 specific gravity, with over 90% of the +21 micron-sized pyrite being removed. The AIA of a raw Illinois No. 6 coal revealed that some 23% of the mineral matter was pyrite, with 56% of the pyrite having size 20 microns or larger (ref 4). AIA of the Illinois coal after being treated by the TRW Gravimelt process showed that 65% of the pyrite had been removed. Specifically, the treatment was reported to have removed some 85% of the +20 micron-sized pyrite.

CCTF sulfur reductions

Coal cleanability tests have been conducted at EPRI Coal Cleaning Test Facility (CCTF) located at Homer City, PA (ref. 5). Laboratory washability tests on a 50/50 blend of raw Upper and Lower Freeport coals resulted in a 56% sulfur reduction for the coal ground to -3/4". Negligible improvement in sulfur reduction was achieved by additional grinding to -¼". CCTF equipment used to test the -3/4" and -¼" fractions of the Freeport blend demonstrated that cleaning via heavy-media cyclone at 1.40 or 1.50, water-only cyclone or concentrating table all achieved higher sulfur reductions (56-64%) than did the laboratory tests. Note, however, that the various prep plant equipment resulted in 79-93% Btu recoveries which were less than the 95% energy recovery achieved in the laboratory.

METHODS

Controlled grinding

Pittsburgh, Ohio No. 6A and No. 8, and Illinois No. 6 coal samples were ground to -8, -28, -50, and -100 mesh fractions using a controlled grinding procedure in order to minimize the amount of fines. Each jaw crushed coal was passed once through a cone crusher followed by roll milling to produce some 90% minus 8 mesh coal. The undersize fraction was screened out while the oversize material was again passed through the roller mill to produce a -8 mesh and a +8 mesh fraction, the oversize again being recycled through the roller mill. This procedure was continued until all the coal had been ground to -8 mesh.

A tighter setting on the roller mill along with the above procedure of screening out the undersize prior to continued grinding of the oversize was used to produce the -28, -50, and -100 mesh fractions for each coal. Each size fraction was screened and plotted on a Rosin-Rammler graph. The value at 50% oversize on the graph determined the mean coal particle size at which 50 percent by weight of the coal is larger or smaller in size, that is d_{50}. This grinding procedure minimized the production of excess fines which is a resulting problem whenever an entire coal fraction is recycled through the grinding equipment.

Float/sink tests

Float/sink tests at 1.4 and 2.0 specific gravities were performed on -8, -28, -50, and -100 mesh fractions of Pittsburgh, Ohio No. 6A and No. 8, and Illinois No. 6 coals, prepared as detailed above. Separations for all size fractions, excluding -100 mesh, were achieved after four hours. The -100 mesh fraction was permitted to separate for six hours. However, it is not clear that a good gravity separation of the finer particles was in fact achieved. The float, sink, and starting coal fractions were analyzed for pyritic sulfur and ash using ASTM procedures.

Petrographic microscope analyses

Each of the -8, -28, -50, and -100 mesh fractions of the Pittsburgh, Ohio and Illinois coals under investigation was microscopically examined as explained below to determine the percent pyrite liberated, i.e., the extent to which the pyrite particles had been freed from the coal due to the controlled grinding.

Pyrite is easily identified under the microscope because it is so much higher in reflectance than the surrounding coal material. It is a simple matter then to classify pyrite as liberated or encased since liberated pyrite is free standing while the encased pyrite is substantially surrounded by coal. To do this, the following microscopic method was developed using the standard one inch, polished pellets employed in coal petrography.

In this technique, only the pyrite within the grid of the microscope's eyepiece was counted as liberated or encased. This permitted the counting of sufficient pyrite particles to produce a statistically relevant population that was also reproducible. This procedure was repeated in 1 mm steps in the vertical direction until the edge of the particle was reached. The pellet was then moved 1 mm horizontally, and the vertical movement was repeated. Two operators each count 500 points of pyrite with one set on the left side of the pellet and another on the right side of the pellet.

DISCUSSION

Results of present study

In general, the experimental results illustrated that more pyrite was liberated and later removed via float/sink tests due to finer grinding of the coal. As much as 75% of the pyrite was liberated with up to 93% pyrite reduction achieved. Both pyrite and ash removals increased when the coal samples were ground from -8 to -28 mesh. Very little improvement in the pyrite/ash reduction was achieved by the additional grinding to -100 mesh.

Statistical analyses of the data showed that the percent pyrite liberated was inversely correlated to the mean grind size for the Pittsburgh seam coals. This was shown by a correlation coefficient of -0.94 at a 99% confidence level for the cleaned Pittsburgh seam coals. The percent pyrite removed from the Pittsburgh coals by float/sink tests was only weakly correlated with the degree of pyrite liberated by grinding as shown by a correlation coefficient value of 0.50 having 90% confidence at 1.40 specific gravity and a correlation value of 0.47 at 88% confidence for 2.0 specific gravity tests.

The raw Ohio coals also showed an inverse correlation between pyrite liberation and mean grind size, correlation coefficient equal to -0.64 at 83% confidence. Here too, the pyrite removed via 1.4 and 2.0 float/sink tests was directly related to the pyrite liberated upon grinding. This was shown by the correlation coefficients of 0.84 having 96% confidence and 0.77 with 92% confidence for the 1.4 and 2.0 tests, respectively.

Consistent with the Pittsburgh and Ohio coal data, the Illinois No. 6 coals showed a significant inverse correlation between the average grind size and the percent pyrite liberated, correlation coefficient equal to -0.88 at 99.8% confidence. The raw Illinois coal data showed a correlation coefficient of 0.80 at 99% confidence between the amount of pyrite that was removed via 1.4 float/sink tests and the initial pyrite content of the coal.

In addition, the amount of pyrite removed through washability tests on the Ohio coal was significantly correlated to the initial ash content of the raw coal, as shown by correlation values of 0.95 and 0.99 (both having 99% confidence) for the 1.4 and 2.0 tests, respectively. The Pittsburgh seam and Illinois coals did not show such a relationship. (The Ohio and Illinois coal samples were raw, i.e., ROM, samples while the Pittsburgh seam samples were cleaned coals.)

Overall, the data showed that more pyrite is liberated upon finer grinding of the coal sample, which thereby permits the pyrite to be removed via float/sink tests. Generally, Figures 1, 2, and 3 show a linear relationship between pyrite liberation and mean grind size of the coal. However, the Ohio No. 8 coal relationship departs substantially from linear form. In this one case, the pyrite liberation is a linear function of the reciprocal of the mean coal size.

These graphs demonstrate that coal samples from different seams possess distinct pyrite liberation characteristics. Over the size range of 100 to 300 microns mean coal size, the Pittsburgh seam (PA and WV) and the Ohio No. 8 seam showed very similar pyrite liberation values. These coals show about 39% pyrite liberation at a mean coal size of 200 microns. Similarly, the Ohio No. 6A samples were very similar to the Illinois No. 6 samples, i.e., at a mean coal size of 200 microns some 66% of the pyrite was liberated for both.

Pyrite reductions up to 93% were achieved through 1.4 gravity washability tests on the raw Ohio coal samples at 53% yield. Up to 80% reduction in pyrite resulted from 1.4 float/sink tests for Pittsburgh seam (85% yield) and Illinois No. 6 coals (36% yield). The pyrite and ash reduction results via float/sink tests for Pittsburgh, Ohio and Illinois coals are shown respectively in Tables 1, 2, and 3.

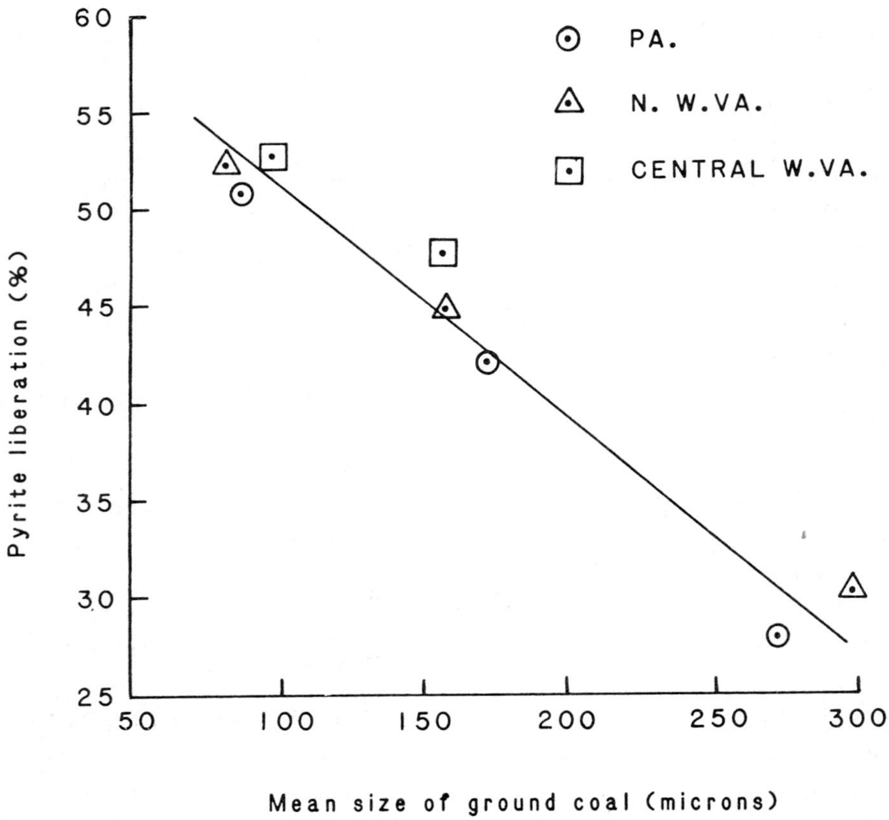

Fig. 1. Inverse relationship between pyrite liberation and mean grind size for Pittsburgh seam coals.

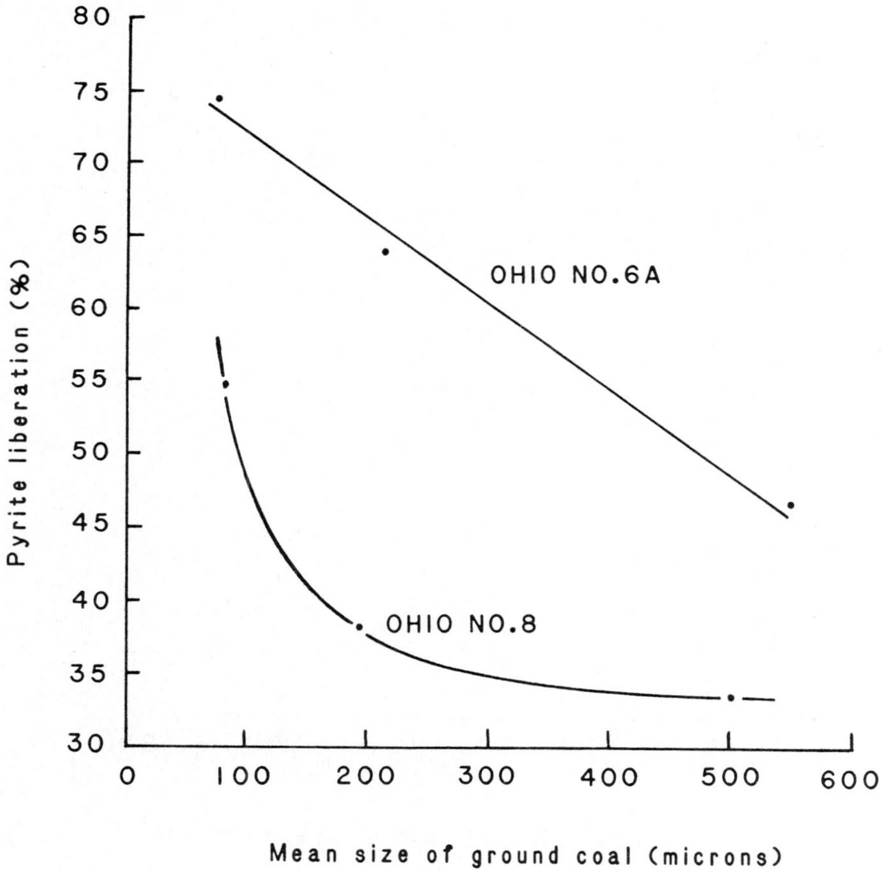

Fig. 2. Inverse relationship between pyrite liberation and mean grind size for Ohio coals.

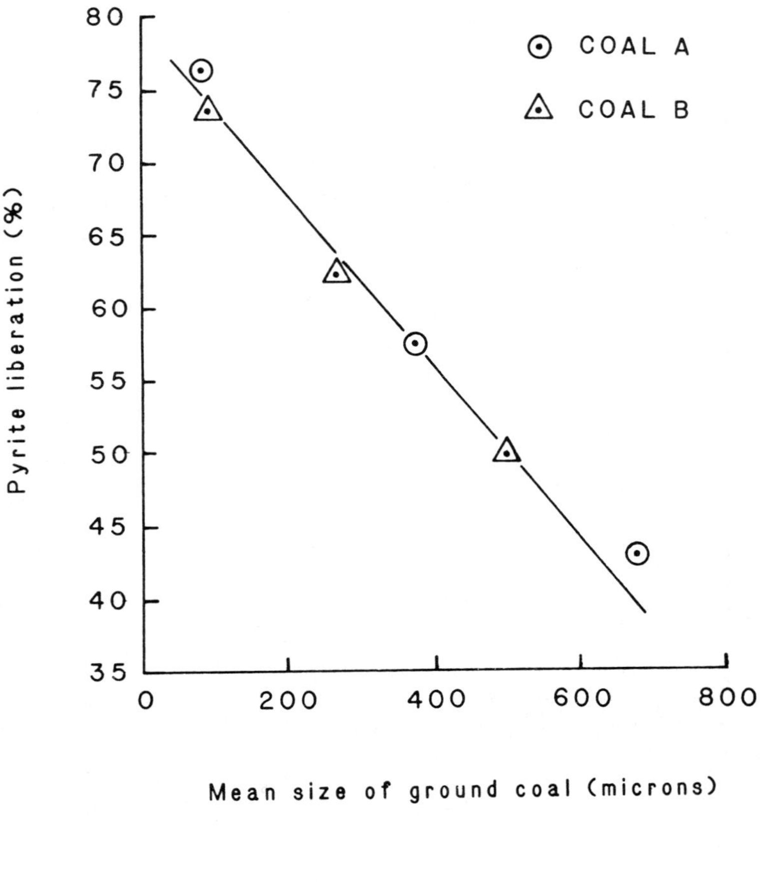

Fig. 3. Inverse relationship between pyrite liberation and mean grind size for Illinois No. 6 coals.

TABLE 1

Pyrite and ash reductions for Pittsburgh seam coals via 1.4 and 2.0 float/sink tests.

Coal	Grind Size, mesh	Mean Size of Coal, microns	Pyrite* Liberation %	Pyrite Reduction %		Ash Reduction %		Yield %	
				@1.4	@2.0	@1.4	@2.0	@1.4	@2.0
PENNSYLVANIA									
Dry Analysis:	8	625	-	35.0	20.0	33.6	13.0	90.5	99.1
1.46 % Total S	28	270	28.0	37.5	17.5	39.3	9.9	85.5	98.2
0.40 % Pyr S	50	170	42.1	37.5	25.0	39.9	13.8	85.9	98.6
7.30 % Ash	100	85	50.8	50.0	27.5	37.7	9.2	81.4	98.6
N. WEST VIRGINIA									
Dry Analysis:	8	900	-	27.3	11.8	49.3	20.7	71.0	73.8
3.87 % Total S	28	297	30.1	41.8	14.6	51.6	21.7	75.4	71.5
1.10 % Pyr S	50	155	44.8	54.6	13.6	59.3	23.8	69.5	96.6
10.97 % Ash	100	80	52.4	62.7	27.3	58.7	13.7	63.7	95.2
CENTRAL W. VA									
Dry Analysis:	8	690	-	51.5	39.4	32.8	17.4	89.8	97.7
2.98% Total S	28	275	-	70.7	48.5	33.9	14.0	87.5	97.3
0.99 % Pyr S	50	155	47.7	73.7	52.5	33.1	17.7	85.2	97.0
7.74 % Ash	100	96	52.7	79.8	59.6	37.2	16.8	85.5	97.4

* Determined via petrographic microscope.
\- Not determined.

TABLE 2

Pyrite and ash reductions for Ohio coals via 1.4 and 2.0 float/sink tests.

Coal	Grind Size, mesh	Mean Size of Coal, microns	Pyrite Liberation* %	Pyrite Reduction %		Ash Reduction %		Yield %	
				@1.4	@2.0	@1.4	@2.0	@1.4	@2.0
OHIO No. 6A									
Dry Analysis:	8	545	46.6	87.1	75.5	85.1	74.6	54.9	67.2
3.33 % Total S	28	210	64.1	93.5	83.8	86.7	79.4	56.9	66.4
2.70 % Pyr S	100	74	74.7	93.2	83.6	81.6	69.2	49.5	68.5
32.92 % Ash									
OHIO No.8									
Dry Analysis:	8	500	33.8	60.3	37.1	73.6	54.0	56.8	80.0
4.20 % Total S	28	190	38.5	60.3	36.0	74.0	52.8	55.2	80.4
2.66 % Pyr S	100	80	54.8	70.4	40.2	73.4	45.4	48.3	81.9
24.37 % Ash									

* Determined via petrographic microscope.

TABLE 3

Pyrite and ash reductions for Illinois No. 6 coals via 1.4 and 2.0 float/sink tests.

Coal	Grind Size, mesh	Mean Size of Coal, microns	Pyrite* Liberation %	Pyrite Reduction %		Ash Reduction %		Yield %	
				@1.4	@2.0	@1.4	@2.0	@1.4	@2.0
COAL A									
Dry Analysis:	8	675	43.0	76.7	66.0	69.2	44.4	56.4	82.8
4.20 % Total S	28	375	57.6	77.3	69.8	74.3	51.0	54.8	81.2
2.33 % Pyr S	100	85	76.4	76.8	58.5	66.2	33.2	36.0	86.4
24.94 % Ash									
COAL B									
Dry Analysis:	8	500	49.6	66.7	54.0	74.8	49.2	54.2	81.4
3.01 % Total S	28	265	62.4	72.7	58.0	76.4	35.6	52.8	83.2
1.37 % Pyr S	100	87	73.6	71.1	57.8	66.7	35.5	44.6	87.5
23.50 % Ash									

* Determined via petrographic microscope.

Percent pyrite and ash removals increased, for all the coals tested, when ground from -8 to -28 mesh, but only little improvement in pyrite/ash reduction was achieved upon further grinding to -100 mesh. The fact that -100 mesh coal particles do not readily separate upon float/sink cleaning may mask to some degree the potential for pyrite and ash reduction for this size range. However, from this study it appears that crushing to a size of -28 mesh allows for substantial reduction in pyrite and ash while limiting the costs due to grinding to finer sizes and the handling of such fine material.

Comparisons with literature data

The raw Ohio and Illinois coal samples of the present study generally show agreement in pyrite liberation values when compared to those of the Bureau of Mines' study (ref 2). This latter study showed 40 to 70% of the pyrite to be liberated at -14 mesh for five distinct Ohio No. 8 coal samples. The -8 mesh sample of Ohio No. 8 coal in the present study showed 34% pyrite liberation. Further, the six Illinois No. 6 coal samples of the Bureau of Mines study showed 21 to 71% pyrite liberation at -14 mesh while the three Illinois No. 6 samples of the present study (-8 mesh) ranged from 31 to 50% pyrite liberation.

If the Bureau of Mines' grinding procedure did not minimize the fines produced as did the present study, then this could account for their somewhat higher range of pyrite liberation values obtained. In addition, the coal size fractions and specific gravities used for cleaning were not exactly the same for the two studies. Some variations in the results may also be due to using coal samples from different mines, as well as to different degrees of weathering.

No comparison is made for the Pittsburgh seam coals since our experimental data (Table 1) are for cleaned coal samples, whereas all coals studied by the Bureau of Mines were raw samples.

REFERENCES

1. D.R. Bomberger and M. Deul, Society of Mining Engineers - Transactions, (March 1964) 65-99.
2. J.T. McCartney, H.J. O'Donnell, and S. Ergun, Bureau of Mines RI 7231, February 1969, pp. 1-18.
3. W.E. Straszheim and R. Markuszewski, ACS Preprint - Division of Fuel Chemistry, 29, 4 (1984) 310-318.
4. W.E. Straszheim and R. Markuszewski, ACS Preprint - Division of Fuel Chemistry, 30, 1 (1985) 47-55.
5. Coal Cleaning Test Facility Update No. 2, EPRI Technical Newsletter, Spring 1985, pp. 2-9.

CHARACTERISTIC PYRITE OCCURRENCES AND DISTRIBUTIONS IN A FINE BITUMINOUS COAL

E. TOMPOS

Dept. of Mineral Dressing, Tech.Univ. of Heavy Industry, H-3515 Miskolc (Hungary)

ABSTRACT

Polished samples of <1.5, 1.5-2.0 and >2.0 specific gravity fractions of various size fractions of a Hungarian fine bituminous coals were microscopically examined. Simultaneously with the petrographic examinations the contents and types of pyrites were also determined. Pyrite distributions were determined by a petrographic method developed in the Department. The pyrite distributions in coal comply with Rosin-Rammler's function with a good correlations.

The adherences of pyrites with the microlithotypes of coal are very characteristic in the specific gravity fraction <1.5 clean coal quality, in the sp gr fraction 1.5-2.0 middling for power plant and in the sp gr fraction >2.0 refuse. Size distributions and average sizes of pyrites are also different in the sp gr fractions.

One of the important conditions of pyrite elimination or concentration is the proper size reduction of coal. This is evident from the washability curves of ash and total sulfur of the <12 mm fine coal, of the comminuted coal <0.5 mm and <0.1 mm. The determination of the size distributions of pyrite in coal can be very useful in the experiments of pyrite elimination.

INTRODUCTION

The Liassic bituminous coal in Hungary is a high quality coking coal. As well known, the high sulphur content is nondesirable in the coking coals. To assess the possibilities of pyrite elimination, pyrite types, contents and distributions were microscopically and coalpetrographically analysed in the sp gr fractions <1.5, 1.5-2.0 and >2.0 of the size fraction 6-12, 0.5-6, 0.037-0.5 and <0.037 mm in the fine coal of Pécs and Komló of Southern Hungary.

METHOD TO DETERMINE THE GRAIN SIZE DISTRIBUTION OF PYRITE IN COAL

According to the method developed in the Department of Mineral Dressing, Technical University of Heavy Industry (ref. 1) fine coal particles are dried, coarser particles dried and fixed with Canadian balsam and embedded into mountain wax or granular gypsum. Then polished samples are prepared and they are systematically analysed with a microscope by observing several hundred fields of vision. As indicated in Fig. 1, the characteristic dimensions of the pyrite particles have to be measured in each field of vision (diameter in the case of circular cross-section and the framing length and width for irregular shapes).

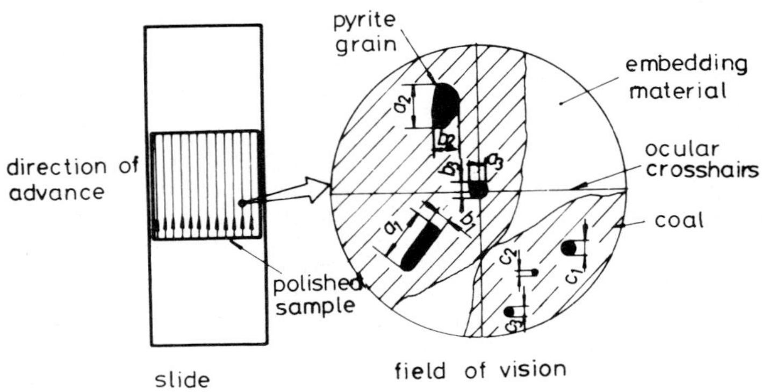

Fig. 1. Field of vision of microscope

Then the real dimensions have to be calculated by taking into account the magnification which provides the real circle diameter d_i or the diameter of equiarial circle d_i'. The particles are grouped into particle size fractions according to their dimensions. With the particle numbers m_i and n_i the average particle size by volume can be determined for each size fraction. The average particle size of the i-th size fraction will be

$$\bar{x}_i = \left[\frac{\sum_{i=1}^{k} \left(m_i d_i^3 + n_i d_i'^3 \right)}{\sum_{i=1}^{k} \left(m_i + n_i \right)} \right]^{1/3} \quad [\mu m] \qquad (1)$$

The mass or weight fraction Δw_i of a size fraction can also be calculated since it is proportional to the particle number

$N_i = m_i + n_i$ and the third power of the average particle size (if the density is identical):

$$\Delta w_i = \frac{N_i \bar{x}_i^3}{\sum_{i=1}^{k} \left(N_i \bar{x}_i^3 \right)} 100 \qquad (2)$$

The calculations have been programmed for a HP 97 pocket calculator. The calculator prints out the mass fractions and cumlative mass percentages of the pyrite particles, data of the distribution according to the numbers and the average particle size by volume of all pyrite size fractions. From the grain size characteristics the important particle sizes can be determined.

It can be proved that the spherical model applied in the evaluation can be used for determining the grain size characteristics with sufficient accuracy. In applying this model, pyrite particles are supposed to be spheres and the expected sizes of the pyrite particles are determined from the diameters of the sphere sections.

If cutting single spheres randomly or analysing sections of particulate materials consisting of balls with identical diameters, the correlation between the expected value of the diameters of the sections M(d) and the diameter of the particle d can be written as

$$M(d) = \frac{1}{2} \int_{x=-R}^{R} \sqrt{R^2 - x^2}\, dx = \frac{R\pi}{2} = 0.79\ d \qquad (3)$$

where R is the radius of a sphere with a diameter d.

The variance of the circle diameters in the sections $D^2(d)$ becomes

$$D^2(d) = M^2(d) - M(d)^2 = 0.20\ R^2 = 0.05\ d^2 \qquad (4)$$

where $M^2(d) = \frac{8}{3} R^2$ is the expected value the square of the chord length.

This result has been checked experimentally by analysing polished samples of beads in synthetic resin and lead balls in mountain wax.

TYPES OF PYRITE IN COAL

Pyrite particles can be found in the coal in strong and fine associations with microlithotypes. In the <1.5 g/cm^3 density fraction of coking coal (ash content 10-11 %) pyrite appears in the vitrinite and exinite in form of finely disseminated syngenetic concretions i.e. globulites. In the density fraction 1.5-2.0 g/cm^3 of middling quality (ash content 35-45 %) pyrite is locked with almost all microlithotypes in from of striped epigenetic and in groups arranged globulites. In the >2.0 g/cm^3 density fraction waste, large singular globulites and disseminated irregular globulite aggregates locked with carbonaceous shale and waste. In the >2.0 g/cm^3 product of the smallest size fraction discrete (free) pyrite particles are also present but their mass remains unimportant. Scanning micrographs of an unlocked pyrite particle in 300, 1000 and 3000 times magnification are shown in Figs. 2, 3 and 4.

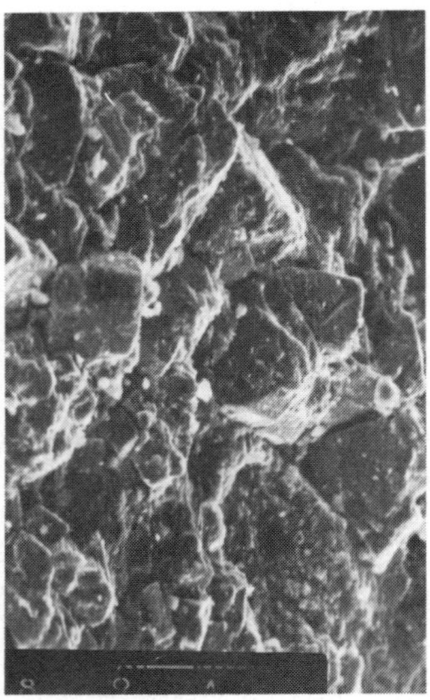

Fig. 2. Scanning micrograph of pyrite in coal, 300 x

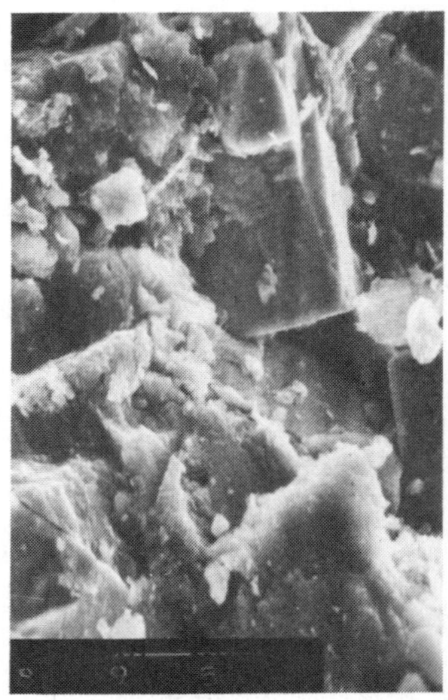

Fig. 3. Scanning micrograph of pyrite in coal, 1000 x

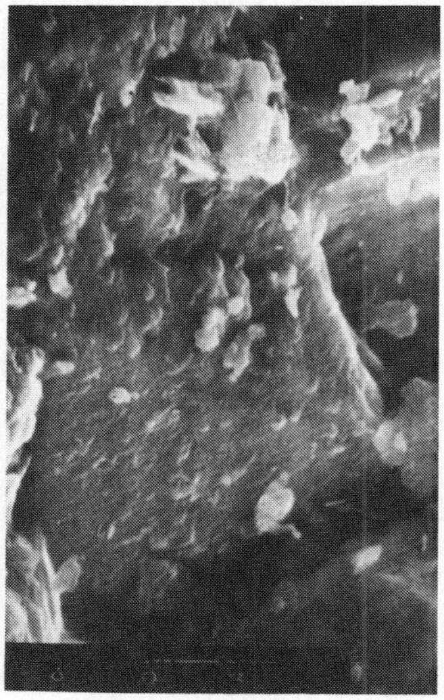

Fig. 4. Scanning micrograph of pyrite in coal, 3000 x

In the sequence of the figures pyrite crystalline aggregate; poorly developed crystalline planes of pyrite particles; pyrite surface contaminated with illite and montmorillonite are illustrated.

From the microcopic pyrite contents and the calculated pyrite recoveries (ref. 2) pyrite balance has also been determined. The pyrite balance of the Pécs fine coal is illustrated in Fig. 5:

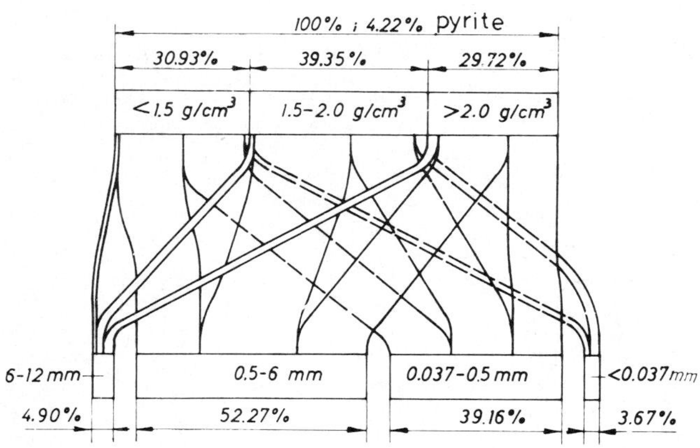

Fig. 5. Pyrite balance of Pécs fine coal

the 4.22 per cent pyrite content is distributed among the coking coal (31 per cent), the middling (39 per cent) and the waste

(30 per cent). The distribution according to the size fractions is also of interest.

GRAIN SIZE DISTRIBUTION OF PYRITE

Grain size distributions of pyrite in the coal determined with the method outlined earlier, are illustrated in Fig. 6

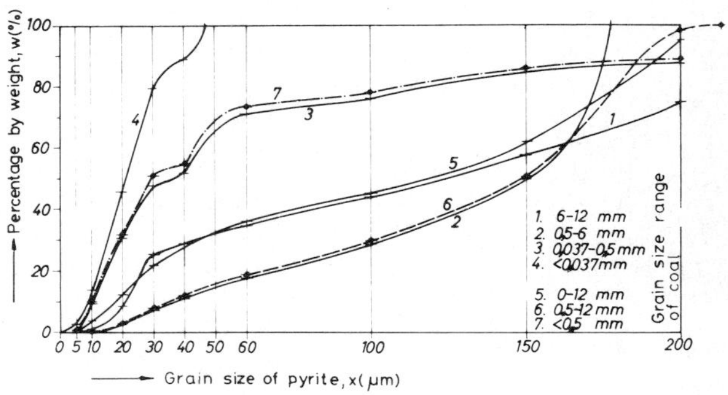

Fig. 6. Pyrite distribution of Pécs fine coal

for the size fractions of the Pécs coal and in Fig. 7 for the size and density fractions of the Komló coal. The average size

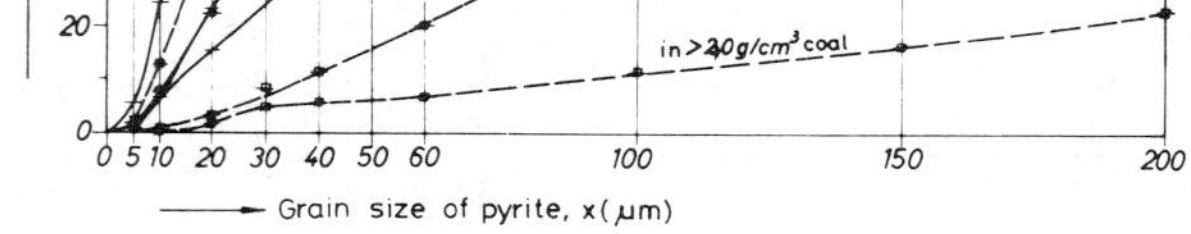

Fig. 7. Pyrite distribution of Komló fine coal

of the pyrite is the smallest in the coking coal ($x_{max} < 100$ μm) it is greater in the middling ($x_{max} < 200$ μm) and the greatest in the tailing ($x_{max} < 350$ μm) for the Komló coal.

Pyrite distributions in the size and density fractions of both the Pécs and Komló fine coal can be described with Rosin--Rammler's function with correlation coefficients in the range of 0.92-0.99. Fig. 8 illustrates the grain size distributions of the density fractions of the technological size fractions <0.5 and 0.5-12 mm in the Pécs fine coal in Rosin-Rammler's diagramm.

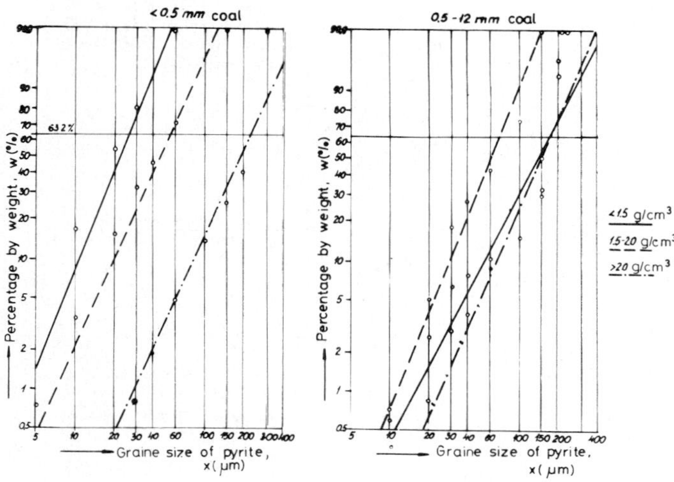

Fig. 8. Pyrite distribution in Rosin - - Rammler's diagram

LIBERATION OF PYRITE

Laboratory comminution of the fine coal (<12 mm) below 0.5 mm and 0.1 mm and a subsequent density analysis proved for both the Pécs and the Komló raw and coking coal that with increasing grinding fineness total sulphur contents and recoveries increase in the density fraction > 2 g/cm^3. Figs. 9, 10 and 11 show the ash and total sulphur washability curves of < 12, <0.5 and < 0.1 mm raw fine coal. The total sulphur content of the waste > 2 g/cm^3 has increased from 1.5 per cent to 3.9 per cent and the total sulphur recovery ($= 100 - m_s$) from 22 per cent to 40 per cent.

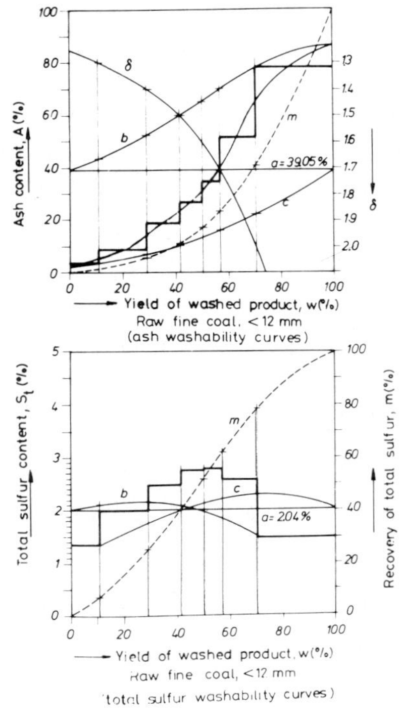

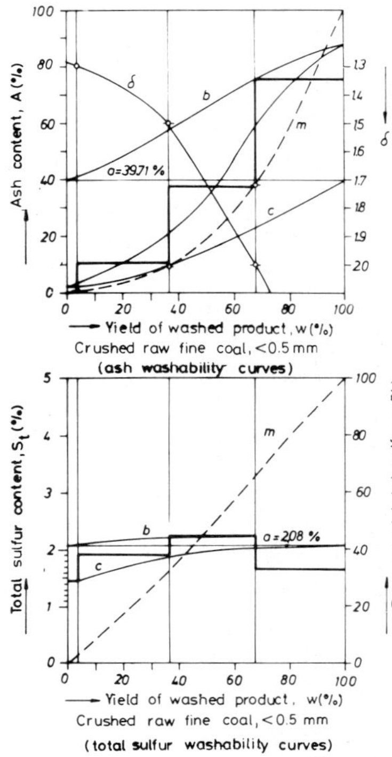

Fig. 9. Washability curves of <12 mm Pécs fine coal

Fig. 10. Washability curves of < 0.5 mm Pécs comminuted coal

a - average content

b - cumulative sink ash

c - cumulative float ash

- specific gravity (yield)

m - recovery

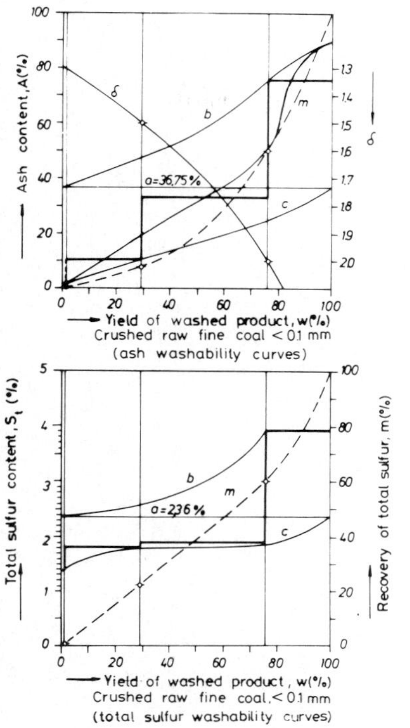

Fig. 11. Washability curves of <0.1 mm Pécs comminuted coal

As regards the elimination of pyrite in the Liassic fine coal, only partial removal of pyrite (reduction of pyrite content) is feasible (ref. 3, 4). The reason of that is multiple: fine distribution of pyrite, its associations with the microlithotypes, the low ratio of the pyrite sulphur to the total sulphur, physical properties of pyrite in the coal differing from those of ore pyrite (magnetic susceptibility, zeta potential, contact angle, low density-increasing effect etc.). Middlings and wastes have to be separated sharply to remove pyrites locked to carbonaceous shale, waste and other microlithotypes as well as in free state.

REFERENCES

1 E. Tompos, Petrográfiai módszer az ásványelőkészitési technológiák optimális tervezéséhez, (Petrographic method to optimise mineral dressing technologies. Ph.D.Thesis.), Miskolc, 1983, (In Hungarian).
2 G. Tarján, Mineral Processing Vol. I. Fundamentals, Comminution, Sizing and Classification, Akadémiai Kiadó, Budapest, 1981.
3 R.C. Eliot, Coal Desulfurization Prior to Combustion, Noyes Data Corporation, Park Ridge, New Jersey, U.S.A, 1978.
4 E. Stach et al, Stach's Textbook of Coal Petrology, Gebrüder Borntraeger, Berlin, Stuttgart, 1975.

SIZE, SHAPE, AND DISTRIBUTION OF MICROSCOPIC PYRITE IN SELECTED OHIO COALS

W. A. KNELLER[1] and G. P. MAXWELL[2]

[1]Geology Dept., University of Toledo, 2801 W. Bancroft St., Toledo, OH 43606

[2]Consulting Geologist, 2508 E. 11th St., Apt. 201, Denver, CO 80206

ABSTRACT

The size, shape and distribution of microscopic pyrite (-840 μm) was determined in 6 Ohio coal beds. The pyrite morphologies noted were classified petrographically into 8 classes: (1) framboidal; (2) euhedral; (3) fracture filling; (4) layered; (5) dendritic; (6) cell filling; (7) bleb and (8) submicron.

The analyzed maximum and minimum dimensions in plane-section of the pyrite were encouraging. Testing the distribution of each set of sample values for normality with a modified version of the Kilomogorov-Smirnov Test revealed that each sample distribution occurred in the limits specified by its Gaussian distribution. At a level of significance of 0.05, it is assumed that these distributions could aid in predicting the occurrence of pyrite at selected size fraction.

Volume percent distribution of microscopic pyrite revealed a marked increase within the -32 to 4 μm size range. Within this range the maximum percent was observed to occur between the -16 to 8 μm size. Below the 4 μm size, the data show a notable decrease in the pyrite particles. This concentration with minor variations existed in all samples. It is, therefore, possible to develop a partition for the optimum washability of a coal bed. The homogeneous distribution permits the efficient economic deep-cleaning of Ohio coals at the -8 to -4 μm size. This uniform distribution minimizes capital investment required for preparation plant equipment and decreases daily operational costs by reducing the flow variables of mineral dressing.

INTRODUCTION

The United States possesses approximately 1.5 trillion net tons of identifiable coal reserves, including those down to 3,000 feet below the surface (ref. 1). Additional estimates place 1.3 trillion short tons of coal in unmapped and unexplored areas throughout the United States. Because it accounts for 80 percent of our fossil fuel reserve, coal is the one resource certain to be available to meet the nation's immediate and future energy demands. Consequently, it is clear that the United States in the future will have a greater reliance on its most abundant fossil fuel, coal.

Although the United States possesses coal resources with a supply potential, based on current demands, of up to 2,000 years (ref. 2), its use is dependent upon the sulfur content. Historically, sulfur in coal has plagued industry and society for many years. Technological and environmental problems created by sulfur in coal include the contamination of coke during the metallurgical process, the formation of hydrogen sulfide in the hydrogenation process, the formation of sulfur oxides harmful to both industry and the ecosystem, the potential contribution to spontaneous combustion of coals in mines and stockpiled coals, and the pollution of ground water flowing from or through coal mines. These effects have led to considerable government legislation which has imposed stringent limits upon coal utilization.

The presence of sulfur in coal, however, is not entirely negative. Tarrer, et al., (ref. 3) illustrates the catalytic effect of various coal mineral matter, including pyritic sulfur, on reaction rates during coal liquefaction. In 1980 Moroni and Fisher (ref. 4) confirmed the use of pyritic sulfur as a disposable mineral catalyst in coal liquefaction.

As a result of the variable nature of sulfur, interest has mounted concerning all aspects of its presence in coal. Unfortunately, the extent of this interest is at most incomplete. In order to provide for the safe and efficient use of our coal reserves, it becomes apparent that the characterization of sulfur's presence in coal is essential.

The forms of sulfur in coal as reported by ASTM chemical analyses are organic sulfur, pyritic sulfur, and sulfatic sulfur (ref. 5). Organically combined sulfur, usually as mercaptans, thiophenes or organically complexed sulfides, and inorganic sulfides comprise essentially all of the sulfur found in most coals. The inorganic sulfides are represented by a variety of mineral species which include pyrite (FeS_2), marcasite (FeS_2), galena (PbS), chalcopyrite ($CuFeS_2$), arsenopyrite (FeAsS), and spalerite (ZnS). Occurrences of sulfatic sulfur are rare in fresh coal samples, although significant amounts of sulfatic sulfur appear in weathered coal.

Iron disulfide minerals, primarily pyrite and marcasite, dominate the inorganic sulfide occurrences in most coals. Pyrite (cubic-FeS_2) and marcasite (orthorhombic-FeS_2) are dimorphs, minerals that are identical in chemical composition but different in crystalline form. Gluskoter (ref. 6) states that these two minerals generally cannot be differentiated in coals except by their crystalline structures, which are usually determined by x-ray diffraction methods.

Pyrite in coal originates in many ways, as indicated by its variations in morphology, size, and mode of occurrence. These origins are characterized by different periods of mineral formation. Some pyrite has had a syngenetic origin; it was formed contemporaneously with coal in the peat swamp during the first stage of coalification. Other pyrite has had an epigenetic origin, which means it was formed within the coal seam during the second stage of coalification. Both origins are in response to the geochemical environment at the time of mineral formation.

Many deposits of pyrite in organic-rich sedimentary material are of organic origin. The formation of syngenetic pyrite in organic deposits depends primarily on the action of bacterial forms which are anerobes capable of reducing sulfates to sulfides (ref. 7). The amount of sulfide produced seems to be primarily a function of the availability of metabolizable organic matter to serve as a source of hydrogen sulfide or energy for the sulfate-reducing bacteria (ref. 8). The process of syngenetic pyrite formation originates as the result of the reaction of dissolved hydrogen sulfide with fine-grained detrital iron minerals. Pyrite is not directly formed by this reaction. Berner (ref. 9) cites Berner (ref. 10 and 11) and Rickard (ref. 12) in noting that the substances which form by the reaction of iron minerals or dissolved ferrous ion with hydrogen sulfide under simulated sedimentary conditions in the laboratory are generally non-crystalline FeS, mackinawite (tetragonal-$Fe_{1=x}S$), and gregite (cubic-Fe_3S_4). None of these three phases is thermodynamically stable under sedimentary conditions. These forms are unstable relative to pyrite and stoichiometric pyrrhotite. The final transformation of these metastable iron sulfides to pyrite is not well documented. It is believed by Berner (ref. 13) and others, however, that reactions of elemental sulfur with the metastable iron sulfides results in pyrite.

The emplacement of epigenetic pyrite occurs in cleat spaces, joint cracks, and fractures resulting during dehydration and consolidation of a coal bed or as a consequence of local or regional tectonic disturbance (ref. 14). Mackowsky (ref. 15) also notes epigenetic mineral deposition along bedding planes of coal. The formation of epigenetic pyrite is believed by many authors to originate as a result of the precipitation process. Reyes-Navarro and Davis (ref. 14) cite several authors (ref. 16, 17 and 18) stating that the precipitation results from the interaction of percolating waters bearing sulfate salts, iron compounds and organic matter in the coal bed. Other authors (ref. 19) contend this mechanism to be thermodynamically untenable. Neavel (ref. 19) suggests the reactants traveled different pathways; for example, the iron could have migrated down through the cleats from the muds above the coal, whereas the sulfidic sulfur could have migrated from the adjacent coal walls of the cleats.

METHODS AND RESULTS

Sample Collection

Ohio's coal resources are located in the eastern third of the state, in part of the unglaciated Allegheny Plateau. This area of Ohio displays Pennsylvanian and Permian rock sequences which are primarily non-marine and contain all of Ohio's coal resources. For this study, six coal beds were sampled from two series of the Pennsylvanian System. The Clarion No. 4A, the Lower Kittanning No. 5, and the Middle Kittanning No. 6 were collected from the Allegheny Series, whereas the Pittsburgh No. 8, the Meig's Creek No. 9, and the Waynesburg No. 11 were sampled from the Monongahela Series. Figure 1 shows the locations of the coal sampling sites selected for this study. These six coal beds were sampled in accordance with ASTM STP 611 (ref. 20). In addition, a preliminary examination of known proximate/ultimate analyses and mine production tonnages of these Ohio coals aided in selection of the sample sites. Bulk (600-1000 lbs.) vertical channel sampling proceeded at the various sites.

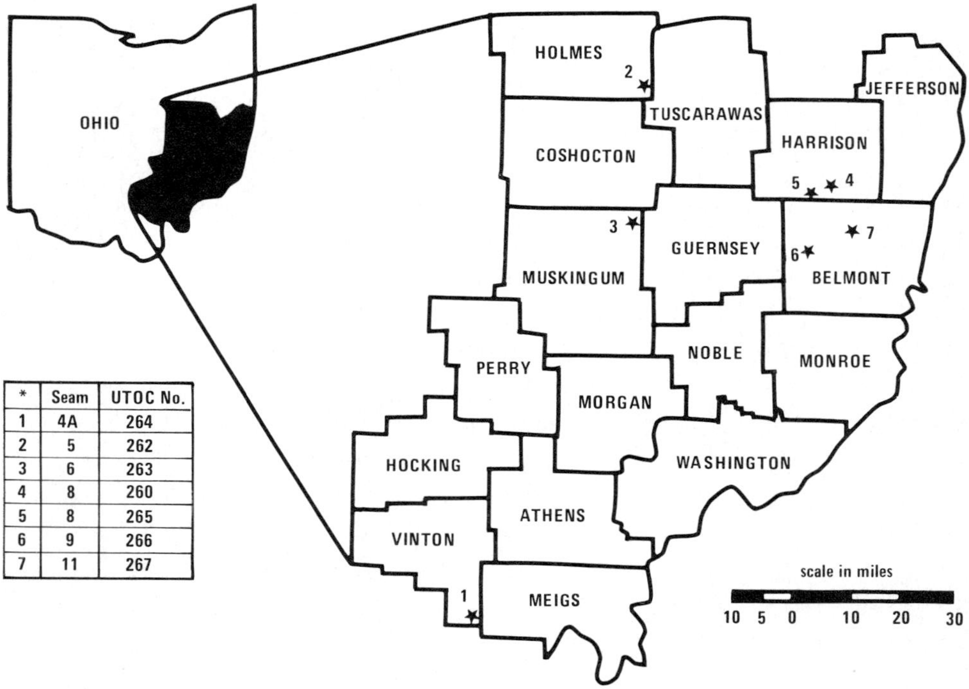

*	Seam	UTOC No.
1	4A	264
2	5	262
3	6	263
4	8	260
5	8	265
6	9	266
7	11	267

Fig. 1. Location map of the collection sites of selected Ohio coals.

Sample Preparation

Initially, the bulk samples were crushed through a Sturtevant Automatic Crusher and Sampler (SACCS) which reduced the size of the raw lump coal to about -10 mesh (2000 mm). Representative samples for the laboratory were taken from the automated splitting device. This split was further reduced in size for various tests in accord with ASTM methods and specifications.

Representative splits of the coal samples were prepared in accordance with ASTM Method D 2797, entitled: Standard Method of Preparing Coal Samples for Microscopical Analysis by Reflected Light (ref. 6). Each coal sample was stage ground using ceramic mortars and pestles until 100 percent passed a No. 20 sieve. A sample of approximately 250 grams was obtained from the sieved coals by using a microsplitter. After each split, a sample was made into a briquette which was polished to the proper degree for microscopic examination.

Maceral Analysis

The maceral analysis of each coal sample was performed in accordance with ASTM Method D 2799, entitled: Standard Method for Microscopical Determination of Volume Percent of Physical Components of Coal (ref. 6). A Zeiss Universal Microscope System MPM-01K was used in conjunction with a power regulated high pressure halogen lamp. A Brinkman Instruments Automatic Point Counter was used for the mechanical advancement of the briquettes. Following ASTM Method D 2799, an interpoint and interline distance of 0.5 mm was employed in the point count. Five hundred counts were made on each briquette and volume maceral percentages were determined. The analytical accuracy of any maceral analysis is dependent upon the number of point counts. If 1,000 counts are made on a briquette, the accuracy for each maceral or maceral group is approximately $\pm$ 3% at the 95% confidence level (ref. 21, 22, and 23). The results of the maceral analysis are summarized in Table 1.

Micro-Morphological Classification of Pyrite

Pyrite occurs in coals in particles from microscopic size to large masses several feet in diameter. These variations in morphology, size, and mode of occurrence arise because of the different geochemical environments in which the minerals were formed and the time of their genesis (ref. 6).

Many studies have been conducted to investigate the microscopic varieties of pyrite in coal. Throughout the course of these studies, numerous attempts have been made to classify pyrite according to its origin, morphological features, such as size and shape, and its distribution within the organic/inorganic constituents of coal.

The following classification of pyrite types, based on the research of Reyes-Navarro and Davis (ref. 14), Neavel (ref. 19) and Grady (ref. 24), was used for the purpose of this study. Eight different classes are defined according to the physical appearance of the pyrite occurrences.

1) Class Framboidal. An occurrence comprised of a spherical aggregation of euhedral crystals, subhedral crystals, and anhedral crystals or blebs.
2) Class Euhedral. An occurrence exhibiting distinctive crystal outlines.
3) Class Fracture Filling. An occurrence filling any kind of fracture, including cleats and dessication cracks.
4) Class Layer. An occurrence exhibiting a tabular or lenticular appearance.
5) Class Dendritic. An occurrence with the appearance of having formed by a branching growth.
6) Class Cell Filling. An occurrence in which the forms of the original cellular structure have remained.

TABLE 1

Volume percent maceral content as received for selected samples of Ohio coals

MACERAL TYPE	CLARION (NO. 4A) UTOC 264 SEAM		L. KITTANNING (NO. 5) UTOC 262 SEAM		M. KITTANNING (NO. 6) UTOC 263 SEAM		PITTSBURGH (NO. 8) UTOC 260 SEAM		PITTSBURGH (NO. 8) UTOC 265 SEAM		MEIG'S CREEK (NO. 9) UTOC 266 SEAM		WAYNESBURG (NO. 11) UTOC 267 SEAM	
	mmcb	mmfb	mmcb	mmfb	mmcb	mmfb	mmcb	mmfb	mmcb	mmfb	mmcb	mmfb	mmcb	mmfb
TELINITE	1.1	1.2	0.0	0.0	0.3	0.3	2.9	3.0	2.2	2.3	0.5	0.5	2.6	2.8
COLLINITE	70.8	77.0	79.8	83.6	81.1	83.8	76.9	80.4	79.7	83.0	72.0	74.7	73.4	77.7
VITRODETRINITE	0.5	0.5	0.1	0.1	0.3	0.3	0.3	0.3	0.2	0.2	3.9	4.0	1.4	1.5
TOTAL VITRINITE	72.4	78.7	79.9	83.8	81.7	84.4	80.1	83.8	82.1	85.5	77.0	79.9	77.4	81.9
SPORINITE	2.5	2.7	2.3	2.4	1.8	1.8	2.5	2.6	1.3	1.4	5.9	6.1	5.5	5.8
RESINITE	2.5	2.7	2.1	2.2	0.9	0.9	2.3	2.4	2.2	2.3	2.7	2.8	1.7	1.8
CUTINITE	0.5	0.5	0.0	0.0	0.2	0.2	0.3	0.3	0.4	0.4	1.4	1.4	0.7	0.7
ALGINITE	0.0	0.0	0.0	0.0	0.0	0.0	0.0	0.0	0.0	0.0	0.0	0.0	0.0	0.0
LIPTODETRINITE	2.7	2.9	2.7	2.8	1.1	1.1	1.1	1.2	2.0	2.1	2.1	2.2	2.1	2.2
TOTAL EXINITE	8.2	8.9	7.1	7.4	4.0	4.1	6.2	6.5	5.9	6.1	12.1	12.6	10.0	10.6
FUSINITE	2.3	2.5	2.1	2.2	2.3	2.4	1.6	1.8	1.8	1.9	1.4	1.4	2.2	2.3
SEMIFUSINITE	3.3	3.6	2.0	2.1	2.9	3.0	2.9	3.0	2.4	2.5	1.2	1.2	1.5	1.6
MACRINITE	0.2	0.2	0.1	0.1	0.3	0.3	0.2	0.2	0.0	0.0	0.4	0.4	0.4	0.4
MICRINITE	3.7	4.0	2.7	2.8	3.4	3.5	3.0	3.1	2.1	2.2	1.0	1.0	1.0	1.0
SCLEROTINITE	0.1	0.1	0.0	0.0	0.1	0.1	0.3	0.3	0.1	0.1	0.2	0.2	0.0	0.0
INERTODETRINITE	1.8	2.0	1.5	1.6	2.1	2.2	1.3	1.4	1.6	1.7	2.1	2.2	2.0	2.1
TOTAL INERTINITE	11.4	12.4	8.4	8.8	11.1	11.5	9.3	9.7	8.0	8.3	7.3	7.6	7.1	7.5
PYRITE	2.8		3.0		2.2		2.3		2.5		2.4		1.7	
CARBOARGILLITE	4.0		1.2		0.7		1.7		1.3		1.5		1.1	
CARBOPYRITE	0.9		0.4		0.2		0.3		0.2		0.5		1.9	
CARBANKERITE	0.3		0.0		0.1		0.1		0.0		0.0		0.0	
CARBOSILICATE	0.0		0.0		0.0		0.0		0.0		0.0		0.3	
CARBOPOLYMINERITE	0.0		0.0		0.0		0.0		0.0		0.0		0.5	
OTHER MINERALS	0.0		0.0		0.0		0.0		0.0		0.0		0.0	
TOTAL MINERAL MATTER	8.0		4.6		3.2		4.4		4.0		3.6		5.5	
COUNTS PER PELLET	1000		1000		1000		1000		1000		1000		1000	

NOTE: mmcb-mineral matter containing basis mmfb-mineral matter free basis

7) Class Bleb. An occurrence possessing an irregular morphology which does not fit into any of the other classes.
8) Class Submicron. An occurrence measuring less than one micron in diameter exhibiting an undistinguishable form.

Pyrite Analysis

The pyrite analysis of each coal sample was performed following a modified version of the ASTM Method D 2799 (ref. 6). A Zeiss Universal POL Microscope having an incident light illuminator was used with an attachable graduated mechanical stage. Two ocular lenses, a Kpl W 12.5x with a 100 point graticule disk (Zeiss Integrations Platte II) and a Kpl W 12.5x with a cross-line micrometer scale disk, were used along with two oil immersion objective lenses, 16x and 40x, to provide the recommended magnification. An immersion oil with an index of refraction (n) of 1.518 at 20°C was used in the analysis.

A square, with diagonal length of 25.4mm, was inscribed on each briquette to produce a point count area uniform to all briquettes. The point count area consisted of 18 linear traverses, with each traverse possessing 18 points. The interpoint and interline distance were set to 1mm to conform with ASTM Method D 2799. The total number of counted points was increased a hundredfold with the use of the aforementioned graticule, instead of the 4 point graticule recommended by ASTM methods. Thirty-two thousand and four hundred points were counted on each briquette. Considering two briquettes were examined for each of the seven samples, volume pyrite percentages were based on 64,800 point counts. Results of this analysis are recorded in Table 2.

In addition, a calibrated cross-line micrometer scale disk was used in measuring the maximum and minimum (normal to maximum) dimensions in a plane-section for each individual pyrite occurrence encountered in the point count. These data were used in determining the size frequency distribution of microscopic pyrite within each set of sample values.

Pyrite Content

Pyrite in the -20 mesh (-840 μm) fraction was observed in all samples from a maximum of 1.3 percent in the Middle Kittanning No. 6 (UTOC 263) to a minimum of 0.56 percent in the Meig's Creek No. 9 (UTOC 266) samples. Table 2 displays the relationship of the various classes of pyrite morphologies with the coal maceral groups, mineral matter, and matrix.

Examples of each of the eight morphological classes of pyrite were observed in each of the samples, with the exception of the cell filling class in the Lower Kittanning No. 5 (UTOC 262) and the layer and dendritic classes in the Waynesburg No. 11 (UTOC 267) coal samples.

Blebs, which exhibited an irregular morphology, are by far the most abundant and diverse class of pyrite occurrence observed in this study. Neavel (ref. 20) states that the most evident feature of blebs is the absence of traces of cell walls. Throughout this study, no trace of plant cell walls was found in the bleb occurrences. Over half of the blebs appear to be syngenetic in origin, because they are associated with various maceral types, principally vitrinite group macerals. Over a third of the bleb occurrences, however, are associated with mineral matter. This could imply a syngenetic or epigenetic origin, depending on the distribution of the mineral matter before sample preparation. Blebs are the predominant pyrite morphology observed in all samples, except the Waynesburg No. 11 (UTOC 267).

Euhedral pyrite particles are those occurrences which display distinctive crystal outlines. These particles were observed to exhibit crystal outlines which are square in plane-section (cube), triangular in plane-section (octahedron or tetrahedron), and pentagonal in plane-section (pyrithohedron). The pyrite particles, however, also exhibit many variations in crystal outline. In the majority of occurrences, euhedral pyrite particles are with a group of other isolated pyrite particles. On occasion, these particles coalesced into an aggregate, which lacked any internal structure or crystalline outline; therefore, this aggregation was classified as a bleb. The overwhelming majority of euhedral occurrences are associated with vitrinite group macerals in each sample, thereby indicating a syngenetic origin.

Tabular or lenticular pyrite was assigned to the layer class. Microscopic layers are present in all samples, except the Waynesburg No. 11 (UTOC 267). The majority of these occurrences are associated with vitrinitic material or present as mineral matter. The layers display a wide variety of granularity and crystallinity. In some occurrences, weakly and moderately coalesced euhedral crystals were noted as comprising the layer. These occurrences can be indicative of a syngenetic or epigenetic origin, depending on the mode of occurrence before sample preparation. The Clarion No. 4A (UTOC 264) contains the greatest observed amount of layer occurrences, exceeding one-eighth of the total pyrite found in the sample.

Occurrences which were observed to fill dessication micro-cracks or micro-cleats were classified as fracture filling pyrite. Fracture filling pyrite is present in varying amounts in all samples, comprising relatively large amounts of the total pyrite observed in the Pittsburgh No. 8 (UTOC 265), the Meig's Creek No. 9 (UTOC 266), and the Waynesburg No. 11 (UTOC 267) coal samples. Reyes-Navarro and Davis (ref. 14) state that fracture filling pyrite is probably epigenetic in origin, but dessication crack fillings may be deposited in the very early stages of coal formation. For example, Neavel (ref. 19) states that cleat filling pyrite may also be emplaced in the early stages of coal formation. Neavel notes the draping of linear elements over the top of the pyrite filled cleats and the imbrication of the pyritic infillings as evidence of compaction. This observation substantiates the early origin of the cleats and their pyritic infillings. Fracture filling pyrite, therefore, can originate syngenetically or epigenetically.

Dendritic pyrite includes those occurrences which give the appearance of having been formed by a branching growth. Neavel further explains this morphology as resulting from the fission and subsequent growth of individuals in a bacterial colony which expands in essentially random directions. Dendritic pyrite is present in six of the seven samples, and except for minor instances, is almost always associated with macerals of the vitrinite group. Neavel also notes that dendritic pyrite apparently consumes or replaces incipient vitrinite rather than pushing it aside, implying the bacteria digested the incipient vitrinite. In addition, he recognizes this morphology as containing small micron-sized holes which he presumed to represent the original position of an individual bacterium. Dendritic pyrite is considered to be representative of a syngenetic origin.

Pyrite that was observed to fill intercellular cavities (lumina) or replace cell walls was classified as cell filling pyrite. Cell filling pyrite represents minor amounts of the total pyrite observed in each sample. Only one sample, the Middle Kittanning No. 6 (UTOC 263), displays an amount that closely approximates a tenth of the total pyrite observed within each sample. Only the Clarion No. 4A (UTOC 264) sample displays cell filling pyrite that replaced cell walls, the other samples contain pyrite that exclusively filled the cell lumina. Pyrite filling cell lumina are unique to macerals of the inertinite group (fusinite and semi-fusinite). The rare occurrences observed replacing cell walls are contained in the vitrinite macerals, specifically telinite. It should be noted that the extent of pyrite which fills the cell lumina is seldom limited to one lumen, nor does the pyrite always fill the lumen totally.

Framboidal pyrite was observed as a spherical aggregation of euhedral crystals, subhedral crystals, and anhedral crystals or blebs. These framboids are present in all samples and reach a maximum in the Meig's Creek No. 9 (UTOC 266) of 12 percent of the total observed pyrite. These aggregates were observed dominantly to be distributed within particles of vitrinite. In all samples, the framboids exhibited varying degrees of void space (porosity) within the aggregate. Grady (ref. 25) states that as a result of diagenetic alteration, pyrite growth has caused the crystals and/or blebs in some framboids to coalesce along common boundaries. He further notes that this process does not increase the size of the framboids, but reduces the void volume between the crystals.

Pyrite particles that in plane-section measured less than 1 μm in dimension were classified as submicron pyrite. In essence, the submicron class was a "wastebasket" class for those occurrences that possess an undistinguishable morphology due to the resolution of the microscope system used. These occur-

TABLE 2

Association of pyrite types with coal maceral groups, mineral matter, and matrix

Pyrite Type	Clarion #4A, UTOC 264, Vinton Co., OH			Lower Kittanning #5, UTOC 262, Holmes Co., OH			M. Kittanning #6, UTOC 263, Muskingum Co., OH			Pittsburgh #8, UTOC 260 Harrison Co., OH		
	Vol. % Total Pyrite	Vol. % Pyrite Type	Vol. % Total Coal	Vol. % Total Pyrite	Vol. % Pyrite Type	Vol. % Total Coal	Vol. % Total Pyrite	Vol. % Pyrite Type	Vol. % Total Coal	Vol. % Total Pyrite	Vol. % Pyrite Type	Vol. % Total Coal
FRAMBOID												
Vitrinite	2.8	94.7	0.03	5.7	96.4	0.04	1.9	59.2	0.02	3.0	62.5	0.02
Exinite	--	--	--	--	--	--	--	--	--	--	--	--
Inertinite	--	--	--	0.2	3.6	tr	--	--	--	1.4	29.2	0.01
Mineral Matter	--	--	--	--	--	--	0.9	29.6	0.01	0.2	4.2	tr
Matrix	0.2	5.3	tr	--	--	--	0.4	11.1	tr	0.2	4.2	tr
Total	3.0	100.0	0.03	5.9	100.0	0.04	3.2	99.9	0.03	4.8	100.1	0.03
EUHEDRAL												
Vitrinite	10.0	72.2	0.10	13.2	78.5	0.10	5.0	77.8	0.06	8.5	84.3	0.07
Exinite	--	--	--	0.4	2.5	tr	0.1	1.8	tr	0.4	3.9	tr
Inertinite	0.3	2.2	tr	0.2	1.3	tr	0.1	1.8	tr	0.2	2.0	tr
Mineral Matter	--	--	--	0.4	2.5	tr	0.2	3.7	tr	--	--	--
Matrix	3.5	25.6	0.04	2.5	15.2	0.02	0.9	14.8	0.01	1.0	9.8	0.01
Total	13.8	100.0	0.14	16.7	100.0	0.12	6.3	99.9	0.07	10.1	100.0	0.08
FRACTURE FILLING												
Vitrinite	9.2	92.3	0.09	1.7	47.0	0.01	10.0	84.8	0.13	4.2	65.6	0.03
Exinite	--	--	--	--	--	--	--	--	--	--	--	--
Inertinite	0.8	7.7	tr	1.9	52.9	0.01	1.8	15.2	0.02	2.2	34.4	0.02
Mineral Matter	--	--	--	--	--	--	--	--	--	--	--	--
Matrix	--	--	--	--	--	--	--	--	--	--	--	--
Total	10.0	100.0	0.09	3.6	99.9	0.02	11.8	100.0	0.15	6.4	100.0	0.05
LAYER												
Vitrinite	--	--	--	1.5	38.9	0.01	2.6	64.7	0.03	6.8	79.1	0.05
Exinite	--	--	--	1.9	50.0	0.01	0.4	8.8	tr	1.8	20.9	0.01
Inertinite	--	--	--	--	--	--	--	--	--	--	--	--
Mineral Matter	13.9	100.0	0.14	--	--	--	0.6	14.7	0.01	--	--	--
Matrix	--	--	--	0.4	11.1	tr	0.5	11.8	0.01	--	--	--
Total	13.9	100.0	0.14	3.8	100.0	0.02	4.1	100.0	0.05	8.6	100.0	0.06
DENDRITIC												
Vitrinite	2.6	100.0	0.03	11.7	98.2	0.08	19.1	100.0	0.25	17.9	98.9	0.14
Exinite	--	--	--	--	--	--	--	--	--	0.2	1.1	tr
Inertinite	--	--	--	--	--	--	--	--	--	--	--	--
Mineral Matter	--	--	--	--	--	--	--	--	--	--	--	--
Matrix	--	--	--	0.2	1.8	tr	--	--	--	--	--	--
Total	2.6	100.0	0.03	11.9	100.0	0.08	19.1	100.0	0.25	18.1	100.0	0.14
CELL FILLING												
Vitrinite	1.2	33.3	0.01	--	--	--	--	--	--	--	--	--
Exinite	--	--	--	--	--	--	--	--	--	--	--	--
Inertinite	2.5	66.7	0.02	0.6	100.0	tr	9.1	100.0	0.12	3.6	100.0	0.03
Mineral Matter	--	--	--	--	--	--	--	--	--	--	--	--
Matrix	--	--	--	--	--	--	--	--	--	--	--	--
Total	3.7	100.0	0.03	0.6	100.0	tr	9.1	100.0	0.12	3.6	100.0	0.03
SUBMICRON												
Vitrinite	17.7	96.6	0.18	19.1	88.2	0.14	8.9	96.2	0.12	5.8	85.3	0.04
Exinite	0.6	3.4	0.01	1.1	4.9	0.01	--	--	--	--	--	--
Inertinite	--	--	--	1.5	6.9	0.01	0.4	3.8	tr	1.0	14.7	0.01
Mineral Matter	--	--	--	--	--	--	--	--	--	--	--	--
Matrix	--	--	--	--	--	--	--	--	--	--	--	--
Total	18.3	100.0	0.19	21.7	100.0	0.16	9.3	100.0	0.12	6.8	100.0	0.05
BLEB												
Vitrinite	12.5	36.2	0.12	18.0	50.6	0.13	14.0	37.7	0.18	20.9	50.0	0.16
Exinite	--	--	--	1.7	4.8	0.01	0.1	0.3	tr	0.4	1.0	tr
Inertinite	--	--	--	--	--	--	0.4	1.0	tr	0.8	1.9	0.01
Mineral Matter	19.4	56.2	0.19	12.3	34.5	0.09	16.8	45.4	0.22	16.7	40.0	0.13
Matrix	2.6	7.6	0.03	3.6	10.1	0.03	5.8	15.6	0.08	3.0	7.1	0.02
Total	34.5	100.0	0.34	35.6	100.0	0.26	37.1	100.0	0.48	41.8	100.0	0.32

tr = trace amount matrix = epoxy binder

TABLE 2 (Continued)

Association of pyrite types with coal maceral groups, mineral matter, and matrix

Pyrite Type	Pittsburgh #8, UTOC 265 Harrison Co., OH			Meig's Creek #9, UTOC 266, Belmont Co., OH			Waynesburg #11, UTOC 267, Belmont Co., OH		
	Vol. % Total Pyrite	Vol. % Pyrite Type	Vol. % Total Coal	Vol. % Total Pyrite	Vol. % Pyrite Type	Vol. % Total Coal	Vol. % Total Pyrite	Vol. % Pyrite Type	Vol. % Total Coal
FRAMBOID									
Vitrinite	4.2	88.5	0.04	11.1	90.9	0.06	6.1	96.7	0.04
Exinite	--	--	--	--	--	--	--	--	--
Inertinite	--	--	--	0.6	4.5	tr	--	--	--
Mineral Matter	0.2	3.8	tr	--	--	--	--	--	--
Matrix	0.4	7.7	tr	0.6	4.5	tr	0.2	3.3	tr
Total	4.8	100.0	0.04	12.3	99.9	0.06	6.3	100.0	0.04
EUHEDRAL									
Vitrinite	2.2	54.5	0.02	4.6	71.4	0.02	12.5	67.0	0.09
Exinite	0.6	13.6	tr	0.3	4.8	tr	0.4	2.3	tr
Inertinite	--	--	--	--	--	--	0.4	2.3	tr
Mineral Matter	0.7	18.2	0.01	--	--	--	--	--	--
Matrix	0.6	13.6	tr	1.4	23.8	0.01	5.3	28.4	0.04
Total	4.1	99.9	0.03	6.3	100.0	0.03	18.6	100.0	0.13
FRACTURE FILLING									
Vitrinite	25.6	87.4	0.21	14.1	75.0	0.08	28.1	89.9	0.20
Exinite	--	--	--	--	--	--	--	--	--
Inertinite	3.8	12.6	0.03	4.7	25.0	0.03	3.2	10.1	0.02
Mineral Matter	--	--	--	--	--	--	--	--	--
Matrix	--	--	--	--	--	--	--	--	--
Total	29.4	100.0	0.24	18.8	100.0	0.11	31.3	100.0	0.22
LAYER									
Vitrinite	0.7	12.1	0.01	0.8	60.0	tr	--	--	--
Exinite	0.6	9.1	tr	0.6	40.0	tr	--	--	--
Inertinite	--	--	--	--	--	--	--	--	--
Mineral Matter	2.9	48.5	0.02	--	--	--	--	--	--
Matrix	1.8	30.3	0.02	--	--	--	--	--	--
Total	6.0	100.0	0.05	1.4	100.0	tr	--	--	--
DENDRITIC									
Vitrinite	7.0	100.0	0.06	3.6	100.0	0.02	--	--	--
Exinite	--	--	--	--	--	--	--	--	--
Inertinite	--	--	--	--	--	--	--	--	--
Mineral Matter	--	--	--	--	--	--	--	--	--
Matrix	--	--	--	--	--	--	--	--	--
Total	7.0	100.0	0.06	3.6	100.0	0.02	--	--	--
CELL FILLING									
Vitrinite	--	--	--	--	--	--	--	--	--
Exinite	--	--	--	--	--	--	--	--	--
Inertinite	2.2	100.0	0.02	1.1	100.0	0.02	7.4	100.0	0.05
Mineral Matter	--	--	--	--	--	--	--	--	--
Matrix	--	--	--	--	--	--	--	--	--
Total	2.2	100.0	0.02	1.1	100.0	0.02	7.4	100.0	0.05
SUBMICRON									
Vitrinite	5.9	97.0	0.05	5.0	90.0	0.04	6.6	79.5	0.05
Exinite	0.2	3.0	tr	0.6	10.0	tr	1.5	17.9	0.01
Inertinite	--	--	--	--	--	--	0.2	2.6	tr
Mineral Matter	--	--	--	--	--	--	--	--	--
Matrix	--	--	--	--	--	--	--	--	--
Total	6.1	100.0	0.05	5.6	100.0	0.04	8.3	100.0	0.06
BLEB									
Vitrinite	16.4	40.3	0.14	36.8	71.5	0.20	15.0	53.4	0.11
Exinite	--	--	--	1.7	3.2	0.01	0.6	2.2	tr
Inertinite	--	--	--	2.5	4.8	0.01	0.8	3.0	0.01
Mineral Matter	17.5	43.0	0.15	7.2	14.0	0.04	7.4	26.3	0.05
Matrix	6.8	16.7	0.06	3.3	6.4	0.02	4.2	15.0	0.03
Total	40.7	100.0	0.35	51.5	99.9	0.28	28.0	99.9	0.20

tr = trace amount   matrix = epoxy binder

rences are primarily associated with the vitrinite material and reached a maximum of 18 percent of the pyrite observed in the Clarion No.4A (UTOC 264) coal.

Size Distribution of Pyrite

It has been noted that plotted size frequency distributions are often highly skewed when the variables are chosen as dimensions in millimeters (ref. 26); therefore, ordinary moment analysis is not applicable. Krumbein (ref. 26) and others, however, have shown that these data are often lognormal and for this reason his method was employed in this study.

It should be noted that the size frequency distribution analysis was based on pyrite particles ranging from 1 to 840 μm. ASTM Method D 2797 served to prescribe the upper limit in requiring 100 percent passage of each sample through a No. 20 (840 μm) U.S. Standard Sieve. The lower limit was a function of the resolution of the microscope and the accompanying lenses. The submicron class was specified for those occurrences less than 1 μm in plane-section dimension, whose form and actual dimensions in plane-section were undiscernable. Size frequency and normal distribution curves pertain to that percentage of the total pyrite observed _not_ occurring in the submicron range. Because the submicron pyrite could not be differentiated into various size fractions, it is suggested that the moment analysis pertain to the discernable size fractions, 1 to 840 μm. The size frequency and theoretical normal distributions, therefore, are open-ended at the 1 μm size, because the actual distribution of the submicron particles is undeterminable. Furthermore, it is difficult to ascertain if these submicron sized particles are artifacts derived from excessive comminution of and subsequent contamination by the +20 mesh (+840 μm) pyrite particles associated with the original sample.

The recorded dimensions (d), in millimeters (mm), were transformed into a new logarithmic variable, phi (ϕ), by means of the following equation.

$$\phi = \frac{\log dmm}{\log 2}$$

The size frequency distributions, in terms of phi, are plotted in Figures 2 and 3 and are examples showing the maximum and minimum (normal to maximum) dimensions.

A normal phi curve, illustrating a theoretical distribution, was constructed for each set of sample values. This hypothetical curve specifies the limits of a normal, or Gaussian, distribution within the sample. The probability equation describing the normal phi curve is as follows:

$$y = \frac{1}{\sigma_\phi \sqrt{2\pi}} e - (\phi - \mu_\phi)^2 / 2\sigma_\phi^{\ 2}$$

where μ_ϕ = arithmetic mean in phi units, σ_ϕ = standard deviation in phi units, ϕ = maximum or minimum dimension of a pyrite particle in phi units. The normal phi curve for each set of sample values were plotted as in Figures 2 and 3 which represent the upper and lower limits of the corresponding size frequency distribution of all samples.

Employing a variation of the Kolomorogov-Smirnov (K-S) test for normality (ref. 27), the sample distributions were tested against their corresponding normal distributions to ascertain whether a significant difference exists between the distributions. This variation of the K-S test allows for testing the distribution of a population when the mean and variance are not specified, but must be estimated from the sample. Frequently, the chi-square test is used to test for the difference between the sample and normal distributions, but this variation of the K-S test was used because it is a more powerful and efficient procedure. Procedural steps of the K-S test follow:

1) Fit a normal distribution to the sample distribution.
2) Plot the two distributions on the same graph as step-wise cummulative curves. Express the cummulative frequencies as percent.
3) Find the maximum vertical difference between the two curves, D_{max}. Compute D, $D = D_{max}/100$. If this value exceeds the critical value of D at a specified α, one rejects the null hypothesis of "no difference between distributions."

where: α = level of significance of the test or risk of a Type I error; the rejection of the null hypothesis when it is true (ref. 28, p. 16).
Crow, _et al._, (ref. 29) note that this level of significance is arbitrary, but should be chosen when the experiment is designed. For this study the value of 0.05 was chosen for the level of significance. This implies that there is a 5 percent risk of error or 5 percent probability that the null hypothesis will be rejected even if the null hypothesis holds true.

The results of the K-S test for the maximum and minimum dimensions of each sample are recorded in Table 3. The calculated value of D did not exceed the critical value prescribed by the chosen level of significance and sample size in any of the seven samples. Therefore, at a significance level of 0.05, the null hypothesis of "no difference between distributions" can be accepted as true for both the maximum and minimum dimensions of all the sample values tested. Table 4 summarizes the volume percent distribution of microscopic pyrite particles at class intervals (-840 μm to 1 μm) as integrated from the theoretical normal distribution curves representing the observed maximum and minimum particle dimensions in plane-section.

TABLE 3

Kolomogorov-Smirnov test statistics for distributions of microscopic pyrite particles (-840 to 1 μm) representing maximum and minimum plane-section dimensions in selected Ohio coals

Sample	Dimension: Maximum Dmax D = 100	Dimension: Minimum Dmax D = 100	Critical Value of D*
Clarion No. 4A (UTOC 264)	0.084	0.086	0.258
L. Kittanning No. 5 (UTOC 262)	0.079	0.063	0.258
M. Kittanning No. 6 (UTOC 263)	0.077	0.072	0.258
Pittsburgh No. 8 (UTOC 260)	0.057	0.066	0.258
Pittsburgh No. 8 (UTOC 265)	0.076	0.047	0.258
Meig's Creek No. 9 (UTOC 266)	0.056	0.036	0.258
Waynesburg No. 11 (UTOC 267)	0.130	0.064	0.258

*Sample Size (N) = 10
Level of Significance (α) = 0.05

Determination of the Forms of Sulfur in Coal

The forms of sulfur (sulfatic, pyritic, and organic) were determined for each of the as received, whole coal samples in accordance with ASTM Methods D2492, entitled: Standard Method of Test for Forms of Sulfur in Coal (ref. 6). This method first determines the sulfatic sulfur and then the pyritic sulfur. The organic sulfur is determined by subtracting the difference from the total sulfur as determined in accordance with ASTM D271, entitled: Standard Method of Laboratory Sampling and Analysis of Coal and Coke. The results of these determinations are listed in Table 5.

COAL WASHABILITY

The major source of sulfur oxide air pollution is coal-fired electric utility plants. Blackmore (ref. 29) states that steam coal for electric utilities is the largest single market for the coal industry, comprising 78 percent of our domestic coal production. In addition, he notes that the latest projections of coal consumption from the Federal government are in the area of 1.2 billion annual tons by 1990. Therefore, it becomes imperative that sulfur oxide emissions from coal combustion must be controlled.

Of the available methods for controlling sulfur oxide emissions, the physical removal (coal cleaning) of pyritic sulfur (pyrite) prior to combustion is the most technologically developed and potentially the lowest in cost (ref. 30). Blackmore (ref. 31) notes that coal cleaning has the potential for giving the coal-burning facility and its customers an economic benefit, and at the same time makes a substantial contribution to sulfur oxide reduction. Figure 4 is

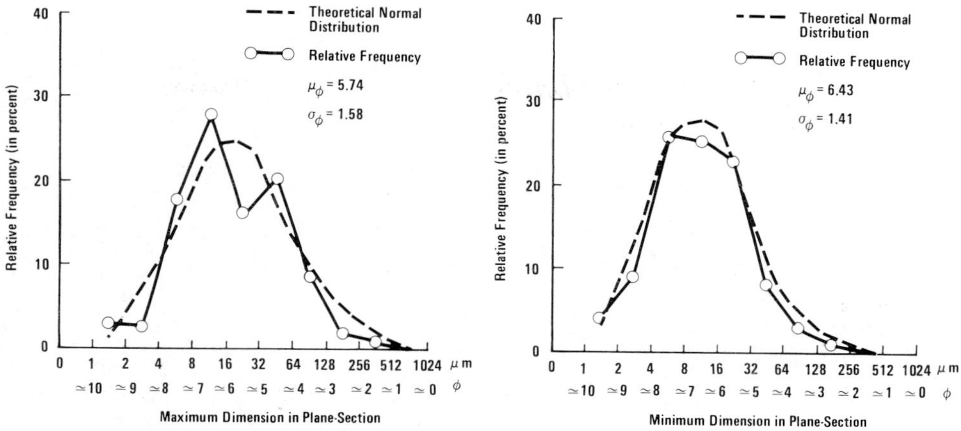

Fig. 2. Plot of the relative frequency and theoretical normal distribution of the observed maximum and minimum dimension in plane-section of pyrite particles in the Pittsburgh No. 8 seam, UTOC 265, Harrison County, Ohio.

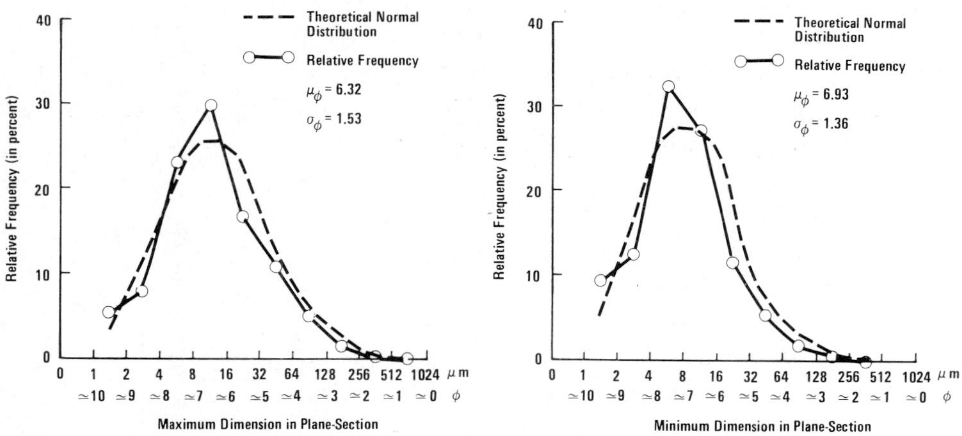

Fig. 3. Plot of the relative frequency and theoretical normal distribution of the observed maximum and minimum dimension in plane-section of pyrite particles in the Lower Kittanning No. 5 seam, UTOC 262, Holmes County, Ohio.

TABLE 4

Table of volume percent distribution of microscopic pyrite particles at class intervals (-840 to 1μm) integrated from theoretical normal distribution curves representing observed maximum and minimum particle dimension in plane-section

Class Intervals (microns)	Clarion No. 4A (UTOC 264) Volume % Distribution		L. Kittanning No. 5 (UTOC 262) Volume % Distribution		M. Kittanning No. 6 (UTOC 263) Volume % Distribution		Pittsburgh No. 8 (UTOC 260) Volume % Distribution		Pittsburgh No. 8 (UTOC 265) Volume % Distribution		Meig's Creek No. 9 (UTOC 266) Volume % Distribution		Waynesburg No. 11 (UTOC 267) Volume % Distribution	
	MXD*	MID**	MXD	MID	MXD	MID	MXD	MID	MXD	MID	MXD	MID	MXD	MID
840-256	1.1	0.0	1.0	0.3	1.5	0.9	1.2	0.0	2.5	0.7	0.9	0.0	0.8	0.0
256-128	2.4	0.4	2.6	0.9	3.8	1.2	3.8	0.6	4.6	1.9	2.9	0.7	2.7	0.7
128-64	5.4	2.6	6.2	2.8	8.6	4.6	7.9	3.7	10.2	4.4	7.4	3.9	7.1	3.2
64-32	11.1	6.9	10.6	6.9	17.3	10.3	15.4	10.1	16.8	10.4	14.1	9.6	14.1	8.0
32-16	20.4	14.4	21.0	17.4	22.4	20.4	22.4	20.5	23.4	20.4	22.2	22.2	22.1	17.5
16-8	24.8	21.2	24.5	24.9	21.8	26.0	22.4	27.8	21.6	25.7	24.8	27.3	24.1	26.1
8-4	20.2	19.9	20.3	23.8	15.3	21.7	16.5	22.2	12.5	21.9	17.0	21.9	17.9	25.2
4-2	10.6	11.8	10.9	16.0	7.1	11.3	8.2	11.7	6.8	11.5	8.2	11.5	9.7	13.8
2-1	4.1	4.5	2.9	6.9	2.0	3.5	2.2	3.3	1.5	3.1	2.4	2.9	1.5	5.5

* MXD = Volume percent distribution integrated from theoretical normal distribution curves representing observed maximum dimension in plane-section.

** MID = Volume percent distribution integrated from theoretical normal distribution curves representing observed minimum dimension in plane-section.

NOTE: Volume percent distribution of submicron particles are not included in the above table.

TABLE 5

Sulfur and pyrite content of selected Ohio coals

SAMPLE	FROM WHOLE SEAM CHANNEL SAMPLE, AS RECEIVED							MICROSCOPIC PYRITE RETAINED @ -840 to -1 μm	
	Total Sulfur %	Sulfatic Sulfur %	Organic Sulfur %	Pyritic Sulfur %	WT. Pyrite Whole Coal %	True Density Whole	Apparent Density Whole	Vol. %*	WT %**
Clarion No. 4A (UTOC 264)	3.48	0.37	1.37	1.74	3.25	1.52	1.33	1.00	3.26
Lower Kittanning No. 5 (UTOC 262)	4.80	0.11	0.77	3.92	7.32	1.40	1.18	0.73	2.58
Middle Kittanning No. 6 (UTOC 263)	4.42	0.33	1.79	2.30	4.30	1.37	1.23	1.30	4.70
Pittsburg No. 8 (UTOC 260)	3.75	0.32	1.73	1.70	3.18	1.43	1.29	0.78	2.70
Pittsburgh No. 8 (UTOC 265)	3.36	0.29	1.55	1.53	2.86	1.42	1.20	0.84	2.93
Meig's Creek No. 9 (UTOC 266)	2.51	0.12	1.11	1.28	2.39	1.42	1.28	0.56	1.95
Waynesburg No. 11 (UTOC 267)	2.49	0.22	0.96	1.31	2.45	1.49	1.32	0.73	2.43

* Derived from the integration of the theoretical normal distribution.
**Weight percent retained between class intervals -840 on -1 μm based upon use of true density, whole coal and a pyrite density of 4.95.

representative of a type of advanced conventional coal washing system.

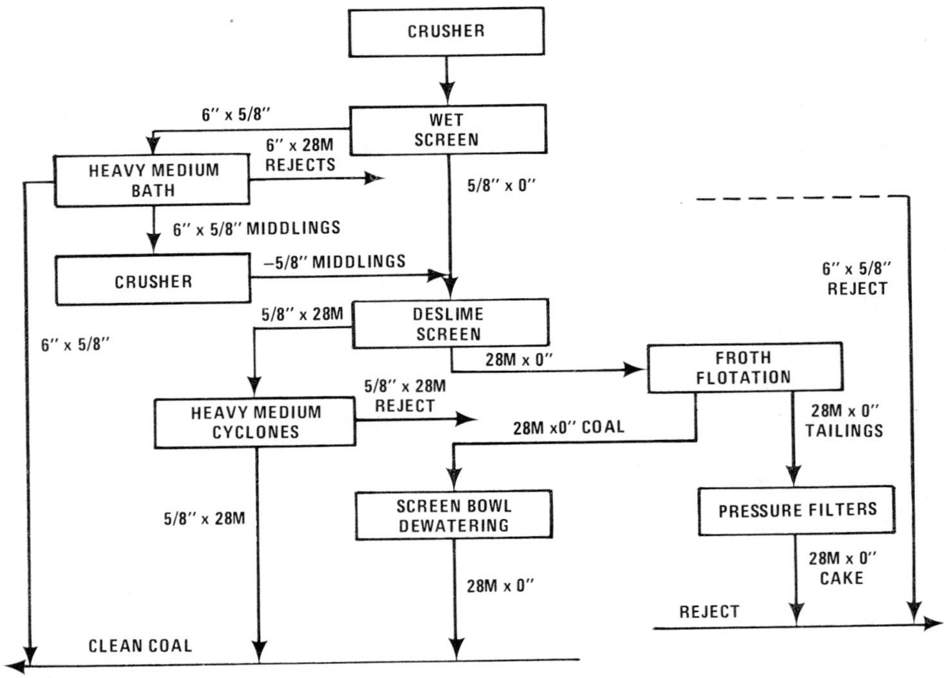

Fig. 4. Schematic of an advanced conventional coal washing system.

The values in Table 6 show the capital and operating costs for washing three different size fractions of coal. Although these figures can vary according to a number of factors, such as present-day costs, these figures clearly show the capital cost increase in washing the finer size fractions of coal. In addition, approximate present-day cost comparison of No. 2 fuel oil, No. 6 fuel oil, crude oil, and unwashed, delivered spot market coal is shown in Table 7. In addition to these figures Table 7 also shows the estimated costs of washed, delivered spot market coal for three different size fractions. (These estimated costs for washed coal were derived by converting the operating (washing) costs per ton fuel, given in Table 6, to operating costs per MBTU (million British Thermal Units) coal. These conversions were accomplished by employing the standard coal contract figure of 12,500 BTU/lb. coal. The resultant costs were then added to the cost for unwashed spot market coal. The cost comparison illustrates the present economic viability of coal utilization versus petroleum utilization. Note that the cost of washed coal, for any of the three size fractions, does not even exceed the cost at 1980 constant dollars for the unrefined crude oil.

These values clearly show the cost benefits associated with coal usage, even with deeply cleaned coal.

In previous years, coal-fired electric utility operations have reaped considerable economic pay-backs in maintaining as coarse a size fraction in coal as possible. This, in turn, retarded the development of the technological methods in beneficiating the finer size fractions of coal. Most of this finer size fraction, usually 100 mesh - 0 (149 μm to 0 μm), was simply thrown away. Unfortunately, this has led to the current high state of inefficiency in the froth flotation method, which is the primary method for deep cleaning the finer fractions of coal. Blackmore (ref. 29) notes that the net result is that the coal operator is confronted with decisions on process equipment and associated conditioning and flotation oils which must, in themselves, satisfy EPA toxic restrictions.

TABLE 6

Coal preparation plant capital and operating cost in coarse, average, and fine size fractions of coal

Size Fraction	Capital Cost Per Ton/Hr. Capacity	Operating Cost Per Ton Product*
Coarse (4" x 1/4")	$ 16,000	$ 1.70
Average (1/4" x 28 mesh)	22,500	3.75
Fine (28 mesh - 0)	30,000	6.10

*Includes interest and depreciation Source: (ref. 29)

TABLE 7

Average cost of selected energy sources

Energy Source		Cost Per Million BTU
No. 2 Fuel Oil		$ 7.50
No. 6 Fuel Oil		4.73
Crude Oil		3.50
Unwashed Spot Market Coal		1.38*
Washed Spot Market Coal -	Coarse (4" x 1/4")	1.45**
	Average (1/4" x 28 mesh)	1.53**
	Fine (28 mesh - 0)	1.62**

* Includes transportation **Estimate including operating plant cost

Washability tests must be conducted to determine the proper method and equipment needed in deep-cleaning. Washability tests involve stage crushing to release the impurities and specific gravity fractionation to show the quality and quantity of the cleaned product (ref. 30). Even though these tests prove extremely valuable in predicting the washability of a coal, they remain as costly commitments for producers and consumers. The petrographic technique employed in this study was anticipated as aiding in the rapid determination of the deep-cleaning of coal.

Importance of this study to the deep-cleaning of Ohio coals

The results of this analytical petrographic technique confirm that conventional washing, size degradation in conjunction with specific gravity fractionation, will not provide satisfactory yields in the finer fractions of coal. This claim is reinforced by the substantial occurrence of pyrite in the size fractions less than the conventional degradation size, 100 mesh (149 μm). In addition, the overwhelming majority of these pyrite occurrences are dispersed within the coal macerals. The underlying premise of coal washability is that the liberation of pyrite is a function of relative specific gravities. If the pyrite is intimately associated with the coal macerals, specific gravity fractionation is not possible. Therefore, additional size degradation and specific gravity fractionation (deep-cleaning) are mandatory in attaining the desired quality and quantity of coal.

In applying the results of this study to the deep cleaning of coal, volume percent distributions of microscopic pyrite were examined at the class intervals -840 to 1 μm (see Table 4). These data were amended to include the particles in the submicron sizes and to calculate the reduction level in percent for pyrite at these class intervals. For brevity only the percent volume and pyrite reduction level for the maximum plane dimension are included in this discussion (see Table 8). Figure 5 illustrates a band which graphically defines variation of the distribution of microscopic pyrite in seven Ohio coals and the pyrite reduction level which can be expected with increased comminution. Both Table 8 and Figure 5 reveal a marked increase in values within the -32 to -4 μm size. In this range, the maximum volume percent of observed microscopic pyrite is contained in the -16 to 8 μm class interval. Below the 4 μm threshold, volume percent values show a notable decrease in the distribution of the pyrite particles. (Only two samples, the Clarion No. 4A (UTOC 264) and the Lower Kittanning No. 5 (UTOC 262), display substantial values in percent distribution below the 4 micron threshold. Each of these samples contains approximately 20 percent by volume submicron pyrite.) This concentration of microscopic pyrite particles exists for each observed dimension in all samples. The importance of this relationship is evident in developing a partition for determining the optimum deep-cleaning of a coal bed.

TABLE 8

Volume* and reduction level percents of microscopic pyrite particles (maximum plane-section dimension) retained at class intervals (-840 to -1 μm) for selected Ohio coals

Sample	Particles Retained at -840 to -1 μm		Particles Retained at -256 to -1 μm		Particles Retained at -128 to -1 μm		Particles Retained at -64 to -1 μm		Particles Retained at -32 to -1 μm		Particles Retained at -16 to -1 μm		Particles Retained at -8 to -1 μm		Particles Retained at -4 to -1 μm		Particles Retained at -2 to -1 μm		Particles Retained at -1 μm	
	% Volume	% Weight	% Volume	% Reduc.	% Volume	% Reduc.	% Volume	% Reduc.	% Volume	% Reduc.	% Volume	% Reduc.	% Volume	% Reduc.	% Volume	% Reduc.	% Volume	% Reduc.	% Volume	% Weight Remaining
Clarion No. 4A (UTOC 264)	1.0	3.26	0.99	1.00	0.97	3.00	0.93	7.00	0.87	13.00	0.67	33.00	0.47	53.00	0.30	70.00	0.21	79.00	0.18	0.59
Lower Kittanning No. 5 (UTOC 262)	0.73	2.58	0.72	1.37	0.71	2.74	0.67	8.22	0.61	16.44	0.49	32.88	0.35	52.05	0.24	67.12	0.17	76.71	0.16	0.57
Middle Kittanning No. 6 (UTOC 263)	1.30	4.70	1.28	1.54	1.25	3.85	1.14	12.31	0.93	28.46	0.67	48.46	0.41	68.46	0.23	82.31	0.15	88.46	0.12	0.43
Pittsburgh No. 8 (UTOC 260)	0.78	2.70	0.77	1.28	0.74	5.13	0.67	14.10	0.57	26.92	0.41	47.44	0.25	67.95	0.13	83.33	0.07	91.10	0.05	0.17
Pittsburgh No. 8 (UTOC 265)	0.84	2.93	0.82	2.38	0.78	7.14	0.70	16.67	0.57	32.14	0.39	53.57	0.22	73.81	0.12	85.71	0.06	92.86	0.05	0.17
Meig's Creek No. 9 (UTOC 266)	0.56	1.95	0.55	1.79	0.54	3.57	0.50	10.71	0.44	21.43	0.31	44.64	0.18	67.86	0.09	83.93	0.04	92.86	0.03	0.10
Waynesburg No. 11 (UTOC 267)	0.73	2.43	0.73	0.0	0.72	1.37	0.70	4.11	0.65	10.96	0.53	27.40	0.36	50.68	0.19	73.97	0.10	86.30	0.06	0.20

* Derived from the integration of the theoretical normal distribution curves representing the maximum dimension in plane-section.

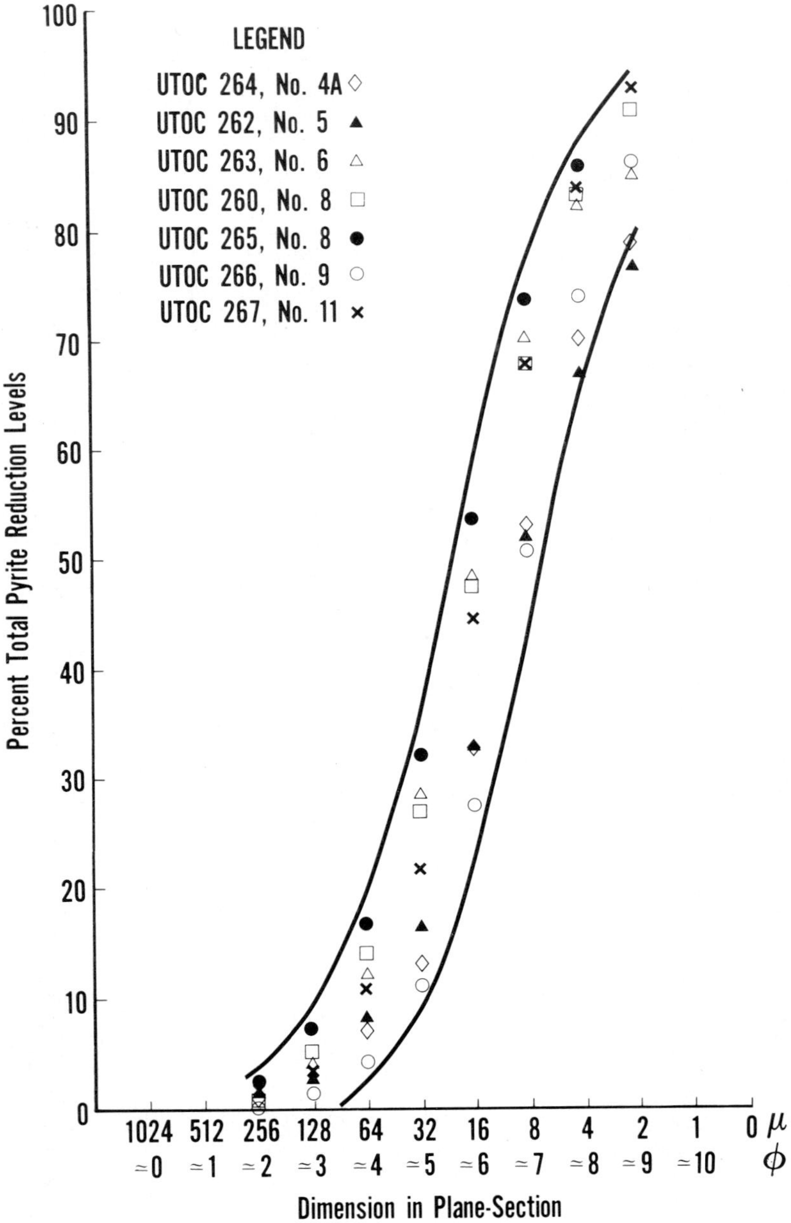

Fig. 5. Percent reduction levels of pyrite vs. size distribution (observed maximum dimension in plane-section) of microscopic pyrite particles in selected Ohio coals.

Of note, however, is the homogeneity of the theoretical normal distribution curves from which the volume percent distribution values were integrated. These representative curves, illustrated in Figures 2 and 3 of this study, display theoretical distributions of microscopic pyrite particles which nearly mirror one another. These results are incongruent with the findings of a similar study by Reyes-Navarro and Davis (ref. 14). Their results show theoretical normal distributions for sample populations of microscopic pyrite which are notably different from one another. The homogeneity of distributions noted in this study, therefore, are unexpected. These findings possibly could be fortuitous—a chance happening, or the homogeneity could be a result of bias introduced by sample processing (splitting and grinding) or simply the result of a different geologic history of the samples.

Initially, bulk vertical channel samples were obtained and crushed through a Sturtevant Automatic Coal Crusher and Sampler (SACCS). Representative samples for laboratory study were taken from an automatic splitting device. In the laboratory, the samples were then prepared and examined according to ASTM standards methods, which are similar to the methods employed in the study by Reyes-Navarro and Davis (ref. 14). Consequently, the methods of sample processing utilized offer no indications of bias which could contradict the validity of the homogeneous nature of the results obtained in this study.

The similarity in environmental conditions required for the uniform distribution of microscopic pyrite over approximately 10-20 million years seems improbable. This, however, does not prove bias, because it is at least possible that conditions of deposition responsible for these coals favored a uniform distribution of microscopic pyrite.

Table 8 shows that beneficiating the coal samples to a proposed optimum size partition of 4 μm would yield coals with less than 0.4 percent by volume and 2 percent by weight pyrite content. In addition, Table 8 and Figure 5 permit the ranking of the coals according to yield at the various size fractions. At the proposed optimum partition of 4 μm, the ranking from most attractive to least attractive yield is as follows: 1) Meig's Creek No. 9 (UTOC 266); 2) Pittsburgh No. 8 (UTOC 260 and 265); 3) Waynesburg No. 11 (UTOC 267); 4) Lower Kittanning No. 5 (UTOC 262) and Middle Kittanning No. 6 (UTOC 263); and 5) Clarion No. 4A (UTOC 264). (These rankings are based on either volume or weight percents.)

If the results of this study are valid and representative of Ohio coals, they have great economic implications. The homogeneous distribution of microscopic pyrite observed in the coal seams, therefore, permit the development of technology geared to more efficient deep-cleaning of Ohio coals at the -8 to 4 μm size range. This homogeneity of the microscopic pyrite distribution minimizes the capital investment required for preparation plant equipment. Furthermore, this pyrite distribution decreases daily operational costs. This is a striking economic incentive for producers and legislators concerned with the development of high sulfur coals as a primary source of energy.

REFERENCES

1 C. E. Bagge, in L. C. Ruedisili and M. W. Firebaugh (Eds.), Perspectives on Energy, Oxford University Press, New York, 1977, pp. 176-180.
2 M. K. Hubbard, Coal as an Energy Resource, Conflict and Concensus, National Academy of Sciences, Washington, D.C., 1977, 326 pp.
3 A. R. Tarrer, J. A. Guin, W. S. Pitts, J. P. Henley, J. W. Prather and G. A. Styles, in R. T. Ellington (Ed.), Liquid Fuels from Coal, Academic Press, New York, 1977, pp. 45-61.
4 E. C. Moroni and R. H. Fischer, Disposable Catalysts for Coal Liquefaction, U.S. Dept. of Energy, Office of Coal Processing, Div. of Coal Liquefaction, Presented at 179th National Mtg., American Chemical Society, Houston, March 24, 1980, 31 pp.
5 Annual Book of ASTM Standards, Part 26, Gaseous Fuels, Coal and Cokes, Atmospheric Analysis, Amer. Soc. for Testing and Materials, Philadelphia, 1982, 480 pp.
6 A. J. Gluskoter, Energy Sources, 3 (1977) 125-131.
7 B. E. Balme, J. Inst. Fuel, 29 (1976) 21-22.
8 C. E. Zobell, in I. R. Breger (Ed.), Organic Geochemistry, Pergamon Press, New York, 1963, pp. 543-578.
9 R. A. Berner, Am. J. Sci., 268 (1970) 1-23.
10 R. A. Berner, J. Geology, 72 (1964) 293-306.
11 R. A. Berner, in G. H. Lauff (Ed.), Estuaries, Am. Assoc. Adv. Sci. Pub. 83, 1967, pp. 268-272.
12 D. T. Rickard, Stockholm Contr. Geology 20 (1969) 67-95.
13 R. A. Berner, Principles of Chemical Sedimentology, McGraw Hill, New York, 1971, 240 pp.
14 J. Reyes-Navarro and A. Davis, Pyrite in Coal: Its Form and Distributions as Related to the Environments of Coal Deposition in Three Selected Coals from Western Pennsylvania, Spl. Report, No. SR-110 Coal Research Section, Pennsylvania State University, State College, 1976, 141 pp.
15 M. Th. Mackowsky, in D. Murchinson and T. S. Westoll (Eds.), Coal and Coal-Bearing Strata, American Elsevier, New York, 1968, pp. 309-321.
16 D. White and R. Thiessen, The Origin of Coal, U.S.B.M. Bull. No. 38, 1913, 390 pp.
17 R. Thiessen, AIME Trans. Soc. Min. Eng. 63 (1920) 913-931.
18 G. Theissen in H. H. Lowry (Ed.), Chemistry of Coal Utilization, Wiley & Sons, New York, 1945, 425 pp.
19 R. C. Neavel, Sulfur in Coal: Its Distribution in the Seam and in Mine Products, Unpubl. Ph.D. Thesis, Pennsylvania State University, 1966, 332 pp.
20 R. R. Dutcher, Field Description of Coal, ASTM Spl. Tech. Pub. 661, 1976, 71 pp.
21 E. Stach, G. H. Taylor, K. Th. Mackowsky, D. Chandra, M. Teichmüller and R. Teichmüller, Stach's Textbook of Coal Petrology, Gebrüder Bortraeger, Berlin and Stuttgart, 1975, 428 pp.
22 D. Chandra, Fuel, 42 (1963) 457-465.
23 International Handbook of Coal Petrology, International Committee for Coal Petrology (ICCP) 2nd ed., Centre National Recherche Scientifique, Paris, 1963, with Suppl. 1971 and 1975.
24 W. C. Grady AIME, Trans. Soc. Min. Eng. 262 (1977) 268-274.
25 W. C. Grady, A. Descriptive Approach to Microscopic Varieties of Pyrite in West Virginia Coals, 51st Ann. Mtg. West Virginia Acad. Sci., Report No. 126, Bethany College, W. Va., 1976, 13 pp.
26 W. C. Krumbein, J. Sed. Petrology 8 (1938) 84-90.
27 H. W. Lilliefors, Amer. Statistical Assoc. J. 62 (1967) 399-402.

28 E. L. Crow, F. A. Davis and M. W. Maxfield, Statistics Manual, Dover, New York, 1960, 288 pp.
29 G. Blackmore, A. Discussion on Coal Economics as it Relates to Electric Utilities and Energy Independence, Consolidation Coal Co., Am. Management Mtg., August 19-20, 1980, 62 pp.
30 J. A. Cavallaro, M. T. Johnston and A. W. Deurbrouck, U.S.B.M. R.I. No. 8118, 1976, 323 pp.
31 G. Blackmore, Coal Cleaning, Steam Raising and Environmental Control, An Economic Interpretation of Systems, Integrated Environmental Control for Coal-Fired Power Plants Symposium, Denver, February 22-25, 1981, 15 pp.

SULFUR CHARACTERIZATION OF EASTERN CANADIAN COAL FOR OPTIMIZATION OF BENEFICIATION PROCESSES

C.W. Angle , W.M. Leung*, R.J. Mikula , and H.A. Hamza
Coal Research Laboratory, CANMET, EMR, P.O. Bag 1280, Devon, Alberta, Canada

ABSTRACT

Several coal characterization techniques were used to investigate the presence of sulfur in two eastern Canadian high sulfur coals before and after cleaning. In addition to the normal ash, sulfur, and petrographic analysis, scanning electron microscopy, differential thermal analysis, Fourier transform infrared spectroscopy using photoacoustics, and electrophoresis were used to study these coals. These techniques identified the composition, and characterized the surface functional groups of the materials. Electrophoresis showed that the surface behaviour of these coals in aqueous medium with varied pH, was very different, although their apparent compositions were similar. These results were rationalized in terms of the effects of incorporated clays and pyrite.

INTRODUCTION

Mineral matter, especially sulphur and its compounds, is considered to be undesirable and detrimental to almost every aspect of coal utilization. Removal of the minerals is necessary, and choosing a process that optimizes the production of a quality product, requires intimate knowledge of the coal, both chemically and physically, as well as its associated minerals.

In this study several techniques were used to characterize two high-sulphur eastern Canadian coals beneficiated in the 10tph pilot plant of Coal Research Laboratory, Devon, Alberta, Canada. The raw and clean coals and their low temperature ashes were studied by scanning electron microscopy (SEM), differential thermal analysis (DTA), Fourier transform infrared spectroscopy using photoacoustics (FTIR-PAS), and electrophoresis. Standard samples of pyrite and clay were used as reference materials for identification of the mineral matter in using DTA and FTIR-PAS techniques. Furthermore, the reject products of each coal as well as the raw and clean coals were studied electrokinetically at various pH values.

SEM provides a qualitative characterization of mineral in terms of their size distribution within the coals and their morphology. The DTA technique which was used exclusively for the study of clay minerals until the early 1950's (1), is a powerful tool in mineral studies especially when used in conjunction with SEM and FTIR. FTIR-PAS provided a means for quick analysis of the the clean and raw coals and furthermore was capable of showing the presence of certain clays that would not have been easily detected by x-ray diffraction. FTIR-PAS also suggests assignments other than carbonyls for the commonly discussed 1600 cm^{-1} coal peak.

Finally, electrophoresis provides information on coal surface properties and uses the characterization results including ash, sulphur, and moisture, and petrographic analysis to rationalize their behaviour in aqueous media. The results were compared to those of other coals and pyrites reported in the literature.

EXPERIMENTAL

Materials

Two high volatile A bituminous coals from eastern Canada were obtained as samples of the beneficiation products from tests in the 10 tph coal washing pilot plant of the Coal Research Laboratory. Each coal underwent density separation in a Batac jig and water-only cyclone (2). The minus 60 mesh fraction of raw, clean and fine reject of each coal were used for the study. The ash, sulfur and moisture contents as determined by ASTM methods are shown in Table 1. Splits of the raw and clean coal samples were exposed to an oxygen plasma ashing using a Brenson/IPC low temperature asher with an oxygen flow of 150-200 cm^3/min and 50 watts rf power per chamber, for 24 hours with frequent stirring. The low temperature ashes (LTA) thus obtained were used to study the 'carbon-free' mineral of each sample; the minerals identified by X-ray diffraction analysis using a Philips diffractometer are shown in Table 2. Petrographic analysis of the coals is presented in Table 3.

The pyrite standard was obtained from Wards Natural Science Establishment Inc., Rochester N.Y.. The illite clay standard was obtained from Alberta Research Council, Geology Section, Edmonton, Alberta, Canada.

TABLE 1
Ash, Sulfur and Moisture Analysis of raw and clean coal samples

	Coal A			Coal B		
	Moisture	Ash% (db)	S% (db)	Moisture	Ash% (db)	S% (db)
Raw Feed	3.58	22.12	4.63	1.27	16.23	7.06
Clean Coal	10.74	8.52	3.55	5.81	12.17	6.02

TABLE 2
X-ray Diffraction Analysis of Coal LTA

Samples	Minerals
Raw Coal A	Pyrite, Marcasite, Quartz, Clay
Clean Coal A	Pyrite
Raw Coal B	Pyrite, Quartz, Clay
Clean Coal B	Quartz, Pyrite

TABLE 3
Petrographic Analysis of Raw and Clean Coals A and B

Maceral Counting Volume Percent	Raw Coal A	Clean Coal A	Raw Coal B	Clean Coal B
Vitrinite	61.8	77.0	56.2	70.8
Semifusinite	2.2	3.0	4.8	5.4
Fusinite	3.8	2.4	4.0	2.2
Macrinite	1.4	1.8	3.4	2.6
Micrinite	3.2	0.6	3.8	3.4
Inertodetrinite	3.2	1.8	2.6	1.4
Sporinite	4.6	4.0	5.0	4.4
Resinite	0.6	1.8	2.0	2.2
Pyrite	3.6	2.6	8.4	3.4
Other Mineral Matter	15.6	5.0	9.8	4.2
Random Reflectance				
Mean	.634	.657	.708	.732
Maximum	.756	.748	.886	.891
Minimum	.535	.567	.541	.576
Standard deviation	.042	.039	.073	.071

Methods

Scanning electron microscopy (SEM) was performed on the clean and raw coals using Hitachi X-650 scanning electron microscope equipped with a Tracor Northern 30 cm^2 Si(Li) energy dispersive X-ray detector. The LTA elemental composition was qualitatively determined using this X-ray detector. Raw and clean coal samples were fractured before examination, thus the photographs represent internal features of the coal that would be exposed during the crushing process. The samples were carbon coated for the x-ray analysis and gold coated to increase the electron signal for the morphology studies.

Thermal analysis of the LTA samples, the pyrite, illite kaolinite and their synthetic mixtures was performed using a Perkin-Elmer DTA-700 High Temperature System, equipped with a System-4 Thermal Analysis Data Station. The temperature was calibrated using a 99.99% pure aluminum standard. Ceramic liners were used in the DTA sample holder cup. All samples were studied in a dynamic nitrogen or carbon dioxide atmosphere at a flow rate of 50 cm^3/min.

Fourier transform infrared photoacoustic spectroscopy was performed on the raw, clean, and LTA samples, as well as on the illite and pyrite standards using a Bruker 113V FTIR , with a photoacoustic cell from EG&G Instruments. After packing the samples in the PAS accessory, the entire system was evacuated for 15 minutes then purged with nitrogen gas prior to data collection. A resolution of 4 cm^{-1} was used at a scanning velocity setting of zero (3). About 2000 scans per sample were acquired. The spectra obtained were ratioed against the spectrum of charcoal black powder prior to plotting.

Difference FTIR spectra between the clean and raw samples were done for both

Coal A and Coal B powders. Spectra of the low temperature ash of the raw coals were compared to those of pyrite and illite clay standards in an attempt to identify the mineral species.

Electrophoretic mobilities were measured according to methods previously developed for coal at CRL using a Pen Kem system 3000 Electrokinetic Analyser (4,5). Coal samples (minus 60 mesh) were weighed and suspended in deionized water (18 M Ω), stirred for 4 hours and left to equilibrate for 12-24 hours. Prior to analysis, the samples were wet screened through a 400 mesh (37 μm) sieve. Adjustments of pH were made with ACS grade hydrochloric acid or sodium hydroxide. The pH of the samples was determined on-line in the System 3000 as well as by using a Radiometer pH meter (5).

Each coal was studied at the equilibrium pH of its slurry in deionized water where several measurements were made. Since all conditions were kept constant, the pH displayed at equilibrium would represent the degree of solvation and the acidity of the surfaces. To explain similarities or differences between the surface properties of these coals, they were studied as they approached the isoelectric region. To locate the isoelectric region, the pH was then adjusted towards the acidic or the basic range depending on whether the particles were naturally electronegative or electropositive respectively (6,7). The average mobilities obtained from several spectra were used for the calculation of the zeta potential using the Smolochowski formula for $ka > 300$ (8).

RESULTS AND DISCUSSION

Scanning Electron Microscopy (SEM)

Scanning electron microscopy showed that pyrites were present in both of these coals in highly crystalline pockets, framboidal crystal agglomerates, or as almost amorphous masses.

Examples of these features are shown in Figures 1 to 3. Figure 1 shows the well defined crystal structure of some of the finely dispersed pyrite pockets. Figure 2 shows the raspberry-like framboidal structure. Figure 3 shows the relatively structureless pyrite (identified by X-ray analysis) which has intruded, forming a separation in the coal matrix. These three features are found in both coal samples although Coal A appears to have a higher proportion of the crystalline pyrite structures than Coal B.

X-ray studies on the coal samples, and on their LTA showed qualitatively that, beneficiation effected much greater removal of clays than of pyrites from Coal A. This is supported by the ash, sulphur and moisture analysis. This could be a result of the finely disseminated nature of the pyrites in this particular coal. From a beneficiation point of view grinding to liberate the pyrite would be required only to expose the pockets of pyrite and not to release individual grains of crystals. This has important implications for deep or intensive cleaning of pyritic coals.

Fig. 1. Scanning electron micrograph of Pyrite crystals (Coal A).

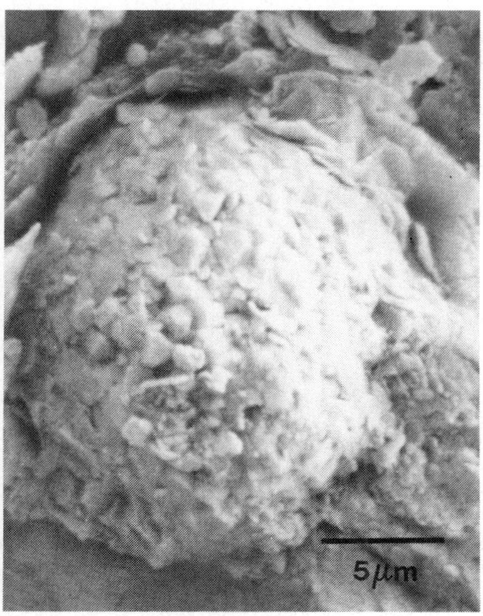

Fig. 2 Scanning electron micrograph of frambodial pyrite (Coal A)

Fig. 3 Scanning electron micrograph of relatively amorphous pyrite (Coal B)

Differential Thermal Analysis (DTA)

Figure 4 illustrates a typical DTA curve of pyrite in a dynamic nitrogen atmosphere. There is a strong endothermic peak at about 650 - 700°C followed by an exothermic peak at 750°C. The strong endotherm has been interpreted as the decomposition of pyrite (9). Any oxygen present or which becomes available at the decomposition temperature, causes rapid oxidation, and is shown by the exothermic peak at 750°C followed by the endothermic decomposition peak. The melting point of pyrite is shown by a small endothermic peak at 1150°C.

The thermal curves of the raw and clean coal LTA samples are shown in Figures 5 and 6. A comparison of these curves with that of pyrite, shows that the LTA samples of raw Coal B and clean Coals A and B consisted mainly of pyrite. The thermal curve of the LTA of raw Coal A was more complicated. The broad endothermic peak with a minimum temperature at 530°C indicates that other minerals were present along with the pyrite. The decomposition peak of pyrite at 630°C can still be identified as a shoulder on the high temperature side of the broad endotherm. The exothermic peak at 750°C and the endothermic peak at 1150°C were identified as the oxidation and melting peaks of the pyrite.

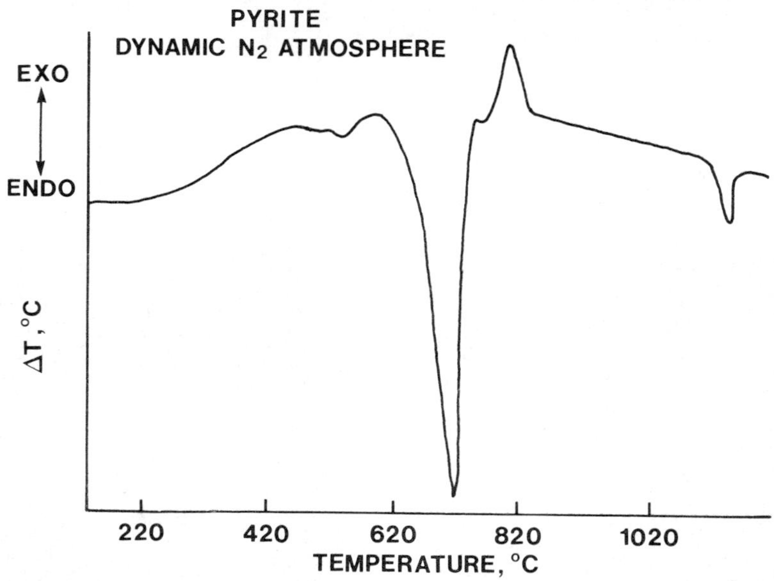

Fig. 4. DTA thermal curve for pyrite using purge gas rate of 50 cm^3 /min and heating rate of 20° min.

A convenient way to determine whether the unknown mineral is carbonate is to heat the sample in a carbon dioxide atmosphere. The increase in CO_2 partial pressure will cause an upward shift in the decomposition temperature of carbonates. For the LTA of raw Coal A, the thermal curves remained essentially unchanged with the switch of the purge gas from N_2 to CO_2. The similarity in the peak position eliminates, within the limits of detectability, the possibility of existence of carbonates, in this coal.

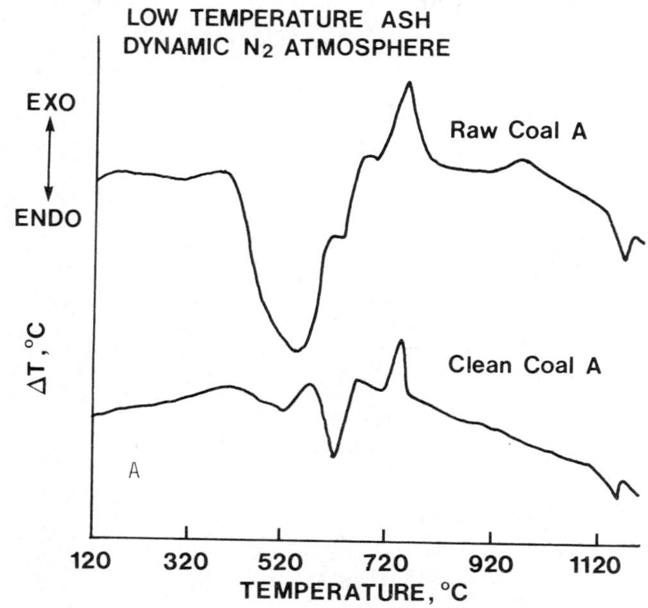

Fig. 5.

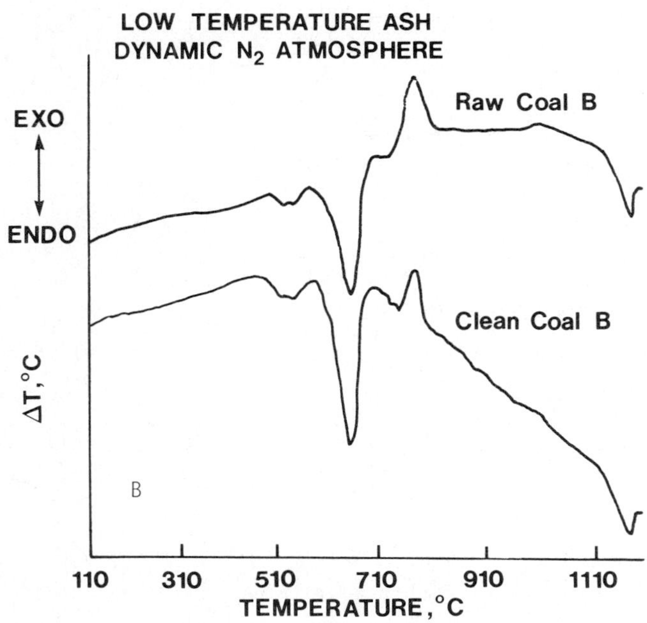

Fig. 6.

Fig. 5 and 6. DTA thermal curves for LTA samples A&B, using purge gas flow rage g 50 cm^3 /min. and heating rate of 20°C/min.

To characterize the unknown mineral in the raw Coal A LTA sample, several synthetic mixtures of minerals (kaolinite, bentonite, illite, quartz and iron(II) sulfate) which are commonly encountered in coal LTA were mixed with pyrite and their thermal behaviours were studied by DTA. The chosen levels of pyrite and clay minerals in the mixtures are believed to be close to those in the LTA sample. Some of the thermal curves of these mixtures are shown in Figure 7.

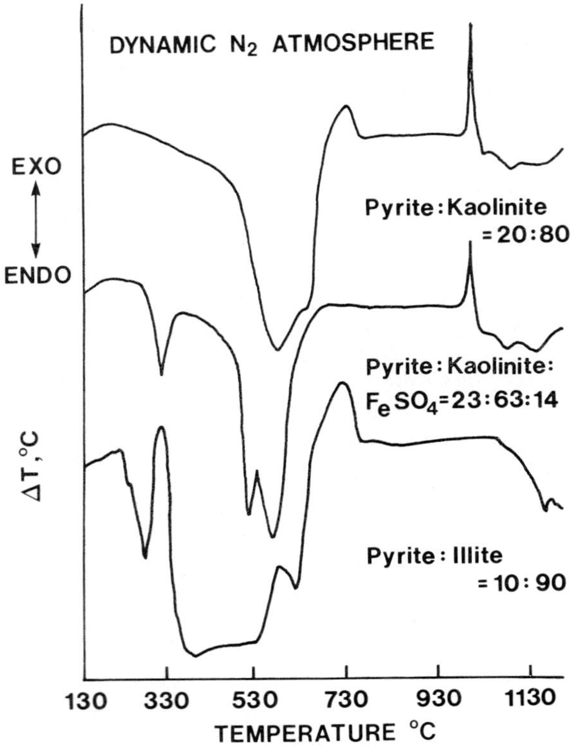

Fig. 7. DTA thermal curves for synthetic mixtures using purge gas flow rate of 50 cm^3 /min heating rate 20°C/min.

In the pyrite/kaolinite and pyrite/kaolinite/$FeSO_4$ mixtures, a sharp exothermic peak is observed at 1010°C. This peak is caused by the crystallization of a spinel phase of kaolinite and can be recognized even if the concentration of kaolinite is as low as 2-5% (10,11). The peaks at 305, 726, and 590°C in the thermal curve for the pyrite/kaolinite/$FeSO_4$ mixture result from the decomposition of $FeSO_4$. The absence of these peaks in the raw Coal A LTA sample indicates that the mineral was neither kaolinite nor iron(II) sulfate. In the bentonite/pyrite mixture, the endothermic decomposition peak of pyrite at 650°C does not overlap with any of the characteristic bentonite peaks.

The pyrite/quartz mixture shows a sharp endothermic peak at 576°C. The peak results from the transformation of alpha quartz to beta quartz. This peak was not observed in the raw Coal A LTA thermal curve. These imply that if bentonite and quartz were present in the sample, their levels were below the detection limits.

In the pyrite/illite thermal curve, a low temperature endothermic peak at 250°C, attributed to the loss of absorbed water in the illite, is followed by a broad dehydration endotherm of illite in the range of 300°C to 600°C. The intensity of this endotherm is so strong that it overlaps the pyrite decomposition peak which appears as a shoulder at 650°C. This is very similar to that shown by the raw Coal A LTA sample. The peaks appearing at higher temperatures are also similar to those found for the raw Coal A. These peaks are contributed by the oxidation and melting of pyrite. Thus it appears most likely that illite was the mineral which was present along with pyrite in the raw Coal A LTA. The absence of the low temperature endothermic peak at 250°C in the LTA sample may have been due to the removal of absorbed water during the low temperature ashing process.

In summary, DTA indicated that the major mineral constituent in raw and clean Coal B was pyrite and that in raw Coal A, pyrite, and probably illite were the two major components. During beneficiation of the coal, effective removal of the clay was achieved. The principal mineral left in clean Coal A was mainly pyrite.

Fourier Transform Infrared Spectroscopy (FTIR)

The difference spectra of the raw and clean coals gave an indication of the effect of processing on the quality of coal products. The difference spectra of A and B raw and clean coals are shown in Figures 8 and 9. No differences in the aromatic and aliphatic C-H stretch regions were observed. However, clean coal A showed higher intensities for the out-of-plane aromatic C-H bending mode (700 - 900 cm^{-1}). There was also some difference in the oxygen-hydrogen bonded O-H stretching bands in the 3300 cm^{-1} region between Coal A and Coal B. The characteristic 1600 cm^{-1} coal peak characteristic of coals was essentially the same in both cases (12). The difference spectra show that the coal was not oxidized during processing. In both cases, the mineral matter signal was more intense in the raw than in the clean coals indicating the removal of mineral matter during the cleaning process. This was the only major difference between the spectra of the raw and clean products. These results were corroborated by ash, sulphur and moisture, (Table 1) DTA (Figs. 5 and 6), and petrographic results (Table 3).

A large part of the coal FTIR literature has been devoted to band assignments of coal FTIR spectra, and there is much focus on the bands arising from oxygen containing functional groups in the 1600-1700 cm^{-1} region (12).

Changes in the 1600 cm^{-1} peak and 1690 cm^{-1} shoulders have been attributed to oxidation of the coal to carbon-oxygen groups. Although changes in intensities of 1600 cm^{-1} band have been related to phenols (12), this assignment is frequently debated (13).

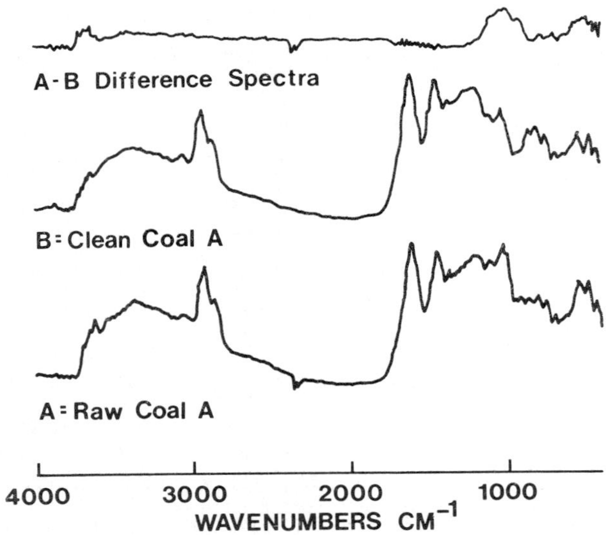

Fig. 8.

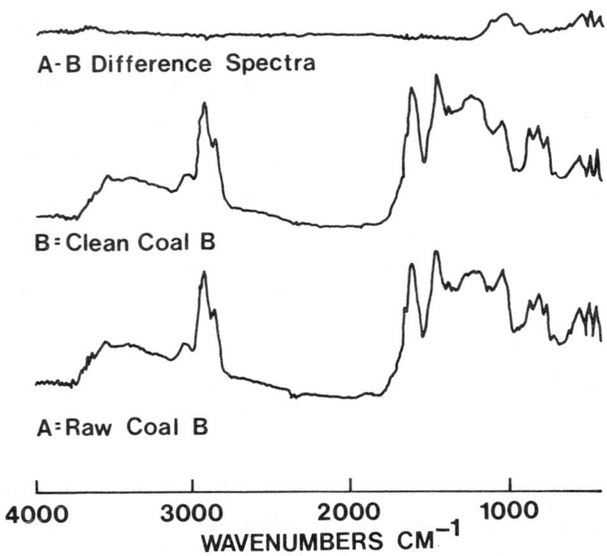

Fig. 9.

Fig. 8. and 9. Difference FTIR-PAS spectra of raw and clean coals.

To identify the mineral matter present in the raw coals the FTIR spectra of LTA samples were compared to those of pyrite and illite standards. Figure 10 compares the spectra of LTA samples of the raw coals A and B with the spectra of the pyrite standard between 2000 and 400 cm^{-1}. The low temperature ash of raw Coal A shows a peak at the 1631 cm^{-1} and a shoulder at 1701 cm^{-1}. The latter correspond to a peak in the standard pyrite spectra which suggests that the existence of the shoulders at 1631 cm^{-1} and 1701 cm^{-1} region for these coalsmay not be exclusively due to the oxidation state of the coal. The 1496 cm^{-1} peak in the pyrite spectrum is obscured by other peaks in the coal spectrum. The major peak of the pyrite spectrum at 1168 cm^{-1} corresponds to a shoulder in the spectrum of the LTA. A comparison of these spectra shows that the broad width of the 1168 cm^{-1} mineral peak may be an indication that the pyrite waspresent in the low temperature ash samples.

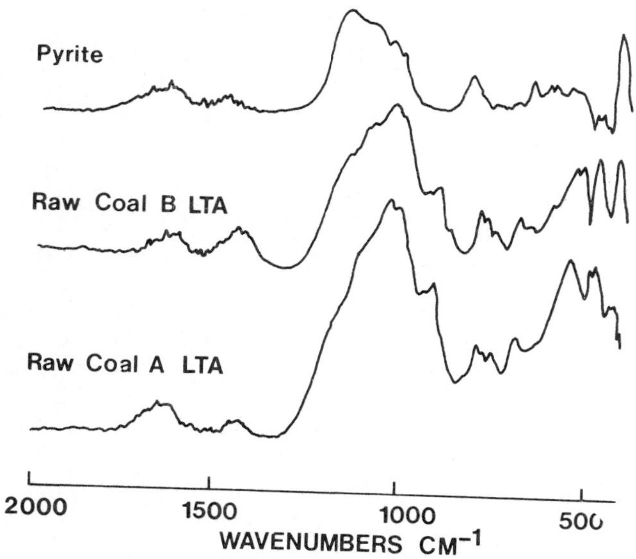

Fig. 10. A comparison of LTA FTIR-PAS spectra with pyrite.

A comparison of the raw Coals A and B LTA spectra in the 700 cm^{-1} to 500 cm^{-1} region shows that there coals A and B have similar bands but of different intensities. This may be an indication of the different amounts of mineral matter present in each coal. This is corroborated by the respective ash, sulphur, X-ray diffraction, and petrographic analyses which indicated that mineral matter composition differed in the two coals.

The X-ray diffraction analysis shows the presence of quartz in raw Coal A and B and clean Coal B. Kuehn noted that in the study of coal structure the mineral matter causes problems of interference for the coal spectroscopist (14). Differences in LTA, and clay mineral FTIR spectra were reported by Snyder, who indicated the contribution of quartz to the LTA spectra (15). Although X-ray diffraction showed the presence of quartz in Coal A, no significant contribution was noted in the LTA FTIR spectrum. The occurrence of the 1400 cm^{-1} peak in LTA is attributed to carbonate in FTIR mineral studies of carbonates (16). A comparison with the FTIR-PAS spectra of illite mineral in Figure 11 shows that this peak appears for both raw Coal A LTA and illite. The clay in this study showed a remarkable resemblance to illite in the 1500 - 400 cm^{-1} region. Although the 3500 cm^{-1} region has some similarity with kaolinite in FTIR-PAS, DTA showed that kaolinite was not present, and as noted earlier, gave strong evidence that the clay mineral was illite. Illite, has been reported by Montano (17) as being one of major clays present in coal.

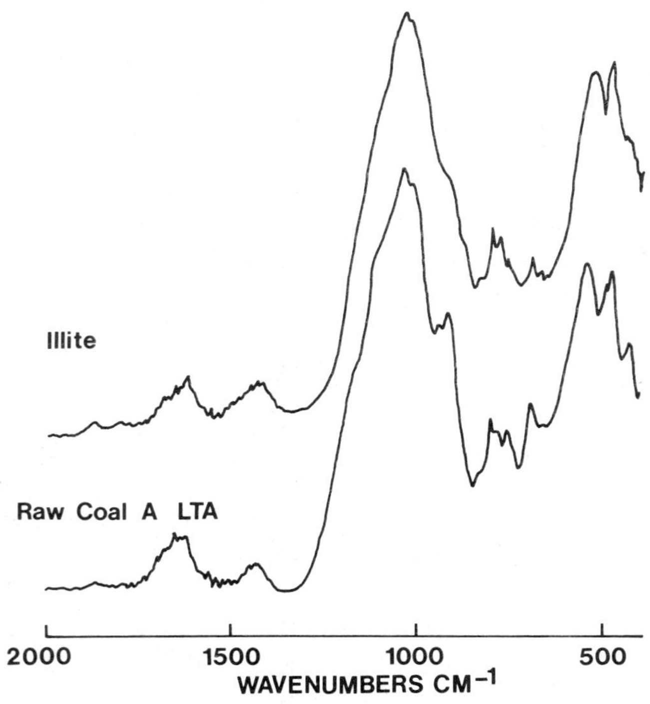

Fig. 11. A comparison of FTIR-PAS spectra of LTA and Illite.

It can be seen that mineral composition has a bearing on the appearance of the coal spectra and that FTIR is useful for quick characterization of coal to be studied by other surface techniques especially electrokinetics.

Electrophoresis

The zeta potential (ζ) results for the raw, clean, and reject samples of each coal at various pH appear in Figures 12 and 13. The detailed electrophoretic spectra prior to, at, and after reaching the isoelectric condition are shown in the order a, b, c, and d in Figures 14 to 18. An exception is Figure 14(d) which represents the pyrite standard at pH 4.3. The x-axis represents electrophoretic mobilities (E.M.) and the y-axis the relative intensity of the scattered particles. The vertical line in each figure marks zero mobility on the x-axis, with negative mobilities located to the left and positive mobilities to the right.

Heterogeneity of each coal's particle population is indicated by the bimodal or trimodal electrophoretic spectra of both coals' raw, clean, and reject fractions at given pH values.

The two raw coals displayed remarkable differences when immersed in theaqueous environment. Although little difference was found in their equilibrium pH, there were large differences in their zeta potentials (Table 4).

TABLE 4
Equilibrium pH and Zeta Potential of Coals A and B

Sample	Equilibrium pH	Zeta Potential ζ (mV)	Electrophoretic Spectra Comments
Raw Coal A	5.83	+19.43 ± 3.1	
Raw Coal B	5.72	-23.63 ± 5.9	
Clean Coal A	4.25	+16.10 ± 4.3	
Clean Coal B	5.52	-23.20 ± 6.8	
Reject A	6.58	-13.87 ± 5.7	Normal distribution (Fig.18)
Reject B	6.34	-22.90 ± 9.9	Bimodal Spectra (Fig.18)

At their equilibrium pH, the ζ of Coal A's raw and clean samples were electropostive (+19.43 and 16.10 mV respectively) in contrast to its reject fraction which was electronegative at -13.8 mV. On the other hand Coal B's samples were all electronegative with similar ζ. It is also apparent from Table 4 that for both coals during the cleaning process, the clean fraction became more acidic and the reject samples were more neutral than the raw samples. The large difference in ζ between Coal A and B was inconsistent with the similarity in composition, rank, oxidation state and mineral makeup. However, in a heterogeneous composite such as coal the influence of the mineral matter would be highly dependent on the experimental conditions. The sensitivity to changes in the concentrations of potential determining ion (p.d.i.) such as H^+ and OH^- would indicate the chemistry of the solids in the electrolyte medium. For solid-oxides and metal-hydroxide complexes Parks noted also that the isoelectric reflects the chemical composition of the solid in the electrolytes in which it is immersed (7).

In the case under study the coals were suspended in deionized water. This ensured that the problems of indifferent supporting electrolytes did not arise in interpretations (6,8). The isoelectric range (i.e.r.), rather than the more familiar isoelectric point (i.e.p.), is reported in this work since it was found that the i.e.p. did not truly represent a heterogeneous material such as coal (6). The i.e.p. in this case is the pH at which the electrical double layer is considered collapsed. Since coal is heterogeneous, various particles of a population may be zero charged, positively charged or negatively charged at a given pH (i.e.p.). A multimodal electrophoretic spectra is most common in heterogeneous cases and usually represents various species of particles which are charged differently under the same condition. This is observed in many cases and is shown in Figs. 14 to 18. This behaviour presents difficulty in reporting a specific i.e.p.. The isoelectric range (i.e.r.) is therefore reported in such cases.

Raw Coal A was electropositive at all pH and was quite similar to clean Coal A at pH = 9.0. Figs. 12,14, and 15 show the respective ζ vs pH curves and electrophoretic spectra. Zero mobilites were especially difficult to achieve with these coals. For instance an adjustment of raw Coal A suspension to pH 9.0 caused a partial change of part of the population to assume negative charges (Fig. 14c) and the pH to drop from 9.01 to 6.71. Similarly, as shown in Fig. 15 (a), the pH of clean Coal A dropped from 9.05 to 6.15 leaving a spectra with a partial zero charge. At pH 10.59 the raw Coal A particles were negatively charged with a ζ of -1.5 mV. This rapid drop in pH was also observed in other studies on hydrophobic low-sulfur coals (6) and may not be a behaviour specific to high sulfur coal. This tendency of the coal to move towards surface neutralization when placed in disequilibrium (18), may account for some of these observations. However the indications of the i.e.r. between pH 10.5 and 6.0 for these two coals suggest that a metallo-hydroxo complex might have been formed; and the system behaved as though it had been doped with some oxide. Alumina was reported to have a higher basic i.e.p. of 9.1 (8) and certain iron oxides,reported by Parks and de Bruyn, to have i.e.p.'s as high as 8.5 depending on the type of complex (19).

Raw Coal B shown in Fig. 16a-16c was electronegative at pH 5.57 and became electropositive at pH 2.74. The i.e.p. in this case was 3.89 as found from the electrophoretic spectra shown in Fig. 16, and 4.0 from the ζ vs. pH curve in Fig. 13. Clean Coal B shown in Fig. 17 was electronegative at all tested pH's except at pH 2.75 where a positive bimodal distribution occured, however at pH 2.62 the particles seemed to shift slightly towards zero mobility. The i.e.r. in this instance lies between 3.62 and 2.62 with an average of 3.12.

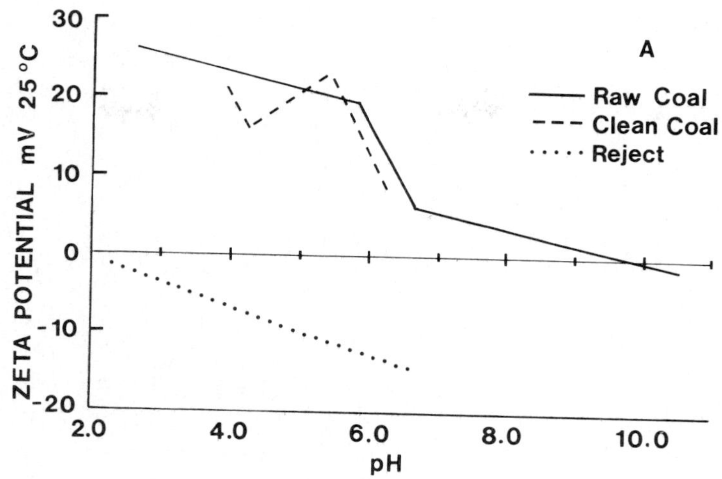

Fig. 12.

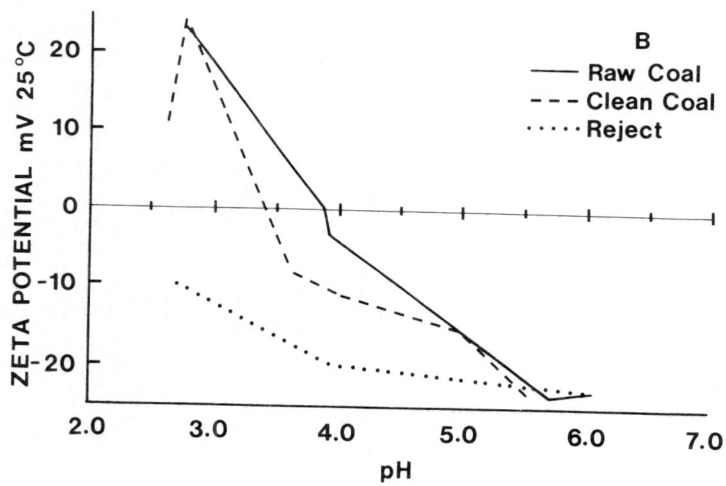

Fig. 13.

Fig. 12. and 13. ζ vs pH curves of Coal A and B samples in deionized water.

The differences found between raw and clean Coals A and B were not marked in their respective rejects. Reducing the pH of Coal A and B rejects from 6.58 and 6.34 to 2.25 and 2.69 respectively changed the ζ from -13.87 mV and -22.9 to zero and 9.7 mV respectively (Figs. 12, 13 and 18). In the case of the rejects it is expected that the majority of particles would behave like the clays and the other negatively-charged minerals that are present since these are in the majority. Evidence of quartz found in raw Coals A and B and clean Coal B indicate that quartz is most likely to be a mineral appearing in the rejects. Quartz has been reported to have an i.e.p. of 2.2 (7). Illite which is a montmorillonite related clay is also electronegative under these conditions (20).

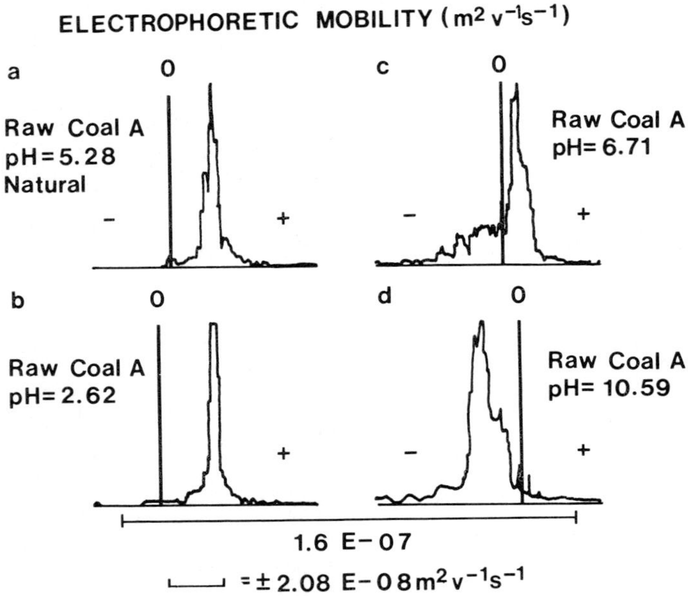

Fig. 14.

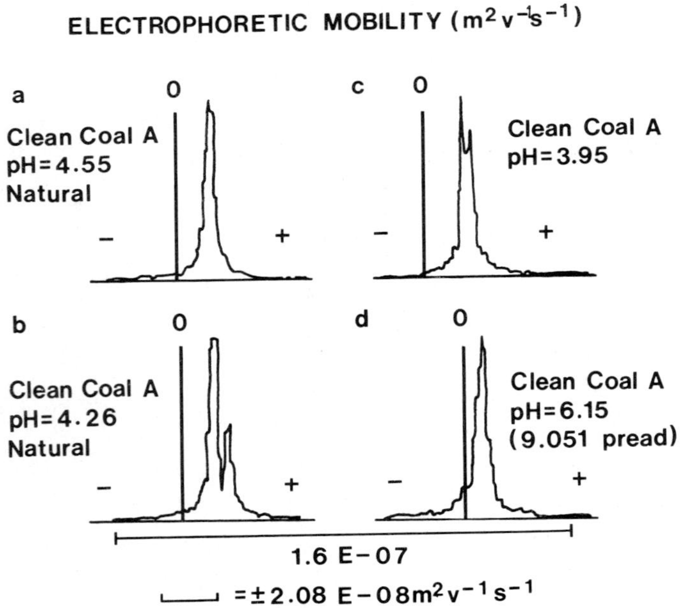

Fig. 15.

Fig. 14. and 15. E.M. spectra of raw and clean Coal A in deionized water.

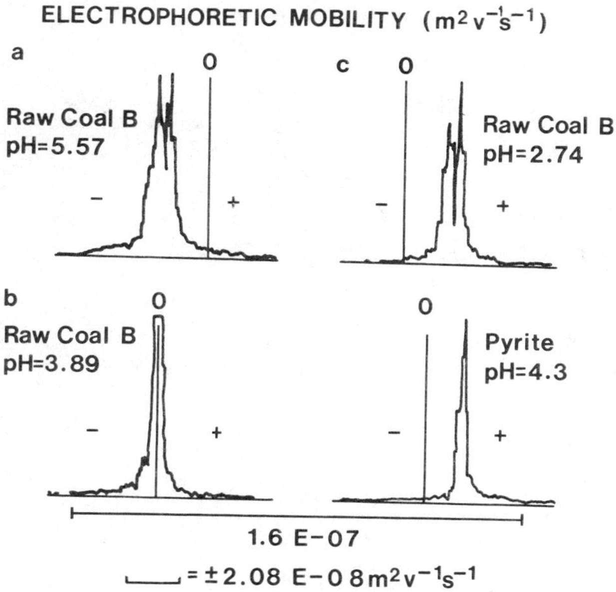

Fig. 16.

Fig. 16. and 17. E.M. spectra of coal samples in deionized water at various pHs
Top: raw Coal B Bottom: clean Coal B

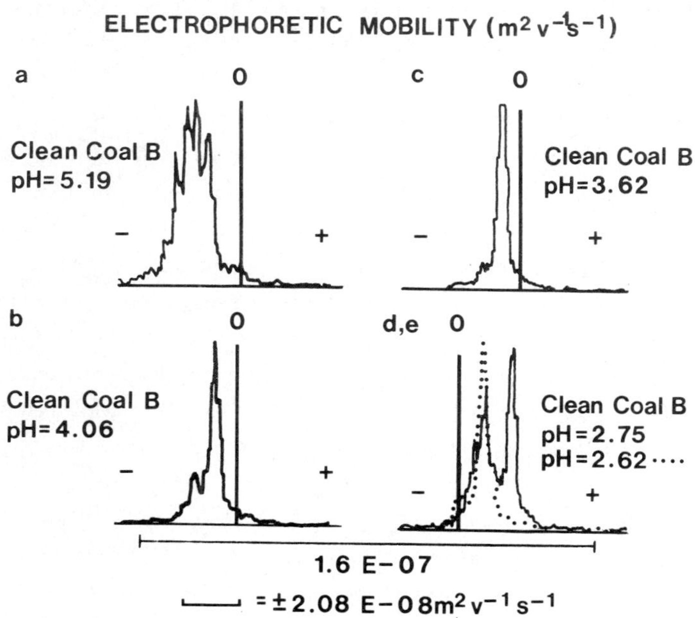

Fig. 17.

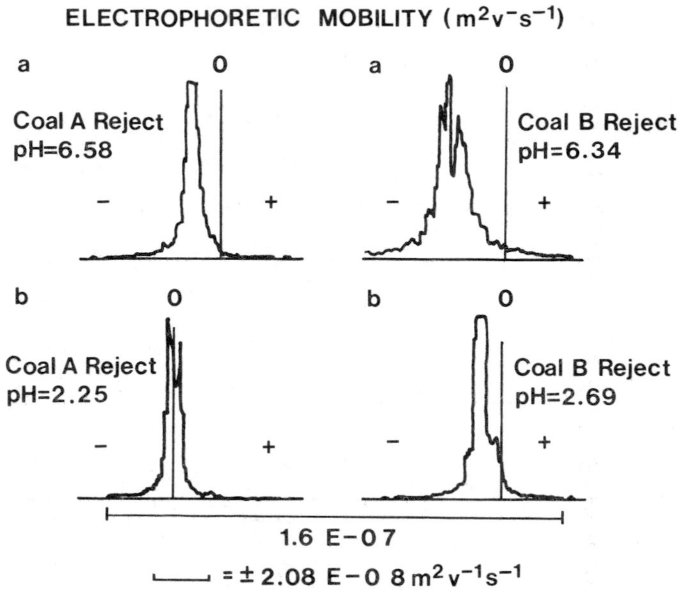

Fig. 18. E.M. spectra of Coals A and B reject samples in deionized water at various pH.

The striking difference between Coals A and B in an aqueous media becomes difficult to understand when all the other characterization results in fact, show little difference in their mineral composition and similarity in rank. These observations may relate to the charging mechanism.

Various reseachers have hypothesized the charging mechanism of coals to be related to the carbon-oxygen groups on the surface being hydrated and dissociated. Other suggested the presence of $SiOH_2{}^+$, $M\text{-}OH^-$ or $MOH_2{}^+$ species where M is the surface of the metal on the surface. Based on Potentiometric Titration and electrophoretic data, the charging mechanism of low-sulfur coals have previously been attributed to a Zwitterionic effect. Here the charges are contributed by many species including the above and also by ion exchange (6) depending on the equilibrium properties (21). For pyrite, charging was reported by de Bruyn and Agar to be from S^{2-}, HS^-, H^+, OH^- (22). Variations in the i.e.p. of pyrite have been reported by many investigators. Gaudin and Sun reported a value of 6.5 which is close to our results on Coal A (23). Fuerstenau et al reported 6.2 and 6.9 for aged pyrite (24). Electrophoretic behaviour of pyrites aged in water is shown to be highly significant (25). Variation in degrees of dissolution of surface-metallo-hydroxy complexes may also account for ionic differences (7,21,25,26).

Coal pyrite was reported by Kalebek et al (27) to have two i.e.p. at 4.5 and 6.5 after aging in water and one at pH = 2.2 immediately after preparation. Iron hydroxide has also been reported to have an i.e.p. of 6.7. Since Fe^{3+} was found in the process water it follows that its hydrolysis could produce complexes as described by Healy (26). These complexes adsorb on the coal surfaces and can render the coal surface similar to a ferrous hydroxide which would account for some of the response observed for Coal A. In the case of Coal B, a different form of pyrite may be presented and perhaps the coal surface itself plays a larger role.

In studies of ζ of coal fines in aqueous suspensions, variations in i.e.p. and charge on the surfaces have also been attributed to chemical or elemental makeup (28). Coca et al suggested that impurities such as clays, salts and ionic adsorption influence the charge and i.e.p. (29). Studies at CRL showed the influence of ionic composition and equilibrium on the electrophoretic mobilities (18,21). Many others have attributed these variations to the oxidation state and coal type (30-34). Campbell and Sun have pointed out that the z.p.c. (=i.e.p.) of the macerals fusain, vitrain, durain and of whole coal studied were 2.8, 3.5, 4.2 and 4.6 respectively, but it was acknowledged that in these materials the mineral matter and sulfur content were low compared to these coals (32-34). Clean Coal B also behaved similarly to vitrain when clean and raw Coal B to durain. However, the petrographic analysis showed little difference between these coals, with vitrinite being the major maceral (Table 3).

The electrophoretic study has shown the complexity of the problems associated with the behaviour of these coals in an aqueous medium. Other workers have recognized the problems of coal heterogeneity (35). Further investigations on these coals are being undertaken using other techniques including NMR analysis.

CONCLUSION

Several techniques used for the characterization of two high sulfur coals which were processed at CRL were shown to compliment one another and contribute to the understanding of the composition and surface behaviour of these coals in an aqueous medium.

Scanning electron microscopy (SEM) and X-ray analysis provided a qualitative view of the pyrite structures in the coals. It was found qualitatively that the pyrite in Coal A was more crystalline than that found in Coal B, and that more clays relative to pyrite were removed from Coal A in the cleaning process.

DTA showed that the clean and raw coals exhibited different thermal behaviour. The main mineral in Coal B was pyrite. The DTA technique was useful for detection of other minerals in addition to pyrite in the raw coals and gave evidence of the presence of illite clays in the raw Coal A.

FTIR confirmed the presence of clays in the cleaned coals and indicated the identification of illite clay. The FTIR-PAS spectrum of pyrite was shown by comparison to the spectrum of the LTA samples to contribute to band broadening. Carbonate bands were indentified in the LTA spectra. The shape of the 1600 cm^{-1} peak not only reflected the oxidation state of the coals, but it also suggested that in these high sulfur coals the shoulders of this peak may not be exclusively related to coal surface oxidation, and can also be related to mineral matter. Most importantly FTIR-PAS showed, through difference spectra the major changes that occur in the coal surface as a result of the cleaning process.

Electrophoresis showed that with the cleaning process, that both coal's equilibrium pH and i.e.r. were shifted towards more acidic values than their raw counterparts. The i.e.r. of Coal B was far more acidic than Coal A. Although their characterization results indicated similar structures, Coal A and B were oppositely charged. The behavioral difference between Coal A and B is explained as a combined response of the coal molecule associated with the pyrite, clay and quartz minerals. Coal B responded similar to a normal low sulphur coal. Coal A behaved similar to a coal coated with a pyrite (metallo-hydroxo) complex. The electrophoretic behaviour of the rejects reflected the presence of a large proportion of minerals than coal. Electrophoresis of these coals demonstrated the heterogeneous nature of the particles and the major changes in their surface properties with pH. Of the four techniques electrophoresis provided the most sensitive probe of the relationship between the aqueous environment and the coal surface. The zeta potential is a powerful measure that can be used in stability control for the preparation of coal.

ACKNOWLEDGEMENTS

The authors wish to express their deep appreciation to all CRL Staff for their support to this work: Ms. L. Svenson, and Mr. V. A. Munoz are acknowledged for the X-ray diffraction and petrographic analyses respectively.

REFERENCES

1 Wendlandt, W.W., Amer. Lab, 9, #1, 59, (1977)

2 Mikula, R.J., Mikhail, M.W., Salama, A.I.A., Humeniuk, O.E., "Some EMR Facilities in Western Canada for R & D on beneficiation of Canadian Coals", CRL Division Report ERP/CRL 82-01 (TR).

3 Bruker, IFS User Manual, Bruker Analytische, Messtechnik GMBH FT-IR Division, Version 3/84.

4 Angle, C.W. and Hamza, H.A. "Exploratory Tests on Sample Preparation Techniques for the Electrophoresis of Unoxidized Coal", ERP/ERL 84-49 (TR)

5 Angle, C.W. and Hamza, H.A. "Evaluation of an Automated Electrokinetic Analyser", ERP/ERL 81-18 (TR)

6 Angle, C.W., Donini, J.C., and Hamza, H.A. "Electrophoretic Behaviour of a Relatively Unoxidized Coal", presented at 57th Colloid and Interface Science Symposium, June/83, Toronto, Canada.

7 Parks, George A.,"The Isoelectric Points of Solid Oxides, Solid Hydroxides and Aqueous Hydroxo Complex Systems", Chemical Reviews, Vol. 65, 1964, p. 177.

8 Hunter, R.J. Zeta Potential in Colloid Science, Academic Press, New York, 1981.

9 Warne, S. St. J., "Differential Thermal Analysis of Coal Minerals" in "Analytical Methods for Coal and Coal Products", Vol IV, Ed. Karr C. Jr. Academic Press (1979)

10 Smykatz-Kloss, W., "Differential Thermal Analysis Application and Results in Mineralogy", Springer-Verlag, Berlin and New York (1974).

11 Warne, S. St. J., Inst. Fuel, 38, 207 (19)

12 Painter P.C., Davis, A. et al. "A Data Base for the Analysis of Compositional Characteristics of Coal Seams and Macerals. FTIR and CMR Studies of Coal Structure." Contract #AC22-80PC-30013 Penn. State University, Technical Information Center U.S. Department of Energy. October, 1983.

13 Starsinic, M.E., Ph.D. Thesis, 1984, Penn State University, Material Science Engineering, United States.

14 Kuehn, D.W. Ph.D. Thesis, 1983, p. 102.

15 Snyder, R.W., Ph.D. Thesis, 1982, Pennsylvania State University, U.S.

16 Friedel, R.A. Ed. Spectrometry of Fuels, Plenum Press, New York, 1970, p. 242-247.

17 Montano, P.A. "Characterization of Iron-Bearing Minerals in Coal" in Coal Structure Gorbaty, M. L. and Ouchi, K. (Eds) Advances in Chemistry Series 192 ACS Washington, D.C. 1981 Ch. 22, pp. 345-351.

18 Angle, C.W. Picard, J.L. and Hamza, H.A. "Short Term Equilibrium Development of an Aqueous Coal Slurry traced by ionic content and pH changes", ERP/CRL report in preparation.

19 Parks, G.A. and de Bruyn, P.L. "The Zero Point of Charge of Oxides", J. Phys. Chem., Vol. 66, 1962, p. 967-73.

20 Angle, C.W. and Hamza, H.A. "The Electrokinetic Behaviour of Kaolinite Montmorillonite Mixture of Coal Associated Clays in Electrolytes", ERP/ERL 81-95 (TR)

21 Angle, C.W. and Hamza, H.A. "Ultrasonication in the Study of Coal. Part II The Effects of Sample Preparation on the Electrophoretic Mobility and Ionic Equilibrium of an Aqueous Coal Suspension", ERP/CRL 55-85 (TR).

22 de Bruyn , P.L. and Agar, G.E. Ch. 5, Froth Flotation - 50th Anniversary Volume, D.W. Fuerstenau, e.d. AIME, New York, 91 1962.

23 Gaudin, A.M. and Sun S.C. Trans AIME, 169, 347 (1946).

24 Fuerstenau, M.C. Kuhn, M.C. and Elgillani, Trans. AIME 241, 148 (1968).

25 Smith, R.W. and Trivedi M.,"Variation of Point of Zero Charge of Oxide Minerals as Function of Aging Time in Water:, SME AIME Trans., Vol. 255, March 1974, p. 69.

26 Healy, T.W. "The Chemistry of Mineral Slurry Processing" A.M.I.R.A. "Handling of Concentrates" HOBART, May 1977.

27 Kelebek, S. Salman, T. and Smith, G.W. "An Electrokinetic Study of Three Coals" Canadian metall. Eng. Quarterly - 1982 21, (2) 205-9.

28 Mori, S., Hara, T., Aso, K., "Zeta Potential of Coal Fine Particles in Aqueous Suspension", Powder Technology, 40(1984), p. 161-165.

29 Coca, J., Bueno, J.L., Sastre H., "The Electrokinetic Behaviour of Coal Particles Suspensions", L. Chem. Tech. Biotechol., 1982, 2., p. 637-642.

30 Campbell, J.A.L. and Sun, S.C. "Bituminous Coal Electrokinetics", Transactions, AIME SME, Vol. 247, 1970, p. 111-114.

31 Sun, S.C. "The Effects of Oxidation of Coals on Their Flotation Properties", Transactions AIME, Vol. 6, Issue 4, 1954, p. 396-401.

32 Sun, S.C. and Campbell, I.A.L. "Anthracite Lithology and Electrokinetic Behaviour", Adv. Chem. Series, Vol. 55, 1965, p. 363-375.

33 Sun, Wen Phd Thesis 11977 Penn State University

34 Campbell, J.A.L., Sun S.C., "Anthracite Coal Electrokinetics" SME AIME Trans., Vol. 247, 1970, p. 120-122.

35 Prasad, M. "Electrokinetics of Coal" J. Inst., Fuel, Vol. 9, 1974, 174.

CHALCOGEN ELEMENTS IN COAL AND THEIR REDISTRIBUTION IN COMBUSTION PRODUCTS

K. Y. CHIOU and O. K. MANUEL

Chemistry Department, University of Missouri, Rolla, MO 65401 (USA)

ABSTRACT

The chalcogen elements (Te, Se and S) have been measured in five coal samples and in three types of coal combustion products from a local power plant. Te and Se were determined by a combination of ion exchange and flameless atomic absorption. A new analytical procedure was developed which takes advantage of the high sensitivity of atomic absorption detection of tellurium after removing interfering elements, such as selenium, arsenic, and copper, by cation exchange. Neutron activation was used to determine S and up to 30 additional elements. The enrichment factor of Te in coal and fly ash is higher than that of any other element. Average concentrations of chalcogen elements in the five coal samples are Te = (0.80 ± 0.14) μg/g, Se = (3.03 ± 0.99) μg/g and S = (2.75 ± 1.10)%. Elemental ratios of the chalcogen elements in fly ash are much closer to those in local aerosols than to those in coal, bottom ash or economizer ash. The enrichment factors of Te, Se, and S in fly ash also show essentially the same pattern as that in local aerosols. These results suggest that coal combustion may be a major source of chalcogen element emission.

INTRODUCTION

Tellurium (Te) and selenium (Se), chemical analogues of sulfur (S), are commonly associated with sulfide minerals. Increased usage of high sulfur coal as a fuel thus enhances the need for practical information on the distribution of Te and Se in the combustion products. Since Te and Se also form independent minerals, their concentrations in sulfide ores are highly variable, depending on the type of deposit and the history of the ore body (ref. 1).

Coal contains relatively large amounts of inorganic elements and is thus considered a dirty fuel. Its combustion probably releases Te, together with Se and large amounts of S to the atmosphere. As potential air pollutants, the concentrations of Se and Te are expected to vary with time and geographic location relative to that of S because of differences in the solubilities and other physical properties of the oxides of S, Se and Te.

Since the oil crisis of 1973-1978, the consumption of fuel oils by industries has decreased, and a program of fuel oil replacement by coal has been initiated. Recognition that this program may lead to an increase in SO_2 emission led to the implementation of new emission controls for coal combustion facilities, but the corresponding increases expected in Se and Te pollution have received little attention.

Although the redistribution of Te, Se and S in coal combustion products has not been previously studied in detail, currently available data on Se suggest that this element is volatilized during combustion and preferentially adsorbed on the surface of coal fly ash. Ruch et al. (ref. 2) report an average Se content of 2.08 ppm in coal, and Andren and Klein (ref. 3) report 2.2 ppm of Se in coal. Wangen (ref. 4) reports 6.0 ± 1.6 ppm of Se in coal ash, and he suggests that Se and As exhibit increasing concentrations as the particle size of the coal ash decreases. Subsequent measurements here (ref. 5) revealed Te, Se and S concentrations in local aerosol particulate matter of (0.45 ± 0.13) ng m^{-3}, (3.63 ± 0.78) ng m^{-3} and (4.45 ± 1.40) µg m^{-3}, respectively, and it was noted that the combustion of coal may be a major emission source for the Te, Se and S in atmospheric particulate matter.

The present study was undertaken to acquire basic concentration data for Te, Se and S with which the concentrations of chalcogen elements in air may be compared. The results may also be used as a basis for possible air pollution emission and air quality control measurements.

METHODS

Apparatus and reagents

Ion exchange columns: Amberlite IR-120, 50-80 mesh (H^+-form) is packed into a column, 1.2 cm I.D. by 10 cm length. The rate of elution for Se and Te was about 1 ml min^{-1}.

Atomic absorption spectrometer: Te and Se were determined with a Perkin-Elmer model 503 spectrometer equipped with a HGA-2200 graphite furnace, deuterium arc background corrector, and hollow cathode lamps.

Activation and counting facilities: Samples were irradiated in a 200 kw swimming pool reactor on campus, and characteristic γ-rays of the activated products were measured with a Ge(Li) detector and a 2000 channel pulse height analyzer.

Standard Te(IV) solution, 1000 ppm: This standard was prepared by dissolving 173.7 mg of analytical grade sodium tellurite (Na_2TeO_3) in 5 ml of 12M analytical grade hydrochloric acid and diluting to 100 ml with distilled water.

Standard Te(VI) solution, 1000 ppm: This standard was prepared by dissolving 186.6 mg of analytical grade sodium tellurate (Na_2TeO_4) in 5 ml of 12M analytical grade hydrochloric acid and diluting to 100 ml with distilled water.

Standard Se(IV) solution, 100 ppm: This standard was prepared by dissolving 142 mg of analytical grade selenium dioxide (SeO_2) in 5 ml of 12M analytical grade hydrochloric acid and diluting to 1000 ml with distilled water.

Analytical Procedure

Coal samples were pulverized and sieved to 80-mesh size. A 1.00 g portion of the sample was weighed into a teflon beaker, and 5 ml each of conc. HF and HNO_3 were added. This was digested and taken to dryness on a hot plate at 100-110°C. Then 5 ml conc. HNO_3 and 10 ml conc. $HClO_4$ were added, the solution heated to complete digestion, and then taken to near dryness. The remainder of the procedure for the determination of Te and Se follows that used in our earlier work for the determination of these elements in aerosols (ref. 6).

The $HClO_4$ and HNO_3 were expelled from the residue by repeated evaporation to near dryness after the addition of 1.5 ml portions of conc. HCl. The final residue was dissolved in 20 ml of 0.05 N HCl and the solution placed on the ion exchange column. The Se was eluted directly with an additional 80 ml of 0.05 N HCl, and then Te was eluted with 150 ml of 0.3 N HCl. For both elements, the flow rate for elution was ~ 1 ml min^{-1}.

Each elutent was evaporated to 1-2 ml, and then diluted to a standard volume. Aliquots of these solutions were placed in sampling cups, and transferred to the graphite furnace in 20λ Eppendorf pipets.

The peak absorbance for Te was measured at 214.3 nm wavelength using a 3 nm slit, and that for Se was measured at 196.0 nm wavelength using a 2 nm slit width. The furnace purge gas was nitrogen.

Neutron activation and γ-ray spectrometry were used to determine the concentrations of up to 35 additional elements, depending on their concentrations relative to the level of interference. Elements determined by activation analysis include Na, Mg, Mn, K, I, Si, F, P, Ti, Rb, Al, Ca, Fe, La, Lu, Ce, Eu, Cr, Dy, Sm, Sc, Cu, Ba, Hf, Ga, Ni, Co, Cs, Ag, Pb, Br, In, Mo, Hg, As and S (ref. 7). The samples were irradiated in the local reactor at a flux of 2×10^{12} n cm^{-2} sec^{-1}. For short-lived activities, the sample and monitors were irradiated for 2 minutes, allowed to cool for 3 minutes, and then counted for 300 seconds. Intermediate-lived activities were measured by counting for 2,000 seconds, about 15 minutes after the irradiation. Long-lived activities were measured in samples that had been irradiated for 2 hours, allowed to cool for 24 hours, and then counted for 4,000 seconds. Longer-lived activities were measured by counting for 6-10 hours about 30 days after the irradiation.

RESULTS AND DISCUSSION

The results of our analysis for Te, Se and S in coal samples are shown in Table 1, together with elemental ratios. The average values obtained for weight ratios of these chalcogen elements in coal, Te/Se = $(2.7 \pm 0.5) \times 10^{-1}$, Te/S = $(3.2 \pm 1.0) \times 10^{-5}$ and Se/S = $(1.2 \pm 0.3) \times 10^{-4}$, show decreasing relative abundances of the heavier elements. This is also the case for average

TABLE 1.
Concentrations and ratios of tellurium, selenium and sulfur in five coal samples

Sample No.	Te µg/g	Se µg/g	S %	Weight Ratios Te/Se x 10	Te/S x 10^4	Se/S x 10^4
1*	0.92	4.30	3.85	2.1	0.24	1.12
2*	0.91	3.81	3.76	2.4	0.24	1.01
3**	0.84	2.75	2.04	3.1	0.41	1.34
4**	0.75	2.25	2.81	3.3	0.27	0.80
5**	0.58	2.04	1.29	2.8	0.45	1.58
Average	0.80 ±0.14	3.03 ±0.99	2.75 ±1.10	2.7 ±0.5	0.32 ±0.10	1.17 ±0.30

*Samples 1 and 2 are from Missouri.
**Samples 3, 4 and 5 are from Illinois.

crustal rock, where the weight ratios are Te/Se = 0.2×10^{-1}, Te/S = 0.6×10^{-5}, and Se/S = 2.9×10^{-4} (ref. 8). Thus, coal has enriched Te relative to crustal Se by about a factor of 10, but relative to crustal S the Te is enriched only by a factor of 5, i.e., the coal has enriched S relative to crustal Se by about a factor of 2.

TABLE 2
Concentrations and ratios of tellurium, selenium and sulfur in coal ash samples

Sample*	Te µg/g	Se µg/g	S %	Weight Ratios Te/Se x 10	Te/S x 10^4	Se/S x 10^4
Bottom Ash	3.00 ±1.24	12.53 ±3.11	0.85 ±0.23	2.4 ±0.6	3.53 ±1.35	14.7 ±2.6
Economizer Hopper Ash	2.85 ±0.72	12.52 ±4.23	0.92 ±0.21	2.3 ±0.7	3.10 ±1.42	13.6 ±2.9
Fly Ash	2.10 ±0.51	11.41 ±2.91	1.47 ±0.35	1.8 ±0.5	1.43 ±0.14	7.8 ±1.7

*All samples are from the university power plant.

The results of our analyses for Te, Se and S in various ash samples collected from the university power plant are shown in Table 2, together with their weight ratios. The nomenclature shown for these ash samples is that used by personnel of the power plant. The bottom ash is the coarse, salt-rich material that remains after relatively complete combustion, the economizer hopper ash is collected and recycled into the burners to complete combustion, and the fly ash is soot-like, low-density material collected in the stacks. Elemental ratios in the first two types of coal ash show about a ten-fold increase in values of the Te/S and Se/S ratios relative to those in coal, but the Te/Se ratio is unchanged. The fly ash displays less enrichment of Te and Se relative to S. This redistribution pattern reflects the greater volatility of S and its oxides relative to those of Te and Se (ref. 9).

TABLE 3
Concentrations and ratios of tellurium, selenium and sulfur in aerosols, fly ash and coal

Sample (#)	Concentrations			Weight Ratios		
	Te*	Se*	S**	Te/Se x 10	Te/S x 10^4	Se/S x 10^4
Aerosol$^+$ (12)	0.45 ±0.13	3.63 ±0.78	4.45 ±1.40	1.3 ±0.3	1.03 ±0.12	8.7 ±2.3
Fly Ash (5)	2.10 ±0.51	11.41 ±2.91	1.47 ±0.35	1.8 ±0.5	1.43 ±0.14	7.8 ±1.7
Coal (5)	0.80 ±0.14	3.03 ±0.99	2.75 ±1.10	2.7 ±0.5	0.32 ±0.10	1.2 ±0.3

#Number of samples analyzed.
*Concentrations of Te and Se are ng m^{-3} for aerosols and ug g^{-1} for fly ash and coal.
**Concentrations of S are ug m^{-3} for aerosols and weight percent for fly ash and coal.
+(ref. 5)

The concentrations and elemental ratios of chalcogen elements in fly ash and coal are compared with those found in atmospheric aerosol (ref.5) in Table 3. The elemental ratios in aerosols are much closer to those in fly ash than to those in coal. These results suggest that the bulk of the chalcogen elements in local aerosols may come from combustion or from other sources in which Te and Se are enriched relative to S. Zoller and Reamer (ref. 10) report, for example, that the enrichment of Se in aerosols from oil combustion exceeds that in aerosols from coal combustion, and that the enrichment of Se in aerosols from volcanic eruptions is sometimes even higher than that in aerosols from oil combustion.

To illustrate the ease with which such sources may dominate the abundance pattern of chalcogen elements in aerosols, we compare the enrichment factors (EF) of elements in coal, fly ash and local aerosol samples.

$$EF = (X/Al)_{sample}/(X/Al)_{crust} \quad (1)$$

The term $(X/Al)_{sample}$ in equation (1) is the ratio of the concentration of element X to that of Al in the sample, and $(X/Al)_{crust}$ is that ratio for elements in igneous rocks of the upper continental crust (ref. 8).

Figure 1 shows the average EF values obtained for chalcogen and other elements in coals, and Figure 2 shows the average EF values obtained for fly ash. Chalcogen elements are highly enriched in coals and fly ash, but the combustion process causes a shift in the relative enrichments of S and Se. This probably reflects the fact that a larger fraction of the S combustion products remain as a gas than do those of Se.

The average EF values obtained for local aerosols are shown in Figure 3. The enrichment pattern of the chalcogen elements in aerosols is very similar to that in fly ash, although their enrichment relative to Al is about an order of magnitude higher in the aerosols than in the fly ash. This probably indicates that Al_2O_3 and other refractory inorganic combustion products are more abundant in the fly ash trapped in the smoke stacks than in the very fine particles that are released, but highly volatile elements such as S and Br are more enriched in the latter than in the fly ash that is trapped in the stacks. Alternately, differences in the EF patterns shown in Figure 2 and Figure 3 may reflect the contribution of other than coal combustion products to the inventory of chalcogen and other elements in local aerosols.

CONCLUSION

Analysis of Te and Se by atomic absorption, after removing interfering elements such as arsenic and copper by cation exchange, appears to be suitable for identifying the pathways of these elements in the combustion of coals.

Our results showed large variations (4-5 orders-of-magnitude) in the enrichment factors of various elements in coal, with the three chalcogen elements showing the largest EF values. After combustion, the fly ash samples displayed large EF values for volatile elements, such as Te, Se, As, S and Mo, that are preferentially released during ashing.

Values of elemental ratios and enrichment factors of chalcogen elements in atmospheric aerosols are much closer to those found in fly ash than to those in coals. These results suggest that coal combustion may be a major emission source of chalcogen elements in local aerosols.

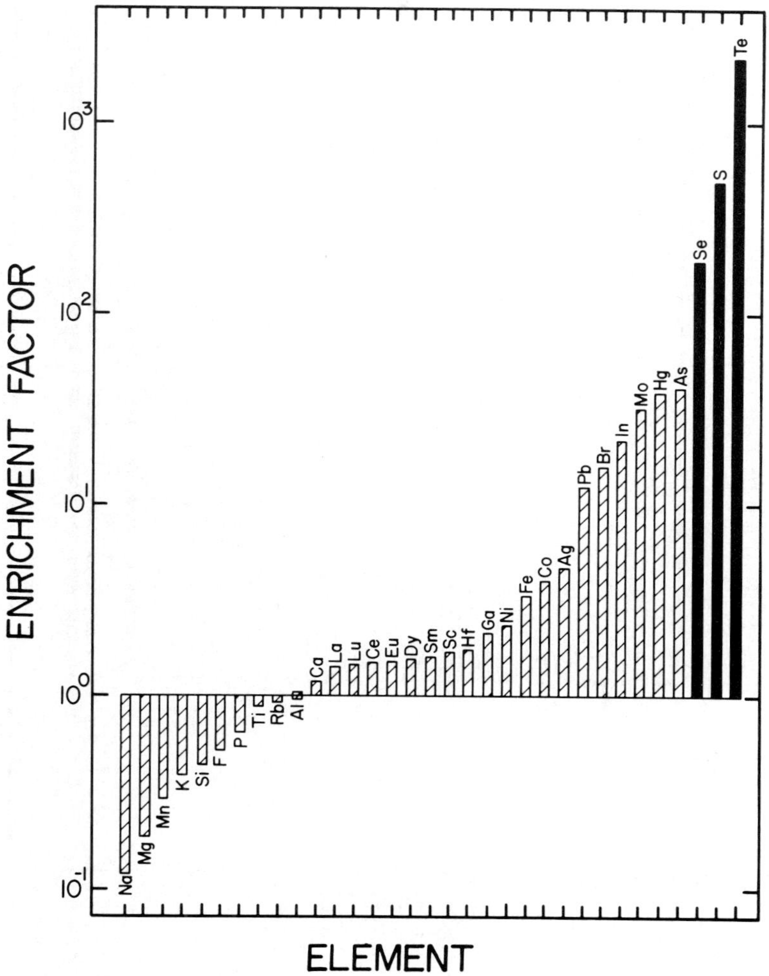

Fig. 1. Average values for the enrichment factor (EF) of chalcogen (filled bars) and other elements in coals, arbitrarily arranged in order of increasing EF values.

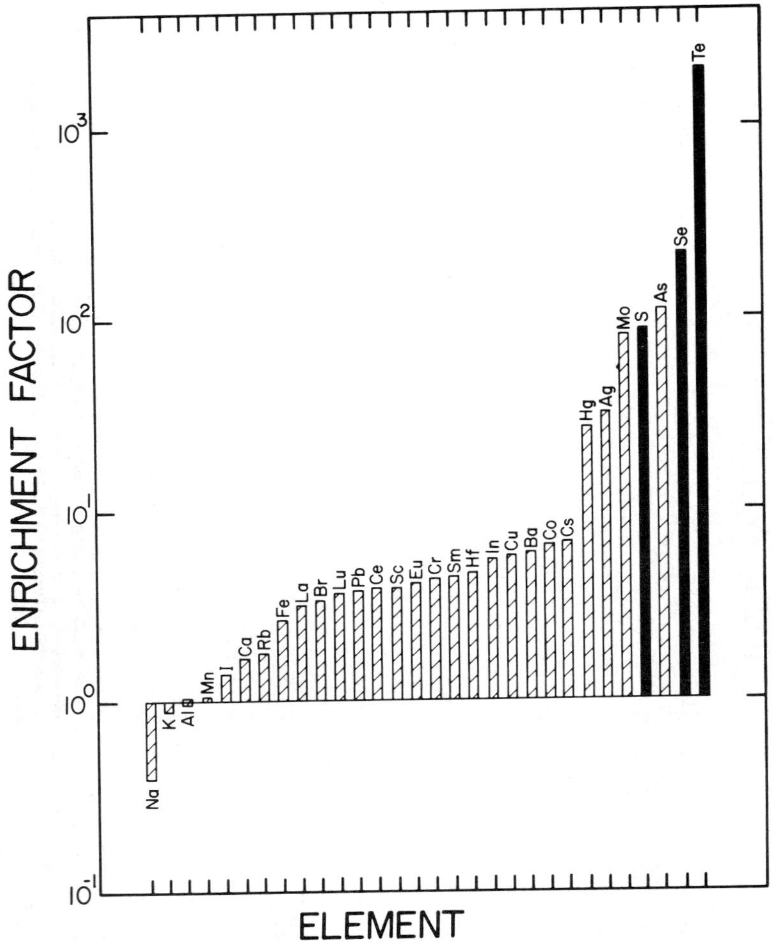

Fig. 2. Average values for the enrichment factor (EF) of chalcogen (filled bars) and other elements in fly ash.

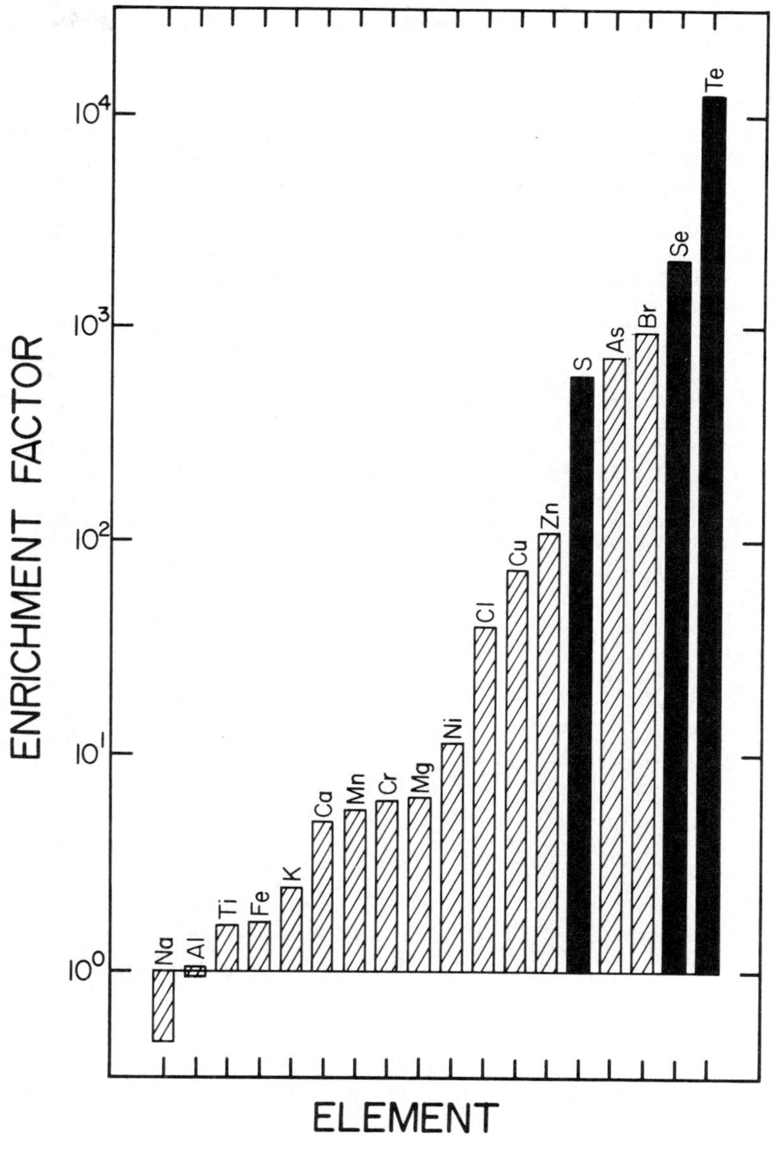

Fig. 3. Average values for the enrichment factor (EF) of chalcogen (filled bars) and other elements in atmospheric particulate material collected over the period of July-October, 1983 at Rolla, Missouri.

ACKNOWLEDGEMENTS

We are grateful to Mr. Larry Matson of the university power plant for providing samples and to Dr. Ernest Bolter of the UMR Department of Geology for his interest and help in making available instrumentation used in this project.

REFERENCES

1. N. D. Sindeeva, Mineralogy and Types of Deposits of Selenium and Tellurium, Intersciences Publishers, New York, 1964, 363 pp.
2. R. R. Ruch, H. J. Gluskoter and N.F. Shimp, Environmental Geology Notes, No. 72, Illinois State Geological Survey, Urbana, IL, 1974, 92 pp.
3. A. W. Andren and D. H. Klein, Envir. Sci. Technol. 9 (1975) 856-858.
4. L. E. Wangen, Envir. Sci. Technol. 15 (1981) 1080-1088.
5. K. Y. Chiou and O. K. Manuel, Chalcogen Elements in Atmospheric Particles, manuscript submitted to Geophysical Research Letters, 1985, 12 pp.
6. K. Y. Chiou and O. K. Manuel, Anal. Chem. 56 (1984) 2721-2723.
7. R. R. Ruch, R. A. Cahill, J. K. Frost, L. R. Camp, and J. J. Gluskoter, J. Radioanal. Chem. 38 (1977) 415-424.
8. K. H. Wedepohl, Geochemistry (translated from German by E. Althans), Holt Rinehart and Winston, Inc., New York, 1971, p. 65.
9. R. C. Weast, M. J. Astle and W. H. Beyer, CRC Handbook of Chemistry and Physics, 64th Edition, CRC Press, Boca Raton, FL, 1983.
10. W. H. Zoller and D. C. Reamer, in Proceedings of the Symposium on Selenium-Tellurium in the Environment, Industrial Health Foundation, Inc., Pittsburgh, 1976, pp. 54-66.

PART 2

ENVIRONMENTAL PROBLEMS AND IMPACT OF HIGH SULFUR ON COAL UTILIZATION

THE RELATIONSHIP OF BIOLOGICAL SULFUR CYCLING AND COAL INDUSTRIES TO ATMOSPHERIC ACID AND ACID DEPOSITION

P.R. DUGAN
Department of Microbiology, The Ohio State University, Columbus Ohio

ABSTRACT

Atmospheric patterns continue to show that acid and sulfur oxides are concentrated regionally and coincide both with mining and combustion of high sulfur coal. In addition to SO_2 the atmosphere receives significant amounts of biogenic H_2S, which is oxidized to SO_x and acid in the atmosphere. Neutral sulfate can be recycled biologically to form acid sulfate at the expense of weak organic acids in anoxic zones, thereby increasing acid deposition with each turn of the sulfur cycle. The primary key to acid deposition in the eastern U.S. is therefore total sulfate. Mining in this context is viewed as a large scale process that exposes reduced sulfur to atmospheric oxygen and moisture which is accompanied by the inevitable production of acid.

INTRODUCTION

Data concerning: (A) The existence of acid deposition, (B) the regional loci of highest acidity and (C) the effects of acid deposition on sensitive aquatic and forrest ecosystems appear to be so convincing that these issues are no longer seriously questioned (4, 7, 16). In some regions acid in the deposition is dominated by nitrous oxides (NO_x); whereas in other regions sulfur oxides (SO_x) dominate. Both NO_x and SO_x have long been known to be extremely deliterious to vegetation (17, 24). In the case of NO_x, photochemical reaction products such as peroxyacetyl nitrate (PAN) are especially phytotoxic and the combined effects of NO_x plus SO_x can be greater than the sum of the individual toxicities. Three environmental factors are important in determining precipitation chemistry: air pollution, agricultural policies and sea spray (21) and air pollution is the dominant source of acid (H^+), sulfate ($SO_4^=$) and nitrate (NO_3^-) in the atmosphere (13).

In the eastern United States, acid in precipitation is correlated much more closely with sulfate than with nitrate or with sulfate plus nitrate. Gorham et. al. (13) reported that H^+ averages 61.7 microequivalents (μeq) per liter (pH 4.21) in a region covering northern Illinois and Indiana, Ohio, West Virginia, Virginia, Pennsylvania and New York. The acid concentration decreased concentrically away from the region. Wet sulfate deposition in the most polluted area was approximately 31 kilograms per hectare per year compared to 15 Kg./ha./yr. in an outer region which included Minnesota, Wisconsin, Upper Michingan, Arkansas, Mississippi, Florida, Georgia, Tennessee, North Carolina and Maine. A pH range of 4.6 to 4.7 is considered to be the lower limit for damage to sensitive ecosystems (11, 14) and wet sulfate loadings of about 15 Kg/ha./yr. produce precipitation acidities at that damage threshold (13).

Consequently the major remaining questions concerning acid deposition are: What is the relative contribution of various sources of acid? and What can be done about it?

In the eastern U.S. the isopleths that dilineate the greatest concentration of atmospheric acid coincide with the eastern coal fields (see 3). Although much has been published concerning SO_2 emissions from the combustion of high sulfur coal in the Appalachian region, there has been a recent controversy over what action should be taken to alleviate the problem of acid precipitation because of the lack of precise correlation between the decrease in SO_2 emissions and the anticipated decrease of acid in precipitation (1, 19, 22, 23). The following cases have been cited: (A) In Virginia acidity in precipitation increased by 74% while measured sulfate dropped by 3%, (B) In Pennsylvania acidity increased by 216% while sulfate dropped by 23% and (C) In Illinois acidity increased by 2700% while sulfate increased by only 22%. The lack of linearity or proportionality between SO_2 emissions and sulfate depositions has also been questioned, expecially in the forrested mountain areas where some evidence points to forrest litter as a source of acidity (19).

FUNDAMENTALS

It is time to return to our fundamental knowledge of acids, bases, salts and buffers. Again, with reference to the eastern U.S. where atmospheric acidity is predominantly associated with sulfate, virtually all the isopleths of the various measurements that track acid deposition have a pattern of roughly concentric rings with the highest values centered over the coal producing region (H^+ in precipitation (3), cumulative wet sulfate deposition (4) weighted mean pH, SO_2 emissions, average sulfate concentrations (7, 16, 22).

pH is a measure of hydrogen ion activity -not total acidity and pH is dependent upon dissociation of H^+ ions from the anions with which they are associated. Both sulfurous acid, H_2SO_3 (i.e. SO_2 dissolved in water) and sulfuric acid H_2SO_4 each have two ionizable H^+ ions and each dissociates from its anion with a different constant (see Table 1). It is possible, for example, to take a dilute solution of sulfuric acid and add sulfate plus other anions and increase the pH while increasing the sulfate concentration.

The reverse is also possible. Divalent (or trivalent) salts in association with their free acids are buffers although the sulfate system will have buffering effect at a much lower pH than the familiar carbonate-bicarbonate or phosphate systems. pH is not a good indicator of total acidity. Titratable acidity must be used to provide accurate values of acid potential when multivalent anions are present. The pH resulting from deposition of sulfate depends on what other anions and cations are also present as well as the amount of each. In addition, nitrogen oxides are known to catalyze the

TABLE 1

Ionization constants (Ka) of some acids in water.

	Ka	
acetic (organic)	1.86×10^{-5}	$pK_a = -\log K_a$
lactic (organic)	1.3×10^{-4}	
H_2S (1) (mineral, weak)	1×10^{-7}	$pH = pK_a + \log \frac{[Anion^-]}{[Acid]}$
H_2S (2) (mineral, weak)	1.26×10^{-13}	
H_2SO_3 (1) (mineral)	1.7×10^{-2}	
H_2SO_3 (2) (mineral)	6.2×10^{-8}	
H_2SO_4 (1) (mineral, strong)	complete dissociation	
H_2SO_4 (2) (mineral, strong)	2×10^{-2}	

oxidation of SO_2 to su furic acid as shown in equation 1, a reaction used in the lead chamber process for making sulfuric acid.

$$NO_2 + SO_2 + H_2O \rightarrow H_2SO_4 + NO \quad (1)$$

Some further consideration of sulfur in the environment is warranted.

ENVIRONMENTAL CYCLING OF SULFUR

Sulfur is the 12th most abundant element in the earths crust. Most of it occurs in combined form as either metallic sulfide or metallic sulfate and most is quite insoluble in water. Some free (elemental) sulfur is found; usually in association with gypsum or adjacent to volcanos.

Over geologic time most sulfur has been sub-surface and therefore under anoxic conditions. It has been placed into contact with atmospheric oxygen and water by either volcanic eruption or slow weathering and the rate of cycling in the environment was determined by these natural processes. More recenty the cycle has shifted toward oxidation because of large scale exposure to the atmosphere due to mining.

On a global scale the natural sulfur resevoir consists primarily of an oxidized form (gypsum, $Ca\ SO_4 \cdot 2H_2O$) and a reduced form (metallic sulfides, predominantly iron pyrite, FeS_2) in approximately a 3:2 ratio with a small amount of free elemental sulfur (S_8^o). Gypsum enters the cycle through solution as Ca^{++} and $SO_4^=$ ions, whereas pyrite is extremely insoluble in water and is dependent upon oxidation in the presence of O_2 for cycling, as shown in equation 2 (see Refs. 12, 20).

$$4\ FeS_2 + 15\ O_2 + 8\ H_2O \rightarrow 2\ Fe_2O_3 + 8\ H_2SO_4 \qquad (2)$$

This process results in the net production of sulfuric acid. The cycling of pyritic sulfur is therefore inseparable from iron and oxygen. When gypsum cycling is considered, the sulur cycle is also inseparable from calcium and carbon. Equation 3 illustrates the formation of gypsum and carbon dioxide from sulfuric acid and calcium carbonate (limestone).

$$H_2\ SO_4 + Ca\ CO_3 \rightarrow Ca\ SO_4 + CO_2 + H_2O \qquad (3)$$

The world's oceans contain an average of about 900 milligrams sulfur per liter as $SO_4^=$ and the world has an estimated 1.2×10^{21}L. water (approximately 10^{15} tons of dissolved sulfur). On average the oceans contain about 10^9 times as much sulfur as the atmosphere and the ocean serves as a net sink for atmospheric sulfur. Estimates of the global atmospheric load of sulfur by source are given in Table 2 and estimates for the atmosphere over the eastern U.S. are in Table 3.

TABLE 2

Estimates of global atmospheric sulfur load. 260 million tons per year.

	Million T./Year
From Ocean	
Biogenic (H_2S, SO_4)	50
Spray (Na_2SO_4, neutral sulfate)	44
90% recycle over ocean	
10% moves over land	(9.4)
From Land	
Biogenic	48
Volcanic	5
Fuel Combustion	104
10% neutral SO_4 from ocean	9.4
Annual Total S over Land	166.4

TABLE 3

Estimates of atmospheric sulfur load over Eastern United States. 17.32 million tons per year.

		Million T./Year
Natural		
From Ocean		0.72
Biogenic		(0.60)
Terrestrial Biogenic		0.10
Flow from West (neutral SO_4)		0.50
Anthropogenic		16.00
	Total	17.32

GEOGRAPHIC CORRELATION OF ACID FORMATION AND MINING

The association of pyritic sulfur with coal deposits is well recognized (2, 9). Coal in the Appalachian region has an especially high average sulfur content of over 3.0% (5). However, averages can be very misleading and the total sulfur content varies with specific location and coal seam within the range of 0.5% to over 10%. Table 4 shows examples of the range in coal quality for several Appalachian and other states.

It has been estimated that 4 million tons of acidity drains into about 10,500 miles of streams in the Appalachian region each year. This polution spans 10,000 square miles across 11 states and also adversely affects 29,000 surface acres of reservoirs and other water impoundments (2).

Acid drainage is a problem associated geographically and geologically with the mining industry and is due to production of sulfuric acid from sulfur containing minerals.

Pyritic minerals and shale are often found geologically in intimate contact with coal or sandwiched between seams of coal. Some of the shale, reject coal, pyrite and other non-marketable minerals are extracted with the coal and left on the surface when the coal is taken to market. These piles of minerals, called **refuse piles**, and sometimes referred to as **gob piles**, may extend to several hundred acres. Several kinds of pyritic minerals are found in nature but iron pyrite (FeS_2), also known as fool's gold, is the one most commonly encountered in association with coal. There is no practical way to segregate pyrite from other minerals and low grade coal in refuse originating from either deep or drift mines.

When a coal seam is sufficiently near the surface, a large shovel or drag line can be used to scoop away the overburden and expose the seam of coal. Consequently, overburden which may contain pyrite is piled on the surface. It is often possible to separate much of the pyrite from other minerals in overburden during a stripping operation and bury it deep in the **spoil bank.** The term **spoil** is used in conjunction with stripped overbuden, in contrast to to the refuse or "gob" from subsurface mines or from coal preparation plants.

Acid formation represents a major environmental problem in the Appalachian region where vast amounts of high sulfur coal are mined. The problem results from exposure of the pyritic minerals to the combined effects of atmospheric oxygen, moisture and a group of acidophilic iron and sulfur oxidizing bacteria (6, 9). These bacteria catalyze the oxidation of pyritic minerals, principally from pyrite, and marcasite to ferric sulfate and sulfuric acid (8, 10) according to equations 4, 5, 6 and 7.

$$Fe^{++} \rightarrow Fe^{+++} + \text{energy} \quad (4)$$

$$2\ S^{-2} + 3O_2 + 2H_2O \rightarrow 2(SO_4^{-2}) + \text{energy} + 4H^+ \quad (5)$$

$$\text{Sum:}\quad FeS_2 + 3O_2 + 2H_2O \rightarrow 2H_2SO_4 + Fe^{+++} \quad (6)$$

The oxidized iron (Fe^{+3}) formed, subsequently reacts with water to produce ferric hydroxide and more acid according to equation 7.

$$Fe^{+++} + 3H_2O \rightarrow Fe(OH)_3 + 3H^+ \quad (7)$$

MICROORGANISMS THAT PRODUCE ACID FROM PYRITE

Relative to the role of microorganisms in pyrite oxidation, a distinction between two categories of microbes is made on the basis of nutritional requirements. Both types have been discussed extensively in the literature.

Autotrophic microorganisms are those organisms which need only carbon dioxide, the oxidation of minerals as their source of energy, and a few trace minerals and/or vitamins as additional nutrients.

Included amongst the known iron and sulfur oxidizing microorganisms are: the acidophilic thiobacilli: **Thiobacillus ferrooxidans** and T. Thiooxidans; the iron oxidizing bacterium: **Leptospirillum ferrooxidans,** the acidophilic thermophilic archaebacteria: **Sulfolobus acidocaldarius** and **S. brierlei** as well as some bacteria whose taxonomic status is presently uncertain. Of the acidophilic microbes known to oxidize pyritic minerals, **Thiobacillus ferrooxidans** is considered to be the primary organism responsible for production of acidic drainage from high sulfur coal, coal refuse and pyritic spoil banks.

Heterotrophic microbes are those which depend upon the oxidation of reduced organic compounds for their energy in addition to their cellular carbon requirements. They also have nutritional requirements for trace amounts of minerals and/or vitamins similar to those of the autotrophic organisms. In general, specific differences among species of heterotrophic microbes are reflected in differences among types of organic compounds required by each species.

Acidophilic bacteria of the **Thiobacillus** group have long been recognized as being associated with pyrite oxidation and acid production in coal mine spoils and refuse and can be readily isolated from acid mine drainage water (8, 9, 18, 19). The bacteria grow optimally in the pH range of 2.0 to 3.5. Maintenance of an adequate supply of Fe^{+2} as an energy source in the absence of high concentrations of organic material requires an environmental pH less than 4.0 because of the rapid auto-oxidation of Fe^{+2} in the presence of O_2 above pH 4.0.

Although iron pyrite will be oxidized chemically in the absence of bacteria, ultimately to $Fe(OH)_3$ and H_2SO_4; the bacteria catalyze the reaction and increase the rate of oxidation up to one million times the chemical rate under acidic conditions (25).

Acidic drainage from exposed high sulfur coal, coal refuse and coal spoils reaches 60,000 milligrams net titratable acidity per liter (6%) with dissolved sulfate reaching 60,000 mg/L (6%) and dissolved iron, 15,000 mg/L (1.5%) (26). The average drainage of one spoil bank was 305 lb./acre/day (as sulfuric acid) with slug doses flushed out as high as 3.3 tons/acre/day. The potential sulfuric acid from coal combustion is summarized in Table 5.

TABLE 4

Examples of the range in ash, sulfur and BTU content of bituminous coal from several states.

State	BTU/LB	% Ash	% Sulfur	
			Pyritic	Total
Alabama	12,676 - 15,092	1.8 - 16.4	.06 - 2.8	.58 - 3.75
Illinois	10,814 - 13,284	7.7 - 24.2	.60 - 5.92	1.15 - 8.2
Indiana	11,550 - 13,477	6.5 - 19.9	.41 - 5.56	.67 - 7.23
Iowa	10,064 - 13,202	11.1 - 23.3	1.80 - 10.55	2.63 - 12.24
E. Kentucky	11,890 - 13,955	5.3 - 19.7	.21 - 1.79	.75 - 3.02
W. Kentucky	8,775 - 13,671	5.2 - 37	.96 - 3.93	2.12 - 5.12
Ohio	8,630 - 13,699	3.5 - 37	.10 - 3.31	.65 - 6.39
Pennsylvania	9,150 - 14,800	3.3 - 37	.04 - 7.0	.58 - 9.21
Utah	12,565 - 13,738	4.1 - 10.2	.06 - .21	.38 - .68
West Va.	9,542 - 14,150	5.3 - 33.3	.01 - 4.53	.56 - 6.13

TABLE 5

Annual potential H_2SO_4 from coal combustion.

	Million Tons	
Production, U.S.	850	
Combustion, U.S. (Avg. 2.2% S)	556	
Utility		(390)
Other		(166)
Combustion, Eastern U.S. (Avg 3.1% S)	372	
Potential SO_2 Generated	22.3	
If all $SO_2 \rightarrow H_2SO_4$	33.5	

If all $SO_2 \rightarrow H_2SO_4 \rightarrow$ Precipitates without neutraliztion
AREA INCL. E. CANADA (1,500,000 sq. mi)
68 lb. conc. H_2SO_4 / acre or
.69 gm./Sq.ft. = 4.3 ml / sq. meter

Actual 1980 Acid Deposition, Approx. 1 lb. / Acre

Mining can be viewed as a very large scale process whereby pre-reduced substances are removed from their anoxic subsurface deposits and exposed to the moisture and oxygen of the atmosphere where they become oxidized. Because oxidations are accompanied by a net release of energy, the natural thermodynamic tendency is in the direction of oxidation; which in the case of sulfur is accompanied by sulfuric acid production. Also, on a thermodynamic basis, chemical reductions require a net input of energy. This is supplied in the environment by either light, concurrent oxidations or pre-reduced organic substances such as fossilized biomass (petroleum, natural gas coal) or contemporary biomass (forrest litter, leaves, plankton, domestic sewage, etc.).

ATMOSPHERIC ACID FROM BIOGENIC H_2S

The net reduction of oxidized sulfur forms (e.g. $SO_4^{=}$) requires energy to drive the reactions which consume acid (equivalent to producing alkali) in the process. In the environment a group of microorganisms, collectively known as "sulfate reducers," is able to utilize sulfate in the absence of oxygen to reduce sulfate to H_2S. These microorganisms must have a nutritional source of energy and that comes from the organic acids found in decomposing biomass on lake bottoms, in forrest litter or other places where the organic content is rich and the oxygen content has been depleted. Equation 8 is an example of this type reaction where the organisms couple the reduction of sulfate to sulfide with the oxidation of lactic acid to acetic acid and CO_2.

$$\underset{\text{lactic acid}}{2\ CH_3-CHOH-COOH} + SO_4^{=} \rightarrow \underset{\text{acetic acid}}{2\ CH_3\ COOH} + 2\ CO_2 + H_2S + 2OH^- \qquad (8)$$

Hydrogen sulfide can be biogenically produced from sulfate in acidic mine drainage in relatively large amounts (15, 27, 28) although quantitation over the coal mine region has not been accomplished. However, the color photograph (projected but not published) clearly illustrates the black precipitate of metallic sulfide that accumulated in the acid drainage when a source of organic nutrient was available (27).

In a study that analysed the relative atmospheric sulfur contributions from three different sources to the Salt Lake City area, Jensen (18) was able to show that copper smelters contributed more than bacteriogenic H_2S from Salt Lake which contributed more than oil refineries plus automobile exhaust in the ratio (20:10:1). The biogenic contribution, ten times greater than oil refineries plus auto exhaust, was significant.

Figure 1 is a schematic drawing which illustrates the overall effect of cycling of sulfur on acid in the atmosphere. Recall that different microorganisms are responsible for the oxidation of pyrite ultimately to sulfuric acid and ferric iron and that iron hydrolyses with water to form ferric hydroxides plus acid. This is equivalent to acidic mine drainage. When that H_2SO_4 or any other $SO_4^=$ finds its way to an anoxic zone where there is a source of organic, the sulfate reducers will reduce the sulfate to H_2S.

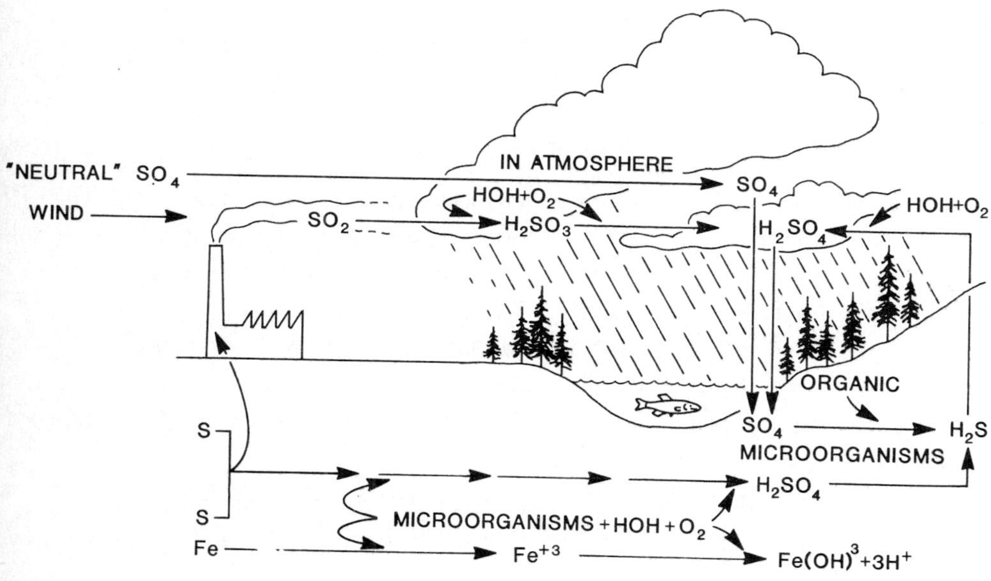

Figure 1. Schematic drawing of the involvement in the cycling of sulfur to produce acid deposition.

H_2S is a volatile gas and it will diffuse into the atmosphere where it will be reoxidized to sulfuric acid as shown in equation 9.

$$2\ H_2S\ +\ 3O_2\ \rightarrow\ 2\ H_2O\ +\ 2\ SO_2\ (2\ H_2SO_3) \qquad (9)$$

H_2S can also react with metals such as iron to produce insoluble metal sulfide precipitates (see equation 10).

$$H_2S\ +\ Fe(OH)_2\ \rightarrow\ FeS\ \text{(black precipitate)}\ +\ 2H_2O \qquad (10)$$

The point is that the source of sulfate for "sulfate reducers" is irrelevant (i.e. it can come from "neutral sulfate from ocean spray, from acid precipitation, from acid mine drainage, gypsum or SO_2 or other) and the microorganisms will convert it to H_2S whereupon it is easily reoxidized in the atmosphere to acid. The question of wet vs dry deposition in this context is also irrelevant. The net effect in the environment is additive and the cycle provides a means where neutral sulfate salts can be converted to strong mineral acid at the expense of consumption of weak organic acid (see Table 1) in anaerobic zones. The key to formation of acid deposition of course is sulfate which correlates with coal mining as well as with combustion of high sulfur fuel and the formation of acid is additive. Each turn of the cycle can produce more acid deposition. Our consideration relative to control measures should focus on how much of the acid deposition is direct from SO_2 and how much is indirect from re-oxidation of H_2S which may come from several sources. Control measures will depend on the source and it will be important to obtain quantitative estimates of the various contributions of each source.

SUMMARY

We are experiencing severe regional dislocations in the dynamics of the natural sulfur cycle because of the increased exposure of sulfide minerals, primarily iron pyrite, to oxygen and moisture due to large seale mining as well as combustion practices.

The oxidation of reduced sulfur is inevitably accompanied by formation of acid. This is also the case with oxidation of iron. These oxidations are catalysed by iron and sulfur oxidizing microorganisms and the rates of reaction are accelerated up to one million fold.

There are several ways in which sulfur forms enter the atmosphere; all of which result directly or indirectly in acid formation. For example:

1. SO_2 from combustion of high sulfur fuel
2. Probable SO_2 directly from high sulfur coal or refuse piles
3. SO_2 and reduced sulfur forms from mine or dump heap fires
4. "neutral sulfate" from ocean spray

5. "acid sulfate" spray or vapor from mine drainage regions
6. microbially generated H_2S gas in local environments where the organic and sulfate content is relatively high and oxygen has been depleted.

Biological recycling of "Neutral" sulfate results in formation of acid sulfate in the atmosphere at the expense of weak acids in anoxic zones.

All of the above correlate with the high sulfur coal mining region and each contribution is additive relative to formation of acid deposition.

The problem of formation of acid deposition is exacerbated by:

uneven distribution of sulfur content in coal.

uneven distribution of fossil energy reserves.

uneven distribution of combustion sources (power plants).

uneven distribution of population use centers.

uneven distribution of precipitation and other atmospheric conditions.

The lack of direct correlation between sulfate deposition and pH of precipitation or receiving water should be anticipated because of the chemistry which governs acidity, buffering, pH and acid-base reactions.

REFERENCES

1. P.H. Abelson, Science. 226 (1984) 1263.
2. Acid Mine Drainage in Appalachia. Appendix C: The Incidence and Formation of Mine Drainage Pollution in Appalachia. (1969) 253 pp. U.S. Corps of Engineeers and U.S. Dept. of Interior.
3. L.A. Barrie, (in) Acid Precipitation, Effects on Ecological Systems. (F.M. D'Itri, ed.), Ann Arbor Press (1982) 141-161.
4. D.W. Bubenick, Acid Rain Information Book 2nd Ed. (1984) Noyes Publ., Park Ridge, New Jersey, pp 397.
5. J.A. Cavallaro, M.T. Johnston and A.W. Duerbrock, U.S. Dept. Interior, Bureau of Mines Rept. of Investigtion, RI 8118 (1976), pp 323.
6. A.R. Colmer, K.L. Temple and M.E. Hinkle, J. Bacteriology 59 (1949) 317-328.
7. F.M. D'Itri (ed.) Acid Precipitation, Effects on Ecological Systems. (1982) Ann Arbor Sci. Publ., Ann Arbor, Michigan, pp 506.
8. P.R. Dugan and D.G. Lundgren, Develop. in Industrial Microbiology. 5(1964):250-257.
9. P.R. Dugan, Biochemistry of Acid Mine Drainage. (In) Biochemical Ecology of Water Pollution. (1972) pp. 123-137. Plenum Publ. Co., N.Y., N.Y.
10. P.R. Dugan, Ohio J. Sci. 75 (1975) 266-279.
11. L.S. Evans, G.R. Hendry, G.S. Stensland, D.W. Johnson, and A.J. Francis, Water Air Soil Pollution. 16 (1981) 469. Nat. Acad. Sci. Press, Wash. D.C.
12. R.M. Garrels and E.A. Perry, (In) The Sea. J. (1974), Wiley Interscience, N.Y.
13. E. Gorham, F.B. Martin and J.T. Litzau, Acid Rain: Ionic Correlations in the Eastern U.S., 1980-1981 Science. 225 (1984) 407-409.
14. A. Henriksen, Nature (London), 278 (1979) 542.
15. A.T. Herlihy and A.L. Mills, Applied Microbiol. 49 (1985) 179-186.
16. R. Herman and A.I. Johnson (eds) Acid Rain, A Water Resources Issue for the 80's. Amer. Water Resources Assoc., Bethesda, MD. 1983, pp 83.
17. J.S. Jacobson and A.C. Hill (eds). 1970. Recognition of Air Pollution Injury to Vegetation: A Pictorial Atlas. Air Pollution Control Assoc., Pittsburgh, PA. pp 103.

18. M.L. Jensen, Science. 177 (1972) 1099-1100.
19. A.W. Katzenstein, Science. 228 (1985) 390-391.
20. W.W. Kellog, R.D. Cadle, E.R. Allen, A.Z. Lazarus and E.C. Martell, Science. 175 (1972) 589-596.
21. J.W. Munger and S.J. Eisenreich, Environ. Sci. Technol. 17 (1983) 32A.
22. National Research Council. Acid Deposition. Atmospheric Processes in Eastern North America (1983) National Academy Press. Washington, D.C.
23. M. Oppenheimer, J.N. Galloway, G.E. Likens and S.A. Norton, Science. 227 (1985) 1154-1155.
24. Photochemical Air Pollution: Formation, Transport Effects. 1975. Rept. NRCC No. 14096, Nat. Res. Council Canada, Ottawa, Canada, pp 224.
25. P.C. Singer and W. Stumm. Science 167 (1970) 1121-1124.
26. Truax-Traer Coal Co. Control of Mine Drainage from Coal Mine Mineral Wastes. Phase I. Hydrology and Related Experiments. Report. U.S. Environmental Protection Agency, Water Pollution Control Research Series 14010ODDH. (1971) 148 pp.
27. J.H. Tuttle, P.R. Dugan and C.I. Randles, Jour. Bacteriol. 97 (1969) 594-602.
28. J.H. Tuttle, P.R. Dugan and C.I. Randles, Applied Microbiology 17 (1969) 297-302.

IMPACT OF HIGH SULFUR ON COAL UTILIZATION

R. B. ENGDAHL

Fuels and Combustion Technology Section, Battelle, Columbus Laboratories

505 King Avenue, Columbus, Ohio 43201

ABSTRACT

The sulfur in coal causes problems at the mine from acid attack on mining machinery and from the discharge of acid minewater. Corrosion and wear of coal processing equipment is increased by higher sulfur content. Boiler superheater corrosion by sulfates is so severe at high temperatures that a cap has been placed on the maximum operating temperature and, therefore, on the efficiency of steam power plants.

Corrosion of scrubbers and chimneys and the cost of sulfate sludge disposal are major burdens of high-sulfur coal.

INTRODUCTION

Although the primary concern about the sulfur content of coal is the resulting air pollutant, sulfur oxides, when coal is burned, there are other problems with sulfur; and generally the higher the sulfur content, the worse the problem.

Acid Minewater

These problems of high sulfur coal begin in the mine, where the inorganic sulfur content, iron pyrites, $FeSO_2$ oxides to form acidic compounds. The first worrysome result is sulfurous and sulfuric acid in mine drainage water.

Very often the higher the pyrites content of the coal, the more acid the mine drainage water. Not only does this cause serious stream pollution when discharged to surface streams, but it is a distinct corrosion hazard when such water is used in coal washing equipment. Mitchell and Leonard (ref. 1) have pointed out that often this acid mine drainage is the only readily available makeup water for the coal washery. To prevent rapid corrosion of the wet coal cleaning equipment water treatment is necessary. Thus, control of acidic corrosion from the sulfur content adds directly to the cost of coal mining and preparation.

Along the way to the coal customer, the pyrites and its acid byproducts cause corrosion of expensive mining machinery and coal handling equipment.

Spontaneous Combustion

In addition, iron pyrites has long been felt to be contributory to the spontaneous combustion of coal in storage, although Lowry (ref. 2) has summarized a long-standing controversy on this point. Fires can occur in piles of even low-sulfur coal, because air flowing through the air spaces in a poorly piled, segregated coal pile becomes heated by oxidation of the coal surface. Also iron pyrites, when it oxidizes, releases considerable heat, so the higher the sulfur content, the greater is the potential for coal pile fires. Of course, fires can be avoided by careful piling of the coal. If it is packed and sealed to keep air from flowing over the moist airfaces of the coal and the pyrites, spontaneous combustion is very unlikely.

Corrosion of Coal-Handling Equipment

As the sulfurous coal is moved along its way to final utilization, coal handling equipment is continually subject to corrosion by the water-soluble acidic sulfur compounds.

Pulverizer Wear

Even when the coal is pulverized at the power plant, sulfur content has a dramatic effect on pulverizer wear. Hardgrove (ref. 3), a pioneer in developing the utilization of pulverized coal, in 1937 showed the following effect of sulfur content on the life of grinding elements in a ball mill when the coal ash content was 10 percent:

Coal Sulfur Content %	Life of Grinding Elements Hours
0.5	27,500
1	15,000
2	9,980
3	6,800
4	5,000

This effect, with element life for high-sulfur coal being only 1/5 to 1/3 that for low sulfur coal, is probably a result of combined abrasion and corrosion. The ash particles are highly abrasive, and the corrosive action of the acidic sulfurous compounds on freshly abraded steel surfaces probably accelerates the wear.

Boiler-Superheater Corrosion

When the high-sulfur pulverized coal burns in large, high temperature furnaces, the discovery was made back in 1950's that alkali pyrosulfates and iron trisulfates deposited on the steel tubes of superheaters caused unacceptable corrosion rates. These corrosive salts can be minimized by furnace design and operation to control the formation of sulfur trioxide which then reacts to form the potentially corrosive sulfates. But SO_3 cannot be eliminated, and in general, the higher the sulfur content of the coal, the more SO_3 is formed in the boiler-furnace.

Reid (ref. 4) has described the research on boiler-superheater corrosion that resulted in a widely agreed-upon design decision to limit the maximum steam temperature in future plants to less than 1100 F (593 C). This placed a limiting cap on the energy conversion efficiency of all future steam electric plants which continues to this day.

Acid Dewpoint Problems

After the sulfurous flue gases leave the boiler, condensation and corrosion problems further limit the amount of useful energy that can be extracted from those gases.

For moist flue gases uncontaminated by any acid gases, the dewpoint is approximately 100 F (38 C). The result below this temperature is condensation. When even a small amount of sulfur in fuel oil, gas or coal results in the formation of sulfur trioxide during combustion, the dewpoint increases to 250 or 270 F (120 C to 130 C). Any equipment used to cool the flue gases below the dewpoints so as to improve boiler efficiency, will be subject to condensation and acidic corrosion.

Electrostatic Precipitation

One slight advantage of high sulfur content is that electrostatic precipitators for capture of flyash from coal-burning furnaces work much better because of the sulfur trioxide content of the gases. SO_3 improves the electrical conductivity of the flyash particles so that the electrically charged particles on the precipitator plates can release their charge and can be cleaned off more easily. With low sulfur fuel, not enough SO_3 is formed to provide high surface conductivity of the ash particles and small amounts of SO_3 added will dramatically improve precipitator performance. Although this may appear to be a distinct advantage for high sulfur coal, it is a relatively minor one compared to all the detrimental aspects of sulfur that have been discussed.

Chimney Deterioration

Even though a plant may be well-designed and operated to avoid acid dewpoint problems, small amounts of flue gas will inevitably invade any cracks in chimney walls. The result is then acid condensation in the cracks and deteroration of the mortar and of any metal supports that may be incorporated in a conventional brick-lined chimney. To combat this problem, very expensive materials are being tried.

Scrubber Corrosion

Corrosion of scrubbers by sulfur compounds is a major factor in first cost and in operating cost and reliability. EPRI reports have detailed those problems and feasible solutions.

Scrubber Sludge Disposal Volume

Almost half of the cost of owning and operating SO_2 scrubbers results from the problem of sludge disposal. The higher the sulfur content of the coal burned, the greater the sludge volume. Until this sludge can be converted to a material of useful value, high-sulfur coal utilization will continue to bear the substantial cost burden of waste sludge disposal.

Summary

Almost all of the points discussed here constitute burdens of cost and operation unique to high sulfur coal utilization. Any advantages of high sulfur content that were covered are quite minor compared to the substantial burdens incurred. Thus, this review calls attention once more to the urgent need for stronger efforts to remove most of the sulfur from coal before it is burned.

REFERENCES

1 D. R. Mitchell, and J. W. Leonard, Coal Preparation, Amer. Inst. Min., Met., Petr., Engrs., New York, 1968.
2 H. H. Lowry, Chemistry of Coal Utilization, Vol. I, John Wiley & Sons, New York, 1945.
3 R. M. Hardgrove, Grindability of Coal from the Producer's Viewpoint, Proc. 17th Fuel Engineer's Meeting, Appalachian Coals, Inc., Cincinnati, Ohio, January 11, 1937.
4 W. T. Reid, External Corrosion and Deposits, Boilers and Gas Turbines, Elsevier, New York, 1971.
5 Battelle, Columbus Laboratories, Construction Materials for Wet Scrubbers, Vol. I and Vol. II, Electric Power Research Inst., Palo Alto, CA, March 1981.

TECHNOLOGICAL OPTIONS FOR COAL DESULFURIZATION

C. W. BULLARD

University of Illinois, Office of Energy Research, 108 Observatory, 901 S. Mathews, Urbana, Illinois 61801 (USA)

ABSTRACT

This paper examines the fundamental physical constraints on technologies for removing sulfur from coal before, during, and after combustion, isolates critical factors affecting their relative economic viability, and identifies key uncertainties to be resolved by R&D. The baseline technology is simply the transportation and substitution of low sulfur coal for high sulfur coal. Many other technologies are considered, despite the fact their theoretical minimum cost may substantially exceed that of coal substitution, recognizing the potential viability of politically mandated efforts that would incur additional costs to minimize disruption of existing coal production patterns.

INTRODUCTION

In 1980, 24.1 million metric tons of sulfur dioxide were emitted due to anthropogenic activities in the United States. Of these emissions, 75% originated in coal-fired boilers, most in the electric utility industry (ref. 1). Because sulfur dioxide is an important precursor of acid precipitation, a variety of regulatory efforts have been directed at its control.

The current approach to regulating SO_2 is directed not so much at forcing development of new clean coal burning technologies, but at forcing the maximum politically and economically feasible use of existing technology. Standards for new sources of air pollution were revised in 1978 from a purely performance basis to a prescriptive standard mandating removal of 70-90% of sulfur dioxide, reflecting the political unacceptability of substituting low sulfur coal. A more stringent percentage removal requirement would not necessarily lead to increased competition among developers of advanced desulfurization technologies; in fact, it may diminish competition because unit SO_2 removal costs of the various technologies diverge exponentially as 100 percent removal is approached.

Recent concern about the effects of acid precipitation have led to a reexamination of the regulatory approach taken in the Clean Air Act of 1980, which left untouched most non-urban stationary sources of SO_2 emissions. Congressional consideration is now focused on two basic approaches to reducing sulfur dioxide emissions from such facilities by 9-11 million metric tons per year. One approach, aimed at minimizing disruption of high sulfur coal producing regions of the East and Midwest, would mandate use of flue gas desulfurization (FGD) equipment at costs exceeding \$500 per metric ton SO_2 removed. The other approach would leave individual states and utilities free to select the compliance technology in the expectation that most would choose to substitute low sulfur coal, at a substantially lower cost (ref. 2). In this latter case, states could utilize transferable discharge permits to facilitate an economically efficient allocation of a fixed amount of emissions among generators who would purchase and sell the marketable permits. Such an optimal allocation would probably include least emissions dispatching of power plants, along with substitution of low sulfur coal at those facilities where it is most economic. However, the net impact on high sulfur coal production is the same, whether least emissions dispatch or coal substitution is the technology of choice.

The five-year Congressional stalemate over acid rain legislation thus exemplifies the tension between two state-of-the-art technologies for SO_2 control: FGD vs rail (or wire) transportation of low sulfur coal. Development of advanced coal desulfurization technologies will be driven by a combination of environmental, political and economic forces, as the system of air pollution control regulations evolves to demand more SO_2 removal at costs lower than today's FGD technology. While it is reasonable to expect future regulations to provide for economically efficient implementation (e.g. transferable discharge permits), it is also likely that they will continue to reflect the strong economic, social and political interests of the high sulfur coal mining industry: mandating use of desulfurization technology.

To date, only wet and dry FGD technologies have seen widespread application as a result of the New Source Performance Standards (NSPS). Another technology, fluidized bed combustion, is not yet commercially competitive with FGD for utility-scale boilers, but is turning out to be the least costly approach to NSPS compliance at smaller industrial scale facilities, where it can also capture the economic benefits of cogenerating electricity and replacing deregulated natural gas.

In contrast to fluidized bed combustion, a closely related technology, LIMB (limestone injection into a multistage burner) has received virtually no stimulus from the new source performance standards because of inherent technological limitations that prevent it from achieving 70-90% removal. The same applies to chemical and physical coal cleaning technologies having low removal rates.

Synthetic fuel technologies, since they produce low sulfur fluid fuels, might also be considered as desulfurization technologies. During the 1970's, synfuels R&D was driven mainly by a national need to develop substitutes for imported oil and gas. Now that conservation and the associated decline in oil prices have made this need less urgent, the economic value associated with desulfurization will be one of the major forces driving innovation. Examples of this new direction include the Cool Water medium Btu gasification project sponsored by EPRI, and the Allis Chalmers low Btu KilnGas technology cosponsored by the State of Illinois and a number of electric utilities.

This paper presents an overview of several of the major technological alternatives to FGD, focusing on the irreducible material and energy requirements that determine the extent to which each may be able to compete with FGD, and with the baseline technology: transportation of low sulfur coal. The next section presents information on the sulfur composition of coal used in utility boilers; sulfur content is a key variable affecting not only the economics of FGD and competing technologies, but also the major political forces driving technological innovation. The third section details the costs associated with the transportation of low sulfur coal, using engineering cost models of coal mining and transportation to determine the cost per ton SO_2 removed. Section 4 presents the same type of analysis for FGD technology, the state-of-the-art technology supported by the existing regulatory structure. Section 5 examines sulfur capture during the combustion process, potentially the least costly technological alternative to FGD. Section 6 describes coal-based synfuels as a precombustion desulfurization technology rather than an oil/gas substitute. Finally, Section 7 presents a discussion of other precombustion desulfurization technologies, identifying key uncertainties to be resolved through R&D. Section 8 summarizes the conclusions and their R&D implications.

SULFUR IN COAL

The sulfur composition of US coal reserves is shown in Figure 1. Despite the pressures from constituencies having an interest in high sulfur coal production, the average sulfur content of coal delivered to electric utilities declined from 1.80% in 1978 to 1.47% in 1983 (ref. 3). Of the sulfur contained in coal that is mined, only a negligible amount is removed by washing, and less than one-fifth of coal fired generating capacity is equipped with FGD. Under current law, therefore, the rate of SO_2 emissions reduction is tied mainly to the lifetimes of existing facilities, a decision largely in the hands of electric utilities.

The current debate over acid rain legislation is therefore focused on a "window" for accelerating the reductions from existing facilities before their nominal retirement dates. Figure 2 shows such a window, derived from a computer simulation for the State of Illinois, assuming a 45 year lifetime for existing power plants. Congressional attention has focused on the fact that nearly two-thirds of the 1980 installed generating capacity was constructed after 1965; most of these plants are larger than 400 MW, and therefore escape the disecon-

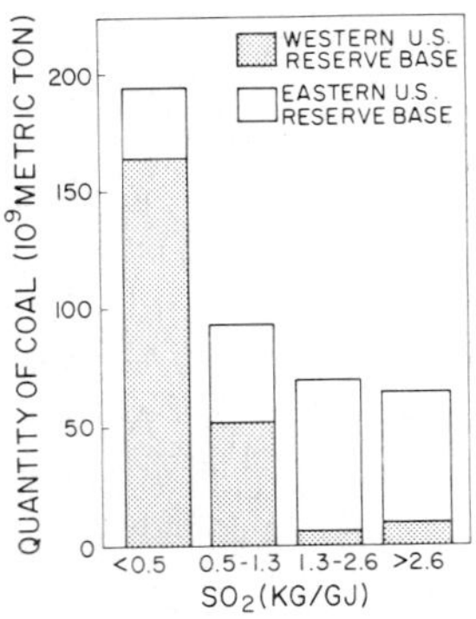

Fig. 1. Sulfur content of US coal reserves (from ref. 2).

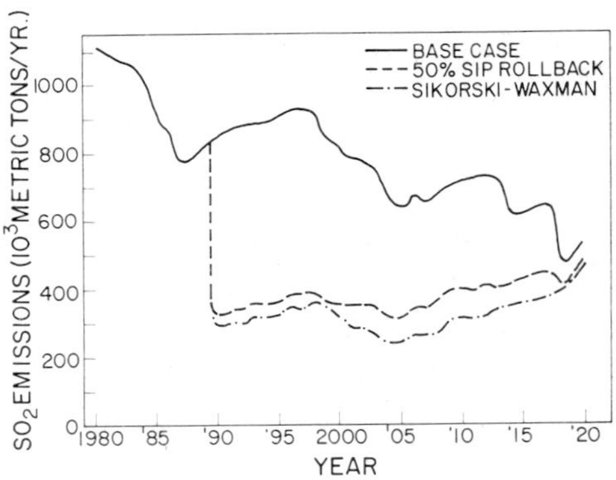

Fig. 2. Sulfur dioxide reduction "window" for Illinois (from ref. 25).

omies of scale associated with FGD retrofits on smaller plants built prior to that time. The average emission rate for plants added between 1965 and 1975 averaged nearly 3 lb SO_2/MBtu (ref. 4).

Some of the technologies discussed in the remainder of this paper may play a significant role in retrofit applications to deal immediately with the acid rain problem. The focus of this analysis, however, is on comparing existing and emerging desulfurization technologies over the long term, and examining the R&D implications.

THE BASELINE TECHNOLOGY: COAL TRANSPORTATION

A detailed econometric analysis (ref. 5) revealed that coal prices in the 1975-80 period did not depend strongly on sulfur content. It could not be determined from that analysis whether low sulfur coals are relatively scarce, or whether their scarcity might have been obscured by such factors as excess capacity in the mining industry, volatility in spot markets, or the vintage of contracts in place in 1980 that were executed during periods dominated by environmental and OPEC-induced uncertainties. Therefore, it was decided to employ a detailed engineering cost model (ref. 6) to identify "ceiling costs" as a function of sulfur content over the long term.

Minemouth costs were first computed for 11 sulfur categories of coal in 35 supply regions, and expressed as a function of seam thickness, depth and size of deposit. This analysis identified 29 price-setting coals having sufficient reserves to meet total US demands for several decades at delivered prices below the cost of more than two-thirds of US coal reserves. In each sulfur category, smaller quantities of even less costly coal were available from other regions, but would be depleted within a few decades. The next step was to plot these delivered (ceiling) costs of the price setting coals as a function of sulfur content in each state. The slope of each curve indicated the long term cost of coal substitution and the convex envelope defined the set of economically optimal choices at the margin.

Figure 3 shows the result of combining these data on substitution choices for the 48 contiguous states, assuming coal demands were distributed by sulfur content as in the 1980 base year. It indicates that significant reductions in SO_2 emissions are possible at relatively modest costs over the long term, using the baseline technology: transportation of low sulfur coal. The curve is surprisingly flat. The displacement of the horizontal axis intercept results from the fact that the long run ceiling costs of many high sulfur coals used in 1980 exceed the ceiling cost of lower sulfur coals. Therefore as the price of these higher sulfur coals rises due to depletion, users can switch to lower sulfur coals at little or no incremental cost, because the costs of the relatively abundant low sulfur coals are less sensitive to depletion.

Due to the long term nature of the analysis, the implicit assumption that boiler modification costs are negligible is justified, because new power plant construction costs do not depend strongly on the rank of coal used (ref. 7). In the short term, coal substitution costs might be greater or less than the long term limits identified here; and possibilities would be limited in existing boilers due to such factors as ash fusion temperatures, and by transportation bottlenecks and boom/ghost town effects in the coal supply system. However, to the extent that such effects can be mitigated by a gradual 20-40 year transition, Figure 3 clearly shows that emerging desulfurization technologies have extremely ambitious cost goals to meet.

Since the amount of coal used by electric utilities is expected to double or triple during the next few decades, these results can also be used to estimate the cost of maintaining utility SO_2 emissions at 50% of 1980 levels over an extended period. Figure 3 indicates that a 50% emission reduction at 1980 levels of coal use could be accomplished at very little cost as prices approach ceiling costs over the long term. If coal use doubled, the marginal cost would increase to more than \$500/ton SO_2. However, Figure 3 shows clearly the infeasibility of reducing emissions to one-sixth of 1980 levels (halving emissions while tripling coal use); the asymptotic limit corresponds to the minimum sulfur content of US coal reserves.

These direct costs of coal substitution would be added to the indirect costs resulting from economic dislocations associated with structural changes in the patterns of coal production and distribution. In the near term, therefore, substitution looks more attractive in terms of direct costs but more costly in terms of the economic dislocations associated with massive structural changes

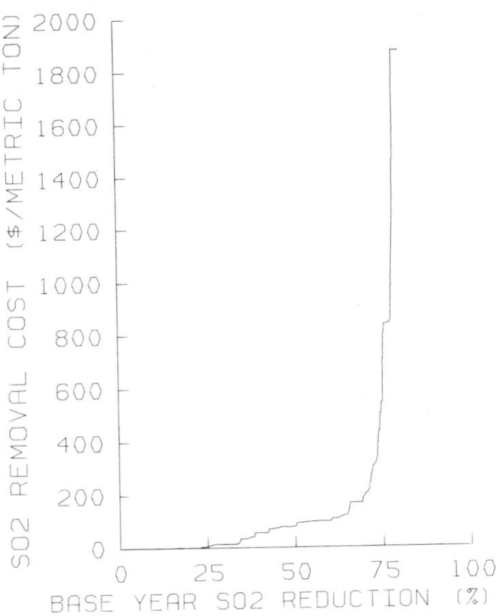

Fig. 3. Estimated cost of switching to low sulfur coal.

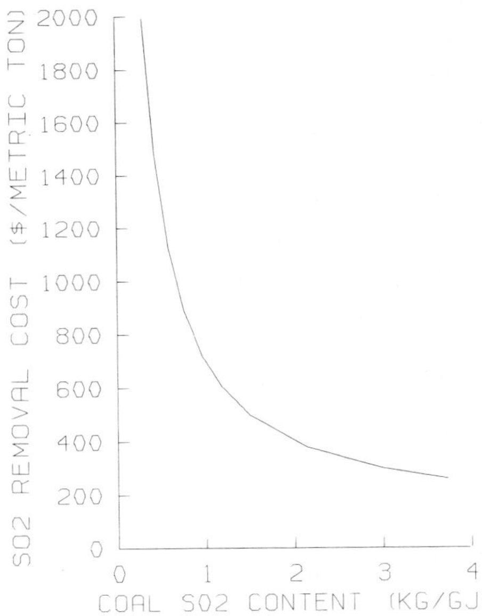

Fig. 4. FGD cost as a function of sulfur content (1.0 kg/GJ=2.33 lb/MBtu)

in the patterns of coal production and use. However, as these indirect costs diminish over the long term, the direct costs of holding emissions constant as coal demand increases could easily exceed the cost of FGD and other desulfurization technologies.

FLUE GAS DESULFURIZATION

The analysis in the previous section indicated the unit cost of SO_2 reductions was highest for coals having low sulfur content. This resulted mostly from the fact that low sulfur coal deposits are located far from the largest demand centers, and only partly from the effects of geologic scarcity on mine-mouth cost. The same effect, only much more exaggerated, characterizes the situation for FGD technologies.

Based on an extensive international review of engineering cost models, and the actual performance of FGD equipment, Rubin (ref. 8) developed a simplified method for estimating capital and operating costs. This method was applied here with minor modifications to estimate the unit cost of SO_2 removal as a function of coal sulfur content. Since FGD system costs depend strongly on the gas volume treated, they are most economically designed to operate at 90% removal efficiency and bypass a fraction of the gas when lower removal rates are required. Using this assumption, the simplified method yields the results shown in Figure 4.

There are two types of irreducible costs associated with FGD technology: the fixed costs of handling large volumes of gas, and the variable costs associated with purchasing and handling reagent. To estimate these costs, consider these two reactions typical of nonregenerable wet FGD systems:

$$SO_2 + CaCO_3 + \frac{1}{2} H_2O \rightarrow CaSO_3 \cdot \frac{1}{2} H_2O + CO_2$$

$$SO_2 + \frac{1}{2} O_2 + CaCO_3 + 2H_2O \rightarrow CaSO_4 \cdot 2H_2O + CO_2$$

These require only slightly more than stoichiometric quantities of limestone at approximately \$13/metric ton (ref. 7), or about \$22/metric ton SO_2 removed. The resulting mixture of calcium sulfite sludge is thixotropic, and must be disposed of at a cost of about \$11/metric ton (ref. 7), or about \$22/metric ton SO_2 removed. Advanced processes that produce calcium sulfate (gypsum) through forced oxidation yield slightly greater quantities of waste due to the higher molecular weight, but contain significantly less water in the filter cake.

These reagent- and waste-related costs, amounting to approximately \$45/metric ton SO_2 removed, are independent of the sulfur content of the coal. It is the other irreducible costs, those depending directly on flue gas volume, that dominate the economics of current FGD systems, and exhibit a strong but unavoidable dependence on coal sulfur content. For example, state-of-the-art processes require 3-5% of the total coal burned to handle and cool flue gas; some regenerable technologies require up to 15%. In terms of volume requirements, FGD systems require approximately 15 m^3 per metric ton per hour of coal burned (480 ft^3/tph) for gas scrubbing during a 5-10 second residence time. An additional 18.5 m^3/metric ton per hour (600 ft^3/tph) is required for the holding tank for the liquids and solids where reactions go to completion over a period of 10-15 minutes (ref. 9).

Due to the large quantity of flue gas to be treated, the volume requirements are unlikely to change significantly unless residence times can be reduced. Since reactions are already near-stoichiometric, reagent-related costs may prove quite difficult to reduce. Some FGD technologies attempt to avoid contaminating the gypsum with flyash, in order to produce a marketable by-product. However, it is not clear that significant credits would continue to be available as gypsum markets became saturated. Other advanced FGD technologies involve sorbent regeneration, but their high capital costs and energy requirements render them economically unattractive (ref. 2).

SULFUR DIOXIDE REMOVAL DURING COMBUSTION

Rather than waiting to capture SO_2 from relatively cool flue gases, another approach is to capture the SO_2 in dry reactions that occur at higher temperatures far upstream. Two approaches are currently under development: limestone injection and fluidized bed combustion. Both of these technologies rely on rather high reaction rates, typical of the 2-4 second gas residence times common in coal combustion equipment. Theoretically, they represent the potentially least costly technologies for desulfurization, because the short residence times avoid the irreducible volume-related costs of other technologies such as FGD.

Limestone Injection (LIMB)

Most research on limestone injection into multistage burners (LIMB) is being funded by EPA. Private sector development is virtually nonexistent because the technology's SO_2 removal rates fall far short of NSPS requirements for new facilities, leaving LIMB in direct competition with inexpensive rail transportation for the market at uncontrolled facilities. The progress and status of the EPA research program has been detailed by Drehmel et al. (ref. 10), and by Martin et al. (ref. 11). Basically, the technology depends on injecting limestone into the high temperature region of the flame where calcination produces fine particles having extremely high surface area, followed by rapid transport of the CaO to the cooler region of the flame where it will not recrystallize. In this cooler region, between 700-1000°C, farther downstream of the flame, SO_2 is captured by the mineral oxide under oxygen rich conditions:

$$CaCO_3 \rightarrow CaO + CO_2$$

$$CaO + SO_2 + \frac{1}{2} O_2 \rightarrow CaSO_4$$

The second reaction must occur in the temperature range bounded above by the decomposition of gypsum, and bounded below by kinetics and residence time. Drehmel and Martin report pilot scale results using ash free (gaseous) fuels showing limestone utilization rates as high as 30-40% are achievable for some temperature profiles not atypical of conventional boiler designs. However, the rates generally drop to 10-15% upon firing with coal. This has led EPA to focus recent research on sorbent-flyash interactions in the hopes of finding alternative sorbents, low cost promoters to improve sorbent utilization, and to identify the specific mineral species in flyash that reduce limestone reactivity. In addition, EPA is investigating the possibility of increasing calcium-sulfur contact under fuel rich conditions to form CaS, which is stable to higher temperatures than gypsum.

Fluidized Bed Combustion (FBC)

Gas temperature in fluidized bed combustors is lower than in conventional boilers, leading to inherently low levels of NOx formation. This temperature regime is also conducive to the formation of gypsum from limestone through a set of reactions similar to those described above. Unlike LIMB research which has focused almost exclusively on retrofit technologies, fluidized bed combustors have been designed from the outset to take advantage of the most favorable thermal histories, slightly longer gas residence times (~4 seconds) in temperature regimes conducive to dry $CaSO_4$ formation, and long solid-solid residence times (~10-40 hours). Moreover, these high residence times are achievable without substantially increasing the overall size of the furnace and boiler, due to the significantly enhanced heat transfer coefficients resulting from immersion of the tubes in the turbulent fluid bed.

Tests conducted at the TVA-EPRI 20 MW pilot plant achieved 90% SO_2 removal using underbed limestone feed and high flyash recycle rates at a calcium/sulfur molar ratio of 2.8 (ref. 12). On low sulfur coals, 70% removal was shown to be achievable at a calcium to sulfur ratio of 1.64 at the EPRI 2 MW research facility. EPRI reports no serious erosion problems in the bed at either test facility, in contrast to experience in the UK and elsewhere (ref. 13).

Cost Limits

If erosion problems can be limited, and overall combustor volume increases held to modest levels, the only irreducible sulfur removal costs associated with FBC or LIMB technologies would be those related to reagent and waste. As research yields fundamental knowledge of the thermal and chemical histories of sulfur and calcium compounds during the combustion process, future technolgical innovation could lead over the long term to construction and maintenance costs for FBC or LIMB systems, approximately the same as those for conventional designs.

Assuming a target stoichiometric ratio of 2 at 90% removal, the incremental cost of SO_2 control could approach the cost of handling reagent and disposing of approximately twice the tonnage of waste generated by FGD systems.

Reagent:

$$\frac{2 \times 100 \text{ ton } CaCO_3}{64 \text{ ton } SO_2} \cdot \frac{\$13}{\text{ton } CaCO_3} = \$41/\text{metric ton } SO_2$$

Disposal of contaminated gypsum:

$$\frac{136 \text{ ton } CaSO_4}{64 \text{ ton } SO_2} \cdot \frac{\$11}{\text{ton } CaSO_4} = \$23/\text{metric ton } SO_2$$

Adding about $15 for disposal of unreacted lime, mixed with flyash to produce caustic waste, yields a total cost of approximately $80/metric ton SO_2 removed.

Since the cost is split evenly between reagent and disposal, this cost could theoretically be reduced by a factor of two through improvements in stoichiometry, and by a similar amount if gypsum contamination could be avoided.

A related technology, pressurized fluidized bed combustion, need not be considered here because it is a separable issue. There is no fundamental difference between AFBC and PFBC in terms of SO_2 removal; PFBC simply involves a tradeoff between the benefits of passing hot gases through a combined Brayton-Rankine power generation cycle, and the additional capital costs associated with turbine blades resistant to erosion by abrasive particles in the uncleaned hot combustion gases. These benefits and costs are essentially independent of SO_2 removal.

GASIFICATION

Most synthetic fuel technologies employ a basic gasification reaction as a first step:

$$C + H_2O \text{ (steam)} \rightarrow CO + H_2$$

and then proceed to obtain additional hydrogen from water to produce liquid and gaseous hydrocarbon fuels. The entire process is quite energy intensive; overall efficiencies between 60-70% are not uncommon. Some process energy is eventually recovered as the enthalpy of formation of H_2O during combustion of the fluid fuels, but the irreversibilities associated with the entire process are quite large. Since many of the irreducible capital and energy costs are incurred to upgrade the systhesis gas to premium fuels valued for their high Btu content, versatility, and transportability, it is possible to compare the benefits and costs of desulfurization by limiting this analysis to the gasification step alone.

Gasification is an endothermic reaction requiring combustion of some of the desulfurized product. The overall efficiency of this step alone is about 85% (ref. 14). In an air-blown gasifier, however, large amounts of nitrogen dilute the carbon monoxide and hydrogen, adding significantly to the volume requirements and flue gas losses after combustion. However, the most significant irreversibilities are associated with excursions to high gasifier temperatures (800-2000^oC, depending on the technology) with subsequent cooling for gas cleanup prior to combustion. Gas cleanup generally requires an additional 6% of the energy input to the gasifier (ref. 14). Additional energy, material and labor costs are incurred to treat the resulting wastewater.

Finally, there are irreducible volume related costs associated with gasification. Sulfur removal, in the form of elemental S, is accomplished by scrubbing H_2S from the synthesis gas and treating it in a Claus process. An air blown Lurgi gasifier can produce a relatively low heating value (120 Btu/scf) power gas at typical exit conditions of 2.4 MPa, 540°C (350 psi, 1000°F). The gas volume at these conditions is about 3.5 m^3 per metric ton per hour of coal processed (123 ft^3/tph) assuming a typical coal heating value of 28 GJ/metric ton and a residence time of 15 seconds (ref. 15). The largest volume requirement is for the large heat exchanges needed with current technologies to cool the gas for H_2S removal; it is defined by the flow rate calculated above, and the residence time. Using design data for a recent EPRI project (ref. 16) the residence time was estimated at about 15 minutes. The corresponding volume required for cleanup of low heating value power gas is therefore about 60 m^3/metric ton coal processed per hour (2000 ft^3/tph). The total volume requirement is therefore about twice that for FGD, and the cost per unit volume is significantly higher due to the materials required for high pressure operation.

The clean fuel gas is well suited for combustion in a combined Braytor-Rankine cycle power generating system. The resulting improvement in power generation efficiency, from about 34% to 39% (ref. 7), offsets a significant fraction of the energy costs of the gasification process.

Compared to FGD, gasification holds the promise of significantly lower reagent costs (except in regions where water is scarce) because sulfur is removed in its elemental form. However, gasification has some significant disadvantages in terms of volume and energy-related costs compared to FGD, even assuming corrosion-resistant material coatings can be developed through future R&D. Even assuming the most optimistic outcome of future R&D efforts, the irreducible costs associated with gasification are not likely to ever reach those associated with fluidized bed combustion. The greatest advantage of gasification as a clean coal combustion technology lies in its ability to approach 100% elimination of both SO_2 and particulate emissions, without the extremely rapid exponential cost experienced by FBC, and to a lesser extent by FGD. However this fundamental benefit is not likely to be realized unless emission standards are made substantially more stringent.

PRECOMBUSTION DESULFURIZATION

While chemical cleaning can theoretically attack both organic and inorganic sulfur in coal, conventional and advanced physical cleaning technologies are limited to removing pyrite, which typically accounts for less than 50% of total sulfur. This section first examines the potential for economically optimal combination of physical cleaning and FGD technologies to achieve high overall sulfur removal rates. With this as background, two promising approaches to chemical cleaning are described. Their fundamental technical and economic limitations are then compared to gasification, another precombustion desulfurization technology.

Physical Cleaning

Since most data on coal cleaning are proprietary, it is not possible to determine the amount of sulfur currently removed from coal mined in the US. The best available statistics (ref. 10) indicate about 48% of raw coal is subjected to some type of physical cleaning. Very little sulfur is removed, however, since most coal is cleaned to eliminate roof and floor material that is mixed with the coal as a result of modern continuous mining techniques. The sulfur content of the cleaned coal is reduced only slightly below that present in the seam. All physical cleaning methods now in use are based on exploiting differences in specific gravity between coal and mineral matter, and are economically limited by the lower yields resulting from grinding the coal finely enough to expose a significant fraction of the pyrite and other minerals. Skea and Rubin (ref. 17) developed a set of engineering cost models of conventional cleaning processes, showing that yields as low as 70-80% are not uncommon for technologies achieving relatively high (30-40%) sulfur removal rates.

Most physical cleaning research is being funded by DOE, and is summarized by Brown and Feeley (ref. 1). One interesting approach uses ultrasound to reduce the energy requirements nearly an order of magnitude below those of mechanical grinding while liberating more mineral matter, because coal tends to fracture preferentially along surfaces where minerals are present. Most approaches being explored by DOE, however, are aimed at taking advantage of differences other than specific gravity, between the mineral matter and the coal. These include using true heavy-media cycloning using fluids such as freons to avoid the problems encountered separating fine magnetite particles from fine coal and mineral particles in existing cleaning technologies. One approach uses high intensity magnetic fields to separate paramagnetic pyrite from diamagnetic coal; conventional low intensity fields are inadequate. Another approach, selective coalescence, exploits differences in surface properties to separate fine coal and mineral particles (ref. 18). Agglomerating agents are added to a coal water slurry causing coal particles to coalesce into a two-phase slurry of immiscible liquids. Related efforts are investigating the use of various surfactants and flocculants to change the fundamental structure of filter cake, thus improving dewatering techniques.

Skea and Rubin's engineering cost models can be used to calculate the incremental cost of cleaning raw coal type k, mined in region i, from sulfur content $m \leq k$ to $n = m - 1$:

$$ICC_{ikmn} = \left[\frac{MC_{ik} + W_{ikn}/Y_{ik}}{a_{ikn}} - \frac{MC_{ik} + W_{ikm}/Y_{ik}}{a_{ikm}}\right] \frac{1}{SO2_m - SO2_n}$$

where MC = minemouth cost ($/seam ton)

W = washing cost ($/raw ton)

Y = basic preparation yield (seam ton/raw ton)

a = cleaning yield (GJ/seam ton)

SO2 = sulfur content (ton SO_2/GJ).

For each of 35 supply regions i and 11 sulfur contents k, these incremental cleaning costs were compared with the cost of FGD (Figure 4) for scrubbing type m coal. This identified a set of "cleaning pathways" (ikn) that are more economical than FGD for small or moderate sulfur removal fractions. There were only a few cases where the deepest level of cleaning turned out to be more costly than FGD. In most cases, physical cleaning was far less costly, although severely limited in terms of the sulfur removal fraction.

Assuming all such ik coals in 1980 had been cleaned to their maximum extent (i.e. to a point where the marginal cost of SO_2 removal equals that of FGD), yields the curve shown in Figure 5. Also shown for comparison is a curve derived from Figure 4 indicating the cost of SO_2 removal using FGD technologies alone, and the low sulfur coal substitution curve from Figure 3. An optimization analysis would be necessary to define a fourth curve, indicating the cost of the economically optimal combination of physical cleaning and FGD needed to achieve a certain level of SO_2 removal. However, the cross on Figure 5 shows the endpoint of that curve, which would begin by following the cleaning curve and then cross the FGD curve as percentage removal rates were increased. It was determined by computing for each (ikm) cleaning pathway less costly than

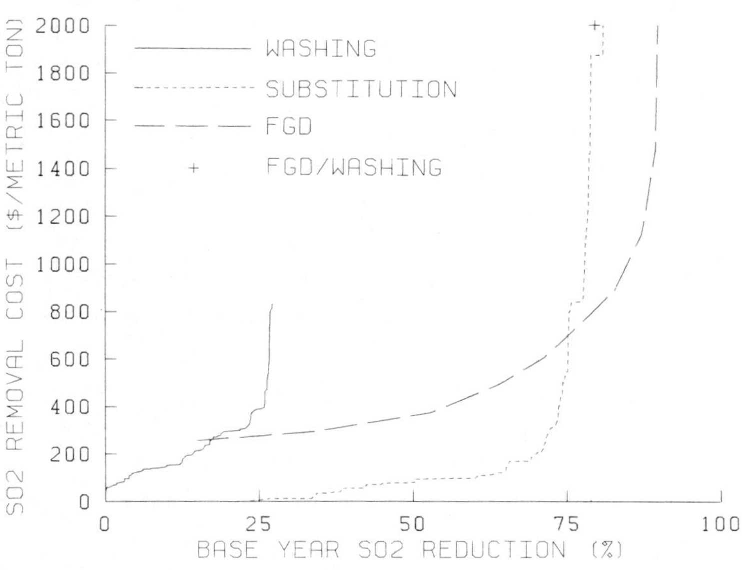

Fig. 5. Comparison of sulfur dioxide removal costs.

FGD, the difference between cleaning and FGD costs. This was used as a budget constraint to determine the breakeven amount of additional SO_2 that could be removed from the cleaned coal through FGD.

It is not necessary to perform the detailed optimization analysis, however, to conclude that the possibilities for combining precombustion cleaning with FGD technology are extremely limited, especially when high (~90%) sulfur removal rates are desired. The steepness of the curve in Figure 4, which results from the irreducible volume-related costs of FGD, significantly increases the unit cost of SO_2 removal from cleaned coal. This higher cost, when averaged with the significantly lower unit costs of precombustion cleaning, yields an average cost of SO_2 removal that is not significantly lower than the cost of scrubbing the raw high sulfur coal.

Chemical Cleaning

Currently the most advanced chemical cleaning technology under development by DOE is the TRW Gravimelt process, which achieves 80-90% sulfur reduction and works nearly as well on organic as on pyritic sulfur (ref. 19). The process occurs at atmospheric pressure and at a relatively low temperature, below the 380^0C point where coal softens. Fine coal is placed in a mixture of molten potassium and sodium hydroxides, producing a waste containing sulfides plus caustic ash products. Most of the sulfides are then carbonated and converted to hydrogen sulfide, from which the sulfur is ultimately removed. Problems have been experienced extracting the fine coal from the ash, and then regenerating the spent caustic. Since ash removal requires an aqueous solution, the evaporation energy requirements for the regeneration process are significant; the most recent target cited by TRW is an 11% loss. This factor, and its tradeoff against regeneration efficiencies less than 100%, define the major irreducible operating costs of the Gravimelt process.

The irreducible fixed costs, on the other hand, are potentially quite small relative to FGD and gasification because the coal is treated in its solid phase. Since the coal is only diluted 2:1 with caustic, the volume requirements for the 2 hour process would be relatively modest: on the order of 4.4 m^3/metric ton per hour of coal processed (70 ft^3/tph) for the reactor (assuming a specific gravity of 1.35). This volume should be at least doubled, however, to account for the additional 2 hour residence time in holding tanks for acid rinses, etc. Additional volume would be required for the regeneration process. The total is therefore less than that required for gasification, and the materials requirements are far less severe.

In summary, the irreducible cost of the Gravimelt process will likely be dominated by its energy losses, in the form of fine coal losses and the evaporation requirements of an efficient process.

A second approach to chemical cleaning involves pyrolysis. Utah Power and Light plans to demonstrate the commercial feasibility of producing, from each ton of bituminous coal, 0.6 tons of char to be used as boiler fuel and 1.1 barrels of marketable liquid petroleum byproducts (ref. 20). The process occurs at atmospheric pressure, and temperatures below 700°C. The energy efficiency, assuming the char need not be cooled, exceeds 80%. Volume-related costs of the reactor are minimal because the 2:1 gas/solid ratio the solid coal requires only 2.2 m^3/metric ton per hour of coal processed (23 ft^3/tph) for each hour residence time, and reaction times less than one hour are possible. The volume required for gas cleanup, however, would be comparable to that for synfuels: about 60 m^3/metric ton per hour of coal processed (~2000 ft^3/tph). Since the volatile matter is at atmospheric pressure, it occupies a large volume during cooling and cleanup, but the unit cost of providing this volume is substantially less than for gasification.

The economic feasibility analysis for the Utah project has been conducted by Bechtel (ref. 21). Although this project involves no desulfurization, its results could have significant implications for the development of chemical coal cleaning technologies. A successful project would demonstrate that the coal could be pyrolyzed in such a way as to produce enough premium hydrocarbon byproducts from the hydrogen naturally occurring in the coal, to essentially pay the complete cost of the pyrolysis process. The same process performed on a high sulfur coal would yield a char containing only a slightly lower percentage sulfur than the original coal. The char would need additional desulfurization. However, fundamental research on the thermal and chemical histories of coal particles undergoing controlled pyrolysis could identify a vast array of chemical and physical states in which the sulfur could be chemically attacked during charring. The possibilities for identifying a low cost chemical cleaning process would therefore be substantially increased.

Even if chemical cleaning of the char proves infeasible, there is reason to believe that the char itself could be superior to coal for use in fluidized bed combustors. Recent fundamental research on coal charring and combustion has revealed that sulfur dioxide formation is not coincident with CO_2 formation. Moffat (ref. 22) noted a bimodal pattern of sulfur dioxide release, the first sulfur leaving the coal quickly during the devolatilization phase, and the remainder at the very end of burnout. He notes that "most sulfur which remains in the char appears to be inorganic or is converted to inorganic sulfur during char burnout." Kruse (ref. 23) and Hill (ref. 24) hypothesized that the sulfur remaining in the char is in the form of pyrrhotite, high molecular weight organic compounds such as thiophenes, or other organo-iron compounds which could remain in the char long enough to benefit greatly from the long solid-solid residence times with limestone in a fluidized bed combustor.

CONCLUSION

For each type of desulfurization technology, three types of irreducible costs were discussed. The first, reagent and waste disposal costs, are limited by R&D target stoichiometric ratios of 1.0 for FGD and 2.0 for FBC and LIMB, respectively, and are theoretically avoidable in precombustion cleaning processes that remove sulfur in its elemental form. The second cost, energy loss, amounts to 2.5-5.0 ton coal/ton SO_2 for losses of 10-20%, respectively, assuming the sulfur removed is 2% of the coal mass for a typical coal. The lost high sulfur coal may be valued at about \$30/ton or more, depending on transportation requirements. Finally, the volume-related costs were found to be dominant for both FGD and for gasification (due to high materials costs), relatively small for precombustion cleaning processes that handle solids and liquids rather than gases, and possibly negligible for FBC/LIMB.

Thus reagent-related costs are relatively small, ranging from about \$45-80/ metric ton SO_2 for FGD and advanced FBC/LIMB technologies, respectively. Energy costs are almost negligible for FBC/LIMB, but can range from about \$40/metric ton SO_2 for FGD, to a few hundred dollars for precombustion cleaning or gasification technologies having efficiencies less than 80%. Volume requirements vary most widely, in terms of m^3/metric ton coal processed per hour: about 60 for gasification and pyrolysis, 30 for FGD, and 5 or less for some types of physical and chemical cleaning. Volume-related costs are significant, but comparisons are not always straightforward. For example the costs of pyrolysis may be offset by the value of liquid byproducts, and the unit costs of pressure vessels for gasification are higher than those for technologies that operate at atmospheric pressure.

Because volume-related costs are significant for pyrolysis, gasification, and FGD, they are likely to be implemented first at large scale; R&D incentives lie mainly with large industry. Scale economies follow approximately a two-thirds power law because materials costs are proportional to area while capacity is proportional to volume. Technologies having small volume-related costs like FBC are already successfully competing with FGD at small scale; R&D is being done by vendors who mass-produce these systems.

The critical policy questions center on the ability of any of these desulfurization technologies to compete economically with the baseline technology, transportation of low sulfur coal. Recall from Figure 3 that the long run marginal cost of SO2 removal through coal switching could remain below the \$200/ metric ton SO_2 range even if SO_2 emissions were held at 1980 levels while coal use doubles or triples. Under these conditions the role of all desulfurization technologies is clearly limited, and the emerging technologies will have to approach their irreducible minimum costs in order to compete effectively with FGD in large-scale applications. On the other hand, substitution costs would increase quickly and dramatically if emissions were reduced and held to 50% of 1980 levels as coal use grew. If mandatory desulfurization policies remain in effect over the long term and require removal rates substantially exceeding today's 70-90% levels, the economic advantage could tilt towards technologies that remove sulfur from flue gas or synthesis gas, rather than finely ground coal particles. Otherwise the long term advantage could lie with technologies that remove sulfur from solid coal, provided that energy losses can be held to less than 20%.

ACKNOWLEDGEMENT

I am indebted to D. Banerjee, C. Kruse and W. Wells for many helpful discussions, and to H. Hottmann and J. Lytle for valuable comments on an earlier draft. Although the information described in this paper has been funded in part by the U.S. Environmental Protection Agency under Assistance Agreement CR-808514 to the University of Illinois, it has not been subjected to the Agency's required peer and administrative review and therefore does not necessarily reflect the views of the Agency. No official endorsement should be inferred.

REFERENCES

1. F. R. Brown and T. J. Feeley, "Status and Progress of the Physical Desulfurization of Coal", Technical Note, U.S. Dept. Energy Technology Center, Pittsburgh, PA, May 1985.
2. U.S. Congress, Office of Technology Assessment, Acid Rain and Transported Air Pollutants, U.S. Govt. Printing Office, Washington D.C., 1984.
3. J. Siegel and S. Spandone, "The Potential Role of Advanced Coal Cleaning Technology in Reduction of SO_2 Emissions from Coal Fired Facilities". Technical Note, U.S. Dept. Energy Office of Fossil Energy, Washington D.C., April 1985.
4. R. Plant, "Survey of Coal Fired Power Plants", interoffice memorandum, Office of Vice Chancellor for Research, University of Illinois at Urbana-Champaign, May 1983.
5. J. W. Green and A. Michelson, "Enconometric Estimation of Regional Coal Price Forecasting Models for the U.S.". U.S. Dept. Agriculture, Economic Research Service, Colorado State Univ., December 1983.
6. C. W. Bullard, "The Coal Supply Module", Chap. 4, The State-Level Advanced Utility Simulation Model: Analytical Documentation, USEPA/IERL, Research Triangle Park, NC, Septemper 1984.
7. Electric Power Research Institute, Technical Assessment Guide, EPRI P-2410-SR, Palo Alto, CA, May 1982.
8. E. S. Rubin, "International Pollution Control Costs of Coal-Fired Power Plants", Env. Sci. and Tech., v. 17, no. 8, 1983, pp. 366A-377A.
9. W. Wells, private communication, Center for Research on Sulfur in Coal, Champaign, IL, May 1985.

10. D. C. Drehmel, G. B. Martin and J. H. Abbott, "Results from EPA's Development of Limestone Injection into a Low NOx Furnace", Technical Note, USEPA/IERL, Research Triangle Park, NC, 1983.

11. G. B. Martin, J. H. Abbott, and F. T. Princiotta, "EPA's LIMB R&D Program Evolution, Status, and Plans", Technical Note, USEPA/IERL, Research Triangle Park, NC, April 1985.

12. T. Moore, "Achieving the Promise of FBC", EPRI Journal, Electric Power Research Institute, Palo Alto, CA, Jan/Feb 1985.

13. E. M. Petrill, private communication, Electric Power Research Institute, Palo Alto, CA, May 1985.

14. R. F. Probstein and H. Gold, Water in Synthetic Fuel Production, MIT Press, Cambridge, MA, 1978.

15. D. Banerjee, private communication, Center for Research on Sulfur in Coal, Champaign, IL, May 1985.

16. N. Holt, private communication, Electric Power Research Institute, Palo Alto, CA. June 1984.

17. J. F. Skea and E. S. Rubin, "Coal Cleaning Methodology", The State-Level Advanced Utility Simulation Model: Analytical Documentation Appendix C, USEPA/IERL, Research Triangle Park, NC, September 1984.

18. R. P. Killmeyer, "A Comparison of Emerging Selective Agglomeration Technologies", Technical Note, U.S. Dept. of Energy Technology Center, Pittsburgh, PA, April 1985.

19. TRW Energy Development Group, Gravimelt Process Development, DOE/PC/42295-T7 (DE84013743), NTIS, Springfield, VA, June 1983.

20. J. R. Thurgood and F. D. Wareham, "Utah Char-Oil Project", Technical Note, Utah Power & Light Co., Salt Lake City, UT, February 1982.

21. Bechtel Group Inc., Coal Pyrolysis Feasibility Study for Utah Power & Light Co., Technical Report Summary, San Francisco, CA, May 1982.

22. A. J. Moffat, "The Chemistry and Mechanism of Sulfur Release During Coal Combustion", Technical Note, Phillips Petroleum Co., Bartlesville, OK, March 1982.

23. C. W. Kruse, private communication, Illinois State Geological Survey, Champaign, IL, May 1982.

24. G. Hill, private communication, University of Utah Dept. Chemical Engineering, Salt Lake City, UT, May 1985.

25. C. W. Bullard and H. Hottman, Strategies for Reducing Acid Emissions from Illinois Electric Utilities: A Preliminary Assessment, IEPA/APC/84-029, Illinois EPA, Springfield, IL, September 1984.

PART 3

SELECTIVE MINING AND BLENDING

AN INTERIM STEP IN THE REDUCTION OF SULFUR EMISSIONS

D.M. Doner[1]

[1]Chemical Engineering Department, West Virginia Institute of Technology, Montgomery, West Virginia 25136 (USA)

ABSTRACT

A first step towards a solution of the sulfur dioxide emission problem resulting from the use of high sulfur coals is developed. The important role that physical cleaning of the coal can play is pointed out. This proposed scheme is designed to "buy time" for the development of a comprehensive solution, while reducing emissions from some high sulfur coals by as much as 60% with no loss in the salable coal. This scheme could result in a substantial reduction in sulfur emissions with minimum initial costs. It also allows new technologies to be developed rather than reacting with a crisis solution.

INTRODUCTION

The causes and effects of acid rain have and are presently being debated with no conclusive answers in sight. At the same time, the use of high sulfur coal is being singled out as the main culprit. Assuming that it is the main contributor to the formation of acid rain then a significant decrease in the sulfur emissions from high sulfur coal usage should reduce or at the very least, delay the long term effects of acid rain. This would allow more time to study the causes and effects of acid rain. Also the new technologies for sulfur removal will have more time for development. Finally, if the above assumption is incorrect, then the high sulfur coal industry is in the position that it can demand action from the real culprits.

The above discussion presents an argument for the industry to make a significant reduction in sulfur emissions. How much of a reduction and how it will be accomplished remains to be defined. This study shows the significant role that physical clean can play in the reduction of sulfur emissions. It also gives some indication of how much reduction is feasible. Once the significance of physical separation has been shown, some implications of how the results of this study can be applied are discussed.

OPTIONS

The options for reducing sulfur emissions that are to be considered in this study are shown in Figures 1, 2 and 3. Figure 1 presents Option 1 in which the sulfur is reduced solely by scrubbing the entire stack gas stream.

Fig. 1. Diagram showing Option 1 in which the entire stack gas stream is scrubbed as the sole means for sulfur reduction.

Figure 2 presents Option 2 in which the sulfur is reduced by scrubbing a fraction of the stack gas and then blending this scrubbed gas with the unscrubbed stack gas.

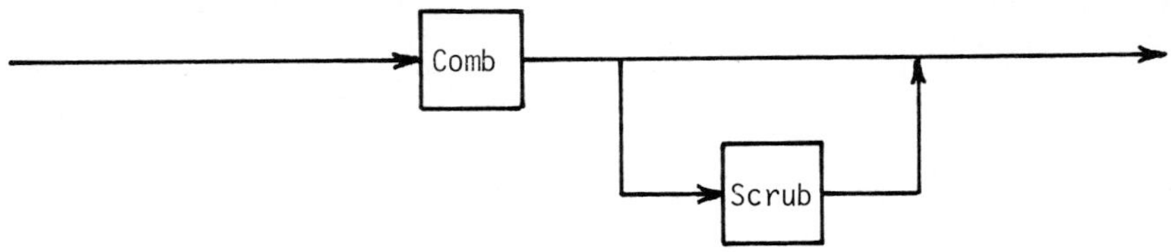

Fig. 2. Diagram showing Option 2 in which a portion of the stack gas is scrubbed and blended with the remaining gas.

Figure 3 presents Option 3 in which the coal is physically separated into two streams. The lighter fraction is utilized with no stack gas clean-up while the heavier fraction is utilized with stack gas clean-up.

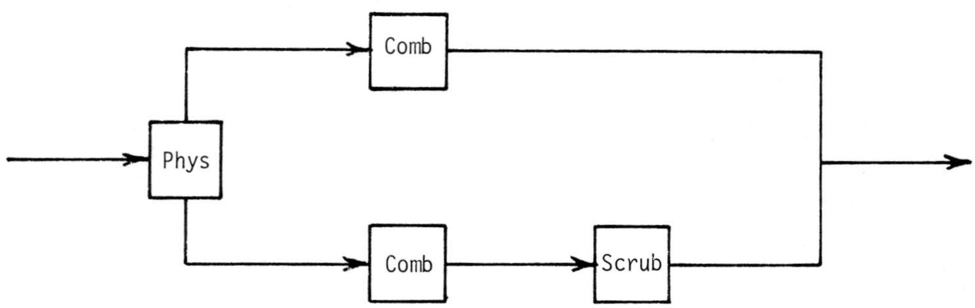

Fig. 3. Diagram showing Option 3 in which the coal is physically cleaned and part of the stack gas is scrubbed.

It is now necessary to develop a method for comparing these options.

METHOD OF COMPARISON

The options presented will be compared by determining the relative scrubber sizes required to produce a given sulfur reduction. Option 1 will be used as the base case for comparison with a sulfur dioxide emission rate of 1.2 lbs. SO_2 per million Btu. This emission rate will be called "compliance" in this study but this is only a base case for comparative purposes. A simple approximation of scrubber size will be used to determine the ratio of scrubber size to that of the base case, since a comparison is all that is desired.

The volume of a scrubber is determined by (ref. 1):

$$V = \frac{N}{K_G a \ (p_{ag} \ p_{al})_{\log \text{mean}}} \tag{1}$$

where V = volume of scrubber; K_G = overall absorption coefficient; a = interfacial area per unit volume; N = moles of material absorbed in the scrubber; p_{ag} = partial pressure of component a in the main body of the gas; and p_{al} = partial pressure of component a if it were in equilibrium with a solution having the same concentration as the main body of liquid. The log mean value is based on the pressure differences of the gas stream entrance and exit.

The ratio of scrubber volume to that of the base case is:

$$R = \frac{V}{V_B} \tag{2}$$

where R = scrubber volume ratio; and the subscript B stands for the base case.

Substituting equation (1) into equation (2) and using the following simplifications and assumptions:

- $(K_G a)_B = K_G a$
- P_{al} is negligible
- The scrubbers are operated at the same pressures.

The ratio of scrubber volumes is then given by:

$$R = \frac{N \quad (Y_{Bi} - Y_{Bo}) \ln \frac{Yi}{Yo}}{N_B (Yi - Yo) \ln \frac{Y_{Bi}}{Y_{Bo}}} \tag{3}$$

where y = sulfur dioxide mole fraction; and the subscripts i and o represent input and output, respectively.

The comparisons are made with each option and the base case utilizing a million Btu of coal. It is also assumed that the stack gas flow rate is 31.5 lb. moles per million Btu.

COMPARISON OF OPTIONS

It is an obvious fact that it is more difficult to achieve sulfur emissions reduction from some types of coal than others. The initial comparison will be made using the published preparation characteristics of a sample of a Pittsburgh seam of coal (ref. 2). This data is given in Table 1.

TABLE 1
Preparation characteristics of a sample of Pittsburgh coal from Waugaman No.1 (ref. 2).

Sp.Gr.	Float		Sink	
	Wt. Fraction	%S	Wt. Fraction	%S
1.30	0.716	1.95	0.284	9.77
1.35	0.852	2.07	0.148	16.27
1.40	0.896	2.18	0.104	21.35
1.45	0.913	2.27	0.087	24.12
1.50	0.922	2.34	0.078	25.81
1.55	0.929	2.41	0.071	27.23
1.58	0.933	2.46	0.067	28.07
Total	1.000	4.17	-	-

Heating value (dry) = 14050 Btu/lb.

Options 1 and 2 are compared in Figure 4. This figure has the fraction of "compliance" as the absicissa and the fraction of the gas stream scrubbed in Option 2 is shown as a parameter.

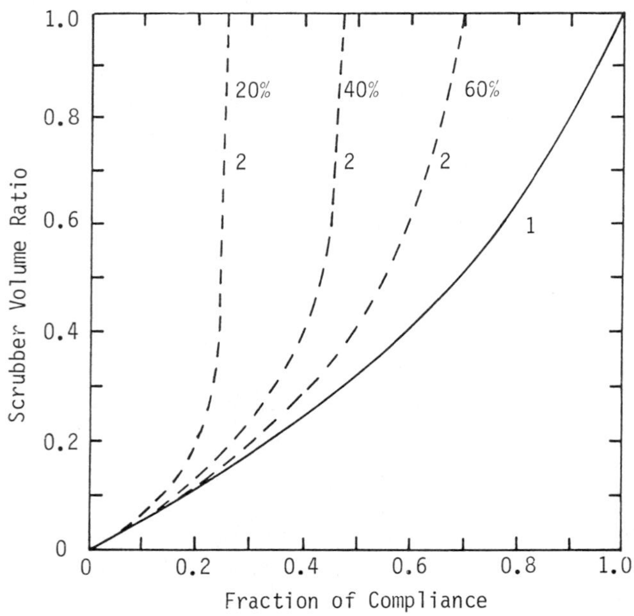

Fig. 4. Comparison of Options 1 and 2 using a sample of a Pittsburgh seam of coal (ref.2).

This comparison shows that to accomplish the same fraction of "compliance," Option 2 requires a larger scrubber than Option 1. Option 2 will always result in a larger scrubber than Option 1 because the portion of gas cleaned in Option 2 must have the sulfur dioxide concentration reduced to lower levels, which is more difficult to accomplish.

Figure 5 compares Options 1 and 3. The parameters shown for Option 3 are the separation specific gravity used in the physical cleaning step. The heavy line shown for Option 3 represents the locus of minimum scrubber sizes.

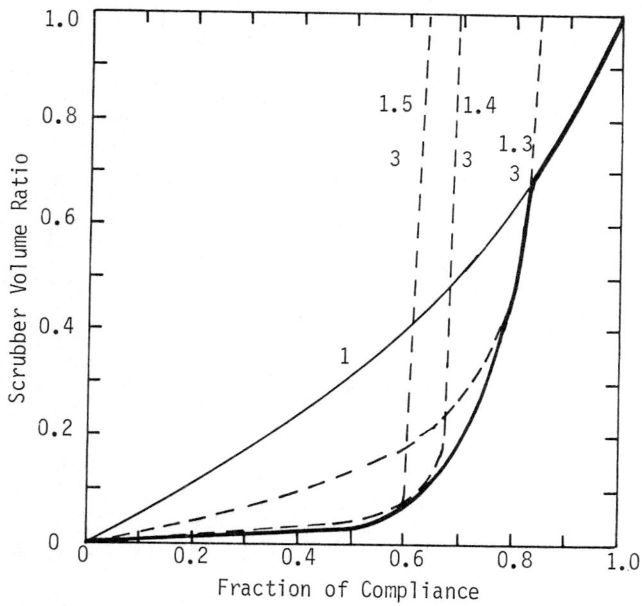

Fig. 5. Comparison of Options 1 and 3 using a sample of a Pittsburgh seam of coal (ref. 2).

This comparison shows that a considerable reduction in the scrubber size is accomplished by using Option 3. The specific case shown here indicates that for a 50% "compliance" a scrubber volume of one tenth of that for Option 1 is required (based on the same percentage of "compliance" for Option 1).

It was pointed out earlier that achieving a reduction in sulfur emissions is more difficult with some coals than other. The preparation characteristics data for three other coal samples (ref. 2, 3) have been compared in a similar manner. The Pittsburgh sample of coal has been used for the base case in all comparisons. Figure 6 compares Option 1 for the various coals. Figure 7 compares Option 3 for the same coals.

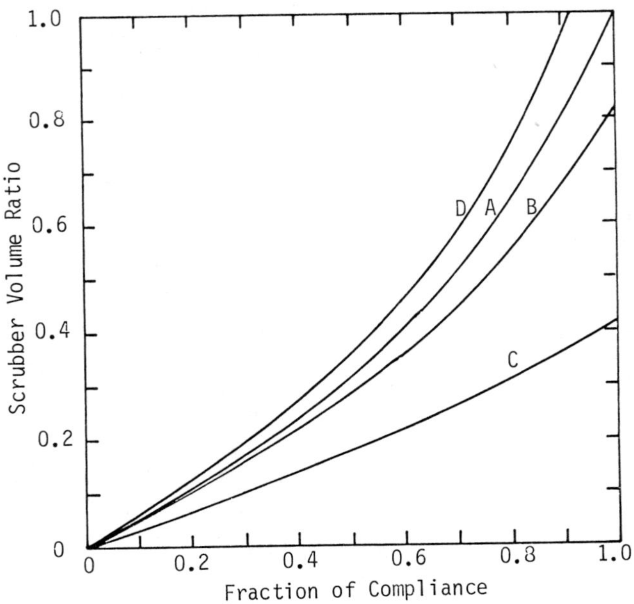

Fig. 6. Comparison of Option 1 for four types of coal. A = Pittsburgh (Waugaman No. 1); B = Middle Kittanning (Badger No. 13); C = Lower Kittanning (Jay No. 1); D = Upper Clarion (McKissick)

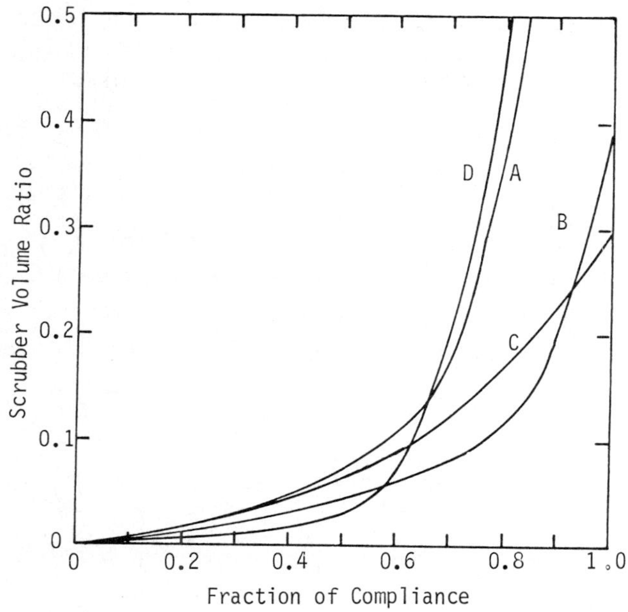

Fig. 7. Comparison of Option 3 for four types of coal. A = Pittsburgh (Waugaman No. 1); B = Middle Kittanning (Badger No. 13); C = Lower Kittanning (Jay No. 1); D = Upper Clarion (McKissick)

These comparisons point out the variations in the coals. They also indicate that by using Option 3 a 40 to 60% "compliance" can be achieved with relatively small scrubber volumes. These comparisons indicate that the physical cleaning of coal can play a significant role in the reduction of sulfur emissions. How this concept might actually be applied is developed in the next section along with some suggested future work that should be considered.

IMPLICATIONS AND SUGGESTED FUTURE WORK

Some of the implications of Option 3 and how it might be applied are now considered. A concept of composite emissions will be considered. Finally, some area for future work are suggested.

Composite Emissions Concept

This concept would allow a group of high sulfur coal users to approach the emissions problem as a group. They would dedicate one or more facilities to the combustion of the heavy or high sulfur fraction of coal. These facilities would have sulfur emissions controls while the majority of the facilities within this group would burn the lighter or low sulfur fraction of coal without emissions controls. This type of approach should be permissible since the problems being attributed to the sulfur emissions do not occur in the region near the combustion facilities. This scheme will require a cooperative atmosphere between industry, government, and environmental groups instead of the antagonistic atmosphere that too often prevails.

The adoption of Option 3 along with the composite emissions concept could provide for a quick reduction in sulfur emissions. Any facility which presently has some form of stack gas sulfur removal can be switched to burning the heavier, high sulfur fraction of coal. At the same time neighboring facilities burn the lower sulfur fraction of coal.

Suggested Future Work

Some of the areas which need future study are:

- A cost analysis of the various options presented along with a detailed scrubber design.
- An evaluation to determine if the present physical cleaning facilitiy can be merged into this concept.
- An evaluation of the modifications that would be necessary to adapt an existing sulfur removal system to this concept.
- The development of a price schedule for the light and heavy fractions of coal so that the burden of sulfur removal is equitably passed to all parties involved.
- Most facilities use a mixture of coals, therefore a method for determining the optimum application of Option 3 should be developed.
- An application of this concept should be made to a "real" composite situation.

CONCLUSIONS

Three conclusions drawn from this study are:

1. There is no benefit in applying stack gas clean-up to a portion of the stack gas using the comparison methods of this study.

2. Physical separation of the coal can play a significant role in the reduction of sulfur.

3. Option 3 as described in this study could reduce the sulfur emissions problem allowing time for the further development of new technologies.

REFERENCES

1. Peters,M.S., "Elementary Chemical Engineering",McGraw-Hill Book Co.,1984, p.224
2. Deurbrouck,A.W., "Preparation Characteristics of Coal from Barbour County, W.Va." US Bur. of Mines-Report Investigation 6707,1965
3. Zeilinger,J.E. and A.W.Deurbrouck, "Preparation Characteristics of Coal from Clarion County, Pa.", US Bur. of Mines-Report Investigation 7174, 1968

SELECTIVE MINING, HOMOGENIZATION AND BLENDING OF CANADA'S HIGHER SULPHUR THERMAL COALS

R.K. SINGHAL[1] and D.G. OSBORNE[2]

[1]CANMET, Calgary Coal Research Laboratory, Calgary, Canada.

[2]Kilborn Engineering (B.C.) Ltd., Vancouver, Canada.

ABSTRACT

Canada is blessed with large reserves of low sulphur (minus 1 percent) thermal coal of bituminous rank, much of which is located close to the eastern and western seaboard to make coal exportation a major and expanding industry.

Contained within these potential export coal reserves, is a relatively small proportion of coal with higher sulphur content (up to about 8 percent). Such coal has other redeeming qualities which have demanded its inclusion in the mineable reserves planned for utilization, although inclusion of such coal in export products is at present, extremely limited.

In eastern Canada, where the majority of the higher sulphur containing coal occurs, mining production capacity is currently about 3 million tonnes of which about 75% is metallurgical coking coal. In western Canada, no higher sulphur coal reserves are at present being mined, although reserves of bituminous coal with 1-3% sulphur exist close enough to the coast to represent a strong future potential export source.

This paper describes the factors which must be considered during the planning of simultaneous extraction of coal seams of variable sulphur content. Similar discussion is included of the coal preparation requirements in terms of blending and homogenization prior to utilization as thermal coal. In the context of both mining and preparation, current research and development work in Canada is also described.

INTRODUCTION

Canada has very large reserves of low sulphur (minus 1%) content coal distributed throughout all ranks from anthracite to lignite. In this regard it does not necessarily need to develop specific strategies for the direct utilization of its relatively small economically recoverable deposits of higher (1-5%) sulphur coal. These deposits are contained only in bituminous rank coal and are located in the three areas shown in Figure 1, i.e. the Sydney coalfield in Nova Scotia; the Minto coalfield in New Brunswick, and the Comox coalfield on Vancouver Island. In all three cases, however, the coal has other valuable properties or attributes encouraging extraction and therefore demanding individual means for overcoming the various specific problems of their utilization.

The three western provinces currently produce 54 million tonnes of coal annually all of which is low in sulphur. This represents about 95% of all coal mined in Canada at the present time. Of this total, slightly less than half is currently exported and most of this goes to Japan. The remainder, 3 million tonnes, is produced by the provinces of Nova Scotia and New Brunswick in the eastern coastal area of Canada about 5500 kms away from the centre of the western coalfield. The majority of western Canadian coal (97%) is at present produced by surface mining methods, where as, on the east coast, mining is predominantly by underground methods and only a relatively small (and decreasing) portion of the total (15%) is produced by surface mining.

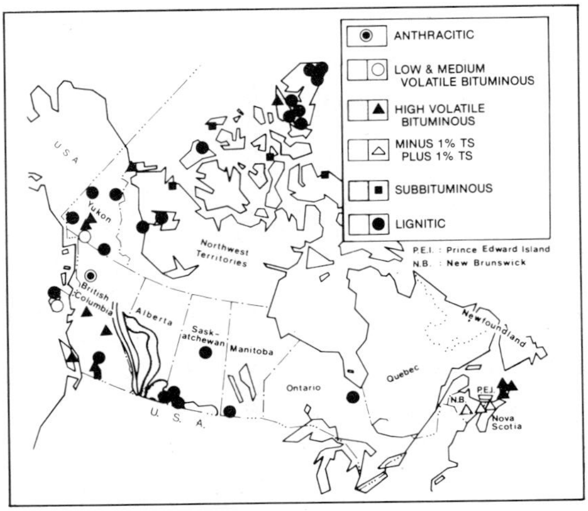

Fig. 1. Coal mining areas of Canada showing areas of high sulphur content

In 1984, Canada exported about 25 million tonnes of bituminous coal of which 20.5 million was high grade low sulphur, metallurgical coking coal. At present, no coal is produced from the Comox coalfield but one project, Quinsam, is well advanced and appears to be the only likely new Canadian export thermal coal mine in the foreseeable future (ref.1). Quinsam's product will average 1% total sulphur compared to the below 0.5% average of other export thermal producers of the western provinces. Mining will initially be by surface methods but eventually underground mining will also be introduced.

The only large scale coal production in eastern Canada occurs at the underground operations on Cape Breton Island in the Sydney coalfield where current capacity is about 2.5 million tonnes per year. Almost all of the entire production of metallurgical coal (1-2% sulphur) and thermal coal (1-3% sulphur) is utilized locally in steel mills and power plants in eastern Canada.

Operations in the Minto coalfield have a capacity of about 0.5 million tonnes per year and a dragline strip mining method is employed. The coal product is used by the New Brunswick Power Commission and is at present burned in the as mined state (6.5-8% sulphur; 18-20% ash).

These eastern Canadian operations therefore represent the only source of higher sulphur coal currently being utilized within Canada. All other coal, of all ranks, currently predominatly utilized for power generation, is of low sulphur content. This situation is unlikely to change within the western provinces due to the large proven reserves which exist of low sulphur anthracite, bituminous, sub-bituminous and lignitic coal. This paper therefore confines attention to the three coalfields described and discusses current problems and practices in the mining and preparation for utilization, of higher sulphur coal. Research and development which is on-going in Canada in the areas of mining and preparation of these coals is also reviewed.

COAL QUALITY OF HIGHER SULPHUR COALS

All of Canada's higher sulphur containing coal is of high volatile bituminous rank and the predominant sulphur component is the mineral pyrites. In eastern Canada, the coal deposits in the Sydney coalfield of Nova Scotia and Minto coalfield of New Brunswick were formed during the Upper Carboniferous (Pennsylvanian) period. In the west, the Comox coalfield was formed much later, in the Cretaceous Period. All three coalfields are known to have originated from marine deposits where a peat swamp is known to have had a marine cover. The change in pH from acid to alkaline is accompanied by bacterialogical action resulting in the formation of sulphide and a consequent nucleation of pyrite (ref.2). These complex processes have resulted in the formation of coals containing significantly higher levels of sulphur than Canada's other major coal deposits which were all formed in in-land non-marine, fresh water environments, and thereby have low sulphur content.

The Table 1 gives quality characteristics of the major coalfields in Canada. Of particular interest is the column giving typical sulphur contents of raw coal. Table 2 shows relative quantities of the three sulphur forms contained in Canada's higher sulphur coals.

COAL MINING METHODS

Large scale underground coal mining operations in eastern Canada are limited to producing collieries on Cape Breton Island, Nova Scotia, in the Sydney coalfield. The mines are operated by the Cape Breton Development Corporation (CBDC), a Government of Canada Corporation (ref.4).

TABLE 1
Quality characteristics of coal in Canada by region and coalfield (ref.3)(continued)

Region/ Coalfield	Rank	Current use (potential use)	Proximate analysis % 1 run-of-mine coal 2 clean coal	M*	A	VM	FC	Sulphur %	Heating value (moist basis) Mj/kg*(kcal/kg)	Remarks
NOVA SCOTIA										
Sydney Coalfield	hvb	70% thermal	1	5-8	10-18	29-32	45-55	1-4	24-31 (5736-7409)	low ash, variable
		30% metallurgical	2	2-8	3-12	29-35	54-62	1-2.5	29-34 (6931-8126)	sulphur, high reactivity
NEW BRUNSWICK										
Minto Coalfield	hvb	thermal	1	1-3	15-20	31-34	47-50	6-8		high sulphur, high ash
SASKATCHEWAN										
Estevan Coalfield	lig	thermal	1	28-35	8-15	26-31	29-34	0.4-0.6	14-17 (3346-4063)	
Willow Bunch Coalfield	lig	thermal	1	36-40	10-13	23-24	26-27	0.5	12-13 (2868-3107)	high moisture
ALBERTA PLAINS										
Battle River Coalfield	sub	thermal	1	23-25	6-12	30-35	36-38	0.3-0.4	18-20 (4302-4780)	
Lethbridge Coalfield	sub-hvb	(thermal)	1	9.5	10-20	35.5	36-44	0.5-0.6	21-24 (5019-5736)	
Sheerness Coalfield	sub	thermal	1	24-27	11-18	25-28	33-37	0.4-0.5	17-18 (4063-4302)	variable quality
South Swan Hills Coalfield	sub	(thermal, chemical feedstock)	1	22-24	21-22	24-26	30-31	0.3	15-16 (3585-3824)	
Wabamun Coalfield	sub	thermal	1	20-22	15-17	26-27	35-38	0.2-0.3	17.5-19.5 (4183-4661)	
Wetaskiwin Coalfield	sub	(thermal)	1	18-20	13-16	27-28	36-38	0.3-0.4	19-20 (4541-4780)	
ALBERTA FOOTHILLS										
Coalspur Coalfield	hvb	thermal	1	11	24-25	28	37	0.4-0.5	21 (5019)	variable quality; clay
			2	8.5-9.5	8-10	33	49	0.3	26 (6214)	bands in coal demand selective mining
Obed Marsh Coalfield	hvb	(thermal, chemical	1	16-20	30-31	33	20	0.5	19 (4541)	
		feedstock)	2	7-8	12-13	36	44	0.5	20 (4780)	
ALBERTA MOUNTAINS										
Cadomin Luscar Coalfield	mvb	metallurgical	1	5	12-20	18-20	55-56	0.3	30-31 (7170-7409)	
		(thermal)	2	8	10	22-24	58-60			
Smoky River Coalfield	lvb	metallurgical	1	4	17-20	16	60-63	0.4	28 (6692)	
		and some thermal	2	6	7	18	65-69	0.4	34 (8126)	

TABLE 1

Quality characteristics of coal in Canada by region and coalfield (ref.3)(concluded)

Region/ Coalfield	Rank	Current use (potential use)	Proximate analysis % 1 run-of-mine coal 2 clean coal	M*	A	VM	FC	Sulphur %	Heating value (moist basis) Mj/kg*(kcal/kg)	Remarks
BRITISH COLUMBIA										
Comox Coalfield	hvb	(thermal)	1	3	12-25	33-37	42-47	0.7-3.5	26 (6214)	Composite from 3 seams in Quinsam Project
			2	9	12	40	45	1.0	(6950)	
Crowsnest Coalfield	lvb-mvb	metallurgical and thermal	1	4-5	15-20	19-23	55-60	0.3-0.4	30-31 (7170-7409)	Clean thermal coal has higher ash and lower heating value
			2	8	9-10	21-22	59-61	0.3-0.4	31-33 (7409-7887)	
Elk Valley Coalfield	lvb-mvb-hvb	metallurigcal and and some thermal	1	3-6	20-35	19-33	40-60	0.3-0.5	20-25 (4780-5975)	several blends of clean coal are produced
			2	8	6.5-10	21-34	50-64	0.3-0.5	30.37 (7170-8843)	
Flathead Coalfield	mvb	(thermal, metallurgical)	1	3-6	25-35	22	40-50	0.5	20-25 (4780-5975)	% VM is an average value
			2	6	16	23	60	NA		
Hat Creek Coalfield	lig-sub	(thermal, chemcial feedstock)	1	23	32	24	21	0.4	11.6 (2772)	values shown are averages for a highly variable coal
Peace River Coalfield	mvb	(metallurgical and thermal)	1	6	20-30	21-27	45-55	0.3-0.5	21-26 (5019-6214)	clean metallurgical coal
			2	6-8	6-9	22-24	63-66	0.3-0.5	33-36 (7887-8604)	clean thermal coal
			2	8	16	20	56	0.3-0.5	28-30 (6692-7170)	

Abbreviations used for rank: lig = lignitic; sub = subbituminous; hvb = high volatile bituminous;
mvb = medium volatile bituminous; lvb = low volatile bituminous

Abbreviations used for proximate analysis: M = moisture; A = ash; VM = volatile matter; FC = fixed carbon
(moisture* given are; as received (1) or product(2))

TABLE 2

Forms of sulphur content in raw coal

		Coalfield	
	Sydney	Minto	Comox
Total sulphur, % (normal value)	1-4	8	1-2.5
Sulphur form, wt% of total: sulphate	1	1.1	0-3
pyrite	70-75	84.3	35-60
organic	29-24	14.6	65-37

The CBDC steadily produced approximately 3 million tonnes per annum of thermal and metallurgical coal until April 1984 when one of the main mining operations was ceased due to a mine fire. This mine remains sealed and its future is still not known. At present only two mines, Lingan and Prince are operating; a third one is under development. Expansion plans are at present under review.

Although room and pillar mining was practised at shallow depths in the past, all the existing mines currently employ the longwall system of extraction. Lingan employs the typical advancing longwall whereas Prince has converted to shortwall retreat mining from a room and pillar operation.

A single seam is extracted by all of the mines and all of the present workings are submarine. Further development in each mine will extend operations even farther from the shore. Mining must adhere to strict stipulations with respect to minimum thickness of cover between the sea and mine working and mining design is based upon a calculated maximum tensile strain of 8.5 millimeters per meter (mm/m) at the seabed.

Future mining calls for the recovery of coal from depths of up to 1200 m and from workings extending up to 13 km from the coastline. As the mine workings extend deeper, many problems associated with deep mining particularly in gassy seams, are being identified. The first and most significant, is the problem of gas outbursts. At No. 26 Colliery in Nova Scotia, prior to its closure, a number of incidents of gas outbursts have occurred. Other depth related problems are expected to arise due to increased ventilation requirements and larger roof pressures. There is no local experience in multiple seam longwalling. Sequence of extraction could help alleviate the problem of higher roof pressures and studies are in progress to develop first-stage guidelines utilizing the prevailing experiences in other countries longwalling under similar geological environment.

Undersea mining places serious constraints on the transportation routes. The workings continue to extend farther from the land based portals/shafts increasing the distance to be travelled by personnel, materials and raw coal. There is also the inherent problem of spontaneous heating to be considered wherever coal transportation is interrupted by the need to store coal en-route to the surface.

The mining operations at Minto, New Brunswick are at present surface methods employing the use of draglines for stripping overburden and loaders and trucks for extracting coal.

Similar surface methods are proposed for the Quinsam mine located near the town of Campbell River on Vancouver Island in British Columbia.

The main difference between these two surface mining operations is that, where the former mines from a single thin seam of up to 0.5 m in thickness and containing between 5-8% sulphur, the latter will extract from three different seams, all containing differing quantities of sulphur; (0.5-3.5%) and varying in thickness from 1.2 m to over 4.5 m.

In underground coal mining methods such as those practised in Nova Scotia, coal is extracted on the face by shearer loaders. The main component of a shearer loader is the cutting drum which is equipped with cutting picks. A single ended ranging drum shearer consists of a motor, the hydraulic haulage unit, the gearhead and the ranging arm mounted on the underframe. If another gearhead and a ranging arm is added, the single drum shearer is converted into a double ended ranging drum shearer. Both types of shearer-loaders are used on longwall faces in the Sydney coalfield. These machines are very powerful ranging from 170 to 1000 kW and are capable of operating in seams from 0.75 to 5 m in thickness. Selective mining is not however, a forte of such machines.

The total sulphur content of coals in all the seams mined in the Sydney coalfield is relatively high. There is a pattern of increased sulphur in the top and bottom parts of the seams. Furthermore, the presence of sandstone channels which exist in the overlying roof strata and also intrude into the coal seams are mined with the coal, tending to increase the sulphur content of the run-of-mine coal in addition to causing unavoidable dilution.

At present no special provision is made in extracting the coal seams on the longwall faces from the point of view of sulphur reduction. Any sulphur removal which occurs is carried out in the washing plant. High horsepower shearer loaders do however serve to break the coal into finer size form causing some liberation of pyritic sulphur but contributing to higher preparation cost due to the generation of larger quantities of fine coal. In order to improve roof stability, a few centimeters of coal is left at the top (roof) and bottom (floor). This practice, reduces the thickness of the mined seam but can also leave some of the high sulphur containing material in the waste. Raw coal containing pyrites may be prone to spontaneous heating (self-heating) and therefore elaborate precautions are usually taken to prevent an occurrence of spontaneous combustion (ref.6).

It is anticipated that when multi-seam mining is introduced in Sydney, Nova Scotia, greater emphasis will be placed on underground mine scheduling such that coal tonnages from various seams and from a number of pits will be mined and blended in a pre-defined sequence. This is expected to provide the means for maintaining the sulphur content in the run-of-the-mine coal to a prescribed limit. Any additional sulphur removal and control will be carried out in the washing plant.

In surface mining, opportunities for selective mining and in-pit blending for quality control are far greater than with underground mining. A variety of mining equipment is available to accomplish these objectives. In a surface mine there is plenty of space for equipment turn around, visibility is good and many of the environmental problems met in underground mines are insignificant or even non-existent in surface operations.

Hence, the success of a selective mining operation is largely dependent on matching the operational characteristics of the excavation equipment with the geology of the deposit. For example; thickness and number of seams; the dip, thickness and nature of interburden; location in the stratigraphic column in relation to the seams to be mined; the thickness and nature of overburden, the quality of coal in seams including variations in sulphur content, spatial variation of coal quality within the seams; the presence of folds, faults and other complexities and sub surface water all influence the choice of equipment, the manner of application and the resultant performance.

Conventional equipment such as draglines, scrapers, dozers, front end loaders have all been used successfully in selective mining. The criterion at all times is to obtain a clean separation of coal from waste thus maximizing resource recovery at the lowest possible cost and in a safe manner. Recent developments in surface mining equipment for application in selective mining are the hydraulic shovels and drum type surface miners (ref.5).

The newly developed hydraulic shovel combines the operating merits of a shovel with a front end loader. The "wrist action" of this type of machine allows efficient separation of waste from coal thus reducing the downstream cost of screening and preparation. When converted to a backhoe it can be used to mine inclined and folded seams. The greater breakout force of these machines makes them ideal for loading compacted materials. The machine has given outstanding performance in selective mining and when excavating dipping and geologically complex seams.

In multiple seam mining where seams are separated by thin bands of overburden (interburden); in cases where coal seams are split and where materials of different qualities must be separated application of a new breed of mining equipment called surface miners is available. Such machines can mine to very narrow limits improving resource recovery and providing an uncontaminated product. By separating materials of various qualities at the source considerable savings in downstream processing and blending can be realized. (Fig. 2).

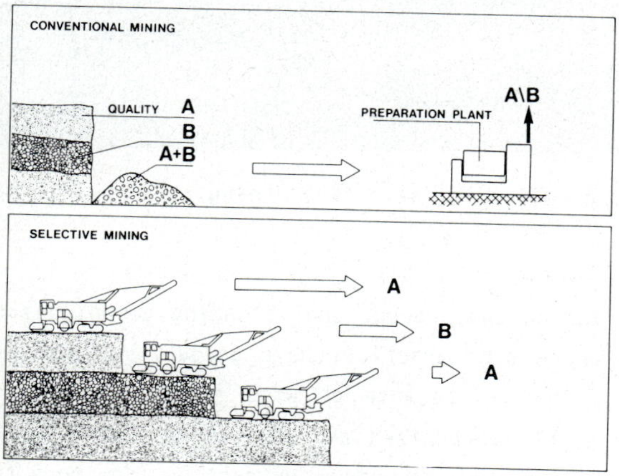

Fig. 2. Selective mining (bottom) permitting separation of quality product at source.

One example of such a miner is the surface miner developed recently by WIRTGEN in Federal Republic of Germany. Such machines are continuous, highly flexible, diesel powered, do not require drilling and blasting and can discharge into either trucks or mobile conveyors. This particular machine moves on three crawlers, one in the front and two behind. It can move at speeds up to 6 km/hr under its own power. Material cutting is accomplished by a rotating drum similar to those seen on underground continuous miners. The drum is raised, lowered or lifted by hydraulic controls. There is no primary crushing requirement as coal sizing is controlled to a maximum particle sizing of 300 mm in any dimension automatically (Fig. 3).

Fig. 3. Wirtgen surface miner for selective mining.

HOMOGENIZATION AND BLENDING

Raw coal homogenization and product coal blending techniques have largely emerged from Europe where such practices have gradually become more or less obligatory (ref.7). This is because of wide fluctuations in run of mine coals produced from highly mechanized and productive mining locations which are often remote from each other and where extraction from more than one seam occurs simultaneously.

In order to maintain independence between mine and plant, storage capacity must be provided to create a buffer between the operations of both. The appropriate capacity for this can usually be readily determined once the operating schedules for both have been established. Homogenization or blending systems are commonly employed in order to cater for variations in size consist, moisture content, coal washability and quality parameters such as ash, heat and sulphur content. These are usually in the form of bedded stockpiles or multi-compartment bunkers. An example of the end result of utilizing such a system is shown in Table 3 and Figures 4,5,6.

Table 3 gives analytical values obtained from samples of raw coal at ten minute intervals for the following:

a) raw coal discharged from a run-of-mine belt conveyor to the stacking conveyor of a coal stockpile blending system

TABLE 3

Analytical values for incremental samples of raw coal.

Time (min)	Unblended				Blended			
	ash		sulphur		ash		sulphur	
	%	Devn.	%	Devn.	%	Devn.	%	Devn.
10	30.3	-3%	1.7	59%	31.3	-1%	1.05	1%
20	34.5	10%	1.51	41%	31.6	0%	1.11	6%
30	27.2	-13%	1.63	53%	31.8	0%	0.99	-5%
40	32.6	4%	0.73	-32%	31.9	1%	1.03	-1%
50	33.3	6%	0.66	-38%	32.7	3%	1.08	4%
60	36.5	17%	1.05	-2%	31.8	0%	1.00	-4%
70	30.3	-3%	1.77	66%	31.9	1%	0.92	-12%
80	29.4	-6%	1.01	-5%	31.1	-2%	1.04	0%
90	32.2	3%	1.22	14%	32.8	3%	1.08	4%
100	32.5	4%	1.68	57%	32.3	2%	1.09	5%
110	31.9	2%	1.01	-5%	32.2	1%	1.01	-3%
120	26.2	-16%	0.77	-28%	32.4	2%	0.99	-5%
130	31.7	1%	1.04	-3%	32.1	1%	0.98	-6%
140	31.5	1%	0.57	-47%	31.6	0%1	1.07	3%
150	27.2	-13%	1.11	4%	32.6	3%	1.03	-1%
160	33.6	7%	1.03	-4%	31.8	0-%	1.10	6%
170	31.4	0%	1.22	14%	31.2	-2%	1.16	11%
180	34.7	11%	1.01	-5%	30.6	-4%	1.02	-2%
190	29.7	-5%	0.75	-30%	31.6	0%	1.01	-3%
200	28.8	-8%	0.88	-18%	31.2	-2%	0.98	-6%
210	27.4	-31%	1.22	14%	31.9	1%	0.99	-5%
220	30.8	-2%	0.93	-13%	31.3	-1%	0.99	-5%
230	29.7	-5%	0.81	-24%	31.7	0%	1.03	-1%
240	31.6	1%	0.77	-28%	31.7	0%	1.08	4%
250	38.3	22%	0.83	-22%	31.9	1%	1.12	7%
260	29.2	-7%	0.72	-33%	32.2	1%	1.08	4%
270	33.7	8%	0.49	-54%	32.3	2%	1.08	4%
280	25.1	-20%	1.20	12%	30.9	-3%	1.06	2%
290	30.9	-1%	1.08	1%	30.8	-3%	1.05	1%
300	37.7	20%	1.66	55%	30.8	-3%	1.05	1%
Average	31.33		1.068		31.73		1.042	
Standard Deviation	3.13	10.0%	0.35	32.7%	0.57	1.8%	0.05	4.9%
Variance		9.78		0.12		0.32		0.002

b) raw coal from the same source as (a) on the reclaiming conveyor after removal by a drum reclaimer from a chevron form of blended stockpile.

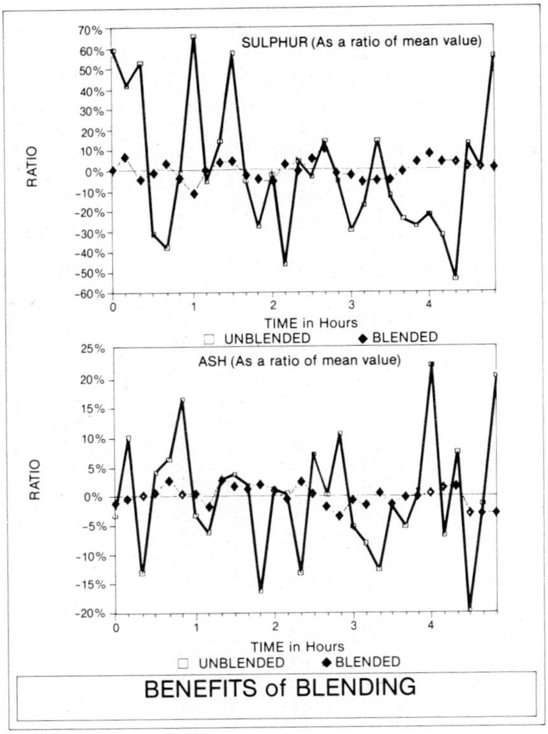

Fig. 4. Graph showing ash and sulphur distribution for blended and unblended coal.

Figure 4 and 5 are graphical representations of the data given in Table 3 which show clearly how this form of blending has resulted in greatly reduced ash and sulphur fluctuation in the subsequent raw coal feed to the washing plant.

Figure 6 shows similar benefits to the washing plant obtained by controlled blending of two different seams. In terms of both ash and sulphur content, the resultant task of cleaning the coal is simplified as a result of blending and the required clean coal specification is more readily achieved.

Pre-cleaning blending therefore serves to "iron-out" inconsistancies in the run-of-mine coal, aiding cleaning and also creating savings in the capital and operating cost of the coal preparation plant. Post-cleaning blending ensures that the required product specification is always achieved with despatched coal. The extent to which both are applied is usually mainly

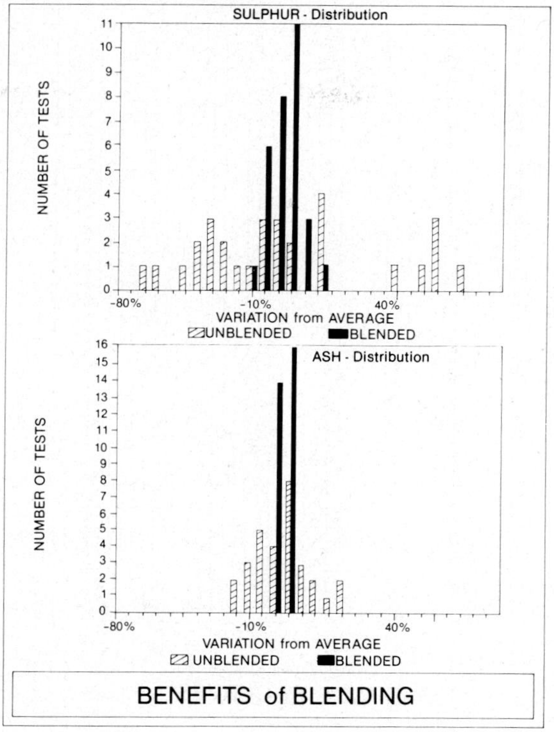

Fig. 5. Graph showing ash and sulphur variation for blended and unblended coal.

determined by the variability of the raw coal. This is because of the relatively high costs of the various levels of blending system that are currently available.

Most of Canada's current large capacity coal mining operations are fortunate in being able to avoid the very costly and sophisticated systems which produce improvements of the type shown in Figure 6. However, the eastern operations of CBDC in Nova Scotia do employ a simplified raw coal homogenization system (ref.8) as shown in Figure 8 and future developments currently planned in this province will probably result in the introduction of the more sophisticated types of blending bed stockpiles in order to provide for the required control of size distribution, washability characterisitcs and fluctuations in ash and sulphur content.

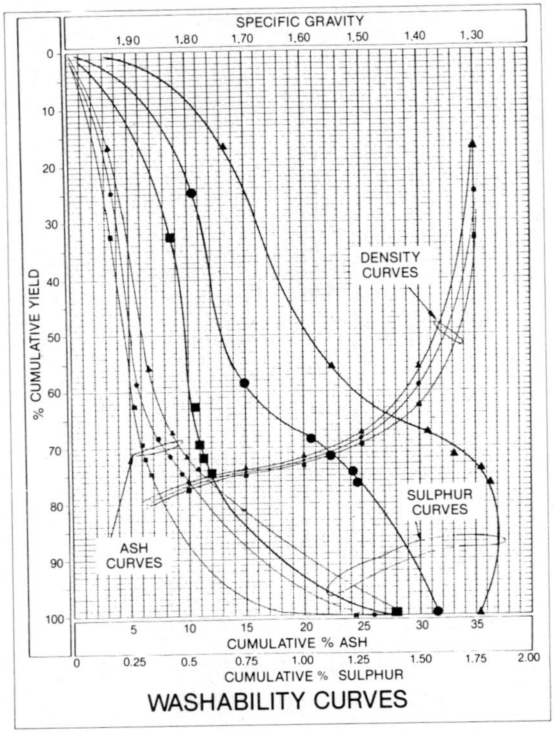

Fig. 6. Graphs of washability curves for individual and blended coal from two seams.

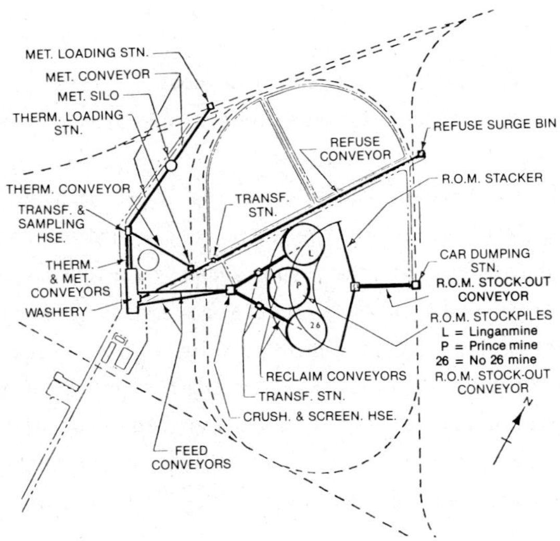

Fig. 7. CBDC Victoria Junction Coal Preparation Plant: surface facilities layout.

COAL CLEANING

All three seams encountered in the Comox coalfield on the Quinsam property are readily cleaned and therefore represent no problems to conventional coal cleaning equipment. The fine coal, (minus 0.5 mm) can be cleaned by froth flotation with good potential for reduction in pyrite sulphur content with careful reagent selection. Providing that the raw feed is blended to eliminate any severe fluctuations, no problems are likely to be encountered with maintaining a good quality export thermal coal specification. By a combination of raw coal and cleaned coal blending and coal washing, sulphur variations from 0.5 to over 3.5% in the run of mine coal can be consistantly maintained at below 1.0% in the clean coal product.

Eastern coal from the Minto coalfield contains 6-8% sulphur and averages 18% ash content. About 75% of the sulphur content is pyritic which is present in the coal in a variety of forms. The iron sulphides, pyrite and to a lesser extent, marcassite, occur over a wide size range from large nodules to finely disseminated in-fillings within the cleat of the coal. The coarser material, sized down to 150 micron could be fairly easily removed by the simpler types of gravity coal cleaning devices ranging from Baum jig and water-only cyclones in the coarser sizes, to tables and spirals for finer coal. It is in the fine grain sizes, intimately mixed with the coal constituents, where the separation becomes more difficult. Sub-sieve particle size analysis has shown than fine grain sulphide crystals contained within the coal can comprise 50% smaller than 30 microns.

CANMET coal laboratories are currently studying alternative methods of effectively cleaning this coal (ref.9). The results given in the following table summarize the performance of several coal cleaning units as determined by means of computer simulation for 9.5 by 0.15 mm coal. This work formed part of the initial examination into the possible methods for upgrading higher sulphur coals carried out by provincial and federal government laboratories in Canada.

Cleaning unit	Separation density	Clean coal quality %ash	Clean coal quality %sulphur	Degree of Reduction %ash	Degree of Reduction %sulphur	Yield %
Dense medium cyclone	1.64	10.6	6.2	48	23	75
	1.45	6.3	-	66	38	51
Water only cyclone	1.65	12.2	6.6	40	19	71
	1.31*	7.3	-	63	37	16
Table	1.65	11.1	6.3	46	22	75
	1.42	6.6	-	68	38	44
(Raw coal) reference	-	21.2	7.9	-	-	-

*not practical

Coal from CBDC's mines in the Sydney coalfield have very high fluidity and unusually low ash content (ref.12). However, sulphur content, mainly pyritic, normally ranges from 1 to 2.5% in the run-of-mine coal. In order to obtain the low and consistant sulphur level required in metallurgical coking coals, preparation at low relative density is required. Consequently, the dense medium cyclone circuits at the Victoria Junction plant are operated at relative densities between 1.30 to 1.35, perhaps the lowest levels anywhere in the world. The thermal product is obtained by rewashing the sinks product from the primary dense medium cyclone stage in a secondary dense medium cyclone circuit. In order to maintain the correct magnetite size distribution for the primary state (95% passing 325 mesh), raw magnetite is milled in the plant and quality is carefully monitored.

RESEARCH AND DEVELOPMENT

Table 4 gives an indication of the various product qualities from the three operations previously described.

Associated with western Canadian mining, research work is directed towards improving resource recovery by application of newly developed mining equipment and cost reduction by efficient use of available resources and greater application of computers to mine planning. The hydraulic excavator has received wide acceptance, and its use is expected to continue to expand as new applications occur and further improvements are made. A surface miner is now operating in Alberta and use of this type of machinery in selective mining is also expected to increase. Computerized scheduling of equipment and coal extraction sequence in the pit will result in tighter cost control and effective smoothing of run-of-mine (ROM) coal quality as well as creating a shorter reaction time to deal with variations in the ROM coal quality.

Spontaneous combustion is another problem in coal mining and stockpiling. Low rank coals are generally more vulnerable to self-heating than high rank coals. The Calgary and Ottawa Coal Research Laboratories of Canada Centre for Mineral and Energy Technology (CANMET) are currently engaged in extensive studies relating to development and control of spontaneous combustion (ref.6). Some tentative recommendations arising from this work are:

- Coal should be stacked in well compacted layers about 0.6 m in thickness. Each layer should be rolled before commencing the next.
- Coals from different ranks and of different sizes should not be stored together. If the stacking of coal of different sizes becomes unavoidable, the risk of heating can be reduced by alternating layers of fine coal and large coal.

TABLE 4

Product quality of higher sulphur coals.

Quality Parameter	CBDC Victoria Junction, NS		New Brunswick Coal Co., NB	Brinco Mining Ltd. Quinsam, BC
Coal Use	Met. coal	Thermal coal	Thermal coal	Thermal coal
Purpose	Local steel mills	Local power utility	Local power utility	Export
Moisture	7.5	7.0	1-3	10.0
Ash (dry basis)	3	10-15	15-20	12.0
Volatiles (dry basis)	36	33	31-34	40.0
(Fixed carbon)	61	52-57	47-50	58.0
Total Sulphur	1.2	2-4	6.8	1.0
Cal. Value as recd kCal/kg	7700	6610-7055	6600	6250
Free Swelling Index	7.5		5.5	0.5
Ash Fusibility reducing °C IT	1080		1077	1250
HT	1150		1200	1310
FT	1280		1227	1320
Ash Analysis				
SiO_3	38		34.4	35
Al_2O_3	20		15.6	25.3
Fe_2O_3	32		45.6	15.7
P_2O_5	0.2		1.0	0.6
TiO_2	1.1		0.5	1.6
Na_2O	0.8			0.2
K_2O	1.3			0.1
MgO	1.1			0.3
CaO	2.1			16.6
Hard Groveability	63 min^m			45 min^m

- Since ROM coal is more susceptible to spontaneous combustion than washed coal, faster turn around of stockpiled ROM coal should be provided.
- Spraying of road tar or similar binding agents provides an effective seal to air leakage and is recommended for coal stacks likely to be left over for a long time.

In the Sydney coalfield in eastern Canada, mining research is in progress in the underground mines aimed at improving safety, productivity, resource recovery and coal quality on the working face. One aspect of particular interest is the horizon control instrumentation for the shearer loaders. With the aid of this instrumentation, the cutting machine can be programmed to extract a specified thickness of the coal seam leaving a pre-set amount of roof and floor coal. Standard instrumentation for such mahcines is available which can distinguish coal from non-coal thus providing control in the mixing of waste with the mined coal and also helping to eliminate highly pyritic elements of roof and floor material in some cases.

The mines in the Sydney coalfield are gassy with a methane release approaching 60 m^3/tonne of coal mined. Consequently, systematic methane drainage is carried out at all existing mines and plans are under active consideration to sell this gas for commercial use.

Research and development in coal beneficiation of Canada's higher sulphur coals has grown in intensity during the past decade. The need for such work has been largely stimulated by greater acid rain awareness and consequent tightening up of environmental legislation of the provincial governments concern to increase coal utilization (ref.13).

In Nova Scotia, the provincially owned electric utility (Nova Scotia Power Corporation) which generates 95% of the electricity in the province, has recently commissioned two new 150 megawatt coal fired units and two more are currently under development (ref.14).

In New Brunswick, coal mining is controlled by the New Brunswick Coal Co., a provincial Crown Corporation, and the majority of the provincial output is consumed by the New Brunswick Electric Power Commission. Both producer and user are actively engaged in research work to improve the quality of the product coal. The beneficiation potential of coals from both eastern Canadian coal producing areas is currently the subject of an intensive study funded by the federal government through CANMET's Energy, Mines and Resources (EMR). The ultimate goal of the program is to help further develop the coal industries in these two provinces and the focus of the work is stated to be; the reduction of high sulphur and mineral matter content, in all economically recoverable coal deposits. Various beneficiation methods have already been examined including; selective agglomeration (ref.15) using a technique developed by the National Research Council (NRC) of Canada; wet and dry high gradient magnetic cleaning (ref.10,11) and various chemical communition processes (ref.16). Research work is on-going for each of these methods together with a number of others. Scotia Liquicoal, a privately owned company, has been involved in process and product development since 1979. Its main interest is in developing a coal-liquid mixture but has worked with the NRC in developing a pilot plant in Dartmouth, Nova Scotia which utilizes the NRC oil agglomeration process. The slurry fuel produced normally contains 60% coal, 20% water and 20% oil, but other combinations aimed at reducing the oil content are currently being developed.

Another coal liquid mixture development program has been underway for a similar period of time, employing a Swedish product called "Carbogel". This product, developed by AB Carbogel contains 70% coal and 29% water and 1% stabilizer. A pilot plant incorporating coal grinding and froth flotation is located at CBDC's Victoria Junction preparation plant and can produce up to 5 tonnes/hr of the fuel. Several thousand tonnes of this product have now been tested at Chatham power station in New Brunswick and very favourable results have apparently been obtained from various trial burns. Other development projects are also underway including research into the coal preparation requirements for deep cleaning eastern Canadian coal to facilitate fluidized bed combustion and coal water mixture production (ref.17,18). Both of these avenues of research are viewed as being potential stepping stones for encouraging and expanding the exportation of eastern Canadian coal and coal technology as well as a stimulation for increasing local utilization. Fine coal beneficiation technology presents formidable and consequently, costly problems of communition, separation, recovery and drying. These are major reasons why the development of coal liquid mixtures for combustion is viewed as being a major prioity by the eastern Canadian coal users. However, when balanced against estimated annual sulphur emission clean up costs of $25-$50 million for Nova Scotia and $15 to $38 million for New Brunswick at current coal consumption levels, the impetus for actively pursuing all reasonable coal beneficiation route is clearly, well justified.

REFERENCES

1 I. Irwin, Quinsam Coal, Mining Review, V. 4, No. 4, pp. 8-13, July 1984.
2 D.E. Pearson, The Quality of Western Canadian Coking Coal, CIM Bulletin, V. 73, No. 813, pp.70-84, January 1980.
3 A.S. Romaniuk, Coal Reserves and Production for Canada 1981, CANMET/EMR, Energy Research Program, Coal Research Laboratories, Division Report ERP/CRL 82-09(TR), November 1982.
4 R.K. Singhal and D.G. Osborne, Coal Mining Developments in Canada, World Mining Equipment, to be published.

5 R.K. Singhal and D.G. Osborne, Steep Seam Coal Mining in Western Canada by Open pit Mining, SME-AIME Annual Meeting 1984, Preprint No. 84-96.
6 R.N. Chakravorty, Spontaneous Combustion Problems Associated with the Mining and Storage of Thermal Coals, Paper 3, Proceedings 2nd CIM Technical Conference on Western Canadian Coals, Edmonton 1982.
7 D.G. Osborne and A.W. Walters, Design Criteria for Canada's Export Coal Preparation Plants, Paper 89, 87th Annual General Meeting, CIM, Vancouver, April 1985.
8 W.H. Lilleker and A. Lambert, The Role of Coal Preparation in the Cape Breton Coal Industry, Paper 52, 77th Annual General Meeting, CIM, 1975.
9 D. Abbott, Limits of Upgrading Eastern Canadian Coals High in Sulphur by Washing, CIM Bulletin, V. 74, No. 827, pp. 153-1, March 1981.
10 G.I. Mathieu and L.L. Sirois, Dry Cleaning of Thermal Coal, Paper 91, 87th Annual General Meeting, CIM, Vancouver, 1985.
11 G.I. Mathieu, Wet and Dry High Gradient Magnetic Cleaning of Eastern Coals, Proceedings 64th CIC Coal Symposium, Ottawa, 1982.
12 B.W. Clyburn, Low-gravity Separation: Part 1, CIM Bulletin, V. 73, No. 823, pp. 145-9, November 1980.
13 W.R. Effer, Acid Rain - A Re-evaluation of Some Issues, Proceedings 64th CIC Coal Symposium C.S.Ch.E., Ottawa, 1982.
14 J.J. Laffin, The Future of Coal Utilization Nova Scotia, Proceedings 64th CIC Coal Symposium, Ottawa, 1982.
15 C.E. Capes, R.D. Coleman and W.L. Thayer, Selective Oil Agglomeration: An Answer to Fine Coal Treatment Problems, Proceedings 64th CIC Coal Symposium C.S.Ch.E., Ottawa, 1982.
16 P.L. Silverston, R.R. Hudgins, B.R. Spink, B. Smith and G.I. Mathieu, Chemical Comminution of Nova Scotia Coals, Proceedings 64th CIC Coal Symposium C.S.Ch.E., Ottawa, 1982.
17 N.S.H. Stover, E.J. Anthony, D.L. Desai and F.D. Friedrich, Combustion Performance and Sulphur Capture during Fluid Bed Combustion, CIE Symposium New Fuels - New Firing, Calgary 1982.
18 D. Abbott, Eastern Canadian Coal Research and Development for a Competitive Energy Market, Proceedings 35th Canadian Conference on Coal, Calgary 1983.

SCHEDULING TO CONTROL COAL QUALITY

PRASHANTA K. CHATTERJEE[1], CINDY WILSON[1], AND E.Y. BAAFI[2]
[1]Department of Industrial and Systems Engineering, Ohio University, Athens, Ohio 45701-2979 (U.S.A.)
[2]Melbourne University, Melbourne, (Australia)

INTRODUCTION

Underground coal mine planning is largely manual and can demand considerable resources to generate a working plan and to perform limited tests of sensitivity to variations in geological and mining conditions. It is possible to employ computer aided mine planning to automate much of the manual procedures and allow the planner to evaluate many layouts, extractions strategies and schedules.

Ohio coals have an inherent high sulfur content which restricts its widespread use. There are limited ways in which this sulfur can be removed, either before or after extraction. Generally, the post removal of sulfur to an acceptable level is considered uneconomical. Thus, renewed emphasis is being placed on optimizing extraction procedures to either selectively produce coal that has little or no sulfur or to blend the production in such a way that an acceptable sulfur level is obtained. This paper examines a mine planning approach which employs a scheduling routine to optimize run of mine coal quality.

OBJECTIVE

The objective of a scheduling approach would be to meet coal quality requirements for a given production and planning period subject to physical, geological, and mining constraints.

METHODOLOGY

The geological characteristics of the deposit are an important consideration. The variation and distribution of coal quality (sulfur, ash, etc.) can be modelled with geostatistical methods, and based on this modelling technique, a predicted sulfur block model can be obtained. The greater the detail in the input data, geology, and other information, the more reliable the prediction. A geostatistics software package is employed to produce a block model (Chatterjee, 1984). A mine plan drafting software is employed to generate an interactive, user friendly center line representation of a detailed mine plan (Chatterjee, 1984). The plan is generated from a series of nodes. These nodes contain the necessary information for a roadway, i.e. width, height of opening, coal quality attributes, etc. The geological block model is internally referenced by these nodes, and geological and related quality information queried at each node can be retrieved from this block model.

Once a layout is defined either by digitizing or through actual screen drafting, simulation and scheduling can be performed. Mine operating results such as productivity and coal quality can be predicted for any given area. If there is more than one area, the simulation and scheduling functions are employed to produce a working schedule. Judicious manipulation of productivity values and shift schedules from different areas will allow maximization of productivity for the entire mine. In the simulation phase, the user must specify production variables based on historical data, past mining experience and expert knowledge. An example of such information is shown in Table 1. Some of these variables, for example, a mineable thickness of the coal seam and the specific gravity of coal, are dependent on the actual geology of the deposit. Other variables are dependent on the particular management procedures followed in the mine such as the average production per shift and the number of working days in a month.

Additional information that will be computed by the program is shown in Tables 2 and 3. The user is furnished with an evaluation in the form of a detailed panel and summary reports. The detailed panel report provides monthly figures for total tons produced, average production per day, and the number of idle shifts per month. These values are derived by using a probability function. Both the panel and summary reports give a description of the attributes of each panel that is being mined, the date on which mining begins and finishes in each panel, and the final summary at the end of the simulation. The user will be able to perform as many simulations as desired and to compare the final evaluations to help him select the optimum mine plan and schedule.

SCHEDULING MODEL FORMULATION

Several recent reports have cited the need for an efficient mine scheduling program. In a recent report (Anon, 1984) the author outlined how run of mine quality can be controlled and scheduled in a large underground mine. The concept of an incremental block system is introduced. All coal quality and geological information is examined over a light table and information from the maps is entered into a formula which produces a run of mine percentage. Armstrong (1985) outlined possible ways of blending run of mine coal. Kim et. al. (1981) in a special report outlined the concepts of operational

TABLE 1

USER SPECIFIED PRODUCTION PARAMETERS

1.	The specific gravity of the coal	84.0
2.	The number of working days in a month	20 days
3.	The number of productive shifts per day	2
4.	The starting date for the simulation	YY, MM, DD 84 07 14
5.	Which panels are to be mined	1,2,3,6,8
6.	The height of the coal seam in each panel	6.5 ft.
7.	The mineable thickness of the coal in each panel	6.4 ft.
8.	The average production per shift	1500 tons
9.	The standard deviation of the normal production rate	100 tons
10.	The probability of having normal production	.90
11.	The face recovery of each panel	.90
12.	The number of days required to move the face equipment from one panel to the next	24 days

TABLE 2

DETAIL PANEL REPORT

YEAR	MONTH	WORK DAYS	TOTAL TONS	AVERAGE PER DAY	IDLE SHIFTS
85	8	13	43.2	3.32	3
85	9	20	65.8	3.29	5
85	10	20	66.7	3.34	4
85	11	20	69.2	3.46	5
85	12	20	65.4	3.27	6
86	1	16	51.5	3.54	2

PANEL 1 BEGINS ON 85 8 12
PANEL FINISHED ON 86 1 25 TOTAL TONS PRODUCED = 361.8

TONS IN THOUSANDS

TABLE 3

SUMMARY REPORT

AREA	PANEL NO.	PANEL LENGTH	FACE LENGTH	PANEL RESERVE	ACCEPT. TONNAGE	START YY MM DD	FINISH YY MM DD	PRODUCED TONS
2	1	3337.1	562.1	470.1	434.0	84 10 16	85 05 04	412.3
2	2	3377.4	562.1	475.8	439.2	85 05 25	85 12 13	417.2
2	3	3417.8	562.1	481.5	444.4	86 01 04	86 07 27	422.2
2	4	3458.1	562.1	487.2	449.7	86 08 18	87 03 13	427.2
2	5	3498.4	562.1	492.9	454.9	87 04 04	87 10 25	432.2
2	6	3538.8	562.1	498.5	460.2	87 11 16	88 06 15	437.2
							TOTAL =	2548.3

scheduling in an underground coal mine. In this scheduling application, a model was developed to consider twelve mining sections using eight continuous miners. The scheduling model was developed for the Helen Coal Mine of Pennsylvania. The mining sections included main and sub-main developments, room and pillar sections and a shortwall section. The purpose of the scheduling model was to be able to determine which working areas in the multi-face mine should be mined during a planning period to meet the coal sulfur quality and tonnage requirements. The working areas were determined for one shift at a time during the planning period.

The authors employed a zero-one integer programming model. The 0-1 model building philosophy was to minimize total sulfur content of coal mined. Constraints of the model are basically:

(i) to keep the most productive section (the shortwall) and its development operating at all times,

(ii) to utilize all available machines during a working period, and

(iii) to tie-in the short term scheduling with the long term mine plans. This constraint prevents selective mining of only the low sulfur sections.

With X_i denoting the i_{th} working area whose predicted sulfur and tonnage block values for a shift (or period) are S_i (%) and T_i (tons) respectively, the objective function of the Integer Programming formulation of reducing total sulfur content is expressed by equation (1) below.

$$\text{Minimize} \quad \sum_{i=1}^{N} X_i \, S_i \, T_i \tag{1}$$

where N is total number of working areas.

The restriction with regard to both quality and tonnage requirements are formulated in the following way. First, ROM composite sulfur must not exceed maximum allowed, S_{max} (%), during a shift.

$$\frac{\sum_{i=1}^{N} X_i \, S_i \, T_i}{\sum_{i=1}^{N} X_i \, T_i} \leq S_{max} \tag{2}$$

Also, all the available crews, NC, must be assigned during a shift. This constraint guarantees a minimum production during a shift.

$$\sum_{i=1}^{N} X_i = NC \qquad (3)$$

This constraint assumes that at any shift, the available crews do not exceed the total number of potential working areas. Thirdly, the most productive section, e.g. the shortwall and its development areas, would be worked at all times. If decision variables X_3 and X_4 represent the shortwall development and production sections respectively, this constraint is given by equation (4) below.

$$X_3 = X_4 = 1 \qquad (4)$$

Fourthly, a decision variable, X_i, must either be 0 or 1. If X_i is zero, then the i_{th} section is not mined.

$$X_i = 0 \text{ or } 1 \qquad i = 1, 2, \ldots, N \qquad (5)$$

The above 0-1 formulation was solved using implicit enumeration method (Balas, 1965). The current approach utilizes an LP model to determine an optimum schedule. The model is formulated as follows:

$$\text{Minimize } Z = C_i\, T_i + G\, D_i\, T_i + K\, (S_i - S_{stand})\, T_i$$

Subject to

$$\sum_{i=1}^{N} T_i \geq TDT$$

$$T_i \leq MT \text{ for all } i = 1$$

$$T_1 = L$$

where C_i = the production costs associated with section i (\$/ton); T_i = the tons removed from section i (tons); G = hauling costs (\$/ton/foot); D_i = the hauling distance from section i to the shaft (feet); K = a penalty/bonus factor associated with the sulfur content (\$/% sulfur/ton); S_i = the actual sulfur content in the ore removed from section i (% sulfur); S_{stand} = the standard (specified) acceptable sulfur content (% sulfur); TDT = total daily tonnage required from room and pillar sections; L = daily tonnage required from longwall section; MT = maximum daily tonnage possible from room-and-pillar section; and N = total number of sections.

The level of acceptable sulfur content has not been included in the model as a constraint. Instead it is taken into account by using a penalty/bonus cost factor in the objective function. If the overall sulfur content is less than the standard, a bonus will be paid because of the higher quality. If the sulfur content exceeds the standard, a penalty is charged for the lower quality coal. The sulfur content in the deposit is assumed to be normally distributed with a given mean and standard deviation. The hauling distance is calculated for each section daily. Once these variables are calculated, the linear programming subroutine is called to optimize the objective function for each day of operation, according to the variables read into it. The optimization is done with respect to the amount of tonnage to be removed from each section.

If the sulfur content of the total tons removed exceeds a user specified maximum value, the main program will find the room-and-pillar section with the highest sulfur content (section k) and then add another constraint to the linear programming model. This constraint is $t_k = 0$, indicating that section k should not be mined at this time. The linear programming subroutine will then be called to re-optimize the objective function with the additional constraint.

Generally, there is not too much freedom of choice in an underground operation. Mining is dictated by physical, geological, and rock mechanics constraints and re-assignment of crews to new areas is rarely permissible. This restriction makes modelling very difficult. It is possible, however, to simulate the extraction process and determine if a given schedule is feasible.

RESULTS AND DISCUSSION

The scheduling simulation algorithm is still in its experimental form. The model has been tested with historical data and has been found to provide meaningful results. However, the algorithm is being incorporated into an interactive mine planning and drafting software package and currently work is underway to complete this additional feature.

The interactive scheduling function will have an extensive data base of all working sections designated by nodes with priorities and coal qualities. The sequence of extraction will follow a given priority schedule. If due to physical, geological or other constraints, this sequence is unable to be completed, alternative sequences will be automatically examined. Crew assignments will be also performed on a nodal basis. The schedule will sequence the extraction from all operational sections to generate daily, weekly or monthly coal production, within given bounds of coal quality represented by sulfur, ash, etc.

CONCLUSION

Coal quality is becoming an increasingly important consideration in mine planning. Current manual methods are unable to accomodate the complex calculations involved to derive an acceptable schedule of production. The present approach attempts to incorporate a heuristic as well as a mathematical approach to mine scheduling. The interactive graphics mode will enable planning for coal quality to be performed quickly and at the same time provide a basis for selecting the optimum schedule from many possible alternatives.

REFERENCES

1 Anon, Knowing R-O-M Quality Pays Off, Coal Age, October, 1984, pp. 72-73.
2 P.K. Chatterjee, C. Xintan, and Y. Jianyu, An Interactive Graphics Approach for the D.sign Simulation and Scheduling of Underground Coal Mines, AME-AIME Symposium, October, 1984, Denver, Colorado.
3 Y.C. Kim, H.P. Knudsen, and E.Y. Baafi, Development of Emission Control Strategies Using Conditionally Simulated Coal Deposits, University of Arizona, Report to Homer City, 1981.
4 M. Armstrong, Ecol Polytechnique, Fontainebleau, France, Personal communications, 1985.
5 E. Balas, An Additive Algorithm for Solving Linar Programs with Zero-One Variables, Operations Research, 13, 1965, pp. 517-546.

OPTIMALLY BLENDED COAL SUPPLY FOR RAILROAD AND BARGE TRANSPORTATION

SLOBODAN LUKOVIC AND JOZSEF GOZON

Mining Engineering Division, The Ohio State University, 148 Fontana Labs, 116 W. 19th Ave., Columbus, Ohio 43210

ABSTRACT

Coal blending is an already available and well established technology for converting high sulfur coals into sellable products. Modern facilities and equipment are providing sufficient capacities at reasonable costs. Together with them, computer programs are written to assist in coal selection, to help in plants' design, and to provide a powerful tool for making decisions towards their maximal overall efficiency.

INTRODUCTION

Coal production in the United States has been hit many times in the past by fluctuations in the energy market. Unfortunately, this is not likely to change in the future. Small coal mine operators, both surface and underground, generally sell their products on the unstable spot market. Large coal companies, enjoying the convenience of long term contracts, usually are less effected by the market behavior. However, the blendable coals, especially those that could be considered for export, offer an opportunity for the small mine operators to become less vulnerable.

Due to the fact that today's coal producers must meet strict quality requirements set by the market and governmental agencies, coal blending has evolved into a very important function in coal marketing and processing. The purpose of this paper is to provide some information about the present state of the coal blending technology and some insight into the existing computer software that is now in reach of every coal producer and could help them in finding the market for their troubled coals.

Stockpile Systems and Equipment

Numerous articles and books on both theoretical and practical aspects of homogenization and blending of raw materials have been published in the last decade [1,2,3 etc.]. They indicate that due to changes in the coal market, many ports and rail-heads had to modernize, or establish new materials handling procedures and equipment. It is not easy to give an exact answer to the question on what type of equipment is to be considered for the particular stockpile system, because all applications are different in regard to various conditions and given locations. In general, criteria for evaluation of coal storage systems can be related to:

- the location of a stockpile,
- the capacity of a stockpile,
- the purpose of the stockpiling system,
- the type of coal transportation involved,
- the capital cost,
- the reliability of the storage facility,
- the environmental impact, and
- the future development.

The coal is usually blended at the transfer point, but the mine site, the preparation plant and the power station are also frequently chosen locations.

Stockpiles are commonly formed by using CHEVRON or WINDROW method [4,5,6]. Both methods have been very successful in practice, with some respective advantages and drawbacks. The CHEVRON method can provide higher property quality in the output flow and the equipment costs can be lower, compared to the equivalent WINDROW system. On the other hand, if the segregation of the grain size is likely to cause problems, the WINDROW stacking method is clearly superior.

In most cases, open storage will be preferred over covered storage system [2]. For less than 40,000 tons of coal being handled, a lowering tower seems to be the best solution. For storage capacities varying between 40,000 and 80,000 tons, a tunnel conveyor is advisable. The bridge-type reclaimer is appropriate for materials handling facilities up to 100,000 tons on open storage stockpiles. Finally, the bucket wheel reclaimer will probably be dominant, if 200,000 tons of storage capacity is required.

If segregation is expected to take place during the stacking operation, and at the same time, high blending effect is required, then the bridge-type reclaimer offers the best solution. Bridge scraper reclaimers are available for pile widths up to 60 meters and with handling capacities ranging from 1,500 to 2,000 tons per hour. For higher capaciites, bucket wheel bridge reclaimers may be applied. In this group, drum reclaimers give the best homogenization effects, allowing for easy control by use of computer-assisted weighing and proportioning devices.

Computer Modeling

One of the major problems in running an efficient blending process is undoubtfully obtaining the right quantities of specified raw materials. In the case of coal blending, the general problem scenario can be presented in the forms of the input (availability) and the output (requirements) matrices, such as shown in Tables 1 and 2.

TABLE 1
Coal Availability

Mine I	Ash A(I)	Sulfur S(I)	Moist. M(I)	BTU B(I)	Available Quantity Q(I)	Cost Per Ton C(I)
1	A(1)	S(1)	M(1)	B(1)	Q(1)	C(1)
2	A(2)	S(2)	M(2)	B(2)	Q(2)	C(2)
3	A(3)	S(3)	M(3)	B(3)	Q(3)	C(3)
"	"	"	"	"	"	"
N	A(N)	S(N)	M(N)	B(N)	Q(N)	C(N)

TABLE 2
Coal Requirements

User J	Tolerated Ash TA(J)	Tolerated Sulfur TS(J)	Tolerated Moisture TM(J)	Required BTU RB(J)	Required Quantity RQ(J)	Price Per Ton P(J)
1	TA(1)	TS(1)	TM(1)	RB(1)	RQ(1)	P(1)
2	TA(2)	TS(2)	TM(2)	RB(2)	RQ(2)	P(2)
3	TA(3)	TS(3)	TM(3)	RB(3)	RQ(3)	P(3)
"	"	"	"	"	"	"
M	TA(M)	TS(M)	TM(M)	RB(M)	RQ(M)	P(M)

By observing these two tables, it is very easy to realize the complexity of the blending problem. Even when only three types of coal are mixed, and only one blend is obtained, the calculations may take a very long time and eventually yield a non-optimal result. In the case of a real feasibility study project for a coal blending facility, many types of coal could be taken into consideration, and this initial step would require tremendous effort from the engineering team. Fortunately, today's availability and low price of computers has made it possible for everyone engaged in this sort of business to use very efficient tools and painlessly arrive at the desired solution. At the moment of writing this paper, the authors had information about four commercial blending programs [7,8], and more of them are likely to exist. Some of the common features of these programs are:

- user friendliness through the interactive input,
- determination of correct blending proportions, and
- economic analysis.

In addition to the above-mentioned programs, the authors have themselves modified a general simplex method program [9] to be used in solving coal blending problems.

One of the most significant applications of linear programming is to maximize profits using available amounts of raw materials; in this case, different types of coal. The assumption was made at the beginning that a linear model can be considered as an adequate approximation to describe the blending process. Specifically, linear programming maximizes or minimizes a linear function of several variables. By use of several restrictions, or constraints in the form of linear equalities or inequalities, the economic function can be solved. The best means in this procedure is the simplex method [10], which leads to statistics that are useful to management in the decision making phase. The program was named after the simplex method and the blending procedure, as SIMPLEBLEND.

The user of the SIMPLEBLEND program should follow four steps:

1) Definition of the economic function.
2) Identification of the decision variables.
3) Construction of a linear objective function.
4) Determination and classification of the constraints.

The economic function should be either maximization or minimization, depending on the nature of analysis. This step is taken care of by the first input prompt in the program.

The decision variables may include the sulfur, ash, and moisture content of the coals available for blending in determined quantities, along with their heat values. An example of these variables is given in Table 3.

TABLE 3
Example List of Decision Variables

Coal Seam	Sulfur (%)	Ash (%)	Moisture (%)	BTU
Upper Freeport No. 7	2.12	10.81	2.80	11,787
Pittsburgh No. 8	3.88	7.23	3.10	12,551
Hazard, KY No. 8	0.80	6.52	4.60	12,800

The objective function takes the economical factors into account. The selling price of coal with defined parameters can be obtained from the market analysis reports. The same information source can be used for production and transportation cost figures. Typical values are listed in Table 4.

TABLE 4
Example List of Economical Factors for Coal ($/Ton)

User Area	Production Cost	Transportation Cost	Selling Price
Ohio River Valley	18	1.73	27
Columbus, Ohio	25	8.11	50
Overseas, Europe	29	21.05	90

The nature of constraints can be associated with either technological requirements, such as type of transportation, capacity limitations, storage and loading facilities, etc., or qualitative specifications, such as coal characteristics. As mentioned earlier, all of the constraints must be expressed in linear form of equalities and/or inequalities. The numerical relationships describing the constraints are built into the program.

Once the problem is solved in regard to desired blends, various decision variables and new constraints may be entered into the program for detailed economical evaluation and cost optimization.

CONCLUSION

In order to sell high sulfur coals, mine operators practically have two options that they can use separately, or combine into one. The first option is coal desulfurization. The other is coal blending in order to obtain a mixture with a tolerable sulfur content. Since stockpiling is an unavoidable part of any coal handling process anyway, it is very important that the coal industry realizes the full potential of coal blending technology. With today's equipment and available decision tools, it is fairly easy to arrive at feasible coal marketing projects and to satisfy both important requirements, the full utilization of natural resources and conservation of our environment. All that is needed is a little bit of joint effort.

REFERENCES

1 R. H. Wohlbier, Stacking, Blending, Reclaiming of Bulk Materials, Trans. Tech. Publications, Clausthal-Zellerfeld, West Germany, 1977.
2 H. Althoff, Comparison of Stockpile Systems: Bucket Wheel Reclaimer versus Underground Reclaimer, Bulk Solids Handling, January, 1981, pp. 87-94.
3 C. G. Schofield, Homogenisation/Blending Systems Design and Control for Minerals Processing, Trans. Techn. Publications, Clausthal-Zellerfeld, West Germany, 1980.

4 G. Fisher, Design and Operation of Coal Storage and Homogenization Systems, Bulk Solids Handling, May, 1981, pp. 317-322.
5 Th. Koch, Comparison of Different Stacking and Reclaiming Methods for Bed Blending Systems by Means of Computer Simulation, Bulk Solids Handling, April, 1985, pp. 301-305.
6 C. G. Schofield, Review of Raw Material Stacking and Reclamation Methods, Bulk Solids Handling, September, 1981, pp. 429-436.
7 A. Sos, "COALBLND" A Computer Program for Blending Coal, Proceedings of the Second Conference on the Use of Computers in the Coal Industry, Tuscaloosa, Alabama, April, 1985, pp. 137-139.
8 Gibbs Associateds: 1985 Directory of Mining Programs, Boulder, Colorado, pp. 4-5.
9 X. T. Bui, Executive Planning with BASIC, Sybex, Berkeley, 1985, pp. 11-22.
10 A. W. Spivey and R. M. Thrall, Linear Optimization, Holt, Rinehart and Winston, Inc., New York, 1970.

PART 4

COAL DESULFURIZATION BY PHYSICAL CLEANING TECHNIQUES

LOW-GRAVITY COARSE COAL SEPARATION

DELBERT I. LILLER*,

Babcock & Wilcox Research and Development Division, Alliance, Ohio 44601

ABSTRACT

Low-gravity (<1.3 specific gravity) coarse coal separation can provide coal for use in coal-water fuels (CWFs) to replace oil, natural gas, and in some cases, dry pulverized coal. With grinding, the coarse (3/8 inch x 200 mesh) low-gravity coal can make the larger particle size end for efficient packing in CWF to obtain 70% to 75% loading. It is difficult to load much above 50% solids content with size distributions of 95% less than 5 microns without using substantial amounts of additives.

Patented improved hydrocyclones (PIH) have achieved <1.3 specific gravity separation cut points with 0.13 and 0.24 probable error (EP) values at the 75 and 90 ordinates. The PIH body diameter was 18 inches. The feedstream consisted of 1-1/4 x 0 raw coal at 23% solids with a throughput rate of 75 TPH. By using a multistage circuit to rewash overflow and underflow, the premium low-gravity coal is concentrated into one stream, while isolating the 1.3 x 1.65 specific gravity coal fraction in another, and rejecting the >1.65 specific gravity refuse via the third stream. Work is continuing to improve the efficiencies of the PIH design and process.

SUMMARY

There is an anticipated growing need for coarse, deep clean coal without modifications to its characteristics (for example, 3/8 inch x 200 mesh, <3% ash, <1% sulfur). This coal can be used to make coal-water fuel a substitute for oil, natural gas, and dry pulverized coal.

Patented improved hydrocyclones (PIH) have achieved <1.3 specific gravity separation cut points (SGSCP) with 0.13 and 0.24 EP values at the 75 and 90 ordinates. Pressures of 20 to 22 psig with the new inlet design produces 18 to 21 feet per second (fps) velocities, which were found to be important in improving the equipment's efficiencies and lowering the SGSCP settings. Pulp densities between 5% and 35% were tested; 23% is the recommended value based on bituminous coal cleaned to date. The vortex finder settings at 1.25, 1.55, and >1.60 SGSCP were determined and used to extrapolate other SGSCP settings.

*The presented text was developed while at Liller Laboratory and Engineering Service P. A., Deer Park, Maryland 21550

The PIH operating variables were used to guide the dimensioning. To improve separation efficiencies and lower the SGSCP, 18-inch and 10-inch units were tested and evaluated.

By knowing the operation characteristics of the PIH, plant circuitry can be designed using stage washing to enhance overall plant performance. Rewashing cleaned coal by another unit, set at the same setting, will give a combined EP value for both stages, which is higher than the value for the unit when operated by itself and at a lower SGSCP. These values are referenced to the plant raw coal feed. Similarly, if the refuse is rewashed the combined effect will be a decrease in EP and an increase in SGSCP. By using both rewashes and by adjusting SGSCP settings at each stage, depending on the feed coal characteristics, plant outputs can be almost pure clean coal, middle gravity coal, and refuse with no carbon value.

BACKGROUND

Low-gravity coarse coal separation is concerned with sorting light-liberated pure coal particles from heavy refuse and heavier contaminated coal particles. A large number of eastern U.S. bituminous coals start liberating pure coal at relatively coarse sizes (ref. 1). In the interest of economics and utilization, equipment and methods are needed to separate this pure coal without further size reduction.

Hydrocyclones designed for making separations by specific gravity were investigated as early as 1950. In the 1960s and early 1970s the units were widely accepted as a rougher type device to separate pyrite and other heavy refuse. Researchers have spent considerable effort improving the design for more efficient performance (ref. 2, 3, 4). Mostly by empirical design, a short stubby cyclone has evolved for specific gravity separations.

Parameters affecting performance

The hydrocyclone parameters, which affect the unit as a separator to sort by specific gravity weight, are numerous and difficult to analyze for weighing each ones importance (ref. 5). According to previous research (ref. 6) determining factors that control particle flow direction through a cyclone are:

- Shape of particles
- Solids concentration
- Feed pressure
- Back pressure

- Cyclone diameter
- Diameters of the feed opening
- Overflow opening
- Apex opening
- Length of the cylindrical part (vortex finder)
- Cone angle
- Specific gravity
- Percent solids of the feed and apex discharge
- Viscosity of the feed suspension and the liquid

Other investigations (ref. 7) have determined the cyclone diameter (D) as a relationship to particle size as follows:

- Maximum particle size of coal processed = D/10
- Recommended maximum particle size (allowing 15% oversize = D/20)
- Gravity cutpoint control is good, over the specific gravity range of 1 to 2.5 for particles of greater diameter than D/300
- The specific gravity cutpoint decreases as particle diameters decrease below D/300

Feed particles should not be larger than one fourth the apex opening to avoid plugging (ref. 8).

PIH operating variables

A PIH was empirically designed and developed through commercial use. Six operating variables observed to be critical for optimum performance (ref. 4) are:

- The inlet feed velocity (ref. 9, 10) and direction to produce the angular acceleration and velocity in the annular space for establishing the required centrifugal force.
- A straight tangential inlet with the non-tangential inlet side compressing the feedstream with a flat deflector located at three angles (ref. 9, 11, 12). This flat deflector directs the incoming slurry at a precise intersection angle with the rotating slurry in the annular space and drives the slurry rotation without intrusive interaction between stratification layers.
- A constant annular volume (ref. 10, 13) to set the revolving slurry at the correct angular velocity, acceleration, and lead to provide the centrifugal and pressure forces necessary in the separation zone.

- The vortex finder sleeve height and diameter (ref. 14) for setting the specific gravity separation range by varying the fluid drag force zone where the light clean coal is withdrawn from the heavy refuse and middle gravity coal.
- Cyclone bottom geometry (ref. 15) to establish the central region for separating the heavy liberated pyrite and refuse from the light-liberated coal through the reverse flow path by centrifugal and fluid drag forces, while maintaining stable mantles between both the inner and outer vortexes carrying light and heavy products to their respective outlets.
- A minumum height between the bottom edge of the vortex finder and the top edge of the cyclone bottom (ref. 16). The particles are accelerated radially outward by the outside swirl causing separation due to specific gravity differences, preventing size classification and allowing sufficient distance for middling spin out.

Cyclone description

The resulting design is shown in Fig. 1 with dimensional values given in terms of the cyclone inside diameter B (Table 1).

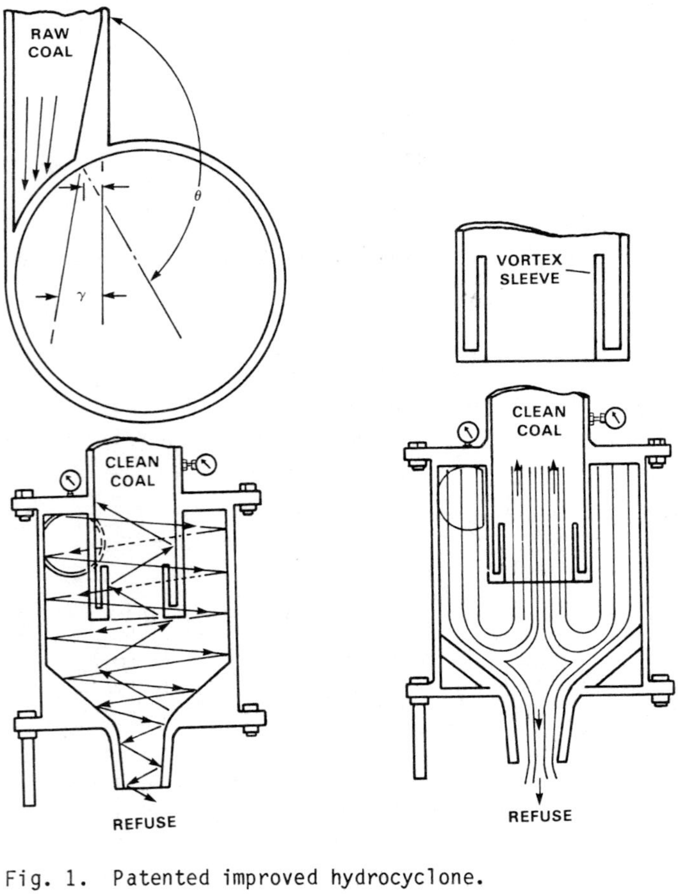

Fig. 1. Patented improved hydrocyclone.

TABLE 1

Best mode and range of selected cyclone dimensions size expressed in terms of bowl diameter (B) and inches. Dimensions of larger and smaller cyclone parts are proportional for each part to "B" product values.

Part	Preferred Size in Inches	Preferred Size Expressed in Terms of B
Bowl Diameter (B)	18.0	1.00
Inlet Tube	6.0	.33
Vortex Finder Depth	4.3	.23
Vortex Finder Sleeve ID	6.5	.36
Orifice Small ID	3.6	.21
Bowl Width	4.2	.23
Θ_4 Included Angle Depth	2.0	.11
Dish-Orifice Unit Height	12.4	.69
Dish Height	5.4	.30
Height of Orifice	7.0	.39
Bowl Height	22.0	1.22
Dish Depth	16.7	.93
Dia. at Throat Entrance	7.1	.39
Radius Connecting Θ_5, Θ_6-Θ_9	3.5	.20
Radius Connecting Θ_4-Θ_5	4.5	.25
Depth or Height of Radius	2.5	.14
Orifice Small ID Height	1.0	.06
Throat Height	6.9	.38
Orifice Bottom Outside Dia.	5.0	.28

Operation parameters

Because the PIHs were used in water-only cyclone wash plants, the 18-inch size was chosen to accommodate 1-1/4 inch x 0 feed size. This exceeds the recommended size to prevent plugging (ref. 8) since the small diameter of the bottom orifice is 3-11/16 inch. Because of the higher throughput rates, increased velocities and the new design, particles as large as 2 inches have passed through the units without plugging. Although operating at 2-inch size is not recommended, it does provide flexibility before plant repairs to crushers and other sizing equipment, without interrupting production. These plants are based on two-stage washing with hydrocyclones. The second stage unit rewashes the first-stage refuse. Although the design limits the number of seams that can be washed effectively, it does provide simple economical washing for certain applications.

The objective of this task (to improve the hydrocyclone design) was to broaden the application of these plants.

The PIH has performed best between 18 and 21 fps inlet feed velocity. The percent solids was varied from 5% to 35% with no apparent change in performance. A 23% pulp density was chosen as the recommended setting because it doubled the plant capacity. No best setting has been established experimentally. Mikhail and Humeniuk (ref. 18) found 8% to be best for 0.6 x 0.15 mm coal sizes in the water-only cyclone work conducted on Western Canadian bituminous coals. The maximum pulp density used in their work was 15%. At 23% solids and 21 fps velocity the inlet pressure is about 22 psig or 50.8 feet of head. Maintaining a constant steady head pressure was found to be important. The siphoning effect mentioned by researchers (ref. 8, 17) is avoided by placing the dewatering sieve bend inlet directly above the PIH and open to the atmosphere.

The head loss between the inlet and vortex finder outlet is 36.9 feet of head (16 psig) when operating at 22 psig inlet pressure. Best performance has been obtained at 20 to 22 psig inlet pressure. Although the performance was not measured at other pressures, operating below 18 psig resulted in a performance reduction.

Performance measurements

The throughput capacity was measured at two inlet pressure settings. At 12 psig inlet pressure the throughput capacity was 140.4 gallons per minute (gpm) in a 10-inch PIH. At 18 psig inlet pressure the throughput was 305.2 gpm. The vortex finder outlet flow was measured at 138.5 and 302.2, respectively. This unit was set at the highest specific gravity setting (estimated at >1.6 specific gravity). The exact specific gravity of separation is being determined.

Fig. 2 presents a graph developed from a series of plant measurement tests, which relate to the inlet pressure of the PIHs with the gpm per square inch of inlet pipe area and the velocity in fps at the PIH bowl entrance. From this graph, pumping rates can be determined for each size PIH. For example, to determine the pumping rate for an 18-inch PIH operating at 22 psig, follow the pressure value over to the intersection of the curve, extend this point down to the horizontal axis and note 45.6 gpm per square inch of inlet pipe.

An 18-inch PIH has a 6-inch inlet pipe having 28.27 in^2 of area. Multiplying 45.6 $gpm/in^2 \times 28.27\ in^2$ equals 1289 gpm pumping rate. The velocity at the bowl entrance can then be determined by knowing the percent inlet reduction made by the patented deflector angles. Normal design of about 30% reduction gives a velocity of 20.9 fps.

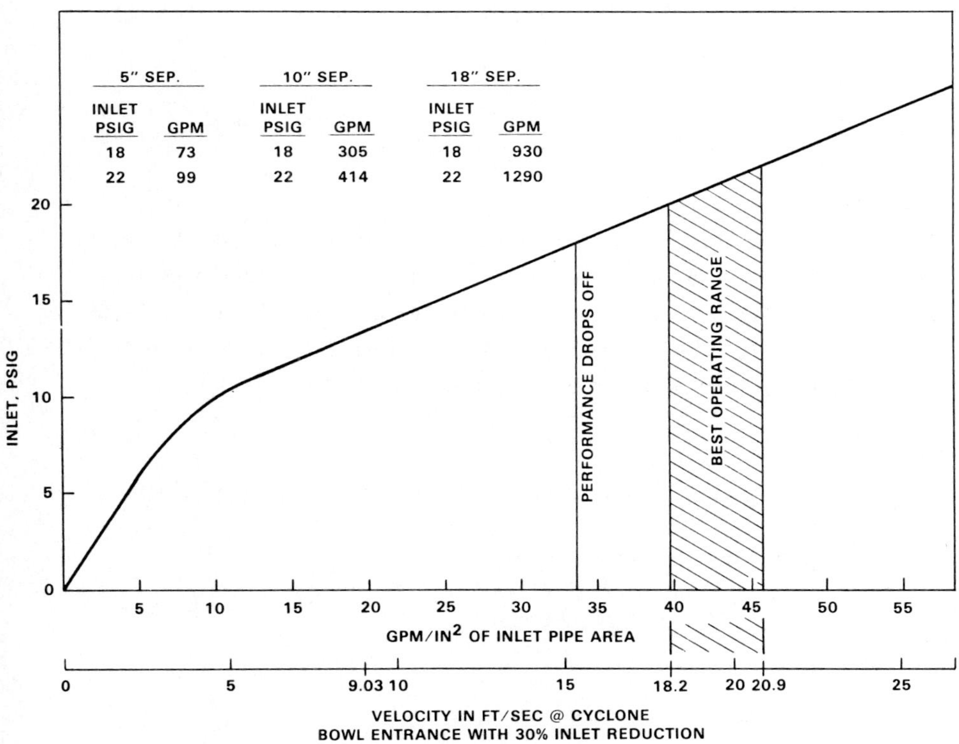

Fig. 2. Cyclone flow measurement.

Plant results

Improvement in selectivity over other conventional coal cleaning hydrocyclones was demonstrated by full-scale comparative tests. 18-inch PIHs were installed and operated alongside original equipment units in a first-stage coal-cleaning circuit. Five coals were cleaned simultaneously with both types of equipment. A common slurry sump with two pumps feed both. Table 2 gives the % ash and % sulfur of the clean coal produced by each.

TABLE 2

Performance comparisons.

Bituminous Coal	Clean Coal From PIH % Ash	PIH % Sulfur	Original Hydrocyclones % Ash	Original Hydrocyclones % Sulfur
#1	10.60	1.32	14.34	1.37
#2	10.69	1.49	12.60	1.71
#3	9.80	1.28	12.07	1.35
#4	9.91	0.99	12.04	1.10
#5	7.41	0.86	11.70	1.39

Specific gravity adjustments

The specific gravity separation cut point (SGSCP) is altered by changing the vortex finder sleeve inside diameter. A larger sleeve increases the SGSCP and vice versa. Table 3 gives the sleeve inside diameter versus the SGSCP for 18-inch PIH (ref. 10) and conventional 18-inch hydrocyclones. The vortex finder height is held constant with respect to the bottom bowl elevation.

TABLE 3

Specific gravity settings; 18-inch body ID vortex inner diameter in inches.

SGSCP	PIH	Conventional Hydrocyclone
1.25	6	
1.30	6-3/8	
1.35	6-3/4	
1.40	7-1/8	6
1.45	7-1/2	6-3/8
1.50	7-3/4	6-3/4
1.55	8-1/16	7-1/8
1.60	8-3/8	7-1/2
1.65	8-11/16	7-3/4
1.70		8-1/16
1.75		8-3/8

Not all of the vortex finder settings have been confirmed by experiment. The low setting of 1.25 specific gravity was confirmed and a Tromp curve developed (see Fig. 3). The 8-inch vortex finder setting for an 18-inch unit was confirmed by comparing the clean coal product with washability data; refer to Tables 4 and 5.

The washability test was performed in 0.10 specific gravity increments. The 1.50 cumulative float yielded a 10.56% ash and 68.99% weight. The PIHs yielded 11.58% and 11.94% ash with 66.21% and 67.04% weight recovery, respectively. Comparing these results and the 1.60 cumulative float yields indicates the 8-inch setting equals approximately 1.55 SGSCP for the 18-inch PIH. The 1.70 SGSCP setting for the conventional hydrocyclone was estimated at 8 inches.

Table 6 gives the results of 8-inch settings for both the PIH and conventional hydrocyclones. Comparing this information with the washability data supports the conclusions.

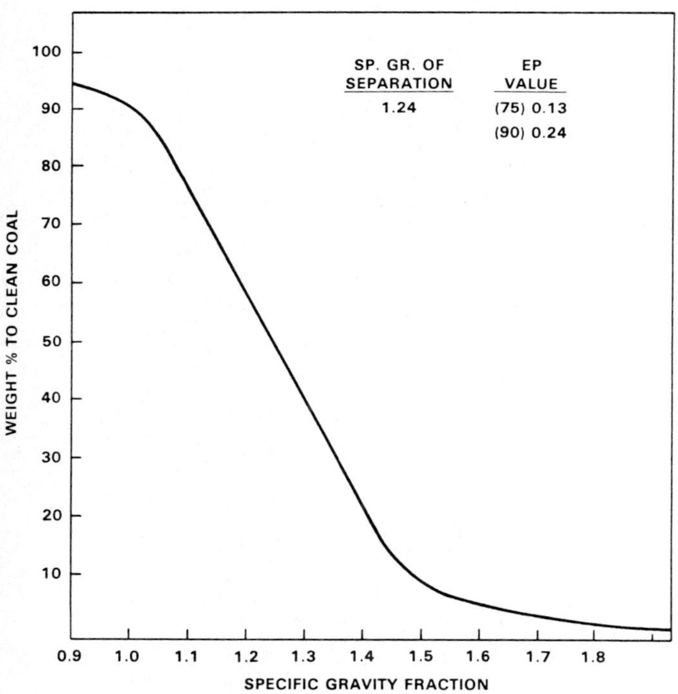

Fig. 3. PIH Tromp curve.

TABLE 4

Results at 1.55 SGSCP for the 18-inch PIH; WSR Bone Shaker coal -- 93 Hardgrove Grindability -- crushed to 3/4 x 0.

Run of 8 Hr	Tons of Clean Coal	% Raw Coal Rec'd	Raw Coal Sulfur			Clean Coal Sulfur			Raw Coal	Clean Coal
			% Pyritic	% Organic	% Total	% Pyritic	% Organic	% Total	% Ash	% Ash
1C	713	67.04	Traces	0.74	0.74	Traces	0.79	0.79	25.72	11.94
2C	688	66.21	Traces	0.70	0.70	Traces	0.73	0.73	26.89	11.58

TABLE 5

Float-and-sink analyses.

Specific Gravity		Dry Basis				Cum. Recovery (Float)				Cum. Reject (Sink)			
Sink	Float	Wt%	% Ash	% Sul	Btu	Wt%	% Ash	% Sul	Btu	Wt%	% Ash	% Sul	Btu
--	1.40	62.53	9.03	0.89	13792	62.53	9.03	0.89	13792	100	27.27	0.69	10078
1.40	1.50	6.46	25.39	0.69	10444	68.99	10.56	0.87	13478	37.47	62.89	0.56	4210
1.50	1.60	2.38	55.47	0.70	5225	71.37	12.06	0.87	13203	31.01	70.65	0.53	2911
1.60	--	28.64	71.97	0.52	2719	100	27.27	0.69	10078	28.64	71.97	0.52	2719

TABLE 6

8-inch setting for PIH and conventional hydrocyclones.

Hydrocyclones	Product Name WSR Bone Pile Shaker Coal Sample ID	Analysis Dry Basis					Tons/hr	Theoretical % Recovery Based on Washability	Plant % Recovery	Plant % Efficiency Based on Washability
		% Ash	% Sul	Btu/lb	M&A Free Btu	FSI Coke Button				
Conventional	Raw Coal	30.81	0.71	10150	14670	1	135			
	Clean Coal	18.86	0.62	12309	15170	1	83	84.17	61.22	72.73
	Refuse	50.24	0.58	7108	14285	0	52	15.83	38.78	
PIH	Raw Coal	26.89	0.70	10736	14685	1/2	130			
	Clean Coal	11.58	0.73	13552	15327	4	86	70.56	66.21	93.84
	Refuse	57.67	0.53	6136	14496	0	44	29.44	33.79	

Equipment setup and laboratory analysis expenses involved in establishing each data point in Table 3 make exact data difficult to obtain. The information is gathered as new operation requirements develop. Table 3 is the best estimation based on data gathered at 1.25, 1.55, and greater than 1.60 specific gravity separation cut points for the PIH and conventional hydrocyclones.

Stage washing

Stage washing enhances the total separation capability of a wash plant by forcing more clean coal to the composite clean coal stream and more refuse to the composite refuse stream, thereby reducing the amount of clean coal lost to refuse. The SGSCP and probable error (EP) of the cleaning circuit can be raised or lowered in relationship to the equipment values by stage washing. For example, if the clean coal from a cleaning device is rewashed by another unit and set the same, the combined EP value for both stages will be higher and the combined SGSCP will be lower. If the refuse is rewashed, the combined effect will be a decrease in EP and an increase in SGSCP (see Fig. 4).

FEED CONSTITUENTS	AFTER STAGE WASHING	
	COMBINED EP	COMBINED SP. GR. OF SEP.
CLEAN COAL	↑	↓
REFUSE	↓	↑

Fig. 4. Effects of stage washing.

Fig. 5 schematically shows rewashing clean coal and refuse via a four-stage circuit. This scheme makes possible the separation of almost pure low specific gravity (<1.3) coal with PIHs even though the EP values at this operation range are above 0.10 ($\cong$.13). A raw coal characterization analysis will give direction to the best choice of circuitry.

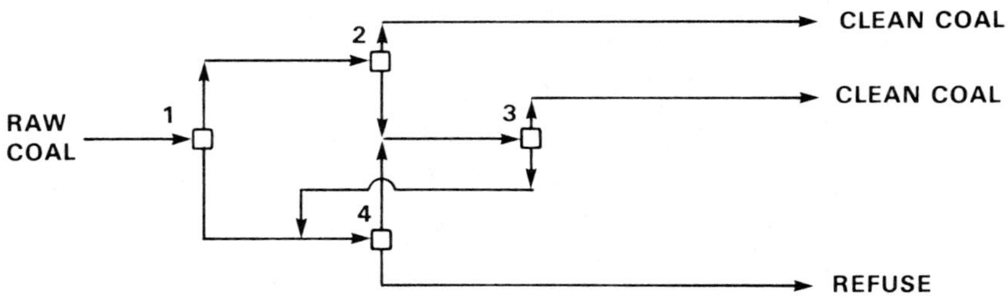

Fig. 5. Four-stage washing.

CONCLUSIONS

An improved hydrocyclone has been designed and patented that can clean coal in the size range of 1-1/4 inch x 0, between 1.25 and 1.65 specific gravity separation cut points with EP values in the range of 0.13 at pulp densities around 23% and with 18 to 21 fps feed velocities. By using stage washing it is possible to separate practically pure coal, isolate middling coal for further cleaning, and reject unusable refuse.

Further study areas

Researchers have established equipment size versus coal particle size relationships for conventional hydrocyclones. This is an area requiring additional testing to determine optimum PIH size for the coal size being cleaned.

It is not certain if the velocities used to date will be the best for cleaning fine coal; for example, 28 x 200 mesh. Researchers (ref. 18, 19) have demonstrated the applicability of using hydrocyclones to clean coal slimes, and have obtained efficiencies greater than with froth flotation. They also claim that recoveries and economics were improved over froth flotation cleaning.

In the near future, testing with 10-inch and 5-inch PIHs are planned to determine the feasibility of fine and ultrafine coal cleaning. It is planned to determine the change in EP value for different SGSCP settings. This information is necessary to more accurately design stage washing circuits. One of the problems encountered in stage washing (without common sumps level control) was trapping a middle gravity fraction within a circuit loop resulting in circuit plugging.

Encouraging results have evolved from the work done to date, but alot of work remains in understanding why hydrocyclones separate according to specific gravity. An understanding of the liberation characteristics of coal, along with extensive practical experience using hydrocyclones as specific gravity separators, holds promise for developing an overlooked economical way of making low-gravity coal separations.

REFERENCES

1 E. W. Albough, et al., "Coal and CWM Selection for the DOE Characterization Project", Proc. of Sixth Int. Symposium on Coal Slurry Combustion and Technology, June 25 - 27 1984, Orlando, Fla., PETC.

2 F. J. Fontein, U.S. Patent 2,573,192, Oct. 1951.

3 J. Visman, "Integrated Water Cyclone Plants for Coal Preparation", Can. Mining and Met. Bul., March 1968.

4 D. I. Liller, "Coal Desulfurization and Demineralization with Low Cost High Capacity Centrifugal Cyclone Separators", Coal Preparation and Utilization Symposium, Coal Conf. and Expo. VI, Louisville, Ky., Oct. 1981.

5 W. E. Foreman and J. R. Lucas, "Current Status of Hydrocyclone Technology", Mining Congress Journal, Dec. 1972.
6 H. H. Dreissen and F. J. Fontein, "Application of Hydrocyclones and Sieve Bends in Wet Treatment of Coal Minerals and Mineral Products", AIME Transactions, March 1963.
7 J. Visman, "Integrated Water Cyclone Plants for Coal Preparation", Dec. 1972.
8 H. F. Trawinski, "Practical Hydrocyclone Operation", Filtration and Separation, Jan./Feb. 1985.
9 D. I. Liller, U.S. Patent 4,164,467, Aug. 1979.
10 Ibid. U.S. Patent 4,341,352, July, 1982.
11 Ibid. U.S. Patent 4,159,409, Aug. 1980.
12 Ibid. U.S. Patent 4,219,409, Aug. 1980.
13 Ibid. U.S. Patent 4,317,716, March 1982.
14 Ibid. U.S. Patent 4,397,071, Aug. 1983.
15 Ibid. U.S. Patent 4,224,143, Sept. 1980.
16 Ibid. U.S. Patent 4,235,363, Nov. 1980.
17 H. F. Trawinski, "About the Practice of Hydrocyclone Operation", 2nd Intl. Conf. on Hydrocyclones, BHRA, The Fluid Engineering Centre, Bath England, Sept. 1984.
18 M. W. Mikhail and O. E. Humeniuk, "Influence of Slimes in Fine Coal on Separation Efficiency of Water-Only Washing Cyclone", Proceedings 64th CIC Coal Symposium, A. M. Al Taweel, Editor, C. S. Ch. E. Ottawa (1982).
19 J. L. Picard, et al., "Improved Compound Water Cyclone for Cleaning Coal Slimes" Canada Center for Mineral and Energy Technology, Energy Research Laboratories, Report ERP/ERL 75-143 (IR), Edmonton, Alberta, Canada.

MAGNETIC DESULFURIZATION OF COAL

R. R. ODER
EXPORTech Company, Inc., P.O. Box 579, Export, PA 15632

ABSTRACT

HGMS desulfurization of a blend of micronized Freeport seam coals from southwestern Pennsylvania using magnetic fields up to 15 Tesla is reported. Measurements show that 56% of the total sulfur can be removed from this coal by the magnetic method. Experimental parameters indicated that more than 94% of the pyritic sulfur is removed. Independent measurements with use of a Frantz Isodynamic Separator determine the magnetic susceptibility of the maf test coal to be -0.6 ± 0.2 *10^-6 cgs/gm and of the mineral fraction removed in magnetic separation to be +0.8 ± 0.4 *10^-6 cgs/gm. Use of this information indicates that mineral impurities are removed in sizes down to 0.8 ± 0.2 microns mean particle diameters. Measurements of maximum sulfur removal for the test coal and of matrix loading characteristics for the HGMS filter bed indicate that magnetic fields of 2 Tesla are adequate for separation of 50% total sulfur at flow velocities up to 0.4 cm/s. Use of magnetic fields up to 6 Tesla extends efficient separator operation to flow velocities of 3 cm/s.

INTRODUCTION

The use of High Gradient Magnetic Separation (HGMS) to desulfurize coal by removal of iron pyrite impurities has been studied extensively [see, for example, refs. 1 and 2]. Adaptation of the technology to commercial coal cleaning for the steaming coal market has not followed, however, mainly because of the high capital cost of conventional iron magnets and because of the poor clean coal yields achieved when processing -28 mesh material characteristic of preparation plant fines. The former is a function of system throughput and the latter is a matter of mineral matter liberation.

There is a new opportunity perceived in the preparation of clean coal for which the magnetic method is perhaps better suited. It is believed that micronized compliance coal of less than .5% ash may be used to replace a portion of the petroleum distillate and natural gas now used in diesel and turbine applications. In this scenario, for example, Class One U. S. railroads alone offer a potential market for 17 million tons ultra clean coal per year to be used in replacing diesel fuel consumed by locomotives. The value added for preparation of ultra low ash compliance coal may be as much as $15 per ton per % ash removed, and the market has a potential value in excess of $2 billion.

The magnetic method has several advantages which make it attractive for preparing ultra clean coal. Extensive coal size reduction will be required to achieve the degree of liberation necessary for efficient separation of mineral impurities required in the new application. HGMS is already used commercially for mechanical separations in the micron particle size range. [ref. 3]

Secondly, the magnetic method is simple and is free from explosive and hazardous materials problems. Water or other transport liquids can be used. No reagents are required other than those normally used for viscosity control in processing pulps of micronized minerals. Relatively broad particle size consists can be processed directly. Added equipment is not necessary for particle size fractionation or for process liquid recovery and recycle.

Thirdly, unlike flotation and other surface activated methods, magnetic separation is insensitive to pretreatment of the coal (surface oxidation). Indeed, some controlled preoxidation can actually improve separator performance through conversion of the pyrite surface to relatively magnetic sulfate forms. [ref. 4]

Fourthly, the magnetic method does not require use of hazardous parting liquids or agglomerating agents such as the fluorochlorocarbons. [ref. 5] Unlike most other gravimetric and body force methods HGMS is very effective in the micron particle size range so that the capillary forces characteristic of agglomeration and coalescence methods employing mixed liquids are not necessary. This avoids potential health and environmental hazards by removing the problem at the source.

The technical feasibility of separating pyrites from coal with use of HGMS is well established. However, the extent to which magnetic methods can be used to remove other mineral matter from coal is not known at this time because not all mineral matter in coal is magnetic. There is a fundamental interrelationship between mineral matter removal and pyrite separation which directly affects the practicability of this technology. Both mineral contaminants and pyrites load the separation matrix and together degrade the efficiency of separation more rapidly than either alone. Further study is necessary to delineate the practical bounds which this effect has on cleaning micronized coal by magnetic methods. This effort is justified because of the availablity of innovative high field magnetic separator designs which, if developed, have the potential to lower the cost of magnetic separation by an order of magnitude. [ref. 6]

The work reported here covers results of measurements of sulfur removal from an Upper Freeport seam Pennsylvania coal using a wet HGMS method. Detailed studies of the magnetic removal of ash from this and other coals are to be reported elsewhere. [ref. 7] Measurements of ash removal by high field and high gradient magnetic separation alone for the coal studied here have been reported earlier. [ref. 8]

EXPERIMENTAL

The experiments in high gradient magnetic separation were carried out using the facilities of the Francis Bitter National Magnet Laboratory at the Massachusetts Institute of Technology. Both a 15-T Bitter solenoid and a 2-in. room temperature access, 15 T superconductive magnet built by IGC of Guilderland, New York, were used. The range of experimental parameters covered in the measurement program which are reported here are given in Table 1.

TABLE 1

Experimental parameters for high field HGMS measurements

Magnets	Bitter and superconductive solenoids
Magnetic field strength	Up to 15 T
Coal	Freeport seam blend
Mass median particle diameter (microns)	9.6
50th percentile (microns)	7.2
Canister dimensions:	
Length (cm)	9.05
Volume (cm^3)	103
Flow velocity (cm/s)	.42 to 3.07
Steel wool strand average diameter (microns) [a]	127
Matrix packing (%) [a]	6
Slurry solids (wt%)	14.8 to 38.4
Slurry viscosity (cP)	3.2 to 31

[a] Medium grade #430 felted stainless steel wool pads.

A Model LB-1 Frantz Isodynamic Magnetic Separator was used for the magnetic susceptibility measurements which were made at the EXPORTech Laboratories in New Kensington, Pennsylvania. The isodynamic separator can be used to measure magnetic susceptibility of materials through the relationship,

$$\frac{\sin(S-S_o)}{X} = \frac{H dH/dx}{g} \tag{1}$$

In Equation 1, S is the side slope of the separator tray, So is the slope setting for equal splitting with no magnetic field and was determined to be +0.6 $\pm$ 0.2 degrees for these measurements, X is the material magnetic susceptibility, g is the acceleration due to gravity, and HdH/dx is the magnetic force term which is, by design, constant over the surface of the separation tray. [ref. 9] The magnetic force term HdH/dx has been calculated from first principles [ref. 10] and is given by,

$$HdH/dx = 0.262\ Hm^2 \qquad gauss^2/cm \qquad (2)$$

Hm is the maximum magnetic field generated in the gap of the separator. The relationship between Hm and magnet current, I, measured for the unit at EXPORTech is shown in Figure 1. The field was calibrated with use of a Model STG3-0404 Bell Hall probe. The uncertainty in field measurement estimated at $\pm$ 5% is due primarily to difficulty in locating the position corresponding to Hm. For current less than .75 amps.

$$Hm = 13{,}900\ I \pm 5\%. \qquad (3)$$

Using Equations 2 and 3, one obtains

$$\frac{Sin(A-Ao)}{X} = 51.6*10^3 *I^2 \pm 10\% \qquad \text{for } I \leq 0.75 \text{ amps} \qquad (4)$$

The coefficient 51.6 used here can be compared with 42.2 measured by separations with 30X50 mesh ferrous ammonium sulfate, magnetic susceptibility = 31.4*10^-6 cgs/gm at 28 deg C, [ref. 11], with the value 48.7 obtained by Nesset and Finch [ref. 10] and by the value 52 obtained by McAndrew some years ago [ref. 12]. The value used for HdH/dx in interpreting experimental data is calculated using Eq. (2) and the magnet calibration data of Figure 1. All data points were generated on the increasing current portion of the magnet hysteresis curve.

The coal used for the HGMS desulfurization experiments, Freeport-83, was a blend of 80% deep mined Upper Freeport seam and 20% deep mined Lower Freeport seam obtained from the middlings circuit of the Homer City Preparation Plant, Homer City, Indiana County, Pennsylvania, in October, 1983. The coal used for the susceptibility measurements, Freeport-84, was substantially the same blend obtained from the same source in October, 1984. Some results are also reported for a deep mined Pittsburgh seam "clean coal," Pittsburgh-84, taken from the Emerald Mine Preparation Plant, Waynesburg, Greene County, Pennsylvania, October, 1984. Characteristics of the as-received-coals are given in Table 2. For the work reported here, standard deviations of 0.35 and 0.08 characterize the variance in ash and sulfur measurements respectively.

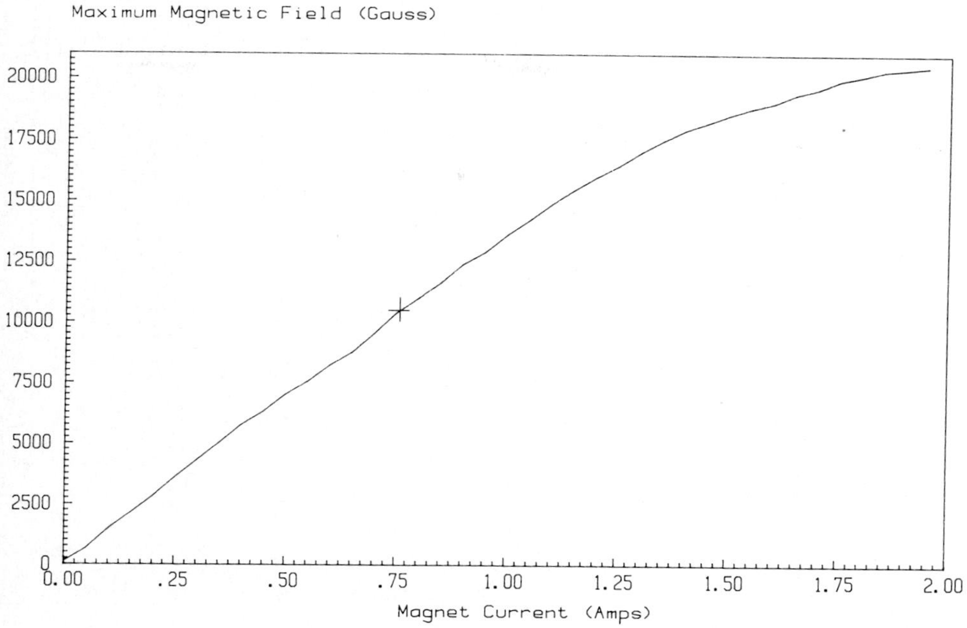

Fig. 1. Magnetic Field Calibration. Maximum magnetic field, Hm, versus magnet current for Frantz Isodynamic Separator

TABLE 2
Analysis of as-received coals

Analysis, Dry Basis	Homer City Freeport-83	Homer City Freeport-84	Emerald Mine Pittsburgh-84
Ash (%)	16.32	15.3	6.36
Btu/lb	12947	13114	14325
Total sulfur (%)	1.88	1.84	1.69
Pyritic sulfur (%)	1.22	1.04	0.59
Sulfate suflur (%)	0.01	0.01	0.00
Organic sulfur (%) [a]	0.65	0.79	1.10

[a] Calculated from difference

The coal was wet size-reduced with the use of a batch-operated bench-scale Perl Mill employing 3/16 in. diameter steel balls. The coal/water slurry was dispersed to minimum viscosity with the use of Diamond Shamrock A22 or A23 dispersant.

The effect of dispersant on the viscosity of a 24.5% solids coal water mixture containing nominal 13.8 micron mass median particle diameter Pittsburgh seam coal can be seen in Figure 2. The viscosity was measured with a Brookfield

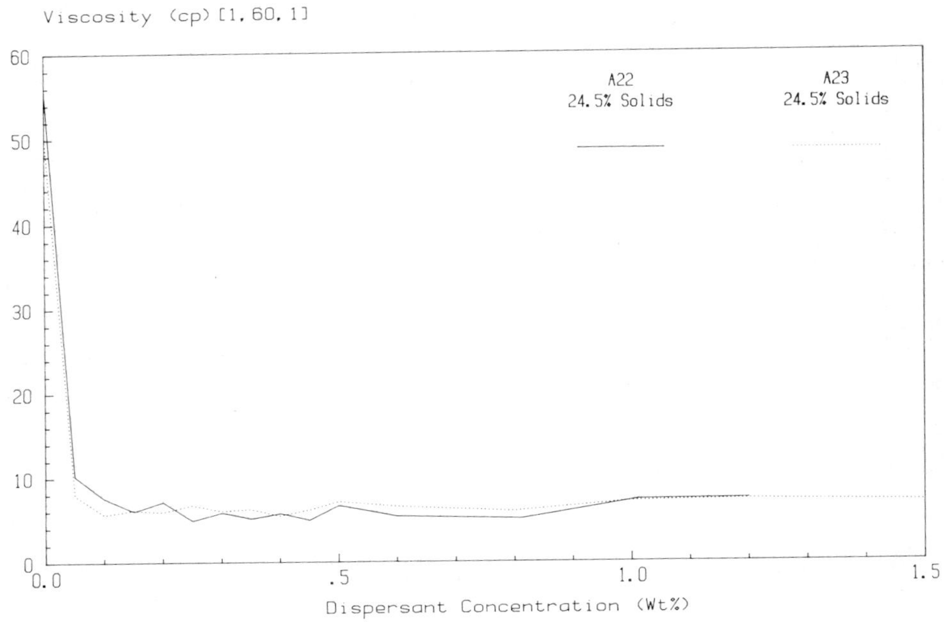

Fig. 2. Viscosity of 13.8 micron MPD Pittsburgh-84 dispersed with Diamond-Shamrock A22 and A23.

viscometer. The measurement configuration of spindle, rpm, and time of measurement in minutes are shown, in that order, within the brackets on the vertical axis label of the figure (e.g., [1,60,1]). Use of dispersants reduces both measured viscosity and associated time and rpm effects.

The flow through the HGMS separator was controlled by a hand-operated manifold and pump arrangement described in Reference 7. The manifold was connected to a canister embedded in a high field magnet. The canister contains a fixed-bed magnetic filter made from ferritic stainless steel wool. Slurry was pumped from the canister bottom with the use of a parastaltic pump. The throughput was regulated between 100 and 3000 cc/min by control of the drive motor dc voltage.

Canisters for the HGMS experiments were made from 2 in o.d. nonmagnetic stainless steel. The maximum o. d. and the maximum magnetic filter length (3 + 9/16 in.) were fixed by magnet considerations.

The HGMS experimental procedure consisted of pumping slurry through the apparatus at predetermined experimental conditions of flow rate, magnetic field strength, slurry solids, etc. Experimental samples of approximately 25 cc were collected from the initial spillover and at time intervals equivalent to processing of 2, 4 and 6 canisters.

More detailed information on the apparatus and experimental procedure can be found in Reference 8.

RESULTS

The measured sulfur values, reported earlier [Ref. 8] and given below in Table 3 have not previously been analyzed for effects of magnetic field, flow velocity and slurry viscosity. That analysis is given in next section of this paper. The analysis is extended and supported by recent measurements of the magnetic susceptibilities of the bare coal and of the mineral matter removed by the HGMS unit. These measurements are discussed in the following section -- Magnetic Susceptibility Measurements. Measurements of weight recovery and of combustible yield for HGMS coal cleaning will be reported elsewhere. [ref. 7]

HGMS Measurements

Total sulfur was measured for the first material to emerge from the HGMS filter and for samples taken at different times during the filtration operation using the Freeport-83 coal. The initial spillover samples are used to characterize the ideal performance of the separator and as such represent the maximum sulfur removal achievable with use of the HGMS technology. The timed samples are used to study the loading characteristic of the HGMS filter. Together, operating parameters such as flow velocity and the batch filter loading time determine the practicability of the batch operated HGMS process.

Maximum Sulfur Removal. Measurements of sulfur and ash in initial spillover samples, for the Freeport-83 coal, are given in Tables 3a, 3b, and 3c as functions of slurry solids, flow velocity and magnetic field strength. The elements of Table 3b indicate that 55 to 56% of the total sulfur can be removed from the test coal with use of the experimental HGMS filter employed. Further, as the measurements at low flow velocity indicate, high fields are not required for effective pyrite removal using this system.

Several trends are discernable in the sulfur levels shown in Table 3. First, at 2 T, increasing flow velocity above 0.42 cm/s decreases the maximum sulfur reduction by a significant amount. This is illustrated in Figure 3 where it can be seen that the adverse effect of increasing velocity can be compensated in part by the use of magnetic fields greater than 2T. The sulfur removals at magnetic fields greater than 6T are equal to within the experimental uncertainty of $\pm$ 10% for velocities up to 3.07 cm/s.

Secondly, at 2T, increasing solids above 14.8% also decreases maximum sulfur removal by a significant amount for the experimental HGMS filter used. Again, as is illustrated in Figure 4, the use of magnetic fields up to 6T can offset

TABLE 3a

Total sulfur in HGMS initial spillover samples for Freeport-83 coal

Head Sample S%	1.66			1.61	1.77
	14.8% Solids			29.4% Solids	38.4% Solids
	Flow Vel, cm/sec				
	0.42	1.14	3.07	1.15	1.15
H(T)	Total Sulfur ± 0.08%				
2.19	0.72	0.77	1.02	0.82	0.99
5.58	0.76	0.74	0.85	0.77	0.81
15.01	NM	0.74	0.83	0.77	0.77

TABLE 3b

Calculated values of sulfur removal by HGMS for the Freeport-83 coal

H(T)	Reduction in Total Sulfur ± 11%				
2.19	56.6	53.6	38.6	49.1	44.1
5.58	54.2	55.4	48.8	52.2	54.2
15.01	NM	55.4	50.0	52.2	56.5

TABLE 3c

Measured values of ash for initial spillover samples

Head Sample Ash%					
Ash ± 0.35%	16.0			16.2	16.4
H(T)					
2.19	7.15	8.03	11.4	8.47	10.5
5.58	5.42	6.33	8.87	6.45	7.13
15.01	NM	4.81	7.24	5.01	5.93

this loss of separation efficiency. For fields greater than 6 T sulfur reductions for initial spillover samples with solids up to 38.4% were equal to within experimental error of ± 10%.

No significant benefit to maximum sulfur removal was observed in this work with the use of fields greater than 6 T for the Freeport-83 middling coal. For flow velocities less than 1 cm/s and slurry solids less than 30%, magnetic fields of 2T are adequate for total sulfur removals up to 50%.

Matrix Loading. Grab samples were taken at uniform time intervals during the test cycle. Sulfur and ash was measured for these samples in order to correlate matrix loading with the loss of separation efficiency. Results for some of these measurements are given in Table 4.

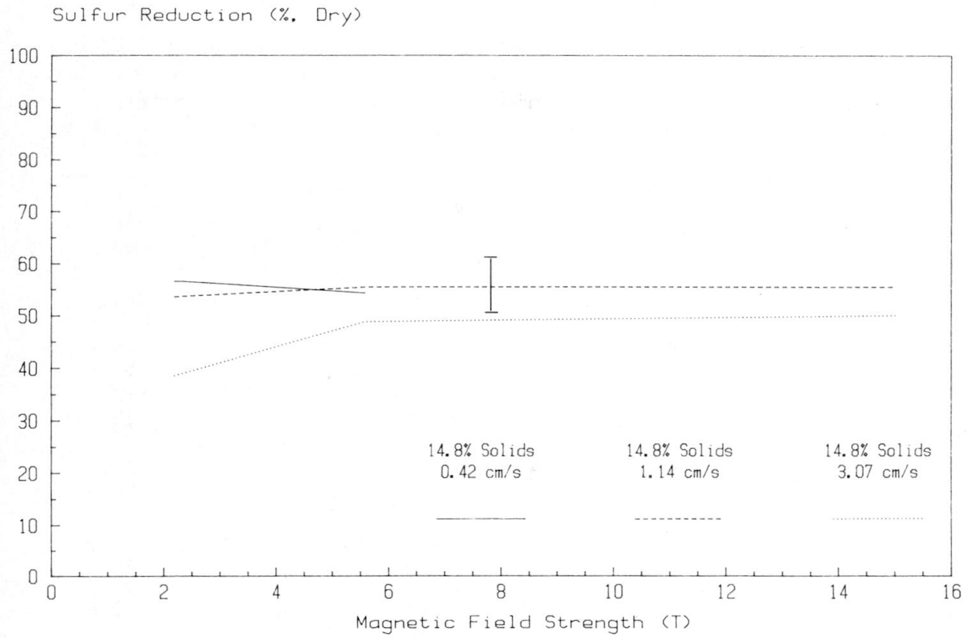

Fig. 3. Magnetic field and flow velocity dependence of maximum HGMS sulfur reduction for micronized 1.66% sulfur Freeport-83 coal

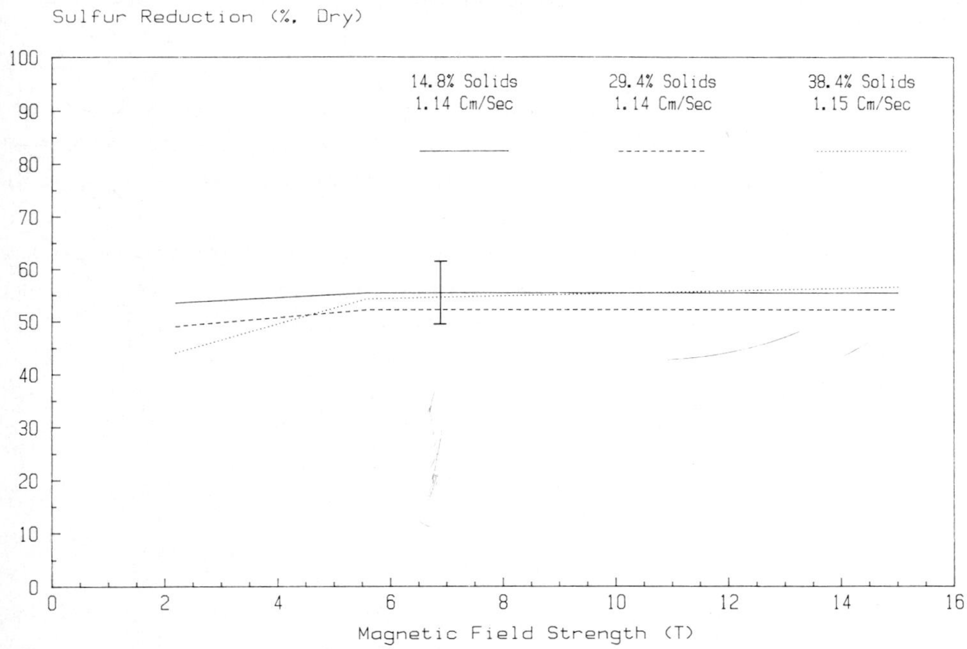

Fig. 4. Effects of magnetic field strength and slurry solids content on the maximum HGMS sulfur reduction for 1.66% sulfur Freeport-83 coal

TABLE 4

Sulfur and ash measurements of grab samples taken throughout the test cycle

H (T)	V (cm/s)	n (cP)	Feed Ash (%)[a]	Feed Sulfur (%)[b]	Can's Passed	Load cc/cc ± 20%	Total Sulfur (%)[b]	Sulfur Reduction (%)
2.19	0.42	3.8	16.0	1.66	0.25	0.20	0.72	56.6
					1.98	1.46	0.90	45.8
					3.95	2.91	1.10	33.7
					5.89	4.33	1.31	21.1
15.01	0.42	3.8	16.0	1.66	0.25	0.19	0.76[c]	54.2
					1.98	1.39	0.81	51.2
					3.95	2.80	1.18	28.9
					5.89	4.14	1.35	18.7
2.2	3.07	3.8	16.0	1.74[c]	0.25	0.10	1.02	41.4
					1.96	0.71	1.57	9.8
					3.92	1.42	1.69	2.9
					5.88	2.12	1.74	0.0
15.01	3.07	3.8	16.0	1.66	0.25	0.10	0.83	50.0
					1.96	0.74	0.97	41.6
					3.92	1.47	1.13	31.9
					5.88	2.19	1.24	25.3

[a] Ash standard deviation 0.35 ash units.
[b] Sulfur standard deviation 0.08 sulfur units.
[c] Estimated

The beneficial effects of high magnetic field strength on matrix loading are illustrated in Figure 5 where sulfur reduction measured for the Freeport-83 Middling coal is plotted versus matrix loading, expressed as volume of magnetic material collected per volume of matrix used.

The matrix loading, L, espressed as grams coal processed per gram matrix was measured in the experiments. The matrix loading expressed as cc magnetics, Vm, per cc matrix, VM, is related to L through the following approximation,

$$Vm/VM = 7.88\ [0.769 - 0.545\ Am/100]\ L\ (1 - Wr/100) \quad (5)$$

The expression of Eq. (5) can be derived from the relation,

$$Vm/VM = SGM/SGm\ (1 - Wr/100)\ L \quad (6)$$

where (1 - Wr/100)L is the weight of the magnetic fraction per weight of matrix and SGM is the specific gravity of the steel wool matrix material, 7.88, and SGm

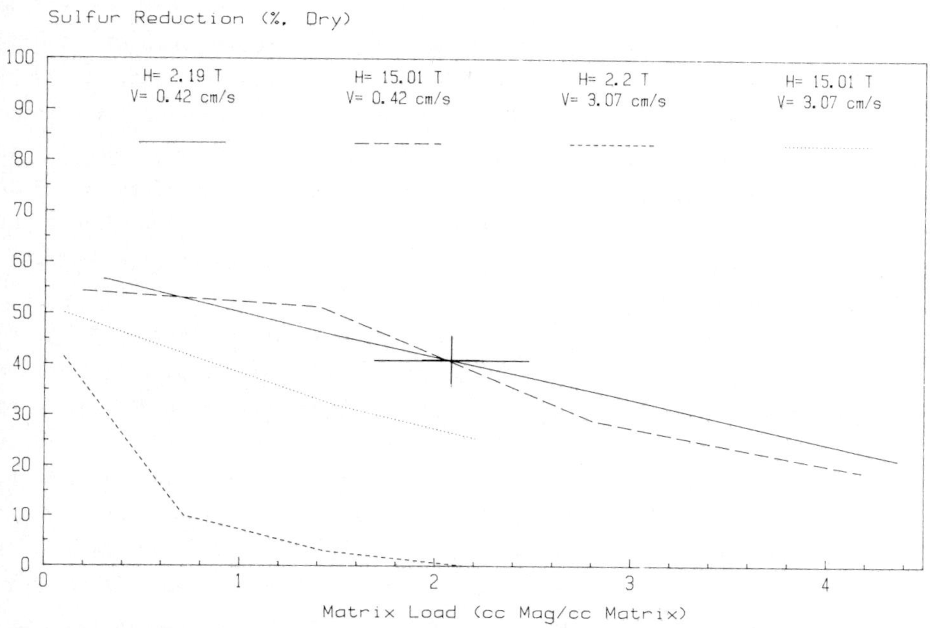

Fig. 5. The effects of magnetic field strength and flow velocity on HGMS matrix loading for 14.8% solids micronized Freeport-83 coal

is the specific gravity of the magnetic fraction. The specific gravity of the magnetics is related to the ash of the magnetics through,

$$\frac{1}{SGm} = \frac{Am/100}{SGmm} + \frac{(1 - Am/100)}{1.3} \tag{7}$$

where SGmm = 4.46 is the specific gravity of the minerals in the magnetics.

It is assumed that the magnetic minerals are 90% pyrite by weight and 10% minerals of typically 2.25 SG. The specific gravity of mineral matter free coal is assumed to be 1.3.

The rather large uncertainty in the loading (cc/cc) is due mainly to the fact that weight balance in the HGMS experiments reported here is 92% with a standard deviation of 17.7 reflecting a 19% uncertainty in weight determinations for HGMS experiments.

The curves of Figure 5 illustrate the observed loss of separation efficiency as the HGMS process proceeds into the separation cycle. This loss is due to loading of the matrix, which diminishes available capture sites, and lessens retention of magnetics through increases in the shear forces at the surface of the loaded matrix.

The upper two curves of Figure 4 illustrate the fact that increasing magnetic field strength at constant flow velocity does not improve the loading relationship observed at low flow velocity. Increasing flow velocity to 3.07 cm/s at 2 T, however, causes significant loss of separations efficiency and results in a severely shortened cycle time. The use of high magnetic fields can, to the extent shown by the curve corresponding to H = 15.01 T and V = 3.07 cm/s, improve separations efficiency and extend cycle time for flow velocities greater than 0.42 cm/s. This distinct advantage can only be realized with use of superconductive magnet technology.

Process volumes may be limited to 2 to 4 canisters for this material. Batch operated HGMS will probably not be practical for direct application to the Freeport-83 coal because of the high ash content. Continuous methods should be effective, however.

Pyritic Sulfur Removal. The total sulfur values shown in Table 4a can be used along with ash values given in Table 4c to calculate approximate pyritic sulfur levels assuming no sulfate and no elemental sulfur. To make this estimate, one subtracts from the measured total sulfur an estimate of organic sulfur in the sample, Op. Assuming that recovery of organic sulfur in the feed, Of, is proportional to combustible yield, Y = (1-Ash Product/100)/(1-Ash Feed/ /100)*Wr where Wr is the weight recovery, one estimates,

$$Op = (Y/Wr)Of \qquad (8)$$

Using this expression for organic sulur the following estimate for pyritic sulfur can be obtained,

$$Sp = St - 0.78(1\text{-Ash Product}/100) \qquad (9)$$

The pyritic sulfur values calculated in this manner are shown in Table 5.

The results shown in Table 5b indicate that better than 96% of the pyritic sulfur in the micronized Freeport-83 coal has been removed by HGMS. HGMS is an effective method for separating pyrite from coal.

TABLE 5a

Calculated values of pyritic sulfur in HGMS products for Freeport-83 coal

Head Sample	1.66% S, 16% Ash			1.61% S, 16.2% Ash	1.77%S, 16.4% Ash
	14.8% Solids			29.4% Solids	38.4% Solids
	Flow Vel, cm/sec				
	0.42	1.14	3.07	1.15	1.15
H(T)	Calculated Pyritic Sulfur, ± 23%				
2.19	-0.001	0.056	0.332	0.109	0.294
5.58	0.025	0.012	0.142	0.043	0.089
15.01	NM	0.000	0.109	0.032	0.039

TABLE 5b

Calculated values of reduction in pyritic sulfur by HGMS for Freeport-83 coal

H(T)	Calculated Reduction in Pyritic Sulfur, ± 46%				
2.19	100.1	94.5	67.1	88.6	73.7
5.58	97.5	98.8	85.9	95.5	92.1
15.01	NM	99.9	89.1	96.6	96.5

Modeling. The relationship of sulfur removal efficiency to field strength, flow velocity and solids discussed above can be interpreted quantitatively through the measured filter transmittance. The transmittance, defined by the ratio of the concentration of the magnetic portion of the feed at the output, Co, to that at the input of the filter, Ci, is expressed in terms of the length of the filter, L, and of the filter absorption length, a, through the relation,

$$Co/Ci = e^{-L/a} \tag{10}$$

For the interpretations to be derived here, a widely accepted model based on capture from flow (trajectory model) is used [ref. 13]. The trajectory model stresses capture by magnetic attraction resulting in motion across flow lines and ignores shear forces in effect on the surface of the steel wool strand. The applicability of the model to separations under conditions of high shear, e. g., high values of nV, is a matter of considerable commercial interest.

For the trajectory model the filter absorption length, a, is given by

$$a = Am/(2RcY) \tag{11}$$

where, Am = cross-section area of a matrix element, sq cm
Y = matrix packing density
Rc = capture radius = C(xDp^2MBoAm/18PInVo)^1/3, cm (12)
C = numerical constant of the order of 1
PI = 3.14159

The elements of Eq. (12) are defined as follows:

x = magnetic susceptibility of the mineral component removed, cgs/cc
Dp = diameter of magnetic particles, cm
M = magnetization of the stainless steel = 1590 gauss
B0 = applied magnetic field, up to 150,000 gauss
Ds = diameter of stainless steel wool strand, 102-152 microns
Am = Cross-section of strand transverse to length, (PiDs^2)/4=1.267*10^-4 sq cm for 127 micron mean strand diameter.
n = slurry viscosity = 3.8 cP @ 14.8% solids, 7.9 cP @ 29.4% solids and 31 cP @ 38.4% solids.
Vo = flow velocity, cm/s

Transmittance measurements for all the experimental runs are plotted in Figure 6 versus the separation parameter, X, derived from the trajectory model,

$$X = [H/(nV)]^{1/3} \qquad (13)$$

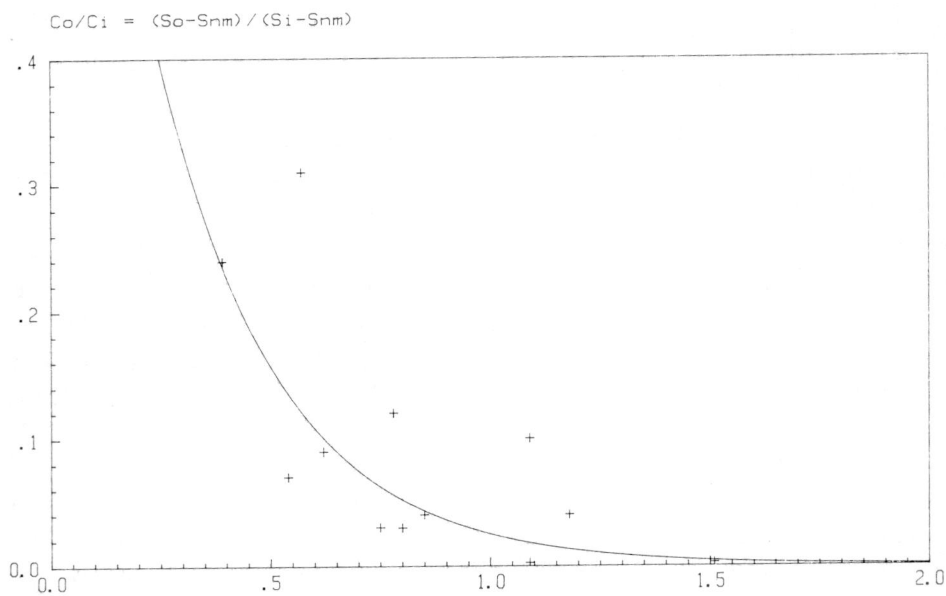

Fig. 6. HGMS sulfur transmittance for micronized Freeport-83 coal

In the figure,

So = total sulfur of product

Snm = total nonmagnetic sulfur

Si = total sulfur of feed

The values of So and Si used in the plot are directly measured and the value of Snm is derived in forcing the exponential fit to the experimental data. Because of the large scatter in the experimental points, the exponential fit has a large degree of uncertainty as evidenced by the small value of the coefficient of determination, r^2, in the least squares fit,

$$Co/Ci = (So-0.738)/(Si-0.738) = e^{-3.7(H/nV)^{1/3}}, \quad r^2 = 0.57 \tag{14}$$

A value of 0.738 has been determined for Snm in forcing the exponential fit of Eq. (14). The average organic sulfur calculated for the three test coals is 0.66% and the average pyritic sulfur is 1.02%. Using these values the estimate for Snm indicates pyritic sulfur removal of 92% for these samples. This estimate is in good agreement with the pyritic sulfur removal of 96% inferred in the above section -- Pyritic Sulfur Removal.

The exponent in Eq. (14), 3.7(H/nV)^1/3 , determined in the statistical fit, can be used to develop information on the size and the magnetic susceptibility of the particles which are removed by the magnetic separator. Using Eq. (12) above and L = 9.05 cm for the canister length, the expression for the exponent,

$$L/a = 3.7\ (H/nV)^{1/3} \tag{15}$$

can be evaluated to obtain

$$xDp^2 = 2.254*10^{-14}\ cm^{-1},\ \pm\ 15\% \tag{16}$$

Measurements of the size of particulates requires sophisticated microscopic evaluations beyond the scope of the effort reported here. Indirect information on particle size can be developed with use of Eq. (16), however, if the magnetic susceptibility is known.

Magnetic Susceptibility Measurements

We have used the Frantz Isodynamic Separator to measure magnetic susceptibility of maf Freeport-84 test coal and of the minerals separated from it by magnetic methods. Use of this information permits determination of the size of particles separated from the test coal with use of HGMS and gives a check for internal consistency of the experimental results.

<u>Magnetic Susceptibility of Extracts from 100X200 Mesh Test Coal.</u> The Freeport-84 coal was size reduced to 100% finer than 30 mesh and screened into five size fractions. The screen distribution is given in Table 6.

TABLE 6
Screen analysis of Freeport-84 test coal

Screen Fraction	Wt.%	Ash %	Sulfur %
-30 Mesh	100.00	14.42	1.61
30X50 Mesh	28.13	15.41	1.41
50X100 Mesh	23.99	13.49	2.00
100X200 Mesh	17.50	13.49	1.98
200X325 Mesh	7.07	13.95	2.21
-325 Mesh	23.31	17.06	1.53
	100.00		
Calculated Composite		14.89	1.74
Difference,%		3.3	7.8

Sulfur and ash values calculated for the composite with use of the component values agree with measured values for the head sample to within 3% and 8% for ash and sulfur respectively.

The 100X200 mesh screen fraction was separated into components of differing magnetic susceptibilities with use of the Frantz Separator. The resulting distribution is shown in Table 7.

TABLE 7
Magnetic fractions of 100X200 mesh component of Freeport-84 test coal

Magnetic Fraction (10^-6 cgs/gm)	Wt.% of Feed	Wt% of Screen Fract	Wt.% of Total Coal	Ash%	Sulfur%
Head	100.00	100.00	17.50	13.49	1.98
>3.7	0.37	0.37	0.07	39.41	2.27
>0.7 <3.7	6.54	6.52	1.14	31.09	3.38
>0.2 <0.7	73.91	68.86	12.04	13.41	2.28
<0.2	100.00	24.29	4.25	9.65	1.62
		100.00	17.50		
Calculated Composite				13.75	2.19
Difference, ± %				1.9	10.7

Table 7 shows that the Frantz separator can remove high ash and high sulfur components from the test coal. The calculated ash and sulfur values for the mathematically reconstructed composite agree with measured values for the head sample to within 2% and 11% respectively.

Assuming the susceptibility of the material less magnetic than 0.2*10^-6 cgs/gm to be diamagnetic and to represent the coal, then the material extracted with susceptibility less than 0.7 10^-6 cgs/gm is composed of a mixture of mineral matter and maf coal and the susceptibililty is given as the sum of components.

$$Xemm = 0.7391\ Xmm + 0.2609\ Xc \quad \text{for } .2 < Xe < .7,\ (10^{-6}\ \text{cgs/gm}) \tag{17}$$

where Xemm is the magnetic susceptibility of the material extracted between 0.2*10^-6 and 0.7*10^-6 cgs/gm.

The diamagnetic fraction of the 100X200 mesh component of the test coal was leached to remove mineral matter. The leaching operation consisted of the following steps:

1) washing in 70% HF at room temperature for one hour,
2) rinsing in alternating hot and cold deionized water,
3) washing in 70% HF for one hour at 100 deg C,
4) repeating step 2),
5) washing in 20% HCl for one hour at 100 deg C,
6) repeating step 2),
7) washing in 12% HNO3 for one half hour at 100 deg C,
8) repeating step 2).

The ash of the diamagnetic product was reduced from 9.65% to 0.45%. This material was fractionated with use of the Frantz separator in a manner similar to that described above for the mineral components. Considerable variablility is observed in measurement of the diamagnetic component. For a total of four separate suites of measurements, the susceptibility is composed of the following components,

$$Xec = 0.9922\ Xc + 0.0045\ Xmm, \quad \text{for } -0.8 < Xec < -0.5,\ (10^{-6}\ \text{cgs/gm}), \tag{18}$$

where Xec is the magnetic susceptibility of the material extracted between -0.8 10^-6 and -0.5 10^-6 cgs/gm.

Magnetic Susceptibilities of maf Coal and Mineral Matter. Estimates of the magnetic susceptibilities of the maf coal and of the mineral matter are compared with measurements and calculations by others in Table 8. The measured values for this work were obtained by solution of the simultaneous equations, Eq, (17) and Eq. (18). Xemm and Xec are mean values for the appropriate intervals.

TABLE 8

Comparison of measured values of the magnetic susceptibility of maf coal and mineral matter extracted from the Freeport-84 test coal

Coal:

Xc (10^-6 cgs/gm) Measured	Reference(14)	Theoretical(16)
-0.6 ± .2	-0.5 to -0.56	-0.57

Mineral Matter:

Xmm (10^-6 cgs/gm) Measured	Reference (15)
+0.85 ± 0.41	0.6

The diamagnetic susceptibility of the test coal studied here is in good agreement with measurements by others [ref. 14]. Further, the measured value of the paramagnetic susceptibility of the mineral component agrees to within experimental uncertainty with that of iron pyrite [ref. 15].

A typical ultimate analysis for a 80/20 Freeport blend characteristic of the test coal studied in this work and corresponding Pascal Constants for the molar susceptibilities of the major atomic constituents of the coal are given in Table 9. The range of values of the magnetic susceptibility of coal measured by others corresponds to coals with carbon and hydrogen concentrations similar to those of the test coal studied here.

TABLE 9

Composition of test coal and associated Pascal constants

	maf	Molar	Pascal's Constant*	L*
C	86.95	7.25	-6.0	+5.5
H	5.76	5.76	-2.9	
O	5.84	0.36	-3.4	
N	1.56	0.10	-5.6	

*10^-6 cgs/gm

The theoretical value of the diamagnetic susceptibility of coal shown in Table 8 has been calculated with use of the empirical Pascal Constants [see page 1778, ref. 16].

Calculation of Dp. Using the measured value of the magnetic susceptibility of the mineral matter extracted for the Freeport-84 test coal one obtains an estimate of the finest particle extracted with use of HGMS, Dp = 0.8 + .2 microns.

The particle size distribution measured for the Freeport-83 test coal with a Micro-Trac particle sizer is shown in Figure 7. Extrapolation of the measurements to 0.8 micron particle size suggests that more than 99% of the pyrites are removed in HGMS processing of this coal.

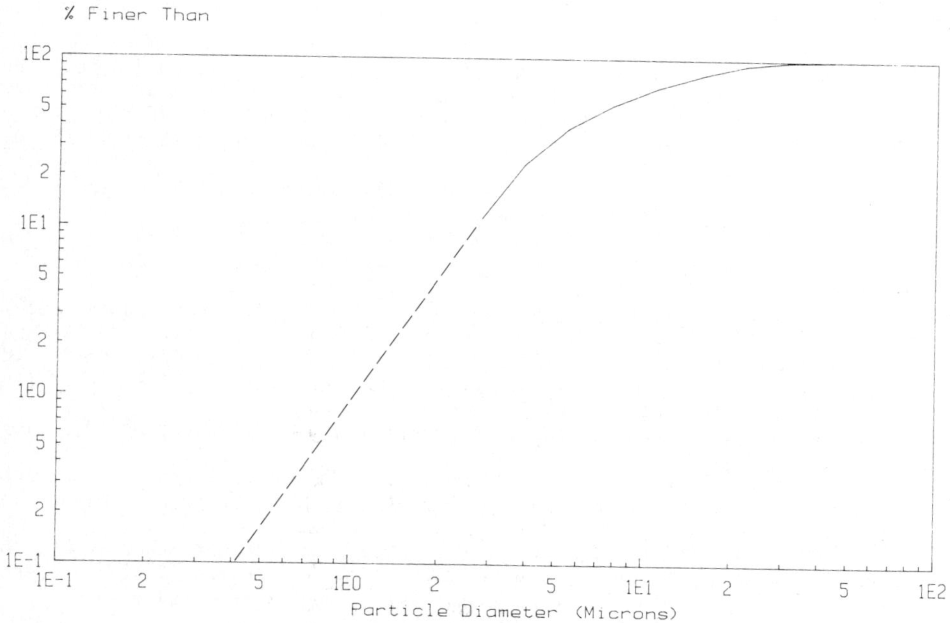

Fig. 7. Particle size distribution for micronized Freeport-83 coal

CONCLUSIONS

Measurements have shown that HGMS can remove up to 56% of the total sulfur from a micronized blend of Upper Freeport coals taken from southwestern Pennsylvania. Measurements of maximum sulfur removal for the test coal and of matrix loading characteristics for the HGMS filter bed indicate that magnetic fields of 20,000 Gauss are adequate for separation of 50% total sulfur at flow velocities up to 0.4 cm/s. Use of magnetic fields up to 60,000 Gauss extends efficient separator operation to flow velocities of 3 cm/s. Continuous HGMS will be required in desulfurizing high sulfur coals.

The maf coal is diamagnetic with a susceptibility of -0.6*10^-6 cgs/gm +30%. The minerals removed are probably pyrite with a susceptibility of +0.8*10^-6 cgs/gm + 50% and are extracted down to sizes of nominally .8 microns. Parameters determined in fitting an HGMS capture model to the experimental data suggest that more than 95% of the pyrites are removed in HGMS processing of this coal.

ACKNOWLEDGEMENTS

This work was supported in part by the Small Business Innovation Research Program of the U. S. Department of Energy under Contract No. DE-AC01-83ER80037.

REFERENCES

1 Proc. of Symposium on Magnetic Desulfurization of Coal, March, 1976, published in IEEE Transaction of Magnetics, Vol. MAG-12, No. 5, September, 1976.
2 Proc. of Conference on Industrial Applications of Magnetic Separation, July, 1978, Ed. by Y. A. Liu, Publication No.78CH1447-2MAG, Institute of Electrical and Electronic Enegineers.
3 R. R. Oder and C. R. Price, "Brightness Beneficiation of Kaolin Clays by Magnetic Treatment," TAPPI, Vol. 56, No. 10, p. 75 (October, 1973).
4 R. R. Oder, B. N. Murthy, and E. L. McGinnis, "Chemical Coal Cleaning Assessments at Gulf," Separation Science and Technology, 18 (12 & 13), pp. 1371 - 1393 (1983).
5 C.D. Smith and D.V. Keller, Jr. "Methods of and Apparatus for Cleaning Coal," U. S. Patent No. 4,173,530 (November 6, 1979); E. C. Tveter, "Sink-Float Separation Processes," U. S. Patent No. 3,348,675 (October 24, 1967); and W. B. Foulke and W. A. Willis, "Process of and Apparatus for Separating Media," U. S. Patent No. 2,150,917 (March 21, 1939).
6 Z. J. J. Stekly, "Magnetic Separating Apparatus," U. S. Patent No. 4,213,854 (July 22, 1980).
7 R. R. Oder, "The Preparation of Low Ash Coal by Magnetic Methods," to be published in the Proceedings of the ACS Symposium on Environmental Science of Fossil Fuels, Miami (April, 1985).
8 R. R. Oder, "The Application of High Field and High Gradient Magnetic Methods to the Magnetic Separation of Mineral Matter from Micronized Coal," Separation Science and Technology, Vol. 19 (11 & 12), pp. 761-781 (1984).
9 S. G. Frantz, U. S. Patent No. 2,056,426 (1936).
10 J. E. Nesset and J. A. Finch, "Determination of Magnetic Parameters for Field Dependent Susceptibility Minerals by Frantz Isodynamic Magnetic Separator," Transaction/Section C of the Institute of Mining and Metallurgy, Vol. 89, pp. C161-C166 (1980).
11 L. C. Jackson, "Investigations on Paramagnetism at Low Temperatures," Phil. Trans. Roy. Soc., London, 224a, No. 1 (1923).
12 J. McAndrew, "Calibration of a Frantz Isodynamic Separator and Its Application to Mineral Separation," Proc. Australas. Inst. Min. Metall. No. 181, pp. 59-73 (1957).
13 Z. J. J. Stekly and J. V. Minervini, "Shape Effect of the Matrix on the Capture Cross Section of Particles in High Gradient Magnetic Separation," IEEE Trans. Magn., MAG-12(5) (September, 1976).
14 H. Honda and K. Ouchi, "Magnetochemistry of Coal I -- Magnetic Susceptibility of Coal," Fuel, Vol. 36, pp. 159-175 (1957).
15 S. Ergun and E. H. Bean, "Magnetic Separation of Pyrite from Coal," U. S. Dept. of Interior, Bureau of Mines, Report of Investigation 7181 (September, 1968).
16 Pascal's Constants and constitutive correction constants taken from: L. N. Mulay, Magnetic Susceptibility, p. 1779 and 1780, Interscience Publishers, John Wylie & Sons (1963).

POLYSACCHARIDES AND XANTHATED POLYSACCHARIDES AS PYRITE DEPRESSANTS DURING COAL FLOTATION

R. W. PERRY[1] and F. F. APLAN[2]

[1]58 Greate Bay Dr., Somers Point, NJ 08244 (U.S.A)

[2]Mineral Processing Section, The Pennsylvania State University, University Park, PA 16802 (U.S.A.)

ABSTRACT

The adsorption of polysaccharides, xanthated polysaccharides and similar compounds onto pyrite was found to correlate with pyrite depression in parallel coal flotation experiments. Unfortunately, some of the pyrite depressants are also good coal depressants. Their use requires a judicious application of the reagent and sometimes a trade-off between coal yield and pyrite rejection. Attachment of these polymers onto pyrite appears to be achieved through some combination of: (1) electrostatic attraction, (2) specific adsorption and (3) hydrogen bonding depending on the system. Superior pyrite depression was found to correspond to the number, location and configuration of the hydroxyl groups in the polymers studied, and for the starches with their amylose content.

INTRODUCTION

One of the more vexing problems facing coal preparation engineers is how to remove the maximum amount of pyritic sulfur from the coal. The problem is more critical for the preparation of coal fines, because the sulfur and ash content of fine coal cleaned by flotation is invariably higher than that of the coarse cleaned coal. This paper will focus only on means of rejection of pyritic sulfur during coal flotation. There are three methods by which this may be accomplished (ref. 1):

1. Flotation circuitry (including froth sprinkling).
2. Coal depression with concomitant flotation of the pyrite.
3. Pyrite depression.

The present study is concerned only with the last of these three alternatives.

The elements of coal flotation are well detailed in the literature (refs. 2-4) and will not be repeated here. Development of pyrite depressants for coal flotation has been less well studied though classic work in this area has been done by Yancey and Taylor (ref. 5) and by Miller (ref. 6). Coal pyrite is known to differ significantly from ore pyrite (ref. 7), and experience in our laboratories has indicated its depression is much more difficult in coal flotation than in ore flotation (using, e.g., cyanide). More recently, Choudhry (ref. 8) has shown that certain polysaccharides and other natural and synthetic polymers are good pyrite depressants for use during coal flotation. Xanthating them also often produces an effective pyrite depressant. This concept was probably first developed in our laboratories (refs. 8-10) but has also been used by others (refs. 11, 12). One of the major problems in the selection of any pyrite depressant is our finding that in almost every case a good pyrite depressant is also a good coal depressant at the same or a higher reagent concentration (ref. 8, 13). Thus, pyrite depression during coal flotation may involve a trade-off between coal yield and the sulfur content of the floated coal and will usually require careful reagent control.

The objective of this study is to evaluate polysaccharides, xanthated polysaccharides and related compounds as pyrite depressants during coal flotation. This will be done through both adsorption and flotation testing and the most favorable conditions and mode of attachment will be studied.

EXPERIMENTAL MATERIALS AND METHODS

Pyrite

The pyrite used was purified from sulfur balls taken from the Pittsburgh seam. It was concentrated by sizing, hindered settling classification and repeated tabling of narrow size/gravity fractions. Two pyrite fractions were used. The first was a rather coarse (10x48 mesh) pyrite of ∿80% purity (40.95% S) used for flotation testing. This was freshly ground to -48 mesh in a mortar and pestle immediately prior to flotation to minimize oxidation.

For adsorption testing a -48 mesh concentrated pyrite fraction was used. This was further purified by hand vanning and repeated treatment in an ultrasonic bath and desliming. It was filtered, quickly dried and stored in a vacuum desiccator. It assayed 42.14% S. Just prior to adsorption testing it was ground in an agate mortar and pestle, and the 100x140 mesh fraction was used for testing. The pyrite so prepared had a Blaine surface area of 0.077 m^2/g.

A few tests were made from ore pyrite from Rico, Colorado (Wards Natural Science Establishment). It assayed 52.87% S. The 100x400 mesh pyrite had a Blaine surface area of 0.025 m^2/g.

Coal

The coal sample used was a heavy media cleaned coal from the Robena mine of U.S. Steel Corporation, Greene County, Pennsylvania. It has a proximate analysis of 7.11% ash, 35.30% volatile matter, 57.60% fixed carbon, 14,153 BTU/lb,

1.14% total sulfur, 0.44% organic sulfur, 0.01% sulfate sulfur and 0.69% pyritic sulfur.

Polymers

The organic polymers used in this study and their source and identifying characteristics are shown in Table 1. They represent natural and modified corn, potato, tapioca and wheat starches, guar gums (natural and modified), gum Arabic, Irish moss (carrageenan), carboxymethyl cellulose (CMC) and hydroxyethyl cellulose (HEC). The basic structure of the various classes of these compounds is shown in Figures 1 and 2. The starches (ref. 14) have two principal constituents, the straight chain amylose, amounting to $\approx$ 25% of the total, and the branch chain amylopectin which makes up the balance. The amylose content of common starches varies from $\sim$17 to $\sim$28% depending on the starch, with corn starch showing, typically, $\sim$25% amylose. Special corn hybrids are available which contain $\sim$100% amylopectin ('waxy maize', typified here by Amioca starch) or are high (50-80%) in the straight chain amylose (typified here by Hylon VII). The starches can be modified by introducing, e.g., anionic or cationic groups or by dextrinization. A dextrin is made from a starch by cooking in the presence of acid. This produces hydrolyzed fragments which are repolymerized (1→6 linkage) to give highly branched dextrin of lower molecular weight than the original starch (ref. 14). The gums (guar, Arabic, carrageenan) are mutations of the basic polysaccharide structure differing from the basic structure by linkage, location of hydroxyl groups and configuration of the 1-4 oxygen configuration (α,β) (ref. 15). Also shown in Figure 2 is the structure of CMC and HEC.

Several of the polymers were xanthated (denoted by the suffix 'X' after the polymer name). This was done by calculating the molecular weight of each basic polysaccharide group per OH group contained in this structure. The stoichiometric amount of carbon disulfide and excess NaOH were then added to produce the desired degree of xanthogenation. This was usually 10% but varying amounts were used in other cases. It was assumed that all of the CS_2 reacted to produce a xanthate.

Adsorption Procedure

One gram of 100x400 mesh pyrite was added to 10 ml of polymer solution containing 150 ppm of the polymer. This amount was determined by separate experiment as sufficient to produce a constant adsorption density on the pyrite. The pH was adjusted immediately before and after the addition of the pyrite. The sample-solution mixture was then shaken on a wrist action shaker for 10 min. and the solution then poured over a 400 mesh stainless steel screen to separate the solids from the polymer. The residual concentration of polymer was determined by the colorimetric phenol-sulfuric acid method of Dubois et al. (ref. 16)

TABLE 1

Natural and synthetic polymers studied.

Ident. No.	Trade Name	Manufacturer	Identifying Characteristics
1	Hylon VII	Nat'l starch	Corn starch - high amylose
2	Amioca	Nat'l starch	Corn starch - high amylopectin
3	CPC 3005	CPC Int'l	Corn starch - unmodified, $\sim$25% amylose
4	Potato starch	Staley	Potato starch - unmodified, $\sim$22% amylose
5	SDU	Ind. Grain Prod.	Wheat starch - MW $\sim$600,000 fraction
6	Supergel 1202	Ind. Grain Prod.	Wheat starch - MW $\sim 10^6$ fraction
7	Tapioca	Staley	Tapioca starch - $\sim$17-22% amylose
8	Sta-Lok 1303	Staley	Corn starch - cationic, $\sim$25% amylose
9	Cato-2	Nat'l. Starch	Corn starch - cationic, $\sim$25% amylose
10	Cato-14	Nat'l. Starch	Corn starch - cationic, high amylopectin
11	Hamaco 196	Staley	Potato starch - anionic, $\sim$22% amylose
12	Jaguar MDD	Celanese	Guar gum - unmodified
13	Jaguar C-13	Celanese	Guar gum - cationic, D.S. $\sim$0.13
14	Gendriv 162	Gen'l. Mills	Guar gum - cationic
15	Gendriv 307	Gen'l. Mills	Guar gum - anionic
16	Gum Arabic	Celanese	Gum Arabic - naturally anionic
17	Irish Moss	Meer Corp.	Irish moss - mostly λ-Carrageenan
18	CPC 8071	CPC Int'l.	Corn Starch dextrin-canary
19	CMC 4H1	Hercules	Carboxymethylcellulose - D.S. 0.2, high viscosity
20	CMC 7H	Hercules	Carboxymethylcellulose - D.S. 0.7, high viscosity
21	CMC 12M31	Hercules	Carboxymethylcellulose - D.S. 1.2, medium viscosity
22	Natrosol 150 GXR	Hercules	Hydroxyethylcellulose - M.S. 1.5
23	Natrosol 250 MR	Hercules	Hydroxyethylcellulose - M.S. 2.5
24	Gum Ethylex 3095	Staley	Corn starch - Hydroxylethylether derivative

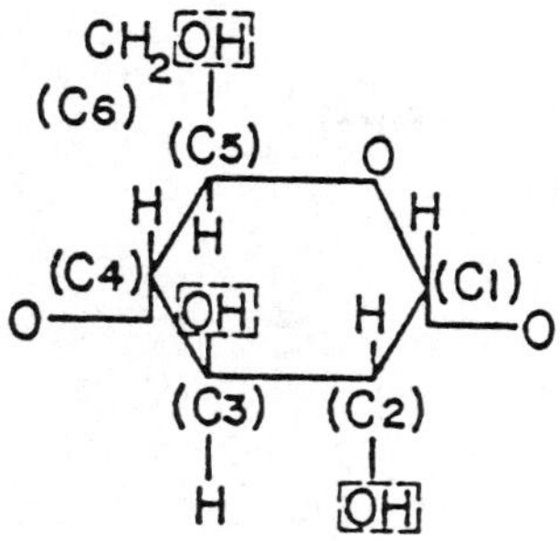

Starch molecule showing reactive hydroxyl sites.

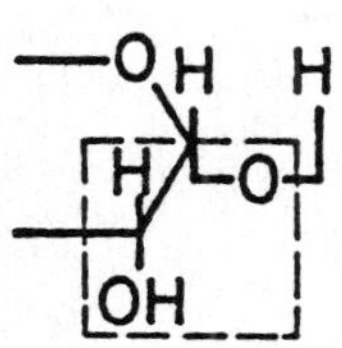

α-configuration

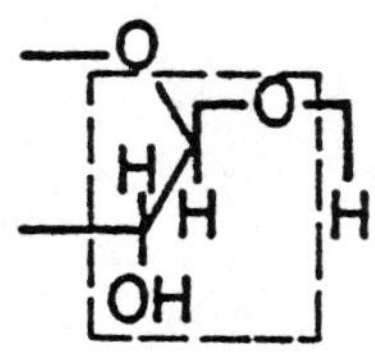

β-configuration

Cis configuration

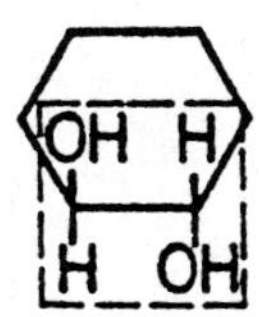

Trans configuration

Fig. 1. Hydroxyl group positioning in polysaccharides.

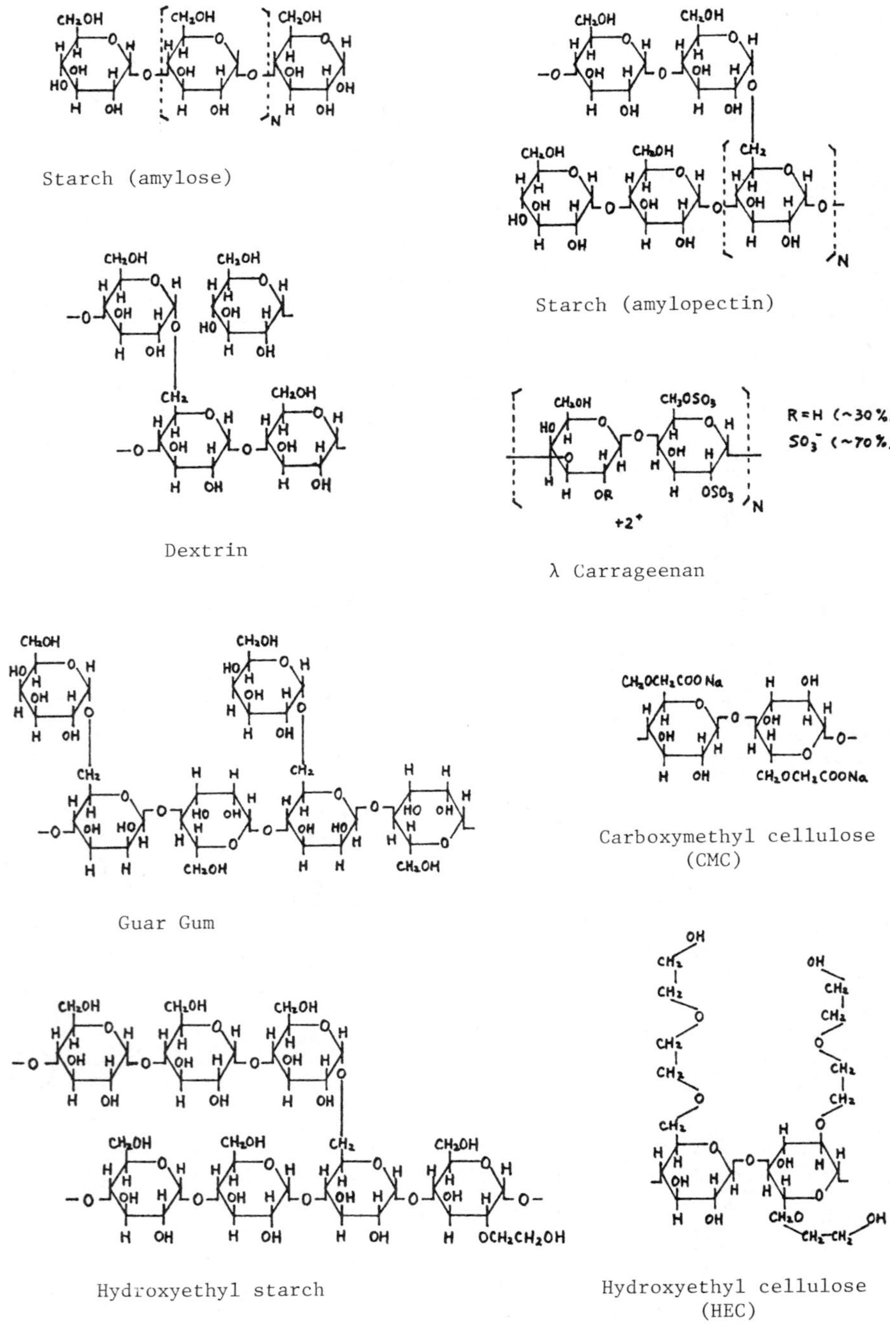

Fig. 2. Chemical structure of polysaccharides and related compounds (refs. 14, 15).

using a Coleman-Hitachi Model 124 double beam spectrophotometer and measuring the adsorbance at λ=486 nm. Adsorption was then determined by difference.

When xanthate solutions were used, the xanthate in solution was determined by measuring the characteristic xanthate adsorbance at λ=301 nm and calculating the adsorption by difference from the starting solution. In this way only the adsorption of that portion of the polymer actually xanthated was measured.

Flotation procedure

Flotation tests were made in the older-style WEMCO-Fagergren laboratory flotation machine with a circular glass cell of ∿2.5 l nominal capacity. Impeller speed was 1750 rpm and an aeration rate of 7.1 lpm. Preliminary testing demonstrated that 60 μl (0.24 kg/t) of MIBC and 10 μl (0.04 kg/t) of No. 2 fuel oil floated this coal in good yield. Pre-prepared solutions of the pyrite depressants were added in the amount of 0.1 kg/t.

A 200 g charge (190 g coal, 10 g pyrite) was added to a small amount of water in the cell and agitated for two minutes to insure wetting. Additional water was then added and the pulp conditioned for five minutes during which time the pH was adjusted. The pyrite depressant was then added and an additional three minutes of conditioning given, and the pH re-checked and re-adjusted, if necessary. Frother and fuel oil were added using a syringe and given an additional one minute of conditioning. Air was turned on and zero time was fixed as that time when froth just started to spill over the top of the cell. Froths were collected at one and two minute intervals and the flotation products filtered, dried, weighed and analyzed for sulfur using a LECO apparatus. Subsequent test evaluation showed that the second minute of flotation resulted in only an additional 5-7% yield and comparable sulfur distribution, so that most tests were evaluated on the basis of the product floated in the first minute.

EXPERIMENTAL RESULTS AND DISCUSSION

Adsorption onto Pittsburgh seam pyrite

The basic adsorption data for the 24 polymers are shown in Figures 3-6. It is immediately apparent that the adsorption of these compounds is pH sensitive; greatly so in some cases. The greatest adsorption is invariably noted at pH 7. Note that the range of adsorption density (Γ, mg/m^2) shown by the ordinate varies from figure to figure. The adsorption of non-ionic starches onto pyrite (Figure 3), while averaging ∿8.5 mg/m^2, shows maximum adsorption at pH 7 and, generally, poorest adsorption at pH 9. There is a great variation in adsorption density, with the straight chain starch, Hylon VII, being most strongly adsorbed while the branch chain, Amioca shows the poorest adsorption. The ionic starches shown in Figure 4 were all cationic compounds. While they show

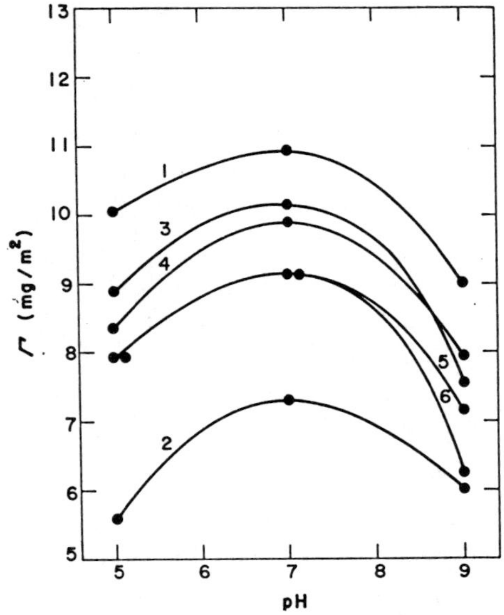

Fig. 3. Adsorption of non-ionic starches onto Pittsburgh seam pyrite.
1. Hylon VII 4. Potato starch
2. Amioca 5. SDU
3. CPC 3005 6. Supergel 1202

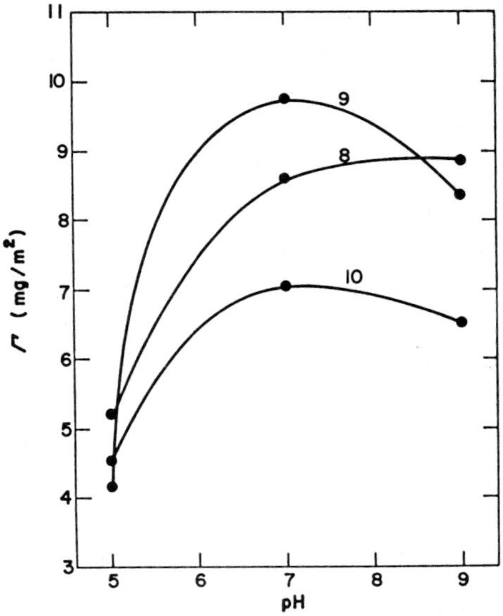

Fig. 4. Adsorption of ionic starches onto Pittsburgh seam pyrite.
8. Sta-Lok 1303 (+) 10. Cato-14 (+)
9. Cato-2 (+)

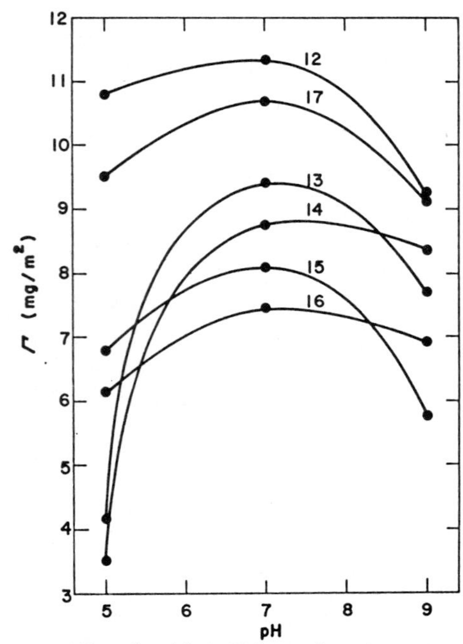

Fig. 5. Adsorption of gums onto Pittsburgh seam pyrite.
12. Jaguar MDD 13. Jaguar C-13 (+)
16. Gum Arabic 14. Gendriv 162 (+)
17. Irish moss 15. Gendriv 307 (-)

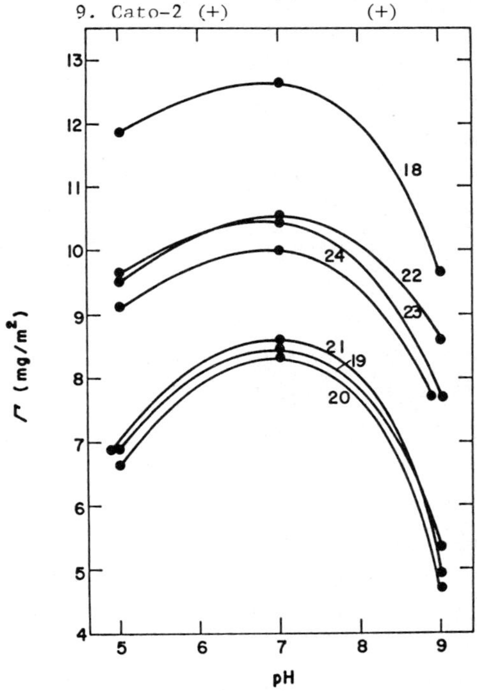

Fig. 6. Adsorption of various polymers onto Pittsburgh seam pyrite.
18. CPC 8071 24. Gum Eth 3095
22. Natrosol 150 GXR 19. CMC 4H1
23. Natrosol 250 MR 20. CMC 7H
21. CMC 12M 31

an average adsorption density similar to the non-ionic starches, the variation of adsorption with pH is very great (more than two-fold between pH 5 and 7 for compounds Nos. 8 and 9). This is likely a manifestation of their cationic charge and will be discussed later. The cationic starches Sta-Lok 1303 and Cato-2 act similarly, and they are both based on ordinary corn starch (∿25% amylose, ∿75% amylopectin) but were probably cationized by different methods (ref. 13). Cato-14 is much less strongly adsorbed, except at pH 5. It is a cationized branched chain waxy maize (∿100% amylopectin). As was noted in Figure 3, compounds containing the straight chain amylose are more strongly adsorbed than are those rich in the branched chain amylopectin.

The adsorption of gums onto pyrite is given in Figure 5. As before, maximum adsorption occurs at pH 7, and there is a wide variation in adsorption from compound to compound. The guar gum (Jaguar MDD) which has only modest branching (Figure 2) is much more strongly adsorbed than the highly branched gum Arabic. As previously noted the charge on the molecule influences its adsorption. The two cationic guars (Nos. 13 and 14) adsorb in a manner almost identical to the cationic starches (Nos. 8 and 9). The anionic guar, Gendriv 307 is much less pH sensitive and is less strongly adsorbed.

The data for organic polymers that did not fall into the previous classes are given in Figure 6. Maximum adsorption occurs at pH 7 with poorest adsorption at pH 9. The corn dextrin, CPC 8071 is the most strongly adsorbed of all the compounds tested. It would appear to be an excellent pyrite depressant, but, unfortunately, it has been found to be the best of all starch products for the depression of coal (ref. 13). Though highly branched, it is structurally different from the other compounds tested and appears to follow a different pattern. The three different carboxymethyl cellulose (CMC) products all act similarly irrespective of their degree of substitution. The hydroxyethylcellulose (HEC, Natrosols) and the starch hydroxyethylether (Gum Ethylex 3095) all respond similarly but are better adsorbed than are the CMC's.

Effect of xanthating. Tests were also made using the xanthated form of several of the starches and gums because of the well-known attachment of xanthates to sulfide minerals (ref. 18) and because of previous success in pyrite depression using this approach (refs. 8-10). Proof that the various starches were indeed xanthated was shown by their UV adsorption spectra using a Varian-Carey 118-CX spectrophotometer. The height and shape of the ±301 nm band changed, however, from compound to compound and between CPC 3005 corn starch xanthated at the 10% and 110% levels (ref. 17). The variation in the xanthates formed is likely a manifestation of the availability of the various polysaccharide OH^- groups for reaction with CS_2 and NaOH.

The adsorption data for the xanthated polysaccharides is given in Figure 7. Highest adsorption density is found at pH 5, decreasing with increasing pH.

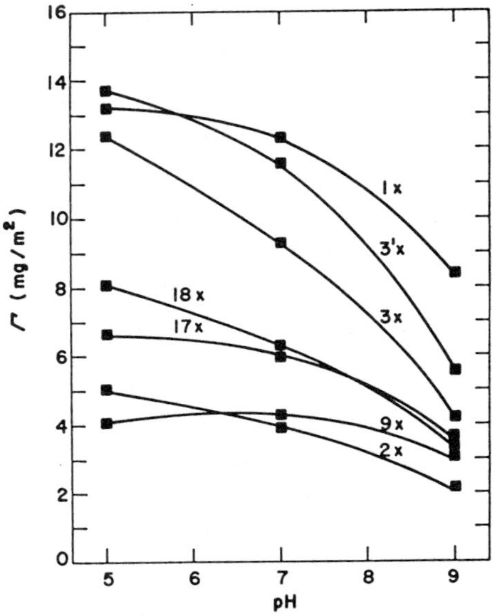

Fig. 7. Adsorption of xanthated polysaccharides onto Pittsburgh seam pyrite.
1X. Hylon VII-X, 10%
2X. Amioca-X, 10%
3X. CPC 3005-X, 10%
3'X. CPC 3005-X, 110%
9X. Cato-2X, 10%
17X. Irish moss-X, 10%
18X. CPC 8071-X, 10%

This is as expected based on Sutherland and Wark's data for the pyrite-xanthate system (ref. 18). Best adsorption occurs when the high amylose content corn starch, Hylon VII, is xanthated. Xanthating ordinary corn starch (∿25% amylose) at either the 10% or 110% levels produces compounds only slightly less well adsorbed. (Note: the authors realize it is not possible to xanthate starch to 110% of theoretical. This was used as a 'device' to indicate sufficient CS_2 was made available to xanthate all the readily reactable hydroxyl groups.) Irish moss and the corn dextrin, CPC 8071, when xanthated at the 10% level, are much less well adsorbed. The fact that both of these xanthates are less well adsorbed than their parent compounds indicates that the adsorption of the parent compound onto pyrite occurs through one or more of the reactive hydroxyl groups (at C_2, C_3, C_4 or C_6, see Figures 1 and 2) which have been partially usurped by the xanthating process.

The Amioca and Cato-2 xanthates are the most poorly adsorbed of all those evaluated. They are also more poorly adsorbed than their parent starches, again, indicating that the xanthating process interfers with adsorption onto pyrite either through diminution of the OH^- groups or through partial electrostatic rejection caused by the presence of the quaternary groups as in Cato-2. As before, the highly branched waxy maize (amylopectin), Amioca, is very poorly adsorbed even with the introduction of the xanthate group. It would not appear to be a good pyrite depressant.

Mechanisms of attachment. The adsorption of these compounds onto pyrite is likely controlled by (1) electrostatic attraction between a charged surface and a charged polymer; (2) specific adsorption through selected active groups (e.g., the xanthates); (3) hydrogen bonding between hydroxyl groups on the surface of the pyrite (obtained through oxidation of the pyrite) and one or more of the hydroxyl groups in the polysaccharide or other polymer.

Electrostatic attraction. The zeta potential of the Pittsburgh seam pyrite used here is shown in Figure 8. It is apparent that the pyrite oxidizes readily and, in fact, other experiments indicated that curves similar to those

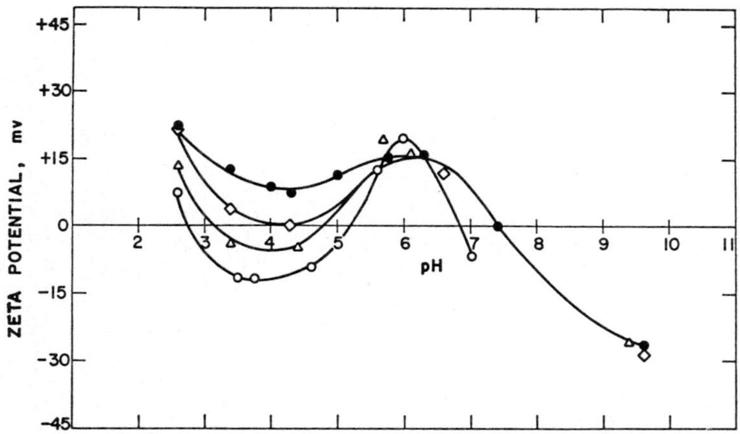

Figure 8. Zeta potential of Pittsburgh seam pyrite. Ground wet, unwashed. Fresh and aged.
Aging Time: ● <5 min, ◇ 10 min, △ 60 min, ○ 24 hrs

obtained with 10 min to 24 hr of aging can be produced by adding $FeSO_4$ externally (ref. 17). In view of the time lapse in performing the adsorption and flotation experiments, a curve similar to that obtained for 10 min or more probably pertains here.

While the effect of electrostatic attachment is not always obvious from Figures 3-6, and structural factors also influence polymer adsorption, Table 2 clearly shows that the adsorption of the charged polymers onto pyrite is definitely influenced by electrostatic attraction between a charged surface and an oppositely charged polymer. At pH 9, where the pyrite is strongly negative, the cationic polymers are much more strongly adsorbed than are the anionic ones. At pH 5, where the pyrite is slightly positive to neutral, the anionic polymers are better adsorbed than are the cationic ones, though neither are particularly well adsorbed.

The high adsorption at pH 7 where the pyrite surface is essentially neutral must be due to reasons other than pure electrostatic attraction. While Figure 8 indicates the slightly oxidized surface is probably neutral at this pH,

TABLE 2

Adsorption density as a function of the charge on pyrite and on polysaccharide.

pH of Solution Charge on Pyrite	5 (slightly +)	7 (∿0)	9 (strongly -)
Adsorption Density Range (mg/M^2)			
9-10	---	Cato-2 (+)*	---
	---	Jaguar C-13 (+)	---
8-9	---	Gendriv 162 (+)	Sta-Lok 1303 (+)
	---	Sta-Lok 1303 (+)	Gendriv 162 (+)
	---	CMC 12M31 (-)	Cato-2 (+)
	---	CMC 4H1 (-)	---
	---	CMC 7H (-)	---
	---	Gendriv 307 (-)	---
7-9	---	---	Jaguar C-13 (+)
6-7	CMC 12M31 (-)	---	---
	CMC 4H1 (-)	---	---
	Gendriv 307 (-)	---	---
	CMC 7H (-)	---	---
5-6	Sta-Lok 1303 (+)	---	Gendriv 307 (-)
	---	---	CMC 4H1 (-)
4-5	Jaguar C-13 (+)	---	CMC 12M31 (-)
	Cato-2 (+)	---	CMC 7H (-)
3-4	Gendriv 162 (+)	---	---

*Value in parenthesis after each compound indicates the charge on the polysaccharide used.

other experiments (ref. 17) indicate it can be slightly negative depending on preparation and treatment. Furthermore, electrostatic attraction cannot be used to explain the strong adsorption of non-ionic polymers especially at pH 7.

Specific adsorption. The specific adsorption of the starch xanthates CPC 3005 and Hylon VII onto pyrite is easy to understand based on ore flotation experience (ref. 18), but the strong adsorption of non-ionic compounds onto pyrite at pH 7 requires another explanation. One strong possibility is the formation of some type of glucose-iron complex at the pyrite surface near pH 7. Dextrin-iron complexes are well known (ref. 19). This concept was evaluated by measuring the turbidity of a 10^{-5} mol l^{-1} $FeSO_4$ solution in the presence and absence of 2 ppm of several polymers (Figure 9). By itself, $FeSO_4$ begins to

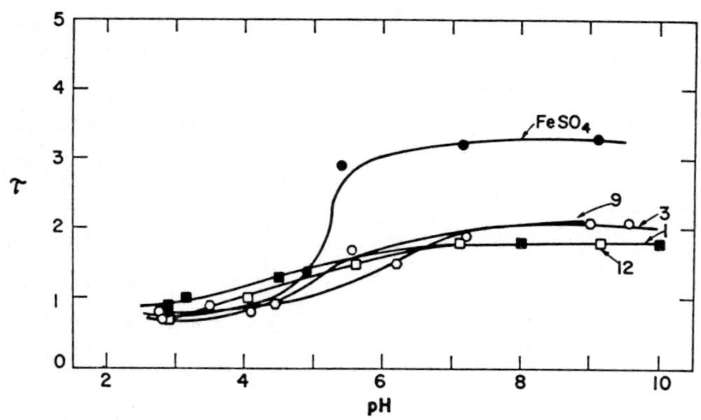

Fig. 9. Effect of various polymers on iron hydroxide precipitation in 10^{-5} mol ℓ^{-1} $FeSO_4$ solutions.
plus: 1. 2 ppm Hylon VII 9. 2 ppm Cato-2
3. 2 ppm CPC 3005 12. 2 ppm Jaguar MDD

precipitate at pH $\sim$5.5, but this precipitation is supressed in the presence of Hylon VII, CPC 3005, Cato-2 and Jaguar MDD. This gives strong support to the concept of a polysaccharide-iron complex being responsible for the strong adsorption of such polymers onto pyrite.

Effect of functional groups. The effect of introducing xanthate groups has previously been discussed as has the introduction of cationic groups which can either influence electrostatic attachment or can interfer with bonding through hydroxyl or xanthate groups.

If hydrogen bonding via hydroxyl groups in the polymer is the primary means of adsorption, the introduction of functional groups would decrease the number of hydroxyl groups available for bonding and, hence, their adsorption. While this is clearly the case for the guars (Figure 5), a different picture emerges for the starches. Comparison of the cationic Cato-2 with the neutral CPC 3005 corn starch (Figures 3 and 4) shows that the addition of the cationic functional group has little effect on its adsorption except for the decrease at pH 5. This difference between guars and starches is probably related to structural differences between them. Guar (Figure 2) can be described as an alternating polymer and the structure should leave the OH^- groups physically unhindered in solution and available for substitution. The starch CPC 3005, contains as its primary component the highly branched amylopectin and since branching occurs at the most reactive site, C-6 (ref. 20), the number of available OH^- sites for substitution will be decreased. Furthermore, the amylopectin molecule is believed to have a cluster form (ref. 20) and this would result in only the OH^- groups on the outer branches readily amenable to substitution.

Differences in the nature and location of the functional groups is also important. While both CMC and HEC are derived from cellulose, the CMC compounds contain carboxylmethyl groups in varying amounts at the C_2, C_3 and C_6 hydroxyl sites while the HEC compounds contain hydroxyethyl groups at these sites (see Figure 2). As shown in Figure 6, the HEC (Natrosol) compounds are more strongly adsorbed under all pH conditions than are the CMC compounds. This infers a hydrogen bonding mechanism since rather than dimishing the number of OH^- groups as for CMC, they have only been replaced in the HEC compounds.

Effect of structure. The degree to which a polymer can be chemically modified or adsorbed depends on its structural characteristics. For the polysaccharides the availability of the very reactive OH^- at the C-6 site will be a function of whether this site was previously used for branching (as in the amylopectins). The superior ability of amylose (Hylon VII) to be xanthated as compared to Amioca, Irish moss or dextrin (CPC 8071) (data not shown here) can be attributed to the unavailability of a hydroxyl at C-6, or blockage by steric hinderance (C-2, C-3) of these sites. The inability of a compound to be effectively xanthated is reflected in the poor adsorption of these branched compounds onto pyrite. As previously stated, poor xanthogenation can also result from prior substitution of the reactive OH^- sites.

The role of branching on adsorption can be seen from Figure 3 where the adsorption is found to increase in proportion to the straight chain amylose content:

Amioca (∿1% amylose)	↓
Potato starch (∿22% amylose)	Increased amylose,
CPC 3005 (∿25% amylose)	increased adsorption
Hylon VII (∿50-80% amylose)	↓

For corn starches, the more highly branched (higher amylopectin content) the starch, the poorer the adsorption.

Effect of hydroxyl group configuration. The availabiltity of an OH^- group to participate in adsorption, either through hydrogen bonding or through formation of a polysaccharide - iron complex, will depend on both the number and configuration of the OH^- groups. Their location at the very reactive C-6 site or the less reactive C-2, C-3 and C-4 sites, as well as their cis or trans form at the C-2 and C-3 sites (see Figs. 1 & 2), will control their availability for reaction. For example, starches exist in the α/trans form, guars in a combination of two β/cis (main chain) to one α/trans (side chain) and cellulose in the β/trans form. No absolute conclusions can be drawn because the configurations cannot be completely isolated, but it should be noted that the guar gums, which show strong adsorption onto pyrite, contain glucose units of the β/cis hydroxyl group configuration.

Adsorption onto ore pyrite. Figure 10 gives the adsorption of several of the polymers onto ore pyrite. The nature and amount of the adsorption is seen

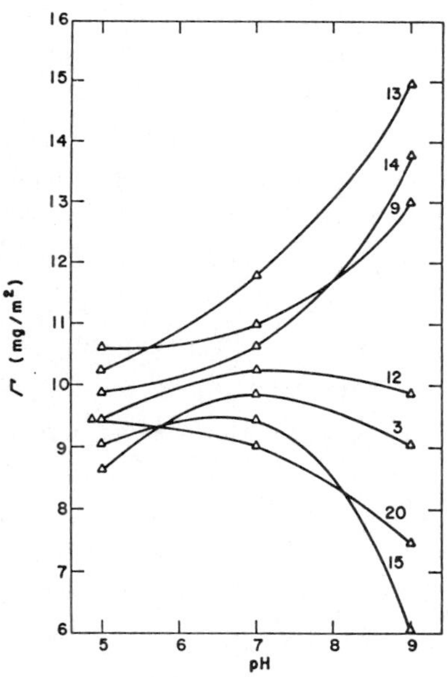

Fig. 10. Adsorption of various polymers onto ore pyrite.
5. CPC 3005 9. Cato-2 (+) 14. Gendriv 162 (+)
12. Jaguar MDD 13. Jaguar C-13 (+) 15. Gendriv 307 (-)
20. CMC 7H

to be quite different from that of the same compounds onto pyrite from a coal source. This indicates the danger of extrapolating flotation response from ore pyrite to coal pyrite. The adsorption of the three cationic polysaccharides is particularly different here and their adsorption at pH 9 onto the strongly negative pyrite seems to be by electrostatic attraction. The adsorption of CPC 3005 and the anionic Gendriv 307 are about the same as on coal pyrite (compare Figures 3 and 5 with 10). The adsorption of Jaguar MDD follows a similar pattern to that on coal pyrite except the adsorption density is less while the adsorption of CMC 7H is a little higher, especially at pH 5, on ore pyrite than on coal pyrite.

Adsorption onto coal. Data for the adsorption of these same polymers onto Pittsburgh seam coal has also been collected and will be reported elsewhere (ref. 21). Their adsorption densities, averaging ~ 5.5 mg/m^2 are significantly lower than the same compounds adsorbed onto pyrite from the same coal seam (averaging ~ 8.5 mg/m^2). This circumstance reflects the problem, previously noted, that good pyrite depressants are also generally good coal depressants.

Desorption. Several water desorption experiments, using both the xanthated and non-xanthated form of several of these polymers, indicated that most of the polymer, after washing, still clings tenaciously to the pyrite. Essentially all of the xanthated material and about 80-85% of non-xanthated polymer remains on the surface after washing with water.

Flotation

Flotation experiments were performed on the Pittsburgh seam coal-pyrite mixture using most of the polymers whose adsorption behavior had previously been studied. Comparison of the adsorption and the flotation data should then allow an estimation to be made as to which features constitute a good pyrite depressant for use in coal flotation. While data are available for both one minute (F_1) and two minutes (F_1+F_2) of flotation, the two were generally found to differ by only about 5-7% yield (an comparable sulfur distribution) and so the data reported here will be based on an analysis of the one minute (F_1) flotation performance. The complete data may be found elsewhere (ref. 17).

The flotation yield (F_1) and total sulfur content (S_T) of the floated coal are shown in Figures 11-15 which may be compared to the adsorption data of Figures 3-7, respectively. The numbering system for the various polymeric compounds follows that of Table 1. For ease of comparison all of the test data are given on the grade-yield curve of Figure 16 with the potentially superior tests identified by test number (Table 3).

Reference to Figures 11-15 show that the pH-S_T curves are essentially a mirror image of the adsorption curves Figures 3-7, for a given class of compounds, i.e., strong adsorption of polymer onto pyrite results in a low sulfur content of the floated coal. The minimum sulfur content usually occurs at pH 7 where maximum adsorption was found to occur. Unfortunately, some of the compounds that produce a low sulfur clean coal also seriously depress coal (i.e., the gums Jaguar MDD, Irish moss, the HEC Natrosol 150 GXR, the dextrin CPC 8071 and many of the compounds at pH 9). This is discussed more fully elsewhere (refs. 13, 21 - 23). In evaluating these data is is suggested that the reader focus first on the S_T data, and after identifying a compound capable

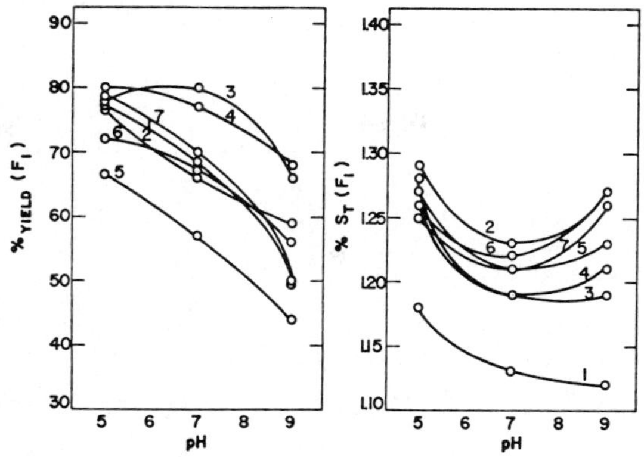

Fig. 11. Flotation results with non-ionic starches.
1. Hylon VII
2. Amioca
3. CPC 3005
4. Potato starch
5. SDU
6. Supergel 1202
7. Tapioca starch

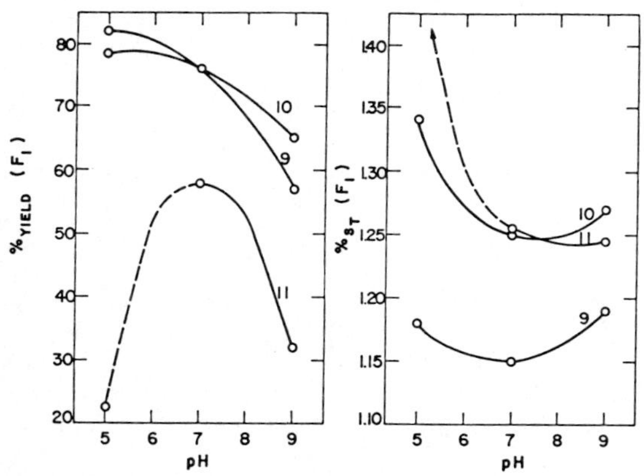

Fig. 12. Flotation results with ionic starches.
9. Cato-2 (+)
10. Cato-14 (+)
11. Hamaco 196 (-)

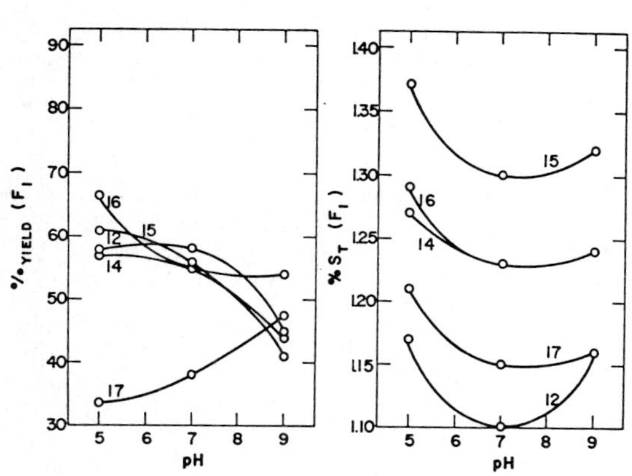

Fig. 13. Flotation results with gums.
12. Jaguar MDD
14. Gendriv 162 (+)
15. Gendriv 307 (-)
16. Gum Arabic
17. Irish moss

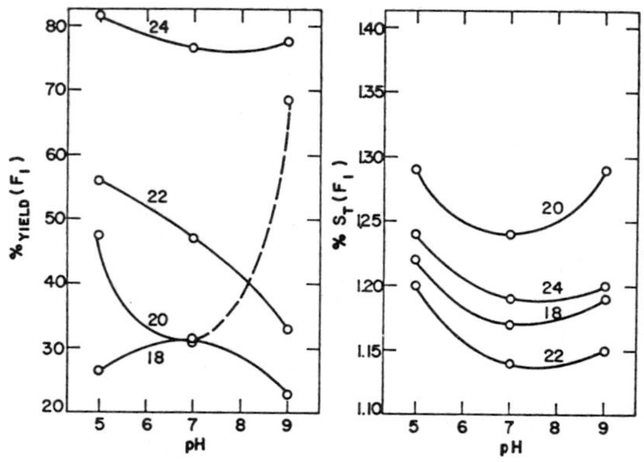

Fig. 14. Flotation results with various polymers.

18. CPC 8071
20. CMC 7H
22. Natrosol 150 GXR
24. Gum Ethylex 3095

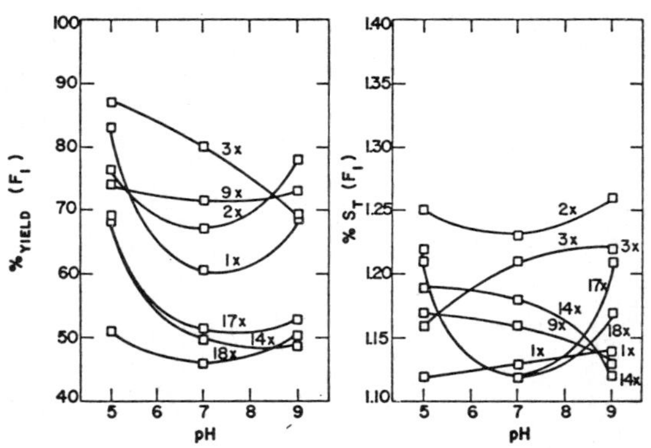

Fig. 15. Flotation results with various xanthates.

1X. Hylon VII-X
2X. Amioca-X
3X. CPC 3005-X
9X. Cato-2X
14X. Gendriv 162-X
17X. Irish moss-X
18X. CPC 8071-X

TABLE 3

Superior pyrite depressants based on grade-yield criterion (listed by test number, not by performance).

Test No.	pH	Yield F_1	Sulfur ST_1	Reagent
101	5	82.1	1.18	Cato-2
102	7	76.2	1.15	Cato-2
113	7	57.5	1.10	Jaguar MDD
151	7	79.7	1.19	CPC 3005
159	5	77.5	1.18	Hylon VII
160	7	68.4	1.13	Hylon VII
179	7	76.8	1.19	Potato starch
182	7	76.5	1.19	Gum Ethylex 3095
183	9	77.4	1.20	Gum Ethylex 3095
201	5	86.9	1.16	CPC-3055 xanthate
205	5	82.8	1.12	Hylon VII xanthate
206	7	60.5	1.13	Hylon VII xanthate
207	9	68.8	1.14	Hylon VII xanthate
209	5	73.9	1.17	Cato-2 xanthate
210	7	71.5	1.16	Cato-2 xanthate
211	9	72.9	1.13	Cato-2 xanthate

of producing a low sulfur clean coal, then focus on those compounds that give respectable yields. This process is greatly facilitated by use of the grade-yield curve of Figure 16 which clearly indicates the superior tests. These

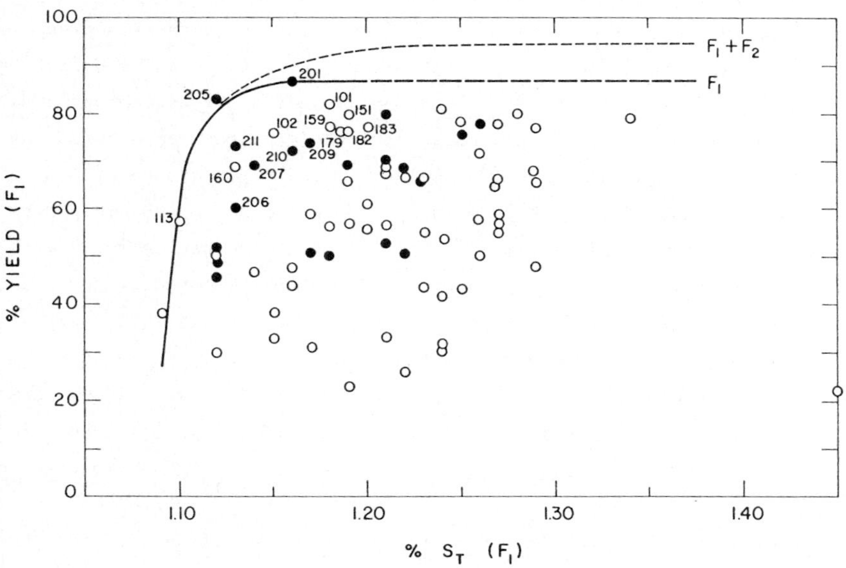

Fig. 16. Grade-yield curve of flotation test results. Dark circles are for xanthates. Test numbers refer to Table 3.

are also indicated by the number on the figure and in Table 3. Even this is not a sufficient evaluation criterion as only one pyrite depressant concentration was used (0.1 kg/t) and compounds that produce a low sulfur content clean coal yet still seriously depress coal may prove to be satisfactory pyrite depressants at some lower depressant concentration. This is typified by Jaguar MDD (Test No. 113, Figure 16) and the three xanthates of Gendriv 162, Irish Moss and (highly unlikely) the strong coal depressant, dextrin, CPC 8071; all of which produce sulfur contents of S_T = ∿1.12 at pH 7 (see Figure 15).

The most promising pyrite depressant systems shown in Table 3 indicate that while pH 7 is the most common value (corresponding to the maximum in adsorption), excellent results are also often obtained at pH 5 and to a lesser extent at pH 9. The problem at pH 9 is often the depression of the coal. The most effective reagents tested here are relatively few in number, with Hylon VII, Cato-2, and to a lesser extent CPC 3005, being effective in both the xanthated and unxanthated form. Gum Ethylex 3095 and potato starch also appear attractive as would the common guar gum, Jaguar MDD, if the coal depression problem could be solved.

Effect of xanthating. The data for the xanthated compounds are shown in Figure 15 and as the filled points in Figure 16. This latter figure clearly demonstrates that while xanthating can produce remarkable results (Tests 201, 205, 211, Table 3) xanthating is no panacea and flotation behavior may vary greatly from compound to compound. Not only are some xanthated polymers not very effective in depressing pyrite, but some of them also greatly depress coal. The xanthates of Hylon VII and Cato-2 are shown to be quite effective at all pH values as is common starch CPC 3005X at pH 5. The other xanthated polymers perform in a manner roughly similar to their parent compounds. The high, straight chain amylose content of Hylon VII and the reasonably high (∿25%) amylose content of the CPC 3005X and Cato-2X appear to enhance pyrite depression. It is especially interesting to note that the cationized starch xanthate Cato-2X works so well especially in view of its relatively poor adsorption at pH 5 (Figure 4). Apparently the presence of its relatively high amylose content, plus xanthated and cationic groups, all serve to effectively attach this compound to the pyrite surface. Figure 15 shows that xanthating a high amylopectin starch (Amioca X) and a 25% amylose starch (CPC 3005X) produces results clearly inferior to the use of the high amylose starch Hylon VII-X.

From Figure 15 there is no clearly identifiable effect of pH. If attachment were only through the xanthate group one would expect superior results at pH 5 (Figure 7 and ref. 18), but since this appears not to be the case, other factors also contribute to the efficiency of these compounds as depressants.

This may be through attachment by more than one mechanism or (less likely) by the ability of the adsorbed polymer surface to reject an air bubble.

On the other hand, flotation with about half of these compounds is favored at pH 5 because of the superior coal yields obtained at this lower pH.

Effect of functional groups and mechanism of adsorption. For the non-xanthated compounds, Figure 11 shows only Hylon VII (amioca) and CPC 3005 produce sulfur contents below S_T=1.20%, and of these Hylon VII is definitely the best. Unfortunately, it also seriously depresses coal at the 0.1 kg/t level at pH 9. Introduction of cationic or anionic groups (Figure 12) results in low sulfur contents only for the Cato-2. As for the non-ionic starches, coal depression is substantial at pH 9. The low moleuclar weight wheat starch fraction, SDU seriously depresses coal (compound No. 5, Figure 11). From Figure 13, two gums, Jaguar MDD and Irish moss look attractive as pyrite depressants, but at least at the concentration tested, both also seriously depress coal as well, especially the Irish moss. Figure 14 shows several potential pyrite depressants with the HEC, Natrosol 150 GXR, dextrin CPC 8071 and the hydroxylethylether starch, Ethylex 3095 all producing clean coal products with S_T<1.20%. As previously noted the corn starch dextrin, CPC 8071, seriously depresses coal as well and, for this reason, is not attractive. The HEC (Natrosol 150 GXR), however, would be an attractive pyrite depressant candidate if its coal depressing capabilities turn out to be substantially less at some lower concentration. The Gum Ethylex 3095 is the most promising of these groups of compounds producing low sulfur at excellent coal yields in the range pH 7-9.

The introduction of functional groups, other than the xanthates (Figure 15) and cationic groups (Figure 12), does not appear to greatly influence pyrite depression except that the introduction of anionic groups into starch (Hamaco 196) and guar gum (Gendriv 307) is counter productive (see Figures 12 and 13). Electrostatic attraction seems definitely to be an important mechanism of polymer attachment (i.e., tests 101, 102, Cato-2, Table 3, though the cationic guar, Gendriv 162, not only seriously depresses coal but is not a very good pyrite depressant either (Figure 13). Certainly, structural factors and hydroxy groups also greatly influence polymer adsorption onto and depression of pyrite. The introduction of hydroxy ethyl groups, $-CH_2CH_2OH$, into corn starch, as in Gum Ethylex 3095, results in the creation of a very attractive pyrite depressant (Figure 14).

It would appear that the mechanisms of electrostatic attraction, specific adsorption and hydrogen bonding are all involved in producing a superior, well adsorbed, pyrite depressant. There seems to be adequate OH^- groups remaining unattached to the surface to allow for creation of the hydrophilic coating on the pyrite to reject an air bubble.

Effect of Structure. The correlation between adsorption and pyrite depression also is seen to be a good one. The high, straight chain amylose content produces a superior pyrite depressant, as in Hylon VII, either in the xanthated or non-xanthated form. Ordinary corn (CPC 3005) or potato starch with a 23-25% amylose content also produces reasonably good pyrite depression while the nearly pure amylopectin, Amioca, is a poor depressant.

Two of the guar gums, Jaguar MDD and Irish moss (carrageenan), are excellent pyrite depressants though they also seriously depress coal (Figure 13). Their effectiveness as pyrite depressants is also likely related to their structure with guar composed as linear chain of 1-4 linked B-D-mannopyranosyl units with single α-D-galactopyranosyl side branches (Figure 2 and ref. 15) while carrageenan is composed of linked groups (Figure 2). Their use as pyrite depressants would depend on mitigating their strong coal depression effect.

Effect of hydroxyl groups availability. Hydroxyl groups perform three functions: (1) they influence depressant attachment through hydrogen bonding, (2) they furnish reactive sites for molecular modification and (3) they strongly reject air bubbles. The depressant action of the compounds tested follows the adsorption trend so item (3), rejection of air bubbles, is not a critical factor here and a sufficient amount of such groups appears always to be available to create the necessary hydrophilic coating. Introduction of functional groups (except in the case of the cationic corn starch, Cato-2) is nearly always counter-productive; likely due to blockage of hydroxyl sites otherwise available for hydrogen bonding to the surface. This is typified by the inferiority of the substituted guar gums to the unmodified form Jaguar MDD (Figure 13). The substitution of the carboxymethyl for hydroxyl groups, as in CMC, hinders adsorption and, accordingly, produces a poor pyrite depressant. On the other hand, their replacement with other hydroxyl containing groups, as in the HEC, Natrosol 150 GXR and Gum Ethylex 3095, results in good pyrite depressants (Figure 14).

It is unfortunate that the β/cis found in the guars not only depresses pyrite but also appears to strongly depress coal. More research is clearly indicated in this area.

As previously discussed the poor response of certain xanthates as pyrite depressants is also directly related to the influence of OH^- group configuration and, hence, availability to be xanthated.

SUMMARY AND CONCLUSIONS

Based on the adsorption and flotation experiments made with a series of natural, modified and synthetic polymers, the following conclusions may be drawn:

1. The adsorption of a polymer onto pyrite generally correlates with its flotation performance as a pyrite depressant, i.e., a strongly adsorbed polymer is usually a superior pyrite depressant. Unfortunately, some of the strongly adsorbed polymers such as the guars, Irish moss and dextrin also seriously depress coal.

2. Generally, pH 7 was found to be the best for adsorption of the polymers studied here onto pyrite. It was also found to be the best pH for coal flotation with accompanying good pyrite rejection in the plurality of the cases, with pH 5 next best and pH 9 the poorest. As a crude generalization, coal was less depressed by these polymers at pH 5 than at higher pH values.

3. The adsorption of these polymers is generally much stronger onto pyrite than onto coal.

4. Xanthating some of the OH^- groups in the polymeric compounds sometimes produces a superior pyrite depressant. However, there is a wide variation in pyrite depressant performance from compound to compound, and xanthating is no panacea.

5. Evidence was found for three different mechanisms for the attachment of the polymers onto coal: (1) electrostatic attraction of charged polymers, (2) specific adsorption through selective active groups and (3) hydrogen bonding through hydroxyl groups. All of these appear to be operating in one or more of the systems studied.

6. Specific adsorption appears to occur between iron at the pyrite surface for both polysaccharides with introduced xanthate groups and for hydroxyl groups naturally present in the polymers. It is believed that specific adsorption of a polysaccharide-iron complex and similar groups account for the superior adsorption and flotation generally found at pH 7.

7. Superior adsorption and pyrite depression by starches is correlated with the straight chain, amylose content of the starch. The high amylose containing starch, Hylon VII was found to be a superior polysaccharide for pyrite depression in either the xanthated or unxanthated form.

8. Superior reagents appear to be those with readily available hydroxyl groups (numerically and sterically). If an OH group in a polysaccharide is replaced by another group (as in carboxymethyl cellulose) it is an inferior depressant whereas if it is replaced by another and more readily accessible OH group (as in hydroxyethyl cellulose or hydroxyethyl ether starches) it may produce a superior pyrite depressant.

9. Fragmentary evidence indicates that the β/cis configuration (as in the guars) is more strongly adsorbed onto pyrite than is the α/trans (starches) or β/trans (cellulose) form. Unfortunately, the guars also seriously depress coal. Because of the pyrite depression potential of these compounds, this should be a fruitful line of research.

10. Ore pyrite responds quite differently in its adsorption of these compounds than does pyrite from a coal source. This factor should be kept in mind when studying pyrite depression in a coal flotation system.

ACKNOWLEDGEMENTS

The authors wish to thank the Penn State Cooperative Program in Coal Research, Dr. William Spackman, Director for providing financial assistance for this work. Thanks are also due to the various companies, named in the text, for supplying the polymers, and to U.S. Steel Corporation for furnishing coal and pyrite samples.

REFERENCES

1 F. F. Aplan, Use of the Flotation Process for Desulfurization of Coal, in T. D. Wheelock (Ed.), Coal Desulfurization, ACS Symposium Series No. 64, ACS, Washington, D.C., 1977.
2 D. J. Brown, Coal Flotation in D. W. Fuerstenau (Ed.), Froth Flotation - 50th Anniversary Volume, AIME, New York, 1962, p. 518.
3 F. F. Aplan, Coal Flotation in M. C. Fuerstenau (Ed.), Flotation - A. M. Gaudin Memorial Volume, AIME, New York, 1976, pp.1235.
4 R. C. Rastogi and F. F. Aplan, Coal Flotation as a Rate Process, Mineral and Metallurgical Processing (Trans. AIME), in press.
5 H. F. Yancey and J. A. Taylor, USBM RI 3263, 1935.
6 F. J. Miller, Trans. AIME, 229 (1964) p. 7.
7 F. J. Chernowsky and F. M. Lyon, Trans. AIME, 252 (1972) p. 11.
8 V. Choudhry, M.S. Thesis, The Pennsylvania State University, 1979.
9 L. C. Newcomer, An Evaluation of Pyrite Depressants in the Froth Flotation of Coal, Senior research project, Metal 410, The Pennsylvania State University, 1977.
10 P. B. Bradley, V. Choudhry, L. C. Newcomer and F. F. Aplan, Pyrite Depression During Coal Flotation - Part II: Polysaccharides, Xanthated Polysaccharides and Related Compounds, manuscript in preparation.
11 P. Vojislov, U.S. Patent 4,211,642, 1980.
12 D. J. Nagaraj, S. S. Wang, P. V. Avotins and E. C. Dowling, "Structure-Activity Relationship for Copper Depressants", submitted to Trans. IMM (London).
13 C. J. Im and F. F. Aplan, Depression of Coal by Starch and Starch Derivatives in A. R. Swanson (Ed.), 1st Australian Coal Prep. Cong., Newly and Beath Printers, Pty, Broadmeadow, N.S.W. Australia, 1981.
14 R. L. Whistler and E. F. Paschall (Eds.), Starch: Chemistry and Technology, Vols. 1 and 2, Academic Press, New York, 1965.
15 R. L. Whistler (Ed.), Industrial Gums, 2nd Ed., Academic Press, New York, 1973.
16 M. Dubois, et al., Analyt. Chem. 27 (1956) p. 196.
17 R. W. Perry, M.S. Thesis, The Pennsylvania State University, 1982.
18 K. L. Sutherland and I. W. Wark, Principles of Flotation, Australasian Inst. Min. and Met., Melbourne, 1955.
19 J. R. Herb, Chem. Abst. 53 (1959) 13517.
20 M. W. Rutenberg, Starch and Its Modifications in R. L. Davidson (Ed.), Handbook of Water Soluble Gums and Resins, McGraw-Hill, New York, 1980.
21 R. W. Perry and F. F. Aplan, Adsorption of Polysaccharides and Related Polymers onto Coal, manuscript in preparation.
22 C. J. Im and F. F. Aplan, Coal Depression by Polysaccharides and Other Polymers, manuscript in preparation.
23 C. J. Im and F. F. Aplan, The Use of Coal Depressants for Pyrite Removal From Coal, manuscript in preparation.

FINE GRINDING AND FLOTATION TO DESULFURIZE COAL

Kenneth J. Miller, Coal Preparation Division, Pittsburgh Energy Technology Center, U.S. Department of Energy

ABSTRACT

Because of the growing need to develop new technology for producing superior quality coals to replace oil and natural gas, the U.S. Department of Energy's Coal Preparation Division is looking at fine grinding and multistage flotation processes to maximize the liberation and removal of sulfur-bearing and ash-forming mineral matter from coal.

The beneficiation of micronized coal by froth flotation has long been considered unsatisfactory because of poor product quality and/or poor recovery. However, our current work suggests that superclean products of high yield can be obtained by a combination of coarse coal cleaning or rougher flotation, followed by wet grinding and recleaning by second-stage coal flotation. For example, by a combination of rougher flotation, wet grinding of the rougher concentrate, and two-stage flotation of the wet-ground product, an Upper Freeport bed coal sample was cleaned from 26% ash and 1.1% pyritic sulfur to 4.5% ash and 0.1% pyritic sulfur.

INTRODUCTION

Despite the fact that coal flotation was first used in a U.S. coal preparation plant more than 50 years ago, very little effort has been expended by the coal industry since then to improve the process. Potential benefits of recleaning by two- and three-stage flotation, classifying and splitting the feed, or desliming have seldom been exploited. However, in defense of this apparent indifference about flotation in commercial preparation plants, it must be remembered that the process was first used to clarify the plant recirculating water--not to recover coal for utilization--and any fine coal recovered adversely affected the moisture content and handling characteristics of the plant product. Furthermore, because of the efficiency and economy of coarse coal specific gravity separation processes, flotation was used to recover only those fines created inadvertantly in mining and handling operations--usually less than 15% of the plant product. Thus there was little economic pressure to optimize the flotation process. This is no longer the case. Mining costs have risen dramatically, and new mining and handling methods are producing more fine coal. To deliberately discard this fraction would not only waste coal but also divert additional tonnage to an already capital- and labor-intensive section of the plant--the fine refuse dewatering and disposal.

In addition to these immediate reasons for improving the efficiency and economy of coal flotation, there is the growing interest in producing superior quality coals that can replace oil and natural gas in both new and retrofit applications, and the need for better desulfurization of all U.S. coals to meet stringent EPA sulfur emission standards. To achieve these latter objectives, it is necessary to look more closely at fine grinding followed by flotation or other processes that utilize the surface/interfacial properties of coal. It is with this objective that we have undertaken the work discussed here.

EXPERIMENTAL WORK

The Upper Freeport bed coal used in this work was obtained from Helvetia Coal Company's Lucerne No. 6 mine in Indiana County, Pennsylvania. The sample was dried in our laboratory and crushed to 28-mesh top size through a jaw crusher and a hammer mill. The crushed coal was split into representative 200-g samples for laboratory flotation and fine grinding work. The 200-g samples were stored under argon to prevent oxidation and removed as needed for experiments. A size analysis of the 28-mesh by 0 coal is given in Table 1, and a washability or float-sink analysis is given in Table 2.

TABLE 1
Wet Screen Size Analysis of the 28-Mesh by 0 Upper Freeport Bed Coal.

Size, Mesh	Direct, %	Cumulative, %
28-48	9.9	9.9
48-100	26.5	36.4
100-200	20.7	57.1
200-400	11.9	69.0
400-0	31.0	100.0

TABLE 2
Centrifugal Float-Sink Analysis of 28-Mesh by 0 Upper Freeport Bed Coal.

	Direct, %				Cumulative, %			
	Wt	Ash	Pyritic Sulfur	Total Sulfur	Wt	Ash	Pyritic Sulfur	Total Sulfur
Float-1.27	23.7	1.2	0.03	0.58	23.7	1.2	0.03	0.58
1.27-1.30	16.0	2.9	0.10	0.65	39.7	1.9	0.06	0.61
1.30-1.40	20.2	9.2	0.26	0.79	59.9	4.4	0.13	0.67
1.40-1.60	11.8	22.4	0.52	0.91	71.7	7.3	0.18	0.71
1.60-Sink	28.3	67.1	3.04	3.34	100.0	24.2	0.99	1.45

Fine Grinding

Recent work aimed at determining the best methods for preparing ultrafine feed for flotation has demonstrated that wet grinding is clearly superior to dry grinding in terms of high-ash mineral matter rejection [1,2]. This difference in flotation properties between dry- and wet-ground coal is difficult to interpret or explain with existing knowledge of the coal flotation system. However, a number of possible reasons have been suggested. Among these are the following:

1. Differences in particle size distribution and mineral matter association.
2. Differences in surface roughness.
3. Different degrees of slime coating caused by dry and wet grinding.
4. Oxidation of coal particle surfaces due to dry grinding.
5. Hydration of coal particle surfaces due to wet grinding.

Because of the demonstrated superiority of wet grinding over dry grinding, the coal used in this investigation was always wet ground. A vibratory ball mill was used for the grinding. The particular mill used, the Megapot Vibratory Pot Mill* (Figure 1), is a small batch unit capable of handling laboratory samples of a few hundred grams or less. The unit consists of a vibrating chassis mounted by means of a flexible suspension system to a rigid steel base. The chassis oscillates at about 3,400 cycles/minute with an amplitude of 2 mm. Polyethylene "pots" (jars) containing the sample and the grinding media are held firmly in place in the chassis by means of two fixed bars on the bottom and by an adjustable bar on the top.

*Reference in this report to any specific commercial product, process, or service is to facilitate understanding and does not necessarily imply its endorsement or favoring by the United States Department of Energy.

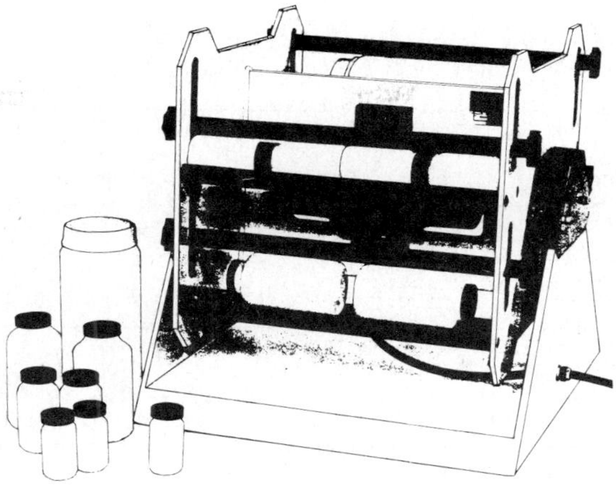

Fig. 1. Megapot Vibratory Pot Mill

For the particular grinding work discussed in this report, 3/8-inch stainless steel balls were used as the grinding medium. Grinding times ranged from 10 minutes to 30 minutes, and grinding was applied to the rougher (first-stage) flotation concentrate only. The raw 28-mesh by 0 coal was not ground prior to first-stage flotation. The size consist of the coal particles obtained by grinding the 28-mesh by 0 concentrate for the various time periods is shown in Table 3.

TABLE 3
Size Analysis of the Vibratory Ball Mill Product After Grinding for Various Lengths of Time.

Grinding Time, Minutes	10	20	30
Size, μm	Percent of Particles Finer		
125	100	100	100
88	99	100	100
62	91	96	100
44	84	94	100
31	69	85	94
22	56	73	84
16	44	60	71
11	32	46	57
5.5	15	23	30
3.9	9	13	18
2.8	3	6	8
Mean Particle Diameter	26 μm	18 μm	13 μm

Table 4 shows the degree of liberation of pyrite particles obtained by grinding for the different lengths of time. The data show a great increase in the number of free pyrite particles as a result of grinding. That is, the number of free particles increased from about 23% in the 28-mesh by 0 material to over 62% in the nominal 400-mesh by 0 material produced by wet grinding for 30 minutes. Conversely, the number of locked pyrite particles decreased from 43% to 5%.

TABLE 4
Degree of Pyrite Particle Liberation Resulting from Wet Grinding for Various Lengths of Time, as Determined by Microscopic Image Analysis.

Grinding Time, Minutes	0	10	20	30
Mode of Occurrence	Percent of Total Particles Counted			
Locked	42.8	10.7	7.7	5.0
Semi-Locked	34.6	42.7	29.7	32.7
Free	22.6	46.6	62.6	62.3
	100.0	100.0	100.0	100.0

It must be pointed out that virtually all of the pyrite particles viewed and counted were smaller than 10 µm in diameter -- 91% in the 28-mesh by 0 sample and over 99% in the 400-mesh by 0 sample. Most coarser pyrite particles in the raw coal were removed in the primary or rougher flotation carried out before cleaner flotation or grinding.

The relatively simple and largely manual approach used here for counting the pyrite particles and determining their mode of occurrence is preliminary in nature. Other more quantitative and sophisticated approaches are being evaluated by us and by others at this time. The purpose of providing these preliminary results is to demonstrate the potential usefulness of microscopic image analysis as an aid in fine grinding and desulfurization of ultrafine coal.

Flotation Procedure

All flotation tests were conducted in a DOE designed [3] laboratory flotation machine fitted with a 4-liter plastic flotation cell. The cell is equipped with a liquid level controller and motor-driven froth scraper paddles that operate through an automatic timing switch. A diagram of the cell and ancillary equipment is shown in Figure 2.

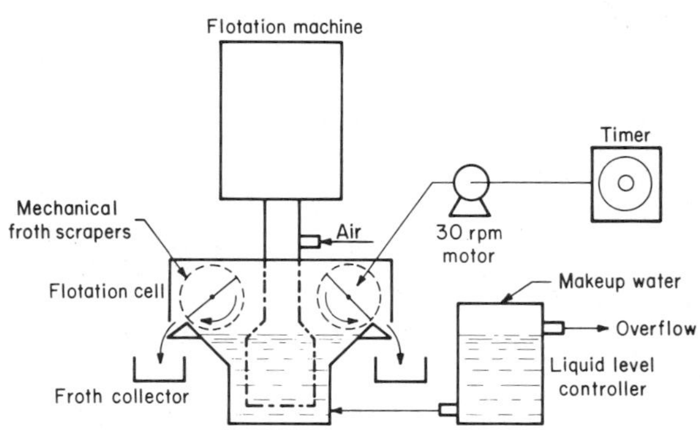

Fig. 2. Laboratory batch flotation unit with mechanical froth scrapers.

In all flotation tests, the initial feed solids concentration was 5% (200 grams of coal in 4 liters of water). The only reagent used was the frother MIBC. The MIBC was conditioned with the slurry in the operating flotation machine for 30 seconds before the cell aeration was turned on and froth scraping was started. The froth scraping time was always set at 1.5 minutes.

Flotation Test Results

Initially, three-stage rougher-and-cleaner type flotation tests were run with the 28-mesh by 0 coal sample at several different MIBC reagent addition levels to obtain a typical grade/recovery relationship that might be expected from more or less conventional flotation. Then for comparison, similar flotation tests were run in which the first-stage froth product was wet ground in the vibratory ball mill for 30 minutes before second- and third-stage flotation. Results of several flotation tests are given in Table 5.

In the first flotation test with the 28-mesh by 0 coal, the wt% recovery at the end of the three stages of flotation was 51.6%, and the ash and pyritic sulfur contents of the product were 6.3% and 0.31%, respectively. In the second test, where no MIBC was added in the third stage, the final wt% recovery was only 30.9%, but the product ash and pyritic sulfur content were down to 5.7% and 0.20%, respectively. While these tests with the 28-mesh by 0 coal show some improvement in product quality (at the expense of yield), when no MIBC frother was used in the third stage, the results are not as good as those obtained with the wet-ground nominal minus-400-mesh material. The greater liberation of mineral matter, afforded by fine grinding for 30 minutes in the vibratory ball mill, apparently provides an opportunity for greater sulfur and ash reduction. For example, at 36.4-wt% yield, the wet-ground coal had 4.5% ash and only 0.12% pyritic sulfur.

Figures 3 and 4 compare the flotation test results with the centrifugal float-sink analysis of the 28-mesh by 0 coal on a pyritic sulfur versus yield and an ash versus yield basis, respectively. In both graphs, the slope of the curve obtained with the minus-400-mesh material suggests that this finely ground coal can be beneficiated to a greater extent than can the minus-28-mesh material.

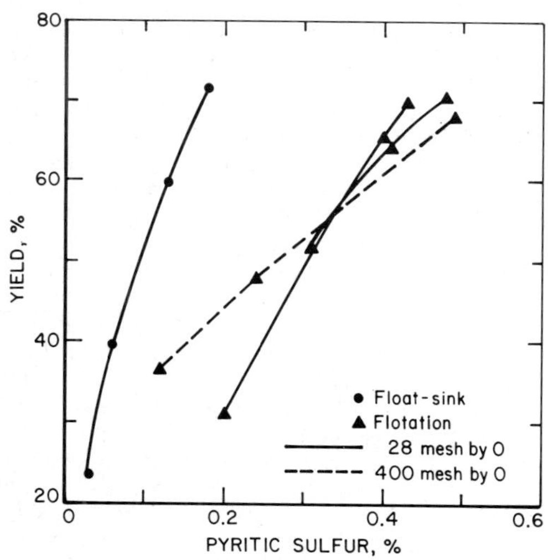

Fig. 3. Comparison of centrifugal float-sink and flotation on product pyritic sulfur content and yield.

SUMMARY AND CONCLUSIONS

Our coal flotation work at this time is directed toward fine grinding to obtain maximum liberation of ash-forming and sulfur-bearing mineral matter, and then applying the flotation procedures deemed most effective for producing the cleanest possible product. The difficulty of floating ultra-fine particles has long been recognized, and much has been written about this subject over the years. In view of the fact that coal flotation selectivity and flotation rate--as well as dewatering--are adversely affected by the presence of extreme fines, it would seem most prudent to grind no finer than

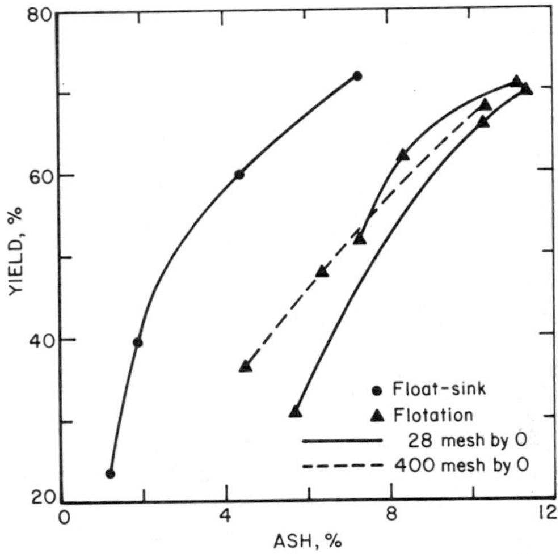

Fig. 4. Comparison of centrifugal float-sink and flotation on product ash content and yield.

is absolutely necessary to obtain adequate liberation of the mineral matter. For this reason, it is most likely that fine grinding and deep cleaning of ultrafine coals will be applied as secondary measures after most extraneous mineral matter is removed by more or less conventional coal-cleaning methods.

Preparation plant flow sheets developed to maximize sulfur reduction will likely have a precleaning circuit where most of the high-ash mineral matter will be removed. It might be possible with some coals to scalp off a high-quality coarse-size clean coal fraction before crushing the remaining coal for better mineral matter liberation. This would reduce the size of the fine coal beneficiation circuit with minimal effect on the final product quality or recovery. Such a flow sheet is shown in Figure 5 and was discussed by Maronde and Deurbrouck in a recent report [4].

Float-sink tests of fine (28-mesh by 0) coal continue to show vast superiority over laboratory flotation in terms of product quality versus yield. Apparently, even if mineral matter liberation is achieved by fine grinding, the flotation process alone is not sensitive enough to take advantage of it. Continued effort must be directed toward improving the efficiency of the flotation process as it applies to ultrafine coal by such means as improved flotation cell design, better air bubble generating and dispersion methods, more selective reagents, and better reagent mixing and conditioning methods. Also, we need a better understanding of the mechanism of coal flotation, especially as it applies to ultrafine wet-ground coals, and a better understanding of the effect of different grinding methods on fine coal surface characteristics.

In addition, work is continuing in an effort to improve the selectivity of the DOE-developed "reverse" coal-pyrite flotation process [5]. As shown in an earlier report [6], the aforementioned process has never been effectively applied to coal with a top size of less than about 35 mesh, nor with a feed containing more than about 30% minus-325-mesh material.

TABLE 5
Laboratory Flotation Test Results with Upper Freeport Bed Coal.

	Direct, %				Cumulative, %				MIBC Addition, lb/ton		
	Wt	Ash	Pyritic Sulfur	Total Sulfur	Wt	Ash	Pyritic Sulfur	Total Sulfur	First Stage	Second Stage	Third Stage
28-Mesh by 0											
Froth Concentrate	51.6	6.3	0.31	0.84	51.6	6.3	0.31	0.84	0.3	0.3	0.1
Tailings 3	12.6	17.2	0.80	1.59	64.2	8.4	0.41	0.99			
Tailings 2	6.3	30.9	1.21	1.97	70.5	10.4	0.48	1.08			
Tailings 1	29.5	58.9	2.20	2.45	100.0	24.7	0.99	1.48			
Froth Concentrate	30.9	5.7	0.20	0.81	30.9	5.7	0.20	0.81	0.3	0.2	---
Tailings 3	34.5	14.4	0.57	1.24	65.4	10.3	0.40	1.04			
Tailings 2	4.4	27.4	0.95	1.63	69.8	11.4	0.43	1.07			
Tailings 1	30.2	57.2	1.86	2.53	100.0	25.2	0.86	1.51			
400-Mesh by 0*											
Froth Concentrate	36.4	4.5	0.12	0.64	36.4	4.5	0.12	0.64	0.3	---	---
Tailings 3	11.4	12.4	0.64	1.09	47.8	6.4	0.24	0.75			
Tailings 2	20.2	20.0	1.07	1.62	68.0	10.4	0.49	1.01			
Tailings 1	32.0	59.6	2.50	2.68	100.0	26.2	1.13	1.54			

*First stage carried out with 28-mesh by 0 coal; first-stage froth concentrate wet ground for second- and third-stage flotation.

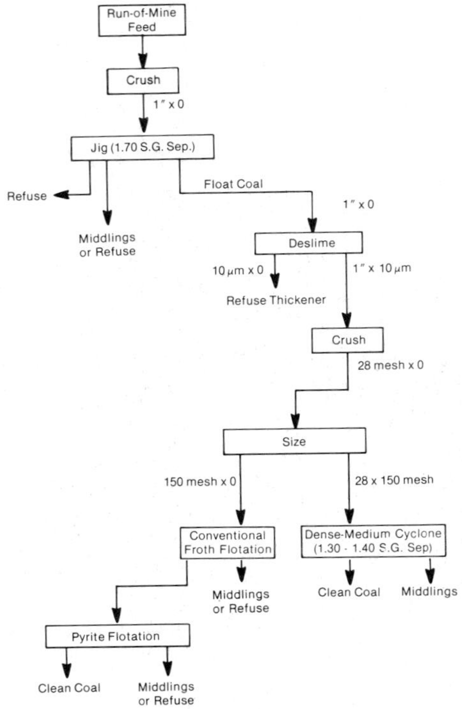

Fig. 5. Advanced Coal Cleaning Circuit

REFERENCES

1. K.J. Miller and R.F. Guzzo, Wet Grinding vs. Dry Grinding for Deep Cleaning Ultrafine Coal by Flotation, DOE/PETC/TR-84/10, September 1984, 11 pp.
2. K.J. Miller, Investigation of Fine Grinding and Flotation to Produce Superclean Coal, 10th Annual Powder and Bulk Solids Conference/Exposition, in Proceedings of the Technical Program, May 7-9, 1985, pp. 861-866.
3. K.J. Miller, Evaluation of Collector Addition in the Flotation of Various U.S. Coals, DOE Report of Investigations PMTC-7(80), January 1980, 29 pp.
4. C.P. Maronde and A.W. Deurbrouck, Sulfur Reduction Potential of the Coals of the Northern Appalachian Region, in Coal Prep 85 Conference Papers, Lexington, Ky., April 30 - May 2, 1985, pp. 16-21.
5. K.J. Miller and A.W. Deurbrouck, Froth Flotation to Desulfurize Coal, in Physical Cleaning of Coal: Present and Developing Methods, Marcel Dekker, Inc., edited by Y.A. Liu, 1982, pp. 255-291.
6. R.E. Hucko and K.J. Miller, A Technical Performance Comparison of Coal-Pyrite Flotation and High-Gradient Magnetic Separation, RI-PMTC-10(80), November 1980, 41 pp.

DESULFURIZING FLOTATION OF EASTERN CANADIAN HIGH-SULFUR COAL

J. LASKOWSKI[1], M. BUSTIN[2], K.S. MOON[3], and L.L. SIROIS[3]

[1]Mining and Mineral Process Engineering Dept., University of British Columbia, 6350 Stores Rd., Vancouver, B.C., V6T 1W5 (Canada)

[2]Geological Sciences Dept., University of British Columbia, Vancouver, B.C. (Canada)

[3]CANMET Mineral Sciences Laboratories, Ottawa, Ontario (Canada)

ABSTRACT

Prince Mine coal contains about 3% pyritic sulfur. The pyrite is very finely disseminated, and very fine grinding (below 400 mesh) is necessary for liberation. A flotation process has been developed which utilizes carboxymethyl cellulose, kerosene and MIBC under alkaline conditions. The process provides clean coal concentrates that contain one-half percent of pyritic sulfur at high coal recoveries. Pyrite retained in the clean coal under such conditions is mainly non-liberated, thus proving the high selectivity of the developed flotation procedure.

INTRODUCTION

In recent years the emphasis in coal preparation has been placed more on reducing the sulfur content of coal than on ash, and on recovering the combustible material. Of the two types of sulfur in coal, inorganic and organic, only the inorganic sulfur, represented by pyrite, may be separated from coal by physical methods.

Large differences in the density of coal (1.2 - 1.6 g/cm^3) and pyrite (about 5.0 g/cm^3) provides the basis for efficient separation of pyrite from coal by gravity separation methods. Although these methods are very efficient in separating coarse coal and pyrite particles, their efficiency decreases drastically with decreasing size of particles (ref.1) and, because sufficient pyrite liberation often requires fine grinding, flotation becomes a major separation process.

The floatability of coal-pyrite is somewhat different from that of ore-pyrite and its separation from coal presents a difficult problem. Thus, research based on the behaviour of ore-pyrite in the flotation process, (namely its poor floatability under alkaline conditions), has not resulted in the development of an effective coal-pyrite selective flotation process.

A two-stage reverse flotation process has proven to be more successful (refs. 2-4). A useful review of the state-of-art in this area was recently published by Miller and Deurbrouck (ref.5).

In the two-stage reverse flotation process, the first stage is conventional flotation with the froth product comprising both coal and pyrite. This product, reconditioned with dextrin which depresses coal in the second stage, is floated with xanthates which allow the removal of pyrite.

The two-stage reverse flotation process is based on the selective adsorption of dextrin on hydrophobic particles through hydrophobic interactions (refs.4,6). Another reagent, carboxymethyl cellulose (CMC), has been extensively tested in the selective flocculation of coal slimes (ref.7,8) and, recently, also in the flotation of inherently hydrophobic solids such as graphite (ref.9) and coal (ref.10). CMC, although chemically similar to dextrin, was shown to behave quite differently at the hydrophobic solid/aqueous solution interface. Its adsorption onto graphite was revealed to proceed through inorganic impurities (Ca, Fe) on graphite which serve as adsorption centres. Therefore, while CMC adsorption on the graphite purified by leaching was very low and did not cause any depression, the floatability of natural Ceylon graphite samples were strongly depressed with CMC (ref.9).

Oil agglomeration, which is a possible alternative to coal flotation in cleaning fine coal, is unable to separate pyrite selectively from coal. Wheelock et al (ref.11) tested pretreatment in 2% Na_2CO_3 solutions at 50-80°C before oil agglomeration to enhance the selectivity of the pyrite removal. They also showed that the selectivity of oil agglomeration is improved at alkaline pH's.

In the 1970's, Krukiewicz and Laskowski (ref.12) described the magnetizing alkali leaching process. They showed that the magnetic susceptibility of siderite particles may be increased by pretreatment in alkaline solutions which facilitates subsequent magnetic separation of siderite ores. The same concept has been further tested by Hall and Finch (ref.13) who have shown that similar pretreatment also improves magnetic separation of pyrite from coal.

During the last three years, CANMET has sponsored a research program at the University of British Columbia in which the use of various modifying agents in the desulfurizing flotation of coal has been studied. The results of the pyrite liberation and flotation tests obtained with a high sulfur coal from the Prince Mine (Nova Scotia, Canada) are presented in this paper.

EXPERIMENTAL

Material

The chemical composition of coal from the Prince Mine used in the experiments is summarized in Table 1.

The preparation of the sample for flotation experiments is depicted in Fig. 1. Wet grinding in a laboratory rod mill for 21 minutes provided material 90% below 38 μm (400 mesh). Fig. 2 shows the effect of longer grinding times on the particle size distribution as determined with ELZONE Apparatus. Mean content of

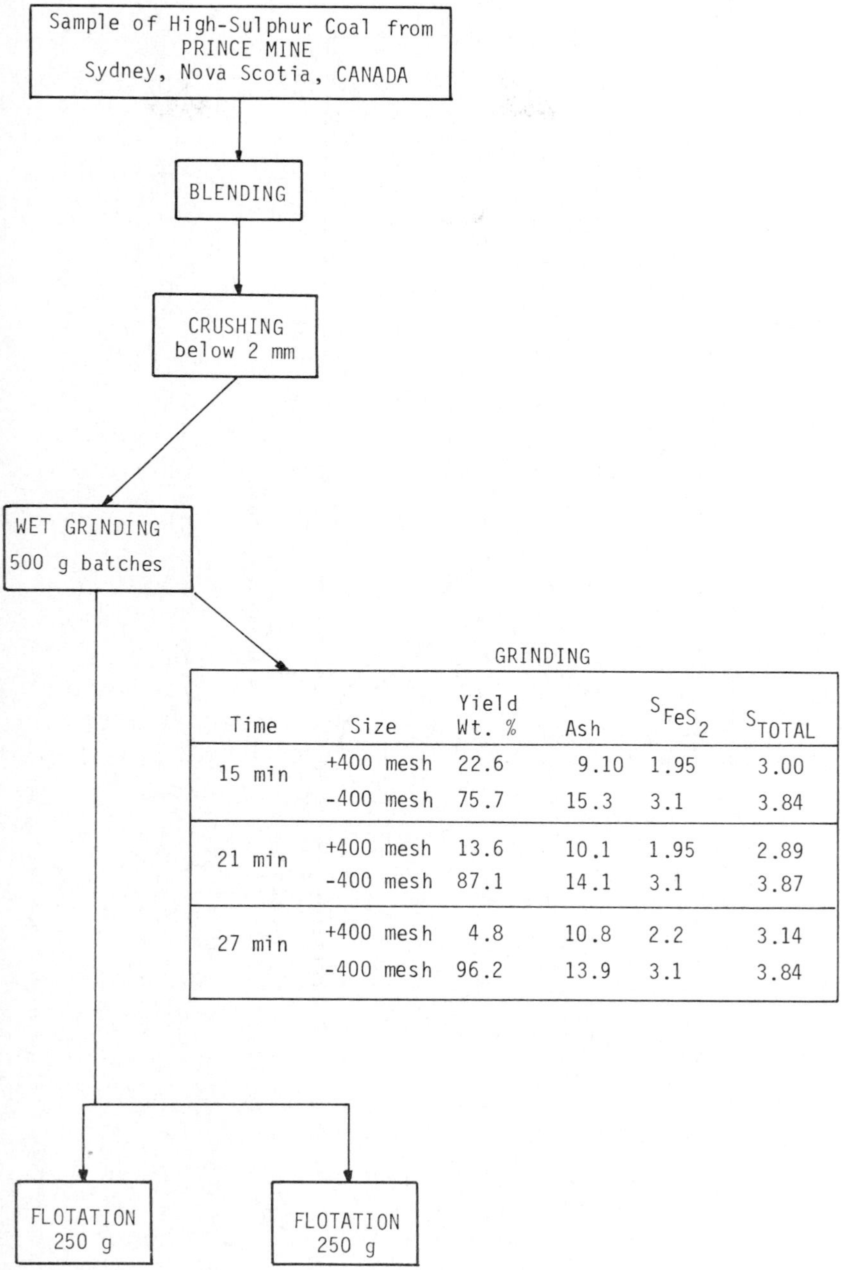

GRINDING

Time	Size	Yield Wt. %	Ash	S_{FeS_2}	S_{TOTAL}
15 min	+400 mesh	22.6	9.10	1.95	3.00
	-400 mesh	75.7	15.3	3.1	3.84
21 min	+400 mesh	13.6	10.1	1.95	2.89
	-400 mesh	87.1	14.1	3.1	3.87
27 min	+400 mesh	4.8	10.8	2.2	3.14
	-400 mesh	96.2	13.9	3.1	3.84

Fig. 1. Coal preparation procedure for flotation experiments.

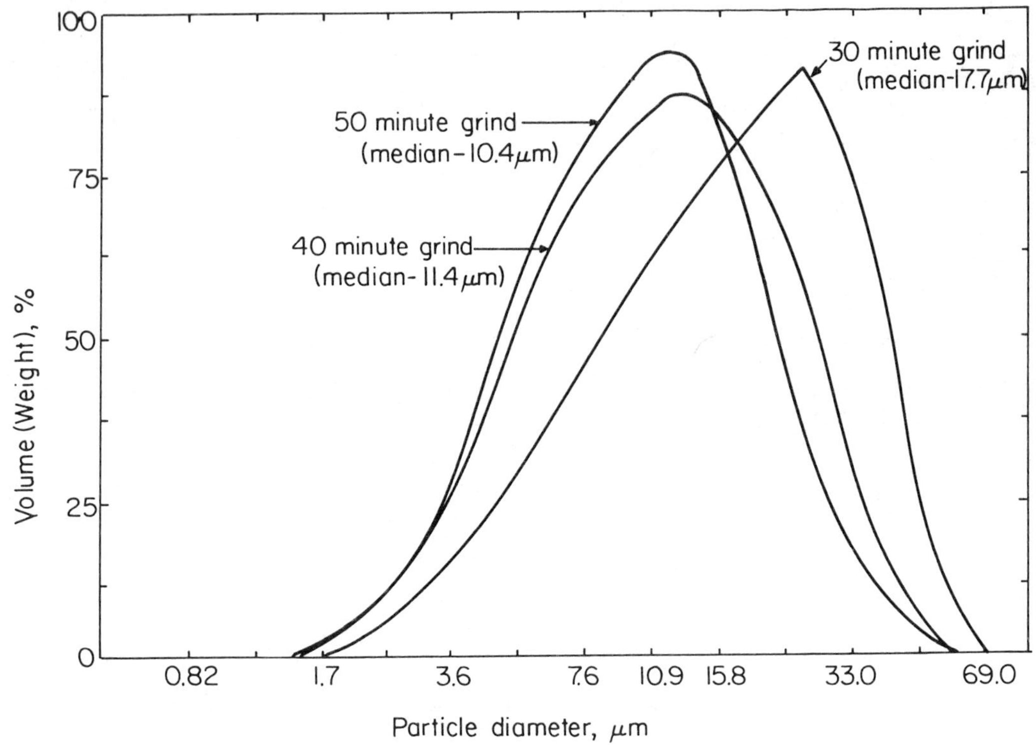

Fig. 2. Particle size distributions of coal sample showing variation with grinding time.

pyritic sulfur in the raw coal calculated from the data given in Fig. 1 is 3% while the total sulfur content is in the range from 3.6 to 3.8%.

TABLE 1

Chemical composition of coal from the Prince Mine, Nova Scotia, Canada

Proximate Analysis:	
Moisture	1.64%
Volatiles (dry)	28.1 %
Ash (dry)	13.2 %
Fixed Carbon (dry)	58.7 %
Volatiles (dmmf)	32.37%
Fixed Carbon (dmmf)	67.62%
Pyritic Sulfur	2.76%
Total Sulfur	3.7 %

Pyritic sulfur was determined according to Standard ASTM procedures. The samples were first extracted with dilute hydrochloric acid, water-washed, and the pyrite was leached from the insoluble residue (remaining after the HCl treatment) with dilute nitric acid. The iron extracted into this portion of the solution was determined by atomic adsorption and the pyritic sulfur was calculated as a stoichiometric combination with the iron.

The Eschka analytical procedure was used to determine the total sulfur of the coal.

Pyrite occurrence and liberation

Methods. Representative raw coal samples and coal samples from each phase of the flotation process were formed into pellets using a transoptic powder and polished using standard procedures with a final grit size of 0.05 μm. The size distribution of pyrite in each sample was determined using image analysis. At least 1,000 particles were measured per sample and population parameters determined. The extent of liberation of the pyrite in each sample was determined by point count analysis of at least 250 pyrite particles per sample using standard coal petrographic techniques (ref.14). For descriptive purposes, four classes of liberation were differentiated:

1. Completely liberated pyrite (pyrite distinctly separated from organics or pyrite comprising greater than 80% of the particle);

2. Greater than 50% liberated, in which the pyrite particle comprises greater than 50% but less than 80% of the particle, and/or the pyrite comprises a significant portion of the surface area of the particle;

3. 20-50% liberated (pyrite is partially bound and/or comprises greater than

20% but less than 50% of particle; and

4. Bound pyrite (pyrite completely encapsulated in organics with the organics comprising greater than 80% of the particle.

The different classes of liberation are illustrated in Fig. 3. The distinction between the classes 20 to 50% liberated, and greater than 50% liberated, are in part subjective but nevertheless within a sequence of samples (such as examined in this study) providing additional data on the success of liberation.

Mode of occurrence of pyrite. The size distribution of the pyrite in the raw coal is shown diagrammatically in Fig. 4. One hundred percent of pyrite in raw coal is non-liberated, as summarized in Table 2.

TABLE 2

Pyrite particles in the raw coal

Sample	Degree of liberation, % (completely bound)	Particle Size, μm		
		Mean	Standard Deviation	Range
Raw 1	100	4.64	5.24	0-21
Raw 2	100	3.38	2.19	0- 9
Raw 3	100	9.50	8.58	0-28
Raw 4	100	1.85	1.36	0- 5
Raw 5	100	7.69	5.80	0-26
Raw 6	100	3.35	1.69	0- 6

The mode of occurrence and size distribution of pyrite in the raw coal (untreated) is highly variable. Analysis of a number of sub-samples indicates that the mean diameter of pyrite in different sub-samples ranges from 1.85 μm to 9.50 μm and the maximum particle size range within a sub-sample is less than 1 μm to 28 μm (Table 2). Some samples have a distinct bimodal distribution with a coarse and fine mode whereas others are unimodal but are distinctly skewed towards the coarser end. The coarser pyrite particles (greater than 5 μm) consist of massive subhedral to anhedral forms. The fine fraction (less than 5 μm) includes two distinct forms:

1. framboidal pyrite that ranges from 1 to 5 μm in diameter; and
2. fine singular particles of pyrite that range in size from less than 1 μm to 5 μm in diameter.

All forms of pyrite are predominantly (greater than 95%) associated with and are intergrown with the various forms of vitrinite. In some vitrinite particles, pyrite infills cell lumens but in most particles the pyrite is randomly distributed. Pyrite is rarely associated with liptinite, inertinite or semi-inertinite macerals.

Grinding the samples for flotation feed had a marked effect on the size distribution and degree of liberation of the pyrite. Some notable observations are:

1. Pyrite in most samples not crushed below 53 μm for the most part has a

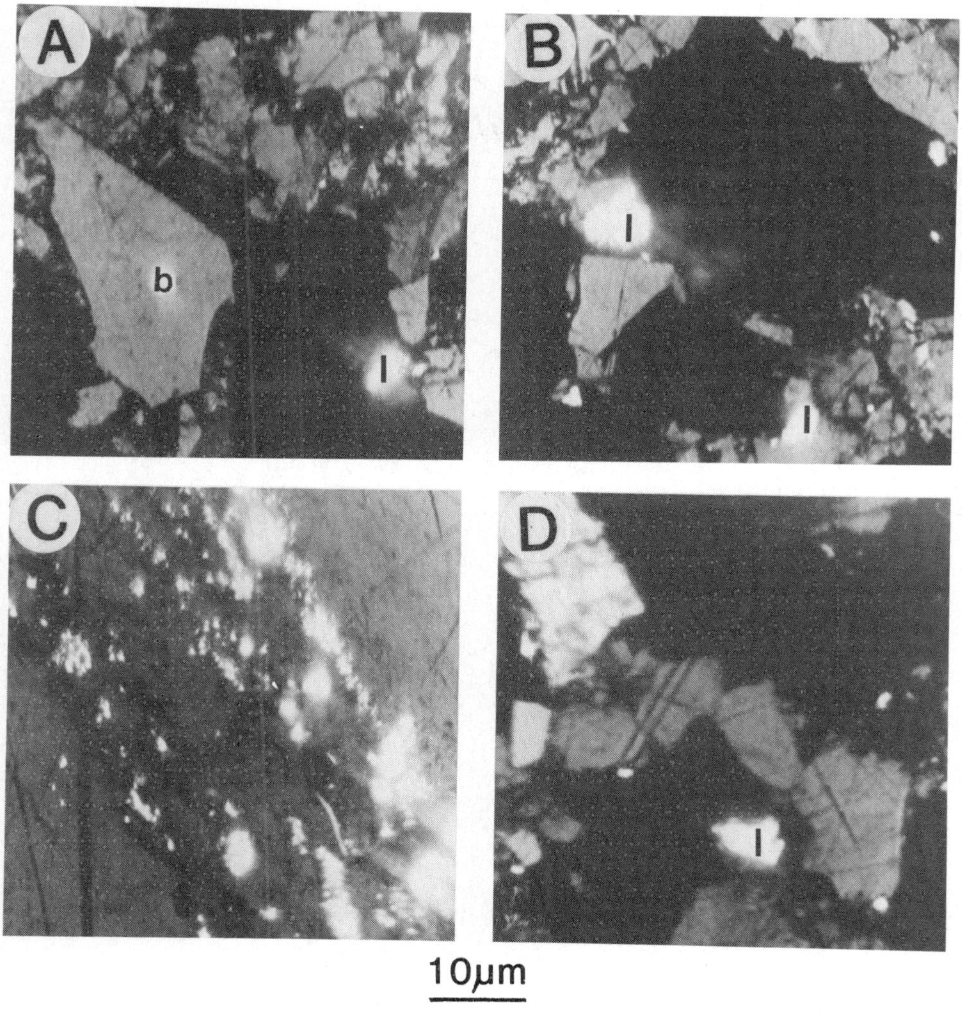

FIGURE 3.
A. Example of completely bound pyrite (b) and completely liberated pyrite (l) from Sample number 6. (See Fig. 9)

B. Examples of partially bound pyrite. In the upper left (l) the pyrite is classified as 20-50% liberated whereas the sample in centre right (l) is greater than 50% liberated. Sample number 6.

C. Framboidal and subhedral pyrite completely bound in vitrinite and liptinite. Note the pyrite occurs along distinct laminae in the organic matter. Sample number 3.

D. Subhedral completely liberated pyrite (l) in Sample number 6.

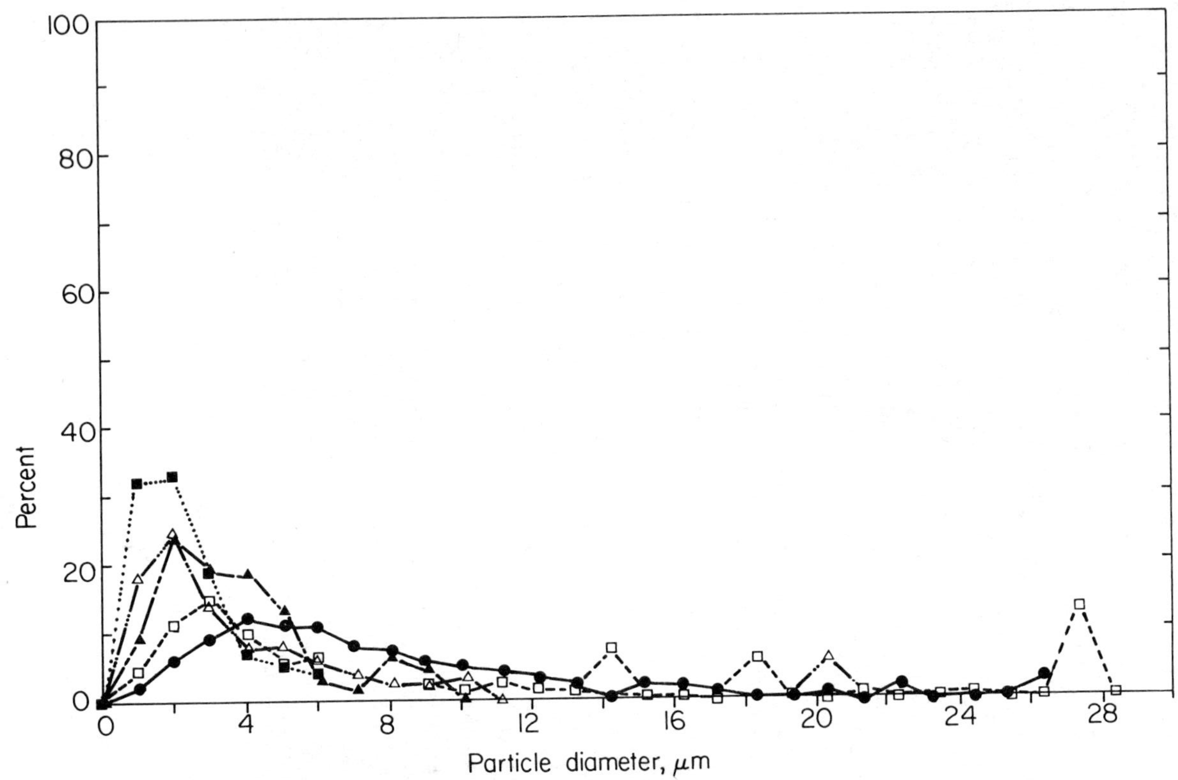

Fig. 4. Pyrite particle size distribution in the raw coal.

bimodal distribution, inherited from the raw coal;

2. Samples crushed below 53 μm for the most part have a unimodal distribution with a mode between 3 and 7 μm. The pyrite is poorly sorted.

3. The amount of liberated pyrite increases with increased time of grinding. The coarse pyrite fraction is liberated more readily (less grinding) than fine pyrite. Even prolonged grinding, however, (50 min.) does not liberate the finest pyrite particles (less than 2 μm).

4. Wet grinding of samples is more efficient at liberation of pyrite than dry grinding.

5. The size distribution of pyrite and the coal particles shows a progressive and predictable shift with increased grinding.

Flotation

Denver lab flotation cell modified by Roberts et al (ref.15) was utilized in our tests.

The main portion of the flotation experiments was conducted with the feed ground in the rod mill for 21 minutes. Such a grinding product contains about 90% of the material below 38 μm (400 mesh).

The initial experiments with the use of kerosene, MIBC and CMC were carried out as shown in Fig. 5. In all the tests, all the kerosene was emulsified with aqueous solution of MIBC (50% of MIBC used in the experiments) for 1 min in a high speed mixer before introducing into the cell. Another part (50%) of the aqueous solution of MIBC was added directly into the cell just before the flotation. The first addition of kerosene and MIBC was usually followed by 2 min flotation and the next addition of the emulsified kerosene (usually 300 g/t) and MIBC after which flotation was continued for another 2 minutes.

The experimental details of the procedure, which also includes cleaning flotation, are depicted in Figs. 6(a) and 6(b).

CMC of molecular weight 80,000, supplied by Polysciences, Inc., was used in our experiments. The CMC was applied in the form of 1 g/l aqueous solution; every 1-2 days a fresh portion of CMC stock solution was prepared.

RESULTS AND DISCUSSION

The results of flotation experiments with MIBC and/or isopropanol are shown in Table 3. Neither the use of CMC as depressing agent nor the change of pH improved flotation selectivity.

The amount of MIBC in the experiment with 50 min grinding (#123) was to be increased in comparison to Exp. #114 (21 min grinding) but sulfur content in concentrate was found to be higher than for 21 min grinding. This is probably due to the excessive frothing which precludes better selectivity after very fine grinding.

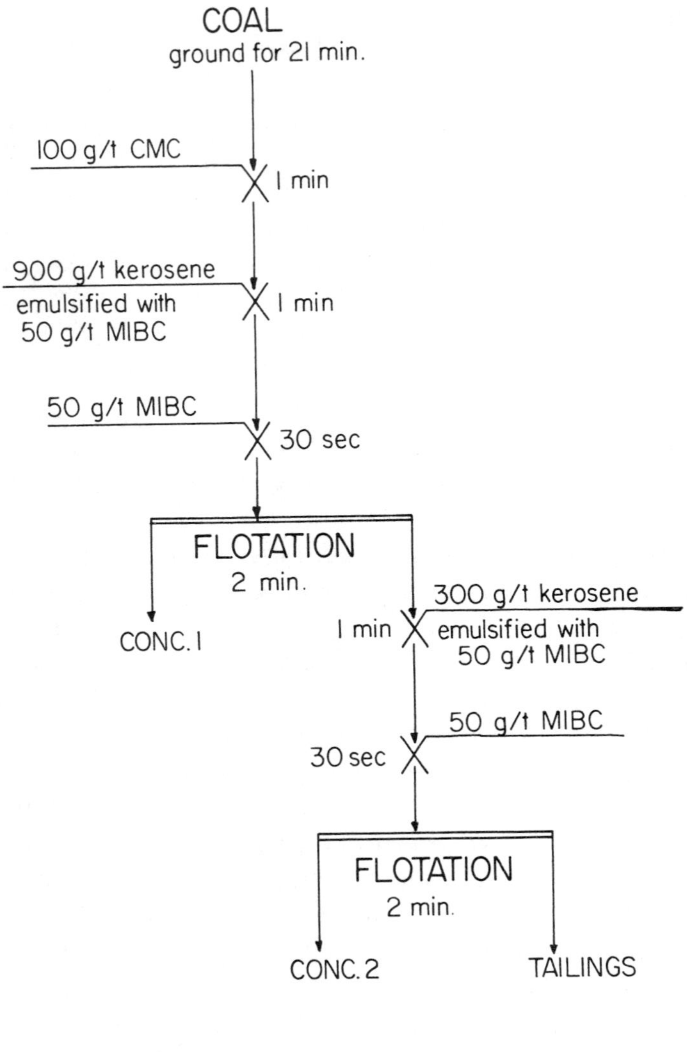

Fig. 5. Schematic illustration of the basic batch flotation procedure (no cleaning).

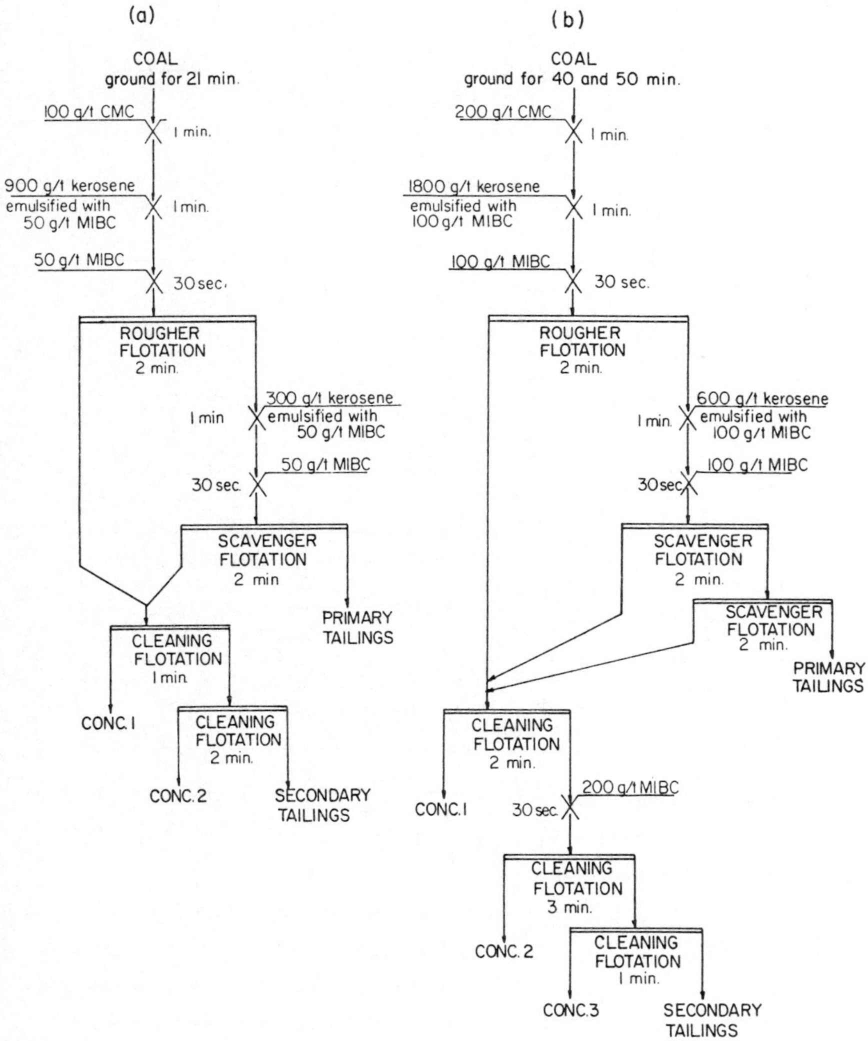

Fig. 6. Schematic illustration of the batch flotation procedures with the use of CMC, kerosene and MIBC and with cleaning stages.

TABLE 3

Results of flotation experiments with the use of frothers only

No.	Products	Yield Wt %	Ash, %	S_{FeS_2}, %	ΣS_{FeS_2}, %	S_{FeS_2} Distribution, %
114	C1	26.7	4.33	0.8	0.8	7.4
	C2	43.2	4.92	1.4	1.16	20.8
	C3	18.6	10.48	3.56	1.66	23.1
	ST	6.5	74.11	11.80		26.7
	PT	5.0	75.70	12.60		22.0
123	C1	48.9	6.23	1.38	1.38	24.8
	C2	25.2	5.45	1.15	1.30	10.6
	C3	12.6	9.33	2.3	1.45	10.6
	ST	9.8	52.70	10.22		36.8
	PT	3.6	71.08	12.97		17.1
115	C1	13.1	5.01	0.9	0.9	4.2
	C2	11.0	4.75	1.03	0.97	4.0
	C3	17.7	4.83	1.03	0.99	6.4
	ST	43.2	12.19	2.65		40.4
	PT	15.0	42.95	8.5		45.0

#114 21 min grinding; 300 g/t MIBC, 200 g/t MIBC, 3 min rougher flotation; and 2 x 50 g/t MIBC in cleaning; pH natural (≃5.5).

#123 50 min grinding; 500 g/t MIBC, 300 g/t MIBC; 100 g/t isopropanol in cleaning; pH natural.

#115 As Exp. 114 but isopropanol was used instead of MIBC. 300 + 200 + 200 + 200 + 200 g/t isopropanol in 10 min rougher flotation; 2 x 100 g/t in cleaning for 6 min; pH natural.

Note: ST = Secondary Tailings; PT = Primary Tailings

The results of Table 3 seem to confirm that MIBC is a very energetic frother which provides high coal recoveries and that isopropanol, as claimed by some authors (ref.16), is more selective.

The flotation experiments with kerosene carried out as outlined in Fig. 5 are more selective. As seen from Fig. 7, the flotation in this case is sensitive to the amount of CMC and the pH. Flotation and selectivity are both enhanced by higher pH (about 9) and this will further be discussed in another publication (ref.10).

The flotation with the use of kerosene and CMC turned out to be amenable to cleaning. Two cleaning stages improve flotation selectivity also after 21 min grinding (Fig. 8). It was found that the flotation carried out with the feed ground for 40 or 50 minutes requires doubled amounts of reagents (Fig. 6(b)). As seen from Fig. 8, the flotation conducted with the feed ground for 40 min is not more selective than after 21 min grinding. However, cleaning changes this situation quite clearly. The same is observed with experiments made after 50 min grinding: flotation cleaning improves these results further and reduces the pyritic sulfur content in the first concentrate down to about one-half per-cent. The selectivity in the flotation experiments with CMC and kerosene and

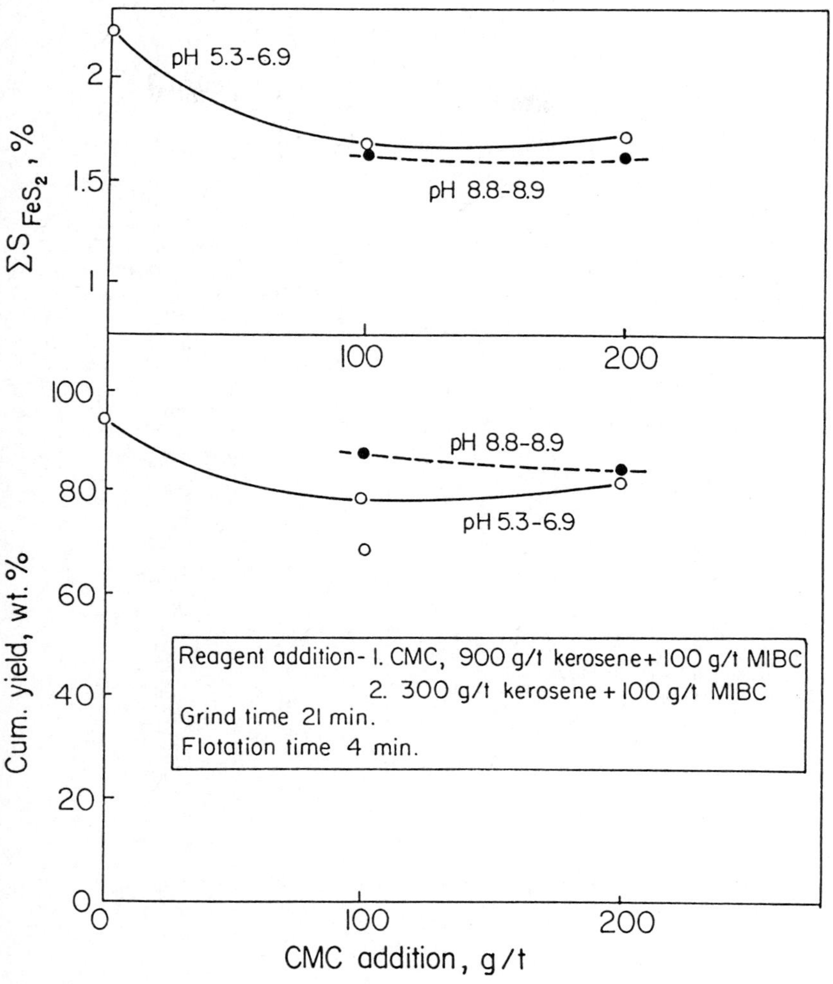

Fig. 7. Effect of pH and CMC on rougher flotation (without cleaning).

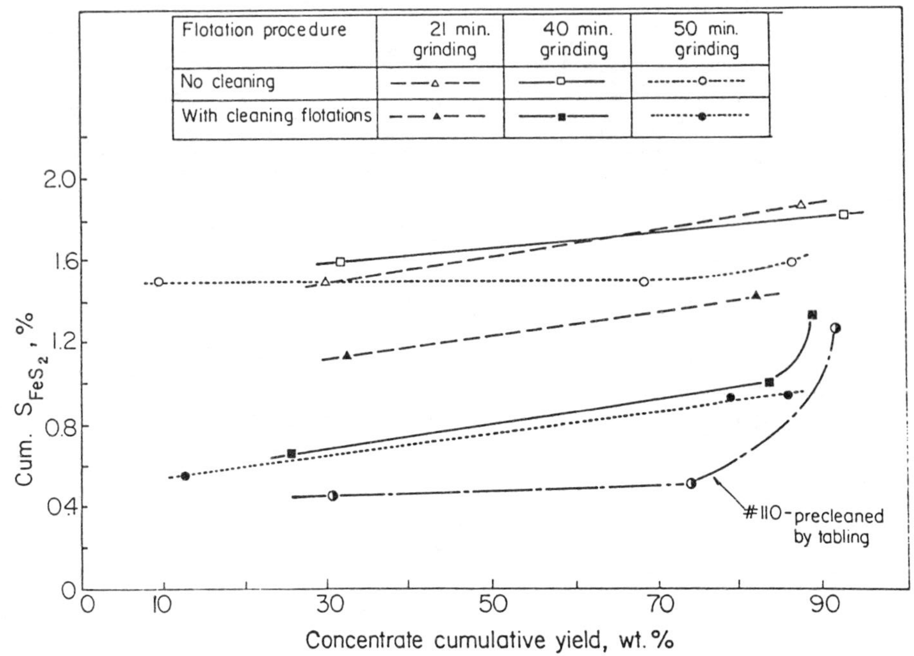

Fig. 8. Effect of grinding time on selectivity of flotation with CMC, kerosene and MIBC. 100 g/t CMC, 900 g/t kerosene + 100 g/t MIBC and 30 g/t kerosene and 100 g/t MIBC, pH 8.9 in the flotation after 21 min grinding; 200 g/t CMC, 1800 g/t kerosene + 200 g/t MIBC and 600 g/t kerosene + 200 g/t MIBC, pH 8.9-9.1 in the flotation after 40 and 50 min grinding. Exp. #110, see Fig. 9 for details.

its change with grinding is clearly seen from Fig. 8.

The experiments so far described were carried out simultaneously with microscopic tests on the pyrite occurrence and liberation, and at that point we learned that the pyrite particle size distribution in the raw coal was in part bimodal and that in the rational flowsheet, the coarser pyrite grains should be removed by gravity methods and only then followed by fine grinding of the precleaned coal and flotation. It also appeared that further improvement in the selectivity might only be achieved by several stages of cleaning flotation.

The use of coal precleaned by tabling enabled us to obtain the concentrate at about 75% yield containing about 0.5% S_{FeS_2}. Complete information on this test is outlined in Fig. 9. These promising results are worth further pilot plant tests.

The pyrite size distribution in the flotation products as numbered in Fig. 9 are shown in Figs. 10 and 11. The degree of liberation of the pyrite, mean size, standard deviation and range of particle size are shown in Table 4. There is no consistent decrease in pyrite particle size with decrease in particle size of the coal, indicating that progressive grinding has not reduced the size of the pyrite; however the degree of liberation generally increases with decrease in coal particle size. The pyrite particle size distribution of samples 1,2, and 3

TABLE 4

Pyrite grains liberation in the separation products (Fig. 9)

	Degree of Liberation, %				Particle Size, µm			Yield Wt. %	S_{FeS_2} %	S_{FeS_2} Distribution, %			$\frac{L}{B}$
PRODUCT NO.	COMPLETELY LIBERATED	MORE THAN 50% LIBERATED	20 - 50% LIBERATED	COMPLETELY BOUND	MEAN	STANDARD DEVIATION	RANGE			TOTAL	COMPLETELY LIBERATED	COMPLETELY BOUND	
A	B	C	D	E	F	G	H	I	J	$K=\frac{[x]}{100}$	$L=\frac{BxK}{100}$	$M=\frac{ExK}{100}$	L/M
1	5	0	4	91	5.96	3.80	0-18	24.8	6.55	61.6	3.1	56.0	0.06
2	17	29	27	27	7.85	5.52	0-24						
3	11	8	4	77	6.72	4.44	0-20						
4	33	0	0	66	4.54	2.21	0- 9	23.2	0.46	4.0	1.32	2.64	0.5
5	5	4	4	89	5.26	3.39	0-15	32.3	0.57	7.0	0.35	6.23	0.056
6	20	31	24	25	3.86	1.85	0- 5	13.5	1.84	9.5	1.8	2.4	0.75
7	36	44	18	2	3.35	2.09	0-11	3.1	7.46	9.8	3.5	0.2	17.5
8	70	25	4	1	4.41	2.80	0-15	2.9	8.38	9.1	6.4	0.09	71.1

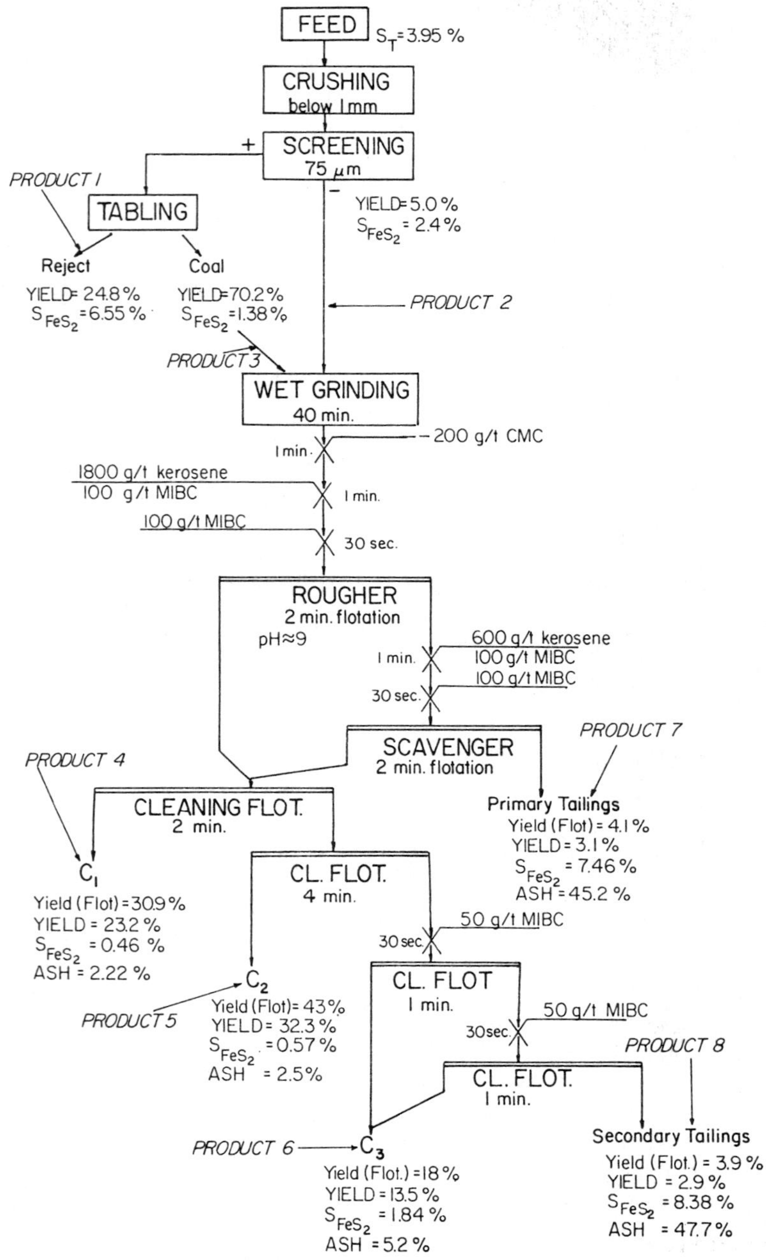

Fig. 9. Final results and flowsheet developed from testwork.

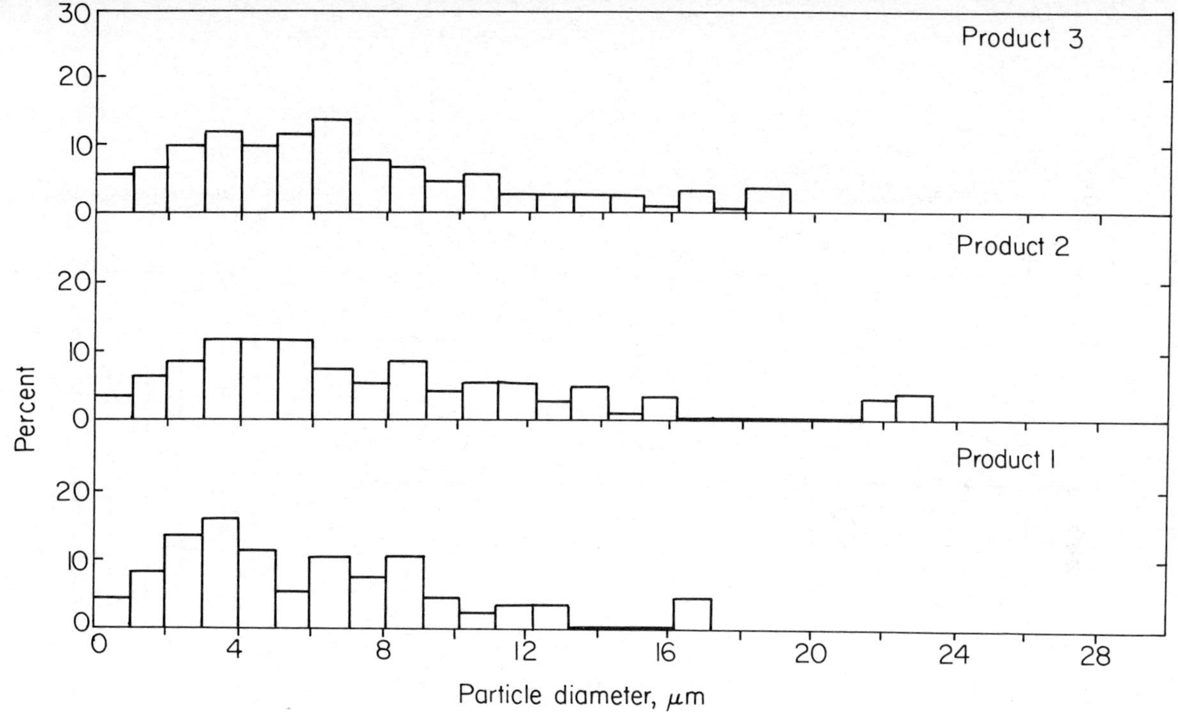

Fig. 10. Pyrite particle size distributions in products: 1,2 and 3.

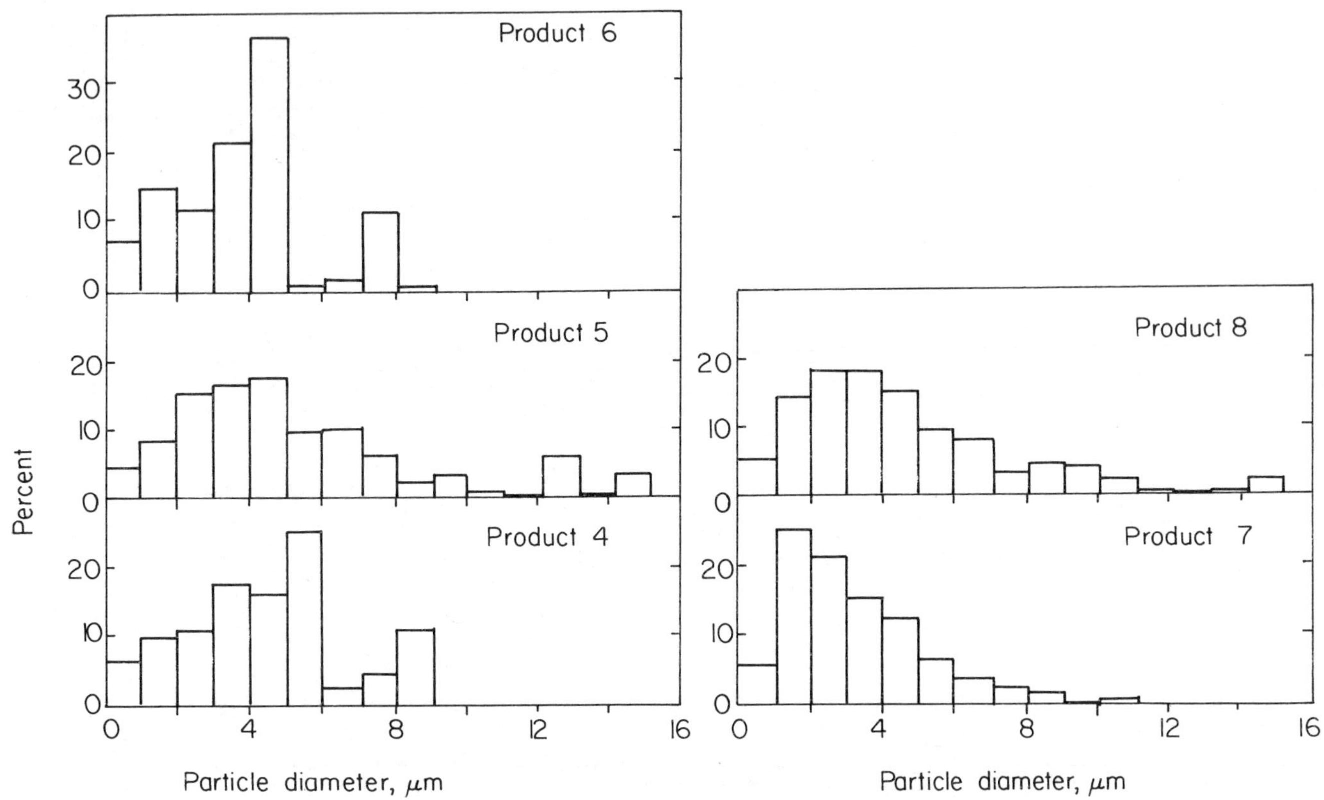

Fig. 11. Pyrite particle size distributions in products: 4,5,6,7 and 8.

which are concentration table products are similar (Fig. 10). Pyrite particle size distribution for samples 4 through 8 (Fig. 11), which are flotation products after an additional 40 minutes grinding, lack the coarse fraction present in samples 1 through 3. Flotation results shown in Fig. 9, recalculated in Table 4, indicate that the ratio of completely liberated pyrite particles to non-liberated pyrite is much higher for flotation tailings than for flotation concentrates. Such results suggest that pyrite retained in the clean coal is mainly non-liberated whereas pyrite reporting to tailings is mostly liberated.

CONCLUSIONS

1. Very fine dissemination of pyrite grains in the Prince Mine coal necessitates very fine grinding to achieve pyrite liberation. Even prolonged grinding (50 min), however, does not liberate the finest pyrite grains (less than 2 μm).
2. Flotation experiments carried out after 21 min (about 90% below 38 μm), 40, and 50 min grinding, did not result in satisfactory selective removal of pyrite when only frothers were used in flotation.
3. Use of kerosene (collector) and MIBC (frother) in the presence of CMC (depressant) under alkaline conditions resulted in much better selectivity. Flotation cleaning stages increase selectivity in this case quite clearly, and finer grinding (40 and 50 min) improves the selectivity further.
4. Coal precleaning by gravity methods followed by grinding and flotation with CMC, kerosene and MIBC allowed the concentrate containing below 0.5% pyritic sulfur to be obtained at 55% clean coal yield from the raw coal which contained about 3% pyritic sulfur.
5. Microscopic observations revealed that in the flotation process developed, pyrite retained in the clean coal is mainly non-liberated whereas pyrite reporting to tailings is mostly liberated, thus proving high selectivity of this flotation procedure.
6. The use of CMC in the one-stage direct flotation with kerosene and MIBC under alkaline conditions seems to offer a simpler and better solution to desulfurizing flotation than the two-stage reverse flotation process.

REFERENCES

1 F.G. Miller, J.M. Podgursky and R.P. Aikman, Trans. AIME, 238 (1967) 276.
2 K.J. Miller and A.F. Baker, Tech. Progr. Rep. U.S. Bur. Mines, 51 (1972) 7.
3 K.J. Miller, U.S. Bur. Mines Rep. Invest., 7822 (1973).
4 H.H. Huang, J.V. Calara, D.L. Bauer and J.D. Miller, in Recent Developments in Separation Science, CRC Press, Florida, Vol.4, 1978, pp. 115-133.
5 K.J. Miller and A.W. Deurbrouck, in Y.A. Liu (Ed.), Physical Cleaning of Coal, Marcel Dekker, New York, 1982, pp. 255-291.
6 J.D. Miller, J. Laskowski and S.S. Chang, Colloids and Surfaces, 8 (1983) 137.
7 Z. Blaschke, Beneficiation of Coal Fines by Selective Flocculation, 7th International Coal Preparation Congress, Sydney, 1976, paper F-2.
8 R. Hucko, U.S. Bur. Mines Rep. Invest., 8234 (1978).
9 J.A. Solari, A.C. de Araujo and J. Laskowski, The Effect of Carboxymethyl Cellulose on the Flotation and Surface Properties of Graphite, 114th Annual AIME Meeting, New York, February, 1985.
10 J. Laskowski, the paper under preparation.
11 W.G. Leonard, R.T. Geer, R. Markuszewski and T.D. Wheelock, Sep. Sci., 16 (1981) 1589.
12 R. Krukiewicz and J. Laskowski, in M.J. Jones (Ed.), Proc. 10th International Mineral Processing Congress, Inst. Min. Met., London, 1974, pp. 391-410.
13 S.T. Hall and J.A. Finch, Minerals and Metallurgical Processing, 1 (1984) 179.
14 R.M. Bustin, A.C. Cameron, D. Grieve and W. Ralkreuth, Coal Petrology, principals, application and methods, Geological Association of Canada Short Course Notes, 1985, 273 pp.
15 T. Roberts, B.A. Firth and S.K. Nicol, Int. J. Mineral Processing, 9 (1982) 191.
16 J.B. Evans, K. Sury, R.O. McElroy and J. Leja, Agglomeration of Ultrafine Coal Flotation Concentrate: The ACOP Project. Presented at International Agglomeration Conference, Toronto, June, 1985.

CLEANING AND DESULFURIZATION OF COAL SUSPENSIONS BY SELECTIVE FLOCCULATION

Y.A. ATTIA

Mining Engineering Division, The Ohio State University, 148 Fontana Labs, 116 W. 19th Avenue, Columbus, Ohio 43210 (U.S.A.)

ABSTRACT

A novel selective flocculation process for the separation-removal of ultrafine pyrite and ash minerals from high sulfur coal slurries has recently been tested and is currently being developed. The process is based on two key innovations. The first was the employment of a selective, xanthate-containing polymeric dispersant for the dispersion-removal of ultrafine pyrite during the selective flocculation of coal. Preliminary tests on high sulfur coal from Kentucky No. 9 seam confirmed this selectivity.

The second key innovation was the preparation and employment of partially hydrophobic polymeric flocculants to flocculate selectively the hydrophobic coal particles while affecting the mainly hydrophilic ash minerals very little. Preliminary tests with various coal samples indicated that this process is capable of producing either steam quality coal from high ash slurries, or superclean coal from medium-ash slurries.

Based on these developments, a selective flocculation process capable of rejecting both the ash and pyrite minerals simultaneously to produce superclean coals is currently being developed for various coals. In addition to laboratory investigations, pilot plant scale up studies of this process are in progress. The economic viability and the near-term commerical potential of this process appear promising.

INTRODUCTION

Advanced ultrafine coal cleaning techniques are needed to treat high sulfur coals. This need stems from the desire to replace oil by very low ash and low sulfur coal-water mixtures, to improve the overall coal (Btu) recovery and lessen costs, and to reduce the environmental concerns in burning high sulfur coals and the disposal of wastes. For coal water mixtures to be a suitable replacement for oil, the ash and sulfur contents must be low (less than 3 percent ash and less than 1 percent sulfur) and the particle size very small (-74 micron). Therefore, deashing and desulfurization of coal by physical cleaning methods will require ultrafine grinding for coal to liberate a high fraction of the ash and sulfur containing minerals.

The notorious problem of acid rain continues to impact coal utilization adversely in many areas. Much of the acid rain is believed to be due to the high sulfur content in coal. The presence of sulfur in coal is generally attributed to two forms, organic sulfur and pyritic sulfur. The proportion of pyritic sulfur to organic sulfur varies significantly from one coal seam to another, but it appears that pyritic sulfur generally represents about 70% or more of the total sulfur (ref. 1). The pyrite is found in coal in a wide size distribution, with a significant proportion in the very fine fraction (less than 25 microns). In fact, there are coal seams such as Kentucky No. 9 where the pyrite particles are often smaller than 10 microns (ref. 2).

Removal of pyritic sulfur or other metallic sulfur could substantially alleviate the problem of sulfur oxide emissions from coal combustion. Efforts to achieve these objectives have been hampered by the lack of adequate and commercially feasible techniques for cleaning and handling fine (-200 mesh or 74 microns) and ultrafine coal slurries (-500 mesh or 25 microns).

Problems of Cleaning Ultrafine Coal Slurries

Any proposed advanced cleaning process must meet certain requirements to be considered effective for the application mentioned. These requirements include 1) treatment of very finely ground coal (necessary for thorough liberation of the ash and sulfur forming minerals; b) rejection of both ash and sulfur from the coal; c) maintenance or improvement of combustion properties, such as ignition temperature, and burning rate; d) treatment in an environmentally acceptable manner; e) suitability to coal-water fuel preparation; and f) performance in a cost-effective manner.

Most techniques commercially available at present generally fail to achieve all the performance criteria mentioned above. For example, froth flotation has been successfully applied to high rank coal fines, but fails in efficiency markedly, when the particle size is much below 50 microns, or when applied to low rank coals. Oil agglomeration has also been successfully used to recover coal fines, but the process cost is often prohibitive. In addition to oil-pollution problems occur accompanied by cleaned coal incompatibility with coal-water mixture preparation. Neither technique adequately removes the very fine pyrite particles which are often the major contributors of sulfur in coal.

The selective flocculation process is an advanced physico-chemical separation technique that appears to address the current need economically. This emerging technique offers much promise as a new separation technology for processing fine and ultrafine coal slurries. Selective flocculation is not used on the entire coal feed which usually has a very wide size distribution (e.g., 28 mesh x 0). A pre-cleaning step using a standard technique such as hydrocyclones or froth flotation should be employed to treat the size fractions above 200 mesh.

This article reports progress made in the selective flocculation process and demonstrates process feasibility for desulfurization and ash reduction of high-sulfur, high-ash coals.

SELECTIVE FLOCCULATION OF COAL

Principles of Selective Flocculation

Flocculation involves the aggregation of finely-divided particles suspended in liquid (usually water) through the utilization of a polymeric flocculant. The particles flocculate if the flocculant is adsorbed on the adjoining particles and bridges the gaps between them. Figure 1 illustrates the current understanding of the bridging mechanism in flocculation. Particles bonded in this way aggregate to form flocs of much larger size so that they settle out of the suspending liquid rapidly.

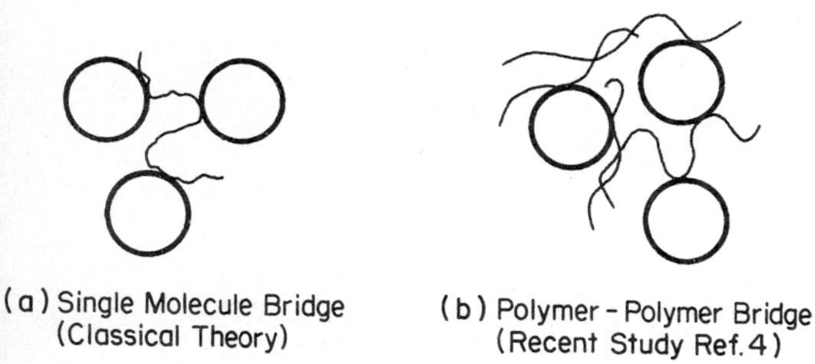

Figure 1. Bridging Mechanisms of Floc Formation with Long Chain Polymers; (a) classical theory showing adsorption of single polymer molecule on several particles (b) recent study (4) showing polymer-polymer bridging in addition to multi-particle adsorption.

Coal preparation plant operators have employed total flocculation of raw coal slurries (black water) for many years to promote the recovery of process water and to aid the disposal of wastes. Natural as well as synthetic flocculants and coagulants have been employed for this purpose. Examples of commercial flocculants are starches, polyethylene oxide, polyacrylamides, and other amine-containing cationic flocculants. In this function, the polymers flocculate coal and shale particles indiscriminately.

Selective flocculation of coal is based on the preferential adsorption of a long-chain organic polymer onto the coal particles in an aqueous slurry. The coal particles flocculate but the mineral matter remains dispersed. The flocculated coal particles can be recovered by preferential gravitational settling or by other appropriate techniques.

Selective flocculation for the separation of coal from fine and ultrafine pyrite and ash minerals is based on two key innovations. The first was the development of a selective, xanthate-containing dispersant for fine and ultrafine pyrite during the selective flocculation of coal.

The second key innovation was the development of partially hydrophobic polymeric flocculants for the selective flocculation of the hydrophobic coal particles. The associated hydrophilic shale minerals remain dispersed. The mechanisms for selective adsorption of dispersant and flocculant are discussed elsewhere (ref. 3,4).

Process for the Selective Flocculation of Coal

The process for the selective flocculation of coal involves four steps. These are:

(1) Total dispersion of raw coal minerals into a slurry. General dispersants for ash minerals and selective dispersants for pyrite are added at this stage.

(2) Addition of polymeric flocculant to effect selective flocculation of coal particles.

(3) Conditioning of the coal flocs to achieve desired performance requirements; and

(4) Separation of coal flocs from the slurry. Separation often involves gravity settling, froth flotation, dissolved air flotation, or a combination of techniques.

Figure 2 illustrates the simplified sequence of events described for the selective flocculation process.

The first step involves dispersion of coal and associated mineral matter in water using a suitable dispersant in a well-stirred tank. A flocculant is then added, in a separate tank, where flocculant adsorption on the coal surface leads to floc formation. The flocculant is typically a water-soluble polymer. The third step involves floc conditioning, which leads to desirable floc characteristics exemplified by minimum mineral matter association and development of sufficient strength to permit separation. Finally,

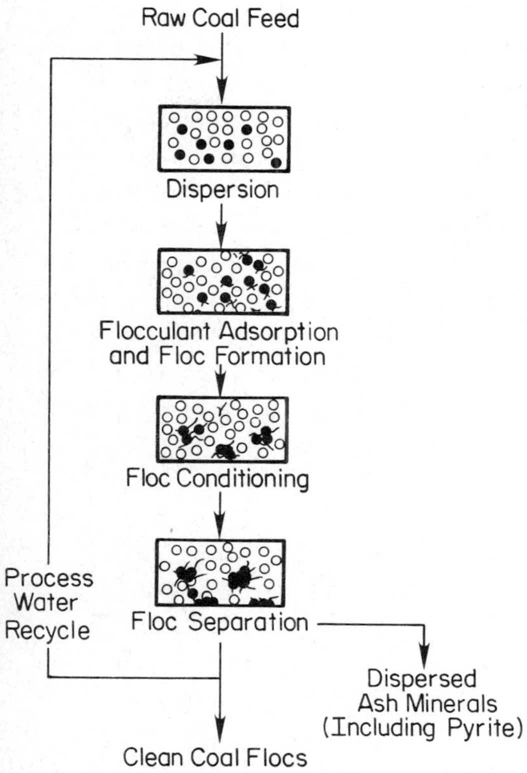

Figure 2. Conceptual Representation of the Selective Flocculation Process for Coal Cleaning and Desulfurization.

the flocs are separated from the dispersed mineral matter (ash constituents). Separation is promoted with minimized floc breakage and maximized mineral matter rejection.

These four steps can be repeated, if necessary, to effect multistage cleaning. The number of stages will depend upon the single-stage process efficiency and the amount of upgrading and recovery desired. For example, if maximum coal values are to be recovered from the fine rejects of a washing plant, then a single-stage may be sufficient. On the other hand, production of very low-ash and sulfur (less than 3 percent ash; less than 1 percent sulfur) coal for a coal-water mixture where the Btu content is maximized (greater than 90 percent) may require two or more stages.

Previous Work on Selective Flocculation of Coal

Several investigators have found empirically that various polymeric flocculants exhibit some degree of selectivity when applied to mixed and raw coal slurries. For example, Blaschke (ref. 5) found that partially hydrolzed polyacrylamide exhibited selective flocculation effects towards coal in mixed suspensions with clay minerals in the presences of a dispersant "Praestor 22K. 21K." Hucko (ref. 6) observed similar effects with a non-ionic polyacrylamide with the inorganic dispersant sodium hexametaphosphate present. Attia and Fuerstenau (ref. 3) studied the adsorption and flocculation properties of an anionic flocculant believed to be sodium polystrene sulfonate, commercially known as Purifloc-A22 (Dow Chemicals, U.S.A.) on coal and shale suspensions. Selective adsorption and flocculation towards coal was observed only at pH greater than 9, where flocculation of shale minerals was negligible. Brookes et. al. (ref. 7) used modified polyacrylamide containing chelating and complexing groups to effect improved selective flocculation of coal from mixtures with clay minerals. On the other hand, Solari and Rubio (ref. 8) did not observe any selective flocculation of raw Brazilian coal using acrylamide flocculants. They attributed the lack of selectivity to the presence of middling particles. The principles and applications of selective flocculation technology for separation of very fine particles have been recently reviewed (rev. 9-12). The commercial feasibility of selective flocculation has already been demonstrated in the iron ore and potash industries (ref. 5,6).

More recently, Attia and Deason (ref. 4,21) and Ishizuka et al. (ref. 15) demonstrated the feasibility of selective flocculation cleaning of ultrafine coal slurries including lignite using partially hydrophobic flocculants. A number of water-soluble polymeric flocculants containing hydrophobic groups have been developed and tested on several coal slurries. The treatment produced both steam quality and superclean coals with high coal (Btu) recovery.

REMOVAL OF PYRITE BY SELECTIVE FLOCCULATION

Basis for the Separation

Selective flocculation separates coal and pyrite based on distinct differences in their surface-chemical properties. Pyrite is a sulfide mineral with an affinity for xanthate-(dithiocarbonate) containing reagents, while coal does not have such an affinity. Therefore, xanthate groups incorporated into the structure of long-chain polymers should selectively flocculate or disperse pyrite, depending on the polymer's chain length. The use of polyxanthates for selective flocculation of sulfide minerals, particularly for copper, was first reported by Attia and Kitchener (ref. 16). The separation of pyrite from coal using polyxanthate dispersants was demonstrated recently by Attia and Fuerstenau (ref. 3). The key to this separation was the design and preparation of the selective dispersant for pyrite (PAAX). The dispersant incorporated xanthate groups into the structure of a low molecular weight acrylic acid polymer (PAA) as described below.

Preparation and Testing of Polyxanthate Dispersants

In a previous article (ref. 3), Attia and Fuerstenau described preparation of polyxanthate dispersants by reacting low molecular weight polyacrylic acid solutions with sodium hydroxide and carbon disulfide. The presence of the xanthate groups (dithiocarbonate) in the polymeric dispersant was ascertained from the UV - spectrum of the purified solution, where the characteristic absorption peak at 305 to 307 mμ was observed. The peak is very close to the UV-absorption peak at 303 mμ for lower xanthate compounds.

The reaction product of polyacrylic acid with xanthate (PAAX) is a dispersant with the possible formula of sodium poly (acrylate-acrylodithiocarbonate) as depicted in Figure 3. The ratio of acrylate to xanthate groups in the polymeric structure depends primarily on a parameter called the degree of xanthation, d_x, which is determined by measuring the concentration of xanthate groups in the polymer solution.

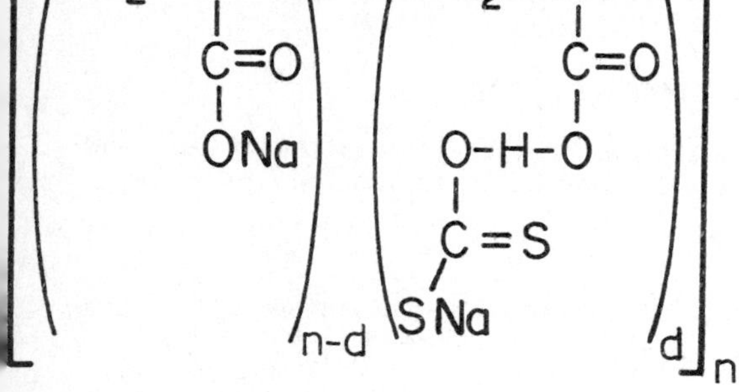

Figure 3. Postulated Structure of Polyacrylic Acid/Xanthate (PAAX) Dispersant: Sodium Poly (Acrylate-Acrylodithiocarbonate).

Selective flocculation of coal from pyrite was investigated using the newly developed polyxanthate dispersant in conjunction with a polystyrene sulfonate (Purifloc A22) flocculant (ref. 3). The flocculation results on the individual mineral suspensions are shown in Figure 4 (A and B). These graphs show the effect of polyacrylic acid dispersant before (PAA) and after xanthation (PAAX) on the flocculation-dispersion behavior of individual suspensions of coal and pyrite with Purifloc-A22 flocculant.

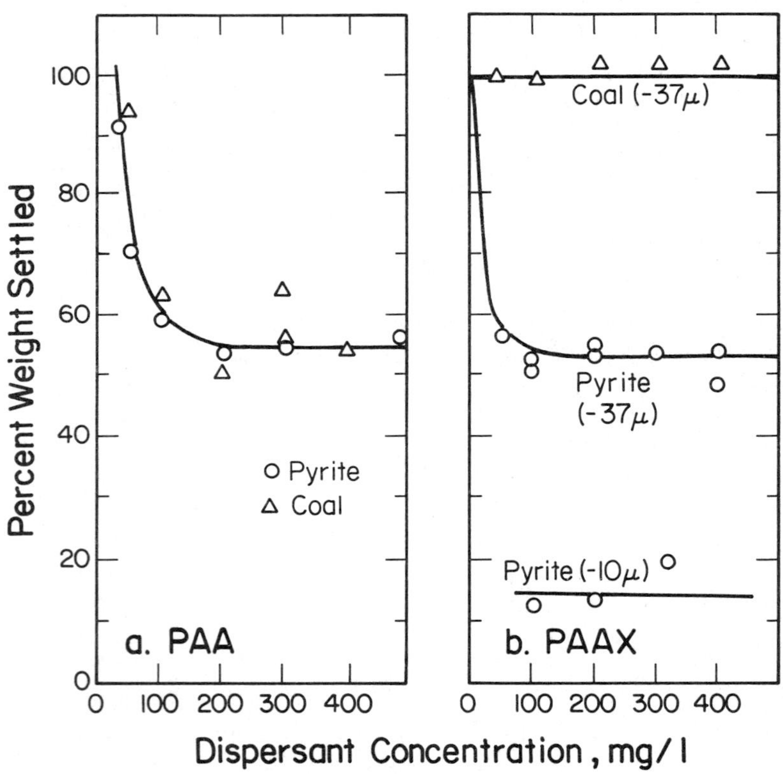

Figure 4. Effect of PAA and PAAX Dispersants on the Flocculation of Coal and Pyrite with Purifloc-A22 (4 mg/l) at pH10, and 2% Solids Content in Suspensions.

From Figure 4 (A), it appeared that PAA inhibited or restrained the flocculation action of Purifloc-A22 on both coal and pyrite suspensions at PAA concentrations of 100 mg/l and above. The dispersive action of PAA in this case was therefore unselective. However, the PAAX crude reaction product in Figure 4 (B) only dispersed the pyrite suspension to the same level as PAA, while the coal suspension was totally flocculated even at high PAAX concentrations. These observations in Figure 4 (B) were repeated and noted several times even with purified PAAX solutions. Selective dispersion of pyrite or selective flocculation of coal from pyrite using PAAX reagent appeared, therefore, possible.

Effect of Particle Size on Pyrite Dispersion

It was suspected that a large proportion of the settled pyrite particles, shown in Figure 4, may have been "coarse" particles between 10 and 37 microns. Pyrite has a specific gravity of about 5.4, while that of coal is around 1.6 to 1.8. Therefore, a pyrite suspension of minus 10 microns and coal suspension of 37 microns were also tested. The results (Fig. 4.8) demonstrated that the coal was almost totally flocculated as before, while the minus 10 micron pyrite suspension remained very stable, with only 10-20% weight of the particle, settled as flocculated. From these observations, it is believed that the selective dispersion of pyrite will be more effective at the smaller particle sizes. Alternatively, a more effective separation technique for the coal flocs such as dissolved air flotation which is more tolerant to the presence of "coarser" pyrite might be utilized instead of the gravitational settling technique.

Selective Flocculation of High Sulfur Coals

Attia (ref. 17) recently described preliminary results for selective flocculation of high sulfur coal with the polyxanthate dispersant mentioned earlier. A sample Kentucky No. 9 coal was obtained from the Occidental Research Corporation. The coal sample was wet ground in a steel ball mill and screened to minus 400 mesh (-37u) fraction. Single-step selective flocculation tests were conducted on dilute suspensions (about 3% wt.) of the minus 37u coal particles, using 2 mg/l, F1029-D flocculant (a partially hydrophobic polymer prepared by Dai-Ichi Kogyo Sieyako, Japan), at pH 10. The results are shown in Table 1.

These preliminary results show that the sulfur content of the dispersed (reject) products was significantly higher than the feed or the flocculated fractions. This was especially evident at the higher PAAX dispersant concentration (500 mg/l) where the sulfur content in the dispersed fraction was 13.74% compared with 2.08% in the flocculated fraction. While these results present a definite evidence for the selectivity of flocculating coal and dispersing pyrite with the reagents mentioned above, the total sulfur removal was low. The low sulfur removal could be due to the fact that pyrite is known to be very finely disseminated (-10 microns) in Kentucky No. 9 seams from which this sample was obtained. In the brief tests mentioned here, no

TABLE 1
Material Balance for One-Step Selective Flocculation of High Sulfur Coal Using PAAX Dispersant and F1029-D Flocculant

Product	Weight %	Total Sulfur %	Total Sulfur Distribution %	Remarks
Dispersed	3.7	13.74	20.3	500 mg/l
Flocculated	96.3	2.08	79.7	PAAX
Feed	100.0	2.52	100.0	
Dispersed	2.7	9.96	9.7	300 mg/l
Flocculated	97.3	2.56	90.3	PAAX
Feed	100.0	2.76	100.0	

attention was paid to the liberation characteristics of pyrite from coal. It was hoped that grinding to minus 400 mesh (37 microns) would liberate a significant amount of pyrite. Also, the results shown in Table 1 did not take into account the presence of organic sulfur. It is believed that Kentucky No. 9 coals contain a very high ratio of organic sulfur (greater than 50%). This selective flocculation process deals only with the pyritic sulfur. Another possible reason for the low sulfur removal was the presence of "coarse" pyrite or pyrite-containing coal particles which would settle at a similar rate as the pure coal flocs, thus interfering in their separation. Other floc separation methods such as flotation might extend the effectiveness of this selective flocculation method to coarser sizes. Alternatively, the coarser pyrite particles (greater than 25 microns, for example) could be separated by other techniques prior to applying selective flocculation. In addition, it is believed that by applying multi-stage selective flocculation (i.e., the flocs from the first separation step are re-dispersed and re-flocculated several times), more of the liberated pyrite particles would be removed.

REMOVAL OF ASH MINERALS BY SELECTIVE FLOCCULATION

The feasibility of cleaning raw coal suspensions by selective flocculatin with partially hydrophobic flocculants has been tested. The objective of this initial study was to obtain a coal product with about 10% ash, but with coal recovery between 80-90% from the fine-ultrafine coal fractions (black water). However, with the development of coal water mixtures as a possible replacement for heavy oils in power generation, the objective of the selective flocculation process was changed to obtain a clean coal with less than 3% ash, and with high coal Btu) recovery from fine and ultrafine raw coal slurries.

Basis for the Separation

For the polymer to function as selective flocculant, it has to first selectively adsorb on the coal particle surfaces and be able to bridge between them. Therefore, the surface chemistry of coal-shale minerals and chemical nature of the flocculant are clearly important.

Shale minerals differ in their surface chemical properties from those of coal in many ways, but the greatest difference that could be exploited for surface-based separations is in the surface affinity towards the solvent water molecules. Fresh surfaces of high to medium rank coals are naturally hydrophobic (i.e., have little or no affinity to bind water molecules), while shale minerals are naturally hydrophilic due to their surface polar groups. The great difference in this hydroprobicity has been the basis for cleaning coal fines by surface techniques such as froth flotation and oil agglomeration.

Selective flocculation using this intrinsic property has only beeen recently recognized. Attia (ref. 12,18) has suggested the use of partially (or totally) hydrophobic polymeric flocculants to selectively flocculate the hydrophobic coal particles from the primarily hydrophilic associated shale minerals. The mechanism of preferential adsorption on coal would be mainly through the hydrophobic (bonding) interactions between the hydrophobic parts of the flocculant and those of the coal surfaces. The nature of the hydrophobic (bonding) interactions and ways of designing and preparation of partially hydrophobic polymeric flocculants are reported elsewhere by Attia and Deason (ref. 4).

SELECTIVE FLOCCULATION WITH PARTIALLY HYDROPHOBIC POLYMERS

A - Individual and Mixed Suspensions

A number of water-soluble polymeric flocculants, containing hydrophobic groups, have been developed and tested for the possible separation of ultrafine coal particles from slurries (ref.4,15). The chemical structure of these partially hydrophobic polymeric flocculants is illustrated in Figure 5. These co-polymers were prepared and supplied by Dai. Ichi. Kogyo Sieyaku (DKS) Company, Tokyo, Japan. The relevant characteristics of the co-polymers tested for selective flocculation are shown in Table 2.

In addition to these co-polymers, another type of flocculant, E-1480, was supplied by Calgon Corporation, Pittsburgh, Pennsylvania, and was tested for selective flocculation of coal. It was believed that E-1480 was also partially hydrophobic polymer, but no detailed chemical or physical property data was provided by the manufacturer.

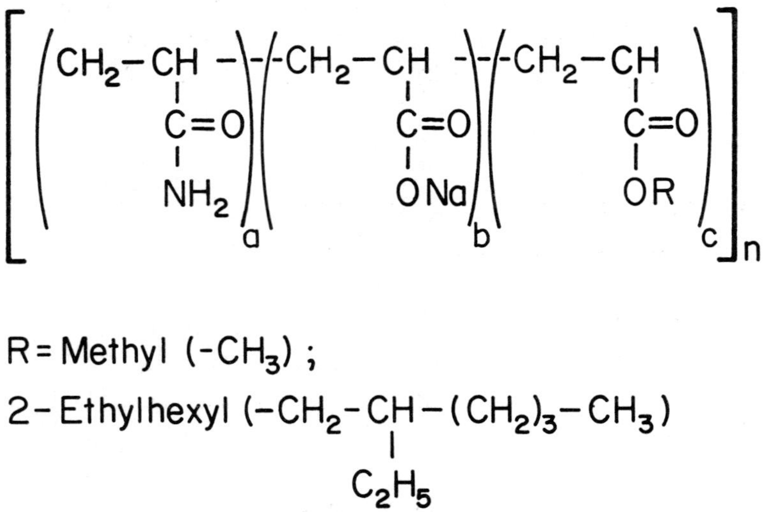

Figure 5. Structure of Partially Hydrophobic Flocculants Series F1029 Co-polymers (DKS Company). The proportions of the various units are as follows: (a) Acrylamide = 55.85%; (b) Sodium Acrylate = 0-15%; (c) Hydrophobic Units R = 10-40%.

TABLE 2

Composition and Molecular Weights of DKS' Partially Hydrophobic Polymers Used in This Study: F1029 Co-polymer Series (ref. 19)

Polymer I.D.	Hydrophilic Unit (A)	Hydrophobic Unit (B)	Molecular Weight	A/B Ratio
F1029-D	acrylamide/ acrylic acid	methyl acrylate	3 million	85+5/10
F1029-I	acrylamide	methyl acrylate	3 million	60/40
F1029-K	acrylamide/ acrylic acid	2-ethyl-hexyl acrylate	3 million	58+13/29
F1029-J	acrylamide/ acryliacid	methyl-acrylate	3 million	60+7.5/32.5

The results of flocculation testing two of these polyers F1029-D and E=1480 on various samples of bituminous coal and shale are shown on Figures 6 and 7, as a function of dispersant concentration (Calgon:Sodium hexameta-phosphate), and the suspension pH, respectively. As indicated by these results, excellent selectivity towards coal was exhibited by both polymers in the acid pH range. In alkaline media, only F1029-D showed marked selectivity still, while E-1480 became ineffective. In any case, the validity of the concept of using partially hydrophobic polymers for the selective flocculation of coal has been confirmed.

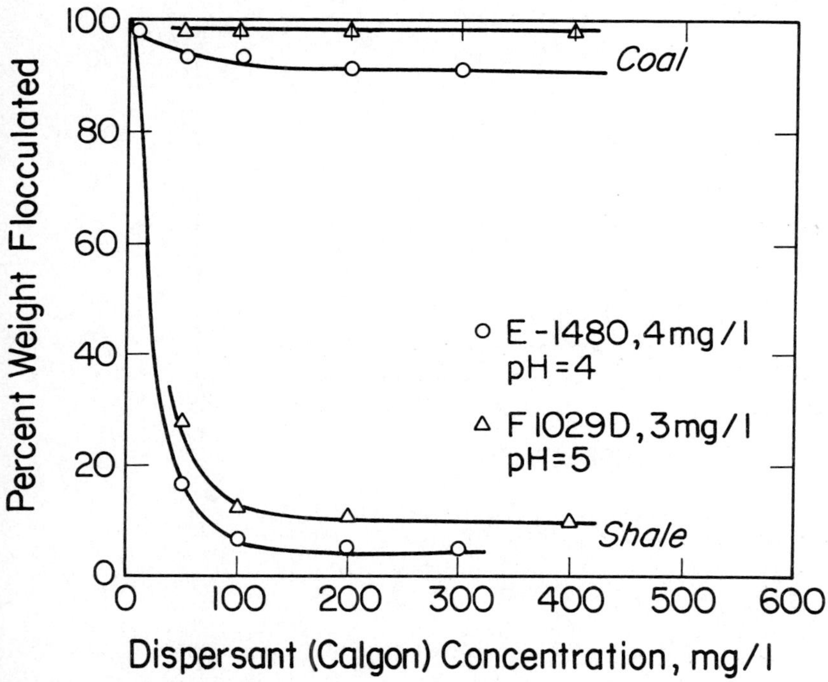

Figure 6. Effect of Dispersant (Calgon) Concentration on the Flocculation of Coal and Shale Suspensions Using F1029D and E-148 Flocculants

Selective flocculation of coal from mixed suspensions with shale minerals simulating a typical raw coal composition was also tested and the selectivity towards coal was confirmed. The four co-polymers described in Table 2 were tested on the mixed suspensions and some relationship between the polymer's hydrophobicity and its selectivity towards coal was found to hold true in general.

On the basis of these tests, a process for cleaning ultrafine coal slurries by selective flocculation with one partially hydrophobic polymer F1029-D was developed and tested. The effects of various process parameters on the performance of selective flocculation were also briefly studied. These included: suspension pH, process water quality, temperature, solids content, and particle size and size distribution of the raw coal slurry. The compatibility of the process with other cleaning techniques and with coal-water mixture preparation was also reviewed and discussed (ref. 4).

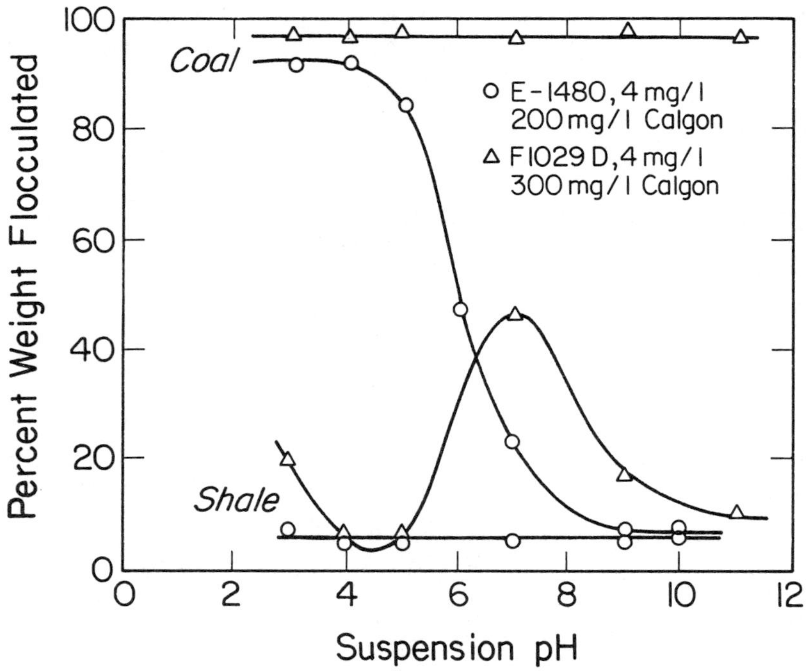

Figure 7. Effect of Suspension pH on the Flocculation of Coal and Shale Suspensions Using E-1480 and F1029D Flocculants

SELECTIVE FLOCCULATION OF RAW COALS

As mentioned earlier, selective flocculation could be used either to produce steam-quality coal (i.e., 8-18% ash) from high ash coal slurries, or to produce superclean coals from medium (about 10%) ash or pre-cleaned coals. Example tests for each case are briefly described here.

Although the objectives are different, the basic steps involved in cleaning the coal slurry by selective flocculation are essentially the same. Figure 8 illustrates the conceptual flowsheet for the selective flocculation circuit used in recent investigations. The cleaning stages can be repeated several times if necessary.

Production of Steam Quality Coal

Preliminary feasibility investigations were conducted on raw coal samples supplied by the National Coal Board, England (U.K.) with an average ash content of 40 to 45%. Typical results are demonstrated here by a few examples shown on Table 3.

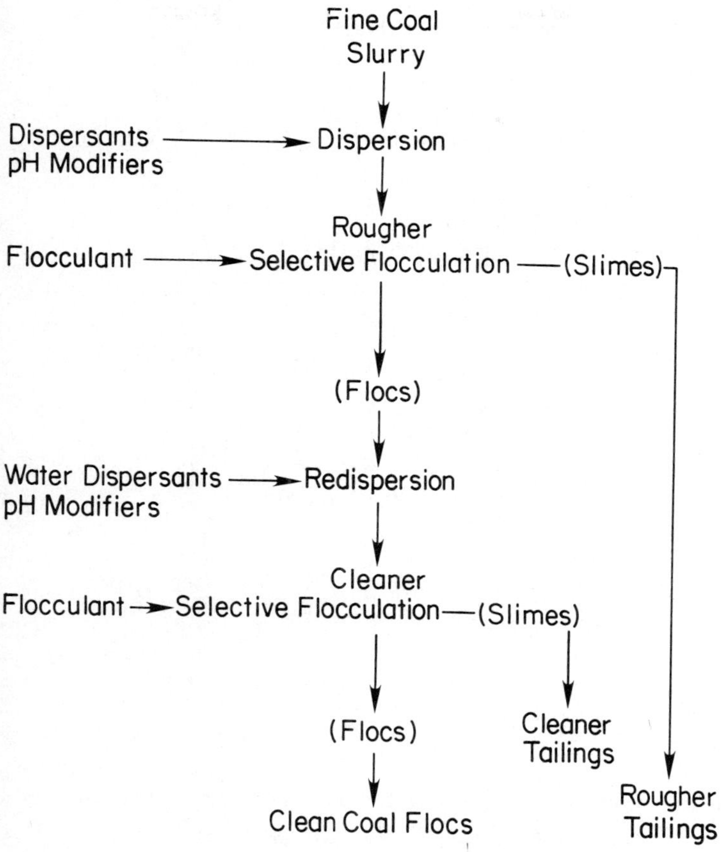

Figure 8. Flow Sheet Used for Selective Flocculation of Coal from Suspensions Containing Shale and Pyrite

In the first example, two flocculation stages were used where the flocculated fraction from the first selective flocculation (rougher) stage was mechanically re-dispersed into a slurry. Selective flocculation was again performed on this slurry in the second (cleaner) stage to produce the final cleaned coal product. Reagent conditions in this test were 5 mg/l flocculant, 500 mg/l sodium hexametaphosphate (SHMP) at pH 11 for the rougher stage and 2.5 mg/l flocculant, 200 mg/l SHMP and pH 11 for the cleaning stage. The results showed that about 64% ash removal with 85% coal (Btu) recovery can be achieved under these conditions. The ash content of the dispersed fractions (tailings) was very high (68-77%), indicating high efficiency for ash rejection.

TABLE 3

Production of Steam Quality Clean Coals by Selective Flocculation

Example	Product	Wt.%	Ash%	Distribution% Ash	Distribution% Coal	Remarks
1	Clean Coal Flocs	65.3	21.5	35.6	84.7	Two Stage Flocculation
	Rougher Tailing	19.1	77.3	37.4	7.2	
	Cleaner Tailing	15.6	68.4	27.0	8.1	
	Feed	100.0	39.5	100.0	100.00	
2	Clean Coal Flocs	48.6	19.9	21.4	71.1	Tap Water, Multi-Stage Flocculation
	Rougher Tailing	18.7	75.0	31.0	8.5	
	Cleaner Tailing	22.8	69.2	35.0	12.8	
	Second Cleaner Tailing	7.3	63.9	10.3	4.8	
	Third Cleaner Tailing	2.6	55.5	3.2	2.1	
	Feed	100.0	45.5	100.0	100.0	
3	Clean Coal Flocs	84.5	10.3	46.1	93.4	Single Stage Flocculation, Pre-Cleaning with Heavy Liquids
	Rougher Tailing	15.5	65.4	53.8	6.6	
	Feed	100.0	18.9	100.0	100.0	

In the second example, Columbus (Ohio) tap water was used rather than distilled water in a multistage test with a rougher and three cleaning stages. Reagent conditions employed in this test were 3 mg/l of flocculant in each stage, pH of 7.5 and 200 mg/l of SHMP in the rougher with 100 mg/l added to each cleaning stage for a 5.6% solids slurry. Polymer dispersion was carried out by adding the flocculant from a 100 mg/l stock solution over a 2 minute period while mixing in the standard baffled mixing vessel. Floc conditioning was conducted using the variable-speed rotating cylinder conditioner for a total of 15 minutes. Results of this test demonstrate that even with the lower levels of reagents and utilizing tap water, a successful separation could be made.

The third example shows that selective flocculation can work well in a flowsheet incorporating a gravity cleaning step at coarser sizes ahead of the selective flocculation circuit. In this test, heavy liquid was used to remove the coarsely liberated ash minerals from the high as coal samples provided by NCB. The "cleaned" coal was then ground to liberate the finely disseminated ash minerals before their separation by selective flocculation. The results indicate even improved ash rejection and coal recovery by a single-stage flocculation.

The foregoing results show that selective flocculation is feasible for high ash coal slurries.

Production of Superclean Coals

Exploratory tests on several coals indicated the possibility of obtaining superclean coal (i.e., less than 3% ash) from ultrafine medium ash coal slurries by selective flocculation. Test results shown on Table 4 confirm this possibility. For example, Attia and co-workers (ref. 20,4) obtained superclean coal with 2.7% ash and about 68% coal (Btu) recovery from an upper Freeport coal slurry containing 9.2% ash. Independent results on bituminous and sub-bituminous coals by Ishizuka and co-workers (ref. 15) also demonstrated the feasibility of obtaining ultraclean coal down to 1.2% ash by selective flocculation. These data show the range of results that could be obtained with various coals. Understandably, it appears that different coals will respond differently to the process.

TABLE 4
Production of Superclean Coal by Selective Flocculation

Coal Type	Feed Ash %	Final Ash %	Ash Removal %	Coal (Btu) Recovery %	Reference
Bituminous (Upper Free-port, PA)	9.2	2.7	81.8	67.9	20
Bituminous	8.5	1.2	87.8	93.2	15
Sub-Bituminous	10.5	2.2	77.9	93.4	15

The results also indicated that up to 88% ash removal can be achieved with coal (Btu) recovery of 68-93%, depending on the coal type, liberation of ash minerals, and the degree of cleaning required. Preliminary economic evaluation indicates that this process has high commercial potentials. Cleaning of fine and ultrafine coals to less than 3% ash appears to be feasible, particularly if the coal is pre-cleaned to less than 10% ash by a suitable technique at coarser sizes to remove loosely associated ash minerals.

SIMULTANEOUS REMOVAL OF ASH AND SULFIDE MINERALS BY SELECTIVE FLOCCULATION

The foregoing results demonstrate the potential for the selective flocculation process in de-ashing and desulfurization (de-pyritization) of fine and ultrafine high-sulfur coal. Further research work is ongoing to improve the process. The combined use of the partially hydrophobic flocculants with the xanthate-containing dispersants should effect the simultaneous removal of ultrafine pyrite and ash minerals during the selective flocculation of coal from slurries. In addition to the laboratory investigations, pilot plant scale-up and studies of this process are in progress. The economic viability and the commercial potential of this process are discussed elsewhere (ref. 20,4).

CONCLUSIONS

From the foregoing discussion, it can be concluded that selective flocculation presents a promising approach for the removal of ultrafine pyrite particles from high sulfur coal suspensions. This is accomplished by the addition of the polyxanthate dispersant (PAAX) during the selective flocculation of coal with F1029-D or Purifloc-A22. It is believed that by applying multistage selective flocculation or use a more adequate means for coal flocs separation, most of the liberated pyrite particles would be removed.

Deashing of ultrafine coal slurries by selective flocculation using partially hydrophobic flocculants appears to be technically and economically feasible. The process can be used to either produce steam quality coals from high ash slurries (e.g., black water from washing plants refuse), or to produce superclean coals from medium ash or pre-cleaned slurries. The process appears to have near-term commercial potential.

REFERENCES

1 "Analysis of Ohio Coal", Information Circular No. 47, State of Ohio Department of Natural Sciences, Division of Geological Survey, 1978.
2 E. Albaugh, Gulf Research and Development, Pittsburgh, Pennsylvania, Private communications with Y. Attia, 1982.
3 Y.A. Attia and D.W. Fuerstenau, "Feasibility of Cleaning High Sulfur Coal Fines by Selective Flocculation", Fourteenth International Mineral Processing Congress, Toronto, Canada (1982).
4 Y.A. Attia and D.M. Deason, "Cleaning Ultrafine Coal Slurries by Selective Flocculation using Partially Hydrophobic Polymers", in press. Also in "Selective Flocculation Technology Group Program", Y.A. Attia, S.V. Krishnan, and D.M. Deason, Final Report, June 10, 1983, Battelle-Columbus, Ohio.
5 Z. Blaschke and S. Sanak, "Separation of Silty Fractions from Coal Slimes by Selective Flocculation", Zesz. Nauk, Akad. Gorn.-Hutn., Cracow, Gron., 66, (1975), 19-30.
6 R. Hucko, "Beneficiation of Coal by Selective Flocculation. A Laboratory Study", U.S.B.M. Report of Investigations 8234, (1977).

7 G.F. Brookes, M.J. Littlefair, L. Spencer, and A.A. Abdelrahman "The Selective Flocculation of Coal/Shale Mixtures Using Commercial and Modified Polyacrylamide Polymers", Fourteenth International Mineral Processing Congress, Toronto, Canada (1982).
8 J.A. Solari and J. Rubio, "Treatment of Brazilian Coal Fines", Fourteenth International Mineral Processing Congress, Toronto, Canada (1982).
9 Y.A. Attia and D.W. Fuerstenau, "Principles of Separation of Ore Minerals by Selective Flocculation", in Recent Developments in Separation Science, Vol. IV (N. Li, ed.), Chapter 5, CRC Press, 1978.
10 G.C. Sresty, A. Raja, and P. Somasundaran, "Selective Flocculation of Mineral Slimes Using Polymers", in Recent Developments in Separation Science, Vol. IV (N. Li, ed.), CRC Press, 1978.
11 Y.A. Attia, "Selective Flocculation - A State-of-the-Art Review", 111th Annual Meeting of AIME, Dallas, Texas, February, 1982.
12 Y.A. Attia, "Fine Particle Separation by Selective Flocculation", Separation Science and Technology, 17(3), pp. 485-493, 1982.
13 A.D. Paananen and W.A. Turcotte, "Factors Influencing Selective Flocculation-Desliming Practice at the Tilden Mine", AIME Annual Meeting, Denver, 1978.
14 A.F. Banks, "Selective Flocculation-Flotation of Slimes from a Sylvinite Ore", in Beneficiation of Mineral Fines, NSF Workshop Report, 1979.
15 T. Ishizuka, H. Hotta, and Y. Nishmura, "Coal-Deashing Process", U.S. Patent #4,437,861, March, 1984.
16 Y.A. Attia and J.A. Kitchener, "Development of Complexing Polymers for the Selective Flocculation of Copper Minerals", in Proceedings of 11th International Mineral Processing Congress, Cagliari, Italy, 1975.
17 Y.A. Attia, "Separation of Ultrafine Pyrite from Coal by Selective Dispersion and Flocculation", Environmental Science of Fossil Fuels Symposium, American Chemical Society Meeting, April 28 - May 3, 1985, Miami Beach, Florida.
18 Y.A. Attia, Private communications with J.A. Kitchener and D.W. Fuerstenau, 1975.
19 T. Ishizuka, Private communications with Y.A. Attia, 1981.
20 Y.A. Attia, H.N. Conkle, and S.V. Krishnan, "Selective Flocculation Coal Cleaning for Coal Slurry Preparation", 6th International Coal Slurry Combustion and Technology Symposium, Orlando, Florida, June, 1984.
21 D.M. Deason and Y.A. Attia, "Selective Flocculation Process for Cleaning of Beulah Lignite", Lignite Utilization and Processing Symposium, National Meeting of AICHE, March 24-28, 1985, Houston, Texas.

A LOW TEMPERATURE TECHNIQUE FOR COAL CLEANING

Shashi B. Lalvani[1], Charles B. Muchmore[2] and Karthik Ramaswami[3]
[1]Department of Mechanical Engineering and Energy Processes, Southern Illinois University, Carbondale, Illinois 62901,
[2]Department of Mechanical Engineerina and Energy Processes, Southern Illinois University, Carbondale, Illinois 62901,
[3]Department of Mechanical Engineering and Energy Processes, Southern Illinois University, Carbondale, Illinois 62901,

ABSTRACT

Coal slurry electrolysis in acidic electrolytes results in production of hydrogen gas at the cathode while coal undergoes desulfurization in the anode compartment. Hydrogen is evolved at cell voltages considerably less than that required for conventional water electrolysis and with Faradaic efficiency of about 95%. Over 50% of the total sulfur from coal can be removed as water soluble sulfate-sulfur under moderate condition of operation. The major anodic reaction involves depolarization of water by pyrite via the following reaction pathway:

Anode: $2FeS_2 + 16H_2O \rightarrow 2Fe^{+3} + 4SO_4^{-2} + 32H^+ + 30e^-$

Cathode: $2H^+ + 2e^- \rightarrow H_2$

Overall: $2FeS_2 + 16H_2O \rightarrow 15H_2 + Fe_2(SO_4)_3 + H_2SO_4$

Electroxidation rate of pyrite, a model compound for sulfur in coal is observed to increase via an autocatalytic reaction pathway. The rates of reaction of coal desulfurization can be enhanced by adding certain multivalent cations of Fe and Ce; these ions transfer charge between the anode (where they are regenerated) and coal particles thus accelerating the rates of reaction. The influence of the electrolysis time, temperature, particle size and the nature of electrolyte on the amount of sulfur removal, reaction rate, energy efficiency and selectivity of the process has been determined.

INTRODUCTION

Presently, physical coal cleaning methods in use remove a major part of the inorganic sulfur while most of the organic sulfur remains intact. The anticipated increase in the use of high sulfur coal and the more stringent restrictions provide enough impetus for investigating novel coal cleaning methods. One such technique is electrochemical desulfurization of coal slurries reported by Lalvani, et al (ref. 1). Anodic oxidation of aqueous coal slurries at potentials of 1.0 V vs. SCE (saturated calomel electrode) causes pyrite in coal to react and dissolve while hydrogen gas is produced at the cathode due to depolarization of water. In addition to inorganic sulfur, the process also accomplishes some organic sulfur removal, both sulfur forms are oxidized to water-soluble sulfate. This coal depolarized water electrolysis occurs at over 90% Faradaic current efficiencies for hydrogen production (ref. 2). A relatively small portion of the carbonaceous part of the coal gets oxidized to CO_2 while solvent-extractable oxygen functionalities are formed on the coal surface (ref. 3). The various reactions that occur during coal oxidation in acidic electrolytes are shown below (ref. 4).

Anodic Reactions

$$FeS_2 + 8H_2O \rightarrow Fe^{+3} + 2SO_4^{-2} + 16H^+ + 15e^- \quad (1)$$

$$FeS_2 \rightarrow Fe^{+3} + 2S + 3e^- \quad (2)$$

$$Fe^{+2} \rightleftarrows Fe^{+3} + e^- \quad (3)$$

thus pyrite (FeS_2) combines with water to produce ferric sulfate and elemental sulfur as co-products of anodic oxidation, whereas ferrous ion (present in the form of $FeSO_4$ impurity in coal) gets oxidized to ferric state. Ferric sulfate further oxidizes FeS_2 by the following mechanism:

$$FeS_2 + Fe_2(SO_4)_3 \rightarrow 3FeSO_4 + 2S \quad (4)$$

However, it is also possible that ferric sulfate at 100-130°C can further oxidize free sulfur to sulfate by the reaction

$$S + 3Fe_2(SO_4)_3 + 4H_2O \rightarrow 6FeSO_4 + 4H_2SO_4 \quad (5)$$

equations (4) and (5) have become the basis for the Meyer's Process (5).

A survey of the literature (ref. 6 - 7) shows that complexes of Fe^{+3} quantitatively oxidize mercaptans to disulfide. This reaction can be written as:

$$2RSH + Fe^{+3} \rightarrow RSSR + 2Fe^{+2} + 2H^+ \quad (6)$$

However, it is also possible to further oxidize disulfides to sulfate. The nature of these reactions is complex and not very well understood. It has recently been shown that thiophene, a model compound for organic sulfur in coal can be oxidized by ferric ions to the sulfate forms (ref. 1). Thus anodic oxidation of various sulfur compounds present in coal leads to formation of non-polluting sulfate and sulfur compounds.

Ferric ions also depolarize the carbonaceous part of the coal to produce various oxygen functionalities on the surface of coal which lead to formation of CO_2 upon decarboxylation (ref. 4):

$$C + Fe^{+3} \rightarrow C_{ox} + Fe^{+2} \quad (7)$$

$$C_{ox} \xrightarrow[Fe^{+3},\ H_2O]{-e^-} CO_2 + H^+ \quad (8)$$

ferrous ions produced via equations (4), (5), (6) and (7) undergo oxidation via equation (3) to form Fe^{+3}. These ferric ions again oxidize sulfur compounds and coal via equations (4) thru (8) in a catalytic manner.

Cathodic Reactions

Protons produced in the anode compartment via equations (1), (6) and (8) get reduced at the cathode to form H_2 gas:

$$2H^+ + 2e^- \rightarrow H_2 \quad (9)$$

The advantages of coal slurry desulfurization by electrochemical pathways are summarized below:

1. Production of H_2 gas at low temperatures (ambient to 90°C) at one-half the applied potentials required for conventional water electrolysis
2. Hydrogen gas is produced separately and free of impurities at high coulometric efficiencies (95-100%) in the cathode compartment.

In a recent paper, Lalvani, et al. (ref. 1) studied the effect of a number of process variables on the rates of electrochemical oxidation of Illinois No. 6 coal, and their preliminary investigations indicate that electrolysis in acidic electrolytes may remove up to 40% of the total sulfur from coal. Here we report even greater sulfur and ash removals form Colchester No. 2 coal slurries oxidized anodically with the help of mediators (for shuttling electrons) of Fe and Ce.

EXPERIMENTAL

The coal electrolysis cell made of Pyrex is shown in Figure 1. It is divided into two parts; I: working electrode (anode) compartment where the coal is kept in suspension in an electrolyte by a magnetic stirrer; this part of the cell is also provided with a Luggin capillary filled with saturated KCl and communicating with a saturated calomel electrode (SCE). A thermometer is also inserted in the electrolyte. The volume of this part of the cell is about 400 ml. The cathode and anode are both made from a platinum electrode of mesh 52 (0.004 in. diameter wire) supplied by Matthey Bishop, Inc. II: The counter electrode (cathode) compartment is separated form the working electrode compartment by a porous glass frit that confines the coal particles to the anode compartment. A hot plate is used to control the temperature of the cell. A potentiostat supplied by ECO Instruments (model 550) is used for applying constant potentials at the working electrode of the electrolysis cell.

Aqueous electrolytes were prepared with Baker analyzed reagents. Coal samples in pulverized form were obtained from the U.S. Department of Energy, Pittsburgh Energy Research Center and were ground to the desired mesh size.

The coal is first contacted with the electrolyte and brought to the desired temperature of reaction. A constant electrode potential is then applied to the reactor.

RESULTS AND DISCUSSION

Electrolysis of coal slurries in acid electrolytes at a constant electrode potential of 1.2 V vs. SCE and 65°C resulted in formation of CO_2 and H_2 at the anode and cathode respectively. Rapid reaction rates during the initial stages

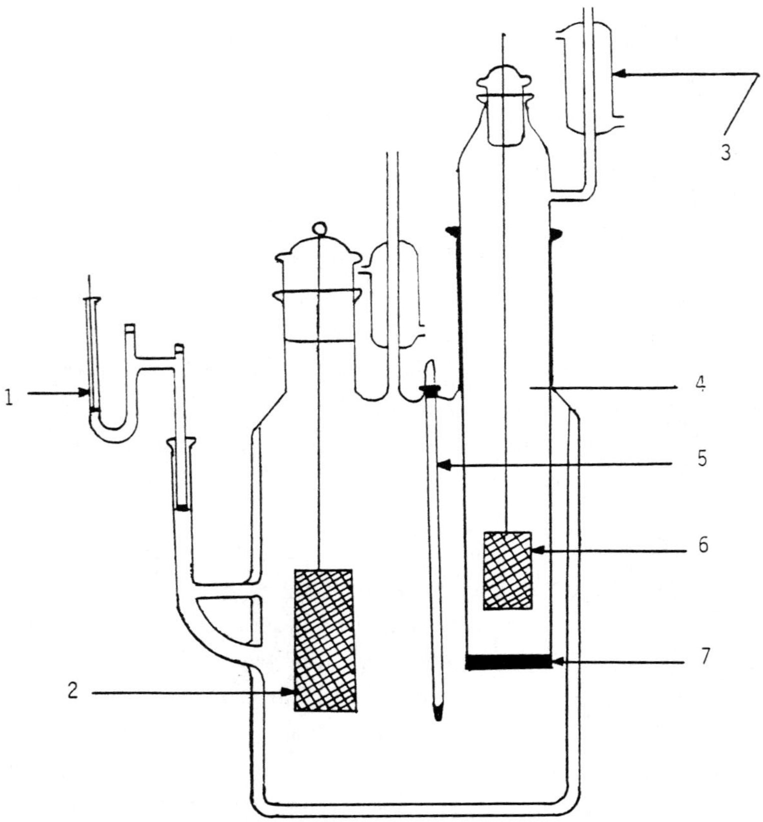

1. CALOMEL REFERENCE ELECTRODE
2. WORKING ELECTRODE
3. CONDENSER
4. CATHODE COMPARTMENT
5. THERMOMETER
6. COUNTER ELECTRODE
7. GLASS FRIT

Figure 1 Coal Electrolysis Cell

of electrolysis were observed to decline after about one hour of reaction time and reach steady state limiting values.

Sulfur Removal From Coal: Addition of Ce(IV)

Electrolysis at 1.2 V vs. SCE

The influence of Ce(IV) on the total amount of sulfur removal and anodic current density is shown in Figure 2. Fifteen grams of Colchester No. 2 coal (63 - 75 μm) were electrolyzed for 6 hours at a Pt mesh anode (19.4 cm^2) and 65°C. A mixture of 5% HCl + 15% H_2SO_4 was used as an electrolyte. Both, the sulfur removal and current density increase with a slight increase in Ce(IV), 1.66 mM Ce(IV) accomplishes a maximum sulfur removal of 34.4% while a current density of 34.4 mA/sq cm is achieved. However, a further increase in the Ce(IV) concentration lowers the sulfur removal and current density. The influence of Ce(IV) on the H_2 production efficiency, sulfur removal and current density is explained below.

Ce(IV) is a better oxidizing agent than Fe(III), it oxidizes Fe(II) to Fe (III) as shown by the following reaction:

$$Ce^{+4} + Fe^{+2} \rightarrow Ce^{+3} + Fe^{+3} \qquad (10)$$

Fe(III) produced by reaction (3) can catalyze reactions (1) and (7), as shown by Lalvain et al. (ref. 1), The relatively large concentrations of Ce(IV) added to the electrolyte leads to corresponding large concentrations of Fe(III). It is shown that although Fe(III) is a desirable agent for leaching pyrite from coal, in large concentrations it undergoes surface adsorption (ref. 5). We believe in our experiments readsorption of ferric sulfate on the surface of the coal results in lower sulfur removals and lower reactivity. It is also possible that chlorine produced in aqueous solutions under electrolytic conditions enhances sulfur removal. No experiments were done to determine the extent of chlorine addition to coal.

Sulfur Removal From Coal: Addition of Fe (III)

Colchester No. 2 coal was electrolyzed in a mixture of 5% HCl + 15% H_2SO_4 at 1.2 V vs. SCE and 65°C. Sulfur removals obtained were lower with Fe(III) as compared to the removal attained with Ce(IV) under identical experimental conditions. Figure 3 displays the total sulfur removal and anodic current density as a function of Fe(III) concentration. Sulfur removal is initially found to increase with Fe(III) concentration and reaches a maximum value of 33.3% when 10 mM of Fe(III) are present in the slurry, however, a further increase in the Fe(III) concentration leads to a steady decline in sulfur removal. This observation is in agreement with the hypothesis that iron sulfate present in large concentrations can readsorb on the coal surface, and thus

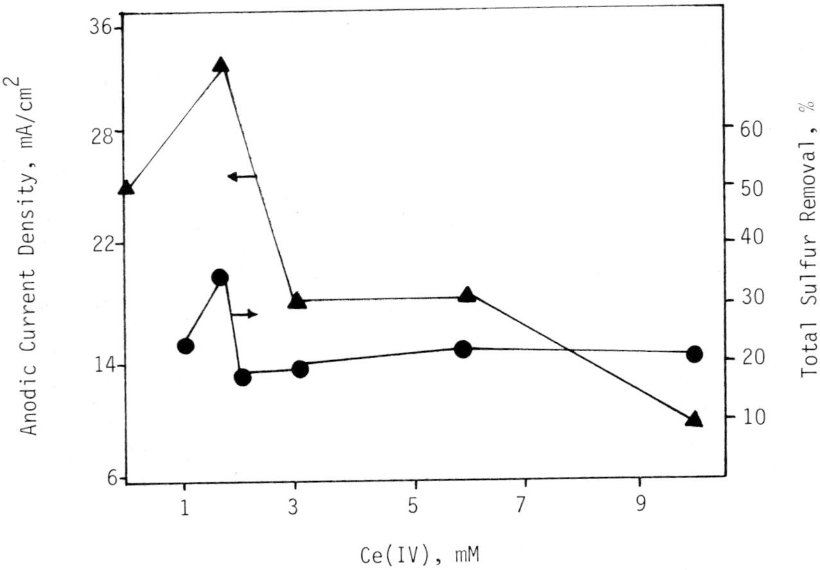

Figure 2 Current Density and Sulfur Removal vs. Ce(IV) Added

15 g of Colchester No. 2 (63-75 μm) coal were electrolyzed in 300 ml of 5% HCl + 15% H_2SO_4 at 1.2 V vs. SCE and 65°C for 6 hours. Anode was a Pt mesh of 19.4 cm^2.

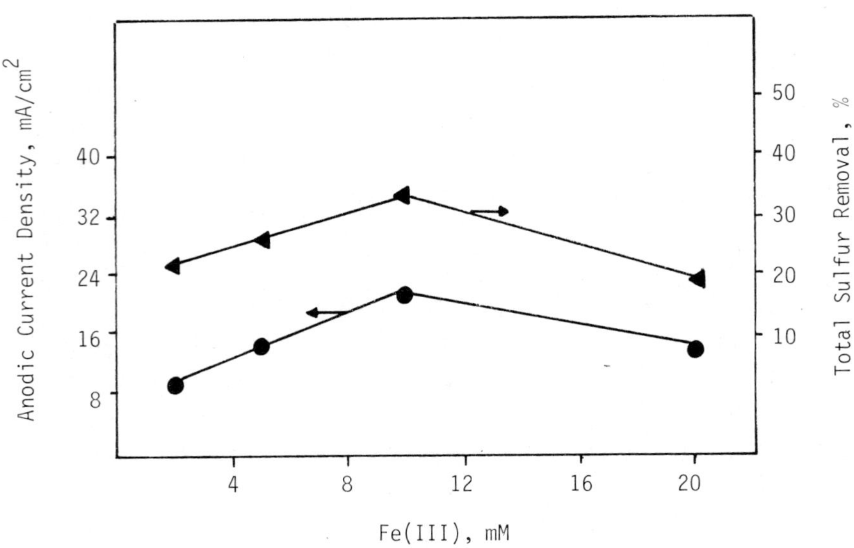

Figure 3 Current Density and Sulfur Removal vs. Fe(III) Added

Reaction conditions as in Figure 2

lower sulfur removals are observed (ref. 5).

Effect of Electrolysis Time

Colchester No. 2 coal was electrolyzed in a mixture of 5% HCl + 15% H_2SO_4 at 1.3 V vs. SCE and 75°C, the experimental data and results are reported in Table I. In one experiment (No. 69), sulfur removal of 57% was achieved while in another (No. 76) the sulfur to heating value ratio of 0.79 lb S/million BTU found is considerably lower than the NFPS (New Federal Performance Standard) ratio of 1.2 lb S/million BTU. The sulfur to heating value ratio as well as the energy efficiency quotient (% S removal x 10^4 per unit watt of electrical energy input) are found to decrease with the electrolysis time. A monotonic relationship between the reduction in sulfur content (percentage basis) of coal and electrolysis time is observed. Up to 80% sulfur reduction is possible for coal slurries electrolyzed for a period of 15 hours.

Effect of Particle Size

The influence of coal particle size on sulfur removal was studied in a preliminary fashion. Under relatively mild oxidizing conditions (i.e., low concentrations of cerium), smaller particles (63 - 75 μm) accomplish greater sulfur removals as shown in Figure 4, however, increase in the Ce(IV) concentration over 3 mM results in higher sulfur removals for larger size (125 - 150 μm) particles. The sulfur removal for 63 - 75 μm coal particles is observed to increase initially with the Ce(IV) concentration and it reaches a limiting value about 44.4% whereas the maximum amount of sulfur removed for 125 - 150 μm particles is observed to be over 53%.

CONCLUSIONS

Significant sulfur removals from coal by electrolysis under relatively mild reaction conditions are possible. An added advantage of this low temperature cleaning process lies in its ability ot accomplish simultaneous hydrogen production at low energy expenditure. More work is needed to elucidate the reaction mechanism and determine the various sulfur forms removed from coal.

ACKNOWLEDGEMENTS

We thank the Department of Energy and the American Chemical Society for financial support.

TABLE I

Effect of Electrolysis Time on Coal Slurry Electrolysis at 1.3 V vs. SCE. 15 g of Colchester No. 2 (63 - 75 μm) coal were electrolyzed in 300 ml of 5% HCl + 15% H_2SO_4 at a Pt mesh anode (19.4 sq cm) and 75°C

Exp. # RR #	Electrolysis Time, hours	Current Density, mA.cm²	Charge, Coulombs	Reduction in S Content, %	Sulfur to Heating Value Ratio, lb.S/ 10^6 BTU	Heating Value Recovery Ratio, %	Energy Efficiency Quotient For Sulfur Removal
74	1.75	44.95	6114	37.18	1.56	93.1	21.7
75	4.0	47.5	13840	45.6			10.7
69	6.0	53.7	22630	60.3	1.97	96.0	7.8
76	10.0	43.9	33555	69.8	0.79		6.3
77	15.0	36.2	42998	79.8			6.6

NOTE: 1. Total sulfur and ash content of raw coal were 2.8% and 6.43% respectively.

2. Heating value of raw coal was found to be 13,325 BTU/lb.

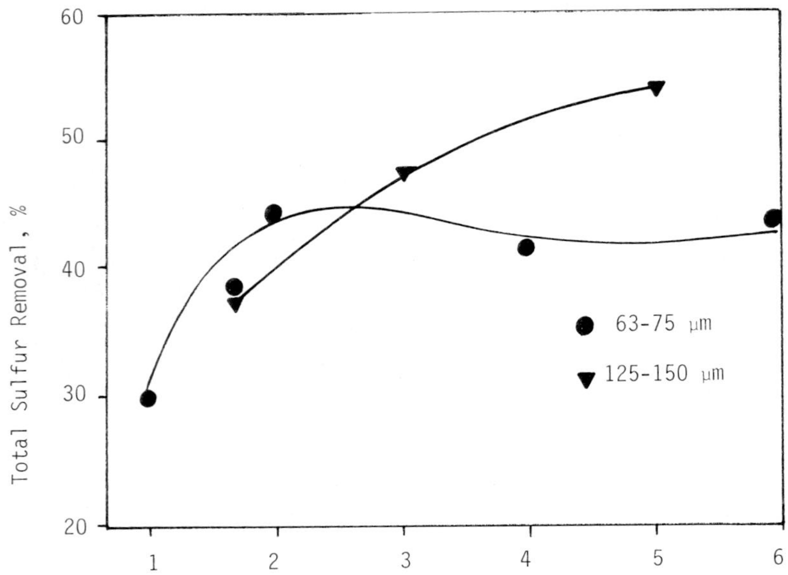

Figure 4 Effect of Particle Size on Sulfur Removal

15 g of Colchester No. 2 coal were electrolyzed in 300 ml of 5% HCl + 15% H_2SO_4 at 1.3 V vs. SCE and 65°C for 6 hours on a Pt mesh anode of 19.4 cm^2.

REFERENCES

1. S. Lalvani, M. Pata and R. W. Coughlin, Fuel, 62 (1983) pp. 427-431.
2. R. W. Coughlin and M. Farooque, Fuel, 48 (1979) pp. 705-708.
3. S. B. Lalvani and R. W. Coughlin, Solubility Enhancement of Electrolyzed Coal Residues, a paper accepted for publication in the Fuel Processing Technology.
4. S. B. Lalvani, Electrochemical Oxidation of Coal in Aqueous Electrolytes, Ph. D. Dissertation, University of Connecticut, 1982, 210 pp.
5. R. A. Meyers, Coal Desulfurization by Electrolysis, Marcel Dekker, Inc., NY, 1977, pp. 52-92.
6. S. Patai, The Chemistry of the Thio Group, Part 2, John Wiley and Sons, London, 1974, 213 pp.
7. A. V. Toholsky, The Chemistry of Sulfides, Interscience Publishers, NY, 1966, 312 pp.

PART 5

COAL DESULFURIZATION BY CHEMICAL CLEANING TECHNIQUES

THE USE OF SOLVENT-EXTRACTION METHODS TO FREE COAL FROM SULFUR CONTAINING MINERAL MATTER

D. DOLLIMORE[1] and A. TURNER[2]
[1]The Chemistry Department, The University of Toledo, Toledo, Ohio
[2]Lanstar, Cadishead, Lancs, U.K.

ABSTRACT

A report of work on coal and related model materials subjected to solvent extraction procedures is presented. It is shown that this process can be used to free coal and carbonaceous matter from its mineral content and from inorganic sulfides in particular. Catalytic hydrogenation followed by fractional distillation produces a gasoline cut. However the practice of precipitating out various coal fractions by using solvent extraction with liquids of low dielectric constant enables a selectivity to be achieved and the production of carbons of high purity tailor made for specific purposes, e.g. adsorbents, catalysts, filter mediums. The characterization of such carbons from adsorption studies is presented.

In laboratory studies the coal used was a medium volatile strongly coking type of coal (N.C.B. Rank Code Number 301). The solvent used was quinoline, and the reprecipitation was achieved using toluene and then petroleum ether (80-100°C fraction). It is shown that oxidation of the residues obtained enhances the adsorption which is an important marketing aspect for speciality carbons. The mechanism of the oxidation process leading to the production of these high area carbons is discussed. The important value of the studies is that the solution formed is free of inorganic minerals, and the subsequent formation of carbons results in high surface area carbon adsorbents.

INTRODUCTION

The sulfur content in coal may be divided into two categories; there is that which has the sulfur combined in organic form and there is that part where the sulfur is present in organic form. The inorganic sulfur is often present as pyrite $Fe\ S_2$, and present also as a separate phase distinct from but mixed in with the coal components. The sulfur present in organic form is incorporated into the organic structures and is thermally resistant. This has been demonstrated by studies where pitch and tar had sulfur containing materials dissolved in them and were then carbonized in an inert atmosphere. The sulfur functional groups were incorporated into the tar structure and then retained until high temperatures had been reached (ref. 1). The process of solvent extraction is therefore going to leave behind as "insoluble" the inorganic sulfur content and indeed most other mineral matter present in the coal. The organic sulfur will pass into the extraction liquors along with other functional groups. The main drawback to the utilization of coal products is the deficiency in hydrogen and in any production or conversion process a stage such as catalytic hydrogenation is necessary (ref. 2). Thus the solvent extraction of British coals in anthracene oil (ref. 3) needs a further hydrogenation with a solid phase catalyst which leads on fractional distillation to a gasoline cut. The practice of liquifaction however can be utilized to separate out various coal fractions with liquids of lower dielectric constant to produce carbons of high purity. These can be processed for various purposes such as adsorbents or electrodes (ref.4). It is this which is the subject of the present report.

Sulfur in Coal

Krevelen (ref.5) points out that chloroform and pyridine extracts give residues with similar sulfur content and concludes that organic sulfur compounds are uniformly distributed throughout the coal. Lissner and Nemes (ref.6) report that groups such as

>C-SH and >C-S-S-C<

were lost between 100°C and 300°C in the presence of superhealed steam as a result of hydrolyte fission. Groups of the kind >CH-SH and >CH-S-CH< evolved hydrogen sulfide with superheated steam. Wnekowska (ref.7) concluded that organic sulfur in coal consisted of thioether and thiophenic groups and that the ratio thio-ether/thio-phenic decreases from 3 in high volatile coals to 1/6 in semi-anthracite. Longanbach et al. (ref.8) find on the basis of short-residence time coal liquefaction that not only is all the inorganic sulfur removed in the solvent proces, i.e. is not dissolved, but an average of 12% of organic sulfur is also removed. Their reaction temperature was high however (413-454°C). The coals were West Kentucky Coal blends with a total of around 3-4% sulfur types (wt %) some 1-2% being pyrotic with the organic sulfur around 1.5%.

The mineral sulfur content is mainly in the form of pyrite. It would seem to be mainly unaffected by the solvent extraction process. However it is often utilized as a catalyst in subsequent hydrogenation. This is discussed in some detail by Bockrath and Schroeder (ref.9). The relationship between pyrite FeS_2, and pyrrhotite FeS_2 is the subject of the study by Richey (ref.10) who notes that pyrroholite samples vary in composition (e.g. FeS troilite, Fe_7S_8 monoclinic pyrrholite; and Fe_9S_{10}-$Fe_{11}S_{12}$ intermediate pyrrholite) and relates this to catalytic hydrogenation of coal solvent extracts.

Solvent Extraction of Coals

Various methods of solvent extraction have been used. Wheeler (ref.11,12) extracted with pyridine at 115°C, followed by residue production using chloroform, petroleum ether, ethyl ether and acetone in succession to give α,β, and γ resins. He showed the γ extraction residue produced a char and not a coke on carbonization. Bone (ref.13) used the following sequence of liquids; benzene, petroleum ether, the solution was evaporated and extracted once more with petroleum ether, and the precipitate was extracted with ethanol.

Types of extraction processes have been classified by Oele et al. (ref.14) as follows:

i. Non-specific extraction: takes out just a few percent of the coal usually below 100°C.

ii. Specific extraction: takes out 20-40% at temperatures below 200°C, using nucleophilic solvents

iii. Extractive disintegration, carried out above 200°C leading to high extraction yields

and

iv. Extractive chemical disintigration; with solvents which are hydrogen donors. The chemical composition of the solvents are affected and temperature is normally over 300°C.

Other schemes of solvent extraction have been put forward by Adam and Sach (ref.15), Demann (ref.16), Mallison (ref.17), Dickenson (ref.18), and McNeil and Wood (ref.19) who have all employed various considerations to determine their choice of solvent. Polarity is one of the main properties having a bearing on solvent power, for example quinoline dissolves more than toluene, which in turn dissolves more than petroleum ether. Table 1 is taken from the Coal Tar Research Association Supplementary Method issued in 1963.

In the present work the solvents selected were quinoline, toluene and petroleum ether (80-100°C fraction).

EXPERIMENTAL

Solvent Extraction: As already noted the solvents selected were quinoline, toluene and petroleum ether (80-100°C fraction). The reasons for their selection were:

(a) Pure grades readily available

(b) Relatively low toxicity

(c) Quinoline and toluene were specified by many of the large-scale industrial uses of electrode binders for quality characterization purposes.

A strict control of experimental technique was important and standardization of the procedure essential. The results are not absolute for the particular solvent used. Mallison (ref.17) and Krenkler (ref.20) extracted first with a powerful solvent before precipitating with a weaker one and the same practice was adopted in this work. The extraction procedure adopted was similar to that of Morgan et al. (ref.21), namely 100 g. of sample were crushed and ground to pass a B.S.S. 60 mesh sieve and used immediately. It was added slowly with stirring to 250 ml of warm (70°-90°C) quinoline and maintained at this temperature for two hours. The mixture was then filtered through a Buchner funnel fitted with Whatman No. 50 filter paper. The insoluble material was then washed with a further 250 ml of warm quinoline a little at a time. The filtrate was retained in a separate container and the precipitate extracted in a Soxhlet with toluene until the returning condensate was almost colorless. The quinoline-insoluble material (α-resin) was then dried in a vacuum oven at 100-110°C for 2 hours.

To the filtrate above was added 250 ml of toluene to precipitate the quinoline-soluble/toluene-insoluble fraction (β-resin). The mixture was allowed to stand overnight at ambient temperature to complete the precipitation, and the precipitate filtered off and dried as above.

To the filtrate was then added 250 ml of petroleum ether (80-100°C fraction) to precipitate the quinoline-soluble, toluene soluble/petroleum ether-insoluble fraction (γ-resin). The procedure was completed as above. The characterization of coal N.C.B. Rank 301 and two coke-oven pitches one of softening point 105°C and the other of softening point 150°C are given in Table 2. The yield of the resin fractions is shown in Table 3.

The various resins could be oxidized at various temperatures. The experimental details of these kind of experiments have been described previously (ref.4, 22-24), and were designed to eliminate diffusion problems.

Figure 1 demonstrates a typical plot of degree of oxidation at 485°C for the Coal 301, α resin, and at 460°C for the β resin. Surface areas are measured by nitrigen adsorption are shown in Figures 2 and 3. It was found that precarbonization at 900°C in nitrogen allowed higher surface areas to be achieved. This is shown in plots of surface area against degree of oxidation for the Coal 301 α and β resins. The elementary analysis for the α and β resins is given in Table 4, where similar analysis data on Pitch 100 and 150 is also given for comparison purposes.

The data for the surface area determinations from the nitrogen adsorption isotherms and the activation energy for the activation process is shown in Table 5. The mass loss on carbonization is shown in Table 6.

DISCUSSION

The sulfur content of the original materials given in Table 2 shows the value for the coal to be 0.79%. This is increased to 1.09% in the α resin, decreased to 0.39% on carbonization and is zero in the β resin. It is tempting to say this is a general behavior pattern but the trend is reversed in the resins

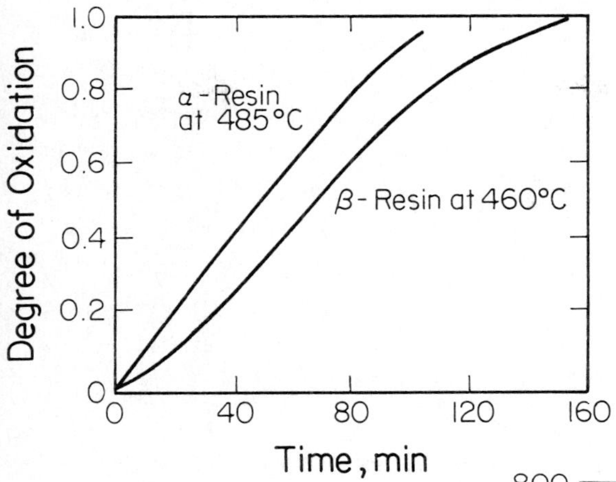

Fig. 1. Typical Oxidation Plots for Coal 301.

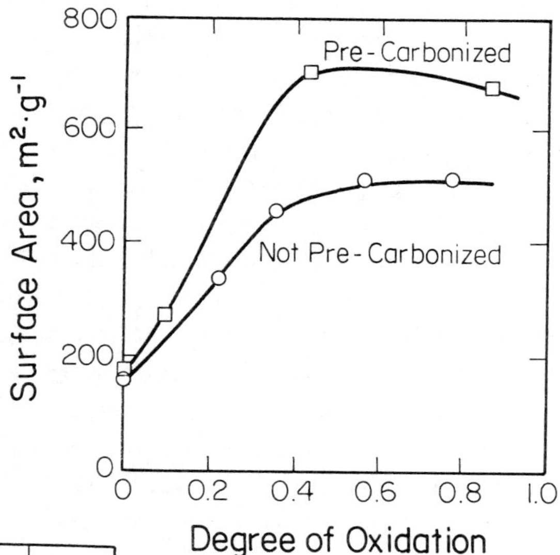

Fig. 2. Surface Area V. Degree of Oxidation for Coal 301 α Resin.

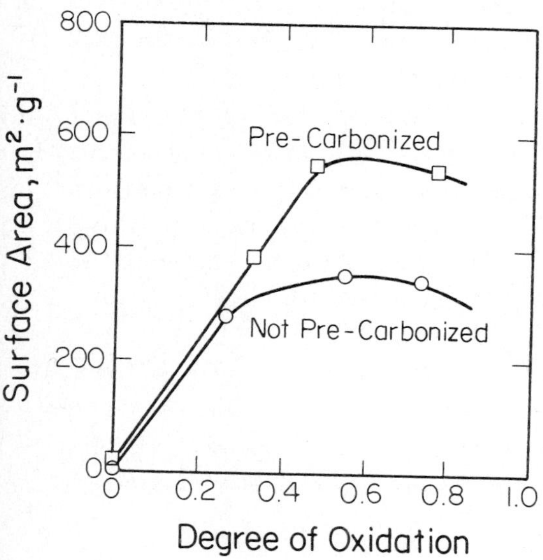

Fig. 3. Surface Area V. Degree of Oxidation for Coal 301 β Resin.

from Pitch 150. It has already been pointed out that the inorganic sulfur is removed by the solvent extraction process.

For the coal it follows that for every sulfur atom present there are 324 carbons. The α resin shows an increased sulfur content, i.e. for every sulfur atom present there are 171 carbons, whilst the position is improved on carbonisation when it is found that for every sulfur atom present there are 567 carbons. No sulfur at all could be detected in the β resin. This variation in the distribution of sulfur atoms with respect to the carbon atoms present in the resins is in contradiction to some statements made by other investigations who suggest on equal distribution among all the solvent extracts (ref.5). However, the values are small and no clear pattern emerges when comparing the various resins from the coal and the pitch resin fractions.

The rate of oxidation of the α and β resins was independent of mesh size over the constant surface area range of the degree of oxidation (see Figure 2 and 3 to observe this range). An average pore size could be calculated from the adsorption isotherms and typical values are shown in Table 7. Here a considerable surface area is to be found as the internal surface of the pores. In all cases the mean pore radii of the activated resin fractions are several times larger than the molecular diameter of oxygen (3.04Å, ref.25). Thus it is concluded that the rate of the oxidation reaction (over the constant surface area range of degree of oxidation) is not controlled by the diffusion of oxygen through the pores to adsorption sites on the carbon surface but is controlled by the rate of the chemical reaction. The fact that the reaction is of zero order is also an indication of the absence of diffusion control. This aspect has been discussed in detail with regard to the α- and β- resin extracts from coal tar pitch (ref.4). It is concluded that during the part of the oxidation in which the surface area remains constant, the molar active site concentration (i.e. the number of sites per unit area associated with the oxidation process) for the α-resin is greater than for the β resins (Table 5).

This difference could be accounted for on the basis of either the presence of a greater amount of metallic impurities in the α-resin which can act catalytically or since at least part of the β-resin is observed to pass through a liquid state during the course of carbonization, the formation of a metaplast could prevent the easy production of labile carbon atoms and render difficult the creation of active sites by subsequent solidification of mobile liquid products.

The behavior of solvolytic resin fractions on heat treatment in an inert atmosphere has been described by other workers (ref.26-29). The general conclusion is that where α-resins are apparently inert and undergo no visible change on heating to form chars, β and α- resins intumesce and pass through a fluid state. These observations have been confirmed in this work, in that all the α-resins were unchanged on carbonization whereas all the β-resins showed evidence of having passed through a liquid state (at least to some extent) and all the γ-resins had melted and fused into a coherent mass. The β-resins on heating did not melt completely, but in some cases sufficient of the resin became liquid to cause the formation of spherical coke particles (0.5∿3mm diameter), presumably as a result of the surface tension of the liquid (lower molecular weight) components gathering up the unmelted (higher molecular weight) components and rolling them into balls which then coked at the elevated temperature. The loss in weight on carbonization (Table 6) and the general trend of the results tabulated here bears out these visual observations. The surface area data based on adsorption of nitrogen (see Figures 2 and Tables 5 and 7) indicate that upon activation these resin fractions of high purity yield good adsorbent materials of a microporous nature.

REFERENCES

1 J. Dollimore, unpublished work.
2 J. Owen, Coal and Modern Coal Processing. An Introduction. G.J. Pitt and G.R. Millward (Eds.), Academic Press, London, 1979, 163.
3 J.W. Clarke, G.M. Kimber, T.D. Rantell and D.E. Shipley, Coal Liquefaction Fundamentals, ACS Symposium Series, 139, (1980), 111.
4 D. Dollimore and A. Turner, Trans. Farad. Soc. 66 (1970) 2655.
5 D. W. van Krevelen, Coal, Elsevier Amsterdam (1961) 171.
6 A. Lissner and A. Nemes, Brennstoff-Chem. 16 (1935) 101.
7 L. Wnekowska, 3rd Intern. Conf. on Coal Science, Volkenburg, 1959.
8 J.R. Longanbach, J.W. Droege and S.P. Chauhan, Coal Liquefaction Fundamentals, ACS Symposium Series, 139 (1980) 165.
9 B.C. Bockrath and K.T. Schroeder, New Approaches in Coal Chemistry, ACS Symposium Series, 169 (1981) 153.
10 W.D. Richey, New Approaches in Coal Chemistry, ACS Symposium Series, 169 (1981) 349.
11 M.J. Burgess and R.V. Wheeler, J. Chem. Soc. 99 (1911) 649.
12 J.E. Ashmore and R.V. Wheeler, J. Chem. Soc. (1933) 1405, (1934) 474.
13 W.A. Bone and L.J. Tei, Proc. Roy. Soc. (London) A 147 (1934) 58.
14 A.P. Oele, H.I. Waterman, M.I. Goedkoor and D.W. van Krevelen, Fuel, 30 (1951) 169.
15 N.K. Adam and J. Sach, J. Soc. Chem. Ind., 48 (1929) 336.
16 K. Demann, Brenn.-Chem., 14 (1933) 121.
17 H. Mallison, Bit. Terre, Asph. Peche., 1 (1950) 313.
18 J. Dickinson, J. Soc. Chem. Ind. 64 (1945) 121.
19 D. McNeil and L.J. Wood, Ind. Carb. and Graphite Conf., (1958) 162.
20 K. Krenkler, Bit. Teere, Asph. Peche., 1 (1950) 234.
21 M.S. Morgan, W.H. Schlag, and M.H. Wilt, J. Chem. Eng. Data, 5 (1960) 81.
22 D. Dollimore and A. Turner, 3rd Conference on Industrial Carbon and Graphite, Soc. of Chem. Industry, 1971, 220.
23 D. Dollimore and A. Turner, Thermal Analysis, Vol. 2, Proc. Third ICTA Davos, 1972, 463.
24 D. Dollimore and A. Turner, Gas Chemistry in Nuclear Reactions and Large Industrial Plant (Editor A. Dyer) Heyden, 1980, 164.
25 J. H. de Boer and S. Kruyer, Proc. Acad. Sci. Amst. 56B (1953) 67.

26 F. Popper, CTRA Report, (1952), Number 0079.
27 A. Darney, Ind. Carb. and Graphites Conf., (1958) 152.
28 E.W. Volkmann, E.O. Rhodes and L.T. Work, Ind. Eng. Chem., 28 (1936) 733.
29 O. Bruckner and H. Huber, Gas U. Wassenfach, 91 (1950) 104.

TABLE 1

Solvent Extraction Method Advocated by Coal Tar Research Association 1,22,63

Solvent	Solvent Action	Insoluble Material (type)
Anthracene Oil Pyridine Quinoline Nitrobenzene	Powerful	Highest Molecular Weight
Benzene Toluene Acetone	Fairly Powerful	Medium Molecular Weight
Petroleum Ether Methanol Hexane Dioxane plus Heptane	Moderate	Low Molecular Weight

TABLE 2

Properties of Coal and Pitches

	Pitch 100	Pitch 150	Coal 301
Softening Point R and B, °C	105	147	--
Density g cm^{-3}	1.30	1.34	1.31
Coking Volume	57.0	62	30.4
Ash Content	0.26	0.34	4.80
Element Analysis (dried mineral free basis) % by wt. carbon	92.46	93.02	85.58
Hydrogen	4.24	4.01	4.42
Nitrogen	0.88	0.81	1.34
Sulfur	0.72	0.79	0.70
Oxygen	1.44	1.03	3.16
Iron	830ppm	1700ppm	--
C/H Ratio	1.82	1.93	1.61

TABLE 3

Yield of Resin Fractions

	Pitch 100	Pitch 150	Coal 301
α Resin % Yield	11.30	19.70	66.04
β Resin % Yield	26.67	30.21	8.22
γ Resin % Yield	5.53	3.94	--

TABLE 4

Elementary Analysis of Solvent Extract Resin Fractions

Sample	Solvent Extract	Pre-Carbonized	Ash %	C %	H %	N %	S %	O %
Pitch 100	α Resin	-	1.70	91.10	2.40	1.75	1.57	1.50
		950°C/N_2	0.34	92.20	3.81	0.91	1.08	1.65
	β Resin	-	0.38	92.45	3.05	1.62	0.74	1.76
		950°C/N_2	0.32	92.39	1.22	0.47	0.93	4.67
	γ Resin	-	0.09	88.65	4.39	3.02	0.37	3.48
Pitch 150	α Resin	-	2.29	92.34	2.50	0.60	0.20	2.07
	β Resin	-	0.30	91.57	4.12	2.22	0.71	1.08
	γ Resin	-	0.01	87.31	4.89	2.96	1.62	3.21
Coal 301	α Resin	-	4.85	70.01	2.46	0.84	1.09	20.45
		950°C/N_2	4.22	81.58	1.32	1.19	0.39	11.30
	β Resin	-	0.09	85.13	5.34	3.53	Nil	5.91

TABLE 5

Surface Area, Activation Energy and Relative Molar Active Site Concentrations for Activated Coal and Pitch Solvent Resin Fractions

	Sample	α Resin		β Resin		γ Resin
Precarbonized			950°C/N_2		950°C/N_2	
Max Surface Area Developed m^2/g	Pitch 100	350	690	330	480	270
	Pitch 150	290	520	320	350	162
	Coal 301	540	760	416	615	-
Activation Energy Kcals $mole^{-1}$	Pitch 100	39.5	-	29.3	-	24.0
	Pitch 150	50.5	24.2	36.4	20.4	22.7
	Coal 301	27.9	23.0	22.4	19.2	-
Relative Molar Active Site Concentration	Pitch 100	10^5	-	10^2	-	1
	Pitch 150	10^{10}	10	10^4	1	1
	Coal 301	10^2	10	1	1	-

TABLE 6

Mass Loss on Carbonization of Coal and Pitch Solvent Resin Fractions

Sample	α-Resin	β-Resin	γ-Resin
Pitch 100	0.9%	9.4%	66.3%
Pitch 150	1.4%	7.6%	73.7%
Coal 301	5.2%	63.8%	-

TABLE 7

Adsorption Data of Typical Resins

Sample	Degree of Oxidation	Specific Surface Area m^2/g	$\bar{r}$ mean Radius in Å
Pitch 100 α Resin precarbonized at 950°C/N_2	0	37	10.3
	25.0	292	9.4
	52.0	543	10.5
	76.0	580	13.6
Coal 301 α-Resin	0	157	9.7
	21.8	331	8.5
	39.1	447	9.1
	58.6	528	9.4
	78.6	545	10.2

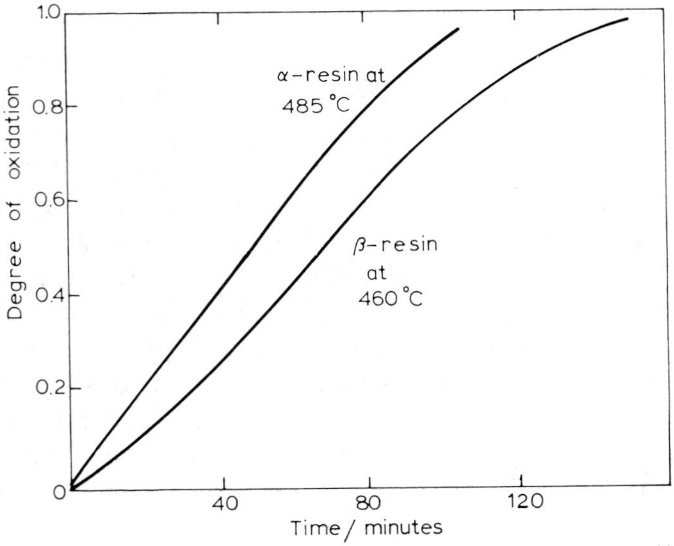

Figure 1. Typical oxidation plots for coal 301

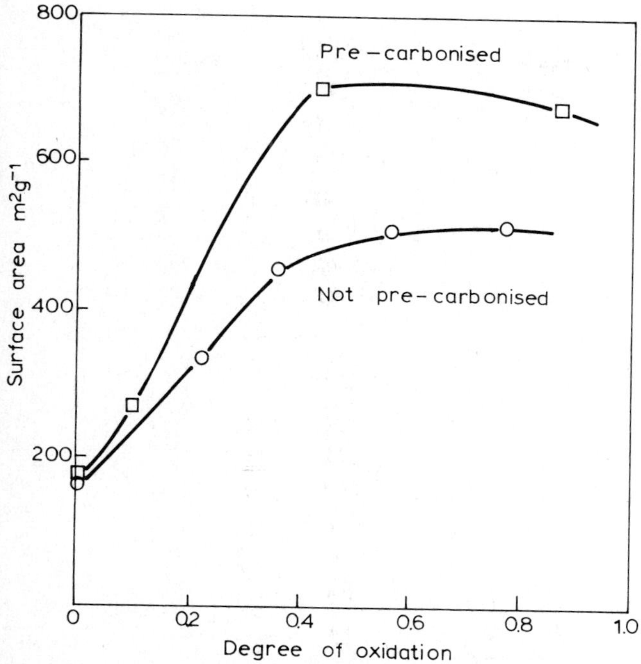

Figure 2. Surface area v. degree of oxidation for coal 301 α-resin

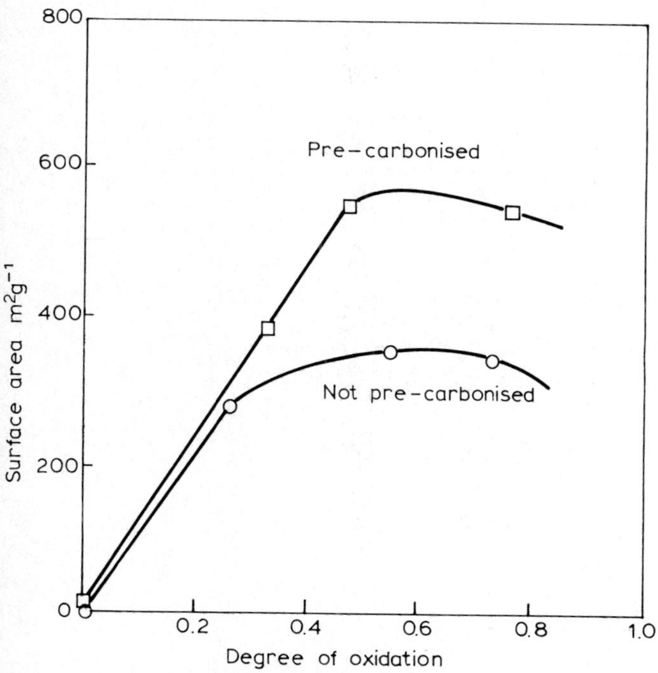

Figure 3. Surface area v. degree of oxidation for coal 301 β-resin

PRECOMBUSTION COAL DESULFURIZATION WITH SODIUM HYPOCHLORITE

I. M. BRUBAKER[1] and T. STOICOS[2]

[1]Separation Science and Technology Department, Signal Research Center, Inc., P.O. Box 5016, 50 East Algonquin Road, Des Plaines, IL 60017-5016 USA

[2]T. Stoicos, Department of Chemical Engineering, School of Chemical Sciences, University of Illinois at Urbana/Champaign, Urbana, Illinois 61801 USA

ABSTRACT

Treatment of coal with sodium hypochlorite removes substantial quantities of sulfur compounds, especially those containing organic sulfur, under mild conditions. The desulfurization efficiency is enhanced when the coal is pretreated with ammonium hydroxide (chemical comminution) and post-treated with sodium carbonate solutions. The initial and final sulfur levels of the three Illinois coals tested (-100 mesh) are 4.7%, 3.6% and 2.6% and 1.0, 0.8 and 0.9%, respectively. After the completion of the desulfurization procedure the heating value recoveries for each sample are 81%, 89% and 95%. The corresponding chlorine contents are only 2.5%, 2.0% and 2.2%. The effects of coal particle size, contact time and concentration of reagents on sulfur removal were studied. Scanning electron microscopy shows that the sodium hypochlorite affects the morphology of the coal particles.

INTRODUCTION

Physical coal cleaning methods have progressed dramatically over the years and currently they are widely practiced by the coal industry. The drawback of these processes is that they can remove only 40-65% of the pyritic sulfur, but none of the organic. The sulfur separation achieved with the majority of these practices is based on the inherent density difference between the pyrite and organic matrix of coal. Recently, high gradient magnetic separation of pyritic sulfur has emerged as a potential physical coal cleaning method, but it is still in the early stages of development.

The increasingly stricter requirements to control the acid rain problem and the inability of the physical routes to remove organic sulfur from coal has led to the development of many chemical desulfurization processes. All these methods can remove most of the pyritic and up to 50% of the organic sulfur, but none has been commercialized. Despite their effectiveness, the capital and operating costs associated with them are still prohibitably high. They use costly and sometimes corrosive chemicals and require special equipment because they are carried out under extreme conditions of high temperature and pressure.

Our desulfurization scheme would be a procedure which includes a combination of physical and chemical methods and removes substantial amounts of both pyritic and organic sulfur from coal in an effective and efficient manner. This research focuses on new chemical desulfurization methods. In order to insure the generality of the desulfurization procedure several coals with varying amounts and proportions of pyritic and organic sulfur have been evaluated.

The process involves successive treatments of the coals with NaOCl (commercial Clorox, common bleach) which oxidizes and selectively removes sulfur compounds, especially those containing organic sulfur. This sequence is preceded by a pretreatment with NH_4OH which acts as a comminution agent. This pretreatment does not cause any desulfurization, but enhances the efficiency of the hypochlorite by making the sulfur sites more accessible and prone to oxidation. After the hypochlorite treatment the coals are leached with hot Na_2CO_3 solution which hydrolyzes and splits off the oxidized sulfur compounds still attached to the hydrocarbon matrix.

EXPERIMENTAL

Three Illinois coal samples were used in this investigation. They were obtained from the State Geological Survey Division of the Illinois Department of Energy and Natural Resources. The analyses and the heating values of these coals are listed in Table 1.

TABLE 1

Analyses of Untreated Illinois Coals (-100 Mesh)

	Coal Samples		
Mass, %	1	2	3
Hydrogen	4.9	5.4	5.0
Carbon	51.1	59.5	66.8
Nitrogen	0.98	1.13	1.47
Ash	19.6	12.3	10.6
Oxygen	18.6	18.0	13.5
Sulfur	4.74	3.62	2.57
Chloride	0.11	0.07	0.01
BTU/lb	8,768	10,254	11,411

The experiments were carried out in batch mode at room temperature and atmospheric pressure. Before use, the coal was sized to pass through the desired mesh screen. The typical procedure was as follows: 20 g sample was slurried in a Pyrex Erlenmeyer flask (with 50 mL ammonium hydroxide for the pretreatment and with 100 mL sodium hypochlorite for the oxidation step) and placed on a Burrell Shaker. The washing cycle with 100 mL sodium carbonate solution was performed under total reflux at 80°C. Between successive treatments the coal was vacuum filtered and the portion which was submitted for analysis was dried overnight at 75°C.

Sulfur contents were determined by the Leco and Eschka methods. The heating value, Q_g, calculations were based on the Dulong formula (ref. 1) which uses the weight percentages of carbon, hydrogen, oxygen and sulfur.

$$Q_g = (145.44C + 620\ (H - \frac{O}{8}) + 41S)\ BTU/lb \qquad (1)$$

This expression represents the gross heating value and assumes that all oxygen which is not part of the ash is associated with H in the proper ratio to form water. The constant volume gross heating value is corrected to a constant pressure net heating value given by:

$$Q_n = Q_g - 92.7H \qquad (2)$$

Chemistry of Desulfurization with OCl^-

The process is expected to use NaOCl to oxidize selectively and remove pyritic and organic sulfur from coal as water soluble sulfates. The iron pyrite can be oxidized to form sulfate as follows:

$$FeS_2 + 8H_2O \rightarrow Fe^{+2} + 2SO_4^{-2} + 16H^+ + 14e^- \quad E = -0.362\ Volts \qquad (3)$$

The hypochlorite ion has an oxidation potential of 0.90V which is more than sufficient to oxidize the pyrite.

$$ClO^- + H_2O + 2e^- \rightarrow Cl^- + 2OH^- \quad E = +0.90\ Volts \qquad (4)$$

The oxidation potentials are guidelines only. The dissolution efficiency will depend not only on the oxidizing agent, but also on the contact time, temperature and oxidant concentration.

Organic sulfur is an integral part of the coal matrix and is found in such compounds as sulfides, mercaptans and thiophenes. These are the same type of sulfur compounds as are present in petroleum distillates on which the action of sodium hypochlorite has been previously described (ref. 2). The probable oxidation pathways are summarized below:

Organic sulfides:

$$R_1SR_2 \rightarrow R_1SO_2R_2 \tag{5}$$

Organic disulfides:

$$R_1S_2R_2 < \begin{array}{l} R_1SO_3H + R_2SO_3H \\ H_2SO_4 \end{array} \tag{6}$$

Mercaptans:

$$RSH < \begin{array}{c} H_2SO_4 \\ \longrightarrow \\ RSO_3H \end{array} > RS_2R \tag{7}$$

Thiophenes:

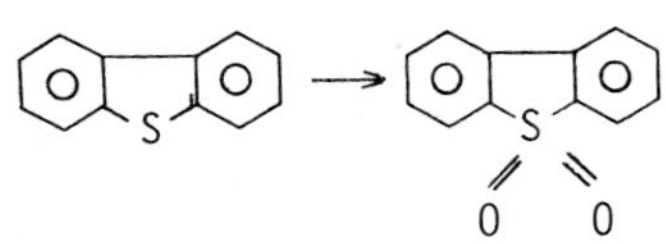

(8)

The oxidation potentials are not generally available for these sulfur compounds. However, diphenyl sulfide can be oxidized at about -0.65V. Thus, it is reasonable to assume that some sulfur compounds are oxidizable by the hypochlorite.

The advantage of using sodium hypochlorite as a desulfurization agent is that it oxidizes the sulfur compounds, but does not affect the rest of the matrix. According to S. K. Chakrabartty (ref. 3), bituminous coals are inert to hypohalite oxidation.

RESULTS AND DISCUSSION

Initially, six 5-hour treatments with NaOCl were performed on all three coal samples. The NaOCl reduced significantly the total amount of sulfur in all coals. For example, the sulfur levels for sample #1 before and after the procedure were 4.74% and 2.06%. The results also indicated that the first treatment removes more sulfur than any other single treatment. Since

5 hours is a long contacting time for a single extraction, a similar procedure was followed to determine the effect of 1-hour treatments. These results appear in Figure 1.

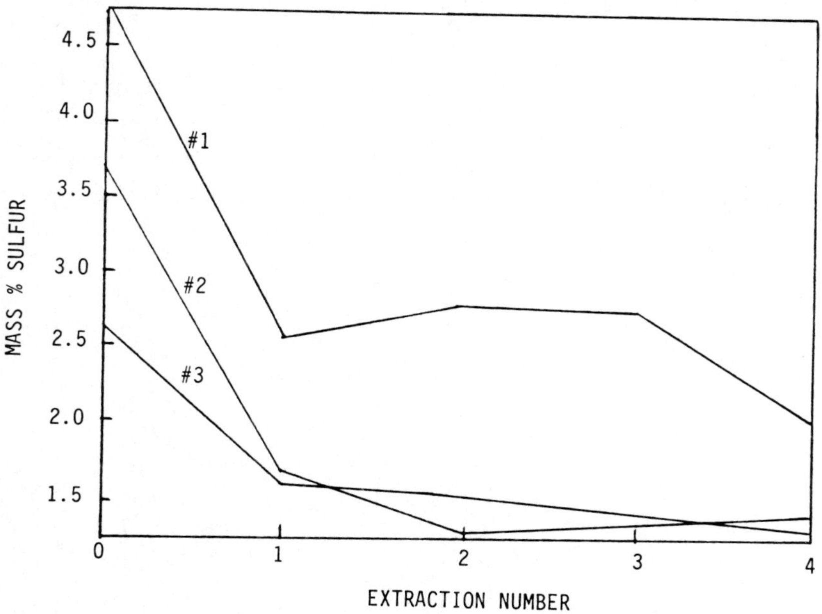

Fig. 1. Diagram showing sulfur reduction as a function of number of extractions with 5.25% sodium hypochlorite; 20 grams coal (-100 mesh) extracted with 100 mL hypochlorite solution for one hour.

As in the previous case, the most significant reduction of sulfur occurs during the first treatment. Subsequently, the sulfur levels decrease at a slower rate.

The activity of NaOCl was also indirectly observed by monitoring the pH change in a treated coal slurry. For example, during a 2-hour treatment of coal sample #3 with 5.25% sodium hypochlorite solution, the pH decreased from 10.2 to 7.2 while at the same time the pH of a water-coal slurry remained acidic at a fairly constant value (pH ~ 4.5). The change in pH can be attributed to a combination of coal dissolution and formation of acid products as was predicted in the desulfurization chemistry section.

A study was done to determine the dependence of the degree of desulfurization on the concentration of NaOCl. After four 4-hour sequential treatments of a coal sample with initial sulfur level of 4.0 wt.%, with 2.62% NaOCl, the sulfur level reached 2.45 wt.%. On the other hand, the same procedure with 5.25% NaOCl resulted in a sulfur level of 1.86 wt.%. Therefore, it appears that higher concentrations enhance the coal desulfurization process.

In evaluating a desulfurization agent it is important to determine the effectiveness not only with respect to the total amount of sulfur removed, but also the type of sulfur removed. Table 2 shows the differential sulfur analysis (Eschka) of the coal samples before any treatment. Table 3 shows the sulfur analysis after water washing and treating with 5.25% NaOCl for one hour.

TABLE 2
Differential sulfur analysis of untreated coals (-40 mesh).

Sulfur Type / Sample	Wt.%			
	Sulfatic	Pyritic	Organic	Total
1	0.11	1.15	2.53	3.79
2	0.31	1.58	1.35	3.25
3	0.15	0.79	1.50	2.44

TABLE 3
Differential sulfur analysis of water washed and NaOCl (5.25 wt.%, 1 hour) treated coals (-40 mesh).

Sulfur Type / Sample	Wt.%			
	Sulfatic	Pyritic	Organic	Total
1	0.03	0.98	2.26	3.27
2	0.04	1.56	0.78	2.38
3	0.06	0.64	1.09	1.79

The results indicate that sodium hypochlorite removed only the sulfatic and organic sulfur. The pyritic sulfur remained almost unchanged. For example, the untreated sample #2 contained 0.31% sulfatic, 1.58% pyritic and 1.36% organic sulfur. After the treatment the corresponding analysis gave 0.04% sulfatic, 1.56% pyritic and 0.78% organic sulfur. A similar trend was observed for all three samples. This sulfur type reduction is supported by the results of SEM studies. Figure 2 shows the SEM images of typical untreated coal particles. The surface of the particles containing organic sulfur is generally smoother than the surface of particles containing pyritic sulfur.

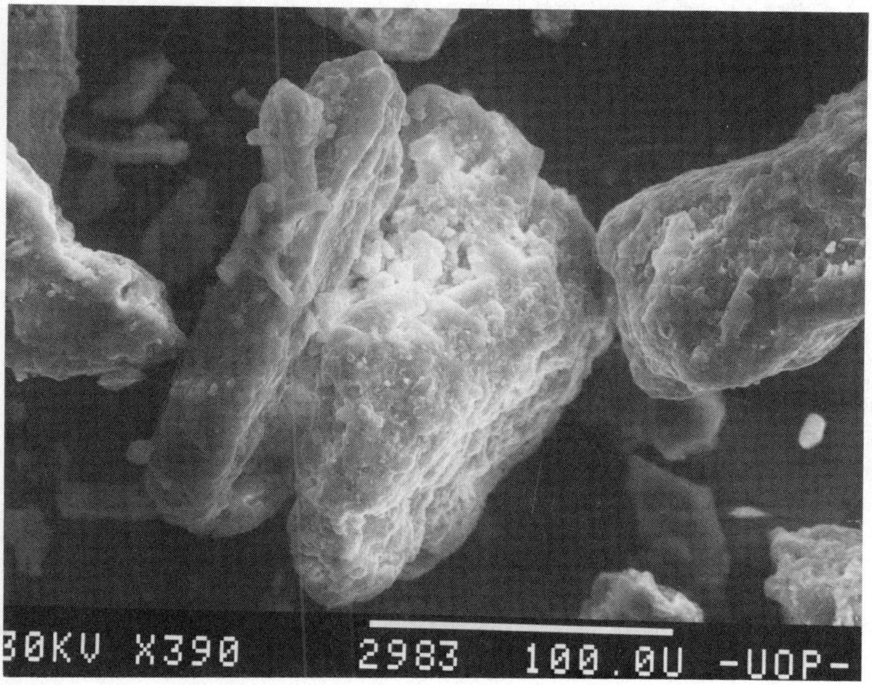

Figure 2. SEM image of untreated coal particles

Treatment with sodium hypochlorite affects the morphology of the particles. It generates multiple channels which transverse the particles in all directions and gives them a sponge-like appearance (see Figure 3). The channeling occurs in all particles except those which contain pyritic sulfur. The pyrite appears to be covered by a material which forms a very

Figure 3. SEM image of coal particles after treatment with sodium hypochlorite.

thin layer (see Figure 4). Differential sulfur analyses have shown that the sodium hypochlorite treatment removes organic sulfur. None or only a small amount of pyritic sulfur is removed. The SEM results support this finding. When sodium hypochlorite comes in contact with the pyrite it initially reacts to form a product which covers the pyritic sulfur and hinders or delays its subsequent removal.

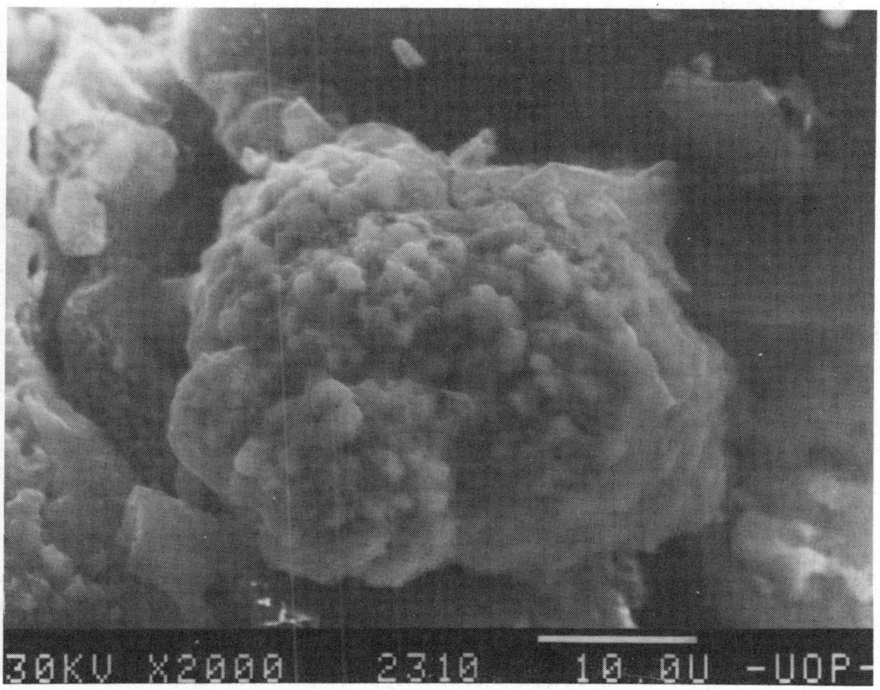

Figure 4. SEM image of pyrite after treatment with sodium hypochlorite.

Chemical comminution is a unique method for coal particle size reduction. Instead of mechanical means a chemical is used to carry out the fragmentation. According to the literature, concentrated NH_4OH is one of the most effective chemicals for comminution (ref. 4). For this reason it was selected to be studied in our desulfurization research program. Analyses showed that it did not result directly in sulfur removal but enhanced the degree of desulfurization in the subsequent NaOCl treatments. Table 4 shows the above effect on coal sample #1.

TABLE 4

Effect of chemical comminution with NH_4OH on sulfur removal; one-hour treatments at room temperature.

Coal Sample	Mesh	Wt.% Sulfur			
		Pretreatment	Untreated	1st Treatment 5.25% NaOCl	2nd Treatment 5.25% NaOCl
1	-40	29% NH_4OH	3.89	2.72	2.26
1	-40	None	3.89	3.13	3.08

SEM studies of coal samples treated with ammonium hydroxide indicated that this treatment does not cause channeling or any other change in the morphology of the coal particles.

A series of experiments showed that sulfur removal could be improved by treating the coal samples with hot sodium carbonate after the sodium hypochlorite oxidation. This trend has been demonstrated by recent studies for other mild oxidizing agents (ref. 5). It appears that the oxidation does not proceed always beyond the sulfone stage, as illustrated in the thiophene case. The additional desulfurization step with sodium carbonate cleaves the organic sulfur compound from coal and enhances the effectiveness of the process. Table 5 shows the effect of sodium carbonate at two concentrations; 0.3 M Na_2CO_3 is more effective than 0.1 M Na_2CO_3.

TABLE 5

Effect of Na_2CO_3 on coal desulfurization; one-hour treatments.

Coal Sample	Mesh	Untreated	1st Treatment 5.25% NaOCl (25°C)	2nd Treatment 5.25% NaOCl (25°C)	3rd Treatment Na_2CO_3 (80°C) 0.1 M	0.3 M
2	-100	3.62	2.70	1.57	1.44	1.30
3	-100	2.57	1.52	1.41	1.35	1.03

The desulfurization improvements with the NH_4OH and Na_2CO_3 treatments indicate that it is feasible to combine these treatments with the NaOCl oxidation to have an overall efficient sulfur removal process. All these steps were included in several schemes and resulted in substantial lowering of the sulfur levels. Table 6 shows the sulfur content of the three samples

(-100 mesh) after they were subjected to the NH_4OH solution and subsequently to two NaOCl (25°C) and two Na_2CO_3 (80°C) treatments in an alternate mode.

TABLE 6

Analyses of treated Illinois coals (-100 mesh)

1-hr. pretreatment with 29% NH_4OH.
2 1-hr. treatments with 5.25% NaOCl (25°C).
2 1-hr. treatments with 0.3 M Na_2CO_3 (80°C, reflux).
NaOCl and Na_2CO_3 treatments performed alternately.

Sample / Mass, %	1	2	3
Hydrogen	5.6	5.8	5.4
Carbon	52.0	54.1	63.5
Nitrogen	0.78	1.17	1.58
Ash	20.5	10.9	9.7
Oxygen	26.6	24.5	16.5
Sulfur	2.00	1.55	1.13
Chloride	2.51	2.01	2.16
BTU/lb	7,082	9,091	10,850
BTU Recovery, %	80.8	88.7	95.1

After the final treatment, the total sulfur in the coals was reduced by 57.8%, 57.2% and 56.0%, respectively. Most of the sulfur reduction was achieved during the first NaOCl extraction (39.0%, 39.2%, 36.6%). Figure 5 shows the overall sulfur reduction and the reduction of sulfur by types for sample 3 after the treatments. All types of sulfur were removed. More organic sulfur was removed than sulfatic and pyritic sulfur.

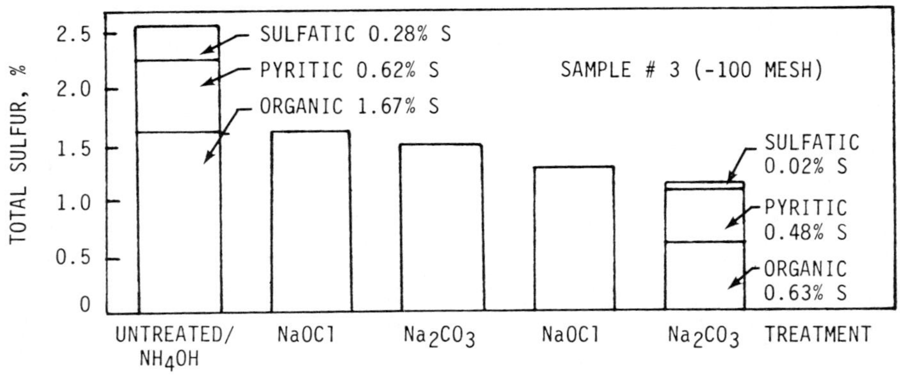

Fig. 5. Reduction of coal sulfur types using desulfurization with $NH_4OH/NaOCl/Na_2CO_3$; all treatments for one hour.

A comparison of the chlorine contents of the starting materials and products from these desulfurization experiments indicates that some chlorine (2.0%-2.5%) has been incorporated in the treated samples.

Under the same experimental conditions, but with the addition of a 1-hour NaOCl treatment an even higher reduction of sulfur levels was observed. Specifically, sulfur was lowered by 78.7%, 78.2%, and 65.4%. The detailed results appear in Table 7.

TABLE 7
Coal (-100 mesh) desulfurization with $NH_4OH/NaOCl/Na_2CO_3$.

1-hr. pretreatment with 29% NH_4OH
3 1-hr. treatments with 5.25% NaOCl (25°C).
2 1-hr. treatments with 0.3M Na_2CO_3 (80°C, reflux).

Sample/ Treatment	Wt.% Total Sulfur		
	1	2	3
Untreated	4.74	3.62	2.57
1st Treatment 5.25% NaOCl	2.07	2.35	1.53
2nd Treatment 5.25% NaOCl	1.84	1.90	1.50
1st Treatment 0.3 M Na_2CO_3	1.52	1.34	1.47
3rd Treatment 5.25% NaOCl	1.26	1.13	0.86
2nd Treatment 0.3M Na_2CO_3	1.01	0.79	0.89

The effect of the desulfurization scheme was studied for larger size coal particles. The results of treatments on -40 mesh samples are given in Table 8.

TABLE 8

Coal (-40 mesh) desulfurization with $NH_4OH/NaOCl/Na_2CO_3$.

1-hr. pretreatment with 29% NH_4OH.
2 1-hr. treatments with 5.25% NaOCl (25°C).
2 1-hr. treatments with 0.3$\underline{M}$ Na_2CO_3 (80°C, reflux).

Sample/ Treatment	Wt.% Total Sulfur		
	1	2	3
Untreated	3.89	3.19	2.29
1st Treatment 5.25% NaOCl	2.86	1.61	1.48
1st Treatment 0.3$\underline{M}$ Na_2CO_3	2.94	1.42	1.58
2nd Treatment 5.52% NaOCl	2.85	1.17	1.42
2nd Treatment 0.3$\underline{M}$ Na_2CO_3	3.06	1.33	1.48

As in the previous cases with the -100 mesh samples the majority of the sulfur was removed during the first NaOCl treatment. Reductions of 26.5%, 49.5%, and 35.4% were achieved for the three Illinois coal samples. In fact, the subsequent steps of the procedure did not remove any additional sulfur. The untreated coal sulfur levels are higher for the smaller particle size samples. This difference is attributed to the relative amounts of pyritic and organic sulfur. Since the organic sulfur is part of the coal matrix then one would expect this type of sulfur to be present in substantially higher levels in the small particles. On the other hand, the pyritic sulfur would tend to go with the larger size coal particles. This phenomenon is clearly demonstrated by the untreated sample #1 which has primarily organic sulfur. Leco analysis has given for the -100 mesh material 4.74% sulfur while for the -40 mesh 3.89%.

Comparison of the initial and final sulfur levels for the -40 mesh and -100 mesh materials indicates that the desulfurization procedure is more effective for the smaller particle size samples. Therefore, it appears that the adopted process steps are particularly successful in removing organic sulfur from coals.

CONCLUSIONS AND RECOMMENDATIONS

A desulfurization scheme has been developed which includes a combination of physical and chemical treatments to achieve a substantial reduction of sulfur levels in coals. This procedure compared to other existing and developing chemical processes can be carried out under considerably milder conditions, yet it can remove comparable amounts of sulfur. Furthermore, it utilizes a chemical (sodium hypochlorite) which is readily available and relatively inexpensive.

The pretreatment with NH_4OH does not cause any desulfurization but renders the sulfur sites more accessible to oxidation by the subsequent hypochlorite treatment. Post-treatment with sodium carbonate enhances the removal of the sulfur containing compounds. The results show that this procedure is more effective in sulfur removal for the smaller particle size coals.

The first sodium hypochlorite treatment is the single step which removes most of the sulfur. Analyses of the different types of sulfur before and after the treatment indicate that the process is especially effective in removing organic sulfur.

Depending on the type of coal treated, the developed method leads to heating value recoveries between 80.8% and 95.1%.

The desulfurization procedure leaves an amount of chlorine in the coal. These residual levels range between 2.0% and 2.7%. This certainly poses a problem and the chlorine must be removed prior to combustion. Chlorine incorporation has been reported previously by Jet Propulsion Laboratory (JPL) researchers (ref. 6) who have developed a coal desulfurization procedure by chlorinolysis under more rigorous conditions. Chlorine uptake levels of 25% were observed, but complete dechlorination was achieved by JPL after treatment with superheated steam. For future work, a similar procedure may be incorporated in our scheme to eliminate the chlorine problem.

The efficiency of the process can be improved by further experimentation on the post-treatment step. It is possible that a different chemical than sodium carbonate will enhance the removal of sulfur containing compounds and shorten the needed treatment time.

Since the sodium hypochlorite treatment removes organic sulfur in relatively short contact times, it is conceivable that this treatment can be combined with a proven physical method to remove the pyritic sulfur. The pyritic sulfur removal and its effect on the overall sulfur removal can be the objective of a future program. It may also be advantageous to complement the precombustion treatment with NaOCl with a post-combustion flue gas desulfurization procedure.

ACKNOWLEDGMENTS

The laboratory work was skillfully carried out by Bill Mukai. The electron microscopist was Dorothy Butkunas.

REFERENCES

1 R. H. Perry and C. H. Chilton, Chemical Engineers' Handbook, 1973, New York, 9-4.
2 S. F. Birch et al., J. Chem. Soc., 127 (1926), p. 1934.
3 S. K. Chakrabartty, in Organic Chemistry of Coal, Ed. J. W. Larsen, A.C.S. Symp. Ser. 71, 1978.
4 P. H. Howard and R. S. Datta, "Chemical Comminution: A Process for Liberating the Mineral Matter in Coal", in Coal Desulfurization, Ed. T. D. Wheelock, Washington, D.C., 1977, p. 58.
5 N. P. Vasilakos and C. S. Clinton, Prepr. Pap. A.C.S., Div. Fuel Chem., (28) 4, 1983.
6 G. C. Hsu et al., "Coal Desulfurization by Low-temperature Chlorinolysis" in Coal Desulfurization, Ed. T. D. Wheelock, Washington, D.C., 1977, p. 206.

COAL DESULFURIZATION USING AQUEOUS CUPRIC CHLORIDE

SUMAN P. N. SINGH, ROBERT L. JOLLEY, and LEONARD S. DICKERSON

Chemical Technology Division, Oak Ridge National Laboratory,* Oak Ridge, Tennessee 37831 (U.S.A.)

ABSTRACT

At present, coal is used primarily to generate electricity. Burning high-sulfur coals within existing statutory constraints requires either beneficiating the coal and/or using flue-gas desulfurization processes to remove the SO_x from the stack gas to meet the sulfur emission limits. Physical coal cleaning is limited in that it can only reduce the mineral sulfur content of the coal. Chemical coal cleaning, on the other hand, promises to reduce not only the inorganic but also the organic sulfur level of coal. However, chemical cleaning processes are still in the development stage. One potentially viable chemical cleaning process uses aqueous cupric chloride ($CuCl_2$) as the reagent to effect deep desulfurization of coal.

An exploratory investigation was undertaken at the Oak Ridge National Laboratory (ORNL) to chemically clean coals using aqueous $CuCl_2$. The results of this investigation suggest that aqueous $CuCl_2$ can be used for deep desulfurization of coals, even those with high organic sulfur contents. However, additional research is necessary to develop the optimum reaction conditions.

INTRODUCTION

Large quantities of coal are used to generate electricity. For example, according to a DOE/EIA report (ref.1), 593 million tons were burned by U.S. utilities to generate electricity in 1983. Of this amount, more than 22% was noncompliance coal even after blending operations. Burning of the noncompliance coal would have required either beneficiation prior to combustion and/or (more likely) the use of flue-gas desulfurization processes to remove the SO_x from the stack gases to meet the statutory SO_x emission levels. Much of the low-sulfur compliance coal is located in the western United States, far from the industrial use centers located in the East or Midwest. In the future:

1. Coal usage for power generation is expected to increase significantly.
2. Regulations involving SO_x emissions are likely to become more stringent due to the concerns about acid precipitation.

---

*Operated by Martin Marietta Energy Systems, Inc., under contract No. DE-AC05-84OR21400 with the U.S. Department of Energy.

3. The availability of compliance coals from the Eastern and Interior basins is expected to decrease dramatically, thereby necessitating increased costs for SO_x emission control.

Increased use of the more readily accessible high-sulfur coals, will require beneficiation or cleaning methods that can sufficiently reduce the coal sulfur content so as to comply with regulations. Physical coal cleaning methods have long been used to beneficiate coal; however, these methods have a severe limitation in that they can only remove the mineral sulfur impurities in coal. The organic sulfur present in the coal is untouched by the physical techniques. Unfortunately, a large fraction of the sulfur content of many coals is organic in nature. Therefore, cleaning methods other than physical beneficiation are needed for these coals. Although chemical beneficiation processes capable of reducing the organic sulfur level in coal are being developed, these processes are likely to be relatively expensive and are far from being commercial. Nonetheless, the impetus for development is that they hold the promise of achieving much greater sulfur reduction and producing a cleaner coal product.

An exploratory project was undertaken at ORNL with the objective of developing a chemical coal beneficiation process using $CuCl_2$ as the desulfurizing agent, based on some results reported by Lompa-Krzymien (ref.2). The results show that coals can be desulfurized to various levels ranging from 6 to 81%, depending on the reagent used and its concentration. However, in this study, we were unable to obtain the deep desulfurization levels reported by Lompa-Krzymien (ref.2). The details of the exploratory study are given below.

EXPERIMENTAL APPARATUS

Figure 1 is a sketch of the experimental setup used in our study. The major equipment consisted of (1) a fluidized-bed sand bath, (2) an air-driven motor, and (3) Pyrex glass reactors. A brief description of each of these units is presented in the paragraphs that follow.

Fluidized-bed sand bath

The fluidized-bed sand bath, which was used to heat the tubing reactors, contained approximately 19 L of 120-mesh aluminum hydroxide sand. It was fluidized with 0.7 L/s (1.5 scfm) of air and was heated with two 4000-W heating elements. Fluidization at an air flow rate of 0.7 L/s ensured a uniform temperature in the top 36 cm (14 in.) of the reactor. A temperature controller was used to maintain the temperature of the sand bath within a narrow range of the desired temperature level. The fluidized-bed sand bath is a TECAM Fluidized

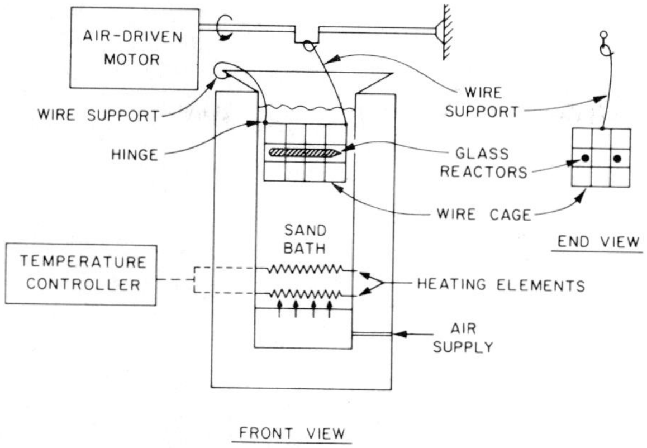

Fig. 1. Simplified sketch of the experimental setup for the coal desulfurization experiment.

Bath Model SBL-2D manufactured by Techne, Ltd., Cambridge, England. The temperature of the sand bath was measured using an ASTM mercury-in-glass thermometer with a 0 to 400°C temperature range.

Air-driven motor

An eccentric camshaft coupled to an air-driven motor was used to oscillate one end of the wire cage (containing the glass reactors) to agitate the reaction mixture. The objective of the agitation was to generate long waves in the reactor so as to thoroughly mix the coal and the reagent solution. The speed of the eccentric camshaft could be adjusted from 70 to 600 rpm; however, at speeds below ~110 rpm, the camshaft rotation became discontinuous. Therefore, for most of the runs the camshaft speed was adjusted to be ~110 rpm to facilitate the creation of a long wave in the horizontally positioned glass reactors. The amplitude of oscillation of the wire cage was adjusted by trial and error to be ~5 cm. The air-driven motor used in the experiments (Model No. 8AM-NRV-5B) was manufactured by the Gast Manufacturing Corporation, Benton Harbor, Michigan.

Pyrex-glass tubing reactors

Pyrex-glass tubing reactors of 25-mm (1-in.) outer diameter (OD) were used for conducting the experiments. The glass reactors were fabricated from No. 7740 Pyrex tubing by the ORNL Glass Shop. Figure 2 is a sketch of a typical reactor before and after sealing one end. New reactors were used for each run. Typical dimensions of the reactors are as follows:

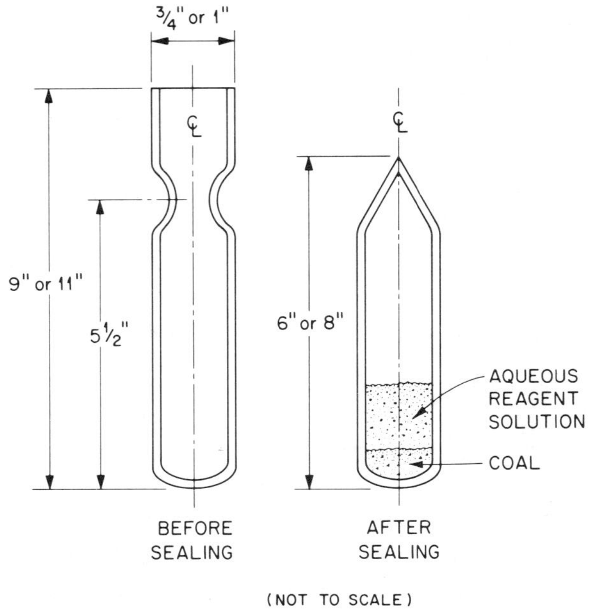

Fig. 2. Sketch of a typical Pyrex-glass tubing reactor used in the coal desulfurization experiment.

Run Nos.	Glass reactor dimensions [cm (in.)] Length	OD	Wall thickness	Av. sealed length
3-9	28 (11)	1.9 (3/4)	0.3 (1/8)	20 (8)
10-17	23 (9)	2.5 (1)	0.4 (5/32)	16 (6)

The first few runs were made using 316 stainless steel microautoclaves. However, further use of the microautoclaves was discontinued because of possible adverse reactions between the metal and the aqueous $CuCl_2$ solutions at reaction conditions.

EXPERIMENTAL PROCEDURE

A brief description of the experimental procedure is given below. The basic procedure consists of the following sequential operations:

1. A mixture of 2 g of oven-dried coal and 10 mL of aqueous $CuCl_2$ solution of the desired concentration ($CuCl_2$ reagent concentrations used ranged from 10 to 50 wt % $CuCl_2$) was placed in each of the two new glass tube reactors.

The reactors were then sealed and emplaced in adjacent sections of the wire cage so that there was no relative motion between the cage and the glass reactor tubes.

2. The temperature of the fluidized sand bath was adjusted to achieve the desired reaction temperature (400°F, or 200°C). When the temperature stabilized at ~200°C, the wire cage with the sealed reactor tubes was lowered into the sand bath so that there was at least 7 to 10 cm (3 to 4 in.) of fluidized sand above the top of the cage. In addition, the cage was placed such that the reactor tubes were oriented in a horizontal plane. Then, the two support wires attached to one end of the cage were secured to the lip of the fluidized-bed bath while the two wires from the opposite end of the cage were secured to the eccentric camshaft attached to the air-driven motor. The air-driven motor was subsequently started, and its speed was adjusted until the eccentric camshaft was rotating steadily at ~110 rpm.
3. When the system was considered to be operating in a stable manner (a matter of a few minutes), the timer was started and the reaction was allowed to continue for the desired time. Although the sand-bath temperature was regulated by the temperature controller, this temperature as well as the eccentric shaft speed were manually checked by the operator every 15 min during the course of the run. Any deviations were corrected promptly.
4. After the system had operated for the desired reaction time, the motor was stopped and the wire cage was withdrawn from the sand bath and allowed to cool to room temperature in the laboratory hood.
5. The two glass reactors were removed from the cage and cut open. The contents of each were carefully poured into a Büchner funnel lined with a preweighed filter paper. The residue on the filter paper was thoroughly washed with distilled water until the filtrate collected was neutral.
6. The filter paper and the residue were then carefully removed and placed in a vacuum oven to be dried overnight.
7. On the following day, the dried filter paper and residue were weighed and the residue was transferred to a sample bottle.
8. The sample was sent to the ORNL Analytical Chemistry Division for a total sulfur analysis using the Leco sulfur analyzer.

EXPERIMENTAL RESULTS

The coals tested in the experimental program are listed in Table 1. A brief description of each of these coals follows.

Kentucky No. 9. This relatively high-sulfur (~3.4 wt % S) coal from western Kentucky was used because sufficient quantities of it were readily

TABLE 1

Coals tested in the experimental program.

Identification of feed coal sample	Description of feed coal	Measured sulfur content[a] (wt % S)
Control 1	Kentucky No. 9	3.42
Control 2[b]	Deashed Kentucky No. 9	2.66 (16.95)[c]
Control 3[b]	Deashed Kentucky No. 9	2.74 (15.69)[c]
Control 4	Kentucky No. 9	3.24
Control 5	Moyer Kentucky No. 9	4.86
Control 6	Kentucky No. 9 fines[d]	3.16
Control 7	Prince coal	4.16

[a]Determined using Leco sulfur analyzer.
[b]The sample of feed coal was float-sink beneficiated using Freon-113. The refuse sulfur content is reported in parentheses.
[c]Sulfur content of refuse.
[d]Particle size, -200 mesh.

available at the time of project initiation. Most of the experimental runs were conducted using this type of coal.

Moyer Kentucky No. 9. This high-sulfur (~4.9 wt % S) coal, also from western Kentucky, is proposed to be used at TVA's Paradise Power Plant. A sample of it was obtained from the Engineering Technology Division at ORNL when the initial runs indicated that deep desulfurization of the Kentucky No. 9 coal was not being obtained. It was thought that deep desulfurization may be dependent on the coal processed.

Prince coal. This high-sulfur (~4.2 wt % S) coal from Nova Scotia, Canada, is the type that Lompa-Krzymien (ref.2) used in experiments where deep desulfurization (product coal containing ~0.07 wt % S) was obtained. Two runs were made using this coal. The results were similar to those obtained from the experiments with Kentucky No. 9 coal.

All the results from the desulfurization experiments are summarized in Table 2.

From the results of runs 1 through 4 (Table 2), it can be seen that although the Kentucky No. 9 coal was desulfurized to some extent using either aqueous $CuSO_4$ or aqueous $CuCl_2$, the level of desulfurization obtained was not as low as reported by Lompa-Krzymien (ref. 2). These results also appear to indicate that aqueous $CuCl_2$ is a better desulfurizing agent than aqueous $CuSO_4$ (runs

TABLE 2

Results of bench-scale coal desulfurization experiments.[a]

Run No.	Desulfurizing reagent	Reaction time (h)	Sulfur content of coal (wt % S)	Coal desulfurization obtained (%)
Control[b]	--	--	3.42	--
1[c]	12% $CuSO_4$	1	3.19	6.7
2[c]	5% $CuSO_4$	1	2.93	14.3
3[d]	10% $CuCl_2$	1	2.12	38.0
4[d]	12% $CuSO_4$	1	2.84	17.0
Control 2	--	--	2.66	--
5	10% $CuCl_2$	1	1.71	35.7
Control 3	--	--	2.74	--
6	20% $CuCl_2$	1	1.60	41.6
Control 4	--	--	3.24	--
7	20% $CuCl_2$-NH_4Cl	1	2.16	33.3
8	4.6% NaOCl	1	2.34	27.8
9	30% $CuCl_2$	1	1.84	43.2
10	30% $CuCl_2$	5	1.58	51.2
11	30% $CuCl_2$ (ammoniated)	1	2.00	38.3
Control 5	--	--	4.86	--
12	20% $CuCl_2$	1	2.29	52.9
Control 6	--	--	3.16	--
13	30% $CuCl_2$	5	1.73	45.3
14	50% $CuCl_2$	6	1.33	57.9
15[e]	50% $CuCl_2$	7.5	0.59	81.3
Control 7	--	--	4.16	--
16	10% $CuCl_2$	1	2.81	32.5
17[f]	50% $CuCl_2$	3.25	2.05	50.7

[a]Unless indicated otherwise, all experimental runs were carried out in Pyrex-glass tubing reactors.
[b]Feed coal sample used in the following experimental runs.
[c]Runs were conducted using 316 stainless steel microautoclaves.
[d]Runs were conducted in glass tubing reactor but with no agitation of the reactor.
[e]Test was conducted using desulfurized coal from run 14.
[f]Test was conducted using only the -200-mesh fines.

3 and 4). Therefore, after the first few runs, the thrust of the project was shifted to scoping the $CuCl_2$ desulfurization process in an effort to identify the operating conditions that may yield deep desulfurization of the coal. To explore these conditions, the following process variables were adjusted: (1) reagent composition and concentration, (2) reaction time, and (3) coal type and particle size. A brief discussion of these variables is given below.

Reagent composition and concentration. Most of the runs were made using aqueous $CuCl_2$ solution as the desulfurizing reagent. However, two runs (Nos. 7 and 11) were made using ammoniated $CuCl_2$ solutions, and one (No. 8) was made using an aqueous sodium hypochlorite solution. The runs with ammoniated $CuCl_2$ solution were made to ascertain if the presence of ammonia would facilitate the desulfurization of the coal. The objective of the run with sodium hypochlorite was to determine whether hypochlorite was the desulfurizing agent. As can be seen from the results of runs 7, 11, and 8, respectively, no major improvement in coal desulfurization was achieved using ammoniated $CuCl_2$ solution or sodium hypochlorite.

In addition to the tests discussed above, the concentration of the aqueous $CuCl_2$ solution was increased from 10% to 20, 30, and 50%. The data in Table 2 show that increasing the concentration of $CuCl_2$ tends to increase the level of coal desulfurization (see data for runs 5, 6, 10, and 14). However, the effects of other variables (e.g., reaction time and particle size) may also be embedded in these improved desulfurization levels.

Reaction time. The reaction time was 1 h for most of the runs; however, it was changed to 5, 5, 6, 7.5, and 3.25 h, respectively, for runs 10, 13, 14, 15, and 17. Apparently, an increase in the reaction time yields a higher level of coal desulfurization (e.g., see runs 9 and 10). This suggests that the desulfurization reaction may be reaction rate controlled.

Coal type and particle size. Examination of the data obtained for the processing of Kentucky No. 9, Moyer Kentucky No. 9, and Prince coals (see runs 9, 12, and 16, respectively) indicates that the level of desulfurization obtained is relatively independent of the type of coal processed or, stated another way, that the coal type appears to have a relatively insignificant impact on the level of coal desulfurization obtained using the proposed methodology.

The effect of particle size on the level of coal desulfurization can be deduced from the results of runs 10 and 13. These results indicate that particle size has little impact on the level of coal desulfurization obtained. Thus, the desulfurization reaction may not be diffusion controlled.

With the above information in hand, an effort was made to identify reaction conditions that would yield deep desulfurization of the coal. Run 15 was made using the partially desulfurized coal from run 14. The data from runs 14 and 15 in Table 2 show that a 3.16 wt % S coal was reduced to ~0.59 wt % S coal (i.e., >80% total sulfur reduction). Therefore, deep desulfurization of coal is possible using a stagewise operation. However, our study indicated that this deep desulfurization also resulted in a significant reduction in the coal's caloric value. Because of time limitations, it was not possible to develop the optimum reaction conditions for achieving deep desulfurization of coal while minimizing caloric value reduction, or to determine whether the coal can be desulfurized even further to levels approaching those reported by Lompa-Krzymien (ref.2).

CONCLUSIONS

The following salient conclusions can be drawn from the exploratory study:

1. Coal can be desulfurized using various concentrations of aqueous $CuCl_2$ solutions at approximately 200°C. It appears that $CuCl_2$ is a more effective desulfurizing agent than $CuSO_4$.
2. Deep desulfurization of coal (even coals with high organic sulfur contents) can be achieved via a two-stage desulfurization process. However, the desulfurization also resulted in a significant loss of the caloric value of the coal.
3. The experimental results reported by Lompa-Krzymien (ref.2) could not be duplicated even for a coal similar to that used by Lompa-Krzymien (i.e., Prince coal from Nova Scotia, Canada).
4. Analysis of the results obtained suggests that the desulfurization reaction may be reaction rate controlled rather than diffusion controlled.
5. The exploratory study demonstrated that coal can be desulfurized to low-sulfur levels using aqueous $CuCl_2$. However, further research is needed to develop the optimum process conditions.

REFERENCES

1. U.S. DOE/EIA, Cost and Quality of Fuels for Electric Utility Plants 1983, DOE/EIA-0191(83), July 1984.

2. L. Lompa-Krzymien, "Complete removal of sulphur from coal using solutions containing cupric ions," Fuel, 61 (September 1982) 871.

AN EFFECTIVE PHYSICAL AND CHEMICAL TREATMENT APPROACH FOR CLEANING HIGH SULFUR AND ASH COALS

DAVID J. BORON [1] AND SCOTT R. TAYLOR, Ph.D.[2]

[1]Senior Project Manager, New York State Energy Research and Development Authority, Two Rockefeller Plaza, Albany, New York 12223

[2]S. R. Taylor & Associates, 843 Colony Drive, Bartlesville, Oklahoma 74006.

ABSTRACT

Laboratory-scale research conducted at the United States Department of Energy Pittsburgh Energy Technology Center has shown that significant sulfur and ash removal from coal can be achieved through specific gravity float/sink treatment followed by mild oxidative chemical processing. A run-of-mine Pittsburgh bed coal from Ohio was first crushed to 8 mesh top-size and floated in a 1.6 specific gravity (s.g.) bath. Representative portions of the float product were then crushed to various topsizes and treated with bleach solutions at ambient pressure and mild temperatures to effect further sulfur and ash removal. The combined treatment steps resulted in over 65 percent ash and nearly 50 percent sulfur dioxide per million Btu removal efficiencies. Organic sulfur removal approached 30 percent and coal Btu recoveries were generally greater than 70 percent.

BACKGROUND

In December, 1971, the Environmental Protection Agency (EPA), in response to the Clean Air Act of 1970, issued "Standards of Performance for New Stationary Sources" (ref. 1). These Standards restricted particulate, sulfur dioxide (SO_2) and nitrogen oxides (NO_x) emissions to 0.10 pounds per million Btu (with an opacity no greater than 20 percent), 1.2 pounds per million Btu and 0.70 pounds per million Btu (except for lignite coal), respectively. In 1979, the EPA issued New Source Performance Standards (NSPS) for utility and industrial boilers in accordance with the amendments to the Clean Air Act of 1977 (ref. 2). These standards reduced particulate emissions to 0.03 pounds per million Btu and NO_x emissions to 0.60 pounds per million Btu (0.50 pounds per million Btu for subbituminous coal). The standards for SO_2 emissions are shown in Table 1.

TABLE 1
EPA NSPS For SO_2 Emission.

Category	SO_2 Content of ROM Coal	SO_2 Reduction Required
1	0-2 lb/M Btu	70%
2	2-6 lb/M Btu	To 0.6 lb/M Btu (70% to 90%)
3	6-12 lb/M Btu	90%
4	12 lb/M Btu	To 1.2 lb/M Btu

Acid rain legislation could further restrict SO_2, NO_x and particulate matter emissions on 1971 and later boilers and possibly impact pre-1971 boilers now exempt from the requirements of the Clean Air Act (but under local and state emission regulations). In a report to the Congress of the United States by the Comptroller General (ref. 3), the General Accounting Office reports that in the year 2000, pre-1980 utility plants will be emitting at 89 percent of their 1980 level and producing 84 percent of all eastern utility SO_2. Recognizing that a vast majority of these plants had begun operation before the 1970 Clean Air Act, it would be of little surprise if pending acid rain legislation or future amendments to the Clean Air Act were to seek SO_2, NO_x and particulate matter reductions in these plants.

INTRODUCTION

With the concern for deteriorating air quality caused by SO_2 emissions from coal-fired combustion sources has come a dramatic increase in research pertaining to sulfur removal from coal. Much of this research has centered on new approaches and novel techniques for removing sulfur from coal prior to combustion. This paper discusses laboratory-scale coal cleaning research conducted at the United States Department of Energy (DOE) Pittsburgh Energy Technology Center (PETC). The approach taken first removes mine dilution mineral matter from coal using a float/sink physical separation process to maximize Btu recovery and then treats the float product through mild chemical oxidation to affect greater sulfur and ash removal. The authors have reported on similar in-house and contract research endeavors in chemical coal cleaning (ref. 4,5,6,7,8). In addition, a number of studies have been conducted that outline and evaluate the merits of various physical and chemical coal cleaning activities and approaches (ref. 9,10,11,12,13,14). The interested reader is referenced to these reports.

Chemical coal cleaning activities at the PETC coal preparation laboratories has focused mainly on approaches for removing the organic sulfur from coal. As an oxidizing agent, sodium hypochlorite (NaOCl) is known to react with various organic sulfur functionalities. It was used for this purpose in research conducted by the petroleum industry early in this century to remove sulfur from crude oil and petroleum distillates (ref. 15,16,17) and in refining petroleum naphtha solutions (ref. 18,19). More recently, NaOCl was used by Chakrabartty (ref. 20,21,22) and Mayo (ref. 23,24) in the oxidative dissolution of coal for the purpose of studying coal structure. These efforts, however, did not monitor the fate of the coal sulfur. The objective of the current study was to determine the effectiveness of NaOCl in the oxidative removal of sulfur (pyritic and organic) from coal.

Using a float 1.6 s.g. Pittsburgh bed coal from Ohio, it was found that commercial grade bleach solutions (containing about 5.25 percent NaOCl) could remove up to 50 percent of the pyritic sulfur and 30 percent of the organic sulfur at ambient temperatures and pressure. Pyritic and organic sulfur removal were seen to vary according to solution pH, coal particle size and reaction temperature. When product coal sulfur levels were compared with those of the run-of-mine (ROM) coal, the combined physical and chemical treatment steps were found to remove over 30 percent of the ROM coal sulfur. On a pounds of SO_2 per million Btu basis, SO_2 reduction was found to approach 50 percent.

Depending on reaction conditions, the hypochlorite treatment did not greatly affect the ash and Btu content of the float 1.6 s.g. coal. Ash reductions ranged from 0 to 34 percent, with the remaining ash content slightly evelated in sodium and chlorine. The Btu content of the coal dropped from 0 to 14 percent and generally corresponded to a similar increase in the coal's oxygen content. The mild test conditions and coal work-up procedures that were used resulted in some coal losses. Coal recovery, however, generally exceeded 95 percent. Coal handling procedures were found to account for the majority of these losses.

EXPERIMENTAL PROCEDURES

Coal sample preparation

A 55-gallon drum of 6 inch x 0 ROM coal from the Pittsburgh bed seam, Monroe County, Ohio, was crushed to 3/8 inch topsize and split into ten representative samples by the method of coning and longpiling outlined in "Coal Sampling, Revision of Technical Paper 133" (ref. 25). Nine of the samples were stored in plastic lined five-gallon pails purged with nitrogen. The remaining sample was crushed in a hammermill to pass 8 mesh topsize and then floated in a 1.6 s.g. bath. The float 1.6 product collected was air-dried and riffled into four fractions. Using a hammermill, one of the four fractions was crushed to pass 14 mesh, the second was crushed to pass 48 mesh while the third and fourth were crushed to pass 100 mesh and 200 mesh, respectively.

The crushed and sized coal samples thus obtained were split into approximate 25-gram samples on a rotary splitter. Each 25-gram sample was then placed in a 4-ounce jar and stored in a light-restricted area. Representative samples of the sized and split float 1.6 s.g. coal and ROM coal were sent to the PETC Coal Analysis Laboratory for analysis. The analysis for these samples are shown in Tables 2 and 3.

TABLE 2
Analysis of samples of float 1.6 s.g. Pittsburgh bed coal from Monroe County, Ohio (analysis reported on an as-received basis).

	14 mesh x 0	48 mesh x 0	100 mesh x 0	200 mesh x 0
Moisture	2.3	2.2	2.4	2.4
Ash	11.3	11.3	11.2	11.3
Hydrogen	5.1	5.1	5.1	5.1
Carbon	69.0	69.1	69.3	69.0
Nitrogen	1.1	1.1	1.1	1.0
Sulfur	3.4	3.4	3.4	3.3
Sulfate	0.2	0.2	0.2	0.2
Pyritic	1.0	1.1	1.0	1.0
Organic	2.2	2.1	2.2	2.1
Oxygen	10.3	10.2	10.5	10.5
Chlorine	0.2	0.2	0.2	0.2
Btu	12,500	12,500	12,490	12,460
SO_2/M Btu	5.4	5.4	5.4	5.3

TABLE 3

Analysis of 8 mesh x 0 ROM Pittsburgh bed coal from Monroe County, Ohio (analysis reported on an as-received basis).

Moisture	2.6	Total Sulfur	3.6
Ash	29.8	Sulfate Sulfur	0.2
Hydrogen	4.3	Pyritic Sulfur	2.1
Carbon	52.9	Organic Sulfur	1.3
Nitrogen	1.0	Btu	9588
Oxygen	8.4	SO_2/M Btu	7.5

Test procedures

A series of NaOCl treatment tests were conducted on these float 1.6 s.g. coal samples evaluating the following variables: coal particle size (14 mesh x 0, 48 mesh x 0, 100 mesh x 0 and 200 mesh x 0); solution pH (12.5, 8.4-Buffered and 8.0) and solution temperature (ambient and 50°C).

Twenty-five grams of coal from one 4-ounce jar was transferred into a 500 ml round-bottom flask (for experiments employing elevated temperatures) or beaker (for experiments employing ambient temperatures) containing 250 ml of commercial grade bleach solution. The flask/beaker was equipped with a thermometer, and mixing was accomplished using a magnetic stirrer. Reaction time was one hour for all tests conducted. At the end of one hour, the reaction mixture was filtered in a Buchner Funnel and washed repeatedly with distilled water until the drippings were neutral to litmus paper. The coal was then dried at 110°F overnight. The treated and dried coal was then weighed and a representative portion (obtained through riffling) analyzed. The remaining sample was sealed in a nitrogen filled jar for future reference. Each test condition was repeated in triplicate.

RESULTS AND DISCUSSION

Effect of hypochlorite treatment on sulfur removal

Several trends were anticipated in the behavior of the organic sulfur in coal relative to particle size, solution pH and solution temperature. First, a greater organic sulfur reduction was expected with smaller coal particle sizes. This reasoning stems from the greater surface area of the smaller coal particles and expected shorter distances with which the reagent would traverse in reaching organically bound sulfur in the coal.

Second, greater organic sulfur reduction was expected with hypochlorite solutions at low pH and with low free alkali content. Several factors are involved in the reactivity and stability of hypochlorite solutions, these are pH, available chlorine and alkali content. The chemistry of hypochlorite solutions is largely dictated by pH (ref. 23,26), with hypochlorite ion dominant at high pH, hypochlorous acid at neutral to slightly acidic or basic pH and chlorine and chlorine dioxide dominant at low pH. The available chlorine content of hypochlorite solutions is also dependent on pH, with low solution pH providing high reactivity and chlorine availability. The alkali content in hypochlorite solutions provide for solution stability. Hypochlorite solutions with high available chlorine and low alkali contents were reported to be particularly effective in oxidizing organic sulfur (ref. 15,16,17,18,19,23).

The effect of solution temperature on sulfur reaction/removal was uncertain. Higher temperatures are expected to increase reagent reactivity but are also expected to reduce reagent selectivity. Thus, at elevated temperatures hypochlorite solutions are expected to be more reactive, but not necessarily more effective in sulfur removal.

Figures 1, 2 and 3 reflect the complex behavior and reactivity of hypochlorite solutions in the removal of sulfur. Organic sulfur values were determined by difference (total sulfur minus pyritic sulfur). All values reflect an average of triplicate tests with the standard deviation of analyses ranging from 0 to 0.2 for total sulfur, 0 to 0.2 for organic sulfur and 0 to 0.1 for pyritic sulfur while averaging 0.1, 0.1 and 0 for total, organic and pyritic sulfur, respectively.

It appears that organic sulfur removal reaches a maximum between the 100 mesh x 0 and 200 mesh x 0 coal fractions. In all but one series, the 200 mesh x 0 coal fraction showed lower organic sulfur removal than the 100 mesh x 0 fraction. This observation runs parallel to an increase in coal oxidation (see Table 4) and decrease in coal Btu (see Table 5) through the 200 mesh x 0 fraction. This suggests that at some point, perhaps after 100 mesh x 0, greater coal surface area may lead to lower reagent selectivity which results in higher carbon oxidation and lower sulfur removal.

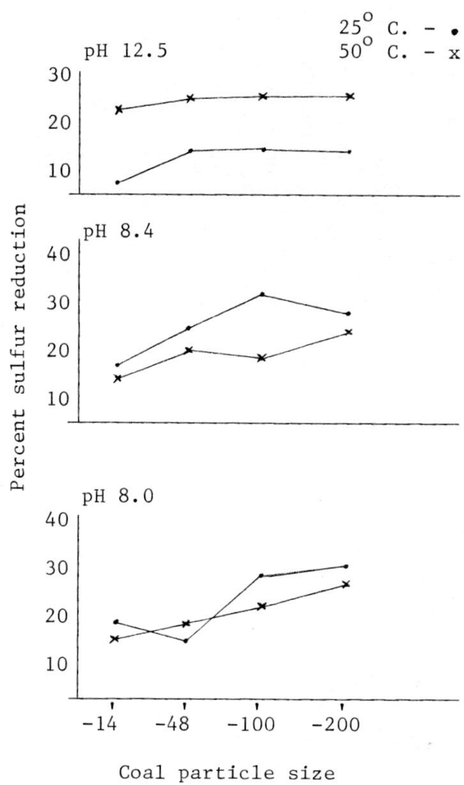

Fig. 1. Total sulfur behavior.

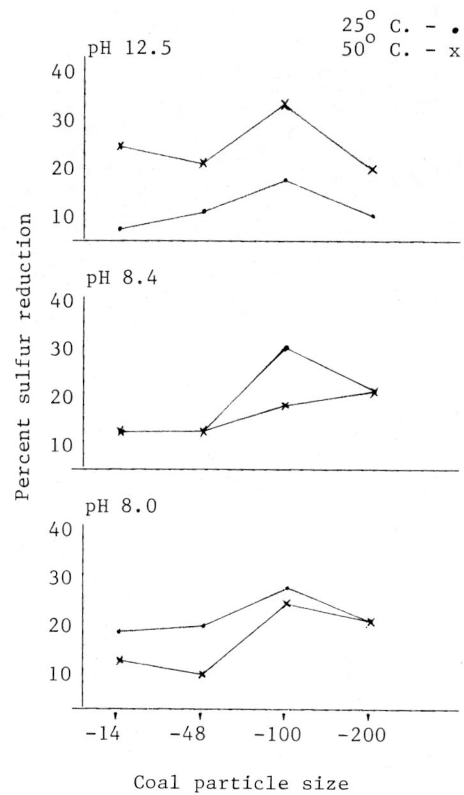

Fig. 2. Organic sulfur behavior.

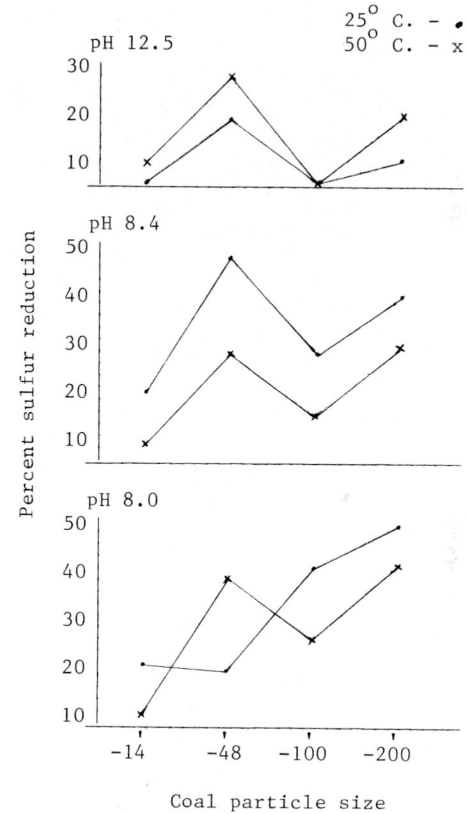

Fig. 3. Pyritic sulfur behavior.

TABLE 4
Oxidation of treated coals as measured by coal oxygen and chlorine content (oxygen determined by difference, chlorine directly).

	pH 8.0		pH 8.4		pH 12.5	
	Ambient	50°C	Ambient	50°C	Ambient	50°C
14 mesh x 0	13.0	12.4	12.9	12.4	10.3	10.6
48 mesh x 0	16.0	14.0	16.2	15.3	10.1	10.5
100 mesh x 0	15.6	14.6	19.7	16.2	9.9	12.3
200 mesh x 0	15.7	16.5	17.9	17.2	11.7	14.3

TABLE 5
Btu values of treated coals.

	pH 8.0		pH 8.4		pH 12.5	
	Ambient	50°C	Ambient	50°C	Ambient	50°C
14 mesh x 0	12,035	12,455	12,240	12,630	12,865	12,660
48 mesh x 0	11,830	12,030	11,620	11,770	12,690	12,305
100 mesh x 0	11,815	11,790	10,920	11,645	12,680	12,060
200 mesh x 0	11,740	11,475	11,480	11,410	12,360	11,905

In contrast, pyritic sulfur removal was observed to plunge for the 100 mesh x 0 fraction and then increase for the 200 mesh x 0 fraction. The exception to this trend was observed in the 100 mesh x 0, pH 8.0, room temperature series where pyritic sulfur removal climbed steadily to a maximum of 50 percent reduction at the 200 mesh x 0 fraction. It is tempting to account for this observation through analytical error. However, all tests were run in triplicate (with a standard deviation between analyses of 0.1) and the pH series for which these observations were made were run several months apart and submitted separately for analytical analysis. Additional tests are required to clarify the behavior of pyritic sulfur reaction in the 100 mesh x 0 fraction.

The effect and relationship of solution pH and temperature on sulfur removal is intriguing. Higher solution temperatures seem to effect greater organic sulfur removal at higher solution pH, whereas, lower solution temperatures seem to be more effective for organic sulfur removal at lower solution pH. Greater pyritic sulfur removal is seen at a low solution pH for all mesh size fractions regardless of solution temperature.

At first glance, these observations seem to part from those expected. However, the pH 8.0 series was observed to oxidize the coal more severely than the pH 12.5 series (see Table 4), with the extent of oxidation increasing with smaller particle size. In view of these coal oxidation and sulfur removal trends, it may be that higher solution temperatures increase the reactivity of the hypochlorite reagent at high solution pH without greatly effecting the selectivity of the reagent. The result is low carbon oxidation and high sulfur removal. In contrast, a lower solution pH may increase the reactivity of the hypochlorite reagent, but with an adverse effect on the selectivity of the reagent (as witnessed by greater coal oxidation). Accordingly, a high solution pH at elevated temperatures appears more effective in treating coal for sulfur removal than a low solution pH regardless of temperature (largely due to near equivalent sulfur removal and lower coal oxidation tendencies).

Table 6 shows the average coal sulfur, ash and Btu values for all of the hypochlorite tests conducted. Also shown are the low and high values found within the individual analyses of the product coals. Table 7 compares these values on a percent basis with the starting float 1.6 s.g. coal.

TABLE 6

Composite average and select sulfur, ash and Btu content of hypochlorite treated coals.

	Average	Low	High
Sulfur	2.5	2.1	3.0
Ash	9.5	7.5	12.1
Btu	12,020	10,660	12,960

TABLE 7

Change in sulfur, ash and Btu content of hypochlorite treated coals and float 1.6 coal on a percent basis.

	Average (% Change)	Low (% Change)	High (% Change)
Sulfur	-26	-38	-12
Ash	-16	-34	+7
Btu	-4	-14	+4

Effect of combined physical and chemical treatment

Sulfur and ash removal efficiencies via specific gravity separation are dependent upon liberation (a function of coal particle size and pyrite and ash distribution) and equipment operation performance (separation efficiency). The theoretical optimum for sulfur and ash removal from coal via specific gravity separation is determined by the washability characteristics of the coal. Table 8 shows the washability data for a select Monroe County, Ohio, Pittsburgh bed coal (ref. 26).

TABLE 8

Washability data of select Monroe County, Ohio, Pittsburgh bed coal.

STATE: OHIO
COUNTY: MONROE

COALBED: PITTSBURGH
RAW COAL MOISTURE: 2.1 %

CUMULATIVE WASHABILITY DATA

SAMPLE CRUSHED TO PASS 1-1/2 INCHES

PRODUCT	RECOVERY,%		BTU/LB	ASH,%	SULFUR,%		LB SO2/M BTU
	WEIGHT	BTU			PYRITIC	TOTAL	
FLOAT-1.30	8.5	12.8	12973	5.3	.59	2.68	4.1
FLOAT-1.40	43.4	62.6	12384	9.6	1.20	3.00	4.8
FLOAT-1.60	59.4	82.4	11918	13.0	1.95	3.51	5.9
FLOAT-1.90	67.3	90.7	11576	15.5	2.25	3.80	6.6
TOTAL	100.0	100.0	8589	37.3	2.70	3.84	8.9
EPA STANDARD	*****	*****	*****	*****	*****	*****	1.20

SAMPLE CRUSHED TO PASS 3/8 INCH

PRODUCT	RECOVERY,%		BTU/LB	ASH,%	SULFUR,%		LB SO2/M BTU
	WEIGHT	BTU			PYRITIC	TOTAL	
FLOAT-1.30	22.3	34.4	13110	4.3	.39	2.45	3.7
FLOAT-1.40	47.3	69.1	12411	9.4	1.02	2.96	4.8
FLOAT-1.60	60.3	85.9	12096	11.7	1.60	3.43	5.7
FLOAT-1.90	66.6	91.0	11603	15.3	1.90	3.70	6.4
TOTAL	100.0	100.0	8493	38.0	2.49	3.75	8.8
EPA STANDARD	*****	*****	*****	*****	*****	*****	1.20

SAMPLE CRUSHED TO PASS 14 MESH

PRODUCT	RECOVERY,%		BTU/LB	ASH,%	SULFUR,%		LB SO2/M BTU
	WEIGHT	BTU			PYRITIC	TOTAL	
FLOAT-1.30	14.6	21.8	12904	5.8	.47	2.35	3.6
FLOAT-1.40	43.3	63.0	12562	8.3	.85	2.57	4.1
FLOAT-1.60	58.7	81.6	12000	12.4	1.51	3.04	5.1
FLOAT-1.90	67.3	88.2	11315	17.4	1.89	3.27	5.8
TOTAL	100.0	100.0	8630	37.0	2.57	3.71	8.6
EPA STANDARD	*****	*****	*****	*****	*****	*****	1.20

Data presented for the float 1.6, 3/8 inch x 0 coal product shows a 60 percent weight recovery and a 86 percent Btu recovery corresponding to a product sulfur of 3.4 percent and product ash of 11.7 percent. This coal product quality is not significantly different than the float 1.6, 8 mesh x 0 coal sample used in the current study (sulfur of 3.4, ash of 11.3 and Btu of 12,500). One noteworthy difference, however, is the cumulative ash (38 percent) and corresponding Btu (8493) values of this coal in comparison to the ROM coal sample used here (29.8 percent ash and 9,600 Btu).

At an equivalent weight recovery of 60 percent, and using Btu values presented in Tables 2 and 3, the float 1.6 coal product used in the current study corresponds to a 78 percent Btu recovery. For the hypochlorite-treated coals, a 96 percent weight recovery (typical recovery for the tests conducted) with an average coal product Btu content of 12,020 provide for a 92 percent Btu recovery. Accordingly, the combined physical and chemical steps correspond to a 58 percent coal weight recovery and a 72 percent Btu recovery.

Table 9 reflects the change in average sulfur, ash and Btu values between hypochlorite-treated coals and the ROM coal on a percent basis. Table 10 shows the SO_2 per million Btu content of the ROM, float 1.6 and treated coals, while the change in SO_2 per million Btu between the coals on a percent basis is shown in Table 11.

TABLE 9

Change in sulfur, ash and Btu content between the hypochlorite treated coals and the ROM coal.

	Average (% Change)	Low (% Change)	High (% Change)
Sulfur	-31	-42	-17
Ash	-68	-75	-59
Btu	+25	+11	+35

TABLE 10

SO_2 per million Btu content of untreated and treated coals.

	ROM	Float 1.6	Treated Coals		
			Average	Low	High
SO_2/M Btu	7.5	5.4	4.2	3.9	4.6

TABLE 11
Change in product coal sulfur on a SO_2 per million Btu basis.

ROM to Float 1.6 (% Change)	Float 1.6 to Treated (% Change)			ROM to Treated (% Changed)		
	Average	Low	High	Average	Low	High
-28	-22	-28	-14	-44	-48	-39

Substantial sulfur reductions are observed for the hypochlorite-treated coals in comparison to the float 1.6 and ROM coal feedstocks. In reference to the washability data in Table 8, only the 1.3 float fraction can provide an equivalent sulfur product. However, weight and Btu recoveries for a product of this quality are less than one-half that expected for the two-step physical and chemical approach.

The reader is cautioned that the comparison of coal product qualities and weight and Btu recoveries has been made between two different coal samples. The washability data for the 8 mesh x 0 coal sample used in the current study was not available. Accordingly, a Pittsburgh bed coal from Monroe County, Ohio, of similar float 1.6 s.g. quality was selected from reference 26 as the basis for comparison.

CONCLUSIONS

In view of the emission restrictions promulgated through the Clean Air Act, and others that could arise through pending acid rain legislation, the coal industry, and the industrial and utility sectors look with interest and hope toward the successful development of new and advanced coal cleaning technologies in order to continue to promote the use of coal in this country. This paper presented a two-step physical and chemical coal cleaning approach for reducing the sulfur in coal. It was found that sulfur reductions generally exceeded 30 percent, while SO_2 reduction, on a pounds per million Btu basis, approached 50 percent. Coal weight recovery for the two-step process was about 58 percent, and Btu recovery about 72 percent. Ash reductions from 12 to 15 percent and Btu losses from 3 to 8 percent were common. The oxygen content of the treated coal was higher than that of the starting coal, and the extent of oxidation incresed with smaller particle sizes and correlated closely with observed Btu losses. Sodium and chlorine contents in the coal were higher after hypochlorite treatment.

The quality of the final coal product, in comparison to the washability data of a similar Pittsburgh Bed Coal from Monroe County, Ohio, indicates that at an equivalent weight and Btu recovery, this two-step process exceeds the quality of coal possible through conventional physical coal cleaning. Only the float 1.3 product from this coal would exceed SO_2 reduction on a SO_2 per million Btu basis, however, weight and Btu recoveries less than one-half that obtained in the two-step process would be expected.

The complex behavior and reactivity of the hypochlorite reagent are reflected in the pyrite and organic sulfur reduction trends presented. Of the 25-plus conditions run, only a few conditions resulted in total sulfur removal below the organic sulfur level in the starting float 1.6 s.g. coal. This might imply that in all but a few instances, the observed sulfur reductions could be accounted for solely by pyritic sulfur removal. However, in related work conducted in this laboratory (ref. 28), a float 1.3 s.g. sample of this same Pittsburgh bed coal when treated with hypochlorite solutions under similar reaction conditions routinely resulted in total sulfur removal below that of the organic sulfur level of the starting float 1.3 s.g. coal. Accordingly, the researchers have concluded that the hypochlorite reagent is effecting organic sulfur removal.

Disclaimer

Reference in this report to any specific commercial product, approach, or service is to facilitate understanding and does not necessarily imply its endorsement or favoring by the United States Department of Energy.

REFERENCES

1. "Standards of Performance for New Stationary Sources", Environmental Protection Agency. Federal Register, Volume 36, Number 247, December 23, 1971.

2. "New Stationary Sources Performance Standards; Electric Utility Steam Generating Units," Environmental Protection Agency. Federal Register, Volume 44, Number 113, June 11, 1979.

3. "An Analysis of Issues Concerning "Acid Rain," Report to the Congress of the United States by the Comptroller General of the General Accounting Office, December 11, 1984.

4. Boron, David J. and Taylor, Scott, R., "Mild Oxidations of Coal. 1. Hydrogen Peroxide Oxidation." Fuel, Volume 64, February 1985.

5. Boron, David J. and Zavitsanos, Peter D., "Microwave Energy as a Means to Clean Coal." Paper Presented at the Conference on Processing of Energy Minerals: Shale, Tar Sands, and Coal. Engineering Foundation, August 1984.

6. Boron, David J. and Kollrack, Reiner, "Prospects for Chemical Coal Cleaning." Paper Presented at the 1983 Society of Mining Engineers Fall Meeting, Salt Lake City, Utah, October 19-21, 1983.

7. Boron, David J., Dietz, Albert G. and Taylor, Scott R., "Sulphur Removal from Coal via Hydrogen Peroxide Oxidation.," Fuel, Volume 60, October 1981.

8. Taylor, Scott R., Miller, K.J., Hucho, R.E. and Deurbrouck, A.W. "New Methods for Coal Desulfurization," Presented at the Eighth International Coal Preparation Congress, Donetsk, USSR, May 1979.

9. "Technical and Economic Evaluation of Advanced Physical Coal Cleaning Processes." Work in progress by Kaiser Engineering Corporation for the New York State Energy Research and Development Authority. Expected Completion date in August, 1985.

10. "Technical and Economic Comparison of Advanced Coal Cleaning Technologies: A Multi-Client Study, "A Report by Norton, Hambleton Inc., Coal Process Engineers. March 1985.

11. "Technical and Ecomomic Study of Chemical Beneficiation Processes" Report submitted to the U.S. Department of Energy by Dr. Reiner Kollrack of the Engineers Societies Commission on Energy, Inc., 1983.

12. "An Analysis of Chemical Coal Cleaning Processes." Final Report. DOE/ET10045-T1. Bechtel National, Inc., June 1980.

13. "Survey and Evaluation of Current and Potential Coal Beneficiation Processes." Final report prepared for the Department of Energy/Fossil Program Control and Support Division Engineering Economic and Standards Section, by Oak Ridge National Laboratory. Contract No. W-7405- eng.-26, March 1979.

14. "Assessment of Coal Cleaning Technology: An Evaluation of Chemical Coal Cleaning Processes." Interagency Energy/Environment R&D Program Report. EPA-600/7-78/173a, VERSAR, Inc., Springfield, Va, August 1978.

15. "Dunstan: The Hypochlorite Process," J. Inst. Pet. Tech., April 1924.

16. Waterman, H.I. and Heimel, J.H., "The Desulphurizing of Crude Oil Distillates by means of Hypochlorite, after the Method of Dunstan and Brooks," Inst. Pet. Tech. J., Volume 46, 1924.

17. Birch, S.F. and Norris, W.S.G.P., "The Chemistry of Petroleum. Part II. The Action of Sodium Hypochlorite on Sulphur Compounds of the Types Found in Petroleum Distillates," J. Chem. Soc., 127, 1925.

18. Wood, A.E., Greene, A.R. and Provine, R.W., "Desulphurizing Effects of Sodium Hypochlorite on Naptha Solutions of Organic Sulfur Compounds," Ind. and Eng. News, Volume. 18, No. 8, August 1926.

19. Wood, A.E., Sheely, Clyde and Trusty, A.W., "Action of Petroleum - Refining Agents on Naphtha Solutions of Pure Organic Sulfur Compounds." Ind. and Eng. Chem., Volume 18, No. 2, February 1926.

20. Chakrabartty, S.K. and Kretschmer, H.O., "Studies on the Structure of Coals: Part 1. The Nature of Aliphatic Groups." Fuel, Volume 51, April 1972.

21. Chakrabartty, Sujit K. and Kretschmer, Hans O., "Studies on the Structure of Coals. 2. The Valence State of Carbon in Coal," Fuel, Volume 53, April 1974.

22. Chakrabartty, Sujit K. and Berkowitz, Norbert, "Studies on the Structure of Coals. 3. Some Inferences about Skeletal Structures." Fuel, Volume 53, October 1974.

23. Mayo, Frank R. and Kirshen, Norman A. "Oxidations of Coal by Aqueous Sodium Hypochlorite." Fuel, Volume 58, October 1979.

24. Mayo, Frank R., "Applications of Sodium Hypochlorite Oxidations to the Structure of Coal." Fuel, Volume 54, October 1975.

25. Snyder, N.H. (revised by S. J. Aresco), "Coal Sampling, a Revision of Technical Paper 133," 1957.

26. "Sulfur Reduction Potential of the Coals of the United States: A Revision of Report of Investigations 7633." Report of Investigations 8118, The United States Department of Interior, 1976.

27. Holst, Gustaf, "The Chemistry of Bleaching and Oxidizing Agents." Chem. Rev., Vol. 54, 1954.

28. Unpublished work conducted by the Authors at the US DOE Coal Preparation Laboratories, PETC, 1977-1980.

A REVIEW OF GAS PHASE DESULFURIZATION OF CHAR

M.D. STEPHENSON, M. ROSTAM-ABADI, L.A. JOHNSON AND C.W. KRUSE
Illinois State Geological Survey, 615 E. Peabody Dr., Champaign, IL 61820

ABSTRACT

It is well-known that many reagents remove sulfur from char or coal undergoing devolatilization. This review focuses only on gas phase treatments with an emphasis on the chemistry involved, conditions which influence desulfurization such as time, temperature, pressure and particle size, and the levels of desulfurization which can be expected. Oxygen, steam and hydrogen have been the most thoroughly researched.

INTRODUCTION

When coal is carbonized, the sulfur in coal is partitioned among all three product phases, i.e. gas, liquid and solid. As a result, high sulfur char is generally produced from a high sulfur coal. Fortunately, some or most of the sulfur in the char can be removed by chemical treatment. Gas phase treatments may have certain advantages over wet methods as the product does not need to be washed, filtered, and dried.

The following review is limited to gas phase desulfurization where the desulfurized product is a coal char. In some cases, the gas is used during pyrolysis; in others, it is used subsequently. Oxygen, steam, and hydrogen have been well-studied. Many other gases have also been examined, but to a lesser extent.

OXYGEN

Chemistry

During carbonization of coal, organic and inorganic sulfur compounds decompose and/or react with products of pyrolysis to form additional sulfur compounds. In an inert atmosphere, pyrite decomposes to ferrous sulfide and elemental sulfur. Decomposition begins at 300°C, is complete at 600°C, and generally reaches its maximum between 450-550°C (ref.1). FeS_x ($1.7 < x < 1.9$) is known to be the intermediate non-stiochiometric sulfide prior to the formation of ferrous sulfide (ref.2,3). The elemental sulfur released from pyrite decomposition can polymerize and be retained in the char or interact with the organic matrix of coal to form carbon-sulfur bonds (ref.4,5). Sulfur fixation reactions can take place between 350 and 500°C.

Various reactions also occur between pyrite, ferrous sulfide, or the sulfur released by pyrite decomposition and calcium or iron-containing minerals to form the corresponding metal sulfides (ref.6,7,8).

Sulfur dioxide is the major product of oxidation of both inorganic and organic sulfur compounds (ref.9-11). The main reactions are:

$$4FeS_2 + 11O_2 \rightarrow 2Fe_2O_3 + 8SO_2 \tag{1}$$

$$2FeS_2 + 7O_2 \rightarrow Fe_2(SO_4)_3 + SO_2 \tag{2}$$

The reaction products from pyrite oxidation depend on temperature, particle size, and oxygen partial pressure. Oxidation of pyrite in excess oxygen proceeds predominantly by reaction (1). Sulfate formation is more favorable at lower temperatures. At 400°C, iron sulfate formation amounted to 12 percent of pyrite conversion and decreased with increasing temperature (ref.12). The conversion of pyrite to sulfate results in an increase in solid molar volume of the mineral which tends to close the pores. Therefore, under the conditions where sulfate formation is favorable, the oxidation rate is limited by oxygen diffusion through the pores of sulfate-oxide layers (ref.10-12).

The reactions between ferrous sulfide and oxygen are:

$$2FeS + 3O_2 \rightarrow 2FeO + 2SO_2 \tag{3}$$

$$FeS + 2O_2 \rightarrow FeSO_4 \tag{4}$$

Sulfate formation, as in the case of pyrite, is favorable at temperatures below 400°C and results in pore closing. The reaction stopped at 45 percent conversion of FeS at 400°C and at 70 percent conversion at 500°C (ref.12).

Calcium sulfide present in char reacts with oxygen to yield oxides (ref.13,14)

$$4CaS + 6O_2 \rightarrow CaS + 3CaSO_4 \rightarrow 4CaO + 4SO_2 \tag{5}$$

This reaction, however, is observed only above 850°C (ref.13). A detailed study of the changes in structures of minerals taking place in lignites during combustion has been reported (ref.13).

The oxydesulfurization of organic sulfur groups found in coal is accompanied by carbon loss. Alkyl sulfide and disulfide bonds exhibit a maximum rate of sulfur dioxide evolution at 230°C (ref.9). The corresponding maximum rate for aryl sulfides and thiophenic substances occurs at 460°C.

Oxygen Desulfurization

Desulfurization of coal and char by oxygen-bearing gases has been the subject of few investigations. Although the results of these studies differ, several common remarks can be made on the effectiveness of oxygen to desulfurize coal and the parameters which influence such processes.

The optimum temperature for desulfurization is between 400-450°C (ref.10,11,15). Below 350°C, the rate and the extent of desulfurization is low. Above 450°C or for residence times longer than 15 minutes at 450°C, desulfurization is accompanied by high weight loss. Below 450°C, an oxygen-bearing gas appears to be more effective for removing pyritic sulfur than hydrogen (ref.11), nitrogen (ref.11,15,16), CO or steam-CO mixtures (ref.16). Removal of inorganic sulfur is due mainly to the oxidation of pyritic sulfur, while the removal of organic sulfur is mainly by pyrolysis and combustion. A major portion (>80%) of the pyritic sulfur was removed from 7 U.S. coals of different ranks by air oxidation at 450°C for 10 minutes (ref.10). Weight loss for low and medium volatile bituminous coals was about 5 percent while the weight loss for the high volatile bituminous coals ranged from 10 to 17 percent. However, desulfurization of 10 U.S. coals of various ranks in air showed that only half of the inorganic sulfur and 20 percent of the organic sulfur could be removed at 450°C for 15 minutes. The weight loss accompanying the treatment ranged from 13-42 percent.

The efficiency of desulfurization is affected by gas flow rate and partial pressure of oxygen. An air velocity of 7.9 cm/sec. was most effective in desulfurizing an Ohio hvAb coal at 450°C for 15 minutes (ref.15). The concentration of oxygen in a fluidizing gas stream to desulfurize a hvCb coal from Iowa appeared to have more effect on the removal of pyritic sulfur than on the removal of organic sulfur (ref.11). At 440°C and for 30 minutes treatment time, a gas containing 10 percent oxygen (in N_2-O_2 mixture) removed 78 percent of the total sulfur (63% wt loss), while a gas containing 4 percent oxygen removed 46 percent (34% wt loss). In both cases the organic sulfur comprised nearly half of the total sulfur removed. On the other hand, mixtures of air, nitrogen, and steam with a 2-10 percent oxygen concentration had a slight effect on the extent of desulfurization (ref.17).

The extent of pyritic sulfur removal is increased by reducing the coal particle size. This suggests that the diffusion of oxygen into the microporous structure of coals (ref.10,17) and through the iron oxide product layer is, in part, a controlling resistance.

Generally, both volatile matter and calorific value decrease following oxidation (ref.10,15). For coals with lower volatile matter, there is a net increase in volatile matter since the chemisorption of oxygen on active sites of char outweighs the loss of volatile matter.

Sulfur dioxide is the primary sulfur-containing product (ref.9,11) and its concentration as high as 10 percent does not affect the extent of desulfurization (ref.10). The release of sulfur dioxide occurs in two stages under isothermal conditions (ref.11,14). The first is associated with oxidation of the volatile sulfur compounds while the second stage is due to the oxidation of pyrite. Under nonisothermal conditions at least four stages of sulfur dioxide release have been observed (ref.9); (1) alkyl sulfides are oxidized in the 225-300°C range, (2) pyritic sulfur in the 350-380°C range, (3) aryl sulfides and thiophenic compounds around 450°C, and (4) inorganic sulfates or other sulfur-containing compounds decompose above 650°C.

Several studies have been conducted on post-oxidation of char as an intermediate desulfurization step toward producing compliance char. A fixed bed reactor was employed to evaluate the effectiveness of a gas sequencing approach in the desulfurization of Illinois No. 6 coal (ref.18). Pyritic sulfur was shown to be effectively removed below 450°C by pyrolysis in nitrogen or hydrogen followed by oxidation of the char. As much as 87 percent of inorganic sulfur and 30 percent of organic sulfur was released when coal was heated in nitrogen for 20 minutes and the resulting char was subjected to 10 percent O_2 - 90 percent N_2 gas for 2 minutes at 450°C. Total sulfur removal was 46 percent. Recent work at the Illinois State Geological Survey (ref.19) has also shown that pyritic sulfur can be released from char by using a dilute oxidizing atmosphere (3 percent O_2 - 97 percent N_2 atmosphere for 15 minutes). During post-charring oxidation at 450°C, 8 to 14 percent of the initial sulfur in coal was released, depending on char preparation temperatures and pre-charring treatment.

In summary, oxygen-containing gases remove mainly pyritic sulfur from coal at temperatures between 400-450°C. At higher temperatures, the removal of additional sulfur occurs at the expense of volatile matter and carbon oxidation. Sulfur dioxide is the major product of pyrite oxidation. The reaction rate is limited by diffusion of oxygen through the iron oxide product layer. This could be the reason for excessive carbon loss at higher oxidation temperatures, since the carbon-oxygen reaction is kinetically controlled below 700°C. Post-pyrolysis oxidation of char below 450°C releases additional inorganic sulfur, but the extent of sulfur removal is not substantial. However, postoxidation treatment results in the development of a more porous and reactive char (ref.20,21,22) which may facilitate the hydrodesulfurization of char.

STEAM

To define the effectiveness of steam as a gas phase reagent for the desulfurization of coal, two points warrant consideration. First, the level of desulfurization achieved through the use of steam must be viewed relative to that which normally occurs during pyrolysis. Secondly, since gasification occurs at temperatures above 500°C, additional carbon loss may accompany desulfurization. To maximize the yield of a solid compliance fuel, steam must exhibit selectivity toward sulfur over carbon. A meaningful evaluation will include the loss of weight or loss of heat content in addition to the percentage of sulfur removed.

Chemistry

Organic sulfur is found as sulfides, disulfides, thiols and thiophenes. The interactions between steam and the organic sulfur in coal are not as well understood as the reactions of inorganic sulfur with steam. Pyrite conversion to iron oxides is thought to occur in two steps. The following reaction has been proposed (ref.23) for the conversion of pyrite to pyrrhotite which is the first step toward complete conversion of pyrite.

$$H_2O + FeS_2 + C \rightarrow H_2S + FeS + CO \qquad (6)$$

This is essentially carbon gasification followed by hydrogen reduction of pyrite. Under typical steam pyrolysis conditions (0.5 atm H_2O, 0.15 atm CO), this reaction is thermodynamically favorable at 400°C and hydrogen sulfide concentrations less than 8000 ppm. Further conversion of the sulfide to iron oxides is more difficult. The following reactions for conversion of iron sulfide to hydrogen sulfide and iron oxides have been proposed (ref.23):

$$FeS + H_2O \rightarrow FeO + H_2S \qquad (7)$$

$$3FeS + 4H_2O \rightarrow Fe_3O_4 + 3H_2S + H_2 \qquad (8)$$

$$2FeS + 3H_2O \rightarrow Fe_2O_3 + 2H_2S + H_2 \qquad (9)$$

The carbon-steam and the water-gas shift reactions also occur during pyrolysis in the presence of steam. The hydrogen produced by both reactions may react with carbon or sulfur compounds in the coal. Therefore, desulfurization or gasification by hydrogen may be involved in the overall chemistry during steam pyrolysis.

General Conclusions

The extent of desulfurization during pyrolysis depends on the ratio of organic to inorganic sulfur, the total sulfur content, coal rank, mineral constituents and physical properties of the coal, and pyrolysis conditions. This makes it difficult to directly compare the results of different studies. Table 1 summarizes the results from many investigations (ref.8,15,17,18,24-29) of steam desulfurization of carbonaceous fuels. These studies support the following conclusions:

1) The optimum temperature range for steam desulfurization is 500-600°C with 600°C most commonly cited as optimal. Since the temperature intervals were often as much as 100°C between runs in the reported determinations of temperature effects, the magnitude of this range is not unusual.

2) Complete conversion of pyrite to pyrrhotite can be achieved at partial pressures of steam above 0.5 atm. Additionally, reduction of pyrrhotite to elemental iron appears to take place, but this is difficult to demonstrate because the sulfidic sulfur, elemental sulfur and organic sulfur are frequently not differentiated in the forms of sulfur analysis. Indirect support for this view shows a large decrease in pyritic sulfur unaccompanied by an increase the organic/sulfide sulfur (ref.15,17). Furthermore, 75±25 percent of the sulfur from an oil shale having 70-85 percent of its sulfur in the inorganic form can be removed (ref.24). Similarly 80-90 percent of the total sulfur was removed from a 12 gal/ton Ohio shale through the use of steam (ref.30). In the same study only 45 percent sulfur removal was achieved when retorting in nitrogen. Clearly steam has the potential to carry the reduction of pyrite further than thermal decomposition alone.

3) Steam is relatively ineffective in the removal of organic sulfur from coal. Again, there is some uncertainly due to the fact that sulfide and organic sulfur are often classified together. However, the results of one study (ref.25), which showed only a 23 percent reduction in organic sulfur in an attempt to desulfurize oil sands coke which was essentially free of pyrite, support this conclusion. There was a significant decrease in the concentration of organic/sulfide sulfur but only after a severe and lengthy treatment (ref.29). Only one additional investigator was able to reduce the level of organic sulfur in char to a level significantly lower than that in the original coal (ref.26).

4) Attempts to increase the severity of the treatment by a longer residence time, higher steam concentration, or higher reactant flow rate have failed to produce a substantially cleaner char (ref.15,17,28). Similarly, the use of steam/air mixtures has not been more effective than the use of steam alone (ref.17,18,25,28).

5) During gasification at temperatures below 850°C, steam shows very little selectivity toward sulfur over carbon. Many studies have shown a correlation between weight loss and desulfurization (ref.8,31). Contrary to this, when extensive weight loss occurs due to the carbon-steam reaction, the percent of total sulfur removed increases, but the weight fraction of sulfur in the char remains nearly constant. For example, when an Ohio No. 8 hvAb coal was treated with steam at 600°C for various lengths of time, the weight loss of the coal varied from 24.6 to 67.4 percent, but the sulfur content of the char remained relatively constant at 2 ± .14 percent (ref.15). At 900°C, a substantial decrease in total sulfur was achieved. Since hydrodesulfurization is effective at 900°C and not at 600°C, it is possible that elemental hydrogen produced by the carbon-steam reaction is responsible for the enhanced sulfur removal. In-situ hydrogen desulfurization has been observed in other processes where coal was gasified at high temperatures (ref.32).

In addition to these general observations, some unique results have been reported by individual researchers. Coals were treated with many gases (H_2, air, H_2O, NH_3) after they had been spiked with radioactive pyrite ($Fe^{35}S_2$) (ref.8). In this direct comparison, steam was shown to be more effective than any of the other gases in the removal of pyrite from seven coals of various rank. An increase in the effectiveness of hydrodesulfurization was observed when the coal was oxidized at 300°C prior to hydrotreatment (ref.15), but another study (ref.28) found that mild oxidation of Texas lignite prior to steam pyrolysis did not increase the degree of sulfur removal. Mixtures of steam and air were also ineffective. In a novel approach (ref.18), steam was used in a series of gas treatments which comprise the overall process. Desulfurization with minimal carbon loss would be difficult to achieve, but recent results have shown 70 percent total sulfur removal from Illinois No. 6 coal (ref.33).

Steam pyrolysis is more effective in removing sulfur than pyrolysis in an inert atmosphere. Steam at 500-600°C can be used to remove pyrite from a variety of coals, but complete desulfurization has not been achieved. Steam has many advantages when compared to other gases. It is effective near 600°C as opposed to the 800-900°C range which is optimal for hydrodesulfurization. The reagent is less expensive than hydrogen and fixed carbon losses are lower compared to direct oxidation using air or oxygen. Finally, off-gases from steam pyrolysis can be condensed to produce a concentrated stream of sulfur compounds to be processed in a Claus plant. Despite these advantages, a weakness is the necessity to develop a complementary process for the removal of organic sulfur. There is some evidence that the use of steam may enhance the yield of oil from pyrolysis (ref.30,34). Thus steam at low temperatures,

Table 1

Desulfurization with steam.

Authors	Reactor type and conditions	Reactant gas	Reaction conditions (optimum T)	Coal	Sulfur [a]			Loss of weight %
					Type	Coal wt %	Char wt %	
Alvin et al.	bench-scale, packed bed; 20 minutes	4.6% steam, 95.4% nitrogen	400°–600°C	Illinois No. 6, -20/+40 mesh; 3 g sample	total pyritic sulfatic $S^{=}$ + Org.	2.10 0.83 0.15 1.12	1.74 0.12 0.06 1.55	ca 9
Block et al.	stainless steel boat in a tube furnace; 3C°/min	steam and nitrogen, compositon not given	450°–900°C 0 psig (500°C)	Pittsburgh No. 8, hvAb, -48/+65 mesh, pretreated with air at 300°C for 10 min;	total	3.59	2.03	27.0
				average of 10 coals	total inorganic organic	3.78 2.24 1.54	1.54 0.31 1.23	27.9
Burnham et al.	packed bed in tubular furnace	50% steam 50% argon	300°–600°C 0 psig (500°C)	western shales (several grades) 0.2-0.55 g samples	total	0.99	0.20	-
Cernic-Simic	packed bed, quartz tube, tubular furnace; 3C°/min	100% steam, 100 ml of water/hour	400°–900°C 0 psig (600°C)	radioactive pyrite added to 7 coals, (Fe35S2); 10 g samples	organic inorganic	3.11 0.23	(% of the 35S added that remained in char = 15.7%)	
Jacobs and Mirkus	fluidized bed, $1\frac{1}{2}$ in. S.S. in electric muffle furnace; 30 min	35% steam 13% air 52% inert gas	510°C	Illinois No. 6 (Peabody No. 10 mine) -1/16 in.; 100 g sample	total pyritic sulfidic organic	4.89 2.53 0.33 2.03	2.40 0.49 0.08 1.83	27

Table 1

Continued.

Authors	Reactor type and conditions	Reactant gas	Reaction conditions (optimum T)	Coal	Sulfur[a] Type	Sulfur[a] Coal wt %	Sulfur[a] Char wt %	Loss of weight %
Parmar and Tollefson	fluidized bed, 2 in. S.S.; up to 2 hrs.	100% steam,	871°C 0 psig	oil sands, petroleum coke, -20/+80 mesh; about 100 g:				
				preoxidized	total	4.80	4.00	–
				not oxidized	total	5.80	4.90	–
Sanigok and Bayramoglu	– 310 minutes	100% steam, 80 g/hr	450°-600°C (500-600°C)	Can lignite from Turkey -8/+14 mesh	total	3.31	0.90	26
					pyritic	1.12	0.40	
					sulfidic	0.98	0.09	
					organic	1.21	0.41	
Soo	fluidized bed, 6 in. diameter at base; 25 minutes	62% steam 31% argon 7% nitrogen	450°-700°C 0 psig (650°C)	Herrin No. 6 and Colchester No. 2 (Illinois) size, 125-177um; 1 kg sample	total	3.40	1.81	35.7
					pyritic	0.83	0.21	
					sulfatic[b]	0.81	0.10	
					sulfidic	–	0.04	
					organic	1.76	1.46	
Stone	fluidized bed, 1 in. S.S.	100% steam, steam/coal ratio varied	404°-454°C 10.3 psig (454°C)	Texas lignite	total	1.03	0.79	22.1
Van Hoesle and Quadri	thin layer at the bottom of a quartz tube; heated by a furnace	100% steam, 6 hours	300°-600°C 0 psig (600°C)	8 coals from West Pakistan; 20 g samples	inorganic	1.54	0.12	56.7
					organic	1.88	1.05	

[a] Average analysis when more than one coal was tested
[b] Unpublished analysis performed at the ISGS coal analysis laboratory

either before or during pyrolysis, coupled with subsequent hydrodesulfurization may be a viable combination of processes. In-situ hydrodesulfurization by steam pyrolysis/gasification at higher temperatures is another option. In any event, development of a supplementary process is necessary to produce a compliance fuel from high sulfur coal.

HYDROGEN

Chemistry

Hydrogen is an effective reducing agent for desulfurizing char. Both organic and inorganic sulfur are removed at high temperatures, usually above 800°C. Generally char contains little pyrite due to thermal decomposition to ferrous sulfide. Any remaining pyrite is quite reactive toward hydrogen, the sulfur being reduced to hydrogen sulfide:

$$FeS_2 + H_2 \rightleftharpoons FeS + H_2S \quad (10)$$

This reaction is thermodynamically favorable. At 380°C, the equilibrium constant has a value of 1, and at 850°C, it has a value of 1000 (ref.35). For further conversion of inorganic sulfur to hydrogen sulfide, ferrous sulfide is reduced to elemental iron:

$$FeS + H_2 \rightleftharpoons Fe + H_2S \quad (11)$$

Thermochemical data (ref.35) indicates the Gibbs free energy is a positive quantity at 850°C. Furthermore, the equilibrium constant has a value of 0.002, indicating that 2000 ppm hydrogen sulfide is sufficient to reverse the reaction.

Inhibition studies with char containing ferrous sulfide, generally confirm that small amounts of hydrogen sulfide reverse the reaction, thus inhibiting desulfurization. In one study (ref.36), char was treated with hydrogen in a small fluidized bed. A 0.5 percent addition of hydrogen sulfide to a 10 percent hydrogen, 90 percent steam mixture was large enough to reverse the reaction. Furthermore, an outlet hydrogen sulfide concentration of 0.2 percent caused a decrease in the relative desulfurization rate on the order of 10 to 40 percent depending on the conditions employed. In another study (ref.37) using a nickel basket suspended in a hot zone of a furnace, a hydrodesulfurized char was subjected to increasing concentrations of hydrogen sulfide at 900°C in a hydrogen treatment gas. A sudden rise in the char sulfur content was observed when the hydrogen sulfide content reached 0.3 percent. The amount of sulfur captured was also in good agreement with the estimated amount required for a quantitative conversion of the iron to ferrous sulfide. Inhibition studies of char from Pittsburgh seam coals were determined at various temperatures (ref.38). For one char, the equilibrium hydrogen sulfide concentration was less than 0.1 percent at 593°C, nearly 0.1 percent at 732°C, and about 0.3 percent at 871°C. These data, obtained in a small fluidized bed reactor, also agree closely with the thermodynamic data for the ferrous sulfide/iron system. Other inhibition studies (ref.39,40) with chars containing ferrous sulfide are in general agreement that the back-reaction of hydrogen sulfide with iron strongly inhibits the desulfurization of char by hydrogen.

The chemistry of organic sulfur removal from char is less understood. When fractions of crude oil are hydrodesulfurized, it is known that mercaptans, sulfides and disulfides contained in distillate fractions are the most readily removed. The cyclic sulfur compounds in residual portions (thiophenes, benzothiophenes and dibenzothiophenes) are less reactive. Similar reactions presumably occur in char (ref.6). The sulfur that is removed is liberated as hydrogen sulfide. As far as char is concerned, a non-isothermal kinetic study (ref.41) reveals at least three types of organic sulfur are removed, each attacked in three different temperature ranges. Organic types I and II are released at relatively low temperatures, whereas organic type III is released around 650°C. In some coals, a fourth release of hydrogen sulfide is seen at 925°C. The release of sulfur from ferrous sulfide is observed to occur at 615°C. Other studies indicate that more than one form of organic sulfur is evolved (ref.42,43). Generally, a certain portion of organic sulfur is even more difficult to remove than ferrous sulfide sulfur. One opinion is that total inhibition isotherms for organic sulfur depend only on the nature and prior history of the char. This indicates that some organic sulfur is thermally fixed into a very stable form.

Hydrodesulfurization studies

Hydrodesulfurization of char has been the subject of many studies (ref.11,15,36-41,43-54). The results of many pertinent studies are summarized in Table 2. Reaction temperatures were generally in the 700-900°C range, and pressures ranged up to 400 psig, but generally were in the 0-50 psig range. In many cases, sulfur contents in the chars were reduced below 1 percent and some of the chars were sufficiently desulfurized to meet the EPA standard of 1.2 lbs. SO_2/MM Btu for power plant fuels.

An optimum temperature apparently exists for hydrodesulfurization of char. According to one study (ref.44), thermal fixation of the sulfur occurs such that some of the sulfur is transformed into a stable form which is difficult

Table 2

Summary of hydrodesulfurization results.

Authors	Reactor type	Reactant gas	Reaction conditions	Coal	Wt % sulfur Coal	Wt % sulfur Char
Bachelor et al.	4 in. fluidized bed (batch)	100% hydrogen	871°C 73.5 psig	Pittsburgh, -35/+100 mesh	1.96	0.33
Belt and Roder	entrained flow, 4 in. ID x 18 in. (continuous)	91% hydrogen 9% nitrogen	815°-1038°C 0-400 psig	Pittsburgh, hvAb, 70% -200 mesh	2.55	0.07
Block et al.	stainless steel boat in tube furnace	100% hydrogen	600°-1000°C	Pittsburgh No. 8 seam, ball-milled	3.59	0.51
		high flow rate	900°C		3.59	0.20
Broodman et al.	fluidized bed, 10 in. S.S. (continuous)	20-40% hydrogen 50-70% methane 50-1000ppm H_2S balance CO_2	750°C 150psig	Illinois No. 6	2-2.5	0.6-0.7
Brothers and Cradlemire	quartz boat in tube furnace	100% hydrogen	25°-900°C 50 psig	Prince coal, -100/+200 mesh	3.8	1.1
Fleming et al.	fluidized bed, 1½ in. S.S. (batch)	100% hydrogen	815°C 0 psig	W. Ky. No. 9 Illinois No. 6 -20/+80mesh	3.74 2.43	0.63 0.62
Gray et al.	fixed bed, 3/4 in. (batch)	100% hydrogen	871°C	Illinois No. 6, +100 mesh	3.4	0.4
Haldipur and Wheelock	fluidized bed, 2 in. S.S. (batch)		395°-400°C	Iowa Jude, hvCb, -20/+40 mesh		
		87% H_2; 13% N_2		ROM	8.06	6.42
		84% H_2; 16% N_2		precleaned	6.35	5.54

Table 2

Continued.

Authors	Reactor type	Reactant gas	Reaction conditions	Coal	Wt % sulfur Coal	Char
Huang and Pulfsifer	thermogravimetric analyzer	100% hydrogen	100°–700°C	Iowa Jude, hvCb; -115 mesh		
				ROM	7.76	3.79
				precleaned	5.75	2.23
Johnson et al.	fluidized bed, 3 in. S.S. (batch)	40% hydrogen 52% methane 8% CO	740°C 125 psig	Illinois No. 6	1.74	0.5
Kor	platinum or nickel basket in furnace	100% hydrogen	800°C; 0 psig	Illinois No. 6, +18 mesh	1.93	0.4
		100% hydrogen	800°C; 60 psig		1.93	0.1
Maa et al.	quartz tube, 0.87 in. (batch)	100% hydrogen	870°C	W. Ky. No. 9, -25/+140 mesh	3.8	1.1
Robinson	fluidized bed, 3/4 in. quartz (batch)	100% hydrogen	860°C 50 psig	Montana Sub-bit	0.92	0.22
Tipton	fluidized bed, 3/4 in. quartz (batch)	100% hydrogen	871°C 50 psig	W. Ky. No. 9, -200 mesh	3.3	0.4
Zielke et al.	S.S. fluidized bed	100% hydrogen	871°C	Disco coal char, -65/+100 mesh	1.9	0.97–1.3 0.36–0.55

to remove. This is much more pronounced at higher temperatures where one would expect increased reaction rates. Differential rate data suggest optimum desulfurization rates occur at 788°C for 1 atm. hydrodesulfurization (ref.55). In another study (ref.40), optimum conversion occurred at 850°C. A reduction in surface area with increased temperatures was noted for several chars. For example, one char had a surface area of 75.0 sq m/g at 760°C, 39.7 sq m/g at 815°C and 13.5 sq m/g at 885°C. The rate of hydrodesulfurization was essentially the same at 760°C as it was at 885°C. There are at least three competing phenomena. Hydrodesulfurization should occur at a faster rate at higher temperatures. However, some sulfur becomes thermally fixed and difficult to remove, and the char has a decreased surface area. As a result, there is a single temperature for optimum desulfurization.

Higher pressures result in enhanced conversions and rates. In a fluidized bed system at 741°C, increasing the total pressure from 15 to 110 psig resulted in a higher level of desulfurization in the char (ref.54). In 30 min, an Illinois No. 6 coal (1.74% sulfur) was desulfurized to 0.77 percent at 15 psig, 0.50 percent at 65 psig and 0.33 percent at 115 psig. Other studies gave similar results (ref.36,38,40,44). Char particle size has little influence on hydrodesulfurization. The reduction in sulfur content on 1/2 in. briquets was roughly the same as -65/+100 mesh char (ref.36). No particle size effects were noted for a char from a Western Kentucky coal (ref.40). Other studies support the idea that particle size effects are minimal (ref.15,43,45).

The rate and extent of sulfur removal can be increased by preoxidizing char before hydrodesulfurization (ref.52). Four bituminous caking coals were preoxidized for approximately 30 minutes at 400°C using air. It was found that 8-12 percent of the coal was consumed, but 25-30 percent of the total sulfur was removed, primarily pyritic sulfur. When treated with hydrogen, 70 percent removal was realized in an untreated Western Kentucky coal whereas 95 percent removal resulted with air pretreatment. The improved sulfur removal was obtained by a more complete removal of organic sulfur. The more complete removal is believed to be due to an increase in pore size, resulting in larger passages for diffusion of hydrogen sulfide. Also, oxidative treatment may have activated the sulfur. In another study (ref.15), four minutes of hydrogen treatment on preoxidized coal was as effective as one hour on untreated coal.

Contrary to the previous studies, U.S. Steel's Clean Coke studies indicated that unoxidized coal apparently gave better desulfurization than preoxidized material (ref.54). However, the desulfurization was carried out at 741°C, 65 psig with a 30 minute residence time and the desulfurizing gas was a mixture

(39% H_2, 52% CH_4, 9% CO). Also, the difference in sulfur contents of the two chars was only 0.02 percent.

Whenever char is treated with hydrogen at elevated temperatures, some gasification occurs according to:

$$C + 2H_2 \rightarrow CH_4 \tag{12}$$

High pressures and low concentrations of methane in the treatment gas lead to higher conversions. To inhibit hydrogasification, a small amount of methane can be added to the treatment gas (ref.36). At 871°C and 6 atm. hydrogen, a 1.92 percent sulfur char was reduced to 0.46 percent sulfur with a 5.5 percent carbon loss (ref.36). Under the same conditions but with 0.6 atm. methane added, a 0.54 percent sulfur char was obtained with only 0.36 percent carbon loss. In another study (ref.45), 50-50 mixtures of methane were compared with pure hydrogen to determine whether the gasification and desulfurization reactions were interrelated. Desulfurization appeared to be about the same regardless of the presence of methane, and the reactions were assumed to be independent. Another study (ref.37) indicated that methane presence inhibits desulfurization. It was difficult to distinguish between the effect of methane and the effect of hydrogen partial pressure. The total system pressure was kept constant at 5 atm., and methane concentrations were varied from 0 to 100 percent. The char sulfur content varied from less than 0.1 percent in 5 atm. hydrogen to about 0.3 percent in 2.5 atm. hydrogen, 2.5 atm. methane. Very little desulfurization occurred in pure methane.

The effectiveness of hydrogen in removing sulfur from char has encouraged the development of a number of processes. Laboratory and bench-scale equipment demonstrated that hydrogen could be used to desulfurize char to a level below the EPA standard of 1.2 lb/MMBtu necessary for direct combustion (ref.53). Approximately 25-30 percent of the sulfur was removed as sulfur dioxide during oxidative pretreatment. Pretreatment also helped to increase the sulfur removal during the subsequent hydrogen step.

Occidental Research Corporation has developed a desulfurization process which handles the back-reaction by acid-leaching the char (ref.40,46). An acid-leached char from a run-of-mine Western Kentucky coal could be treated with hydrogen containing as much as 1.5 percent hydrogen sulfide and still meet EPA standards. Coals precleaned by a float-sink test also responded well to hydrodesulfurization. Occidental also developed a potentially less expensive, two-step process involving hydrogen treatment (ref.47). The first step, called the convert step, involved reacting char with hydrogen at a low flow rate. Because a low flow rate was used, hydrogen sulfide was not swept out. Instead, it reacted with in-situ sulfur scavengers:

$$FeO + H_2S \rightleftharpoons FeS + H_2O \qquad (13)$$

The net result is to convert organically bound sulfur to inorganic sulfur with the total sulfur remaining essentially constant. In the second step, the remove step, the sulfur scavengers were regenerated by steam displacement, the reverse reaction (ref.28). While high flow rates of steam are required to sweep out hydrogen sulfide, very little water is actually consumed. The repetition of several short cycles of the two treatment steps was more effective than a single two-step cycle with longer treatment times. With a proper combination of convert and remove steps, it was possible to produce a low sulfur product meeting EPA regulations for direct combustion.

The U.S. Steel Clean Coke Process (ref.37,55) converts high sulfur coal to low sulfur metallurgical coke by fluidized bed pyrolysis and hydrodesulfurization. In essence, pyrolysis gases are recycled to carbonizers at 440°C and 760°C at pressures up to 150 psig. The gas, rich in methane and hydrogen, is scrubbed at 45°C to keep the hydrogen sulfide concentration low. Normally, hydrogen sulfide in the inlet gas is maintained in the 50-100 ppm range, sufficient to initially prevent the back-reaction with iron. Temperature exerted the greatest effect on desulfurization of the char. A 40 minute residence time at 760°C was equivalent to 157 minute at 718°C in producing a 0.6 percent sulfur char from a coal containing 1.7 percent sulfur. The effect of hydrogen sulfide concentration in the recycle gas on the extent of desulfurization was pronounced. A 0.2 percent sulfur char was obtained with 50-100 ppm hydrogen sulfide. At 500 and 1000 ppm hydrogen sulfide, the product contained 0.7 percent sulfur.

The FMC corporation developed a bench-scale process for hydrodesulfurizing char in the presence of a calcium oxide acceptor (ref.45). Hydrogen sulfide was captured rapidly by the acceptor to form calcium sulfide. The acceptor was successful in keeping hydrogen sulfide concentrations low so that the back-reaction was kept from occurring. Several high sulfur chars were desulfurized to below one percent sulfur.

In summary, the following statements concerning char hydrodesulfurization are well-supported by the voluminous literature data:

1) Hydrogen is very effective in removing both inorganic and organic sulfur.

2) The back-reaction of hydrogen sulfide with iron in char must be prevented for high sulfur conversions. In actual practice this means keeping the hydrogen sulfide concentration below 2000-3000 ppm.
3) Hydrodesulfurization of char occurs at an optimum temperature around 750-850°C. Higher temperatures favor formation of a highly stable sulfur compound in a char of reduced surface area.
4) Higher pressures of hydrogen favor desulfurization. Generally, the highest pressures employed are 75 psig.
5) Char particle size has little effect on hyrodesulfurization.
6) An oxidative pretreatment of coal produces a char which appears to be more reactive towards hydrogen.

OTHER GASES

Researchers have investigated the effectiveness of a number of reactive gases such as chlorine, carbon monoxide, carbon dioxide, ammonia, nitric oxide, nitrogen dioxide, water, hydrogen, oxygen, methane, and ethane for desulfurizing coal. Gases other than hydrogen, water, and oxygen have not been completely effective, but they do possess unique characteristics for removal of certain sulfur compounds under specific conditions. Chlorine gas was used at 80°C to achieve about 30 percent desulfurization of an Ohio No. 8 and an Illinois No. 6 coal (ref.56). Further treatment with hydrogen at temperatures as low as 600°C increased the extent of desulfurization to about 75 percent. It appears gases other than water, oxygen, and hydrogen may be useful, but none have the potential to react with most forms of sulfur in coal.

The acid-base reaction of ammonia and hydrogen sulfide has been shown to be of little importance. Like steam, ammonia probably acts as an in-situ hydrogen source, and hydrodesulfurization is the main sulfur removal mechanism. Up to 700°C, an Illinois No. 6 and a number of Turkish coals showed the same sulfur evolution pattern whether they were pyrolyzed in an inert or an ammonia atmosphere. Above 700°C, the desulfurization increased to a level well above the inert environment (ref.8,15,57). Since hydrodesulfurization is relatively ineffective until 700-800°C and ammonia is 65-85 percent dissociated to nitrogen and hydrogen at 700°C, it appears that in-situ hydrodesulfurization is occurring.

Below 500°C, carbon monoxide reacts with sulfur in coal to form carbonyl sulfide and increases the quantity of sulfur volatilized during pyrolysis. At higher temperatures, the carbon monoxide has no effect (ref.48). Because the extent of carbonyl sulfide liberation is limited, attempts to enhance the sulfur removal with additional gases or stages have been made. Fifteen coals of rank ranging from subbituminous to anthracite were treated with carbon monoxide. The sulfur content of the char averaged 33 percent lower than the parent coal when the char was produced in the presence of carbon monoxide at 450°C compared to the 23 percent reduction in sulfur content observed when the chars were produced in an inert atmosphere. For chars treated with air at 450°C prior to carbon monoxide treatment, the desulfurization due to air oxidation was predominant. Also, the level of desulfurization when carbon monoxide/steam mixtures were used was only slightly higher than the level achieved with carbon monoxide alone. Carbon monoxide can be used to convert pyrite in coal to a ($FeS/Fe_{16}S_{17}$) mixture (ref.58). This mixture then catalyzes the ethanol extraction of organic sulfur which is carried out in the second step of the process. After the char is leached with acid to remove the iron sulfide catalyst, an 80-90 percent reduction in the sulfur content of the solid is achieved. Thus, carbon monoxide may have some utility if it is employed in a planned sequence of stages.

CONCLUSIONS

Desulfurization of char from the pyrolysis of coal has been attempted using various gases. Desulfurization using oxygen or air, steam, and hydrogen has generally received the most attention. Oxygen or air and steam are moderately effective in removing sulfur but suffer the disadvantage of excessive carbon loss. Hydrogen, however, is the most effective desulfurizing gas, and a high percentage of the heating value is conserved except at the highest pressures. In order to achieve adequate desulfurization with hydrogen, the back-reaction of hydrogen sulfide with iron must be prevented.

ACKNOWLEDGMENT

This work was funded in part by the Illinois Coal Research Board through the Center for Research on Sulfur in Coal.

REFERENCES

1 A.R. Powell, 1920, J. Ind. Eng. Chem., 12(11), 1069-1077.
2 N.M. Karavaev, V.N. Amagaeva, 1966, Akad. Nauk SSSR, Inst. Goryuck-Iskop, 164-176, Chem. Abst. 676-1177 43-Y.
3 A. Attar, 1978, Fuel, 57, 201-212.
4 P.H. Given and J.R. Jones, 1966, Fuel, 45, 151-158.
5 P.J. Cleyle, W.F. Caley, I. Stewart and S.M. Whiteway, 1984, Fuel, 63(11), 1579-1582.
6 L. Robinson, 1978, Hydrocarbon Processing, 57(11), 213-216.
7 J.A. Brothers and J.J. Strazonski, 1976, Fuel, 55, 105-108.
8 S. Cernic-Simic, 1961, Fuel 41, 141-151.
9 R.B. LaCount, D.K. Hapen, W.P. King, D.A. Dell, F.W. Simpson and C.A. Helms, 1981, in B.D. Blaustein, B.C. Backrath, S. Friedman (eds), New approaches in coal chemistry, ACS Symposium Series 169, 415-426, Washington, D.C.

10 R.K. Sinha and P.L. Walker, 1972, Fuel, 51, 125-129.
11 G.B. Haldipur and T.D. Wheelock, 1977, in T. D. Wheelock, ed., Coal desulfurization, ACS Symposium Series 64, 305-320, Washington, D.C.
12 G.M. Schwab, J. Philinis, 1947, J. Am. Chem. Soc., 69, 2588-2596.
13 J.C. Mankervis, R.B. Furlong, 1980, Fuel, 59, 425-430.
14 A.J. Moffat, 1982, Paper presented at 1982 Spring Technical Meeting of the Central States Section of the Combustion Institute, Columbus, OH, March 22-23. CCS CI 82/08
15 S.S. Block, J.B. Sharp and L.J. Darlage, 1975, Fuel, 54(4), 113-119.
16 R.K. Sinha and P.L. Walker, 1972, Fuel, 51, 329-331.
17 J.K. Jacobs and J.D. Mirkus, 1958, Ind Eng. Chem., 50(1), 24-26
18 M.A. Alvin, R.A. Newby and D.H. Archer, 1982, Paper presented at the EPRI Coal Gasification Contractor's Conference, Palo Alto, CA.
19 D.D. Coleman, 1984, Illinois State Geological Survey, Annual Center for Research on Sulfur in Coal Technical Report, Champaign, IL.
20 M.M. Ludvig, G.L. Gard and P.H. Emmett, 1984, Fuel 62(12), 1393-1396.
21 S. Leon, M. Klotzkin, G.L. Gard, P.H. Emmitt, 1981, Fuel, 60, 673-676.
22 K. Tomkow, A. Jankowska, F. Czechowski and T. Siemieniewska, 1977, Fuel, 56, 101-105.
23 A.K. Burnham, N.K. Bey, and G.J. Koskinas, 1981, ACS Symposium Series 163, 61-78.
24 A.K. Burnham and R.W. Taylor, 1982, 15th Oil Shale Symposium, Golden, CO. UCRL 87052, preprint.
25 B.S. Parmar and E.L. Tollefson, 1977, Can. J. Chem. Eng., 55, 185-191.
26 U. Sanigok and M. Bayramoglu, 1982, Chimica Acta Turcica, 10, 257-264.
27 S.L. Soo and K.C. Cheung, 1984, Report to the Illinois Coal Research Board, Champaign, IL.
28 R.R. Stone, R.G. Anthony, and J.A. Bullin, 1981, Fuel Processing Technology, 5, 65.
29 H.H. Von Hoessle and S.A. Quadri, 1962, Pakistan J. Sci. Ind. Res., 5, 119-123.
30 S.D. Carter and D.N. Taulbee, 1984, Paper presented at the 1984 Eastern Oil Shale Symposium, Lexington, KY, 267.
31 R.M. Felder, 1981, Report to the E.P.A., Washington, D.C.
32 G. Curran, W. Clark, M. Pell and E. Gorin, 1972, EPA-R2-72-060, 16.
33 M.A. Alvin, 1984, Westinghouse Electric Corporation, Princeton, N.J., private communication.
34 R.A. Graff and S.D. Brandes, 1984, ACS Division of Fuel Chemistry Preprints, 29(2), 104.
35 O. Kubaschewski, E.L. Evans, and C.B. Alcock, Metallurgical Thermochemistry, 4th ed., Pergamon, London, 1967.
36 C.W. Zielke, G.P. Curran, E. Gorin, and G.E. Goring, 1954, Ind. Eng. Chem., 46(1), 53-56.
37 G.J.W. Kor, 1977, in T. D. Wheelock, ed., Coal desulfurization, ACS Symposium Series 64, 221-247, Washington, D.C.
38 J.D. Batchelor, E. Gorin, and C. W. Zielke, 1960, Ind. Eng. Chem., 52(2), 161-164.
39 P.S. Maa, C.R. Lewis, and C.E. Hamrin, 1975, Fuel, 54(1), 62-69.
40 L. Robinson, 1976, Fuel, 55(7), 193-201.
41 A.L. Yergey, F.W. Lampe, M.L. Vestal, A.G. Day, G.J. Fergusson, W.H. Johnston, J.S. Snyderman, R.H. Essenhigh, and J.E. Hudson, 1974, Ind. Eng. Chem. Proc. Des. Dev., 13(13), 233-240.
42 D.K. Fleming, personal communication, May 1985.
43 C.A. Gray, M.E. Sacks, and R.T. Eddinger, 1970, Ind. Eng. Chem. Prod. Res. Dev., 9(3), 357-361.
44 J.D. Batchelor, E. Gorin, and C.W. Zielke, 1960, Ind. Eng. Chem., 52(2), 164-168.
45 C.A. Gray and M.E. Sacks, 1966, OCR-532, 78pp.
46 A.B. Tipton, 1977, in T. D. Wheelock, ed., Coal desulfurization, ACS Symposium Series 64, 280-289, Washington, D.C.

47 A.B. Tipton, 1981, Coal Processing Technology, A.I.Ch.E., 7, 94-97.
48 R.D. Snow, 1932, Ind. Eng. Chem., 24(8), 903-909.
49 R.J. Belt and M.M. Roder, 1973, in R.F. Gould, ed., Pollution control and energy needs, ACS Advances in Chemistry Series 27, 121-134, Washington, D.C.
50 M.T.K. Huang and A.H. Pulsifer, 1977, in T.D. Wheelock, ed., Coal desulfurization, ACS Symposium Series 64, 290-304, Washington, D.C.
51 J.A. Brothers, and A.L. Crandlemire, 1983, Paper presented at 1983 International Conference on Coal Science, Pittsburgh, PA, August 15-19, 1983.
52 N.S. Boodman, T.F. Johnson and K.C. Krupinski, 1977, in T. D. Wheelock, ed., Coal desulfurization, ACS Symposium Series 64, 248-266, Washington, D.C.
53 D.K. Fleming, R. Smith and M.R.Y. Aquino, 1977, in T. D. Wheelock, ed., Coal desulfurization, ACS Symposium Series 64, 267-279, Washington, D.C.
54 T.F. Johnson, K.C. Krupinski, and R.J. Osterholm, 1975, American Chemical Society, Div Fuel Chem. Preprints, 20(4), 33-45.
55 K.C. Channabasappa and H.R. Linden, 1958, Ind. Eng. Chem., 50, 637-644.
56 R. Maddury and J. Kalvinskas, 1984, JPL Invention Report NPO-15924/5604.
57 R.E. Brewer and J.K. Ghosh, 1949, J. Ind. Eng. Chem., 41, 2044-2052.
58 R.H. Shiley, R. Hughes, K. Konopka, C. Hinckley, G. Smith, T. Nishitawa, N. Yoshida, and M. Saporoschenko, Catalytic Desulfurization of Coal Prior to Combustion. I. Process Design Principles, to be published.

AN ECONOMIC EVALUATION OF COAL PYROLYSIS WITH CHAR HYDRODESULFURIZATION

M.D. Stephenson and C.W. Kruse

Illinois State Geological Survey, 615 E. Peabody Dr., Champaign, IL 61820

ABSTRACT

The economics were examined for pyrolyzing high sulfur Illinois coal to produce synthetic oil and gas and desulfurizing the product char to a compliance level by hydrogen gas treatment. A survey of oil and gas forecasts for the period beginning in 1990 was used as a basis to evaluate probable pricing scenarios for the project life. Profitability determined by the discounted cash flow return on investment method (DCF/ROI) ranged from 16.3 percent to 33.3 percent. The DCF/ROI was most sensitive to the oil and gas pricing scenario employed but was also sensitive to the initial capital investment. To a lesser degree, the coal, char, and sulfur prices and oil and gas inflation rates affected the profitability level. Provided char is marketable at coal value or higher, it was concluded that oil from a pyrolysis plant can be sold at a lower price than oil from a liquefaction plant. Hydrodesulfurization of char appears to be particularly well-suited as an adjunct to a pyrolysis process from both an engineering and economic standpoint.

INTRODUCTION

Increasing public concern for protecting the environment from the deleterious effects of sulfur dioxide emissions has prompted the drafting of stringent controls which if enacted would threaten established markets for high sulfur coal produced by the Illinois coal industry. The Illinois State Geological Survey (ISGS), a division of the Illinois Department of Energy and Natural Resources, has a strong interest in developing technologies which have the potential for increasing the utilization of high sulfur coals and minimize the harm to the environment.

Processes to utilize high sulfur coals in an environmentally acceptable manner are at various stages of commercial development. Wet scrubbing is widely considered as the most commercially-proven technology for burning high sulfur coal. Although effective, the high cost of operating scrubbers when added to the high capital costs of installing them either on new plants or as retrofits has resulted in switching to low sulfur coal from other regions by many former users of Illinois coal. To reverse this trend, a clean solid fuel produced locally from Illinois coals and delivered at a price competitive with out-of-state low sulfur coal is needed.

Physical cleaning alone cannnot clean Illinois coals sufficiently because the mineral matter is often highly disseminated and frequently there is as much organic as pyritic sulfur. Clean coal produced by emerging chemical methods which remove both forms of sulfur is not yet economically competitive. The potential exists, however, to combine chemical cleaning with pyrolysis technology to produce more than one barrel of oil equivalent (BOE) of liquid and gaseous fuels and 50-60 weight percent yield of clean solid fuel (char) per ton of coal. Revenues from sale of the higher value synthetic crude oil could subsidize char desulfurization. It is this multiple phase products approach which is the basis of current low temperature charring research at the ISGS. The economics of this approach as it relates to the regulatory framework was reported in 1984 by Bhagwat and Johnson (ref.1). Because the nature of equipment and the required dollar investment on desulfurization of char were not specified, they analyzed the regulatory and economic environment as it existed in 1984, estimated costs for a pyrolysis plant, and determined the limit of spending on desulfurization for the given economic conditions. In the present study, costs for both pyrolysis and char desulfurization were estimated for a plant start-up in 1990. Product prices were estimated based on present projections for 1990-2010.

PYROLYSIS PROCESS

Description

The pyrolysis section of the overall process is outlined in block diagram form in Figure 1. Before pyrolysis, coal may be physically cleaned to remove some ash-forming mineral matter. Stagewise heating in fluidized bed reactors allows caking coals to be successfully processed without agglomeration. The devolatilized coal, or char, is desulfurized by a hydrodesulfurization process.

The off-gas from the pyrolyzer is condensed and the pyrolysis oil is separated from the gas. Methane, ethane, hydrogen, carbon monoxide, and hydrogen sulfide are the main constituents of the gas. Hydrogen sulfide is removed by reaction with an aqueous stream of monoethanolamine. Steam stripping of the amine solvent releases a concentrated stream of hydrogen sulfide for subsequent conversion to elemental sulfur.

A portion of the cleaned gas is used as a feedstock for a hydrogen plant. Methane and other gaseous hydrocarbons are reacted with steam at 1200°-1800°F to form a hydrogen-rich synthesis gas which is reformed to produce high purity hydrogen. It is necessary to hydrotreat the heavy, sour pyrolysis oil to produce a low sulfur, light crude oil. Hydrotreating of pyrolysis oil by the H-Oil or LC-Fining processes produces a synthetic crude oil using an

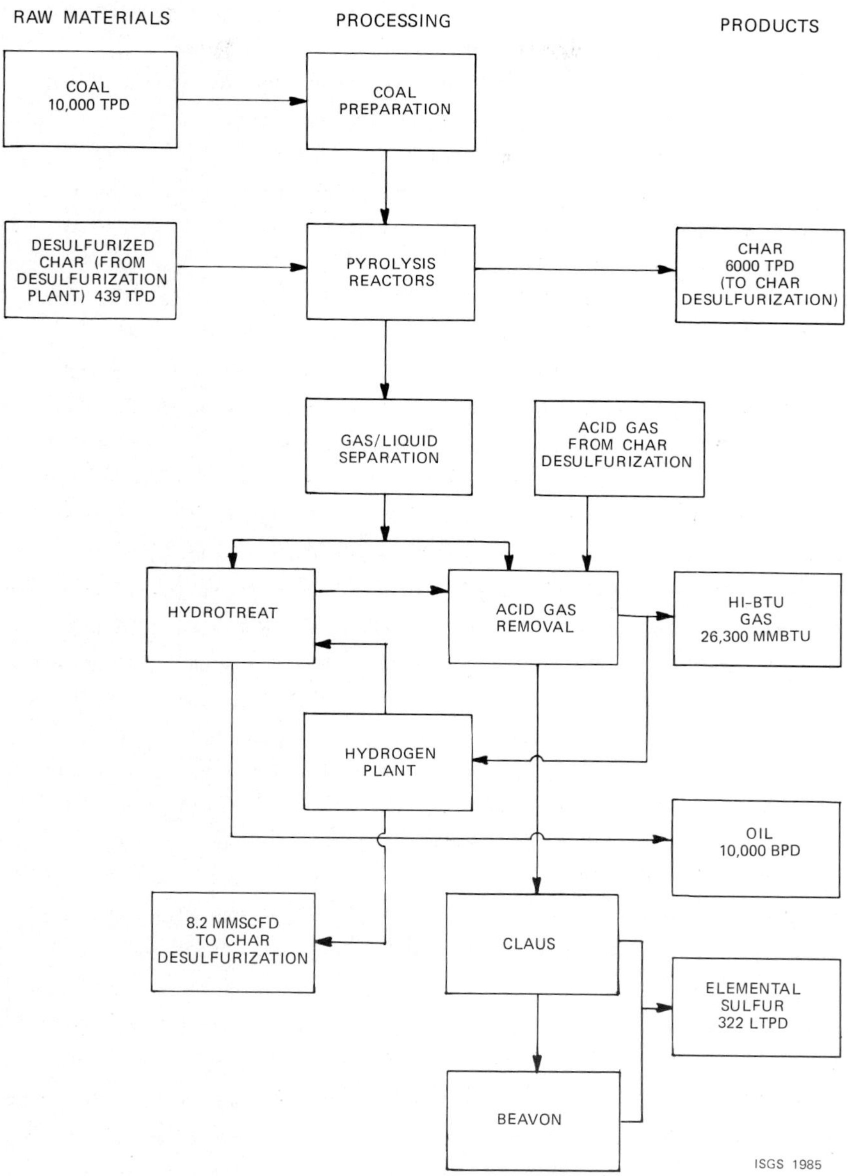

Fig. 1. Block diagram of coal pyrolysis.

ebullating bed reactor. Pyrolysis oils are catalytically hydrogenated and sulfur, nitrogen, oxygen and metals are removed.

Hydrogen sulfide from the hydrodesulfurization of the pyrolysis oil and from the pyrolysis gas is converted to saleable elemental sulfur by the Claus process. One-third of the hydrogen sulfide is burned to sulfur dioxide and the resulting mixture is catalytically disproportionated to form elemental sulfur. The tail gas contains appreciable quantities of sulfur which are recovered by a Beavon sulfur recovery unit.

Pyrolysis products

Oil and gas yields from pyrolysis of caking Illinois coals vary considerably. Because agglomeration of these coals is a major problem, many processes use a preoxidation step to change the surface properties of the coal particles. Unfortunately, the oil yields from the oxidized coal are always less. Oil yields are higher if coal is pyrolyzed in stages such that the temperature at a given stage is below the agglomerating temperature of the partially devolatilized coal. Tar yields from processes using Illinois Herrin (No. 6) coals vary from 0.6 bbl/ton in the Toscoal process where the feed is preoxidized to 1.9 bbl/ton in the Occidental Pyrolysis Process where a special reactor design prevents agglomeration. Approximately 1 bbl/ton is typical for the COED process which utilizes multistaging. Because the process employed in this study utilizes multistaging, an oil yield of 1 bbl/ton was assumed for a high sulfur caking Illinois coal. The hydrogen plant and H-Oil plant are based on a yield of 10,000 bbl/day (1 bbl/ton) of hydrotreated (34 °API) synthetic crude oil.

The oil produced from pyrolysis is hydrogen deficient but high in sulfur, oxygen, and nitrogen. Hydrotreating this oil by a heavy oil process, such as the H-Oil or LC-Fining processes, removes sulfur, oxygen, and some nitrogen and produces a marketable synthetic crude oil which is a suitable refinery feedstock.

A hydrotreated pyrolysis oil should become a more attractive feedstock for refineries as the U.S. crude charge becomes heavier and higher in sulfur. For example, Mexican crude exports to the U.S. have increased significantly in the last five years and a typical analysis of Mexican Maya crude is 22 °API with 3.3 percent sulfur. The API gravity of the U.S. crude slate is expected to average 31 °API by the year 1990 and the sulfur content is predicted to average to 1.18 percent. Also, the crude fraction boiling above 1000°F should reach 20 percent by 1990. In contrast, the fraction of hydrotreated pyrolysis oils which boil above 1000°F is typically 10 percent or less. Therefore, it is even conceivable that a hydrotreated oil could sell at a slight premium.

Installed equipment costs for all process units, shown in Table 1, were determined from literature data, scaled to a 10,000 tpd plant, and adjusted to January 1984 dollars. The fixed and total capital investments shown in Table 2 were calculated using a ratio factor method. For a 10,000 tpd feed, the installed equipment costs were found to be 142.2 million dollars and the total capital investment was determined to be 498 million dollars.

TABLE 1

Installed equipment costs[a]

Equipment	Cost $MM
H-Oil Unit	49.4
Acid gas amine unit	16.2
Hydrogen plant	17.0
Claus sulfur unit	16.3
Beavon tail gas unit	12.6
Pyrolyzers	23.5
Conveyors, storage, crushing	4.6
Waste treatment	2.6
Total	142.2

[a]January 1984 dollars

TABLE 2

Estimate of capital investment for pyrolysis process[a]

Item	$MM
Direct costs	
Installed equipment	142.2
Instrumentation, controls, piping electrical, buildings, service facilities, land	180.0
Total direct cost	322.2
Indirect costs	
Engineering, supervision, construction	71.4
Total direct and indirect costs	393.6
Contractor's fee	19.7
Project contingency	39.4
Fixed capital	452.7
Working capital	45.3
Total capital investment	498.0

[a] January 1984 dollars

CHAR HYDRODESULFURIZATION

The back-reaction of hydrogen sulfide with iron contained in char limits the extent of desulfurization. One means of reducing the back-reaction is to use a large ratio of hydrogen to char. This is not practical from the standpoint of a commercial operation. Removal of hydrogen sulfide from a recycled gas stream by conventional means would require large volumes of gas to be cooled to near room temperature, scrubbed, and then reheated to 1600°F. Use of a hydrogen sulfide acceptor appears to be the most practical means at present of dealing with the back-reaction. In this way, hydrogen sulfide concentrations are kept low so that desulfurization is not hindered. This is the basis of the char hydrodesulfurization process developed by the FMC Corporation (ref.2,3). Since this process has some advantages over other hydrodesulfurization processes, it will be the basis for the analysis which follows.

The process consists of two major sections, char desulfurization and acceptor regeneration, shown in Figure 2. Char is desulfurized in a shaft kiln whereby hydrogen from the pyrolysis plant flows countercurrently to the descending solids bed. Hydrogen reacts with sulfur in the char to produce hydrogen sulfide. The hydrogen sulfide then reacts rapidly with the calcium oxide acceptor to form calcium sulfide and steam. Hydrogen is regenerated by char gasification (carbon-steam reaction) which is controlled by the rate of steam evolution. The reactions taking place inside the desulfurizer are overall, essentially thermoneutral.

Desulfurized char and spent acceptor are pneumatically conveyed to a two stage fluidized bed cooler. A preliminary separation occurs in the second stage where an appropriate fluidization gas velocity causes the acceptor pellets to sink to the bottom of the bed while char is fluidized above the acceptor. Cooled solids are screened for a further separation and an electrostatic separator completes the separation. The spent acceptor is ground to minus 48 mesh and slurried with water and pumped to a regenerator where the calcium carbonate is regenerated by reacting the calcium sulfide with steam and carbon dioxide. The calcium carbonate is then calcined to calcium oxide. The carbon dioxide can be supplied from the pyrolysis plant where it is produced at the hydrogen plant. The noncondensibles in the off-gas, carbon dioxide and hydrogen sulfide, flow to a Claus converter for by-product sulfur recovery. The regenerated slurry is spray-dried and extruded to form 1/8-inch pellets and is conveyed back to the desulfurizer.

FMC evaluated preliminary economics of this process in 1970 based on data obtained from their bench-scale unit (ref.3). Although this work was discontinued a number of years ago, it serves as a basis to evaluate char hydrodesulfurization costs. Updating the FMC study reveals that hydrodesulfurization of 6000 tpd of char has a capital requirement of 38.0 million dollars (Jan. 1984 dollars). When this is added to the cost of the pyrolysis process, the total capital investment increased to 536 million dollars. Annual operating costs have also been calculated using FMC data. Raw materials include char, hydrogen, and limestone.

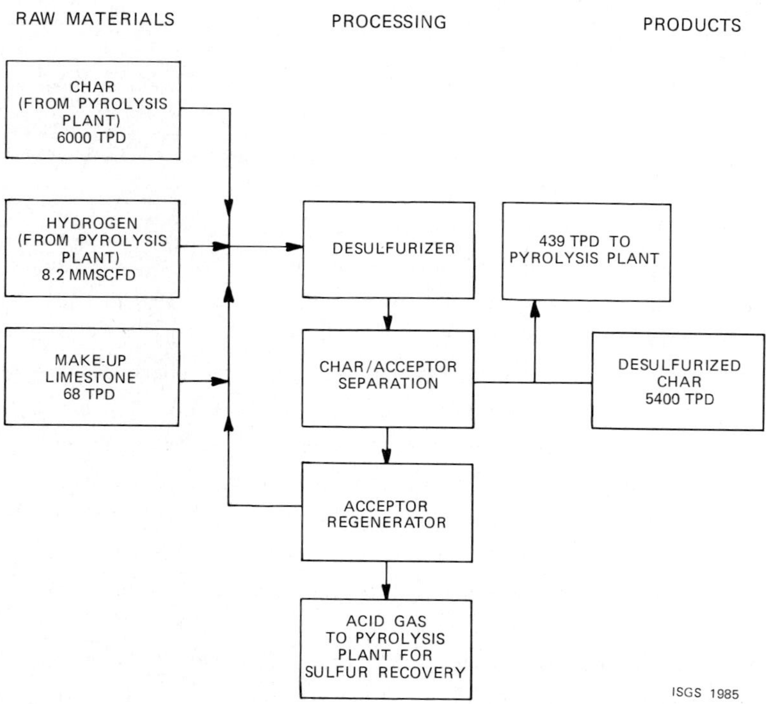

Fig. 2. Block diagram of char desulfurization.

Hydrogen is already manufactured in the pyrolysis plant and a portion of the cleaned char is used as a fuel to drive the pyrolysis and desulfurization processes. Therefore, limestone is the only external raw material required, and since limestone is regenerated, only make-up quantities are required. Regenerating the limestone cuts down on raw materials, reduces waste disposal requirements and produces elemental sulfur as a by-product. The remaining operating costs include utilities and labor. Annual non-fuel operating costs for the combined processes are listed in 1984 dollars in Table 3. At plant start-up, these costs equal 64.79 million dollars.

TABLE 3

Non-fuel annual operting cost estimate for pyrolysis with char desulfurization[1]

Item	$MM, 1984 $
Fixed charges	
Labor overhead, insurance, supplies	36.14
Variable charges	
Utilities	12.21
	48.35

[1] does not include cost of coal which is varied in sensitivity analysis

PRODUCT VALUES

Pyrolysis transforms low priced coal into higher priced oil and high Btu gas. An uncleaned char is certainly worth no more than the parent coal - in fact, it may be worth less since it burns differently and there is little experience in burning it. However, a clean char could be worth appreciably more in eastern and midwestern markets where the alternatives for existing installations are FGD and switching to Western coals. It has been estimated that a desulfurized char which meets New Source Performance Standards could command up to $46/ton in the market place and still retain a 15 percent cost advantage over FGD or switching to Western coal (ref.1). Therefore, pyrolysis with a char desulfurization process would convert inexpensive coal into clean, higher priced liquid, gaseous and solid fuels.

With 1990 being the earliest conceivable start-up date, it is prudent to examine projected oil and gas prices for 1990 and beyond. Forecasters agree that the oil market will, once again, be one of short supplies and continued price increases. The underlying reasons remain the same. OPEC's share in market economies output will expand from its current 45 percent to 51 percent by 1990 enabling OPEC to regain market power and a greater influence over price (ref.4). A recent study by Conoco predicts that OPEC will supply about 54 percent of non-Communist world oil in the year 2000, compared to 41 percent at present. Conoco also foresees U.S. petroleum liquids production (includes natural gas liquids, synthetics, and hydrocarbons from other sources as well as crude oil) will fall far short of demand as output will drop 2 MMbpd by the year 2000. Total U.S. crude oil production will fall to about 7 MMbpd at the turn of the century, down from 8.5 MMbpd currently. As a result, the United States will probably become increasingly dependent on imported oil. Imports will rise from 4 MMbpd in 1983 to 7.5 MMbpd in the year 2000, or about 45 percent of U.S. demand. Bankers Trust Company energy economists (ref.5) predict similar declines in U.S. crude oil and natural gas liquids production from 10.2 MMbpd in 1983 to just over 8 MMbpd by 1995. When this is combined with the anticipated increase in oil demand, imports should increase to 8 MMbpd by 1995. These underlying trends together with the present complacency concerning energy issues portend that the 1990s and beyond could be a period of shortages and price increases.

Oil

Price forecasts for crude oil differ greatly for the period 1984-2000. Generally, there is agreement that oil prices will be soft for the next few years. Beyond that, there is great variation in the forecasts but all see real prices climbing during the 1990-2000 period. For 1990, forecasts in 1984 dollars range from $25.38-40.11/bbl indicating in one case, a 12 percent real decrease to a 38 percent increase. All forecasters predict increasing prices with crude prices in 1984 dollars for the year 2000 ranging from $30.00-61.79/bbl. These forecasts are summarized in Table 4. Due to the variations in the forecasted crude prices, an average price from these studies will be used in the base case analysis. For 1990, a price of $32.95/bbl in 1984 dollars will be used.

TABLE 4

Crude oil price projections (1984 dollars)

Source (Year)	1990	1995	2000
Gas Research Institute (1983)	-	-	43.36
U.S. Dept. of Energy (1983)	34.69	-	61.79
American Gas Association (1983)	36.86	-	45.53
Standard Oil of California (1982)	-	-	32.52-59.62
Energy Information Agency (1983)	40.11	52.20	-
Chemprice Corporation (1984)	27.73	-	34.38
Chemical Data, Inc. (1984)	25.38	27.45	30.00
Sherman H. Clark Associates (1984)	29.13	33.93	41.76

Gas

It is difficult to predict with certainty what level gas prices will reach in a decontrolled environment. A gas surplus presently exists in the U.S., although the general consensus is for supply and demand to be roughly in balance by 1986. According to Bankers Trust energy economists (ref.6), a major challenge to the gas industry in 1986 or 1987 will be to maintain production capacity high enough to meet domestic demand. Prices will remain relatively flat through 1986. Beyond 1986 or 1987, Bankers Trust Company foresees a moderate escalation in gas price and estimates that the average gas price could increase some 25 percent in inflation-adjusted terms by 1990. It is believed that these increases are necessary to provide producers with the incentive and cash flow to make necessary investments.

Paine Webber forecasts a similar gas market (ref.7). According to their study, real wellhead gas prices are not expected to increase until early 1986. At that point, gas supply and demand most likely will be in balance. Real prices will begin to increase in 1986 due to a drop in natural gas deliverability. Beyond 1986, a steady increase is projected to 1990 with the average wellhead price increasing 8.8 percent annually. New gas will reach an average wellhead price of $6.35/Mscf in 1990 or an inflation adjusted value of $4.73/Mscf in 1984 dollars.

An estimate of U.S. natural gas prices through the year 2000 by Sherman H. Clark Associates (ref.8) predicts a slight drop in real prices from 1984 to 1985 and real increases to commence in 1987. Again, the reason given for this turnaround is the disappearance of the present gas surplus. The average field price will show an annual escalation rate of 6.3 percent from 1984 to 1990. For the 1990-2000 period, prices are expected to increase at a real rate of 4.3 percent.

The U.S. Energy Information Administration predicts there will be no significant flyup this year with the decontrol of new gas (ref.9). Gas prices are now driven mainly by competition from oil, rather than by the regulatory process. Gas and fuel oil are in direct competition in the large industrial and electric utility market. The projection states that prices will rise gradually to 1990 in line with the price of oil. More uncertainty prevails for the 1990s. Competition between natural gas and fuel oil for the same market will intensify and gas prices should resume their climb when oil prices increase.

Examining these studies shows agreement for the period 1984-1990. For example, both Paine Webber and Sherman Clark Associates predict 6.3 percent annual real increases from 1984 to 1990. This represents a 44 percent real increase in prices. Bankers Trust's estimates more moderate increases over this period amounting to a real 25 percent price increase. However, this estimate differs only in magnitude of the upward movement. All three studies predict the increase to occur in 1986 or 1987.

Although gas price predictions are subject to appreciable error, the present state of reserves precludes the chances of price declines. Also, if prices do not climb in the late 1980s, they should climb in the 1990s. Since two-thirds of 1995 gas has yet to be discovered, high gas prices will be required to provide incentives and sufficient cash flow. If the prices do not climb by 1990, the resulting gas shortage from a lack of drilling could cause an increase in prices. Consequently, there is a high probability for much higher gas prices in the 1990s.

Although the average wellhead price in 1983 was \$2.59/Mscf, new gas averaged \$3.75/Mscf, high cost gas averaged \$5.50/Mscf and for imports, major pipeline companies paid \$4.51/Mscf. Needless to say, the natural gas pricing system is complex. New gas has now been decontrolled as of January 1, 1985 and although no immediate price fly-up is expected, higher prices are expected in the long run. Gas from a pyrolysis plant would be considered new gas and would command the prevailing wellhead price for new gas. However, the principal competing gas in Illinois is Louisiana gas transported by interstate pipeline. Therefore, it is reasonable to assume that pyrolysis gas used close to its point of origin could command a slight premium for its wellhead price to reflect the reduced transportation costs that a pipeline company would incur. In the present analysis, pyrolysis gas will be priced at the wellhead level for new gas with an added 10 percent premium. For a 1990 start-up year, gas will be priced at \$5.20/Mscf in 1984 dollars. Ultimately, gas prices are expected to approach petroleum market parity. Much of this gain should occur by 1990. At this point, gas prices should track changes in petroleum prices. Therefore, gas prices will be assumed to escalate at a rate equal to oil prices at plant start-up in 1990.

Sulfur

By-product sulfur is most useful in the elemental form. It is easily stored and handled and is highly marketable for sulfuric acid production. The impending decline of Frasch reserves indicates that recovered sulfur from high sulfur petroleum, sour gas, and coal will become increasingly more important. Even so, the Fertilizer Economic Studies Limited forecasts a supply shortage of almost five million tons by 1990. Despite the worldwide recession which saw a lower demand for sulfur in 1981, demand exceeded supply by 2 million tons. The Free World's sulfur market remained in balance by releasing large quantities of producer's inventories (ref.10). Prior to that, the price of Frasch sulfur skyrocketed from \$45/long ton in 1978 to over \$140/long ton in the summer of 1981. At the same time, the U.S. sulfur inventory was reduced at record rates even though Frasch production was at capacity. This portends a long-term tight sulfur market with an increasing U.S. dependence on imports.

Domestic sulfur demand is expected to increase 3 percent annually through the year 2000 (ref.11). It is difficult to predict what effect this will have on long-term prices. For instance, a few years ago, synfuels were expected to supply 1.3 million metric tons of sulfur by 1990 and 3 million metric tons by the year 2000. However, for this to occur the synthetic fuel industry, if exclusively using 3 percent sulfur coal, would need to be producing the equivalent of 430,000 bpd of synthetic crude oil by 1990 and 1 MMbpd by 2000. This now seems an unlikely scenario. On the other hand, sour gas is expected to contribute significantly to the supply picture. Increasing reliance on foreign sources and steadily increasing demand indicate prices could increase substantially by the year 2000. A reasonable assumption, therefore, is that sulfur prices will show a 2 percent real increase in price.

RESULTS AND DISCUSSION

In the following venture analysis, the after-tax discounted cash flow return on investment (DCF/ROI) has been calculated for predetermined product selling prices. An alternate method sets the DCF/ROI at a predetermined level and calculates product selling prices. Since a pyrolysis plant produces multiple products, it is impossible to determine a product selling price without establishing price relationships between the products or price levels for all but one of the products.

Pyrolysis and char desulfurization is a greater-than-average risk venture and requires a commensurate DCF/ROI. A recent study (ref.1) calculated the necessary DCF/ROI to be 20.7 percent. Under today's conditions of historically high real interest rates (the prevailing interest rate minus the inflation rate), a low risk venture would still require a 15 percent DCF/ROI even though the inflation rate is less than 5 percent. A 20 percent DCF/ROI would be expected for a venture with higher risk, such as pyrolysis with char desulfurization. If inflation remains low and interest rates decline, the minimum acceptable return would likewise decline. For this study it is assumed that the federal deficits will remain high and government borrowing will continue to exert upward pressure on interest rates. For a 1990 start-up date, the minimum acceptable DCF/ROI will be considered as 20 percent. Economic parameters for the base case are listed in Table 5 and the product and raw material prices along with their inflation rates are listed in Table 6.

TABLE 5

Economic parameters, base case

1.	Plant size	10,000 tpd Illinois No. 6 coal
2.	Start-up date	1990
3.	On-stream factor	90%, 328.5 days/year
4.	Project life	20 yrs.
5.	Debt/Equity	50/50
6.	Interest rate on debt	16%
7.	Depreciation	Sum-of-the-years-digits, 20 yrs.
8.	Investment tax credit	10%
9.	Inflation rate	5%
10.	Operation cost escalation	5%
11.	Annual operating cost (1990)	64,790,000
12.	Total capital investment (1990)	718,300,000

TABLE 6

1990 Product and raw material prices, base case

Material	1990 Price	Inflation rate for project %
Coal	38.86/ton	7%
Char	61.64/ton	7%
Oil	44.16/bbl	8%
Gas	6.97/MMbtu	8%
Elemental sulfur	210/long ton	7%
Limestone	46.90/ton	5%

An order-of-magnitude capital cost estimate is accurate to ±40 percent. In the present estimate, the char desulfurization process is subject to the greatest error as these results are based on an estimate of a bench-scale process. Examination of the effect on the DCF/ROI of both a gross underestimation and overestimation of the total capital investment is shown in Figure 3. For the base case, a 22.8 percent DCF/ROI is achieved. If the plant could be built for 400 million dollars (1990 dollars), the DCF/ROI increases markedly to 33.3 percent. However, a 1990 plant cost of 1 billion dollars (+40 percent) lowers the return to 18.1 percent. Increasing the capital cost to this upper bound does not make pyrolysis with char desulfurization an unprofitable venture. Instead, the return is close to that required for a higher-than-average risk venture.

In the base case analysis, it has been assumed that a compliance char can command \$46/ton (1984 dollars). However, char burns differently than coal. A hydrodesulfurized char would be deficient in volatile matter having been treated at 850°C. Utilities may be unable to use char without a greater price incentive. Therefore, the value of desulfurized char has been varied to a low value equal to the price of coal. This has been assumed to be the worst case. Under these conditions, the DCF/ROI is 18.7 - close to the desired profitability. At the upper-end, the char price was only increased 13.5 percent above the base case and a 24.1 percent DCF/ROI was achieved. It is unlikely char would command much higher prices as the cost advantages to switch to char would be lost. However, a clean char would interest the steel industry for formed coke. In this situation, the upper-end case is a very reasonable char price even today. The effect of char value is shown in Figure 4. The price paid for coal also influences the profitability. A 20 percent increase in the price paid for coal will lower the profitability from 22.8 percent to 20.2 percent. On the other hand, a 13 percent discount on the price paid for coal increases profitability to 24.3 percent. This effect is shown in Figure 5.

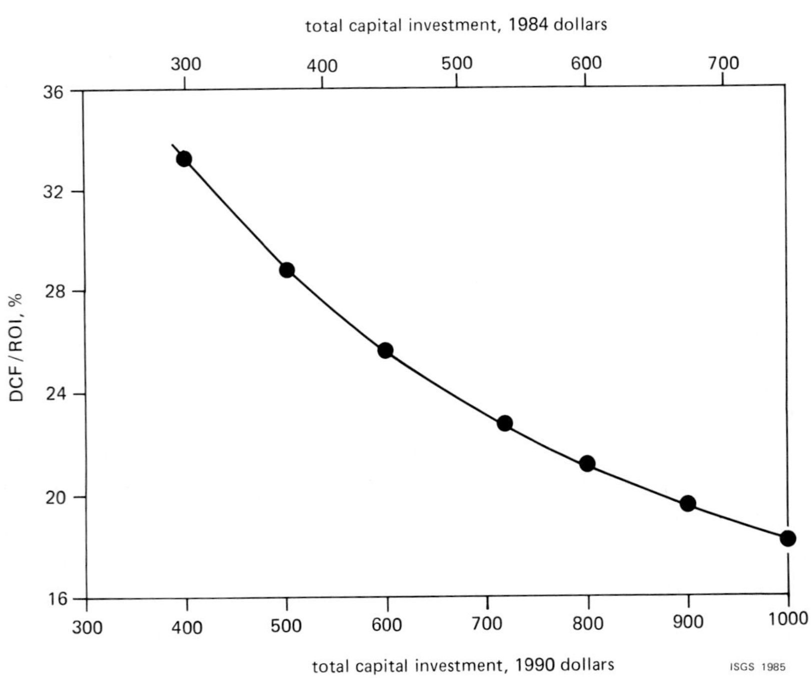

Fig. 3. Effect of total capital investment on DCF/ROI.

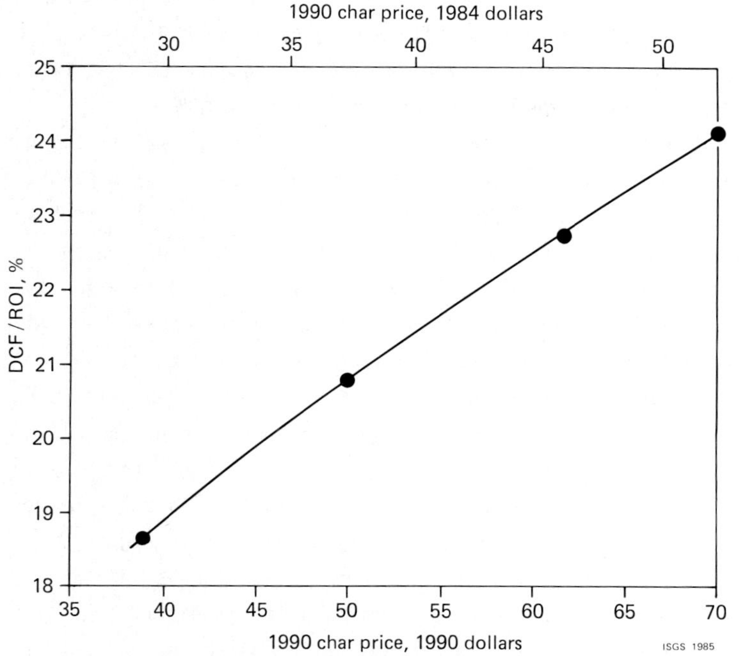

Fig. 4. Effect of char price on DCF/ROI.

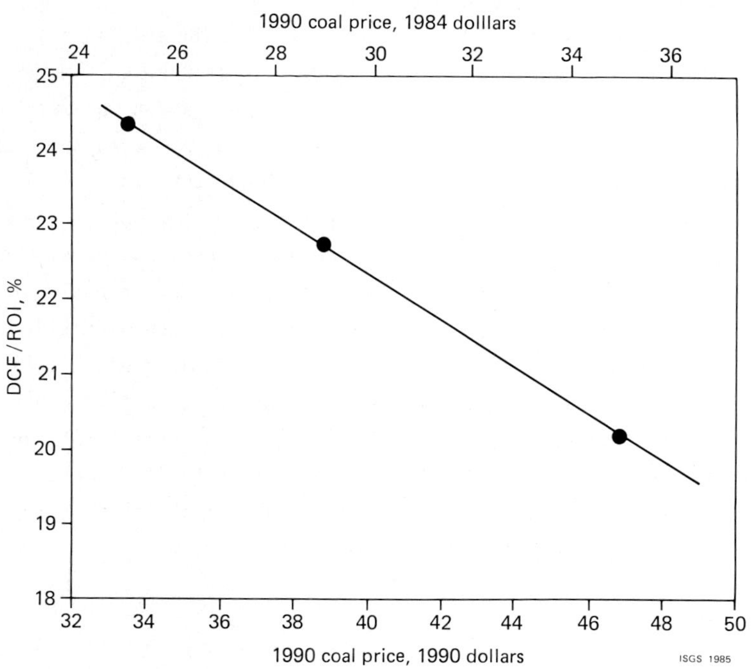

Fig. 5. Effect of coal price on DCF/ROI.

A small but noticeable effect on profitability is due to the price received for elemental sulfur. At \$100/long ton (1990 dollars), the DCF/ROI is 21.6 percent. However, sulfur selling at \$400/long ton raises the DCF/ROI to 24.7 percent. The effect of sulfur price is shown in Figure 6.

As might be expected, profitability is quite sensitive to the oil and gas environment. For the low case, based on the Chemical Data, Inc. forecast, a DCF/ROI of 16.3 percent is achieved. At the upper end based on the Energy Information Agency report, a return of 29.2 percent is achieved. Consequently, the uncertainty in the future oil market is reflected in a fairly wide profitability band. An average oil price from the present studies gives the base case DCF/ROI of 22.8 percent. As expected, oil and gas inflation rates also influence profitability, although not as much as 1990 oil and gas prices. For instance, when oil and gas prices increase 5 percent annually, i.e., no real increase in prices, the profitability is 19.8 percent. As this inflation rate is increased to 10 percent, the profitability shows a commensurate increase to 24.7 percent. This effect is summarized in Figure 7.

Under most scenarios, pyrolysis with char desulfurization is an economically sound venture. The level of profitability is highly sensitive to the oil and gas environment at plant start-up. Other parameters affect profitability to some degree. The DCF/ROI, for instance, is quite sensitive to capital investment, coal and char prices.

Pyrolysis is the least expensive means of producing liquid and gaseous fuels from coal. It is easy to understand why. Pyrolysis involves simple heating to concentrate coal-derived hydrogen in liquid and gaseous phases while carbon concentrates in the solid char. Some external hydrogen that is steam-derived is required to upgrade the pyrolysis oil but this is minor in amount when compared to gasification and liquefaction where copious amounts of steam and hydrogen are consumed. This brute force approach of adding hydrogen to the coal is reflected in the price of the product fuels. For example, design and construction of the Great Plains Gasification Project near Beulah, N. D. was predicated on gas prices of \$9.50 per thousand cubic feet. It is estimated that SASOL III indirect liquefaction has products priced at greater than \$8.00/ MMBtu, and this with low labor and raw material requirements and extensive government subsidy. Indirect liquefaction to methanol is considered to be greater than \$9.00/MMBtu and current estimates for synthetic crude by direct liquefaction are greater than \$60/bbl or \$10.00/MMBtu.

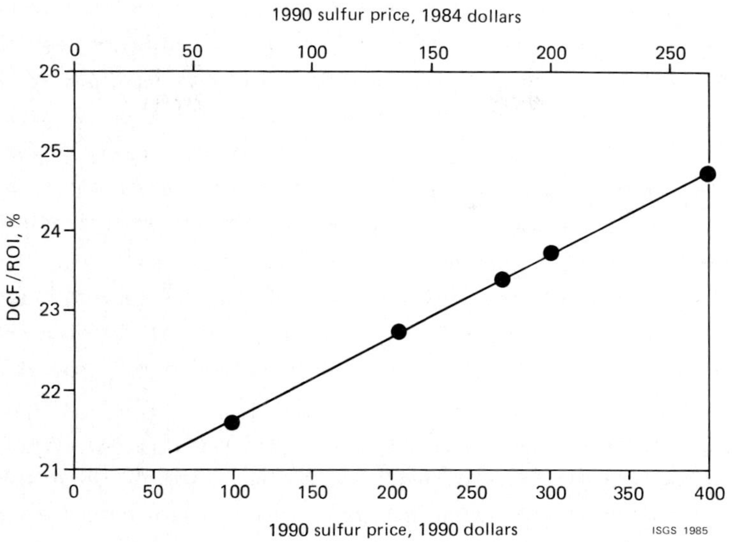

Fig. 6. Effect of sulfur price on DCF/ROI.

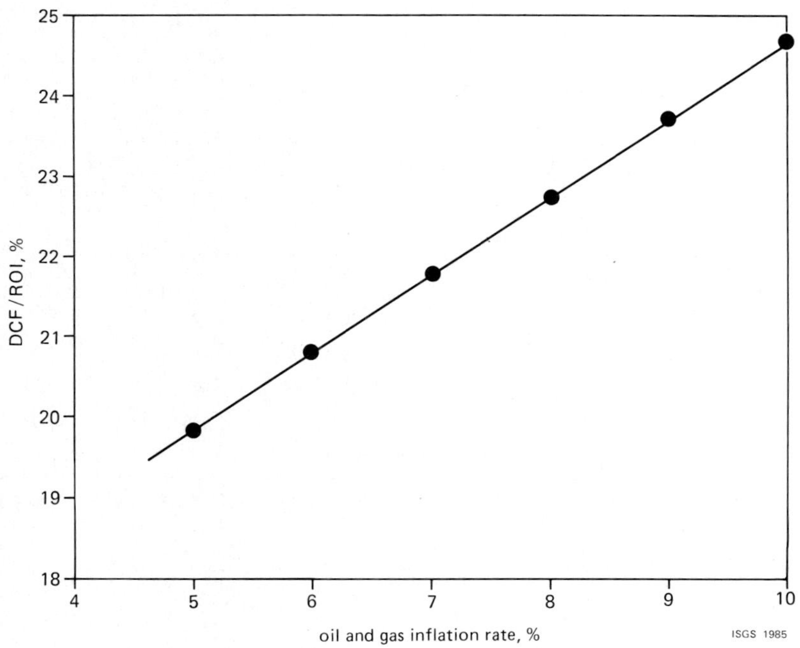

Fig. 7. Effect of oil and gas inflation rate on DCF/ROI.

Part of the problem with gasification and liquefaction is the much lower thermal efficiencies. Direct liquefaction processes range anywhere from 50-65 percent thermally efficient (ref.12-14). Indirect liquefaction processes are even more inefficient with thermal efficiencies ranging from 40-45 percent (ref.15) and thermal efficiencies of gasification processes ranging anywhere from 60-70 percent (ref.14,16). On the other hand, thermal efficiencies of coal pyrolysis processes are much higher and can approach 85-90 percent (ref.17). Thermal efficiencies can be useful in comparing similar processes. Higher thermal efficiencies in coal processes can lead to lower capital and operating costs. In pyrolysis, essentially all of the feed coal is converted to saleable fuels so that coal throughput per Btu of product is minimized.

Recent economic studies support the view that coal pyrolysis has lower capital investments than gasification and liquefaction. Most of these studies were completed in the 1980-1982 period and are generally based on processes which have reached the pilot plant scale. For the sake of comparison these estimates have been adjusted to January, 1984, dollars using the Nelson Refinery Index and have been scaled to a 10,000 tpd operation using the six-tenths rule. The results of these studies are summarized in Table 7.

TABLE 7

Capital cost estimates of pyrolysis processes

Process	Capacity tpd	Capital cost (year) $MM	Adjusted capital cost[a] $MM
Toscoal	10,000	137 (1980)	175
ORNL	44,000	1,309 (1980)	656
Occidental	20,000	1,000 (1981)	768
Union Carbide	21,000	780 (1980)	639
Cochar	36,325	1,300 (1980)	767
Lurgi	20,000	600-700 (1981)	461-537

[a] capacity, normalized to a 10,000 tpd operation using 6/10 rule and inflation adjustment to January 1984 dollars.

As one can see, coal pyrolysis at a 10,000 tpd capacity, regardless of the process, has capital investments ranging from $175-768 million and are modest when compared to the billion or more required for gasification or liquefaction. The Toscoal process economics were evaluated for several scenarios (ref.18). The $137 million capital cost is for a plant pyrolyzing caking Illinois Herrin (No. 6) coal. An economic analysis by ORNL was based on their own hydrocarbonization studies and data from the U. S. Steel Clean Coke process (ref.19). Like the Tosco study, the ORNL analysis examined pyrolysis of Illinois Herrin (No. 6) coal. The Occidental study of their pyrolysis process was evaluated using a caking Kentucky bituminous coal (ref.20). Other studies include a Utah Power and Light study of the Lurgi-Ruhrgas process (ref.21), a Union Carbide study of their hydrocarbonization process (ref.22), and an FMC study of the Cochar process (ref.23). These last three economic evaluations examined pyrolysis of Utah bituminous coals.

CONCLUSIONS

Pyrolysis is the least expensive means of producing liquid and gaseous hydrocarbons from coal. The required capital investment is significantly lower than gasification or liquefaction and the products have a lower required selling price. Conditions likely to be present in the 1990s provide an economic environment that is sufficiently attractive to make pyrolysis with char desulfurization profitable. The reasons are simple. Pyrolysis by itself transforms coal into higher valued liquid and gaseous fuels. In addition, desulfurization of the char produces another higher valued product. This is not meant to say that coal pyrolysis with char desulfurization is a panacea to the domestic energy quandary. It isn't. Instead, pyrolysis and char desulfurization are often overlooked technologies which could add to the nation's oil and gas base and at the same time utilize high sulfur coals in harmony with the environment. Accelerated research and development spending is apparently justified to develop these processes so that the industrial Midwest has viable options for the responsible use of its most abundant natural resource.

ACKNOWLEDGMENT

The authors acknowledge helpful suggestions made by S.B. Bhagwat, L.A. Johnson, R. Boyda, J. Zollars, G.W. Land and W.W. Dodge who read early drafts of this paper. Low temperature charring research at the ISGS is funded in part by contracts from the Illinois Coal Research Board through the Center for Research on Sulfur in Coal.

REFERENCES

1 S.B. Bhagwat and L.A. Johnson, 1984, Economic and regulatory framework for production of solid compliance fuels from medium to high sulfur coal: Proceedings First Annual Pittsburgh Coal Conference, Pittsburgh, PA., September 17-21, 1984.
2 C.A. Gray and M.E. Sacks, 1966, The desulfurization of COED char: PCR-532, 78 pp.
3 M.E. Sacks, 1970, The desulfurization of COED Char-Part III: Office of coal research, U.S. Department of the Interior, OCR report 56, Washington, D.C., 149 pp.
4 Staff, 1984, OPEC oil market share 51 percent by 1990: Hydrocarbon Processing, 63(11), 17.
5 Staff, 1984, U.S. energy crisis not yet resolved according to Bankers Trust analysis: Hydrocarbon Processing, 63(7), 15.
6 Staff, 1984, Natural gas outlook evaluated in report by Bankers Trust: Hydrocarbon Processing, 63(12), 21.
7 Staff, 1984, Paine Webber: No real increase for gas until 1986: Oil and Gas J. 82(37), 88.
8 S.H. Clark, 1984, U.S. Natural gas price forecasting: in AIChE 1984 Summer National Meeting, Philadelphia, Pa., August 19-22, 1984.
9 Staff, 1984, EIA: Oil prices won't rebound quickly: Oil and Gas J., 82(23) 48-49.
10 R. Kiyoura, 1983, Sulfur recovered by desulfurization increased in importance: Chemical Economy and Engineering Review, 15(1-2), 12-21.
11 D.G. Rangnow and P A. Fasullo, 1981, Rapid growth is outlook for recovered sulfur: Oil and Gas J., 79(38), 242-246.
12 J.C. Tao and J.P. Jones III, 1983, SRC-I technology and its economics: Energy Progress, 3(1), 1-6.
13 A.M. Goldstein, C.W. DeGeorge, G.A. Melin, J.S. Morrison and F.N. Nazario, 1982, Exxon donor solvent plants design update: Chemical Engineering Progress, 78(4), 76-80.
14 R. Shinnar, 1978, Gasoline from coal: Chemtech, 8(11), 686-693.
15 P.D. Sunavala, 1982, Energy proficiency: Journal of the Institute of Energy, 55(424), 153-159.
16 K.H. Eisenlohr and H. Gaensslen, 1981, Motor fuels from coal - technology and economics: Fuel Processing Technology, 4, 43-61.
17 H.W. Parker, 1982, Liquid synfuels via pyrolysis of coal in association with electric power generation: Energy Progress, 2(1), 4-8.
18 D.H. Cortez and C. J. LaDelfa, 1981, Coal pyrolysis looks good: Hydrocarbon Processing, 60(2), 111-117.
19 E.C. Fox, 1981, An economic perspective on coal pyrolysis: in Proceedings of the EPRI Conference on Coal Pyrolysis, ch. 19, 22 pp., Palo Alto, Ca., February 25-26, 1981.
20 K. Durai-Swami and W. Deslate, 1981, Occidental's flash pyrolysis coal liquefaction process and its application to power generation: in Proceedings of the EPRI Conference on Coal Pyrolysis, ch. 2, 24 pp., Palo Alto, Ca., February 25-26, 1981.
21 J.R. Thurgood and F.D. Wareham, 1982, Utah char-oil project: in AIChE 1982 Winter National Meeting, Orlando, Fl., Feb. 28-March 3, 1982.
22 J.J. Potter, 1981, Hydrocarbonization-power plant integration: in Proceedings of the EPRI Conference on Coal Pyrolysis, ch 3, 31 pp., Palo Alto, Ca., February 25-26, 1981.
23 R. Bloom, 1981, The Cochar process for coal pyrolysis: in Proceedings of the EPRI Conference on Coal Pyrolysis, ch. 7, 22 pp., Palo Alto, CA, February 25-26, 1981.

PART 6

HANDLING OF COAL FINES

DEWATERING OF COAL-CLAY WASTE AT PARADISE STEAM COMPLEX

B. J. SCHEINER

U.S. Department of the Interior, Bureau of Mines, Tuscaloosa Research Center, P.O. Box L, University, AL 35486 USA

ABSTRACT

The Bureau of Mines has investigated a dewatering technique for coal-clay waste that uses a flocculant, polyethylene oxide (PEO). This flocculant forms strong stable flocs that can be dewatered on either a hydrosieve screen or a combination of hydrosieve and trommel screens. A field test unit (FTU) was operated at a rate of 80 to 160 gal/min at a coal mine in western Kentucky at which coal-clay waste was dewatered first on a static screen, then on a trommel screen, to 50 pct solids using a polymer dosage of 0.12 to 0.15 lb/ton. The dewatered material continued to dewater, reaching 68 to 70 pct solids in 7 days when placed in a pond. The dewatered material was shown to be stable even when subjected to a heavy rainfall. Experiments were also made in which dewatered material was mixed with coarse waste from the preparation plant. A stackable mixture was obtained when four parts coarse waste was mixed with one part dewatered material.

INTRODUCTION

In the washing of coal to produce a low-ash material, two wastes are usually generated: a coarse fraction containing material ranging in size from plus 28 mesh to several inches, and a slurry containing fine-grained materials such as clay. The particle size of these waste slurries usually ranges from minus 28 mesh to several micrometers. Disposal of the coarse fraction is easily accomplished by stacking the waste on an available terrain at the mine site. The method for disposal of the fine coal-clay slurry depends on land availability and local environmental regulations.

The simplest method for disposal of fine coal-clay slurries is by impoundment behind an earthen dam or by ponding in an available mine cut. These slurries usually contain 1 to 3 pct solids and settle naturally to greater than 50 pct solids in several months. Over the lifetime of a typical mine, several ponds and/or impoundments will be required. If land is available, this technique may satisfy the operator's disposal requirement.

One disadvantage of the ponding technique is that disposition of slurries behind dams poses a potential health and safety hazard because of possible dam failure. Also, the recovery and recycling of water contained in the slurries is not efficient, since the settling process is slow and in practice the pond is often located a considerable distance from the preparation plant, thus requiring pumping of the recovered water to the plant (refs.1-5).

Another method for disposing of the fine coal-clay waste slurries is to add flocculant to the slurry and thicken the resulting flocs in a thickener. The underflow from the thickener is then pumped to a pond or impoundment. This technique provides a method for rapidly recovering a majority of the water contained in the waste slurry stream. This technique also requires the availability of enough land environmentally suitable for construction of impoundments.

A method of disposal that minimizes the land requirement is the technique whereby the underflow from the thickener is treated with flocculant and dewatered on a mechanical device such as a filter or belt press. Such treatment yields a material that can be mixed with the coarse waste fraction and disposed of by stacking. This approach is very successful during the dry season, but during the wet season the filtered material has a tendency to reslurry and causes a potentially dangerous mud flow situation.

To overcome the problem of reslurrying dewatered material, the Bureau of Mines has investigated a dewatering technique in which coal-clay waste is flocculated with polyethylene oxide (PEO), and the resulting flocs are dewatered on a hydrosieve screen (ref.6). In field test studies, coal-clay waste from a preparation plant located in Alabama was dewatered from 20 to 60 pct solids using 0.10 lb of PEO per ton of contained solids in the slurry (ref.6). The dewatered material did not reslurry when contacted with water such as rainfall. Tests were also conducted in which the dewatered material was mixed with coarse refuse and stacked at the mine site. The material was stable even though it was subjected to heavy rainfall during the testing program.

Further testing of the PEO-dewatering technique has been conducted in field tests at TVA's Paradise Steam Complex located in western Kentucky. This paper describes the results obtained during that study.

MATERIALS, EQUIPMENT AND PROCEDURES

The screen for the test unit is shown in Fig. 1 and the flowsheet for the unit is shown in Fig. 2. Coal-clay slurry was supplied to the test unit from the preparation plant waste slurry lines. A typical screen analysis of the coal-clay waste is shown in Table 1. The slurry flowed to a 1,000-gal holding tank equipped with a stirrer, to which lime was added to produce a pH of 9 or greater. Previous research had shown that a pH of 9 or greater was required to produce strong flocs. The lime-treated slurry then overflowed into a second 1,000-gal stirred tank to provide time for the lime to react with the waste slurry; previous research has shown that a retention time of at least 6 min was required.

FIGURE 1. - Field test unit dewatering coal-clay waste.

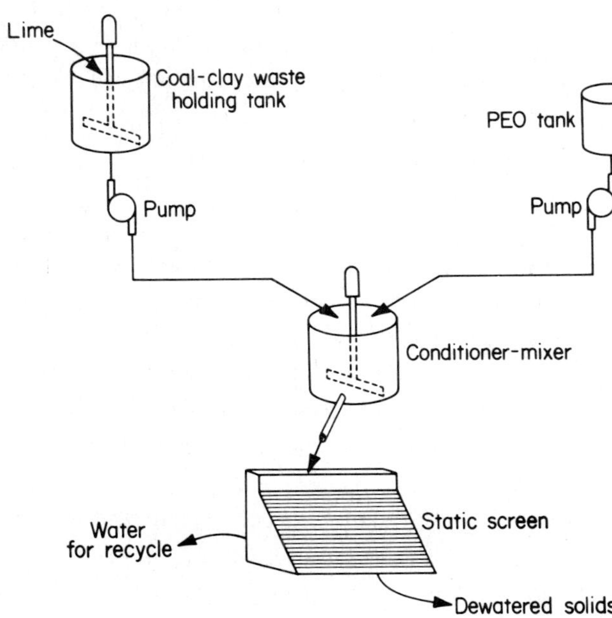

FIGURE 2. - Flowsheet for test unit.

The lime-treated waste slurry was then pumped to a mix tank where PEO was added, and the resulting flocculated material was passed over a set of static screens (hydrosieve) for removal of water that had been released during the flocculation step.

TABLE 1

Screen analysis of coal-clay waste

Screen size mesh	Wt pct
Plus 28.............	2.0
Minus 28, plus 48....	6.0
Minus 48, plus 100...	16.83
Minus 100, plus 200..	14.64
Minus 200, plus 325..	8.60
Minus 325............	51.93
Total	100.00

The hydrosieve is 8 ft wide and 8 ft long and was built in two sections of stainless steel wedge-wire screen; this permits independent angles of inclination for each section. The upper section of screen (8 ft wide, 4 ft long) has slot openings 2.75 in long and 0.02 in wide. The lower section of the screen has slot openings 2.75 in long and 0.01 in wide. Dewatered material from the hydrosieve was conveyed to a pit, where further dewatering occurred.

The flocculant used in all dewatering tests was PEO, a commercially available polymer having a nominal molecular weight of 8 million. PEO is a linear, nonionic, water-soluble molecule composed of repeating units of CH_2-CH_2-O. A stock solution of 0.25 pct PEO was prepared using recycled water from the dewatering unit. More dilute solutions of PEO were prepared by dilution of the stock solution with water recovered from the dewatering unit.

TEST RESULTS AND DISCUSSION

Initial tests indicated that the solids content of the feed for the dewatering unit varied over a wide range within a short period of time. Because of this variation, it was difficult to optimize the mixing of the PEO with the coal-clay waste. The mixing of the PEO with the waste is the most critical step in the flocculation process. The effectiveness of mixing is controlled by the stirrer speed. If the speed is too high, the flocs formed are broken apart; if it is too low, strong flocs are not formed.

Previous research has shown that the solids content of the feed determines the optimum speed. For example, at Jim Walter Resources No. 7 Mine, it had been determined that at 20 pct solids, the optimum speed of the stirrer that mixed the flocculant with the coal-clay waste was 30 rev/min; whereas, for slurries containing 1 to 3 pct solids, 40 rev/min was optimum. Since a constant solids content feed could not be supplied to the unit, the stirrer was operated for several weeks at different stirring rates. These tests provided enough data to conclude that 38 to 42 pct solids would be typically produced during the dewatering sequence.

In previous field tests conducted in Alabama, the dewatered material had typically ranged between 55 and 60 pct solids. The major difference between the wastes from Alabama and Paradise is in the particle size. The waste from the Alabama plant was typically 28.2 pct plus 48 mesh and 43.6 pct minus 325 mesh, whereas the TVA material was 8 pct plus 48 mesh and 52.9 pct minus 325 mesh.

EXPERIMENTS USING BOTH STATIC AND ROTARY SCREENS

To increase the solids content of the dewatered solids obtained in earlier experiments, a trommel screen constructed of 10-mesh stainless steel, 20-in-diam by 8-ft long, was added to the unit for the remainder of the test program. The unit with the trommel screen is shown in figures 3 and 4. In this sequence, dewatered material leaving the static screen was further dewatered in the trommel screen. In initial stages of the testing program it was determined that the feed solids usually ranged between 10 and 20 pct. For this range of feed solids, a speed of 30 rev/min resulted in strong flocs, so a majority of the experiments were conducted using a stirrer speed of 30 rev/min. A summary of test results is shown in Table 2.

FIGURE 3. - Field test unit equipped with trommel screen, with the disposal area shown in foreground.

FIGURE 4. - Coal-clay waste being dewatered in trommel screen.

TABLE 2
Summary of dewatering results obtained from experiments using both a static and a trommel screen

Test[1]	Feed solids (pct)	Feed rate (gal/min)	Screen solids (pct)	Trommel solids (pct)	PEO dosage (lb/ton)	Trommel speed (rev/min)	Trommel angle (°)
1.....	12.0	150	37.8	41.7	0.22	6-10	9
2.....	11.5	150	36.8	42.1	.32	4	6-9
3.....	10.2	135	42.4	46.4	.26	4	6
4.....	4.5	135	40.9	46.5	.38	4	3
5.....	17.2	135	39.0	47.3	.19	4	3
6.....	11.2	135	41.9	51.2	.21	1	3
7.....	14.4	135	38.6	47.9	.22	1	3
8.....	8.5	110	41.6	50.2	.26	.8	3
9.....	15.7	85	40.1	49.9	.20	1	3
10....	20.5	85	36.7	50.0	.12	1	3

[1]Each test represents an average value for each given day.

In test 1, the trommel removed 3.9 pct more water than with dewatering on the static screen only. It was observed that the material moved through the trommel rapidly and still contained free water, indicating that greater retention time was required in the trommel. Two methods of increasing retention time are (1) to decrease the angle on the trommel and (2) to decrease the revolutions per minute of the trommel. So in test 2, the rotation rate of the trommel was reduced to the lowest possible speed, and the angle of the trommel was decreased. As a result, the amount of water removed in the trommel increased to 5.9 pct.

Lowering the rate of material entering the trommel was another method of increasing the retention time in the trommel. Increasing the retention time increased the size of the roll and, generally, a larger roll of material resulted in more water being removed.

In test 3, the angle of the trommel was set at 6° and the feed rate was lowered to 135 gal/min. For this experiment, the solids content of the material coming off the static and trommel screens increased, but the retention in the trommel was still too short, with only a small roll of material being formed in the trommel. In tests 4 and 5, the angle of the trommel was decreased to 3°, but little improvement in the solids content of the dewatered material was achieved.

From these tests it was decided that the trommel speed would have to be reduced even more. A new motor was installed on the trommel so that a trommel speed of 1 rev/min could be achieved. In test 6, where the lower trommel speed was used, dewatered solids of 51.2 pct were obtained; this represents an increase in dewatering of 9.3 pct when comparing static screen solids with trommel screen solids. Results similar to those obtained in test 6 were obtained in test 7. In tests 8 through 10, the feed rate and trommel speed were adjusted, but similar results were obtained. From the data it is apparent that to further increase the percent solids, a larger trommel would be required.

An indication of how a larger trommel would improve solids content was obtained by turning the trommel off and allowing the roll in the trommel to drain for several minutes before starting the trommel again. In a majority of the tests where the trommel was stopped an increase of 2 to 3 pct solids was obtained. In several of the tests, a greater increase in solids content was observed, with 59 pct being the maximum.

FLOCCULANT DOSAGE

A 0.25-pct-PEO solution was made and then diluted to 0.10 pct with recycled water from the small water storage pit at the test site. Water from the static screen was used to fill this pit each morning and then used as needed during the day. At the beginning of each day, 225 to 300 gal of 0.10 pct PEO was made up and used throughout the day. On several occasions, more PEO solution was required.

During the testing program, it was observed that as the day progressed more PEO per ton was often required. Then, if a new batch was made, the PEO requirements would decrease, which indicated that chemical species such as HS^- in the recycle water may have been attacking the polymer. To overcome this problem, starting with test 10, $MnCl_2$ was added to the fresh PEO solution in the amount of 0.003 lb/lb PEO. This reducing agent inhibited breakdown of the PEO, as indicated in test 10, in which the PEO dosage was only 0.12 lb/ton.

LIME DOSAGE

Lime was added to the incoming coal-clay waste as a slurry containing 1 lb of lime per gallon. The pH of the slurry prior to PEO addition varied considerably during the testing program, with most of the pH values ranging from 9 to 9.3. The amount of lime required for this range was 1.3 lb/ton of dry coal-clay waste.

CONSOLIDATION IN DISPOSAL AREA

The dewatered material produced in the test unit was placed in a disposal area approximately 100 ft by 100 ft by 6 ft deep. The material when placed in the area continued to dewater. The released water flowed to one corner of the area and was pumped off periodically. The area was also sampled periodically to determine the extent of consolidation (see Fig. 5)

FIGURE 5. - Sampling of the dewatered material in the disposal area.

During the period between the operating campaigns, the disposal area filled with rainwater. This water was pumped off and the area sampled at two sites, one near the point where the dewatered material entered the area and one in the middle of the area. Results of this sampling are shown in Table 3. Even though this was under water several days prior to sampling, the material was approximately 70 pct solids.

TABLE 3
Consolidation of dewatered material

Sample site	Solids content, pct	
	When ponded	After 1 week
Entry point:		
1 ft........	71.5	68.6
2 ft........	68.8	68.9
3 ft........	64.5	68.4
4 ft........	69.3	69.4
Average...	68.5	68.8
Middle:		
1 ft........	71.1	NS
2 ft........	[1]85.4	NS

NS Not sampled.
[1]Possible error made in this measurement.

After 1 week the disposal area was sampled again at the entry point and the results are also shown in Table 3. It should be noted that heavy rains had fallen during the 2 days prior to sampling. The material averaged 68.8 pct solids. At the end of the testing program the height of the dewatered material in the area was 6 ft at the point where the dewatered material was discharged from the dewatering unit and 4 ft at the wall opposite the dewatering unit.

The consolidated material was removed from the area using a front end loader. The loader removed the material by entering the area at one end and cutting a channel the width of the loader through the material. The side of the channel remained vertical and did not sluff in, thus indicating good stability.

MIXING OF DEWATERED MATERIAL WITH COARSE REFUSE FROM THE PREPARATION PLANT

The next phase of the research was to determine the stability of dewatered material and preparation plant coarse refuse when mixed in a ratio of 2.6, the same ratio of coarse to fine material produced in the preparation plant. Coarse refuse was fed from a hopper onto a variable speed belt that delivered the feed onto a second conveyor belt where the dewatered material exiting the trommel was placed on top of the coarse refuse. This conveyor then deposited the mixture into a pit for removal and stacking by a front-end loader. The test was conducted over a 2-day period. The solids discharged from the trommel, and the operating conditions were monitored during the test. Results are shown in Table 4.

TABLE 4
Results of dewatering experiments of coal-clay waste prior to mixing with coarse refuse (Feed rate, 85 gal/min)

Feed solids (pct)	Trommel discharge (pct solids)	Underflow (pct solids)	PEO dosage (lb/ton)
9.04	46.0	1.10	0.12
8.94	46.7	1.90	.13
8.93	49.0	1.67	.14
16.80	48.2	1.74	.20
17.20	45.9	.85	.22
22.00	NST	NST	NST
18.90	47.6	.30	.10
18.30	47.2	1.06	.12
17.10	50.8	1.63	.13
17.20	49.8	1.46	.13
17.40	49.5	1.11	.13
17.00	47.7	1.34	.19
15.7[1]	48.0	1.29	.15

NST No sample taken.
[1]Average.

For the 2 days of testing, the coarse to dewatered fine ratio average d 2.9. The ratio was varied during the test period from 1.5 to 6.3. It was observed that to get a stackable material, a ratio of at least 3.5 to 4.0 was required. At a 1.5 ratio the material would not stack but flowed much like concrete. To make a stack of coarse and dewatered fines, a dike of stackable material was made, and the nonstackable material was deposited in the diked area. The stack had the approximate dimensions of 18 ft square by 3.5 ft in height. After 2 days, the middle of the stack could be walked on, even though considerable rainfall had been encountered the previous day. The stack was sampled in the middle and the average solids content was 87.2 pct. Four days later the stack was again sampled. The outer portion of the stack was quite hard. A sample was taken at the middle of the stack by breaking through the outer crust and sampling the interior of the stack. The solids content of this material was 84.7 pct.

CONCLUSIONS

The test program at the Paradise Steam Complex has led to the following conclusions:

(1) Coal-clay waste was dewatered on a static screen followed by a trommel screen to 50 pct solids using 0.12 to 0.15 lb of PEO per ton of waste.

(2) For the pH range of 9.0 to 9.3, 1.3 lb of lime per ton of waste solids was required.

(3) The dewatered material continued to dewater when ponded and reached approximately 69 pct solids in 7 days. This material was quite stable and could be handled by heavy equipment.

(4) Mixing the dewatered material with the coarse refuse at a ratio of 1 to 4 produces a stackable product, whereas at a lower ratio of 1 to 2.6, a waiting period of 2 days was required before the material could be stacked.

(5) The capacity of the static screen is in excess of 150 gal/min if a trommel screen is used in conjunction with the static screen.

REFERENCES

1 Zimmerman, R. E. Preparation Headache, Disposal of Refuse Fines. Coal World, April 1978, p. 22.
2 Hucko, R. E. Beneficiation of Coal by Selective Flocculation, A Laboratory Study. BuMines RI 8234, 1977, 13 pp.
3 Milliken, J. C. Coal Preparation--1977. Min. Cong. J., v. 64, No. 2, Feb. 1978, pp. 53-55.
4 Roberts, J. R. and H. J. Geisler. Coal Waste Disposal at National Pocahontas Mine. Min. Cong. J., v. 64, No. 12, Dec. 1978, pp. 39-43.
5 Backer, R. R. and R. A. Busch. Fine Coal-Refuse Slurry Dewatering. BuMines RI 8581, 1981, 18 pp.
6 Brown, P. M. and B. J. Scheiner. Dewatering of Coal-Clay Waste Slurries from Preparation Plants. BuMines RI 8824, 1983, 8 pp.

ON THE DEVOLATILIZATION OF OVERBED SPREADERS FOR ATMOSPHERIC FLUIDIZED BED COMBUSTORS.

F. J. C. Rademacher[1], A. W. Roberts[2], and H. J. Rietman[1]
[1]M.E. Dept., Twente University, Twente, The Netherlands
[2]M.E. Dept., University of New Castle NSW, Australia, visiting

SUMMARY

A feasibility study regarding the use of overbed spreader stokers for atmospheric fluidized bed combustors (AFBC) has been initiated at the University of Twente, The Netherlands. As part of that study the devolatilization in the freeboard has been estimated as part of future design requirements. A computerprogramme has been developed for this purpose, suitable for coal of any particle size distribution and composition, emphasizing SO_2 development.

INTRODUCTION.

Several boiler manufacturers have kept themselves busy during the recent years with building small and medium size industrial fluidized bed combustors. Apparently under time pressure there has been insufficient opportunity, during the scaling-up process from laboratory-, via pilot-, to industrial plants, to carry out component tests regarding alternative feed systems and their evaluation. As a result of that one can see combustors provided with pneumatic in-bed feeding from underneath on one location, and almost identical boilers provided with overbed feed systems on another, often for no clear reasons.

Also in-bed feeding from the side by means of injection screws has been applied. From literature it follows readily that most of the existing feed systems are far from ideal.

This paper aims to produce some relevant design requirements and practical considerations for overbed spreaders. In a preceding paper [2] the required feeding capacity, the ballistics of the coal and constructional needs were already emphasized, whereas this paper focussus on the devolatilization itself of the fresh incoming coal during its flight through the freeboard. An integrated treatment of both aspects will appear soon elswhere [1].

TEMPERATURE INCREASE.

The devolatilization of the coal during its residence time in the freeboard, can be determined from the composition of the coal and its temperature increase with time. In this section first the temperature increase will be paid some attention to, followed by an estimation of the devolatilization focussed on SO_2-forming.

On the heat impact of coal particles in the freeboard.

In this work we confine ourselves to the treatment of sphere shaped particles. Carslaw and Jaeger [17] have obtained a closed-form solution for this case at constant heat flux (Q_c) at the surface when the initial temperature within the sphere is zero throughout. The model for this problem is given by

$$\frac{\delta T}{\delta t} = \zeta \left[\frac{\delta^2 T}{\delta r^2} + \frac{2}{r}\frac{\delta T}{\delta r}\right] \tag{1}$$

for $0 \leqslant r \leqslant r_o$

The solution given by Carslaw and Jaeger is

$$T(r,t) = \left(\frac{3\,Q_c}{\rho_p\, C_p\, r_o}\right) t + \frac{Q_c\,(5\,r^2 - 3\,r_o^2)}{10\,\lambda_p\, r_o} -$$

$$\frac{2\,Q_c\, r_o^2}{\lambda_p\, r} \sum_{n=1}^{\infty} \frac{\sin\left(\frac{r\,\beta_n}{r_o}\right) e^{-\zeta\,\beta_n^2\, t/r_o^2}}{(\beta_n)^2 \sin(\beta_n)} \tag{2}$$

wherein the β_n terms are the positive roots of tan $\beta_n - \beta_n = 0$
In reality the, non constant, heat flux amounts

$$Q = \eta_c\, \alpha\, (T_g - T_{r_o}) + \eta_r\, \sigma\, \varepsilon\, (T_w^4 - T_{r_o}^4) \tag{3}$$

where η_c and η_r are, respectively, coefficients of values $\leqslant 1.0$ to account for the dimination of heat flux reaching each particle due to the total density of particles as a function of the heating zone volume of the furnace. In this explorative study it holds $\eta_c = \eta_r = 1$. The convection coefficient α is a function of the Reynolds number for the gas and is given by

$$\frac{\alpha\, d_o}{\lambda_g} = 0.37 \left(\frac{v_r\, \rho_p\, d_o}{\mu}\right)^{0.6} \tag{4}$$

for Reynolds numbers such that 25 ⩽ Re ⩽ 100,000.

In using a computer a large number of time increments Δt can be considered during which the heat flux as expressed by (3) can be considered constant an during which all the temperature changes can be calculated (and so the new temperature profile through the particles).

Fig. 1 shows some results for the special case that no devolatilization occurs at all, and serves only to demonstrate the high skin temperatures that smaller coal particles of say 6 mm and below will obtain. This indicates first that devolatilization has to be quantified and second that it is not allowed to assume a constant temperature throughout the particles but that the real temperature profile inside the coal particles of different size have to be known at any instant.

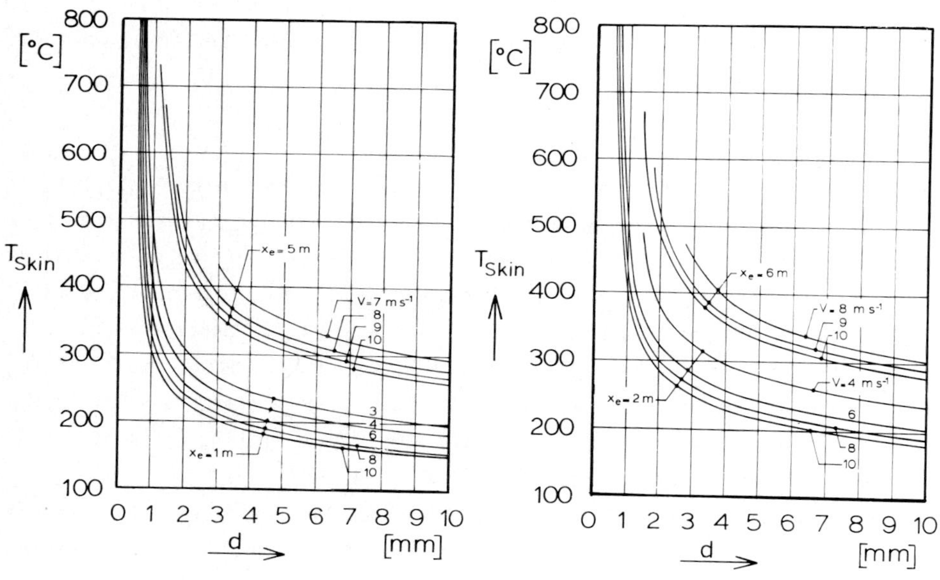

Fig. 1. Surface temperature of single particles as a function of particle diameter. Initial velocity (v) and max. throw distance (x_e) being parameter.

v_g = 2.5 m s^{-1}; ρ_g = 0.311 kg m^{-3}; ν = 0.000142 m^2 s^{-1};
λ_g = 0.074 J m^{-1} s^{-1} K^{-1}; gas temp. 850 ^{0}C;
σ = 5.67 x 10^{-8} J s^{-1} m^{-2} K^{-4}; $\rho_p \simeq$ 1250 kg m^{-3} ;
$C_p \simeq$ 1260 J kg^{-1} $^0C^{-1}$; λ_p= 0.252 J m^{-1} s^{-1} K^{-1}; ε = 1;
initial coal temp. 27 ^{0}C; $\eta_c = \eta_r$= 1; y_o = 0.8 m.

For the common size distribution of fig. 4 of 18 mm minus, the amount of 3 mm minus represents 47% by weight. If particles down to say 1.5 mm would have been permitted, at the cost of some SO_2 emission though, the part to be sieved out could have been reduced from 47% to 19%.

ON THE DEVOLATILIZATION OF COAL PARTICLES IN THE FREEBOARD

Since devolatilization depends on the composition of the coal and of the surrounding gases, the temperature, the rate of temperature increase and consequently on the particle size, a great number of parameters are involved. Therefore a complete kinetic treatment, is precluded [19]. Instead the parameters involved will be considered in perspective of the possible relevance they might have for the freeboard problem.

Heating rate

As shown by many investigators among which Jüntgen and Van Heek [16, 6], the devolatilization process will shift towards higher temperatures when the heat rate increases.

Particle size

In the case of coarser particles, the temperature in the core stays lower compared to the surface temperature, leading to diffusion or heat transfer-controlled pyrolysis. The rate at which gases leave the surface of such large particles is controlled by a physical resistance to the passage of the volatiles through the particle. For melting coals, where outer pores are gradually being closed, considerable pressure build-up may occur inside the particle, possibly causing particle fragmentation. Jüntgen reports grain bursting in the case of 1 to 2 mm coal particles containing 19% VM, at heating rates of 1000 $^{o}C\ s^{-1}$ in an oxygen rich atmosphere [16].
However this did not occur with a 37% VM coal called "Fürst Leopold". In the same paper he reports on an oxygen-free devolatilization experiment at heating rates of 500 $^{o}C\ s^{-1}$ where no bursting occured. Therefore in our study, which deals with heating rates in the order of approximately 500 $^{o}C\ s^{-1}$ at the surface only, and $\leq$ 5% oxygen in the freeboard, bursting is not included.

Modelling of the particle's mass loss.

From the literature [22], it is well known that the non-isothermic release of a gas from a decomposing solid can be described theoretically by the following equation:

$$\frac{dV}{dT} = \frac{k_o V_o}{m'} \exp\{ - \frac{E}{R\,T} - \frac{k_o\,R}{m'\,E} T^2 \exp(- \frac{E}{R\,T}) \} \qquad (5)$$

for a first order reaction that is, and a linear temperature increase denoted by m', (m' = dT/dt). The frequency factor (k_o) and the activation energy (E) are normally deduced from the experimentally obtained devolatilization curve. The latter represents the devolatilization rate as a function of the temperature. In reality there are more different components taking part in the pyrolysis; most often they do overlap each other, fig. 2.

In that case, the total devolatilization rate is represented by the summation of the rates of the single reactions:

$$\frac{dV}{dT} = \int_0^\infty \int_0^\infty \frac{dV_s}{dT} f(E)\, f(k_o)\, dE\, dk_o \qquad (6)$$

V_s being the volume of gas released by one single reaction.

In this equation, two distribution functions have been introduced, one for the activation energy (E) and one for the frequency factor (k_o). Although the total devolatilization can be estimated very closely by assuming Gaussian distribution functions for E and k_o, the method is rather cumbersome.

Instead of proceeding this way it is assumed for the purpose of the drag correction that the total degasification curve, exclusive of sulphur components, will have a form similar to those for single reactions, fig. 2 [19]. It simply requires the determination of two constants (a and b) related to the approximated beginning and end temperatures of the degasification. Besides, the mass loss is only one factor that influences the drag. The other, being swell, may contribute in the same order depending on the coal. The mass decrease of a single volatile component, obeying the first order reaction, can be expressed as follows [20]:

$$-\frac{dm}{dt} = k_o\, e^{-\frac{E}{R\,T}}\, m \qquad (7)$$

Integration leads to the remaining amount at any instant being

$$m = m_o \exp\left\{-\frac{k_o}{m'}\frac{R\,T^2}{E}\exp\left(-\frac{E}{R\,T}\right)\right\} \qquad (8)$$

only when $\frac{E}{R\,T} >> 0$.

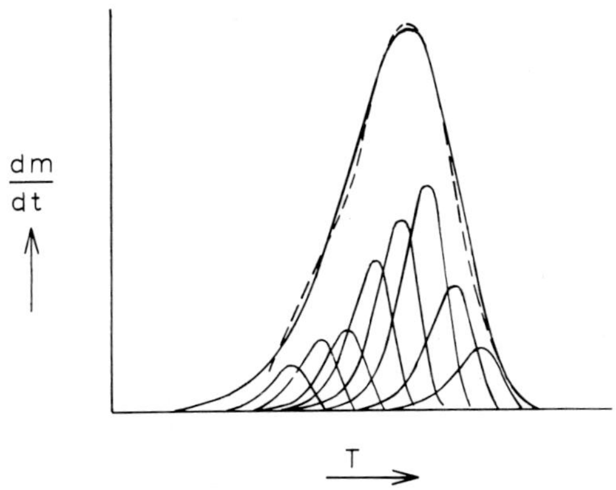

Fig. 2. The total devolatilization as a summation of single reactions overlapping each other [20].

Application of this expression to the whole conglomerate of volatiles, exclusive of the amounts of sulphur 1 and sulphur 2, fig. 2, gives:

$$\frac{m_v}{m_{o_v}} = \exp\{ - \frac{a}{m'} T^2 \exp(- \frac{b}{T})\} \qquad (9)$$

Here a/m' and b can be estimated from experimental results as discussed in [1].

Comparing equations (8) and (9) we learn that a b = k_o, so that (7) can be rewritten as:

$$\frac{\Delta m_v}{\Delta t} = m_v \; a \; b \; e^{-\frac{b}{T}} \; , \qquad \frac{b}{T} >> 0 \qquad (10a)$$

or

$$\Delta m_v = m_v \frac{a}{m'} b \; e^{-\frac{b}{T}} \Delta T \qquad (10b)$$

being the basic incremental form used in the computer programme, Δm_v standing for the decrease of mass over a period Δt, corresponding with a temperature rise ΔT . We first have to determine the original amount of volatiles in each shell. Since the total amount in the grain is known from the proximate analysis in a percentage by weight, it follows readily that

$$m_{o_v} = (m\,a\,v\,p - m\,o\,s\,1\,p \,.\, f_{s_1})\; m_o/100 \qquad (11)$$

The total mass of "volatiles" present in the jth shell equals:

$$m_{o_{v_j}} = m_{o_v} \{(j + 0.5/m)^3 - (j - 0.5/m)^3\} \tag{12}$$

For the jth shell, eq. (10a) takes the form:

$$\Delta m_{v_j} = m_{v_j} \cdot a_v \cdot b_v \cdot e^{-\frac{b_v}{T_j}} \cdot \Delta t \tag{13}$$

The temperature T_j being known, the course of the mass loss can be established for each shell.

Modelling of SO_2 losses.

Many investigators have been studied the evolution of sulphur during devolatilization and hydrodesulphurization of coal [4, 7, 8, 9, 10, 11, 13, 14, 15, 18, 19, 21]. The process is complicated because of the existence of competing reaction processes among several forms of sulphur, that is sulphate, iron sulphides and several organic sulphur forms. After Solomon [13], it can generally be believed that the following reactions occur:

1. The thermal decomposition of pyrite (FeS) at fairly low temperatures (400 ^oC to form sulphide FeS with the release of sulphur, H_2S and more complicated organic sulphur compounds. The sulphur released may combine with the char or be released with the volatiles.

2. Decomposition of FeS at high temperatures (900 ^oC) to form Fe with the release of sulphur and sulphur compounds.

3. Thermal decomposition of various organic forms of sulphur termed by some investigators as organic I, II and III. Organic I and II are removed at low temperature (400 ^oC) while organic III is removed at temperatures above 500 ^oC. Some of the organic III is produced by incorporation of sulphur from pyrite into the organic structure of the char.

For the purpose of our modelling, based on the findings of Cernic and Solomon [21, 13], only a fraction of the organic sulphur has to be accounted for since it is semi-volatile. Although, depending on coal rank, as a rule of thumb for the bituminous coals, it holds that $\leqslant 40\%$ of the organic sulphur forms the easy devolatilizing portion. The remaining sulphur is presumably in thiophenic

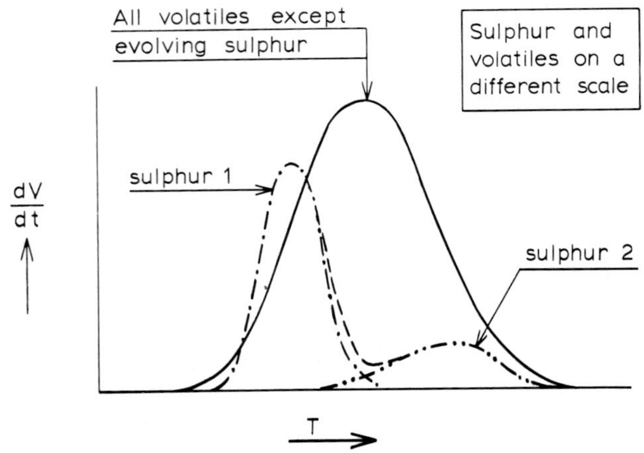

Fig. 3. Approximating devolatilization curves.

compounds.

Depending on the rate of heating and the coal rank, the organic sulphur evolves at somewhere between 200 and 600 ^oC. A part of the pyritic bound sulphur evolves in the form of H_2S, at substantially higher temperatures, ranging up to 1100 ^oC, which however may partly run parallel with the organic freed sulphur. For the purpose of our investigation it is assumed that the released hydrogen bound sulphur will eventually change to SO_2.

Based on these considerations the sulphur devolatilization below 850 ^oC, which is approximately the freeboard temperature, can be represented by the interrupted curve of fig. 3. For the purpose of modelling only, this curve is supposed to be the summation of the two other curves indicated by sulphur 1 and sulphur 2. Each of these (sub) curves are supposed to obey Arrhenius law.

Some investigators, for example Vestal [19], emphasize the influence of back reactions. These influences tend to delay the pyrolysis to somewhat higher temperatures and are neglected in this study. In the case of sulphur rich coal, most of the sulphur is released in the lower temperature range; the amount evolving in the sulphur 2-zone being of minor interest. Both desulphurization-zones are quantified as follows:

Sulphur 1

Original mass that will devolatilize:

$$m_{o_{s_1}} = mos1p \cdot m_o \cdot f_{s_1}/100 \qquad (14)$$

Also for the jth shell:

$$m_{s_{1_j}} = m_{o_{s_1}} \{ (j + 0.5/m)^3 - (j - 0.5/m)^3 \} \tag{15}$$

Mass loss for the jth shell:

$$\Delta m_{s_{1_j}} = m_{s_{1_j}} \cdot a_{s_1} \cdot b_{s_1} \cdot e^{\frac{-b_{s_1}}{T_j}} \cdot \Delta t \tag{16}$$

Sulphur 2

Similarly it holds:

$$m_{o_{s_2}} = m\,o\,s\,2\,p \cdot m_o \cdot f_{s_2}/100 \tag{17}$$

$$m_{s_{2_j}} = m_{o_{s_2}} \{ (j + 0.5/m\,)^3 - (j - 0.5/m)^3 \} \tag{18}$$

$$\Delta m_{s_{2_j}} = m_{s_{2_j}} \cdot a_{s_2} \cdot b_{s_2} \cdot e^{\frac{-b_{s_2}}{T_j}} \cdot \Delta t \tag{19}$$

In the computer programme the masses of devolatilized sulphur 1 and sulphur 2 with time are summed.

DETERMINATION OF THE SMALLEST ALLOWABLE PARTICLE SIZE; INTEGRATED RESULTS

Modelling the coal particle size distribution.

Fig. 4 shows an accumulative weight/coal size distribution for a typical stoker fired boiler [12]. For the solution of the spreader stoker problem however, the percentage contribution of each separate range of particle size is required as represented by fig. 5. The next distribution satifies:

$$m_p = \frac{p}{e^{q|d_o - \bar{d}|^u}} \tag{20}$$

the constants p and q and the exponent u to be determined from the given curve.

For the curve of fig. 5, it holds in particular

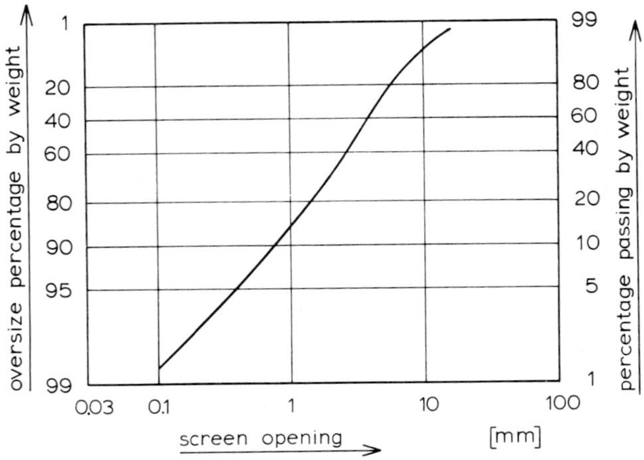

Fig. 4. Graphical representation of the distribution of sizes of broken coal [12].

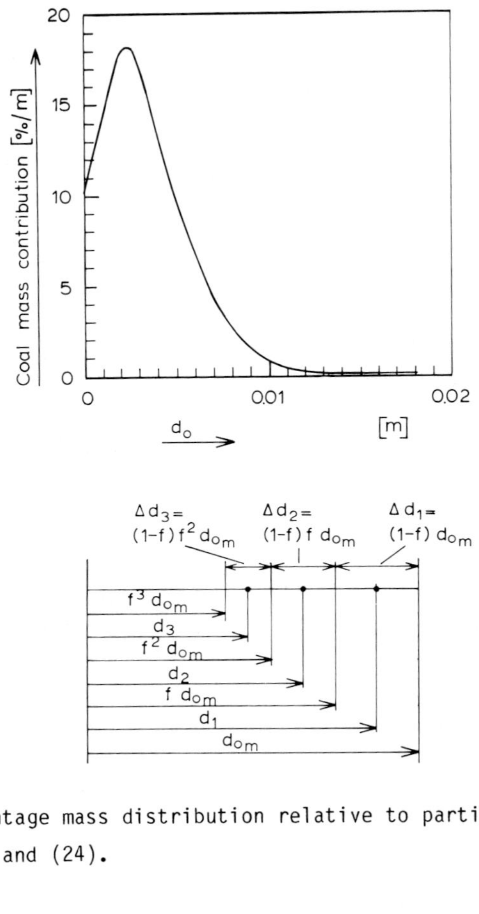

Fig. 5. The percentage mass distribution relative to particle size; see also eqs. (22) and (24).

$$m_p = \frac{18.25}{e^{0.15|d_o - 2.5|^{1.5}}} \tag{21a}$$

when d for practical reasons is in [mm], or

$$m_p = \frac{18.25}{e^{4743|d_o - 0.0025|^{1.5}}} \tag{21b}$$

when d_o in [m].

We will replace the real particle size distribution by a hypothetical one, consisting of a finite number of monosize fractions. If we assign d_{oo} to the diameter of the particles in the smallest monofraction that will not be blown out, and we subdivide into K monofractions each indicated by the subscript k, when $1 \leqslant k \leqslant K$, it holds:

$$m_{t_k} = 10 \cdot m_{p_k} \cdot M \cdot \Delta d_k = 10 \cdot m_{p_k} \cdot M \cdot (1-f)\, f^{k-1} \cdot d_{o_m} \tag{22}$$

wherein, see (20):

$$m_{p_k} = \frac{p}{e^{q|d_{o_k} - \bar{d}|^u}} \tag{23}$$

in which

$$d_{o_k} = \frac{1}{2}(1 + f)\, f^{k-1} \cdot d_{o_m} \tag{24}$$

It holds

$$n_{d_{o_k}} = \frac{m_{t_k}}{m_k} \tag{25}$$

wherein

$$m_k = \frac{\pi}{6} \rho_p d_{o_k}^{\ 3} \tag{26}$$

It can readily be shown with reference to the lower section of fig. 5 that

$$f \simeq F^{\frac{1}{K}} \quad , K >> 1 \tag{27}$$

wherein

$$F = \frac{d_{oo}}{d_{o_m}} \tag{28}$$

The total SO_2-emission in the freeboard.

The equations of the preceding paragraph allow a coal characterisation as illustrated in fig. 4 to be split into a number of K monofractions, the diameter of the particles (d_k) in each fraction being known, as well as the number of particles that each fraction contains. The total mass loss of sulphur for a particle of a specific diameter, during its residence in the freeboard, is calculated in the computer programme by utilising eqs. (16) and (19) repeatedly. Applied to a monofraction of diameter d_k it holds:

$$m\ell ps(d_{o_k}) = m\ell ps1(d_{o_k}) + m\ell ps2(d_{o_k}) \tag{29}$$

The total sulphur release from the kth monofraction amounts to $n_{d_k} \cdot m\ell ps(d_{o_k})$.

So the mass release of sulphur (ΔMS) of all the monofractions in the range (R_k) down to k, and denoted by $\Delta MS(R_k) \equiv (d_k \leqslant d \leqslant d_m)$ equals:

$$\Delta MS(R_k) = \sum_{k=1}^{k=k} n_{d_k} \cdot m\ell ps(d_k) \tag{30}$$

recalling that each single smaller particle will emit relatively more sulphur compounds than each single larger particle. We can expect that the sulphur emission per primary GJ of energy will behave like the dotted curve in fig. 6.

It holds

$$SO_2(FB) \leqslant SO_2(LW) \cdot FBL \tag{31}$$

in which both the developed SO_2 in the freeboard, $SO_2(FB)$, and the amount allowed by law, $SO_2(LW)$, are expressed in grams of SO_2 per GJ of primar energy. For a specific energy (S.E.) in Joules per kg of coal and a total mass of M kg of coal, it apparently also holds

$$SO_2(FB) = \frac{2\ \Delta MS(R_k)}{(S.E.)\ M}\ 10^{12} \leqslant SO_2(LW) \cdot FBL \tag{32}$$

in which $\Delta MS(R_k)$ is given by (30).

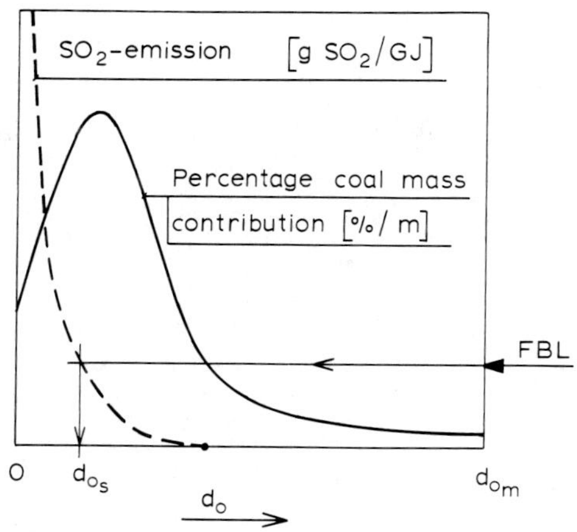

Fig. 6. Qualitative presentation of the expected SO_2 emission related to particle size (– – – – –). The preset FBL (Free Board Limit) determines the smallest allowable particle (d_{o_s}) to be fed in.

Integrated results.

Fig. 7 shows the total production of grams of SO_2 per GJ of primar energy as a function of the smallest particle fraction that one allows to enter the freeboard. It concerns the typical coal of fig. 4. The graph holds for the longest possible shots of the largest particles.

Example

Assume an initial coal velocity of $v = 6\ m\ s^{-1}$ and a gas velocity of $v_g = 3\ m\ s^{-1}$. If we allow all particles of 0.8 mm and up to enter the freeboard then approximately 50 grams of SO_2 per GJ of primair energy will develop. It can clearly be observed how greatly this amount is reduced by increasing the smallest allowable diameter from 0.8 to 1.0 mm. According to fig. 4, these particle sizes do correspond with 11 and 14% of the total weight of the coal that has to be sieved out. Apart from the technological question as how this can be achieved there is evidently an economic optimum to keep in mind. For practical design purposes the graph will be used in the opposite way; starting from a pre-determined allowable amount of SO_2-production by the spreader, the smallest allowable coal particle size follows.

Fig. 8 shows a few examples of the coalmass- and fraction distribution with horizontal distance from the spreader, inclination angle and gas velocity being parameter. This kind of information will serve to determine how the inclination angle, and the initial velocity of the coal stream have to change with time in order to achieve a possible good coalmass distribution across the bed.

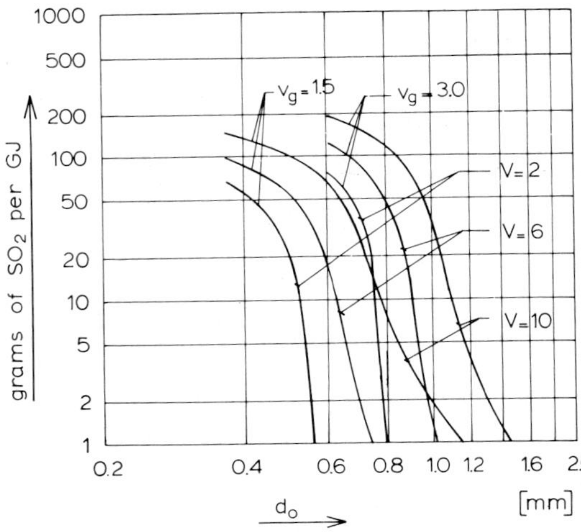

Fig. 7. Development of grams of SO_2 per GJ in the freeboard related to the smallest coal particle (dia. d_0) of the coal of fig. 4 that one allows to enter the freeboard.

v_g = superficial gas velocity [m s^{-1}]

v = initial coal velocity [m s^{-1}].

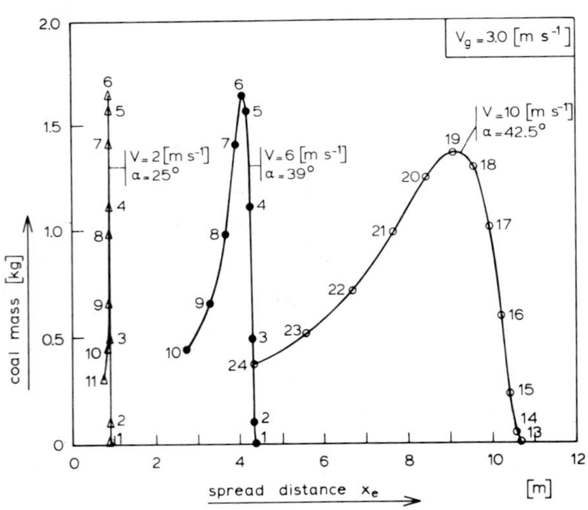

Monofraction number	Particle diam [mm]
1	15.67
2	11.61
3	8.60
4	6.37
5	4.72
6	3.50
7	2.59
8	1.92
9	1.42
10	1.05
11	0.78
12	0.58
13	16.04
14	12.54
15	9.81
16	7.67
17	6.00
18	4.69
19	3.67
20	2.87
21	2.24
22	1.75
23	1.37
24	1.07

Fig. 8. Mass distribution of the coal when it reaches the bed surface. The initial velocities of 2, 6 and 10 m s^{-1} are curve parameters. Inclination is such that the largest particles reach the longest shot; the corresponding inclination angles (α) are indicated. Each monofraction has been given a number, the corresponding diameters to be found in the accompanying table.

PRELIMINARY TESTS WITH AN EXPERIMENTAL MECHANICAL OVERBED SPREADER.

Considering the objections to pneumatic feeding into the freeboard on the one hand, and the damage of the coal by conventional spreader stokers on the other, attempts have been made to demonstrate to what extent the coal particles can be smoothly accelerated by mechanical means. Although this has been reported earlier [2] the relevant obtained results in using a modified impeller for peening installations, fig. 9, are implemented here in the conclusions.

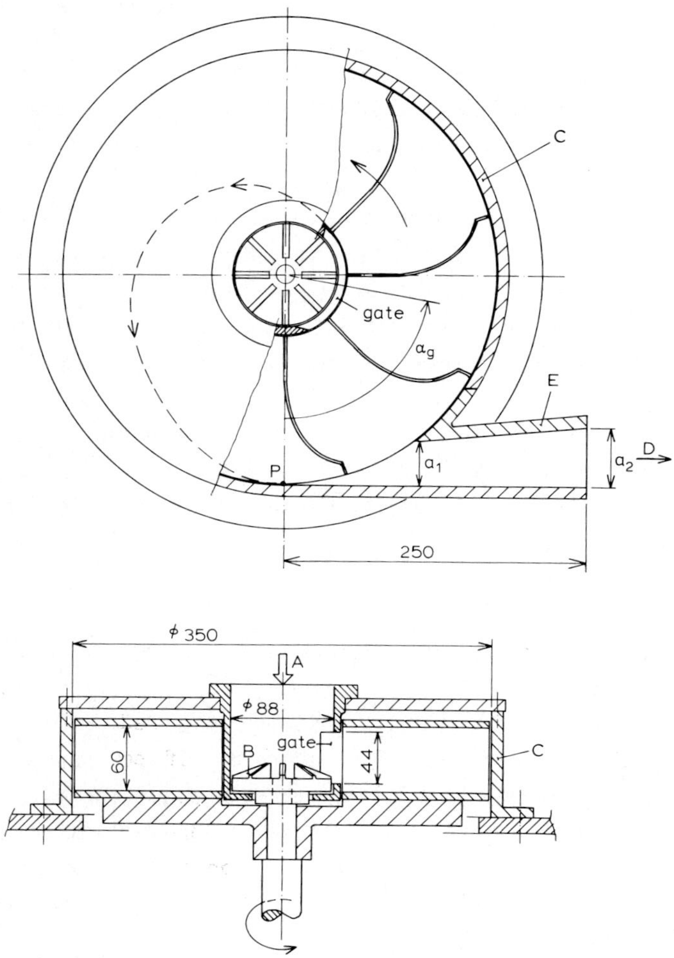

Fig. 9. Sketch of the experimental device being part of the feasibility study on a mechanical overbed feeder that causes a low, total amount of coal damage.

CONCLUSIONS.

Conclusions on what has been achieved.

1. A design tool in the form of a computerprogramme has been obtained by which it is possible to predict, the devolatization of the incoming coal during its residence time in the freeboard, starting from a known coal composition-, particle size distribution-, gas composition-, gastemperature-, superficial gas velocity-, required capacity and ballistics. In particular the SO_2-forming is focussed on.
2. The programme can be used in two different ways. Either to predict the total devolatilization for any given coal, or to prescribe which particles should be sieved off in order not to exceed a preset amount of SO_2 development, fig. 7.
3. Overbed feeding by a mechanical spreader can certainly fulfill the requirements for atmospheric fluid bed combustors, particularly in case of (larger) power generation plants. It has been proved that the SO_2-production of a well designed spreader will stay within negligible-, and therefore acceptable amounts, even for sulphur-rich coals of typical size distribution.
4. Mechanical overbed spreaders have the following practical advantages over mechanical or pneumatical in-bed feeding:
   - Near to ideal coal mass distribution can be realised by standard control devices.
   - Causes no hotspots.
   - Relatively few units required.
   - Easy and almost independent adjustable to changing process parameters.
   - High reliability.
   - Easy maintenance and overhauling jobs due to good accessibility; units can be totally replaced during operation of the combustor when required.
5. An amount of fines may have to be separated before the main bulk of coal enters the spreader. This amount has to be injected either pneumatically in-bed, or via the spreader after pelletizing. For small(er) industrial combustors this extra step is doubtless a disadvantage which can be avoided by using low-sulphur coal.
6. Since, during the preliminary tests with the device of fig. 9, even 90% of the total mass flux stays within a spread angle of 8 degrees, it will be possible to programme the direction of the coal stream such that a triangular bed section in front of the spreader can be fed with an equal load

for every unit of bed area. Basically the following possibilities exist:

- Constant impeller speed; control by both programmed elevation and sweep angle.
- Constant elevation angle; control by both variable impeller speed and sweep angle (not recommended).
- Impeller speed, elevation and sweep angle all controlled.

The feed rate has to be controlled before entering the spreader, either in continuous or batch mode. The latter being of advantage at low power loads. See further [2].

Conclusions on what has to be done.

1. The developed theory has to be refined such that it also covers the case of swelling coals.
2. The same holds with respect to possible fragmentation of coal particles from thermoshock just after having been injected.
3. A prototype of mechanical spreader has to be designed for try-outs in an experimental combustor of possible large size. This need not to be based at all on the idea as exposed by fig. 9.

Acknowledgement.

The authors are very grateful to Mrs. D. Johnstone, president of the Johnstone Foundation, Newcastle University Australia, for her donation that has made the completion of the first phase of this investigation possible.

REFERENCES.

1 Rademacher, F.J.C., Roberts, A.W. and Rietman, H.J., Required characteristics for overbed feed systems for atmospheric fluidized bed combustors. To be published in VDI-Forschungsheft, febr. 1986 (in English).

2 Rademacher, F.J.C., Roberts, A.W. and Rietman, H.J., Considerations concerning coal spreader stokers for atmospheric fluidized bed combustors. Eighth intern. conf. on fluidized-bed combustion, Houston, Texas, March 18-21,1985.

3 Ministerie van Volksgezondheid en Milieuhygiëne, Tweede circulaire inzake eisen met betrekking tot de uitworp van luchtverontreinigende stoffen door kolengestookte installaties. Nr. 162238A/DGMH/L, 3 Sept. (1982), 3 p.

4 Morrison, G.F., Chemical desulphurisation of coal. Report ICTIS/TR15 (1981). IEA Coal Research, London.

5 Wendt, J.O.L., Fundamental coal combustion mechanisms and pollutant formation in furnaces. Prog. Energy Combust. Sci., Vol. 6 (1980), 201-222.

6 Jüntgen, H. and Van Heek, K.H., An update of german non-isothermal coal pyrolysis work. Fuel Processing Technology, 2 (1979), 261-293.

7 Attar, A., Chemistry, thermodynamics and kinetics of reactions of sulphur in coal-gas reactions: a review. Fuel (1978), Vol. 57, 201-212.

8 Maloney, K.L. et al., Sulfur retention in coal ash. EPA Report 600/7-78-153b, Washington, DC20460, Nov. (1978), 59 pages.
9 Hauk, R., Entschwefelung von Steinkohlenstäuben durch Schnellentgasung und Teilvergasung. Diss., Karlsruhe-Univ., Dept. of Chem. Eng., Jan. (1978).
10 Tiwari, D.N. et al., Sulphur in coal and its transformation during pyrolysis. FRI New Institute. Col. 27 (1977), No. 1, 19-22.
11 Hauk, R., Hedden, K. and Huber, L., Entschwefelung von Kohle durch Schnellentgasung und Teilvergasung. Gwf-gas/Erdgas, 118 (1977), H10, 427-430.
12 Biswas, B.K. and Baley, J.U., An investigation of alternative feed systems for utility-scale fluidized-bed steam generators/combustors (200 MWe or larger units). Paper, 5th International Conference on Fluidized Bed Combustion. Washington, D.C., (1977), Vol. II, 248-265.
13 Solomon, P.R., The evolution of pollutants during the rapid devolatilization of coal. United Technologies Research Center. East Hartford, Connecticut 06108, Report Nr. R76-95288-2, April (1977).
14 Baumann, M. et al., Nicht-isotherme Entschwefelung von Steinkohlen mit Wasserstoff. Erdöl und Kohle. Band 30, Heft 4, April (1977).
15 Robinson, L., Hydrodesulphurization of char. Fuel, (1976), Vol. 55, 193-201.
16 Jüntgen, H., The course of combustion of the individual coal particle. EPE, Vol. XII, nr. 3-4 (1976-77), 11-23.
17 Carslaw, H.S. and Jaeger, J.C., Conduction of heat in solids. Second ed. repr. Oxford Clarendon Press (1976), ISBN 0-19-853303-9.
18 Langhoff, J. and Ratajczak, E.A., Das Verhalten organischer Schwefelverbindungen in Kohle und Koks gegenüber Salzsäure und naszierendem Wasserstoff. Glückauf- Forschungshefte, April (1972), 51-54.
19 Vestal, M.L. et al., Kinetic studies on the pyrolysis, desulfurization and gasification of coals with emphasis on the non-isothermal kinetic method. Report NR SRIC 68-13, Scientific Research Instruments Corporation, Baltimore, Maryland, U.S.A. (1969).
20 Jüntgen, H., and Van Heek, K.H., Fortschritte der Forschung auf dem Gebiet der Steinkohlenpyrolyse. Brennstoff-Chemie, Nr. 6, Band 50 (1969), 172-178.
21 Cernic-Simic, S., A study of factors that influence the behaviour of coal sulphur during carbonization. International Conf. on Coal Science, Le Touquet, 30 May - 2 June (1961), 141-151.
22 Krevelen, D.W. van, Coal (1961), Elsevier, Amsterdam.

SYMBOLS USED.*

a	auxiliary constant	-
b	auxiliary constant	-
C_p	Specific heat of c.p.	J kg^{-1} ^{o}C^{-1}
d	arbitrary diameter inside c.p.	m
$\bar{d}$	imaginary average diameter	m
Δd_k	band width taken by the kth monofraction of c.p.	m
d_o	outer diameter of c.p.	m
d_{o_k}	outer diameter of the c.p. of the kth monofraction	m
d_{o_m}	outer diameter of the largest c.p.	m
d_{o_s}	outer diameter of the smallest c.p. allowed to be injected	m

* c.p. stands for (spherical) coal particle

d_{oo}	lower bound diameter of monofractions	m
E	activation energy	Nm kg^{-1}
FBL	Free Board Limit, being a fraction of SO_2 (LW)	-
f	fraction of d_{o_m}	-
f_{s_i}	fraction of the amount of sulphur i that devolatiles	-
j	shell number of a coal particle ($1 \leqslant j \leqslant m$)	-
K	chosen amount of monofractions in the range $d_{oo} \leqslant d_o \leqslant d_{o_m}$	-
k	number of monofraction under consideration	-
k_o	frequency factor	s^{-1}
M	total coal mass	kg
$\Delta MS(R_k)$	total sulphur mass in the coal in range R_k	kg
m	mass of coal particle	kg
m'	heat rate, being dT/dt	K s^{-1}
mavp	percentage of volatile mass in c.p. by proximate analysis	-
m_k	mass of a c.p. of the kth monofraction	kg
$m\ell ps(d_{o_k})$	total sulphur mass loss of a c.p. of monofraction d_{o_k}	kg
$m\ell psi(d_{o_k})$	mass loss of sulphur i from a c.p. of monofraction d_{o_k}	kg
m_o	original total mass of c.p.	kg
m_{o_s}	mass of sulphur i that will devolatile	kg
mosip	percentage of sulphur i in c.p. by proximate analysis	-
m_{o_v}	original mass of volatiles in c.p.	kg
m_{t_k}	total mass of the kth monofraction of coal	kg
m_p	percentage of mass contribution of c.p. of a specific diameter	m^{-1}
m_v	mass of volatiles in c.p. during devolatilization	kg
$n_{d_{o_k}}$	number of c.p. of diameter d_o in the kth monofraction	-
p	modelling constant	-
Q_c	constant heat flux	J m^{-2} s^{-1}
q	modelling constant	-
R	resultant force on a flying coal particle	N
R_e	Reynolds number	-
R_k	coal range from maximum particle diameter down to the kth monofraction	-
r	arbitrary radius inside a c.p.	m

r_o	outer radius of c.p.	m
S.E.	specific energy, or combustion value of coal	$J\ kg^{-1}$
$SO_2(FB)$	SO_2 development in the freeboard from the coal particles injected by the feeder	$kg\ J^{-1}$
$SO_2(LW)$	allowed SO_2-emission by law	$kg\ J^{-1}$
T	temperature	K
T_g	temperature of the combustion gases	K
T_{r_o}	temperature of the surface of the c.p.	K
T_w	temperature of furnace wall	K
t	time	s
u	modelling exponent	-
V	volume of devolatized components	$m^3\ kg^{-1}$
v	initial velocity of coal particle	$m\ s^{-1}$
v_g	velocity of combustion gases	$m\ s^{-1}$
v_r	gas velocity relative to coal particle	$m\ s^{-1}$
y_o	elevation of feeder relative to the bed	m
α	inclination angle of v	rad
β_n	auxiliary angle	rad
ε	emissivity	-
ζ	diffusivity	$m^2\ s^{-1}$
η_c	fraction of convectional heat flux	-
η_r	fraction of radial heat flux	-
λ_p	thermal conductance of c.p.	$J\ m^{-1}\ s^{-1}\ K^{-1}$
λ_g	thermal conductance of combustion gas	$J\ m^{-1}\ s^{-1}\ K^{-1}$
μ	dynamic viscosity of the combustion gas	$kg\ m^{-1}\ s^{-1}$
ν	kinematic viscosity of the combustion gas	$m^2\ s^{-1}$
ρ_p	particle density	$kg\ m^{-3}$
σ	Boltzman constant	$J\ s^{-1}\ m^{-2}\ K^{-4}$

THE HANDLING CHARACTERISTICS OF SULFUR OBTAINED FROM WASTE GASES CONTAINING HYDROGEN SULFIDE

D. DOLLIMORE[1], J.A. GODDARD[2], and T. NICKLIN[3]

[1]The Chemistry Dept., The University of Toledo, Toledo, Ohio

[2]The Department of Chemistry and Applied Chemistry, The University of Salford, M54WT, Lancs, U.K.

[3]North Western Gas, British Gas Corporation

ABSTRACT

Hydrogen sulfide can be removed from waste gas streams by the Stretford process. This produces elemental sulfur suspended in solution and the methods and factors influencing its removal are discussed in this study. The sulfur needs flocculation to achieve an economical fast separation rate. The effect of various added materials in the sedimentation is described. The effect of pH on the filtration and sedimentation of the sulfur is outlined. A method of assessing floc size from direct observation of such experiments is established. However, an alternative technique of froth flotation is also reported leading to a determination of the characteristics and mechanism of flotation agents on a sulfur suspension. The studies should assist the handling and treatment of sulfur slurries in an aqueous environment.

INTRODUCTION

Various methods can be used to remove hydrogen from waste gas streams. This report deals with certain aspects of handling sulfur produced from the Stretford process for the removal of hydrogen sulfide from such gas streams. The data would, however, be applicable wherever sulfur is processed in an aqueous environment. The Stretford process uses an alkaline solution of vanadium and anthraquinone disulfonic acid which absorbs and oxidises the hydrogen sulfide producing sulfur. The vanadium which is reduced in this first stage is then reoxidized by air with the anthraquinone disulfonic acid acting as a catalyst.

The process is, therefore, regenerative and the elemental sulfur produced is carried in the solution until its removal can be facilitated. The purpose of this study has been to examine the methods and factors influencing the removal process. The effect of various added materials such as polyacrylamides, m-cresol, p-cresol, 2-5-xylenol and n-decylamine on the sedimentation was therefore investigated. Another effect studied was the influence of pH on the sedimentation and filtration of the sulfur. An alternative technique of separation by flotation is also reported.

HISTORICAL BACKGROUND OF THE PROCESS

Until about 1960, the conventional method of removing hydrogen sulfide was the dry box process in which the gas was passed through a bed of hydrated ferric oxide. Within the iron oxide bed, the hydrogen sulfide reacted with a small amount of oxygen added to the stream and the interstices of the oxide bed became filled with sulfur. When sufficient sulfur had accumulated the oxide bed was used in a pyrite burning plant for the manufacture of sulfuric acid. Other purification processes replaced this labor intensive technique.

One of the early liquid purification systems was the Manchester process (ref. 1), which used an alkaline suspension of ferric hydroxide to absorb the hydroen sulfide. Plant operation revealed several disadvantages, a low loading factor, high power requirements to circulate the heavily laden suspension and blockages owing to the solids produced by the reaction. Power requirements would be minimized in a truly liquid system capable of a high loading factor. The redox cycle of certain dyes and intermediates was known to be suitable and Evans and Stanier (ref. 2) recommended the use of indigo carmine dye. The failure of this and other systems was due to the instability of the dye to prolonged regenerative cycles.

The Stretford process uses anthraquinone disulfonic acid because it was stable and had a suitable regenerative cycle. Worldwide research into the removal of hydrogen sulfide has produced numerous processes some of which such as the Thylox (ref. 3) and the Giammarco-Vetrocoke (ref. 4) systems also produce elemental sulfur. Other liquid systems such as the use of ethanolamine washers absorb the hydrogen sulfide from the gas stream and it is then subsequently desorbed. Further treatment is then necessary to produce elemental sulfur. This is usually performed by the Claus process. A source of sulfur dioxide, is however also required and when streams contaminated only with hydrogen sulfide are to be cleaned they are passed through a combustion chamber with one third of the stoichiometric air required for complete combustion, when one third of the hydrogen sulfide is converted to sulfur dioxide;

$$H_2S + 3/2\ O_2 \rightarrow H_2O + SO_2$$

This sulfur dioxide then reacts with the unconverted hydrogen sulfide to produce sulfur;

$$2H_2S + SO_2 \rightarrow H_2O + S$$

Unfortunately the Claus process allows up to 10% of the gas to pass through untreated and this may contain up to 8% hydrogen sulfide. A process such as the Stretford process may be needed to clean these tail gases.

THE SULPHUR REMOVAL PROCESS

The Stretford process consists of an absorber, oxidiser and some method of sulfur recovery. The plant solution contains principally sodium carbonate/ bicarbonate, sodium metavanadate and anthraquinone 2,7 disulfonic acid. Sodium potassium tartrate is also added to form a loose complex with the vanadium anions eliminating the loss of quadrivalent vanadium which may occur under overload conditions. The hydrogen sulfide is absorbed in the absorber and sulfur is formed in the oxidizer. In practice, the build-up of sulfur in the system is avoided by flotation at the oxidiser. The froth is normally run off into settling tanks where it is allowed to breakdown leaving a slurry of sulfur. This slurry is filtered, typically a rotary vacuum filter could be used. The precipitated sulfur may be water washed to a desired purity. The efficiency of these removal steps is controlled chiefly by the particle size of the sulfur, the larger the particles, the faster any sedimentation or filtration will proceed. The first stage, the flotation of the sulfur is also governed by the charge and adsorbed species. The material investigated here was from the Stretford process, but the findings would apply to any concentrated suspensions of sulfur.

SEDIMENTATION BEHAVIOR

Using Stokes Law relating resistance to motion exerted on a single sedimenting particle in a fluid, the rate of sedimentation (V_s) in a gravitational field is related to the particle radius r and viscosity of the medium η of by:

$$V_s = \frac{2\, r^2\, (P_s\ P_1) g}{9\eta} \tag{1}$$

where P_s and P_1 are the densities of the particle and medium respectively, and g is the acceleration due to gravity. It is generally considered that this equation is accurate provided that the Reynolds number is less than 0.1. Slurries do not obey Stokes Law due to so called concentration effects. Jarret and Haywood (ref. 5) in a series of comparative tests, found that interference between particles caused departure from Stokes Law behavior at particle concentrations greater than 1% by volume, provided the powder was of a non-flocculating character. Burgers (ref. 6) suggested that the motion of a typical particle in a slurry should be influenced by both the motion and the presence of other particles. He obtained the equation (for spherical particles)

$$Q = V_s[1 + (T_1 + T_2)\ C]^{-1} \tag{2}$$

where Q is the rate of sedimentation taken as the rate of fall of the interface between the settling slurry and the supernatent liquid. The constants T_1 and T_2 are valued at 15/8 and 5, respectively, for a dispersion of random particle size distribution and C represents the volume concentration of the solid.

Hawksley (ref. 7) on the basis of theoretical work by Vand derived the equation;

$$Q = V_s(1-C)^2 \exp \frac{2.5C}{1-0.0609C} \tag{3}$$

This derivation only allowed for hydronamic interactions between particles and assumed collisions did not occur.

Loeffler and Roth (ref. 8) formulated the expression;

$$\frac{Q}{V_s} = \left[\frac{1}{1-C} + \frac{5.7C}{(1-C)^3}\right]^{-1} \tag{4}$$

The term $(1-C)^{-1}$ was arbitrarily introduced so that Q/V_s should approach unity as C approached zero which is a logical limiting condition to which many of the equations do not adhere.

Steinour (ref. 9) assumed that the flow space between particles varied but that at different concentrations the particles kept to the same configuration and his final equation takes the form;

$$Q = V_s\ \varepsilon^2\ \phi\ (\varepsilon) \tag{5}$$

or

$$Q = V_s \frac{\varepsilon^3\phi(\varepsilon)}{1-\varepsilon} \tag{6}$$

where

$$\phi(\varepsilon) = \frac{\varepsilon}{1-\varepsilon} \cdot\phi(\varepsilon) \tag{7}$$

In these equations, ε is the initial porosity. The function $\phi(\varepsilon)$ is the modifying function which takes into account the hindered setting. The $\phi(\varepsilon)$ function is directly related to $\phi(\varepsilon)$ by equation (7) and the $(1-\varepsilon)^{-1}$ satisfies the limiting condition. The function $\phi(\varepsilon)$ was supposed by Steinour to represent those effects which are related to the shape of the particles. In papers by Dollimore et al (ref. 10,11) it has been pointed out that $\phi(\varepsilon)$ depends on other factors than shape and an alternative name "hindrance factor" has been proposed. Steinour also observed that an empirical equation of the type;

$$Q = V_s\varepsilon^2\ 10^{-A(1-\ \varepsilon)} \tag{8}$$

fitted his experimental results with the advantage that a single constant A then characterized the system under study.

A further equation proposed by Richardson and Zaki (ref. 12) takes the form;

$$Q = V_s\ \varepsilon^n \tag{9}$$

where n is dimensionless and this is mathematically more tractable then Steinours equation.

An empirical equation used by Dollimore and McBride (ref. 13) is;

$$Q = V_s\ 10^{-b}\ Ps^{(1-\varepsilon)} \tag{10}$$

where b is a constant and P_s the density of the solid.

The limiting constraint that as ε tends to one, that Q should tend to V_s is not such a problem as might appear. It has been demonstrated generally and for sulfur in particular that sedimentation with a clearly defined settling interface is impracticable at some high value of ε where particles fall unhindered (ref. 10, 14). This ε value for sulfur is around 0.97. Furthermore, in the same publications it was shown that a lower limit of ε exists in these systems because the densest regular packing of spherical particles gives a value of $\varepsilon = 0.2595$ (with a coordination number of 12) and of $\varepsilon = 0.4764$ (with a coordination number of 6 representing primitive cubical packing). Thus, experimental results can be expected only in the range 0.476 to 0.970. In practice, these limits are found to be even more restricted.

EXPERIMENTAL WORK ON STRETFORD SULFUR

Two sulfur systems were used in this series of experiments. Sublimed flowers of sulfur previously dried at 60°C were used as the solid phase in propan-2-ol, dried over a molecular sieve. The sulfur from the Stretford process was supplied as an aqueous slurry containing about 20% sulfur. The liquid was residual plant liquor at a pH of 9.2. The particle size distribution of the sublimed sulfur was obtained from electronic impulse counter measurements, the sample having first been wetted with ethanol and then dispersed in sodium chloride solution containing a small quantity of surfactant. The mean diameter of the sulfur was found to be 76 μm and the sample to have a number of large particles (15% 160μm). To obtain a narrower size range, it was decided to grind the smaple in a ball mill which reduced the mean diameter to 39 μm and the large particles to an acceptable level. Figure 1 shows the particle size distribution for the sublimed sample obtained by the electronic impulse counter method. The density of the sublimed sulfur using propan-2-ol as displacement fluid was 1.86g cm^{-3}. The density of the Stretford sulfur was found to be 1.92 g cm^{-3} which is higher than the quoted value because of occluded plant material.

The general technique for performing the sedimentation experiments has been reported in detail (ref. 15). The sedimentations were carried out at 20 ± 0.1°C in a 250 ml stoppered measuring cylinder using in this case 250 ml of total suspension. The suspension was prepared by adding a known mass of solid to the measuring cylinder containing 50 ml of suspension fluid. This was then made up to 250 ml by further addition of fluid. Figure 2 show typical sedimentation curves for the sublimed sulfur. The data for the two sulfurs is best summarized as plots of linear sedimentation rate against the values of ε,

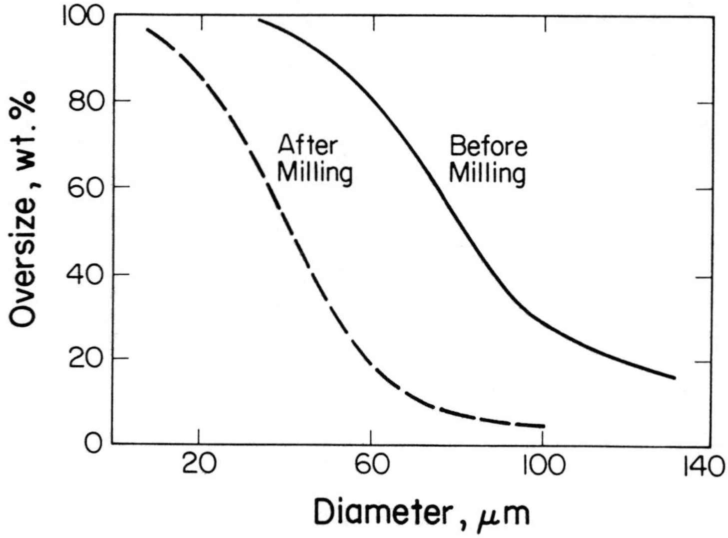

Fig. 1. Particle Size Distribution of Sublimed Sulfur

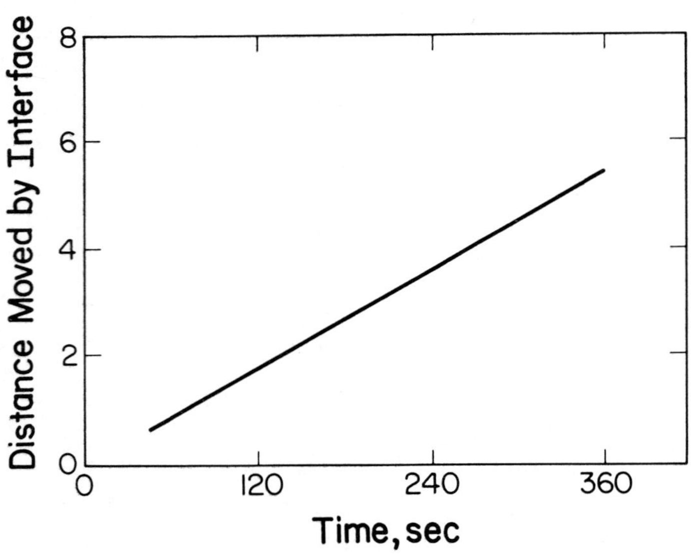

Fig. 2. Sedimentation of Sublimed Sulfur

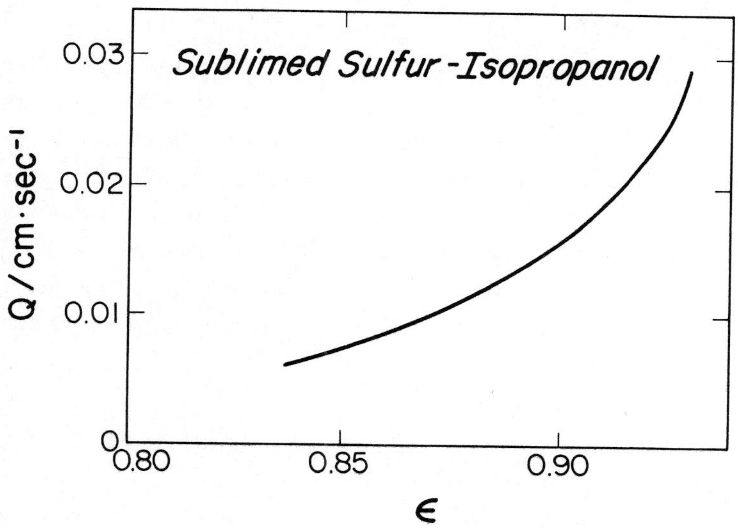

Fig. 3. Sedimentation Rate (Q) as a Function of Voidage (ε)

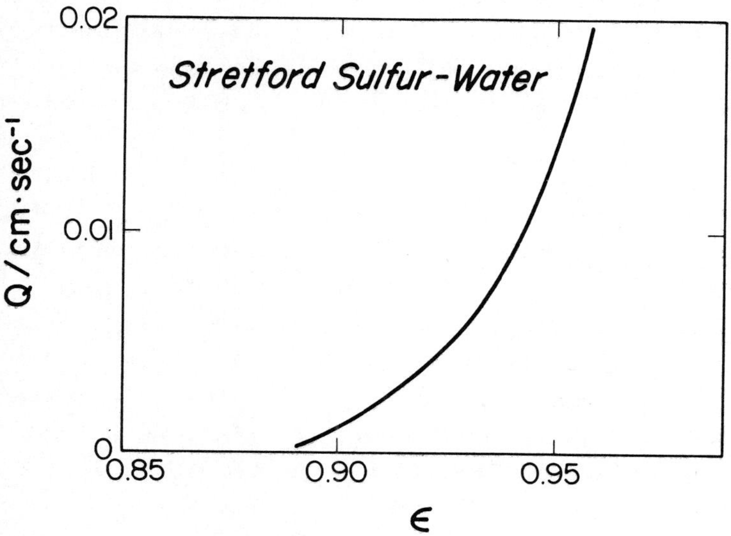

Fig. 4. Sedimentation Rate (Q) as a Function of Voidage (ε)

the initial porosity and these are shown in Figures 3 and 4.

INTERPRETATION OF HINDERED SETTLING DATA ON SULFUR

The hindered settling data used here may be interpreted in terms of modified Stokes Law relationships already developed but can be compared with permeability equations describing the flow of fluid through packed beds. Allison and Muray (ref. 16) described sedimentation behavior by this method using equations developed by Lea and Nurse (ref. 17) and Rigden (ref. 18) but with only partial success. Here a treatment outlined by Davies and Dollimore (ref. 19) is developed further and applied specifically to sulfur sedimentation. The permeability equation utilised is that of Kozeny-Carmen which can be quoted in the form;

$$U = \frac{1}{k}\,\frac{\varepsilon^3}{(1-\varepsilon)}\;\frac{gr^2(\rho s-\rho_1 e)}{9\zeta} \tag{11}$$

where U = apparent fluid flow velocity through the motionless bed and

k = Kozenys constant; originally given a value of 5.0.

In hindered settling, the particles move with a measured velocity Q relative to a fixed horizontal plane through a liquid volume, as compared with the flow of liquid at velocity U through a motionless bed; hence U=Q. Making use of this equality and using equation (1) for the Stokes Limiting velocity allows equation (11) to be written as;

$$Q = \frac{1\ \varepsilon^3}{2k\ (1-\varepsilon)}\ .\ V_s \tag{12}$$

This is the form of the Kozeny-Carmen equation most directly applicable to hindered settling (ref. 20). It has been shown that ε, the initial porosity of the sedimentation experiments can be used in these equations as the actual solid concentraiton of the settling plug does not vary in any single experiment (ref. 21,22,23).

Rearrangement of equation (15) gives;

$$Q(1-\varepsilon) = \frac{1}{2k} \bullet V_s\,\varepsilon^3 \tag{13}$$

The function $Q(1-\varepsilon)$ is of especial importance in plant design as the quantity $Q(1-\varepsilon)\rho$ known as solid flux indicates the mass transfer of solid per unit cross section per unit time down the sedimentation column. Since $1-\varepsilon$ decreases to zero as ε increases to one, a plot of the function $Q(1-\varepsilon)$ against ε will pass through a maximum at some initial porosity ε_1. Figure 5 shows a schematic plot of $Q(1-\varepsilon)$ against ε inducating this feature. This indicates a maximum solid flux at some porosity ε_1. This behavior has been seen in some systems, e.g., glass ballotini in aqueous glycerol (ref. 10) but in many systems only the increasing slope is produced and the maximum of $Q_{(1-\varepsilon)}$ is not observed (ref. 11). This is the case with the sulfur systems reported in this study. The condition for the maximum is found by differentiation and use of the Richardson and Zaki

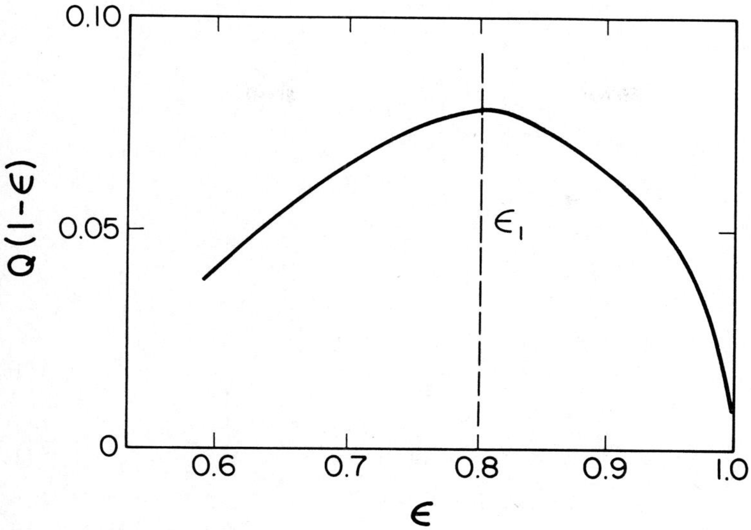

Fig. 5. Schematic Plot of Q (1-ε) Against ε.

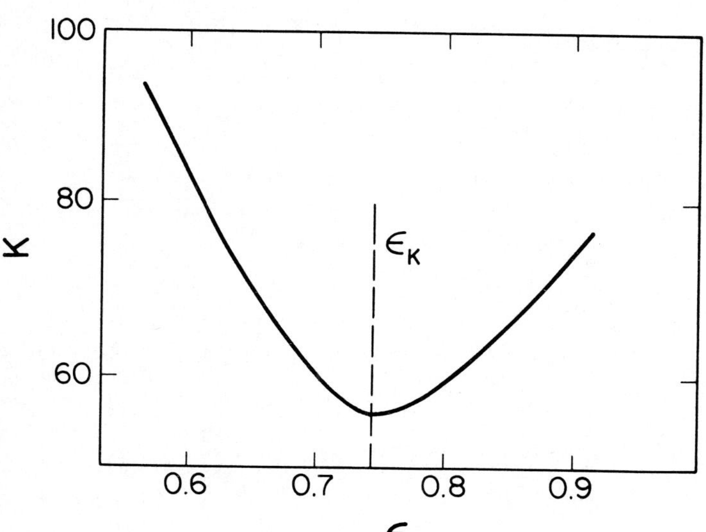

Fig. 6. Schematic Plot of K Against ε

equation;

$$\frac{d(Q(1-\varepsilon)}{d\varepsilon} = \frac{d(Vs\varepsilon^n(1-\varepsilon))}{d\varepsilon}$$

$$= -Vs\varepsilon^n + nVs\varepsilon^{n-1}(1-\varepsilon)$$

At the maximum value of solid flux, $\varepsilon = \varepsilon_1$ and

$$\frac{d(Q(1-\varepsilon))}{d\varepsilon} = 0$$

when

$$\varepsilon_1^{\,n} = n\varepsilon_1^{\,n-1}(1-\varepsilon_1)$$

or

$$n = \frac{\varepsilon 1}{1-\varepsilon_1} \qquad (14)$$

and

$$\varepsilon_1 = \frac{n}{n+1} \qquad (15)$$

which gives a significance to the Richardson and Zaki exponent n not previously realized. Reports by Wason et al (ref. 24) that k decreases as porosity decreases probably result because in their systems the maximum value in $Q(1-\varepsilon)$ was never achieved. From equation (9), due to Richardson and Zaki, we have

$$\frac{Q}{V_s} = \varepsilon n$$

and from equation (12), due to Kozeny-Carmen, we have

$$\frac{Q}{V_s} = \frac{1}{2k}\,\frac{\varepsilon^3}{(1-\varepsilon)}$$

Comparison then gives;

$$k = \frac{1}{2.\varepsilon^{-3}(1-\varepsilon)} \qquad (16)$$

Using various values of n and ε, it can be shown that k in fact goes through a minimum value at a particular value of ε denoted by ε_k for values of n above 3, and that ε_k does not coincide with ε_1. Figure 6 shows a schematic plot of the variation of k with ε.

Differentiation of (16) gives;

$$\frac{dK}{d} = (0.5\varepsilon^{3-n})(1-\varepsilon)^{-2} - 0.5(1-\varepsilon)^{-1}(n-3)\,\varepsilon^{2-n} \qquad (17)$$

The condition for a minimum in k denoted by ε_k is $V_{dk}/d\varepsilon = 0_1$ when

$$\varepsilon_k = \frac{n-3}{1-\varepsilon_k} \qquad (18)$$

Using the relationship from equation (14) gives;

$$n = \frac{\varepsilon_1}{(1-\varepsilon_1)} = \frac{\varepsilon_k}{(1-\varepsilon_k)} + 3 \qquad (19)$$

Rearranging equation (18) gives;

$$\varepsilon_k = \frac{n-3}{n-2} \tag{20}$$

and combining this with equation (14) gives;

$$\varepsilon_k = \frac{4\varepsilon_1 - 3}{3\varepsilon_1 - 2} \tag{21}$$

Table lists values of ε_1, ε_k and n computed from the various equations. The Table shows that as n gets larger the values of ε_k and ε_1 converge. Thus, for values of $\varepsilon_1 > 0.95$, $\varepsilon_1 = \varepsilon_k$. Such systems are highly hindered and k will then vary inversely with ε^3 virtually all the way up to ε_1. Table 2 shows the data for the sulfur systems.

Large values of n represent greatly hindered systems. Other systems are also shown in Table 2 for comparison purposes. It should be noted that for the real systems, ε_1 is higher than 0.75 (see also Table 1). The combination of the Kozeny-Carman and Richardson-Zaki indicates that values of ε_1 below 0.75 are not to be expected. This leads to the prediction that maximum sedimentation mass-transfer rate can never occur at solids concentrations above 25% by volume. The value of k_{min} can be calculated from equation (16) when writing $\varepsilon = \varepsilon_k$)

$$k_{min} = [2\varepsilon_k^{\,n-3}(1-\varepsilon_k)]^{-1} \tag{22}$$

but ε_k is dependent upon n, (equation 20) so k_{min} may be expensed in terms of n only;

$$k_{min} = \left(2\left(\left(\frac{n-3}{n-2}\right)^{-3} - \left(\frac{n-3}{n-2}\right)^{n-2}\right)\right)^{-1} \tag{23}$$

and these values of k_{min} are included in Table 2. It can be seen that the variable value of k can be much greater than the values (~5) commonly quoted. The value for maximum solid flux, ε_1, ε_k and k_{min} are important parameters in plant design and the two sulfurs showed different behavior on sedimentation one showing great hindrence (the Stretford sulfur), the other much less hindrence. Thus in plant design, the actual precipitate formed shows settling behavior dependent upon its pre-history.

OTHER PARAMETERS AFFECTING SULFUR SLURRIES

A brief survey is now made of the other parameters which affect the behavior of sulfur slurries.

Flocculation can be induced by the addition of various additives. With 160mls of stock Stretford slurry containing 20.8g of solid made up to a total of 250 mls at a pH of 9.2 and aged in the presence of flocculant for 24 hours, it was found that an optimum addition for 20% anionic polyacrylamide was around 10 MgL^{-1} and at this concentration an optimum in the filtration rate and the final settled volume was also noted. The flocculation induced by 10% cationic and neutral polyacrylamide was more marked and also occured at a polyacrylamide concentration of $10MgL^{-1}$. The addition of p and o cresols steadily increased the sedimentation rate, the filtration rate and the final settled volume leveling out at a concentraiton of around 1.8g L^{-1}. The addition of 2,5-Xylenol caused similar effects. N-Decylamine, however, showed an optimum flocculation effect at 1g L^{-1}. The effect of suspension pH on the sulfur slurries was to give a maximum sedimentation rate at pH 7 coinciding with a minimum on the filtration rate and a maximum in the settled volume.

The theories for sedimentation outlined in the earlier section can be used to calculate the extent of flocculation expressed as a "floc size." With anionic polymer, the floc size on M varied between 86 and 100 at polymer/solid ratios of polymer/solid ratios of between $0.08x10^{-3}$ to $1.00x10^{-3}$ with the largest floc size at $0.27x10^{-3}$. With cationic polymer, the results were more sharply defined with a floc size of around 190 μM, in other regions of polymer/solid ratio the floc size was around 35 μM. The enhanced effect of the cationic polymer is due to the negative surface charge on the sulfur.

The effect of pH on vacuum flotation experiments showed that p Cresol, m-Cresol and O-Cresol all showed similar effects giving vacuum flotation plots showing a lower critical limit around a concentration value of 0.2g L^{-1} at a pH of about 5 but no upper critical value could be found. The flotation area plot for 2.5 xylenol was broader with a lower critical limit of just below 0.1g L^{-1} extending from a pH of 4 to 7. Again, no apparent critical limit could be found.

REFERENCES

1. J. Powdrill, Inst. Gas Eng. Com., (1959), 551.
2. E.V. Evans and H. Stainer, Brit. Patent (1921) 168504.
3. H.A. Thylox-Gollmar, U.S. Patent, (1929) 1, 177, 719, 777.
4. G. Giammarco, Italian Patent, (1955), 537564.
5. P. Jarret and H. Heywood, Brit. J. Appl. Phys. Sup., 3 (1954) 21.
6. J. Burgers, Verk Akad. Wets. Ams., 44 (1941) 1045, 1177, 44 (1942) 9.
7. P.G.W. Hawksley, Aspects of Fluid Flow, Arnold, London, 1951, 114 pp.
8. A.L. Loeffler and B.F. Ruth, Amer. Inst. Chem. Eng. J., 3 (1957) 293.
9. H.H. Steinour, Indust. Engng. Chem. 36 (1944) 618, 840, 901.
10. J.L. Bhatty, L. Davies, D. Dollimore and A.H. Zahedi, Surface Technology, 15 (1982) 323.
11. L. Davies, D. Dollimore and J.H. Sharp, Powder Technology, 13 (1976) 123.
12. J.F. Richardson and W.N. Zaki, Trans. Inst. Chem. Eng., 32 (1954) 35.
13. D. Dollimore and G.B. McBride, J. Appl. Chem., 18 (1968) 136.
14. D. Dollimore, J.A. Goddard and T. Nicklin, Trans. Instn. Chem. Engrs, 51, (1973) 309.

15. D. Dollimore and T.A. Horridge, Water Research, 6 (1972) 703.
16. E.B. Allison and P. Murray, Trans. Brit. Ceram. Soc., 52 (1953) 204.
17. R.M. Lea and R.W. Nurse, J. Soc. Chem. Ind., 52 (1939) 279.
18. P.J. Rigden, J. Soc. chem. Ind., 66 (1947) 130.
19. L. Davies and D. Dollimroe, J. Phys. Appl. Phys., 13 (1980) 2013.
20. J. Happel, Al Ch. E.J. 4 (1958) 197.
21. A.M. Gavdin and M.C. Fuerstenau, Engng. Mining J., 159 (1958) 110.
22. A.M. Gaudin, M.C. Fuerstenau and S.R. Mitchell, Mining Engng., 11 (1959), 613.
23. J.I. Bhatty, Ph.D. Thesis, University of Salford, 1978.
24. D.T. Wasan, W. Waek, R. Davies, M. Jackson and B.H. Kaye, Powder Tech., 14 (1976) 209.
25. D. Dollimore and R. Kariman, Surface Tech., 17 (1982) 239.
26. J.H. Bhatty, L. Davies and D. Dollimore, Surface Tech., 11 (1980) 269.
27. D. Dollimore and D.L. Griffiths, Proc. Third. Int. Congress of Surface Activity, Cologne, 11B (1960) 674.

Table 1

Values of ε_1, ε_k and n computed from various equations in the text.

ε_1	ε_k	n
0.99	0.99	99
0.95	0.94	19
0.90	0.86	9
0.80	0.50	4
0.75	0.00	3
0.70	-2.00^+	2.33
0.60	-3.00^+	1.50
0.50	-2.00^+	1.00

+ These negative values are a reflection of the fact that below n=3, a k_{min} is not observed (see text).

Table 2

Values of ε_1, ε_k, n and k_{min} for sulfur and other systems

Real System	Ref	n	ε_1	ε_k	k_{min}
Sublimed Sulfur/propanol	this study	15.0	0.94	0.92	17.0
Stretford Sulfur/aqueous	this study	56.0	0.98	0.98	72.9
alumina/water	25	18.2	0.95	0.94	21.3
kaolinite/water	26	88.0	0.99	0.98	46.5
glass-ballotini/ water	26	5.2	0.84	0.69	3.7
coal powder/ water	27	10.1	0.91	0.88	10.3
silica gel/ water	27	16.7	0.94	0.93	19.3

ACOUSTIC STUDY OF COMMERCIAL COAL-WATER FUELS

M. A. BARRETT-GULTEPE, M. E. GULTEPE and E. B. YEAGER

Ultrasonic Research Laboratory, Department of Chemistry, Case Western Reserve University, Cleveland, OH. 44106

ABSTRACT

This paper reports ultrasonic absorption and velocity measurements on two commercial coal-water fuels of average particle diameter 50 and 5 microns. Ultrasonic absorption measurements were made in the frequency range of 1 to 25 MHz for 1°C and 25°C. Velocity measurements were made at 2 MHz in the temperature range of 1°C and 80°C. The general trend found experimentally agrees with that calculated, although the absorption is consistently higher than predicted, even allowing for the averaging effect of a wide particle size distribution.

INTRODUCTION

The use of ultrasonic absorption measurements as a diagnostic tool for monitoring the mass flow, particle size and temperature of coal-water slurries has been suggested by Davis [1],[2]. Although he did not present any experimental data, calculations given by Davis have indicated the promise of this technique.

In the present paper we present some experimental data on ultrasonic absorption and velocity measurements on commercial coal-water fuels. The results are compared with various explicit expressions for heat conduction, viscous drag losses and scattering of sound by particles. Our experimental results also indicate that velocity measurements in a coal-water slurry have the potential of being another possible diagnostic tool along with attenuation measurements.

THEORY

The propagation of sound waves in a medium is attended by periodic variations of pressure and temperature. The propagation of sound waves in a coal-water dispersion results in a shift in the phase and amplitude of the temperature variations within the coal particles and in absorption, due to the fact that heat conduction coefficients of coal and water phase are different. These heat conduction losses are somewhat smaller than that associated with the relative-motion losses caused by differences in the amplitude and phase of the coal particle movement relative to the surrounding medium. The redistribution of the acoustic energy due to scattering and by multiple scattering effects must be taken into account. There are also other relaxation effects involving the adsorbed surface active layer during the propagation of sound through the coal-water slurry. The sound absorption from these effects is believed to be generally small. Further, in coal-water slurries there is formation of surface complexes (i.e., specific adsorption of an ion of the supporting electrolyte) in addition to the association-dissociation of counter ions with the surface sites. We believe that these processes will be accompanied by uncompensated volume changes and therefore will be sensitive to the periodic pressure variations in the sound field. On the basis of the anticipated volume changes, one expects some excess absorption in a coal-water slurry. However, the magnitude of this excess absorption is not known yet. This is partially due to the fact that the surface of coal particles is very complex, and it is difficult to identify a suitable model.

The absorption and dispersion of sound waves due to heat conduction and viscous drag losses are very sensitive to the diameter of coal particles. Measurements of the absorption coefficient and the velocity of sound in a concentrated slurry as a function of frequency will provide information about its particle size distribution. The time dependence study of the particle size distribution will give the flocculation rate of the coal particles.

The sound velocity measurements may also provide the "surface bulk" compressibility of the adsorbed surface active agent at the coal-water interface. The compressibility of an adsorbed surface active agent can provide important information towards a better understanding of the stability of the interface when its value is compared with the compressibility of the molecule in the bulk solution. For example, strongly adsorbed surface active agents tend to be stiffer than weakly adsorbed molecules [3].

Since the density difference between coal particles and water is considerable and the particle size distribution of a commercial coal-water fuel is broad, we primarily consider the viscous drag losses and scattering of sound by particles. The heat conduction losses was calculated using an explicit approximate expression.

The complex and frequency dependent propagation constant k_s of sound waves in a dispersion of particles is given [4] by the following relation:

$$(k_s/k)^2 = \left[1 + \sum_j \left(\frac{4\pi n_j}{ik^3}\right) * \sum_{n=o}^{\infty} (-1)^n a_{2n,j} \right] * \left[1 - \sum_j \left(\frac{4\pi n_j}{k^3}\right) * \sum_{n=o}^{\infty} (-1)^n a_{2n+1,j} \right] \quad (1)$$

where n_j is the number of particles of type j per unit volume, and $k = \omega/c$

For compressible particles the coefficient a_l is given by [5]:

$$a_1 = -\frac{1}{3}k^3R^3\frac{(\rho'/\rho - 1)(1+\tau+is)}{(\rho'/\rho+\tau+is)} \tag{2}$$

where τ and s are given by

$$\tau = \frac{1}{2} + \frac{\delta}{4R}\frac{(2\eta+3\eta')^2}{(\eta+\eta'+R\eta/3\delta)^2+(R\eta/3\delta)^2} \tag{3}$$

and

$$s = \frac{3}{4}(2\eta+3\eta')(\delta/R)^2 * \frac{(1+R/\delta)[(\eta+\eta')+R\eta/3\delta] + 1/3\eta(R+\delta)^2}{(\eta+\eta'+R\eta/3\delta)^2+(R\eta/3\delta)^2} \tag{4}$$

where $\delta[=(2\eta/\rho\omega)^{1/2}]$ is the depth of penetration for shear waves, η is shear viscosity, ρ is the density, ω is the angular velocity and R is the radius of the particle. Primes symbols refer to particles. For $\eta'/\eta \to 0$, the eqs. (3) and (4) reduce to those for rigid particles, originally given by Lamb [6].

The thermal conduction losses are accounted for by the coefficient, a_o. The coefficient a_o can be calculated using the following expression [7]. For a particular case, for which the ratio of the thermal conductivities of the water and particles $\sigma/\sigma' \to 0$, approaches to zero.

$$a_o = -\frac{1}{3}ik^3R^3(1-\beta'/\beta) - \frac{1}{2}k^3R^3(\gamma-1)*(1-\alpha'\rho C_p/\alpha\rho' C_p')^2*(\delta_t/R)*[(1+\delta_t/R)-i] \tag{5}$$

where β is the compressibility, C_p is the specific heat at constant pressure, γ is the adiabatic constant, α is the coefficient of thermal expansion, $\delta_t = (2\sigma/\rho C_p\omega)^{1/2}$ is the thermal depth of penetration and σ is the thermal conductivity.

The attenuation due to scattering can be calculated using the following expression:

$$\alpha_s = \frac{1}{2}k_c^4 R^3\phi[\frac{1}{3}(\frac{\beta-\beta'}{\beta})^2 + (\frac{\rho'-\rho}{2\rho'+\rho})^2] \tag{6}$$

This expression given by Allegra and Hawley [8] for fluid particles was used by Davis [1] for his theoretical calculations for the prediction of scattering losses in coal-water slurries, (see ref. [1], eq. (2)).

Finally writing the following equation for the propagation constant k_s,

$$k_s = \frac{\omega}{C_s} + i\alpha \tag{7}$$

where C_s is the velocity in dispersion and α is the absorption coefficient.

Comparing eq. (7) with eq. (1) and substituting the values of a_0 [eq.(5)] and a_1 [eq.(2)] we can calculate the absorption coefficient and velocity of sound.

EXPERIMENTAL

Ultrasonic absorption measurements were made using the two-transducer VHF pulse echo technique in the frequency range of 1 to 25 MHz with apparatus in our laboratory. A set of matched 1 MHz X-cut quartz crystals were used as transmitting and receiving transducers. The direct received pulse, after being amplified and demodulated is digitized and displayed by the HP5180A waveform recorder. The first cursor is located on the base line of the trace. The second cursor is located on the middle of the flat top of the direct pulse. With the aid of a computer program, a set of voltage readings were obtained on both sides of this particular point. The waveform digitizer records the voltage differences between these two cursor points with 50 ns increments of time. Generally a total of 50 readings of 10.5 bit length were taken and averaged for each transmitting and receiving crystal separation. The amplitude and the distance information is collected in the HP85 microcomputer. The natural logarithm of the amplitude is plotted against distance in centimeters. From the slope of this straight line, the absorption coefficient is calculated. Since the attenuation is very high, the correction for diffraction and waveguide losses were not necessary. Thermostated water from a temperature bath was circulated through the thermostating jacket of the measurement cell and provided control to 0.05^{o}C.

Velocity measurements were made at 2 MHz with an ultrasonic interferometer. The driving signal was obtained from a crystal-controlled oscillator. Standing waves were established between the driving transducer and the traveling reflector in a cylindrical cell holding 10 ml sample. The wavelength of the sound waves was determined by observing the distance of travel of the reflector between a number of successive minima of the driving current of the transducer, up to a separation of 2.5 cm. The interferometer was thermostated so that during the measurements the temperature was kept constant to within 0.02^{o}C over the temperature range 1 to 25°C and within 0.1^{o}C up to 80°C. Due to the very high absorption of sound in coal slurries, most of the velocity measurements were made in the near field region of the present transducer. No diffraction and waveguide corrections were made to the measured velocity values since they would be insignificant.

The coal-water fuel samples were provided by Babcock and Wilcox Research and Development Division, Alliance, OH. The physical properties of coal-water fuels used in this work is tabulated in Table I.

TABLE 1

Physical properties of commercial coal-water fuels.

Coal-water slurry	A (Lower Cedar Grove)	B (Yard Coal)
Mass fraction	0.675	0.50
Viscosity	620 cp at 100 s^{-1}	---
pH	8.3	9.71
Mean diameter (mass)	49.6×10^{-4} cm	5.26×10^{-4} cm
Calc. surface area	1.25 m^2/cm^3	2.96 m^2/cm^3

RESULTS AND DISCUSSION

Sound attenuation measurements were made on two coal-water slurries with differing particle size distribution, designated A and B (Lower Cedar Grove and Yard Coal respectively) over the frequency range 1 to 25 MHz. for two temperatures. The results are plotted in Figs. (1) and (2) for 1°C along with the explicit theoretical expressions for viscous drag [eq. (2)], heat conduction losses [eq. (5)] and scattering contributions [eq. (6)]. The sum of these three equations yields a minimum determined largely by the radius of the particles.

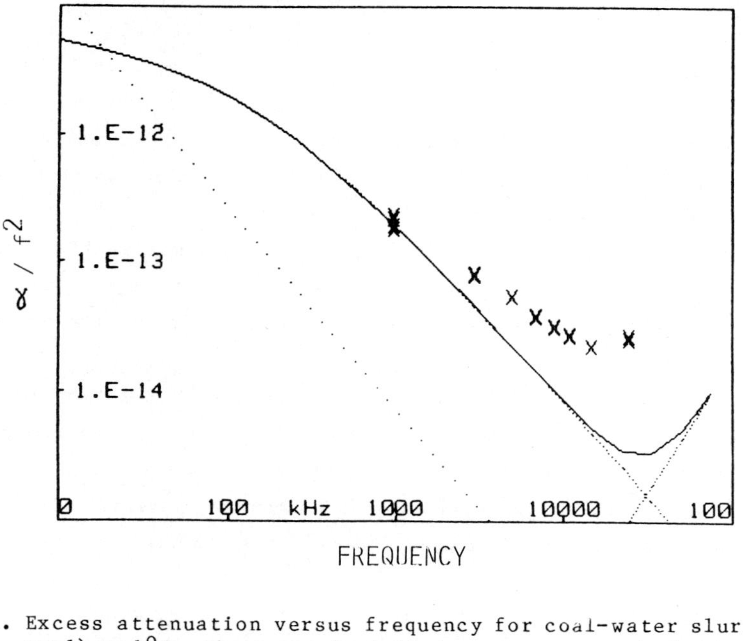

Fig. 1. Excess attenuation versus frequency for coal-water slurry B (yard coal) at 1°C. The sparsely dotted curve represents heat conduction losses. The observed excess attenuation is denoted by the symbol X.

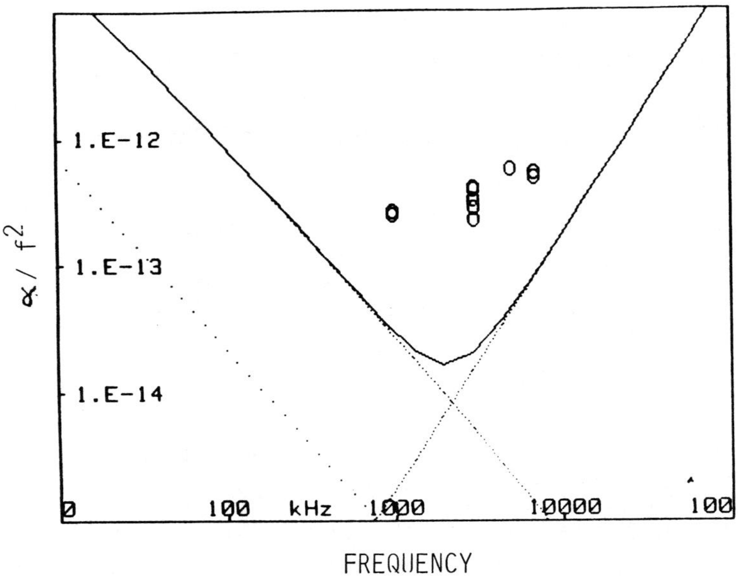

Fig. 2. Excess attenuation versus frequency for coal-water slurry A (Lower Cedar Grove) at 1°C. The sparsely dotted curve represents heat conduction losses. The observed excess attenuation is denoted by the symbol 0.

For a polydispersed system this minimum is rounded and the shape of this curve is determined by the volume average of various particle sizes in the slurry. This is indeed the case for the slurry A. The theoretical curve shown is calculated for the average particle size for this slurry, (50μ). Although the experimental curve is considerably flattened in comparison with the theoretical curve, it is clear that the minimum lies below 1 MHz as predicted.

The attenuation coefficient of ultrasound in slurry A was so high that no meaningful measurements were possible above 7 MHz.

The slurry B with average particle size 5 μ and a considerably narrower particle size distribution was studied in a wider frequency range, 1 to 25 MHz, at 1°C. The minimum predicted by the eqs. (6) and (2) is in fair agreement with the experimental points.

The velocity of sound at 2 MHz in a coal slurry A and B together with that for water as function of temperature is plotted in Fig. (3). Although the velocity of sound varies with the mass fraction of the slurry, the large velocity difference between two coal samples cannot be explained due to this effect. We believe that this is due to the particle size difference between two slurries.

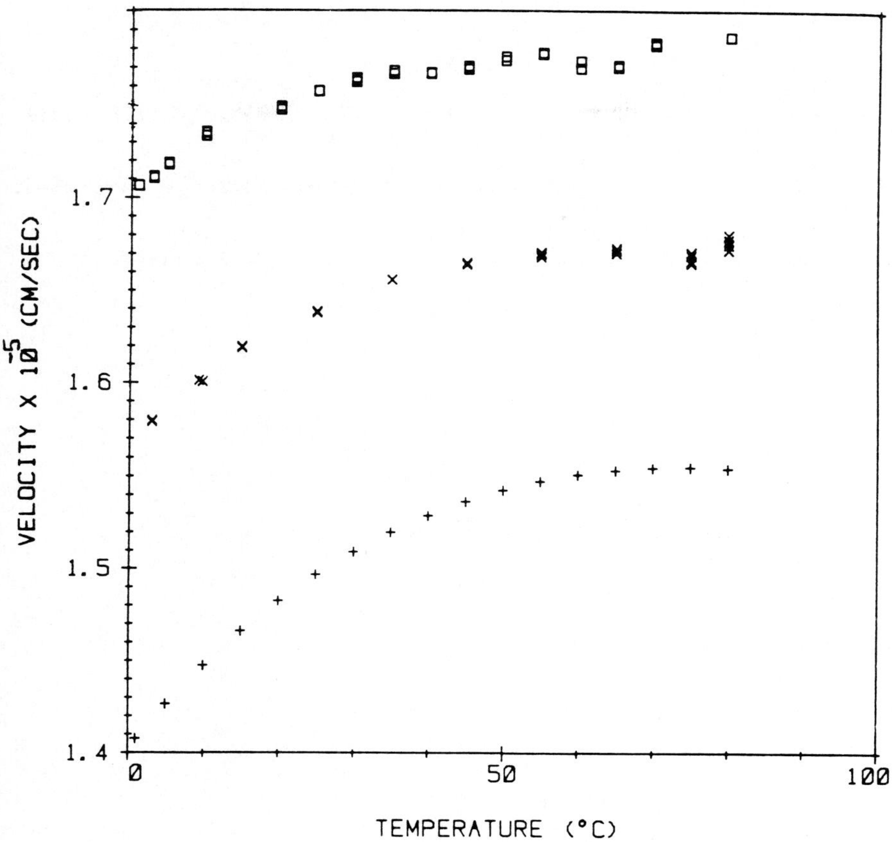

Fig. 3. The plot of the velocity of sound at 2 MHz in coal-water fuels A and B along with the water as function of temperature.

□ - coal-water fuel A
X - coal-water fuel B
+ - water

In conclusion the general trends found experimentally agree with those calculated. The absorption is consistently higher than predicted, even allowing for the averaging effect of a wide particle size distribution.

ACKNOWLEDGEMENT

The authors are indebted to S. J. Vecci, Fuel Technology Group Supervisor, Research and Development Division of Babcock and Wilcox, Alliance, OH, for supplying coal-water fuel samples and the information pertaining to them.

REFERENCES

1 M.C. Davis, J. Acoust. Soc. Am., 64 (1978) 406.
2 M.C. Davis, ibid 65 (1979) 387.
3 M.A. Gultepe, M.E. Gultepe and E. B. Yeager, J. Phys. Chem., 87 (1983) 1039.
4 R.J. Urick and W.S. Ament, J. Acoust. Soc. Am., 21 (1949) 115.
5 A.S. Ahuja, ibid 51 (1970) 182.
6 H. Lamb, "Hydrodynamics", 6th ed., Cambridge University Press, Cambridge, 1932, 657 pp.
7 A.S. Ahuja, J. Appl. Phys. 44 (1973) 4863.
8 J. R. Allegra and S. A. Hawley, J. Acoust. Soc. Am., 51 (1972) 1545.

PART 7

SULPHUR CAPTURE DURING COAL COMBUSTION

COMPARISON OF SULFUR ABSORPTION CAPABILITIES OF PURE LIME AND LIMESTONE IN COAL COMBUSTION ZONE AND FLUE GAS DESULFURIZATION SYSTEMS

M. H. ERTEN

Mining Engineering Dept., University of Missouri at Rolla,

Rolla, MO 65401 (USA)

ABSTRACT

In this research, the sulfur absorption capabilities of pure lime and limestone are investigated when used in combustion zone and flue gas desulfurization methods.

The combustion zone desulfurization tests conducted with coals containing 2.10 and 4.90 percent total sulfur showed that at combustion temperatures ranging between 800°C and 1100°C, lime (CaO) is a more effective sulfur absorbing agent than limestone ($CaCO_3$) when used with the same stoichiometric ratio of lime or limestone to sulfur in the coal.

The flue gas desulfurization of combustion gases given off from the above coal samples showed a higher overall sulfur absorption efficiency with lime than with limestone when the same sorbent-water suspension concentrations were used.

The tests also showed that the flue gas desulfurization method has a greater effectiveness than the combustion zone desulfurization when lime and limestone are used as SO_2 absorbing agents.

INTRODUCTION

The aim of this research was to compare the sulfur absorption capabilities of pure lime and limestone used in the combustion zone desulfurization and flue gas desulfurization systems. Industrially, these two methods are applied in the fluidized bed combustion and pulverized coal burning furnaces, respectively. However, it should be noted that in industrial furnaces there is vigorous particulate motion (Ref. 6), whereas during our experiments the coal and sorbent particles were motionless, except that they were thoroughly premixed before weighing into the charge crucibles.

The high and medium sulfur coal samples used in the experiments were taken from the Bevier seam (ORC-19) mined in the Randolph County of the State of Missouri, and from a core sample (DH-1) obtained from an unidentified coal seam found in the Dade County of the same state. The samples had the proximate analyses given in Table 1 (Refs. 1 and 2). As can be seen from Table 1, the samples chosen contained 2.10 and 4.90 percent total sulfurs. Thus, they represented medium to high sulfur coals.

TABLE 1

Proximate analysis of the coal samples tested.

Sample No	Name of Seam	% Moist. (Total)	% Ash (Dry)	% S (Dry)	% Vol. Matter (Dry)	Btu Value (Dry)
DH-1	Unknown	1.69	7.76	2.10	34.02	13,735
ORC-19	Bevier	12.67	13.72	4.90	37.35	11,349

The forms of sulfur in these two samples are given in Table 2 (Refs. 1 and 2)

TABLE 2

Forms of sulfur in samples tested.

Sample No	% Total S	% Sulfate S	% Pyritic S	% Organic S
DH-1	2.10	0.01	1.20	0.89
ORC-19	4.90	0.45	2.54	1.91

As it is seen from Table 2, both samples contained high organic sulfurs that cannot be separated from these coals by physical cleaning methods.

The limestone used in the tests consisted of laboratory grade $CaCO_3$ powder. The lime (CaO) was obtained by calcining the laboratory grade $CaCO_3$ at 950°C. The yield of CaO from $CaCO_3$ was 58%. The average particle size of $CaCO_3$ powder was 4.2 microns, determined with a Fisher Sub-sieve Sizer.

A modified Leco sulfur analyzer was used to conduct the combustion zone and flue gas desulfurization tests. Figure 1 shows the test setup used during the flue gas desulfurization. For combustion zone desulfurization tests, the Leco combustion tube was directly connected to the titrator with a flexible nylon tubing. The titrator contained 1.15% H_2O solution (by volume) which reacted with the SO_2 gas coming from the combustion tube to form H_2SO_4.

The H_2SO_4 produced in the titrator is titrated with 0.05N NaOH solution in the presence of 10 drops of methyl purple indicator solution. The percentage of sulfur contained in the coal is read directly from the buret of the titrator.

EXPERIMENTAL PROCEDURES

Combustion zone desulfurization

Leco sulfur analyzer shown in Figure 1 was used not only to burn the coal and $CaCO_3$ or CaO mixtures in an oxygen atmosphere but also to determine the SO_2 emitted with the combustion gases. Since the Leco sulfur analyzer is designed to determine directly the percentage of total sulfur present in 0.500 gram coal sample, this weight of coal was used as standart for all the combustion tests.

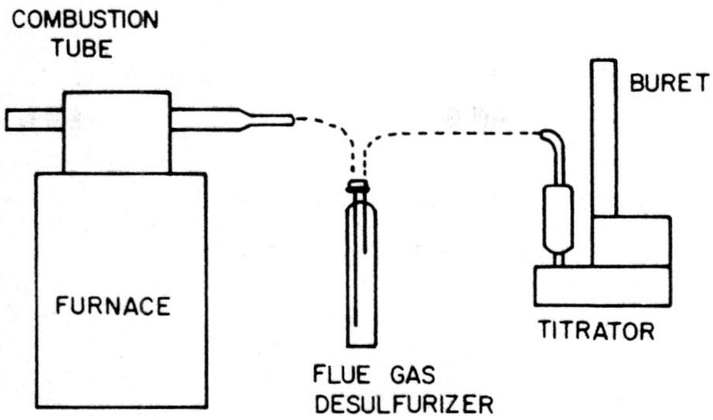

Fig. 1. Modified Leco sulfur analyzer.

The temperatures of combustion were chosen to cover the ranges used in the fluidized bed combustion (Ref. 4) and the lower temperature limits of conventional pulverized coal combustion systems (Ref. 3). The percentage of $CaCO_3$ or CaO addition to the half-gram coal charge depended on the total sulfur content of the coal tested and the choice of the stoichiometric ratio of sorbent to sulfur in the coal.

The stoichiometric ratio of $CaCO_3$ or CaO to total sulfur in the coal was determined as follows. Assuming all the sulfur in a coal sample is converted into SO_2 gas during the combustion of the coal, the stoichiometric ratio of pure limestone or lime to sulfur in the coal charge can be calculated by the following equations.

$$\underset{100}{CaCO_3} + \underset{32\ (S)}{SO_2} + \tfrac{1}{2}O_2 = CaSO_4 + CO_2; \quad \frac{CaCO_3}{S} = \frac{100}{32} = 3.125 \qquad (1)$$

$$\underset{56}{CaO} + \underset{32(S)}{SO_2} + \tfrac{1}{2}O_2 = CaSO_4; \quad \frac{CaO}{S} = \frac{56}{32} = 1.75 \qquad (2)$$

These calculations meant that for a stoichiometric ratio of one, we need to use 3.125 milligrams of $CaCO_3$ or 1.75 milligrams of CaO for each milligram of total sulfur contained in the coal. To find the $CaCO_3$ or CaO requirements for any other stoichiometric ratio, these factors are multiplied by the desired numbers. Table 3 shows the relations between the stoichiometric ratios and the percentages of $CaCO_3$ or CaO in the combustion test mixtures of the ORC-19 coal sample containing 4.90% or 49.0 milligram of sulfur per gram of coal.

TABLE 3

Stoichiometric ratios vs. percentages of $CaCO_3$ and CaO in the combustion zone test mixtures of ORC-19 coal sample containing 4.90 percent sulfur.

Stoichiometric Ratios	Milligrams of $CaCO_3$/0.5g Coal	% $CaCO_3$ in charge	Milligrams of CaO/0.5g Coal	% CaO in charge
1.0	76.56	13.28	42.88	7.90
2.0	153.13	23.45	85.75	14.64
3.0	229.69	31.48	128.63	20.46
4.0	306.25	37.98	171.50	25.54
5.0	382.81	43.36	214.38	30.00

For the medium suflur DH-1 sample, similar calculations were performed in order to determine the percentages of $CaCO_3$ and CaO for charges with varying stoichiometric ratios. For the same stoichiometric ratios, the percentages of $CaCO_3$ or CaO were increased as the sulfur content of the sample increased.

Before deciding the percentage of $CaCO_3$ or CaO in a coal charge to be tested at a given temperature, the maximum uncontrolled SO_2 emission per million Btu of the coal was taken into consideration. For example, the ORC-19 sample contained 4.90% sulfur and had a calorific value of 11,349 Btu/lb (dry basis) determined with a Parr Oxygen Bomb Calorimeter. According to this calorific value, to generate 1,000,000 Btu, it was necessary to use 1,000,000 ÷ 11,349 = 88.11 pounds of this coal that would contain 88.11x0.049 = 4.32 pounds of total sulfur. Since each pound of sulfur produced two pounds of SO_2 upon complete burning, then this coal would produce 4.32x2 = 8.64 pounds of SO_2 per million Btu. This SO_2 emission is 7.2 times greater than the amount presently allowed by the EPA Emissions Standards for a coal of such sulfur content.

During the combustion zone desulfurization tests, the SO_2 not absorbed by the sorbents used reported to the titrator of the Leco sulfur analyzer and was analyzed as the percentage of sulfur emission from the coal tested. This percentage of sulfur can be converted into pounds of SO_2 emission per million Btu by the following simple calculation. For example, if burning of 0.500 gram of ORC-19 sample with a known percentage of CaO at a given temperature resulted in 1.0% sulfur determination in the titrator of the Leco sulfur analyzer, then the equivalent pounds of SO_2 emission per million Btu was found to be (8.64÷4.90)100 = 1.76, which is still in excess of the 1.20 lb. $SO_2/10^6$ Btu limit set by the EPA Emissions Standards for a coal of 4.90% sulfur.

According to its sulfur and Btu contents, the maximum uncontrolled emission rate from the medium sulfur DH-1 sample used in the tests was calculated to be 3.06 lb. $SO_2/10^6$ Btu.

In the combustion zone desulfurization tests, both samples, i.e., ORC-19 and DH-1 containing 4.90 and 2.10 percent sulfurs respectively, were tested at temperatures ranging from 800°C to 1400°C with the use of stoichiometric ratios of $CaCO_3$ and CaO to sulfur in the coals between one and five. A stoichiometric ratio of one represents the theoretical amount of $CaCO_3$ or CaO required in order to arrest any SO_2 generated from the sulfur during the combustion process. However, some excess $CaCO_3$ or CaO is necessary in the combustion zone because the $CaSO_4$ formed during the reaction of SO_2 and CaO plugs the pores of unreacted sorbent, thus making it inaccessible to react with all the available SO_2 for the reaction (Ref. 3&5). The use of stoichiometric ratios other than one showed the effects of excessive $CaCO_3$ and CaO in arresting SO_2 emission from the combustion zone.

Flue gas desulfurization (FGD)

Figure 1 shows a line diagram of the apparatus used for flue gas desulfurization tests. In this figure, "the flue gas desulfurizer" bottle contained a 0.25% $CaCO_3$ or CaO and distilled water suspension that absorbed most of the SO_2 coming out of the combustion tube during the burning of the coal sample. The suspension concentration was chosen arbitrarily at such a level that the number of ½-gram high and medium coal samples during the FGD tests would not be excessive, yet it would allow enough number of test runs during each test series. The 0.25% solid-water suspensions contained 500 mg of CaO or $CaCO_3$ per 200 ml of distilled water.

During the FGD tests, each coal sample was burned at 1400°C for 6 minutes in an oxygen atmosphere with a flow rate of 1.5 liters/minute, and there was plenty of oxygen for the formation of calcium sulfate during the bubbling of combustion gases through the desulfurizing suspension. Any SO_2 or SO_3 that is not arrested by the flue gas desulfurizer was recorded by the "titrator" of the system as percentage of sulfur emitted from the original coal burned. Then, this percentage was converted into pounds of $SO_2/10^6$ Btu emission in the same manner as was explained in the previous section.

Each suspension containing 0.25% solids was used until a sample showed an SO_2 emission rate equal or above the maximum allowable by the EPA Emissions Standards. For ORC-19 sample containing 4.90% sulfur and having an uncontrolled free emission rate of 8.64 lb. $SO_2/10^6$ Btu, the maximum allowable emission rate was 1.20 lb. $SO_2/10^6$ Btu. On the other hand, for DH-1 sample containing 2.10% sulfur and having an uncontrolled free emission rate of 3.06 lb. $SO_2/10^6$ Btu, the maximum allowable emission rate was 0.60 lb. $SO_2/10^6$ Btu (Ref. 3). When these figures were translated into percentages of sulfur emissions recorded at the titrator, the maximum allowable limits were found to be 0.68% and 0.41% for any individual test runs for ORC-19 and DH-1 samples, respectively.

The Leco sulfur analyzer is designed to read directly the percentage of sulfur contained in 0.500 gram of coal when burned at 1350-1400°C. Therefore, in order not to exceed the design limit of the apparatus, each series of FGD tests were conducted with a number of 0.500 gram samples containing medium and high sulfur coals. For example, in order to determine the desulfurizing power of 0.25% CaO and water suspension, first, a number of 0.500 gram high sulfur coal samples (4.90%S) were burned in the combustion tube and the gaseous products were swept through the FGD suspension and the titrator of the analyzer. The SO_2 content of the gases leaving the titrator was recorded as percentage of sulfur emission from the original 0.500 gram coal sample. As the number of high sulfur coal samples burned increased, the SO_2 absorbing power of the 0.25% CaO suspension started to decrease and, therefore, the sulfur percentage reading in the titrator began to increase. For ORC-19 sample, when this reading exceeded 0.68% sulfur, it was the end of FGD test with the suspension used. In calculating the total sulfur absorbed by the 0.25% CaO suspension, any sample exceeding the 0.68% limit was not included. For DH-1 sample, the end of FGD test was reached when 0.41% sulfur was recorded by the titrator.

The same procedure was also followed for determining the effectiveness of 0.25% $CaCO_3$ suspension in absorbing the SO_2 produced by burning the same two coal samples in the manner described.

The total sulfur absorbed by each suspension was calculated by subtracting the total sulfur recorded in the titrator from the total sulfur present in the original coal samples that were used for each series of tests. The sorbent utilization efficiency was determined by calculating the sorbent required for combining with the amount of SO_2 absorbed and dividing the result by the total weight of CaO or $CaCO_3$ in the 0.25% suspension. For each unit weight of sulfur present in the coal, the minimum weights of CaO and $CaCO_3$ were 1.75 and 3.125, respectively. However, as the test results show, the CaO or $CaCO_3$ utilization was never 100%. This may be due to the inaccessibility of part of the sorbent to react with SO_2 because of the formation of $CaSO_4$ over the particles.

TEST RESULTS

Combustion zone desulfurization test results

Use of $CaCO_3$. Laboratory scale tests showed that combustion zone temperature and the stoichiometric ratios of $CaCO_3$ to sulfur in the coal are two of the important factors in combustion zone desulfurization. The accepted reaction for sulfate formation in the combustion zone is given below.

$$CaCO_3 + SO_2 + \tfrac{1}{2}O_2 \rightleftarrows CaCO_4 + CO_2.$$

It is generally believed that (Ref. 3) at temperatures beyond 1000°C, reverse reaction will dominate, leaving much of the SO_2 unreacted and free to escape. At lower temperatures, the retention of SO_2 by the $CaCO_3$ is favored. However, as can be seen from Table 4, for ORC-19 sample, between the temperatures of 800°C and 1400°C, as the stoichiometric ratio of $CaCO_3$ to sulfur in the coal increased, the percentages of sulfur emissions from the same coal decreased. At every temperature tested, the decrease in sulfur emission between the stoichiometric ratio of one and two was much greater than between the ratios of two and three, three and four, or four and five. In other words, after 100 percent excess $CaCO_3$ was used, further increase was not very effective in stopping the sulfur emission from high sulfur coals.

The information given in Table 4 represents the percentages of sulfur emissions from ORC-19 sample (4.90%S) at different temperatures and stoichiometric ratios. In Table 5, this data is converted into pounds of SO_2 emissions per million Btu of the same coal. Each test result given in Table 4 and in all the succeeding tables are the averages of at least two tests under the same test conditions.

TABLE 4
Effects of $CaCO_3$ addition and temperature on the combustion zone desulfurization of sample ORC-19 containing 4.90% total sulfur.

$CaCO_3$:S	% Sulfur emission at various temperatures (°C)							
Ratio	800	840	900	1000	1100	1200	1300	1400
0	4.44	4.62	4.66	4.64	4.69	4.81	4.85	4.90
1	2.37	2.72	2.54	2.52	3.66	4.02	4.50	4.62
2	1.75	1.80	1.58	1.48	1.66	1.86	2.66	3.56
3	1.21	1.48	1.34	1.24	1.34	1.40	1.86	2.84
4	1.18	1.20	1.16	1.02	1.10	1.16	1.56	2.58
5	0.98	1.08	1.06	0.88	0.90	1.00	1.30	2.54

The data in Table 4 show that between 900°C and 1000°C the percent sulfur emissions show a general decrease for most stoichiometric ratios. This may be due to better calcination of $CaCO_3$ and the availability of more CaO for sulfate formation at this temperature range.

TABLE 5
Lb. $SO_2/10^6$ Btu emissions from sample ORC-19 at various temperatures and stoichiometric ratios of $CaCO_3$ to total sulfur in the coal.

$CaCO_3$:S	Pounds of SO_2 emissions per million Btu at various temperatures (°C)							
Ratio	800	840	900	1000	1100	1200	1300	1400
0	7.83	8.15	8.22	8.18	8.27	8.48	8.55	8.64
1	5.62	4.80	4.48	4.44	6.45	7.09	7.93	8.15
2	3.06	3.17	2.79	2.61	2.93	3.28	4.69	6.28
3	2.13	2.61	2.36	2.19	2.36	2.47	3.28	5.01
4	2.08	2.12	2.05	1.80	1.94	2.05	2.43	4.55
5	1.73	1.90	1.87	1.55	1.59	1.76	2.29	4.48

As can be seen from Table 5, in every case the lb. $SO_2/10^6$ Btu emission is higher than the EPA limit of 1.20 lb. SO_2/Btu for a coal containing 4.90% sulfur. As was mentioned before, this coal would have an SO_2 emission of 8.64 lb./10^6 Btu when burned freely without any sorbent addition.

To see the effect of lower sulfur content of the coal, the combustion zone experiments with the use of $CaCO_3$ were also conducted with DH-1 sample containing 2.10% sulfur. The test results are given in Tables 6 and 7.

TABLE 6
Effect of $CaCO_3$ addition and temperature on the combustion zone desulfurization of sample DH-1 containing 2.10% total sulfur.

$CaCO_3$:S	% Sulfur emissions at various temperatures (°C)							
Ratio	800	840	900	1000	1100	1200	1300	1400
0	1.80	1.82	1.80	1.82	1.84	1.98	2.00	2.10
1	1.17	1.11	1.18	1.28	1.60	1.96	2.00	1.92
2	0.76	0.79	0.78	0.76	1.44	1.48	1.88	1.68
3	0.61	0.63	0.64	0.68	0.88	1.12	1.30	1.34
4	0.48	0.51	0.56	0.54	0.64	1.04	1.16	1.22
5	0.45	0.43	0.48	0.58	0.72	0.80	1.24	1.08

TABLE 7
Lb. $SO_2/10^6$ Btu emissions from sample DH-1 at various temperatures and stoichiometric ratios of $CaCO_3$ to total sulfur in the coal.

$CaCO_3$:S	Pounds of SO_2 emissions per million Btu at various temperatures (°C)							
Ratio	800	840	900	1000	1100	1200	1300	1400
0	2.62	2.65	2.62	2.65	2.68	2.88	2.91	3.06
1	1.70	1.62	1.72	1.86	2.33	2.85	2.91	2.80
2	1.11	1.15	1.14	1.11	2.10	2.15	2.74	2.45
3	0.89	0.92	0.93	0.99	1.28	1.63	1.89	1.95
4	0.70	0.74	0.82	0.77	0.93	1.51	1.69	1.78
5	0.66	0.63	0.70	0.84	1.05	1.16	1.81	1.57

According to New Source Performance Standards (NSPS) (1977), for coals which would give an uncontrolled emission rate between 2 and 6 lb. $SO_2/10^6$ Btu, the maximum allowable controlled emission of 0.60 lb. $SO_2/10^6$ Btu must be achieved in large coal-fired plants (Ref. 3). As can be seen from Tables 6 and 7, with a coal like sample DH-1 which contains 2.10% sulfur and will give an uncontrolled emission rate of 3.06 lb. $SO_2/10^6$ Btu, even with a stoichiometric ratio of 5 $CaCO_3$ to one sulfur, the limitations of NSPS are exceeded at all the temperatures used in the tests.

Use of CaO. To compare the effectiveness of CaO with that of $CaCO_3$, a series of combustion zone desulfurization tests were performed with the use of ORC-19 (4.90%S) and DH-1 (2.10%S) samples at temperatures ranging from 800°C to 1400°C. The test results are presented in the following tables.

TABLE 8

Effect of CaO addition and temperature on the combustion zone desulfurization of sample ORC-19 containing 4.90% total sulfur.

CaO:S	% Sulfur emissions at various temperatures (°C)							
Ratio	800	840	900	1000	1100	1200	1300	1400
0	4.44	4.62	4.66	4.64	4.69	4.81	4.85	4.90
1	2.47	2.34	2.52	2.92	3.26	3.89	3.94	4.16
2	1.17	1.04	1.20	1.18	1.23	2.22	3.50	4.02
3	0.56	0.52	0.54	0.62	0.58	0.94	1.94	3.18
4	0.37	0.42	0.42	0.46	0.49	0.67	1.24	2.24
5	0.32	0.36	0.34	0.34	0.41	0.58	1.08	1.64

TABLE 9

Lb. $SO_2/10^6$ Btu emissions from sample ORC-19 at various temperatures and stoichiometric ratios of CaO to total sulfur in the coal.

CaO:S	Pounds of SO_2 emissions per million But at various temperatures (°C)							
Ratio	800	840	900	1000	1100	1200	1300	1400
0	7.83	8.15	8.22	8.18	8.27	8.48	8.55	8.64
1	4.36	4.12	4.44	5.15	5.75	6.86	6.95	7.34
2	2.06	1.83	2.12	2.08	2.17	3.91	6.17	7.09
3	0.99	0.92	0.95	1.09	1.02	1.66	3.42	5.61
4	0.65	0.74	0.74	0.81	0.86	1.18	2.19	3.95
5	0.56	0.63	0.60	0.60	0.72	1.02	1.90	2.89

A study of Tables 8 and 9 shows that for a coal containing 4.90% total sulfur, it is possible to keep the SO_2 emission under 1.2 lb./10^6 Btu with the use of CaO if the stoichiometric ratio of CaO to total sulfur in the coal is maintained above three and the combustion temperature remains below 1100°C. On the other hand, instead of lime (CaO) if limestone ($CaCO_3$) is used for combustion zone desulfurization, the controlled SO_2 emission is always higher than 1.2 lb SO_2/ 10^6 Btu. A comparison of test results in Tables 5 to 9 shows the advantages of CaO over $CaCO_3$ in desulfurizing a high sulfur coal.

To see the effect of sulfur content of the coal on the combustion zone desulfurization power of lime (CaO), the test results obtained with the use of DH-1 sample containing 2.10% total sulfur are presented in Tables 10 and 11.

TABLE 10

Effect of CaO addition and temperature on the combustion zone desulfurization of sample DH-1 containing 2.10% total sulfur.

CaO:S	% Sulfur emissions at various temperatures (°C)							
Ratio	800	840	900	1000	1100	1200	1300	1400
0	1.80	1.82	1.80	1.82	1.84	1.98	2.00	2.10
1	1.20	1.16	1.24	1.28	1.76	1.80	1.84	2.00
2	0.68	0.76	0.80	0.88	1.24	1.72	1.72	1.96
3	0.48	0.56	0.56	0.58	1.00	1.28	1.48	1.80
4	0.44	0.38	0.44	0.46	0.60	0.96	1.22	1.64
5	0.42	0.38	0.40	0.36	0.52	0.68	1.12	1.52

TABLE 11

Lb. $SO_2/10^6$ Btu emissions from samples DH-1 at various temperatures and stoichiometric ratios of CaO to total sulfur in the coal.

CaO:S	Pounds of SO_2 emissions per million Btu at various temperatures (°C)							
Ratio	800	840	900	1000	1100	1200	1300	1400
0	2.62	2.65	2.62	2.65	2.68	2.89	2.91	3.06
1	1.75	1.69	1.81	1.87	2.56	2.65	2.68	2.91
2	0.99	1.11	1.16	1.28	1.81	2.51	2.51	2.86
3	0.70	0.80	0.82	0.85	1.46	1.87	2.16	2.65
4	0.64	0.55	0.64	0.67	0.87	1.40	1.78	2.39
5	0.61	0.55	0.58	0.52	0.76	0.99	1.63	2.21

The results in Tables 10 and 11 show that only with the use of lime with a stochiometric ratio of five CaO to one sulfur in the coal, the NSPS requirement of 0.60 lb $SO_2/10^6$ Btu can be maintained under a combustion temperature of 1000°C.

The results in Tables 4 to 11 showed that for maximum sulfur retention during the combustion zone desulfurization, a stoichiometric ratio of at least four CaO or CaO_3 to one sulfur in the coal should be used. For this ratio, the percentages of CaO and CaO_3 added to the samples will be as follows:

ORC-19; 37.98% $CaCO_3$ or 25.54% CaO

DH-1; 20.79% $CaCO_3$ or 12.82% CaO.

In Figure 2, the lb. $SO_2/10^6$ Btu emissions for a stoichiometric ratio of four CaO or $CaCO_3$ to one sulfur in the coal samples of ORC-19 and DH-1 are plotted against the combustion zone temperatures. This figure clearly indicates the advantage of CaO over $CaCO_3$ when both are used at the same stoichiometric

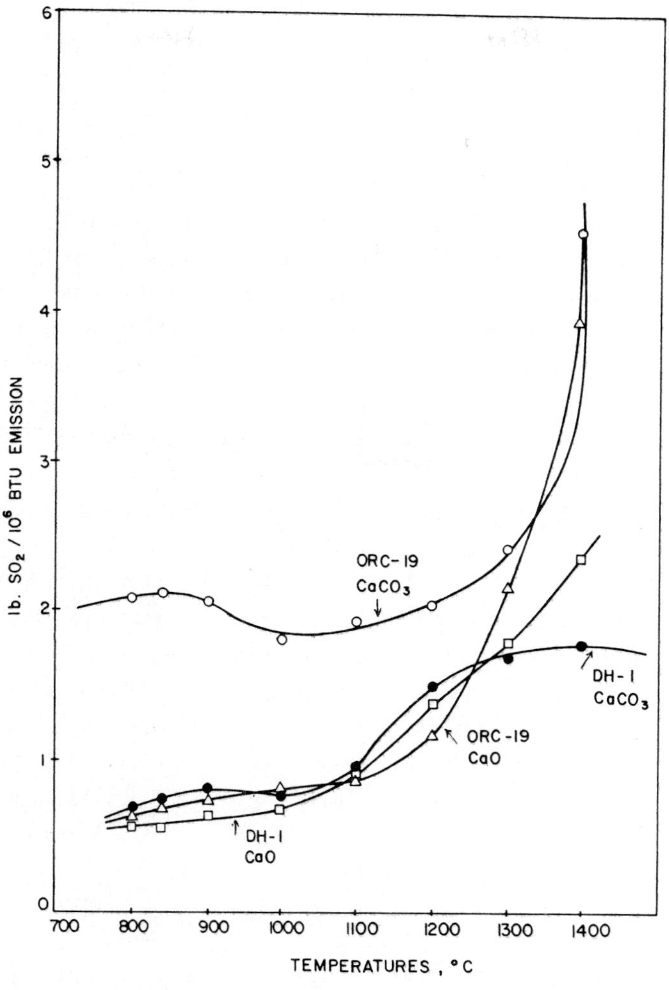

Fig. 2. Comparison of sulfur emissions from ORC-19 and DH-1 samples at various temperatures with the use of $CaCO_3$ or CaO with a stoichiometric ratio of four.

ratio of four sorbent to one sulfur with the ORC-19 sample containing 4.90% sulfur. However, in the case of DH-1 sample of medium sulfur content, the effects of CaO and CaO_3 are not greatly different within the temperature range of 800 to 1100°C.

Flue gas desulfurization test results

Use of $CaCO_3$. Table 12 shows the results of FGD tests conducted with the use of 0.25% $CaCO_3$-water suspensions on ORC-19 and DH-1 samples burned at 1400°C for 6 minutes. The number of test runs of each sample and the total sulfurs absorbed by the suspensions during each test series are also shown in this table. As it was mentioned in the section on Experimental Procedures, during each FGD test-series, several ½-gram coal samples were burned at 1400°C and the combustion gases produced from each sample were first bubbled through the $CaCO_3$-water suspension and then were analyzed for sulfur in the titrator of the Leco sulfur analyzer. Since the EPA Emissions Standards allowed only 0.68% and 0.41% sulfur emissions, which are equivalent to 1.20 lb $SO_2/10^6$ Btu and 0.60 lb $SO_2/10^6$ Btu, from coals like ORC-19 and DH-1, respectively, when these limiting sulfur percentages were read in the titrator of the analyzer, the tests were discontinued. Before and after each series of tests, the pH of the suspension was measured with a pH meter. For example, the pH of 0.25% $CaCO_3$-water suspension was 9.65 at the start of FGD tests, and after 5 half-gram runs of ORC-19 samples, the pH of the suspension fell down to 4.30. Whereas in the case of 11 DH-1 runs, the end pH of the same $CaCO_3$-water suspension fell only to 5.60. This is due to the fact that more SO_2 is absorbed by the suspension from the ORC-19 series of test samples than the DH-1 series of test samples.

TABLE 12
Flue gas desulfurization of ORC-19 and DH-1 samples with the use of 0.25% $CaCO_3$-water suspensions.

Sample No	# of ½g test runs	Total mg of S in original samples	Total mg of S emission	Total mg of S absorbed by suspension	% of original S absorbed
ORC-19	5	122.50	9.80	112.70	92.00
DH-1	11	115.50	9.35	106.15	91.90

The 0.25% $CaCO_3$-water suspension contained 500 mg of $CaCO_3$. During the absorption reaction, the ratio of $CaCO_3$ to sulfur absorbed must be at least $\frac{CaCO_3}{S} = \frac{100}{32} = 3.125$, which corresponds to a stoichiometric ratio of one. The 115.54 mg of sulfur absorbed from ORC-19 samples will need to tie up

115.54x3.125 = 361.06 mg of $CaCO_3$. Since the original suspension contained 500 mg of $CaCO_3$, the sorbent utilization efficiency was (361÷500)100 = 72.20%. Or, with respect to the total sulfur in the original coal, the stoichiometric ratio was (500÷122.50x3.125) = 1.31.

Table 12 also shows that 106.15 mg of sulfur was absorbed by the 0.25% $CaCO_3$-water suspension from 11 test runs of DH-1 sample. For a stoichiometric ratio of one, this amount of sulfur must have used up at least 106.15x3.125 = 331.72 mg of $CaCO_3$ from the suspension. In this case the sorbent utilization efficiency was (331.72÷500)100 = 66.34%. Or, again with respect to the total sulfur in the original coal, the stoichiometric ratio was (500÷115.50x3.125) = 1.39.

The better sorbent utilization in the case of ORC-19 sample is due to the fact that more total sulfur could be passed through the $CaCO_3$-water suspension before reaching EPA emissions limits for high sulfur coals than the medium sulfur DH-1 sample.

Use of CaO. Table 13 shows the results of FGD tests conducted with the use of 0.25% CaO-water suspensions on the same two test samples.

TABLE 13
Flue gas desulfurization of ORC-19 and DH-1 samples with the use of 0.25% CaO-water suspensions.

Sample No	# of ½g test runs	Total mg of S in original samples	Total mg of S emission	Total mg of S absorbed by suspension	% of original S absorbed
ORC-19	10	245.00	8.20	236.80	96.65
DH-1	17	178.50	9.35	169.15	94.76

The 0.25% CaO-water suspension had a pH of 12.65 at the beginning of the tests. After 7 runs of ORC-19 samples, the pH of the suspension fell to 7.0, but reached to 3.0 after the 10th run. The 10th run of this sample was not included in the calculations because it showed an emission rate of 1.46 lb. $SO_2/10^6$ Btu. With 17 runs of DH-1 sample, the final pH of the suspension was measured at 6.2.

The CaO utilization efficiencies during the test runs were 82.83% and 59.20% for ORC-19 and DH-1 samples, respectively. Compared to the total sulfurs in the original coal samples the stoichiometric ratios of CaO in the suspensions were 1.17 and 1.60. Again, the lower utilization of CaO in the CaO-water suspension is due to lower SO_2 emission rate allowed for the medium sulfur DH-1 sample. Actually during the tests, 17th run of DH-1 sample showed 0.26% sulfur emission which was equivalent to 0.38 lb. $SO_2/10^6$ Btu, and was included in the calculations. However, the 18th run of the same sample indicated a sulfur emission of

0.45%, which was equivalent to 0.66 lb. $SO_2/10^6$ Btu, and could not be included in the calculations.

A comparison of the test results in Tables 12 and 13 shows that during the FGD tests of high and medium sulfur coals, a unit weight of CaO absorbed 1.59 (DH-1) and 2.10 (ORC-19) times more sulfur than the same weight of $CaCO_3$.

Comparison of test results

During the combustion zone desulfurization of a high sulfur coal (4.90%S) at temperatures varying between 800 and 1100°C, pure lime (CaO) showed a greater desulfurization ability than the limestone ($CaCO_3$) when both were used with stoichiometric ratios ranging from 1 to 5 (Fig. 2). Actually, with the use of $CaCO_3$, the EPA emission limit of 1.20 lb. $SO_2/10^6$ Btu could not be realized at all the temperatures and stoichiometric ratios tested. Yet, with the use of pure lime with a stoichiometric ratio of 3 or more, the EPA limit was easily realized below a temperature of 1100°C for this high sulfur coal (Tables 5 and 9).

When a medium sulfur coal (2.10%S) was used in the combustion zone desulfurization tests, the performances of CaO and $CaCO_3$ were very close up to 1100°C and with a stoichiometric ratio of 4 sorbent to one sulfur in the coal. However, the only case where the EPA emission limit of 0.60 lb. $SO_2/10^6$ Btu could be realized was the use of a stoichiometric ratio of 5 CaO to one sulfur in the coal at a temperature of 1000°C or lower (Tables 7 and 11).

The FGD tests of the same high and medium sulfur coals showed a greater effectiveness than the combustion zone desulfurization tests with the use of either 0.25% $CaCO_3$ or CaO-water suspensions. When 0.25% $CaCO_3$-water suspensions were used, it was possible to absorb 92.00% and 91.90% of SO_2 coming from the combustion of high and medium sulfur coals, respectively, before the EPA emission rates for these coals were reached (Table 12). The stoichiometric ratios of $CaCO_3$ used with respect to total sulfurs in the high and medium sulfur coals were 1.31 and 1.39.

With the 0.25% CaO-water suspensions, 96.65% and 94.76% of the SO_2 originated from the high and medium sulfur coals were absorbed before the EPA emissions rates were surpassed (Table 13). The stoichiometric ratios of CaO compared to total sulfurs in the high and medium sulfur coals were 1.17 and 1.60.

Table 14 summarizes the stoichiometric ratios of sorbents used for the combustion zone and FGD tests. As can be seen from this table, the FGD method actually used lower stoichiometric ratios of lime and limestone than the combustion zone desulfurization method yet it can comply with the existing EPA emissions standards for high and medium sulfur coals.

TABLE 14

Summary of stoichiometric ratios of sorbents for combustion zone desulfurization and FGD of ORC-19 and DH-1 samples.

Sample No	Desulfurization method	Sorbent used	Stoichiometric ratio of sorbent to sulfur
ORC-19 (4.90%S)	a) Combustion zone	$CaCO_3$	Not possible with 5
	b) FGD	"	1.31
	c) Combustion zone	CaO	3.00
	d) FGD	"	1.17
DH-1 (2.10%S)	a) Combustion zone	$CaCO_3$	Not possible with 5
	b) FGD	"	1.39
	c) Combustion zone	CaO	5.00
	d) FGD	"	1.60

SUMMARY OF CONCLUSIONS

From the studies conducted on the combustion zone desulfurization and flue gas desulfurization of high and medium sulfur coals, the following conclusions were drawn.

1. When $CaCO_3$ is used as sorbent in the combustion zone desulfurization of ORC-19 coal sample containing 4.90% S, the EPA limit of 1.20 lb $SO_2/10^6$ Btu could not be achieved at any temperature or stoichiometric ratio of $CaCO_3$ to sulfur in the coal (Tables 4 and 5).
2. If a coal containing 2.10% sulfur like DH-1 sample is subjected to combustion zone desulfurization with the use of $CaCO_3$ as sorbent of SO_2, then with a sotichiometric ratio of greater than 2, the 1.20 lb $SO_2/10^6$ Btu emission rate is not exceeded if the combustion temperature is maintained under 1000°C. However, the NSPS limit of 0.60 lb. $SO_2/10^6$ Btu is not achieved at any temperature and stoichiometric ratio of $CaCO_3$ used in the tests (Tables 6 and 7).
3. In desulfurizing a coal like ORC-19 containing 4.90% S, if CaO is used as sorbent in the combustion zone, it is possible to have under 1.20 lb. $SO_2/10^6$ Btu emission if a stoichiometric ratio of at least 3 is used and the combustion temperature is maintained under 1100°C (Tables 8 and 9).
4. For medium sulfur DH-1 sample, only with the use of CaO as sorbent with a stoichiometric ratio of 5, the NSPS limit of 0.60 lb $SO_2/10^6$ Btu could be achieved during the combustion zone desulfurization at temperatures below 1000°C (Tables 10 and 11).
5. During the combustion zone desulfurization, the CaO was a more effective sulfur absorbing agent than $CaCO_3$ when used with the same stoichiometric ratio of CaO or CaO_3 to sulfur in the high sulfur coals (Figure 2).

6. In the flue gas desulfurization of high sulfur ORC-19 coal with the use of 0.25% $CaCO_3$-water suspension, 92.00% of SO_2 passing through the suspension was arrested before the emission from individual test runs reached a rate of 1.20 lb. $SO_2/10^6$ Btu. The stoichiometric ratio of $CaCO_3$ to sulfur in original samples was 1.31 (Table 12).
7. The FGD test of DH-1 sample with 0.25% $CaCO_3$ suspension showed 91.90% sulfur absorption before the emission rate from individual test runs exceeded 0.60 lb $SO_2/10^6$ Btu. The stoichiometric ratio of $CaCO_3$ to sulfur in original samples was 1.39 (Table 12).
8. The use of 0.25% CaO-water suspension in FGD tests of ORC-19 and DH-1 samples indicated 96.65% and 94.76% sulfur absorption efficiencies when the EPA emission rates for individual test runs were observed (Table 13). The stoichiometric ratios of CaO to sulfurs in the original coal samples were 1.17 (ORC-19) and 1.60 (DH-1).
9. The test results in Tables 12 and 13 showed that during the FGD tests of coals containing 4.90% and 2.10% sulfurs, a unit weight of CaO absorbed 2.10 and 1.59 times more sulfur than the same weight of $CaCO_3$.
10. The tests also showed that the FGD has a greater effectiveness than the combustion zone desulfurization when CaO and $CaCO_3$ are used as SO_2 absorbing agents.

REFERENCES

1 M.H. Erten, "Desulfurization of Missouri Coals", Process Mineralogy II, The Metallurgical Society of AIME, New York, NY 10017, 1982, pp. 355-372.

2 M.H. Erten, "Beneficiation of Missouri Coals With Emphasis on Desulfurization", Research Report to be Published by the Missouri Department of Natural Resources, Rolla, MO 65401 (USA), 1985, 28 pp.

3 T. F. Edgar, Coal Processing and Pollution Control, Gulf Publishing Co., Houston, Texas, 1983, pp. 221-264.

4 L. Yaverbaum, Fluidized Bed Combustion of Coal and Waste Materials, Noyes Data Corporation, Park Ridge, N.J., 1977, p. 14.

5 L. Grainger and J. Gibson, Coal Utilization, Graham and Trotman, 14 Clifford St., Lond, England, 1981, pp. 76-108.

6 L. D. Smoot and P. J. Smith, Coal Combustion and Gasification, Plenum Press, New York and London, 1985, pp. 184-206.

THE USE OF METALS AND METAL OXIDES FOR SULFUR CAPTURE IN COMBUSTION SYSTEMS

S.F. PROTASIO AND G.R. ST.PIERRE

The Ohio State University, Department of Metallurgical Engineering, 141A Metal. Eng., 116 W. 19th Ave., Columbus, Ohio 43210 (U.S.A.)

ABSTRACT

The thermodynamic stabilities of metal sulfides and the potential for the use of certain metals for sulfur capture are reviewed. Both iron and copper may be used to capture sulfur in situ in hot combustion gases. In both cases, it is necessary to maintain a reduced chemical potential of oxygen in order to favor the formation of the metal sulfide rather than the metal oxide. In the case of copper, the oxygen potential must be maintained at a level corresponding approximately to a CO_2/CO ratio of three at about 1150°C. This necessitates partial coal combustion during sulfur collection by copper followed by secondary combustion for full fuel recovery. At CO_2/CO ratios substantially less than three, H_2S and COS become predominant species while, at higher oxidizing conditions, SO_2 predominates. Hence, the optimum condition is obtained at an intermediate oxygen pressure.

Proposals for the use of iron, iron oxide and zinc oxide are similarly reviewed. In order for these systems to be economic, there must be full recycling of the metals through a closed process loop. A complete system for the use of copper in this manner is discussed.

INTRODUCTION

The increasing demand for electrical power, the inability of nuclear power plants to respond to that demand, and the large availability of solid fossil fuels versus oil and natural gas, have reinforced the trend to burn coal for power generation in North America. Alternative methods of generating electricity (wind, sun, geothermal, and wave power) are also being developed, but they do not generate significant amounts of energy to substitute for traditional fuels and it is not likely for them to be used as more than local energy suppliers during the next decade. Although most of the existing coal consumption contains low levels of sulfur, the electrical utility industry represents the major source of sulfur oxides pollution because of the large amounts of coal used. Standards for sulfur dioxide emission have become very strict and power plants have been preferentially consuming the lower sulfur coal resources. A large number of studies have been conducted on processes able to provide necessary pollution control for use of coal with higher sulfur content.

Most of the processes so far implemented are far from being a final solution as they cannot completely solve the problem or are only effective for coals containing low levels of sulfur. Therefore, research continues on relatively new technologies that, being economical, might also be able to control the emission of higher levels of sulfur species. Based on the overall economics of the processes, solutions for the treatment of the stack gases have ranged from regenerable to throwaway processes, from wet to dry gas desulfurization, and to a number of other processes that do not even fit within the limits of these definitions. It not being the scope of this paper to cover all the desulfurization techniques, we should still note that the removal of sulfur species from flue gases after combustion is not the only way of solving the problem. The removal of sulfur from coal before combustion or the modification of the combustion processes to include desulfurization as part of it, like in the cases of coal gasification or liquefaction, are exciting and promising alternatives. The fine-particle coal slurry systems come closest to fulfilling all technical objectives. These processes may offer the best future solution as they allow the displacement of the fuel as an easily transportable fluid to the place where the electrical power will be used without the need for building desulfurization units at many inconvenient locations.

Although most of the solutions are not yet very encouraging, the overall problem of pollution versus energy needs, coupled with the ethics involved, is pertinent enough to alter our present definition of "economically viable". Also, the continuing development of new materials, e.g., non-metallic erosion-resistant materials and metallic alloys resistant to corrosion at high temperature (ref. 1), and the improvement in the initial design have been able to solve several of the initial difficulties found in some processes, reinforcing the need to continue studying "impossible" solutions that thermodynamically and kinetically seem promising. Because of their thermodynamic affinity for sulfur, metals have been used widely as "sulfur-getters" in laboratory experiments. Both metals and metal oxides offer promise as commercial sulfur collection agents. In the present paper, several possible methods will be analyzed.

METAL OXIDES AND METALS

In this presentation, we will concentrate on the use of metals and metal oxides as desulfurizers, without making any special reference to processes involving metal oxides in slurries, alkali metals and their salts or catalytic oxidation systems. We will not discuss the carbon adsorption either, although it is in essence similar to the processes presented, where the chemisorption is followed by regeneration through washing or heating.

Dry processes were developed as a response to one of the major drawbacks of wet processes, i.e., they require the cooling of the gasified product, which immediately cancels the possibility of using them in low-Btu coal gasification systems due to the loss in sensible heat involved. The need for higher temperature desulfurization methods is also important because low-Btu gas cannot be transported and has to be used immediately for power plant firing.

The choice for a fixating material for a dry process is generally based on its ability to perform desulfurization forming stable solids and other waste materials at the appropriate temperatures. Also, the stable product formed should be completely recyclable in order not to create a new waste disposal problem. The material should be passible through cycles of sorption and regeneration at about the same temperature and period of time as the cycles to provide economic processing.

One of the main advantages of the metal oxides is that they do not generally involve any waste material. This is such an important factor that some studies have already been done on the possibility of using wastes from existing processes to collect more hazardous wastes from other operations. For example, metallurgical slag wastes, under the right circumstances, have the potential to capture large quantities of sulfur from coal combustion.

Dry calcium-based waste has been proposed as a good fixation agent, when mixed with fly-ash for cleaning oil sludge, plating waste, sedimentation waste from aluminum reclamation and other metal processing industries (ref. 2). This, far from solving the problem of waste disposal, may at least concentrate it in a single product.

Although there are several metal oxides that possess the utilization requisites, only a restricted number have been studied so far as most of them raise other toxicity or radioactivity problems, or are too expensive or unavailable in large quantities. The main problem of dry metal oxide processes is their lack of mechanical characteristics to go through several cycles of sorption and regeneration without big dust losses and major reduction of reactivity. Frequently, they are very corrosive to the equipment, although this characteristic is sometimes the basis for the desulfurization process itself and recent improvements on the materials have, as mentioned, attenuated this aspect.

Since the final product in dry systems is generally a sulfate of the metal used, the regeneration is not done simply as in the case of sulfites formed during wet scrubbing. When the sulfates have refractory properties, e.g., alkalis or alkaline earths, their oxides are not convenient for these kinds of dry processes, unless chemical regeneration is economical and then it has to be done with the aid of reducing agents.

To develop these processes, kinetic factors of both sorption and regeneration should also be considered, which is not generally an easy task as gas-solid kinetics are difficult to study. Kinetic studies have shown the reactions to behave as first order with the sorption activity to increase following the sequence: ZnO, Fe_2O_3, CuO, CoO, MnO (ref. 3). These studies based on thermogravimetric measurements of the change of weight during the complete sulfidation of the oxide, at temperatures ranging from 300 to 800°C, have suggested that the rate of reaction for manganese oxide is one order of magnitude greater than for zinc oxide or calcium oxide.

These data obviously only apply for the experimental conditions of the studies and cannot be trusted if other mechanisms are to be considered as, in these cases, no product layer diffusion was considered, i.e., the reaction was considered completed as the first mono-layer had formed around the metal oxide, and also external mass transfer was rendered unimportant by setting an ideal relation between the surface area of the reacting material and the gas flow rate.

In conclusion, it seems that the major drawback for the utilization of metal oxides is their physical and/or chemical inability to go through several cycles of regeneration, their promotion of other pollution problems, and also an eventual disposal of the waste products from the overall operation. The main advantages seem to be the avoidance of gas reheating and the possibility of obtaining a saleable product as the sulfur dioxide can be converted into sulfuric acid, to elemental sulfur through chemical reduction or be purified and liquified as a final product. Some specific examples follow.

Magnesium Oxide. Slurries of magnesium oxide have been used as flue gas desulfurizers in the last ten years, first in oil-fired boilers and more recently in coal-fired ones. This process was easily developed as its chemistry was already known from the paper pulp industry. Its major drawback is the large amount of energy needed for the recycle of MgO through the calcination of the magnesium sulfite formed (ref. 4). Magnesium oxide has also been used as an additive for lime and limestone slurry systems as it increases the alkalinity and sulfite/sulfate solubility of the liquid phase (ref. 5). Still, magnesium oxide is not generally used in dry porcesses as the regeneration of $MgSO_3$ is characterized by losses of up to 20% and dust emission from the dryers containing Mg products are also toxic (ref. 6).

Calcium Oxide. For much the same reasons, calcium, which is widely used for wet scrubbing, is not a good solution in terms of dry processes as the regeneration of its sulfate is not done easily and, as discussed by Bartlett (ref. 7), the Mg present in the initial limestone can form magnesium sulfate during the sorption operation and, being soluble in water, may cause similar pollution problems. Injection of Ca-based materials into the flue gas stream has also been tried but this is still a non-regenerable system and it gives good results only when the gases are previously humidified which causes a decrease in temperature.

Manganese Oxide. Manganese oxide, which has the highest sulfur potential for temperatures above 400°C and up to 1000°C, has been used in industrial processes, especially in Japan, but its regeneration, done with ammonia, also includes too many drawbacks; the main one being the lack of a market for ammonium sulfate (ref. 6). Recently, the Japanese had to abandon this kind of solution as the manganous products formed by reduction during the sorption of sulfur dioxide were also found to be toxic.

Iron Oxide and Iron. Iron oxide seems to be a less problematic fixator in terms of the chemistry involved, but there are also some difficulties in the regeneration step due to mechanical resistance. Some processes have already been developed where liquid iron is used in order to solve this mechanical resistance problem. In this case, the sulfur reacts with the molten materials and is later regenerated. Kloeckner, for instance, has developed a process where the gasification of coal is done in a traditional steel converter using pig iron as the sulfur fixating agent. Although good desulfurization is achieved at a very low cost (extra converters and some extra pig iron are often available in steel plants), this process can only be used on a small scale and for local use of the heat generated. It has to be combined with the gasification of coal and with a good dust collecting system.

In its solid form, iron can be a good desulfurizer at temperatures up to 700°C, while magnetite is the fixating agent. For higher temperatures, "FeO", Wüstite, becomes the stable form and a rapid decrease in fractional desulfurization is observed. In Figure 1 (ref. 7), the stabilities of the principal compound phases in Fe-O-S system are presented at 723°C. The family of diagonal lines with a negative slope of two, are lines of constant partial pressure of SO_2 in atm. In the region of coexistence of metallic iron, "FeO", and "FeS", the equilibrium partial pressure of SO_2 is of the order of 10^{-12} atm. Several processes have been proposed using iron oxide particles in either a fixed or a fluidized bed. In some cases, the fly ash itself has been proposed as a desulfurizing agent since it contains hematite that can be easily converted to magnetite, in an initial quick reduction step

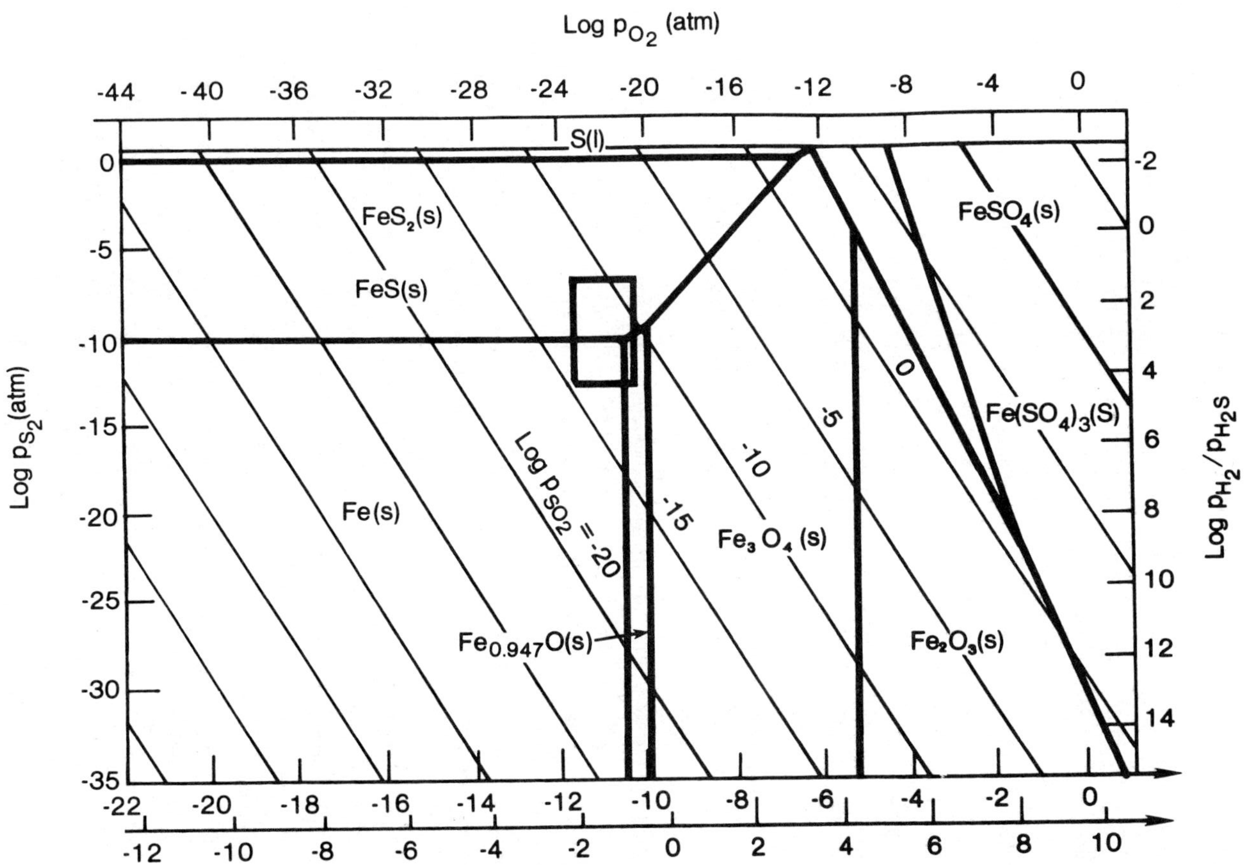

Figure 1. The Thermochemistry of the Fe-O-S System at 1000° K (723°C), After Rocazella and Rapp (ref. 7)

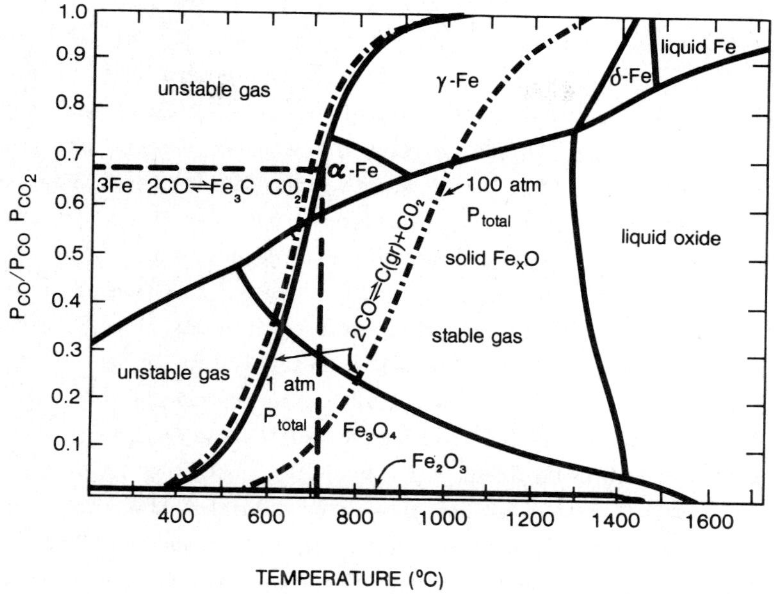

Figure 2. Gas Compositions in Equilibrium with Condensed Phases in the Fe-C-O System. After Bartlett and Shumaker (ref. 8).

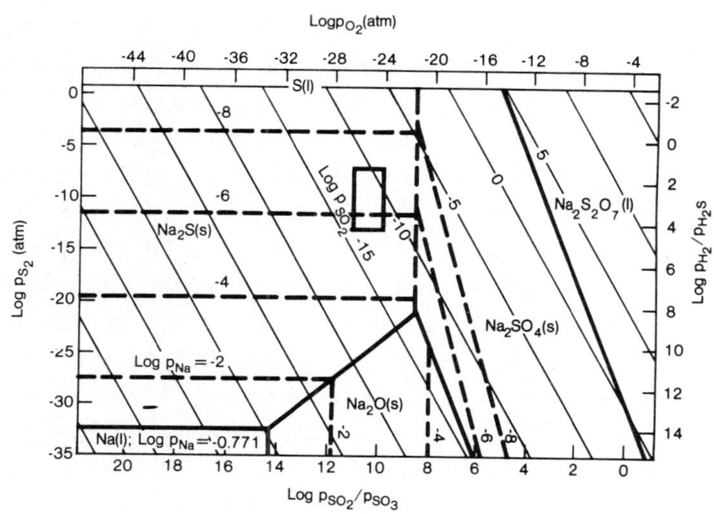

Figure 3. The Thermochemistry of the Na-O-S Sytem at 1000°K (723°C), After Rocazella and Rapp (ref. 7).

(ref. 8). The relative stabilities of the oxides of iron are shown in Figure 2 in terms of CO-CO_2 gas mixtures. Note that weak reducing gas mixtures are sufficient to convert hematite to magnetite at all temperatures. Fly ash has also been used as a matrix, enriched with iron oxide, thus increasing the desulfurizing ability of the bed and reducing the mechanical resistance problem of iron oxide by itself (ref. 9). In this case, a nonstoichiometric ferrous sulfide forms around the particles. The oxides may be regenerated with a dilute oxygen stream. The product stream of this regeneration contains gaseous sulfur and its dioxide, in a suitable relation for either the production of sulfuric acid or elemental sulfur. In this process, no deposition of carbon was found due to the decomposition of carbonous materials in the ash or gas phases in the presence of hydrogen or water vapor, but carbon deposits were found when tar was injected due to its cracking, suggesting that this and similar injections should be avoided.

Much of the technical discussion is centered on choosing between a fixed or a fluidized bed. For materials with low mechanical resistance like iron oxide, attrition seems to be one of the main drawbacks thus suggesting that the fixed bed could be the better option.

For the cases where an alumina or fly ash matrix has been proposed, increased porosity is possible so that it is feasible to work with bigger particles and still have an optimized reaction surface. This way, the need for fluidization and/or very small particles is avoided and plugging or fouling of the system is not as likely to occur, nor will there be a large carry-over of absorbent, which accounts for most of the losses in these types of processes.

Another process has been proposed based on the hot corrosion of iron-carbon pellets coated with a fused alkali sulfate. The coating would be responsible for the removal function and iron would be the sacrificial element in the corrosion process (ref. 7). The thermochemistry of the Na-O-S system is shown in Figure 3. This corrosion process is based on the thermodynamic instability of metals in the presence of sodium sulfate and it is characterized by an acid-base fluxing mechanism. Models to explain the apparent linear kinetics and the self- sustaining characteristics of this mechanism are summarized by Rocazella (ref. 7). In the presence of a reducing species like carbon, the process becomes particularly effective as carbon promotes the formation of sulfite instead of sulfate, avoiding the formation of a protective layer that would block the reaction. The concept involved in this proposal is very interesting, but the production of the pellets, their regeneration, and the disposal of waste products containing corrosive salts do not suggest its industrial application without further improvement. Also, the process has not been tried for flow rates close to practical ones yet, and there is no evidence that it would still give satisfactory removal efficiencies.

Zinc Oxide. Slurries of zinc oxide have been used for flue gas desulfurization as ZnO becomes very reactive in the presence of water vapor. The sulfate obtained through this wet process is easily decomposed upon heating, under a slightly oxidizing atmosphere, into oxide and sulfite which can be directly used in a sulfuric acid plant without the need for sulfur dioxide oxidation (ref. 8).

As the humidification of the gas with water vapor implies big heat losses, zinc oxide has also been proposed as a dry gas desulfurization agent due to its thermodynamic tendency for sulfidation. Also, zinc does not form a high temperature stable sulfate and ZnS can be easily roasted to ZnO (ref. 8). The relative stabilities of ZnO and ZnS are shown in Figure 4 (ref. 8). Note that the partial pressure of SO_2 and the other important sulfur species are low for gases in equilibrium coexistence with ZnO and ZnS. However, the partial pressure of zinc vapor becomes a serious problem at low oxygen pressures. The partial pressures of the principal equilibrium gas species in contact with zinc oxide at 1100°K (823°C) are shown in Figure 5. Note that the combustion products are derived from representative coals. For slightly less than theoretical combustion air, the partial pressures of all sulfur species are reduced substantially by reaction with zinc oxide to form zinc sulfide.

A remaining problem lies in the fact that during oxide regeneration, the particles tend to sinter and the final reactive surface is significantly reduced. Zinc is also very volatile and zinc vapor becomes a problem at operating temperatures in excess of 700°C.

Copper Oxide and Copper. As in the case of iron oxide, the application of copper in dry desulfurization systems may be done with fixed or fluidized bed technologies. The fluidized bed process is still under improvement as it seems to involve complicated technology. The one proposed by Mendoza and Laguerie (ref. 10) allowed a continuous operation, but to reach the same yield as with a typical fixed bed operation it needed a four-stage operation for good control of the solid flow and retention time. If these were not controlled and only the first level was actually operating, the efficiency dropped from 90% to as low as 60% (ref. 10).

Strakey, Bauer and Haynes (ref. 11) proposed some improvement of the fluidized bed method developed by the Bureau of Mines by the injection of ammonia that promotes the selective catalytic reduction of the nitrogen oxide. In this case, the copper oxide sorbent is supported on an alumina matrix and

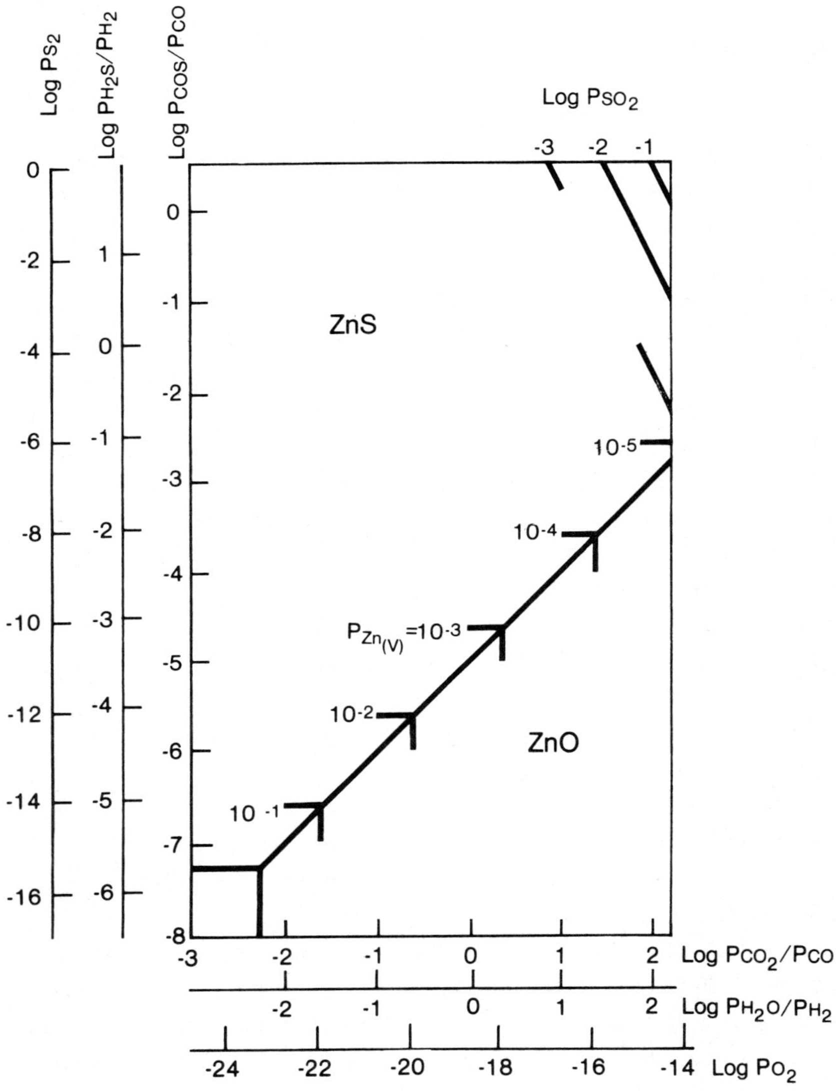

Figure 4. The Phase Stability Diagram of Zn-O-S at 1100°K (823°C), After Bartlett and Shumaker (ref. 8).

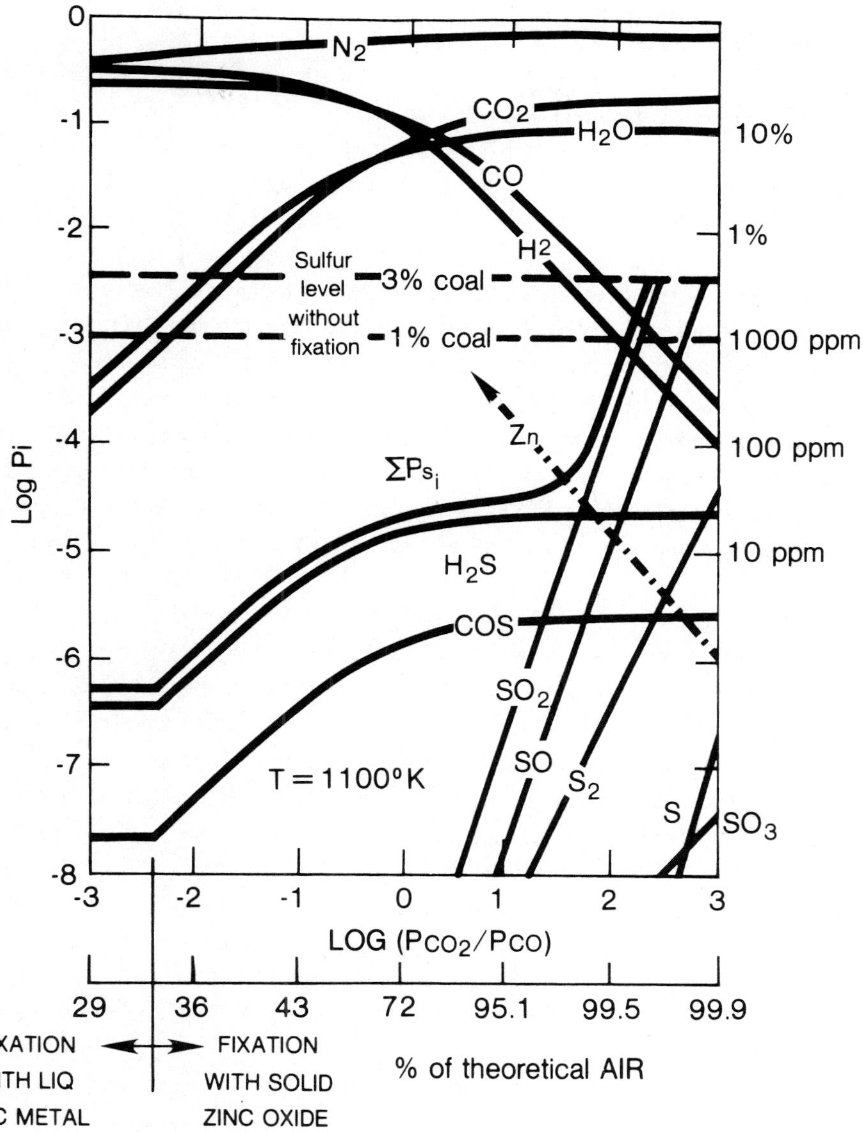

Figure 5. Partial Pressures of Gas Species in Equilibrium with Zinc Oxide at 1100 K (based on the partial combustion of a 60% C, 5% H, 10% O, and 4% moisture coal), After Bartlett and Shumaker (ref. 8).

the operation is made continuous by a second step of regeneration with methane in a gravitating bed (ref. 11).

The fixed bed has its application in the process developed by Shell which consists of parallel passage reactors also made of alumina granules impregnated with copper oxide where both sorption and regeneration with a reducing gas take place cyclically. This process has been improved recently to also control nitrogeneous emissions through the injection of ammonia into the flue gas.

This process, although efficient, requires strict control, because mixing of the stack gas and the reducing gas used for the two different operations creates a serious danger of explosion. Also, during the injection of ammonia, the ratio to nitrogen oxide must be controlled to avoid conditions that plug the system by the formation of ammonia salts. The chemistry of the process must also be correctly controlled if the content of sulfur in the particles drops to less than 4%. In that case, other reaction products may form and sulfur dioxide may even react with the alumina (ref. 4,10).

Whatever the choice, the main problem with copper oxide remains its poor mechanical characteristics. Several materials have been used to provide a porous strong support, e.g., alumina, with impregnated copper sulfite, cloride, bromide, nitrate, etc., all amounting to a content in copper ranging from 4 to 6% (ref. 6). Not much improvement was attained, perhaps because of maldistribution of copper in the pores, plugging of the pores, build-up of aluminum sulfate, or recrystallization of the alumina base during the several operation cycles.

These mechanical stability problems and the known affinity of sulfur for metallic copper have suggested the use of liquid copper as a sulfur collector. In order to create a large surface for absorption as in the case of particles forming a fluidized bed, contacting with a liquid copper spray has been proposed (ref. 12,13).

From the Cu-Cu_2 S phase diagram shown in Figure 6 (ref. 14), it is seen that molten copper dissolves about 1.0 percent sulfur by weight at about 1100°C. St.Pierre (ref. 12,13) has shown that molten copper saturated with Cu_2S at temperatures near 1150°C is highly promising as a sulfur capture agent from thermodynamic considerations. In Figure 7, the partial pressures of the principal sulfur species are shown as a function of oxygen partial pressure (expressed in terms of CO_2/CO ratio). It should be noted that the conditions correspond to gaseous products for combustion of coal with less than stoichiometric air. At low CO_2/CO ratios, the principal sulfur species are H_2S and COS, while at high CO_2/CO ratios, SO_2 predominates. At CO_2/CO ratios near

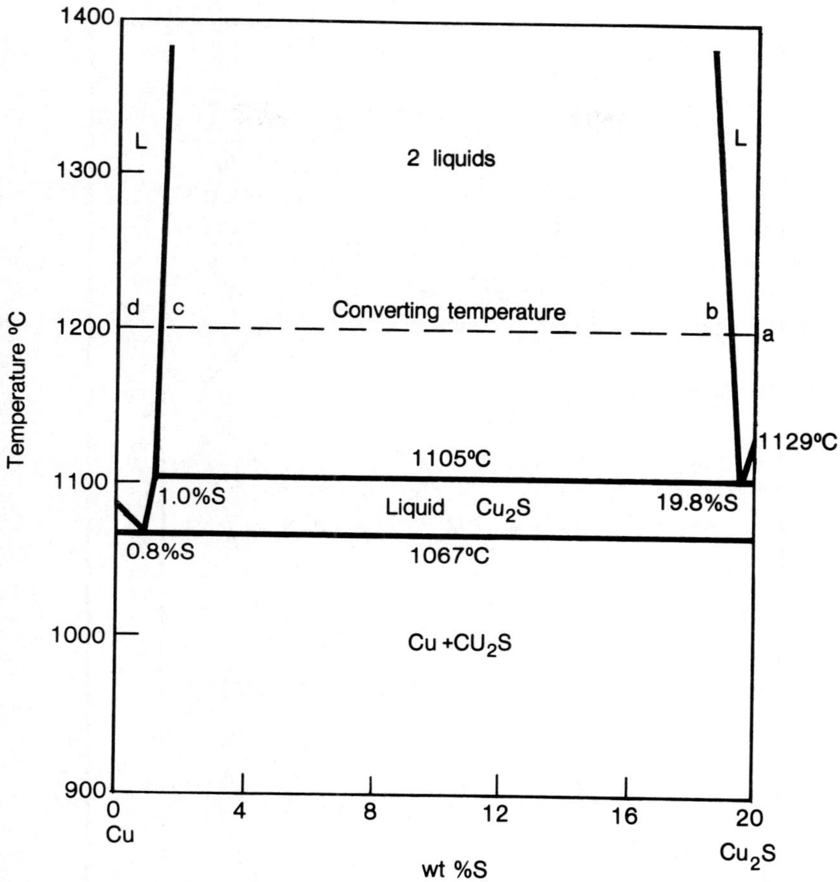

Figure 6. The Cu-Cu_2S Phase Diagram, After Biswas and Davenport (ref. 14).

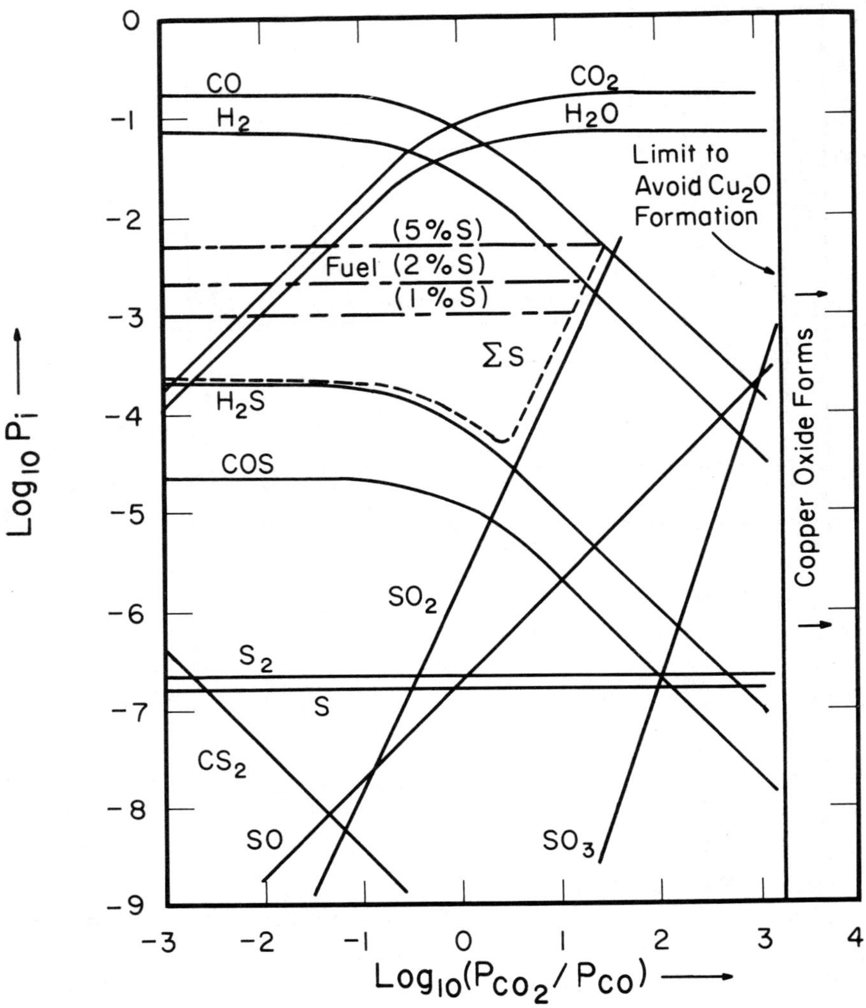

Figure 7. Partial Pressures of Gas Species in Equilibrium with Copper/Copper Sulphide at 1127°C and One Atmosphere Total Pressure, After St.Pierre (ref. 12,13).

three, the total content of sulfur species has a minimum value at 1127°C. Similar results are obtained at higher temperatures. In Figure 8, the potential for sulfur capture by molten copper is illustrated schematically. Three questions remain: can the copper be recycled easily; can the full calorific value of the fuel be realized; and can the operation be put together in a technically sound and economic manner. As is so often the case, only partial answers are available. Sulfur can be removed from Cu/Cu_2S by injection with oxygen, e.g., copper converting. By leaving high residual sulfur contents (approximately 0.2%), little Cu_2O is lost. After contacting the combustion products with a copper spray, secondary air can be admitted for full combustion. Finally, while the coupling of a special copper convertor with the combustion system appears to be technically sound in that the copper remains in the liquid state continuously, the system is considerably more complex than alternative schemes particularly when plants are not technically designed for the addition of a pyrometallurgical operation. Any significant copper loss would render the system economically unsound.

The use of droplets for sulfur removal is not new, and is the basis of most spray drying processes. On a pilot-plant basis, several advantages, namely absence of scaling or plugging and low capital investment, have been demonstrated (ref. 15).

This is particularly relevant if we compare dry with wet scrubbing solutions as these involve a large amount of equipment to handle the wet sludge (thickeners, centrifuges, filters and mixers) and high maintenance. Also, a large part of the pumping operations may be eliminated.

The proposed liquid-copper scrubbing process is based on the chemistry of the Cu-S-O system since the sulfidation of copper is competitive with its oxidation. By a correct control of the oxidizing potential, i.e., the amount of air available for primary combustion, it is possible to favor the reaction with sulfur.

This process, working with a reducing flame, demands the use of a two-stage combustion system. Alone, this does not seem very complicated if we compare it to the technological solutions proposed for fixed or fluidized beds with solid impregnated particles. Still, care must be taken to avoid carryover of copper into the second stage and the sytem has to be buffered to allow only the passage of gases for secondary combustion.

Most of the economical viability of this process has to be discussed in terms of the recycling operation. The recycle of a Cu/Cu_2S mixture should not be a very complicated operation, since the chemistry of this process is well-known from the copper industry itself (ref. 14). It would correspond to the stage of a copper matte conversion during which the so-called "oxidation of

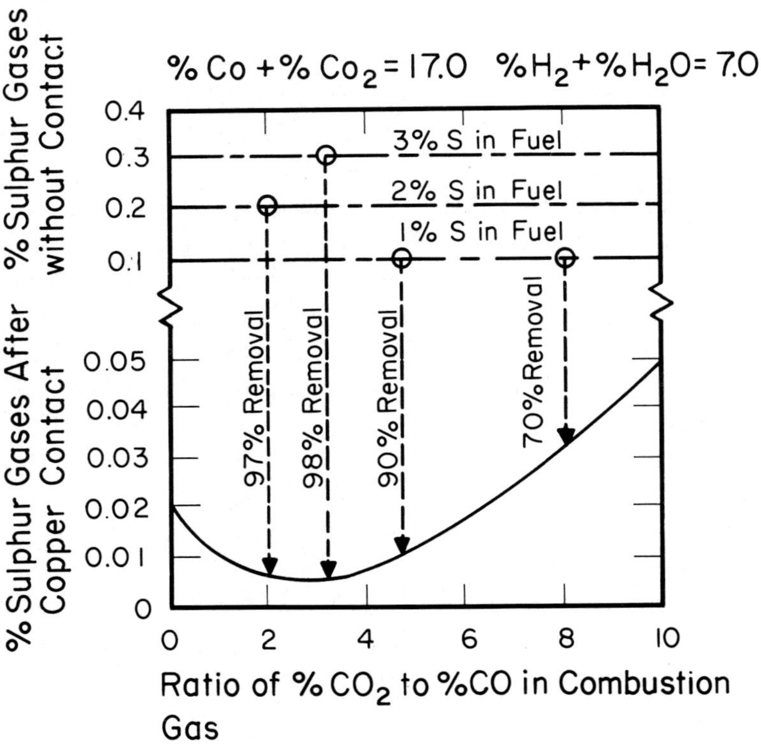

Figure 8. Sulphur Removal from Combustion Gases Contacted with Copper (Cu/Cu_2S) at 1127°C (2061°F) and Atmospheric Pressure, After St. Pierre (ref. 12,13).

white metal" occurs when the amount of sulfur is low enough to allow the formation of metallic copper. For this to be possible, the amount of sulfur in the mixture should not exceed a value of around 1.2%, which is the range of sulfur in blister copper and approaches the saturation of copper with its sulfide at 1100°C. A good control of this oxidation step will minimize copper losses, thus reducing the amount of fresh copper to be introduced in the recirculation loop.

Slags. As mentioned earlier, waste products may be used as sulfur collection agents. For example, molten iron and steel-making slags have sulfide capacities sufficient to offer attractive potential for sulfur capture under certain circumstances, particularly reducing conditions. In this instance, a waste slag would be utilized for an additional purpose before final disposal or byproduct use. The sulfide capacities of typical iron and steel-making slages are well-known (ref. 16).

CONCLUSIONS

In this brief communication, it has been shown that metals and metal oxides offer promise as sulfur capture agents from thermodynamic and kinetic considerations. The oxides of calcium, iron, zinc, and copper and metallic iron and copper are promising for this purpose. However, the future commercial potential for these systems is entirely unknown. It is suggested that the use of a waste product, e.g., molten iron-making slag,for the additional purpose of sulfur capture, may be an attractive possibility.

REFERENCES

1 R.W. Kirchner, Materials of Construction for Flue-Gas Desulfurization Systems, Chemical Engineering, 90, 19 (1983), 81.
2 W.W. Dennis, Hazardous Waste Fixation Using Dry Flue Gas Desulfurization Waste, Resources and Conservation, 9 (1982), 143-148.
3 P.R. Westmoreland, J.B. Gibson, and P.H. Douglas, Comparative Kinetics of High-Temperature Reaction Between H2S and Selected Metal Oxides, Environmental Science and Technology, 11, 5 (1977), 488-91.
4 M.J. Satriana, New Developments in Flue Gas Desulfurization Technology, Park Ridge, N.J., Noyes Data Corp., 1981.
5 J.T. Schrodt, T.K. Eung, and S.V. Bala, Design and Cost Optimization of a Hot Fuel Gas Desulfurization Process, Fuel Processing Technology, 6 (1982), 255-267.
6 A.V. Slack and G.A. Hollinden, Sulfur Dioxide Removal from Waste Gases, Noyes Ed., New Jersey, 1975.
7 M.A. Rocazella, The Feasibility of a Sulfur Dioxide Scrubber System Based on a Hot Corrosion Mechanism, M.S. Thesis, The Ohio State University, 1981. (Adviser: R.A. Rapp).
8 R.D. Bartlett, The Sulfidation of Zinc Oxide, M.S. Thesis, The Ohio State University, 1982. (Adviser: C.B. Shumaker).

9 E.C. Oldaker, A.M. Poston Jr., and W.L. Farrior, Laboratory Evaluation of Properties of Fly-Ash Iron Oxide Absorbents for H2S Removal from Hot Low-Btu Gas, Preprint of Papers, American Chemical Society 170th National Meeting, Division of Fuel Chemistry, Illinois, August 24-29, 1975.

10 M. Vizcarra-Mendoza and C. Laguerie, Etude Experimentale D'Un Procede Continu De Desulfurization D'Un Gas Par Sorption Du SO2 Sur L'Oxide De Cuivre En Lit Fluidise, Environmental Technology Letters, 2, 5 (1981), 214-224.

11 J.P. Strakey, E.R. Bauer, and W.P. Haynes, Removal of SO2 and NOx from Flue Gas in a Fluidized Bed Copper Oxide Process, Energy and Environment, Proceedings of the Sixth National Conference, Pittsburgh, May, (1979), 196-201.

12 G.R. St.Pierre, Method and Apparatus for Sulfur Dioxide Emission Control in Combustion Systems, The Ohio State University, U.S. Patent No. 3,690,808, September 12, 1972.

13 G.R. St.Pierre, Application of Thermodynamics to the Development of Collection Systems for the Control of Sulfur Dioxide Emissions by the Liquid Copper Scrubbing Process, Proceedings of Western Resource Conference, Colorado School of Mines, 1971, pp. 242-62.

14 A.K. Biswas and W.G. Davenport, Extractive Metallurgy of Copper, Pergamon Press, 1980.

15 W.S. Kyte, Some Chemical and Chemical Engineering Aspects of Flue Gas Desulfurization, Trans. Institute of Chemical Engineering, 59, 4 (1981), 219.

16 A. Bronson and G.R. St.Pierre, The Sulfide Capacities of CaO-SiO_2 Melts Containing MgO, FeO, TiO_2, and Al_2O_3, Metallurgical Transactions B, 1981, Vol. 12B, pp. 729-31.

IN SITU SULFUR CAPTURE BY BATTELLE TREATED COAL UNDER SIMULATED COMBUSTION TURBINE CONDITIONS

J.J. REUTHER, H.N. CONKLE, P.R. WEBB, and H.F. FELDMANN

Battelle Columbus Laboratories, 505 King Avenue, Columbus, OH 43201 USA

ABSTRACT

This paper presents the results of an on-going program, sponsored by the United States Department of Energy, designed to determine (1) the technical feasibility of reliably and cleanly firing pulverized coal in advanced combustion turbines, and (2) the temperature and pressure dependencies of the in situ sulfur capture efficiency of a selection of coals under simulated combustion turbine conditions. The selection of pulverized fuels studied consisted of (1) raw high-sulfur Illinois No. 6 coal, (2) the raw coal converted into Battelle Treated Coal (BTC) via a proprietary calcium impregnation process, and (3) the raw coal physically mixed with lime. The simulated combustion turbine conditions consisted of super-atmospheric pressure, ultra-high excess air, moderate average firing temperature, and high carbon conversion.

The analytical and experimental analyses conducted for the program indicate that BTC is favored as a fuel for advanced, coal-capable combustion turbines. Not only do the BTC fuel and flame possess the properties necessary to make coal firing technologically feasible in turbines, they also allow for such combustion to be conducted in an environmentally acceptable manner without the need for troublesome flue gas desulfurization or more expensive chemical coal cleaning.

INTRODUCTION

For the last several years, Battelle Columbus Laboratories has been actively conducting research on the development of a practical, cost-effective process for improving both the fixed-bed gasification (1) and the boiler combustion (2) characteristics of high-sulfur, agglomerating, low-reactivity bituminous coal. The product of this research is a calcium-impregnated coal called Battelle Treated Coal, given the acronym BTC. The properties of BTC are much different from those of the raw coal from which it is processed (ref.1,2). Specifically, these differences are:

- Increased ASTM volatile matter content.
- Higher char reactivity.
- Reduced tar formation tendency.
- Hotter ash fusion temperatures.
- Diminished coal particle agglomeration tendency.
- Enhanced sulfur retention by the ash.

As near-term markets for gasifier and boiler coal diminish or stabilize, respectively, Battelle has expanded the search for other old or new applications for bituminous coal that would also benefit from the alterations that could be given it by the Battelle calcium impregnation process. Recent planning studies by the United States Department of Energy (U.S. DOE) have forecast a need for coal-fired combustion turbines for the repowering of electric utilities when they become capacity-limited or deficient in the early-to-mid 1990s (ref.3). The purpose of the U.S. DOE-sponsored program reported here is to demonstrate the promise of utilizing BTC as a fuel in advanced, coal-capable combustion turbines.

STATEMENT OF THE PROBLEM

Because of differences in combustion hardware and operating conditions, the fuel specifications for candidate combustion turbine coals are much more stringent and quite different from those required for utility boilers (ref.3,4). The principal problem to be solved in this research program was how to demonstrate why BTC would be favored as a fuel for coal-fired combustion turbines. Two approaches were taken, one analytical and the other experimental.

In the analytical approach, the plan was first to understand how the characteristics of and requirements for acceptable pulverized coal combustion under turbine conditions differed from combustion in either conventional coal-fired utility boilers or conventional oil-fired turbines. The next step was to explain how and to what extent the new properties given a raw bituminous coal by the Battelle calcium impregnation process improved the technical feasibility of reliably firing that pulverized coal in a combustion turbine.

In the experimental approach, the plan was to fire raw bituminous coal and BTC produced from it under simulated combustion turbine conditions. Data from such tests would then be used to attempt to demonstrate directly how well the Battelle calcium impregnation process converted the bituminous coal feedstock from a boiler fuel to a turbine fuel. The effect of method of calcium impregnation was also studied.

In addition to obtaining direct information on the feasibility of firing BTC in a combustion turbine, data on the in situ sulfur capture efficiency of BTC were sought as a function of pressure and firing temperature.

EXPERIMENTAL

A coal-fired combustion turbine simulator was constructed at Battelle by converting a continuous high-pressure reactor from a gasifier to a combustor (ref.5). A schematic of the combustion turbine simulator is shown in Figure 1. This device is capable of simulating (ref.3,4) the following combustion turbine conditions:

- Super-atmospheric pressure: 4 atmospheres.
- Maximum coal flame temperature: 2600 F.
- Maximum firing temperature: 1900-2100 F.
- Maximum liner temperature: 1000-1500 F.
- Primary fuel/air ratio: 20 percent excess air.
- Overall fuel/air ratio: 50-100 percent excess air.
- Overall residence time: 150-300 milliseconds.
- Carbon conversion efficiency: 99+ percent.

A premixed methane/air pilot flame was first used to ignite the down-flowing pulverized coal feed at atmospheric pressure. The coal firing rate was then adjusted to about 40,000 Btu/hour. Primary air amounted to 20-30 percent of the total air required for combustion at 20 percent excess air. Secondary air made up the remainder. Temperatures were continuously monitored down the vertical axis of the combustor at locations in the pre-to-post flame zones using Inconel-sheathed Chromel-Alumel thermocouples. Firing temperature was adjustable over a 200 F range by either varying the amount of tertiary dilution air introduced or by oxygen enriching the secondary air to various levels. After combustion was completed within the 12-foot long, 6-inch diameter combustion chamber, samples of gaseous and solid products were collected using an uncooled stainless steel probe and a cyclone mounted in the exhaust, respectively. Flue gases were analyzed for CO_2, CO, O_2, and SO_2 using conventional analytical instrumentation (ref.4). Once the diagnosis of atmospheric-pressure combustion conditions was completed, the system was pressurized to 65 pounds per square inch absolute (psia) and the gas/solid sampling procedure repeated.

The combustion test plan consisted of experiments designed to determine the comparative in situ sulfur capture efficiency, under simulated combustion turbine conditions of: (a) raw pulverized bituminous coal, (b) pulverized BTC prepared from this raw coal, and (c) a physical mixture of the raw pulverized coal and pulverized lime with the same amount of calcium added as for the BTC.

The raw bituminous coal used was from the Illinois No. 6 seam and had a sulfur content of 4 weight percent and a natural calcium-to-sulfur molar ratio (Ca/S) of 0.1. The fuel-bound sulfur content of this high-volatile bituminous coal was split near-equally between inorganic and organic fractions. It was dried to 2 percent moisture, pulverized to 85-90 percent minus 200 mesh (75 microns), and then fired. To produce BTC, coarse raw coal, 100 percent minus 20 mesh (850 microns), was first treated via the proprietary

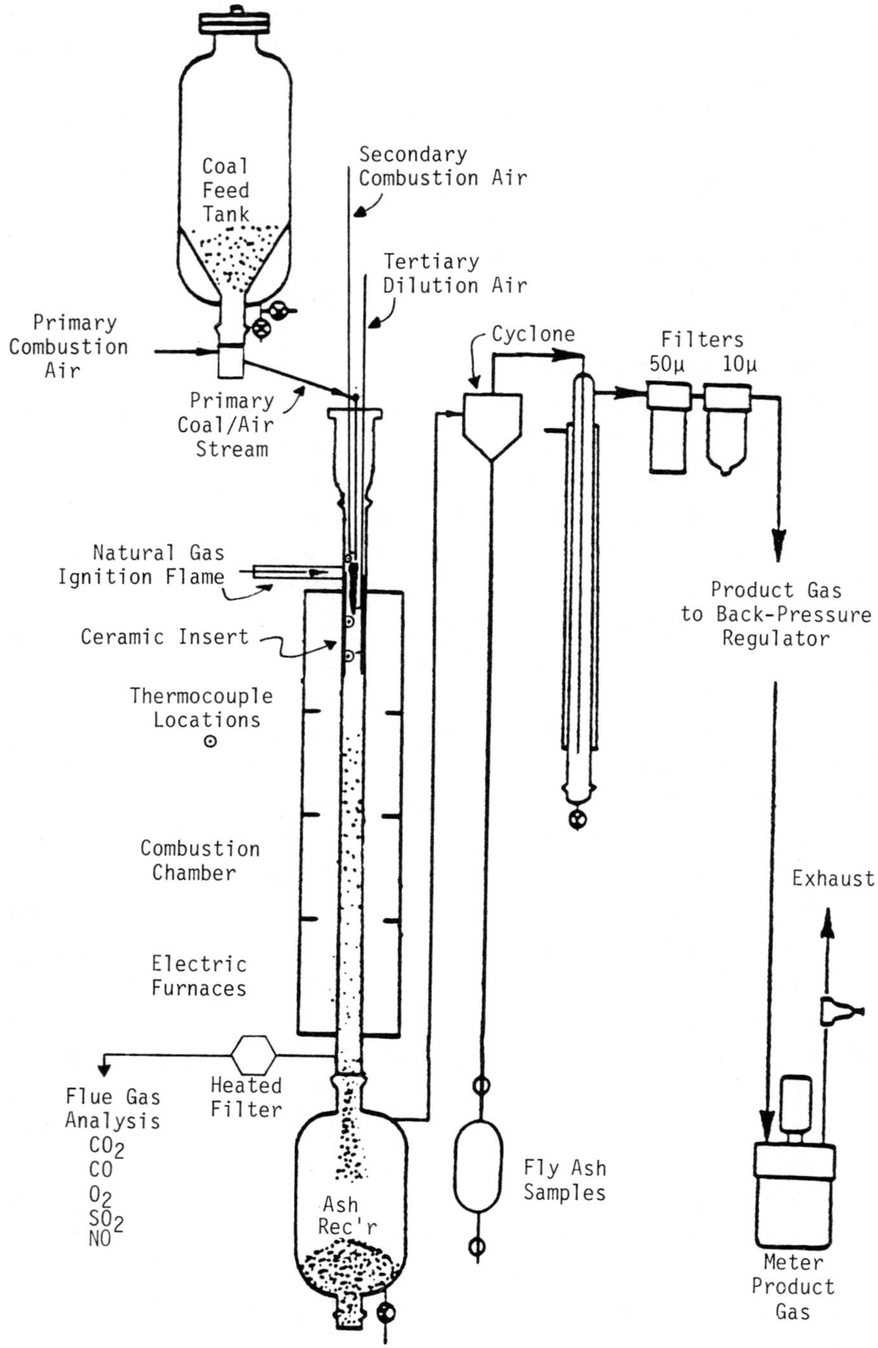

Fig. 1. Battelle combustion turbine simulator.

Battelle calcium impregnation process (ref.1,4) to increase its Ca/S ratio to some prescribed level, in this case 3.0. The BTC was then dried to 2-3 percent moisture, pulverized to 85-90 percent minus 200 mesh, and then fired. For the third fuel, finely-pulverized (85-90 percent minus 200 mesh) raw coal was physically mixed with lime (calcium oxide) having a particle size of 100 percent minus 325 mesh (45 microns); the mix's Ca/S ratio was adjusted to be equivalent to that of the BTC.

RESULTS AND DISCUSSION

The results of the analytical investigation as to why BTC would be favored as a fuel for advanced coal-capable combustion turbines will be discussed first, with some experimental results included where appropriate. The direct firing of BTC rather than either raw coal or chemically-cleaned coal would be technologically favored for the five following annotated reasons.

1. The stable ignition of a pulverized coal flame in a turbine combustor is expected to be more difficult than in a utility boiler because: (a) the chamber pressures are super-atmospheric, 5-10 atmospheres versus 1 atmosphere, and (b) the combustor walls in the near-flame region are cooler, 1000 F versus 1900 F (ref.3,6). The higher pressures in a turbine would reduce the amount of volatile matter than the high-volatile bituminous coal would release if it were under boiler conditions (ref.7,8). The evolution of sufficient volatiles is critical to stable ignition (ref.9). At the rapid-heating rates typical of both boiler and turbine conditions (10^5 F/second), the ASTM volatile yield from high-volatile ($\sim$40 percent) Illinois No. 6 coal would be expected to decrease by about 10-15 percent as the chamber pressure was increased from 1 to 5 atmospheres (ref.7,8). This loss in volatile matter would have an adverse effect on ignition (ref.9). Moreover, the cooler turbine liners would not be capable of thermally assisting ignition to the same extent as the hot refractory walls do in boilers.

   Problems involving the stable ignition of pulverized coal in combustion turbines would be expected to be overcome if the increase in volatile matter yield as a result of calcium impregnation offset the decrease because of higher operating pressures. This appears to be the case as the ASTM volatile matter content of the raw Illinois No. 6 coal is increased by over 30 percent, on a moisture, ash-free basis, upon being impregnated with calcium to a Ca/S ratio of 3 via the Battelle process. This increase would permit pulverized coal to be fired in a turbine at pressures up to about 10 atmospheres before the ease of ignition would be less than that at atmospheric pressure.

It should be noted that a generic result of chemical beneficiation is a lowering of the volatile matter content of a coal (ref.10). Thus, when attempting to fire turbines on coal, ignition problems would be expected to be more severe for the beneficiated coal than for the raw coal. The opposite is the case for BTC.

2. The residence time available for carbon burnout in a combustion turbine is much shorter than that available in a pulverized coal boiler, 0.1-0.4 versus 1-2 seconds (ref.3,6,11,12). The burnout time required for pulverized coal can be shortened by finer grinding (ref.13), but at added expense. Because BTC has a 5-10-fold higher reactivity than that of the raw coal from which it is derived (ref.5), target residence times for carbon burnout in coal-fired turbines would be achievable if BTC were the fuel. This also means either that: (a) the capital cost of coal-capable turbine could be reduced when fine BTC is fired because the flame is shortened and the required combustor volume reduced or (b) the operating cost for a coal-fired turbine could be reduced because the BTC could be pulverized to coarser sizes while retaining the same burnout time required for raw coal of a smaller size.

   Moreover, with coal particle agglomeration eliminated (ref.1,5), the burnout time for BTC should be shorter than that required for the raw coal because the BTC remains as individual particles while the raw coal would agglomerate, forming aggregates that would require longer residence times to burn out than individual raw coal particles.

3. Substitution of pulverized coal for petroleum oils in combustion turbines will probably require the liners to be capable of withstanding higher temperatures (ref.3,6,11,12). This is because coal flames are more radiative than oil flames. Because BTC releases little or no tar (ref.1,5), its radiative heat transfer rate just after ignition should be less than that for raw coal combustion. This effect was observed in simulated combustion turbine experiments at Battelle in the form of measured near-field wall temperatures, which were lower for BTC flames than for raw coal ones at equivalent operating conditions. Thus, increases in liner temperatures and their associated materials problems may not be as severe as first anticipated (ref.3,6). However, the loss of radiative feedback from the flame to the fuel

at early times will also tend to increase ignition instability (ref.12). However again, because the increase in ASTM volatile matter yield upon calcium impregnation is more than the loss upon pressurization up to 10 atmospheres, the ignition instability caused by deficiencies in radiative feedback should be offset.

4. Ash deposition will be a difficult problem to overcome when pulverized coal is fired in a combustion turbine. The use of BTC instead of raw coal can alleviate this problem in two ways. First, by reducing coal particle agglomeration, the particle size of resulting flyash after complete BTC combustion will be smaller than that of the as-fired BTC. Second, the fine flyash particles resulting from BTC combustion are less likely to be sticky or molten because of their higher ash fusion temperatures (ref.1,5). The smaller and more solid the flyash, the more likely it will follow gas streamlines and the less likely it will deposit and/or erode downstream turbine components (ref.3,6). Preliminary results at Battelle indicate that the flyash from BTC flames is smaller in particle size than that from raw coal flames. This is probably because the initial deformation-to-fluid temperature range under reducing conditions for the raw Illinois No. 6 coal was 1995-2165 F, whereas it was 2700+ F for the BTC with a Ca/S ratio of 3 prepared from it.

   It should be noted that while physical or chemical cleaning can greatly reduce the total mineral matter content of a coal, that which is left remaining after beneficiation can have ash fusion temperatures that are equal to or even lower than that of the raw coal (ref.14). Hence, slag deposition problems can actually be made more severe upon beneficiation, rather than less. This is not the case for BTC.

5. Finally, and most importantly, coal contains much more fuel-bound sulfur than petroleum fuels (ref.10). Thus, the direct firing of pulverized raw coal in a combustion turbine will undoubtedly result in non-compliance SO_2 emissions externally (ref.10,15) and severe SO_2 corrosion problems internally (ref.6). Because BTC-firing has been shown to greatly reduce the evolution of SO_2 <u>during</u> simulated boiler combustion (ref.2), the need for expensive pre- or post-combustion SO_2 control technology (3) could be rendered unnecessary if the sulfur capture efficiency were high enough. Subsequently, the capital costs associated with the conversion of turbines from

oil to coal would be reduced. Operating costs would also be reduced because less internal corrosion by SO_2 would mean less maintenance.

This last logical reason for promoting the firing of BTC in advanced combustion turbines was the rationale for the experimental combustion research program, the results of which are now discussed.

Recent research conducted at Battelle has indicated that a high-sulfur (3 percent by weight) Illinois No. 6 bituminous coal impregnated with calcium via the Battelle process efficiently retained its fuel-bound sulfur in its ash during combustion under simulated pulverized coal boiler conditions (ref.2). This retention of sulfur in the ash was much higher than that which is normally observed for high-volatile bituminous coal, which is typically about 15-20 percent (ref.16,17). This enhanced retention of sulfur in the ash exhibited a strong inverse temperature dependence at exhaust temperatures greater than 2200 F. As the coal flame exhaust temperature was decreased to 2200 F, the sulfur capture efficiency increased to approximately 70 percent, when the coal was impregnated to a Ca/S ratio of 3 (ref.2). The peak and exhaust temperatures in combustion turbines are several hundred degrees cooler than those in utility boilers because of the excessive levels of tertiary dilution air (ref.3,6). Because of this, it was suspected at the outset of this program that the enhancement of in situ sulfur capture efficiency afforded raw coal by the Battelle calcium impregnation process would be much greater during turbine-like combustion than during boiler-like combustion. An experimental program was conducted to verify and document this suspicion.

Figures 2 and 3 present the major findings of the experimental program to date. In all cases, the carbon conversion efficiency was 98+ percent.

Figure 2 compares the in situ sulfur capture efficiency of the three test coals as a function of pressure. The Ca/S ratio of the BTC and the physical mixture of raw coal and lime were equivalent at 3; that for the raw coal was 0.1. The excess oxygen and firing temperature for each coal flame averaged about 12 percent and 2000 F, respectively. It should be stressed that straight lines have been drawn in Figure 2 only to assist the eye. Only for the case of BTC does the line accurately reflect the pressure dependence defined by the uncertainty in the data.

Almost all of the data required for the ordinate was obtained by measuring the SO_2 content of the flue gas at a given pressure, normalizing these data to zero percent excess O_2, dividing these corrected SO_2 concentrations by the theoretical maximum amount of SO_2 that the fuel would produce, and then subtracting the result from 100. These data are plotted for the raw coal, the BTC, or the physical mixture as either closed circles, open squares, or open circles, respectively. The SO_2-derived capture efficiency data were

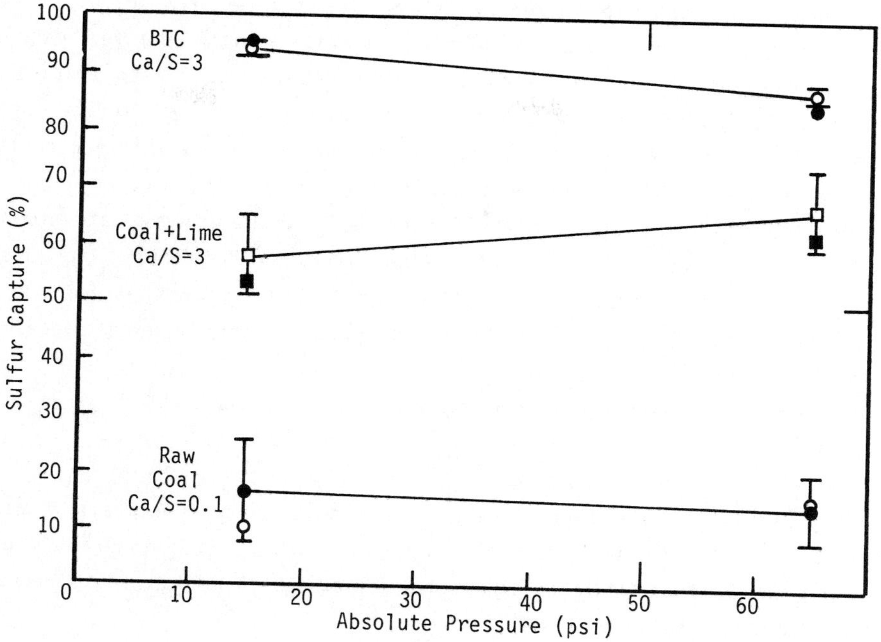

Fig. 2. Comparative sulfur capture efficiency versus pressure.

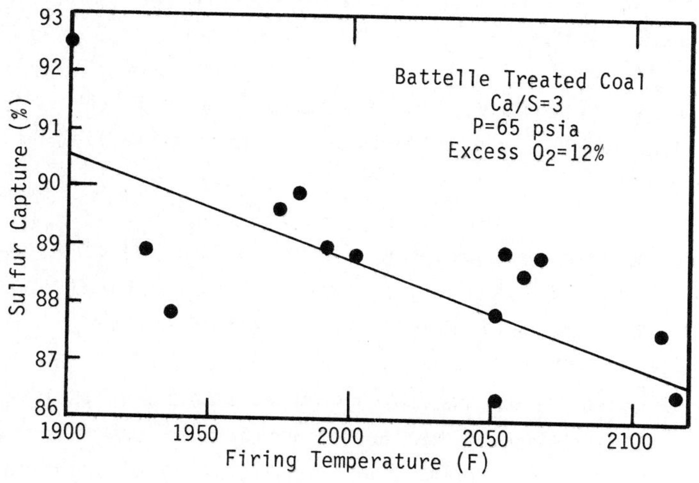

Fig. 3. Temperature dependence of sulfur capture by BTC under simulated combustion turbine conditions.

independently validated by directly determining the percent sulfur retained in the flyash after combustion. The particulate-sulfur derived data are plotted in Figure 2 as open circles, closed squares, or closed circles for the raw coal, the BTC, or the physical mix, respectively. As Figure 2 shows, these complementary measurements produced comparable results within experimental uncertainty. Thus, a reasonable sulfur balance was achieved.

Figure 3 plots the in situ sulfur capture efficiency as determined by SO_2 measurements as a function of firing temperature for BTC with a Ca/S of 3. Pressure and excess oxygen content were held constant at 65 psia and 12 percent, respectively. The line represents a best-squares linear fit of the data.

The following observations can be made from the data in Figures 2 and 3:

- At atmospheric pressure, the raw Illinois No. 6 coal naturally captured about 17 ±9 percent of its fuel-bound sulfur. This inherent capture is comparable to that determined elsewhere for high-sulfur eastern bituminous coals (ref.16,17).

- At 4 atmospheres, the inherent in situ sulfur capture of the raw coal was 14 ±6 percent. This is the first reported measurement of in situ sulfur capture efficiency for pulverized coal combustion as a function of pressure (ref.4).

- Within the uncertainty of the data, the inherent sulfur capture efficiency of the raw Illinois No. 6 coal appears to be independent of pressure.

- At atmospheric pressure, the sulfur capture efficiency of BTC with a Ca/S ratio of 3 was 94 ±1 percent. This is almost 6 times greater than that of the raw coal under similar conditions.

- This enhanced sulfur reduction by BTC at atmospheric pressure is greater than that required by New Source Performance Standards for SO_2 from coal-fired boilers (ref.15), which will probably be the basis for SO_2 compliance limits for future coal-fired turbines.

- The high level of sulfur capture by BTC at atmospheric pressure implies that the mechanism for sulfur removal is independent of the form of sulfur in the fuel. It should be noted that physical beneficiation can only remove part of the inorganic sulfur fraction, which can constitute no more than 70 weight percent of the total sulfur in Illinois No. 6 coals (ref.18).

- At 4 atmospheres, the sulfur capture efficiency of the BTC studied appeared to decrease to 87 ±1 percent.

- The sulfur capture efficiency of BTC with a Ca/S ratio of 3 appears to exhibit a small negative pressure dependence.

- Sulfur capture by the BTC fired at 4 atmospheres and 12 percent excess oxygen exhibited a negative temperature dependence over the 1900-2100 F range. Sulfur capture efficiency was reduced by approximately 2 percentage points for every 100 F increase in firing temperature.

- At atmospheric pressure, the physical mixture of raw pulverized coal and lime with a Ca/S ratio of 3 captured 58 ±7 percent of its fuel-bound sulfur. This level of sulfur retention is comparable to that observed by others elsewhere (ref.19) when cofiring pulverized coal and limestone under atmospheric pressure, staged-combustion conditions.

- At 4 atmospheres, the sulfur capture efficiency of the physical mix of coal and lime was 67 ±7 percent.

- Within the uncertainty of the data, the sulfur capture efficiency of the physical mix appears to be independent of pressure.

CONCLUSIONS

The following are the major conclusions that can be drawn from the data obtained thus far in the program:

- BTC appears to be a technologically and environmentally superior fuel for advanced, coal-capable combustion turbines.

- BTC is 1.2-1.5 times more effective at in situ sulfur capture than a physical mixture of raw coal and lime at an equivalent Ca/S ratio of 3. This demonstrates that the Battelle treatment process impregnates calcium into raw pulverized coal in a different, more effective way than physical mixing.

- The existence of a pressure dependence (slightly negative) for sulfur capture during BTC combustion and the apparent lack of one for either raw coal or raw coal/lime combustion provides indirect evidence that the mechanisms for sulfur capture for these fuels are different, and justifies why the absolute sulfur capture efficiencies observed for each are different.

SUMMARY

Battelle has operated a pulverized coal-fired facility that essentially simulates combustion turbine conditions. This simulator has proven to be of practical value in research concerned with the screening of coal fuels for advanced combustion turbines. Specifically, it has demonstrated that (1) calcium impregnation via the Battelle treatment process improves ignition, burnout, and deposition characteristics of bituminous coal to the point where it can be reliably fired in a turbine, and (2) the production of a low-SO_2, fine-particulate laden flue gas conducive to long turbine life and low maintenance is achievable via BTC-firing. With this singular device, in situ capture efficiency has been measured as a function of super-atmospheric pressure and firing temperature for three different coals. The sulfur capture efficiency of BTC has been compared with that of physically-mixed raw coal and lime at an equivalent molar calcium/sulfur ratio under these simulated combustion turbine conditions. BTC is clearly more effective at in situ sulfur capture. Lastly, although not elaborated upon, the effectiveness of the Battelle calcium impregnation process has been determined using coarse, rather than pulverized, raw coal as a feedstock; this information indicates that the economics of the Battelle treatment process are probably more favorable than originally determined (ref.1).

Future research in this program will consider the effect of lower Ca/S ratio on in situ sulfur capture by both BTC and physical mixtures of coal and lime. This information will permit a more complete determination to be made of the comparative economics of BTC, flue gas desulfurization, and ultra-coal cleaning in terms of SO_2 control in coal-fired combustion turbines.

ACKNOWLEDGEMENTS

The authors gratefully acknowledge the financial support of this research by the U.S. Department of Energy under Contract No. DE-AC21-84MC21289, which is monitored by Mr. Nelson F. Rekos, Jr. of the Morgantown Energy Technology Center.

REFERENCES

1 H.F. Feldmann and H.N. Conkle, Coal Catalysis Expands Gasifier Application Options, Energy Progress, 3:2 (1983) 105-109.

2 E.L. Merryman and A. Levy, In-Situ Capture of Sulfur in Combustion, Central States Section Combustion Institute, CSS/CI-82-10, 1982, 23 pp.

3 F.W. Crouse, Chairman, Proceedings of the First Annual Heat Engines Contractors Review Meeting, Morgantown Energy Technology Center, United States Department of Energy, DOE/METC/84/31, November 1984, 369 pp.

4 J.J. Reuther, H.N. Conkle, and H.F. Feldmann, Evaluation of Calcium Impregnated Coal as a Fuel for Turbine Combustors, Proceedings of the Second Annual Heat Engines Contractors Review Meeting, Morgantown Energy Technology Center, United States Department of Energy, April 1985 (in press).

5 H.N. Conkle, H.F. Feldmann, D.R. Hopper, and E.L. Merryman, Low Sulfur Fuel Gas by the Two-Stage Fluidized-Bed Partial Oxidation/Combustion of Battelle Treated Coal, American Institute of Chemical Engineers Winter National Meeting, March 1984, 23 pp.

6 F.L. McCay, Chairman, The Coal Burning Gas Turbine Project, Report of the Interdepartmental Gas Turbine Steering Committee, Departments of Minerals and Energy and of Supply, Australian Government Publishing Service, Canberra, 1973, 228 pp.

7 J.B. Howard, W.A. Peters, and M.A. Serio, Coal Devolatilization Information for Reactor Modeling, Final Report from MIT on Research Project RP986-5 to the Electric Power Research Institute, April 1981, 279 pp.

8 T.X. Phouc and P. Durbetaki, Heat and Mass Transfer Analysis of Coal Particle Undergoing Pyrolysis, American Flame Research Committee/International Flame Research Foundation International Symposium on Alternative Fuels and Hazardous Wastes, Paper 3, 1984, 26 pp.

9 J.J. Reuther, Ignition, Burnout, and Pollution Characteristics of Various Ranks of Pulverized Coal and Their Slurries, Transactions of the Society of Mining Engineers of the American Institute of Mining Engineers, 276 (1985) 1930-1934.

10 J.J. Reuther, The Fate of Sulfur in Coal Combustion, Earth and Mineral Sciences, The Pennsylvania State University, 51 (1982) 49-53.

11 H.R. Hazard and F.D. Buckley, Experimental Combustion of Pulverized Coal at Atmospheric and Elevated Pressures, Transactions of the American Society of Mechanical Engineers, 10 (1948) 729-737.

12 J.M. Beer and G. Vermes, Gas Turbine Combustors for Coal Water Mixtures, Engineering Foundation Conference on Combustion of Tomorrow's Fuels-1 (in press, 1985).

13 J.J. Reuther, G.G. Karsner, S.T. Jack, Plane Flame Furnace Combustion Studies of Pulverized Wood, Fundamentals of Thermochemical Biomass Conversion, Elsevier Applied Science Publishers, 1985, pp. 793-810.

14 J. Singer, Editor, Combustion: Fossil Power Systems, Combustion Engineering Publishers, 1981, pp. 3-1-33.

15 New Source Performance Standards for Electric Utility Steam Generating Units, Federal Register, Vol. 44, No. 113, United States Environmental Protection Agency, June 11, 1979, pp. 33580-33624.

16 S.C. Hunter and P.K. Engel, Sulfur Oxides Emissions from Boilers, Turbines, and Industrial Combustion Equipment, Workshop on Measurement Technology and Characterization of Primary Sulfur Oxides Emissions from Combustion Sources, P-201, United States Environmental Protection Agency, April 1978, 54 pp.

17 S.L. Chen, M.P. Heap, D.W. Pershing, R.K. Nihart, and D.P. Rees, The Influence of Fuel Composition on the Formation and Control of NO_x in Pulverized Coal Flames, Western States Section Combustion Institute Paper WSS/CI-80-13, 1980, 38 pp.

18 J. Cavallaro, M. Johnson, and A. Deurbrouck, Sulfur Reduction Potential of the Coals of the United States, United States Bureau of Mines Report of Investigation 8118, 1976, 323 pp.

19 P.L. Case, M.P. Heap, R. Payne, and D.W. Pershing, LIMB Testing: The Use of Dry Sorbents to Reduce Sulfur Oxide Emissions from Pulverized-Coal Flames Under Low-NO_x Conditions, Proceedings of the Joint Symposium on Stationary Combustion NO_x Control, United States Environmental Protection Agency and Electric Power Research Institute, EPRI CS-3182, 1, 1980, 25 pp.

GASIFICATION OF COAL-LIME PELLETS

HUGH L. CAMPBELL

Dravo Wellman Company, 4800 Grand Avenue, Pittsburgh, Pennsylvania 15205, U.S.A.

ABSTRACT

During late August, 1984, gasification tests were conducted in the three foot diameter fixed bed pilot gasifier at the Dravo Technology Center on Neville Island. The tests were run on 4.9% sulfur coal, and two coal-lime pellet formulations made from the same coal. Hydrated lime was added to the finely ground coal at a 1:1 and a 2:1 calcium-sulfur ratio. Coke was used to start up the gasifier, and was also used to separate each of the three materials, coal, 20% coal-lime pellets (2:1 calcium to sulfur), and 10% coal-lime pellets (1:1 calcium to sulfur).

Sulfur in the gas was reduced by 77 percent with the 10% coal-lime pellets, and 82 percent when the 20% coal-lime pellets were gasified. Tar and oil loadings in the gas stream were also reduced. The 10% coal-lime mixture produced a 59.4 percent reduction, the 20% coal-lime formulation reduced the tar by 61.4 percent. The higher heating value of the cold, detarred gas also improved.

INTRODUCTION

There are several converging trends which require the conversion of high sulfur coal fines into useable energy. These trends include:

The existence of the fines.

Throughout Western Pennsylvania, Ohio, Indiana and Illinois, coal cleaning plants associated with major utilities are generating high sulfur coal fine refuse which requires disposal. These fines leach into ground water, streams and rivers, and constitute a serious environmental hazard. Conversion of these wastes into useable energy is highly desireable.

Cost of Electricity

During the '70s, electrical utilities responded to increasing demand for power by constructing large coal and nuclear facilities which are now coming on-line. The cost, and cost overruns due to regulatory requirements, are just now being incorporated into the rate base. The direct result of this construction and incorporation of these costs into the rate base will be a dramatic increase in the cost of electricity. In turn, utilities are expected to turn away from construction of large generating plants, and turn towards the construction of smaller facilities, using cogeneration when possible.

Acid Rain

The major concern confronted in the construction of new coal-fired generating stations is the production of still more sulfur dioxide, thus excaserbating an already bad environmental situation.

The convergence of these trends demand a means whereby the accumulating piles of high sulfur coal fines can be converted to useable electrical energy without contaminating the atmosphere with sulfur.

This paper describes a means whereby, using currently available proven technology, coal fines, properly prepared, can be gasified in a fixed bed gasifier, and the gas produced used in a high efficiency engine to generate electricity. The sulfur in the coal is captured during the gasification process and is discharged harmlessly in the ash, not in the atmosphere.

The testwork described in this report follows earlier work done by the McDowell Wellman Company in Cleveland, and by Battelle Laboratories in Columbus. Our tests confirm the earlier results, and continued work on this process will be reported by Battelle during this Conference.

TEST PURPOSES

During the last week in August 1984, the first test run of the Pilot Gasifier at the Dravo Technology Center was conducted. The purpose of the test was to analyze the gas produced in the fixed bed gasifier operating on a high sulfur (4.5% S) bituminous coal, and compare the results of this analysis with analyses obtained from the results of gasifying the same coal when finely ground and blended with lime. The latter analyses were obtained from two coal-lime formulations, the first formulation consisting of a 1:1 calcium to sulfur ratio, the second from a 2:1 calcium ratio.

Earlier work by the McDowell-Wellman Engineering Company has shown that the addition of lime to finely ground coal would result in sulfur capture by the lime, producing calcium/sulfur compounds in the ash. In the McDowell-

Wellman process the pellets were sintered prior to gasification. This process is expensive and reduces the heating value of the pellet fuel.

Subsequent work done by Battelle suggests that the addition of lime to high sulfur bituminous coal would not only reduce sulfur in the gas stream but would also reduce the tar produced. The Battelle experiments were conducted on a small scale in a bench-sized 8 inch diameter gasifier. Verification of the Battelle work requires full scale operation in a commercial size gasifier.

The purpose of this test was to verify the earlier results obtained. To be commercially significant this verification must be done in a commercial gasifier using pellets which are air dried only, not sintered.

In addition, there is concern that the calcium/sulfur compounds present in the ash might contain significant concentrations of calcium sulfides, rather than calcium sulfates, thereby constituting a potential ash disposal problem. An additional purpose of this test, therefore, was to analyze the ash from coal-lime pellets to determine the form of the sulfur species contained herein.

TEST PROCEDURES

COAL PREPARATION

The coal to be gasified was sampled and an analysis was made of the composition. This analysis is included in the following section. The coal was screened and the fraction under 1/4" as separated and used for the coal-lime pellets. Some of the screened fraction of coal 1/4" and over was set aside for the coal gasification part of the test.

The coal for pelletizing was ground in a ball mill to reduce the particle size to 80% minus 200 mesh. The pulverized coal was divided into two lots of equal weight for subsequent addition of lime.

PELLET PREPARATION

Two coal-lime formulations were made; one consisting of 10% hydrate of lime by weight (1:1 calcium to sulfur ratio on a mole basis), and 20% lime to coal (2:1 calcium to sulfur). The pulverized coal was blended with lime, then formed into pellets ranging in size from 3/8" to 3/4" in diameter. The pellets were allowed to air dry for several days. Then the pellets were blended to assure a consistent size mix for each of the two formulations.

COKE PREPARATION

To ensure that there would be a stable bed for gasification of the high sulfur coal and each of the pellet formulations, it was decided that the gasifier be started on coke and that coke would be reintroduced into the gasifier between each of the test runs. The coke was run through the grinder to reduce the top size to 1", but no further sizing was done. Following sizing, the coke was also blended to provide a uniform fuel bed for each of the interim test runs.

FUEL FEED TO GASIFIER

Prior to light off of the gasifier, the gasifier vessel and the lower coal bin were filled with a pre-weighed charge of fuel (coke). The upper bin was also filled with coke, weighed, then elevated into position over the lower bin and dumped. This process was repeated until the gasifier and the lower bin were fully charged. The upper bin was refilled, and placed in its normal position above the lower bin to provide a continuous supply of fuel following light off.

During the test, the upper bin was discharged into the lower bin as required, removed, lowered to ground level, weighed, and refilled. This process was repeated as necessary until the lower bin was completely filled. The weight of the fuel added in each case was divided by the interval between refuelings to obtain an average feed rate.

The various fuels used in this test were batch fed, first coke, then high sulfur coal followed by coke, 10% lime pellets, coke again, 20% lime pellets, and ending with coke. The purpose of interjecting coke between the other fuels was to ensure that there would not be intermixing of gases from the coal and coal-lime mixtures. To ensure that the baseline data on the raw coal was accurate, a subsequent test was run on the coal to permit obtaining several additional samples of tar and sulfur.

GASIFIER OPERATION

Following lightoff, the gasifier was run at low load until a stable bed was obtained. For the first full day of operation, the air flow to the gasifier was raised to establish a fire load loss of 6 inches. (Fire load loss is the pressure drop through the bed). During the first refuel, the average feed rate at 6 inch load was calculated to be 250 pounds per hour. This is more than was desired for this test, so air flow was reduced to provide a 5 inch fire load loss, and this load was maintained for the remainder of the test. This fire load loss corresponds to a feed rate of approximately 175 pounds per hour.

In the commercial Wellman-Galusha gasifier, the saturation of the blast air is accomplished by allowing the blast air to flow over the top of the water in the water jacket. Normally, the temperature of the water leaving the jacket is only 10 to 20 degrees higher than the saturation temperature.

In the pilot gasifier, blast air saturation is obtained in the same manner; that is, by allowing blast air to flow over the top of the jacket water. During the first day, jacket water flow was adjusted to reach a saturation temperature of 135 degrees F (which, from experience, is a good starting place for coke). It was found that jacket temperatures in excess of 190 degrees F were required to obtain proper saturation temperatures. The nominal offtake temperature for the remainder of the test was held at 130 to 140 degrees.

The several changes of feed material during the five day test interval did not provide sufficient time for optimizing saturation temperature for any of the fuels tested. This will be done in subsequent runs. For this test, saturation temperatures were set to minimize clinker formation, thus allowing for continuous operation on each of the test fuels. This procedure ensured that meaningful tar, sulfur, and gas samples could be obtained.

Gasifier operation on coke began on 8/21 and stable operation on this fuel was established. Early on the morning of 8/22 high sulfur coal was introduced into the lower bin. As the coal entered the gasifier vessel later that day, the offtake temperature increased, as expected, from 400 to 900 degrees F. The agitator was turned on at this time, and the offtake temperature dropped to approximately 700 degrees. The reduction in offtake temperature is consistent with past experience at the Bureau of Mines gasifier and is attributed to improved consolidation of the bed.

Also noted and expected, as the coal entered the gasifier vessel, there was an increase in methane and C2+ gases. This occurs as a result of heating the volatile matter in the coal. An advantage of the marked difference in the methane and higher hydrocarbons in the gas stream is that it clearly identifies that the coal or pellets are in the gasifier, and that the tar and oil samples are taken at the correct time.

On 8/23, following operation on coal for approximately 12 hours, coke was introduced into the gasifier to return to a reference condition. Some coke entered the gasifier before sulfur and tar samples were taken and the results of the tar and sulfur analyses, therefore, represent a conservative condition; that is, the sulfur and particularly the tar reported may be less than the amount produced when the retort is filled with coal. A subsequent run was made on the same coal to verify the tar and sulfur content in the gas stream. The data included in this report is based upon the average of the tar and sulfur samples taken during the second test.

The gasifier was filled with coke and the bed stabilized before coal-lime pellets, with 20% lime by weight, were introduced into the feed bins. The 20% lime represents a 2:1 ratio of calcium to sulfur.

On 8/25 the gasifier was completely filled with pellets. Tar and sulfur samples were again taken. As routine practice, the samples were taken to coincide with the gas chromatograph sampling period.

Next, the gasifier was refilled with coke to establish base conditions, then coal-lime pellets with 10% lime (1:1 calcium to sulfur) were loaded. When the gasifier was completely filled with this material, samples were taken once more. The gasifier was filled with coke again and then shut down - completing the test.

On October 3rd a second test was started in the pilot gasifier to affirm the baseline data for the raw coal. The gasifier was started up on coke as before and the bed was stabilized. Operation on coke was continued through October 4th, then the same high sulfur coal used during the first test was loaded into the gasifier. By October 6th the gasifier was fully loaded with coal. Gas, sulfur and tar analyses were made at periodic intervals to serve as an accurate baseline for the coal-lime pellet test.

SAMPLING PROCEDURES

Gas samples of the raw gas were fed directly to an on-stream Hewlett-Packard gas chromatograph and were analyzed for hydrogen, carbon monoxide, carbon dioxide, nitrogen, methane, and hydrocarbons up through C6.

Sulfur samples were taken in sampling bottles and analyzed on a Varian chromatograph for hydrogen sulfide, sulfur dioxide, carbonyl sulfide and methyl mercaptans.

Tar samples were collected in sampling bottles placed in an ice bath. The offtake gas containing tar was aspirated through the sample train by a metering pump so that tar loading as a function of gas volume could be obtained. To the extent possible, the tar and sulfur samples were taken at the same time gas samples were being drawn.

A sample of fuel was taken from each hopper as fuel was being loaded. Composite samples of each fuel were prepared from these individual samples. The same procedure was used to obtain composite ash samples.

TEST DATA

In this section, the data obtained during the coal, coal-lime pellet test is summarized. The data is presented in tabular or chart form for each of the fuels tested, and from a comparison of these tables the material for the next section, "Results and Conclusions", is drawn.

CHARTS AND TABLES

FUEL LOAD DATA

DATE	TIME	FUEL	POUNDS
8/17	1030	COKE	3520
(Producer banked on coke until 8/21)			
8/21	1420	COKE	1420
8/22	0035	COAL	1675
8/22	0830	COAL	1380
8/24	0130	COKE	740
8/24	1550	COKE	775
8/24	2305	10% P	1300
8/25	0330	COKE	1560
8/25	0710	COKE	150
8/25	0710	10% P	1080
8/25	1230	10% P	1130
8/25	1445	10% P	520
8/25	1730	COKE	960
8/25	2330	20% P	1060
8/26	1330	20% P	1020
8/26	1750	20% P	1320
8/26	1850	20% P	800
8/26	2210	20% P	100
8/26	2210	COKE	1410
END OF TEST			

SUMMARY OF FUEL USED:

HIGH SULFUR COAL	3055 LBS
10% LIME PELLETS	4030 LBS
20% LIME PELLETS	4300 LBS
COKE	10535 LBS
TOTAL	21920 LBS

COAL, COKE AND PELLET ANALYSES

Shown below are the results of analyses made on the composite fuel samples.

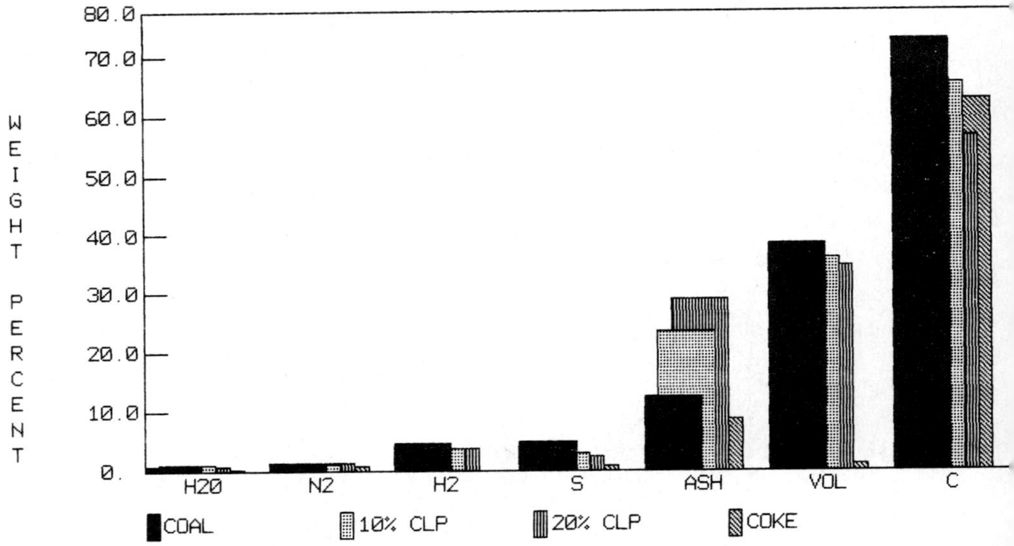

COMPOSITE FUEL SAMPLES

GAS COMPOSITIONS

The two charts below show average gas compositions for each of the four fuels. the 0 to 3% chart scale in the first chart was enlarged for the second chart to show the relationship of trace gases.

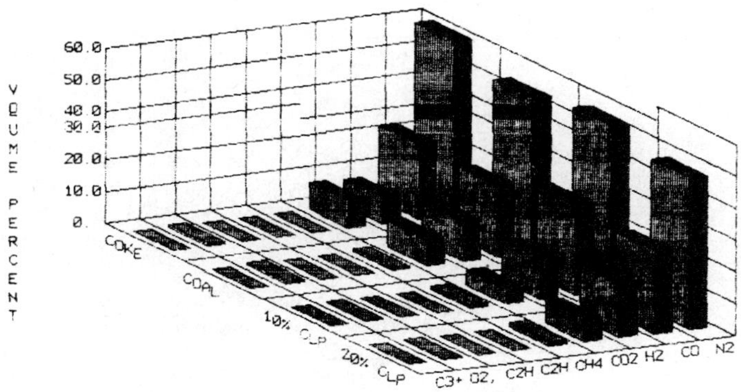

CHART 1 - OVERALL GAS COMPOSITION

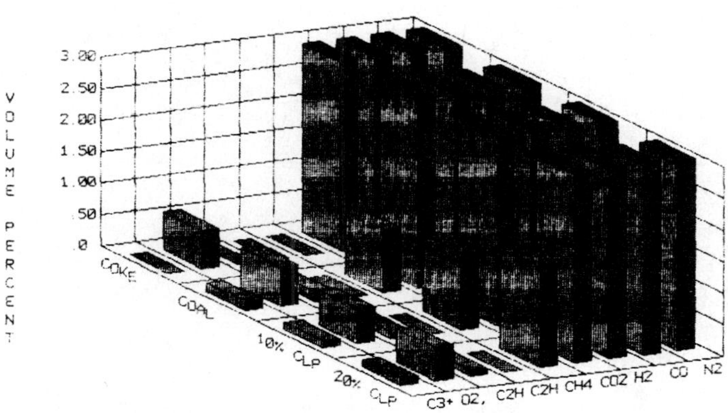

CHART 2 - TRACE GASES

SULFUR ANALYSES

The following graph shows the average sulfur content of the gas for each of the three fuels, coal, 10% coal-lime pellets and 20% coal-lime pellets. No sulfur tests were run on coke.

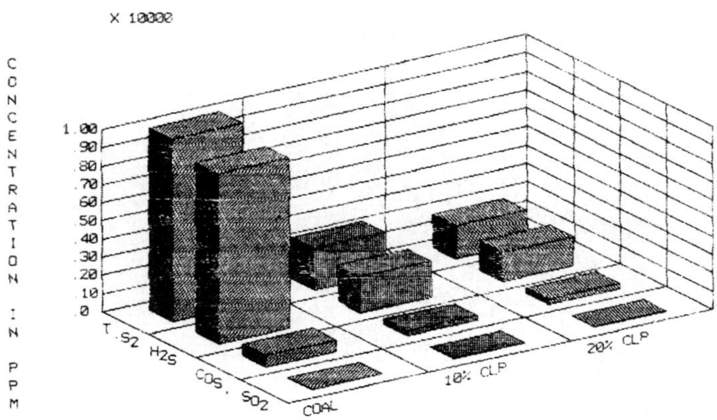

SULFUR SPECIES IN HOT GAS STREAM

TAR CONTENT

Shown below is the average tar content of the gas produced from each of the four fuels. Coal data for the next two charts were obtained from the second test.

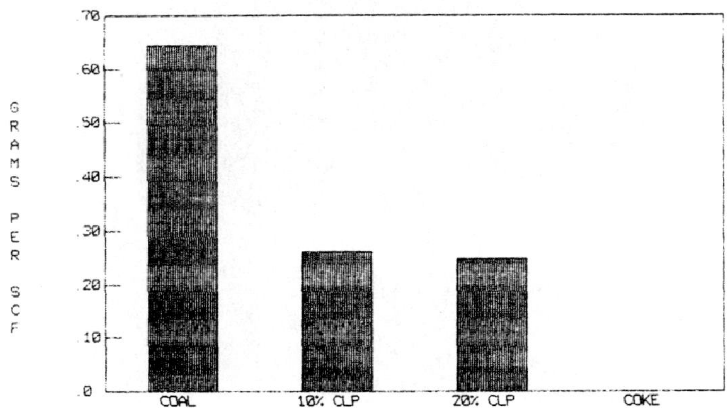

CHART 1 - TAR LOADING IN GAS

TAR COMPOSITION

The major constituents of the tars obtained from raw coal and coal-lime pellets are shown below.

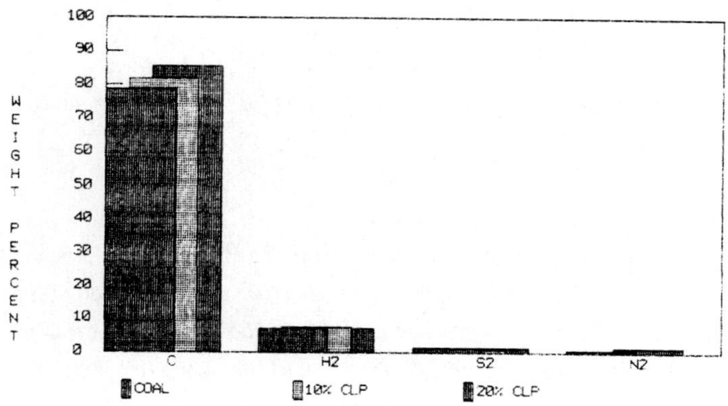

COAL AND PELLET TAR COMPOSITION

RESULTS AND CONCLUSIONS

The objective of this test was to confirm that the addition of lime to high sulfur coal using finely ground material would result in a reduction of sulfur in the gas due to capture of the sulfur in the ash. A further objective of the test was to examine the effect of lime on the production of tar, and the final objective of the test was to determine the chemical characteristics of the ash, particularly with respect to sulfur chemistry resulting from the anticipated sulfur capture.

The results of this test are presented in the order of each of the above objectives.

SULFUR CAPTURE

The sulfur analysis made during the gasification run on high sulfur coal serves as the base case for this test. The sulfur sample taken during both of the tests on coal shows an average hydrogen sulfide concentration of 9325 ppm, and a total sulfur content of 9978 ppm. Sulfur samples taken during the pellet test runs averaged 1983 ppm hydrogen sulfide and 2290 ppm total sulfur for 10% lime pellets. For the 20% lime pellets, hydrogen sulfide was 1593 ppm and total sulfur averaged 1816 ppm. The overall sulfur capture was 77% for the 10 percent coal-lime pellets and 82% for the 20 percent formulation.

TAR REDUCTION

A major consideration in running this test was to observe the extent to which lime would reduce the tar formation in the gas stream when gasifying bituminous coals. Earlier work by Battelle suggested that tar formation is reduced and the production of higher hydrocarbons in the gas stream is enhanced. The results of this test confirm that suggestion, although not to the extent predicted. During the gasification of high sulfur coal, the tar content of the gas was 0.645 g/SCF. When 10 percent coal lime pellets were gasified the tar concentration dropped to 0.262 g/SCF. For 20% lime pellets, the tar concentration was further reduced to 0.249 g/SCF. The tar reduction for 10 percent coal-lime pellets is, therefore, 59.4%. For the 20 percent coal-lime mixture, the reduction is 61.4%.

ASH ANALYSIS

A major concern of this test is the composition of the sulfur forms which are produced in the ash by the gasification of coal-lime pellets (or briquettes). The formation of calcium sulfates is considered innocuous, as these compounds are considered stable and will not produce deleterious results if the ash is used as a landfill material. Production of calcium sulfides or sulfites, however, might produce an ash which could produce hydrogen sulfide when water is added, and therefore become a problem material in a landfill operation.

Ash samples taken during periods when the coal-lime pellets were being gasified were analyzed for sulfur compounds with the following results:

The total sulfur in the coal was 4.96%, all of which is assumed to report to the gas stream or the ash.

The total sulfur content of the ash from the 10% coal-lime pellets was 0.45% which appeared in the ash as 0.39% sulfides, 0.01% sulfites, and 0.01% sulfates.

The ash from the 20% coal-lime formulation contained a total sulfur of 0.74% ash consisting of 0.47% sulfides, 0.03% sulfites, and 0.04% sulfates.

Independent tests were performed on the ash samples by an outside testing laboratory. Their tests showed a slightly different ratio of sulfites to sulfides, but confirmed the fact that most of the sulfur in the ash is in the form of sulfides and sulfites, not sulfates.

The ash analyses (insofar as sulfur species is concerned) are not unlike those obtained from analyses of sludge from wet lime scrubbers, and it is therefore possible to stabilize the gasifier ash with a material such as Calcilox to make it suitable for use in a landfill.

CONCLUSIONS

This test confirms the earlier work of the McDowell-Wellman Company and that done by Battelle. Moreover, it confirms the fact that coal-lime pellets made with high sulfur coal can be gasified in a commercial gasifier, producing compliance gas. Based upon these tests, there is no need to sinter the pellets prior to gasification.

Equally important is the fact that the higher heating value of the gas is increased due to increased production of higher hydrocarbons, especially methane.

The primary conclusion of this test is that there is a significant potential for the gasification of pellets or briquettes made from high sulfur coal fines which are abundant, and inexpensive, both in this country and abroad.

RECOMMENDATIONS

It was noted that the short duration of this test did not allow for optimizing operating conditions for each of the fuels. Such factors as optimum saturation temperature and throughput were not obtained, nor could they be obtained in such a short period.

In addition, only two coal-lime formulations were tested. Battelle's results suggest that increasing the lime content of the fuel pellet will improve sulfur capture and tar reduction. Their data, as do our tests, indicate that the increase in sulfur capture and reduction of tar is not proportional. In our tests, increasing the lime by 100% produced only a 5% reduction in sulfur, and a 2% reduction in tar. Longer test runs and varied calcium/sulfur ratios are needed to determine optimum conditions.

For these reasons, and in light of the promising results of these first two tests, the following recommendation is made:

> Tests of longer duration be run, both on the same pellet formulations used in this test, and also on pellet formulations with higher lime content. Tests should also be run on calcium oxide rather than hydrate. These tests should be of sufficient duration to determine optimum operating conditions. A special effort is needed to determine throughput of the gasifier operating on pellets as these appear to be more reactive than raw coal.

PRESSURIZED FLUID BED COMBUSTION FOR HIGH SULFUR COAL UTILIZATION: AEP'S DEVELOPMENT PROGRAM

R. I. PAWLIGER and M. J. MUDD

American Electric Power Service Corporation, Mechanical Engineering Division, 1 Riverside Plaza, P. O. Box 16631 Columbus, Ohio 43216

ABSTRACT

American Electric Power (AEP) has been a leader in the development of Pressurized Fluidized Bed Combustion (PFBC) since 1976. AEP and its partners have developed a combined cycle PFBC concept with a deep, tapered bed, operating at 1580°F and pressure up to 235 psia to capture sulfur in the coal during combustion.

In this paper, AEP's PFBC development program is discussed including the 15 MW_t Component Test Facility PFBC pilot plant, the 70 MW_e Tidd Demonstration Plant to be installed in Brilliant, Ohio, and future advantages of PFBC are also discussed and compared to conventional coal-fired processes with flue-gas desulfurization to indicate promising future of utilizing PFBC for electric power generation from high sulfur coal.

INTRODUCTION

The American Electric Power (AEP) system provides electricity to its customers in seven states in the midwestern United States. With an installed capacity of almost 23 GW, it is one of the nation's largest investor-owned electric utilities. Almost 90% of AEP's electricity is generated from coal. Because AEP is America's largest single coal consumer -- in 1984 AEP burned over 38 million tons of coal -- and because the AEP system is located in the heart of America's high sulfur coal reserves, AEP is committed to the development of new technologies to generate electricity from high sulfur coal in an economic and environmentally compatible manner.

During combustion, sulfur in coal reacts with oxygen to form SO_2. In order to minimize SO_2 emissions, sulfur must be removed from the process either prior to combustion (coal cleaning), after combustion, (flue gas desulfurization) or during combustion (fluidized bed).

In PFBC, sulfur is removed during combustion by adding a sorbent, such as dolomite, to the process. In this paper, AEP's PFBC development program is presented outlining nine years of privately funded research and development leading to a combined cycle power plant design which is projected to have economic, environmental, and efficiency advantages over conventional coal-based power generating technologies.

THE PFBC PROCESS

The PFBC process applies the fluidization principle -- where an airstream flowing upward through particles in a container maintains the particles in a highly turbulent suspended state, to the sulfidation principle where a sorbent removes sulfur in the coal during the combustion process.

In a fluidized bed, crushed coal and dolomite are continuously injected into a vessel. Air injected through nozzles in the bottom of the vessel maintain these particles in a highly turbulent suspended state, called "fluidization". This turbulence permits excellent surface contact between the air and between the particles themselves, allowing for efficient combustion as well as excellent absorption of the SO_2 by the dolomite.

Boiler tubes submerged in a fluidized bed can be used to generate steam to drive a steam turbine-generator. Overall heat transfer characteristics in these tubes are four to five times greater than tubes in a conventional boiler. Therefore, the total amount of boiler surface is significantly reduced compared to a conventional boiler.

Pressurizing the process results is sufficient energy in the exhaust gases to drive a gas turbine, while the steam generated in the in-bed tubes drives a steam turbine. This configuration allows for an efficient combined cycle power generating plant design.

In PFBC, higher operating pressure allows for the use of deep beds which permit long in-bed residence times -- improving combustion efficiency and sorbent utilization.

Combustion in a fluidized bed occurs at the relatively low temperature of 1580°F. This temperature is below the ash fusion point of virtually all coals. Therefore, slag formation does not occur, and hence the process is not sensitive to the type of coal burned. In addition, this low combustion temperature minimizes the formation of thermal NO_x -- to less than 50% that of a conventional furnace.

The waste products of fluidized bed combustion are dry. The products consist of ash, magnesium oxide, calcium sulfate, and partially calcined dolomite. Unlike scrubbers, no sludge is produced.

These inherent advantages of sulfur removal during combustion for the PFBC process have motivated AEP to develop PFBC technology to the application of combined cycle power generation.

HISTORY

Fluidized beds date back to the 1920's when they were introduced in the chemical industry in processes involving the reaction of solids and gases. Applications for exothermic chemical processes with the concomitant heat removal from the bed raised the possibilities of the application of fluidized beds for coal combustion for generating steam. That came about in the 1950's but not much advance was made in that application because of difficulties in controlling the process and the lack of incentive to develop the concept. In the 1960's when environmental concerns regarding sulfur emissions evolved, the possibility became apparent of utilizing a fluidized bed of a mixture of coal and limestone to minimize sulfur emissions.

Test work in England, at the National Coal Board's Grimethorpe test facility and at the British National Coal Board's Coal Utilization Research Laboratory in Leatherhead, in the mid-1970's, reinforced the concept of the feasibility of burning coal in a fluidized bed under pressure. The test work addressed the concerns about fluidization, combustion efficiencies, sulfur removal, gas clean-up and erosion/corrosion.

Encouraged by these tests, a decision was made in the late 1970's to build an integrated pilot plant to test PFBC hardware and systems. Such a Component Test Facility (CTF) was built and put into operation in 1982 as a precursor to a 70 MW demonstration plant.

COMPONENT TEST FACILITY

The Component Test Facility (CTF) is a 15 MW_t integrated pilot plant located in Malmo, Sweden. It is owned and operated by ASEA PFBC. The CTF contains a representative segment of a full scale commercial bed with the necessary systems to apply PFBC technology to electric power generation. The bed area is approximately 1/36 of the 800 MW_t commercial plant design. An artist's rendition of the CTF is shown in Figure 1.

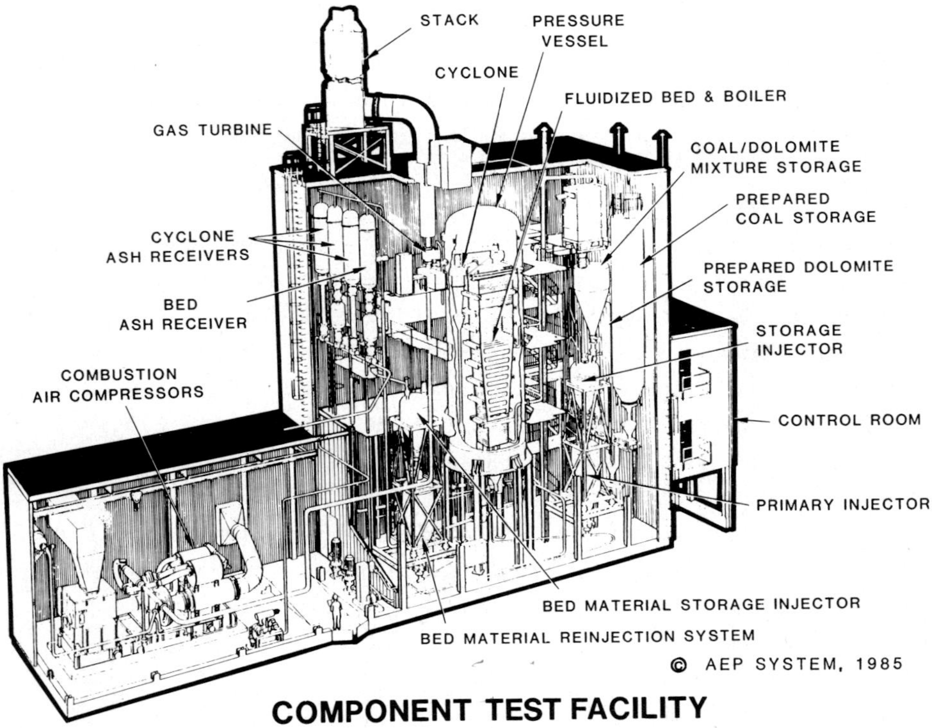

Figure 1

The combustor and cyclones are enclosed within a cylindrical pressure vessel. The combustor vessel is a rectangular tapered vessel enclosed in membrane wall tubes. The boiler is a once through design comprising waterwall, economizer evaporator, and superheater tubes.

There is a single train of three cyclones (which will be replaced with two cyclones in 1985) approximately 3 feet in diameter. They are made of austinitic steel and externally insulated.

A motor driven compressor pressurizes the system to 16 bar, and a free spinning test turbine which simulates commercial plant gas flow, temperature, and velocity is installed in the gas exhaust.

The result of 2000 hours of testing with Ohio and Indiana coal and dolomite are summarized in Table 1. In addition to these tests, the CTF now has approximately 3000 hours of operation on various coals from throughout the world.

OPERATIONAL HOURS (THROUGH DECEMBER 1984)	**2000**
SULFUR RETENTION	
@ 16 bar Ca/S = 1.6	**> 90%**
@ 12 bar Ca/S = 1.6	**90%**
NO_x EMISSION	**<0.3 1b/106 Btu**
COMUSTION EFFICIENCY	
FULL LOAD	**99.5%**
HALF LOAD	**99.0%**
DUST LOADING TO GAS TURBINE	
DUST CONCENTRATION	**80-200 ppm**
MEAN PARTICLE SIZE	**1.5 TO 2.5 MICRONS**
CYCLONE EFFICIENCY	**> 99%**

CTF TEST PROGRAM

PROCESS PARAMETER RESULTS

Table 1

The CTF has also led to extensive developments in the necessary auxiliary system required to apply PFBC to power generation. Key developments include new designs of fuel feed, ash removal, and hot gas clean-up systems.

TIDD PFBC COMMERCIAL DEMONSTRATION PLANT

Paramount to a utility investing upward of a billion dollars in a new commercial PFBC combined cycle electric generating plant, the technology must be demonstrated in a moderate scale facility. Toward this end, AEP has completed the initial design and cost estimate for a 70 MW PFBC Commercial Demonstration Plant. A new PFBC combustor and gas turbine will be used to repower an existing steam turbine cycle at the Tidd Plant in Brilliant, Ohio. If financing can be arranged, it is planned to have this demonstration plant in operation by the end of this decade.

Figure 2 provides a schematic of the plant cycle. An ASEA STAL gas turbine pressurizes the air in two stages, with intercooling, to 12 bar. Air flows through the outer casing of a coaxial pipe and into the combustor vessel shown in Figure 3. Combustion occurs at 1580°F in an 11.5 feet deep tapered zone where all boiler tubes are submerged.

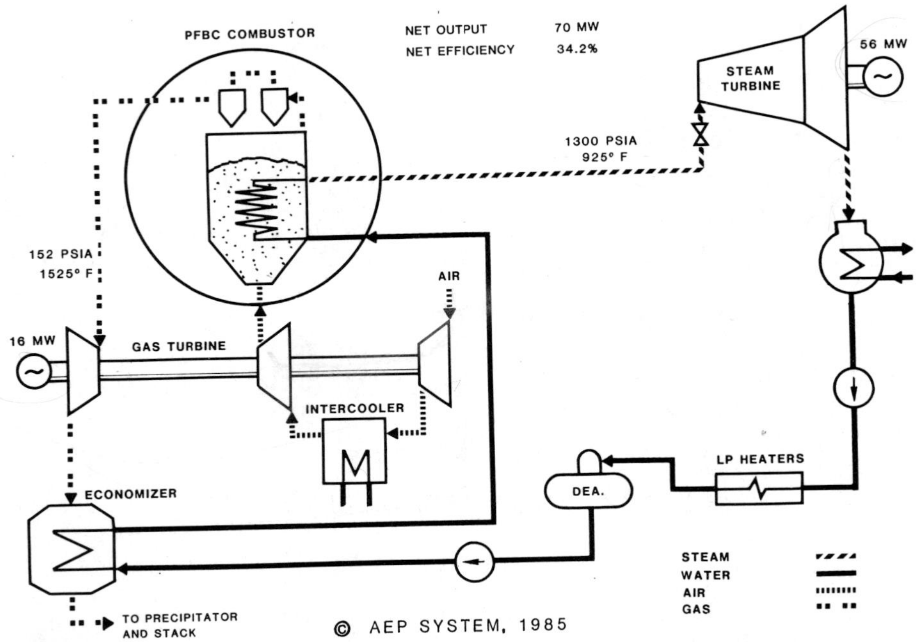

TIDD PFBC COMMERCIAL DEMONSTRATION PLANT CYCLE

Figure 2

Coal is fed through multiple feed points as a slurry containing less than 25% water by weight. Dolomite is fed pneumatically.

The gases from the bed travel through the freeboard into a collecting manifold and through seven parallel strings of cyclones, each with two stages. Approximately 50% of the total ash eleutriates from the bed and is carried to the cyclones, which have a design efficiency of 99%. The balance of the ash is removed from below the bed. The ash is removed via a system which cools the ash with combustion air and depressurizes the ash without the use of valves.

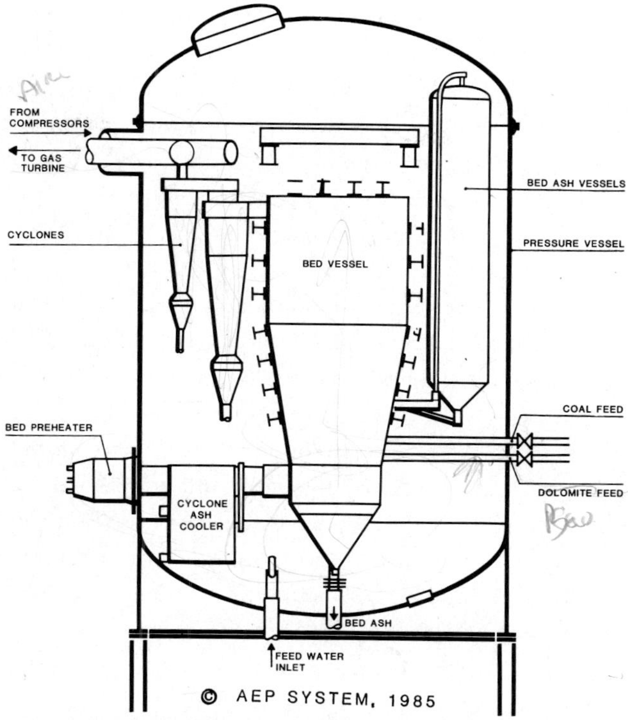

PFBC COMMERCIAL DEMONSTRATION PLANT COMBUSTOR VESSEL ASSEMBLY

Figure 3

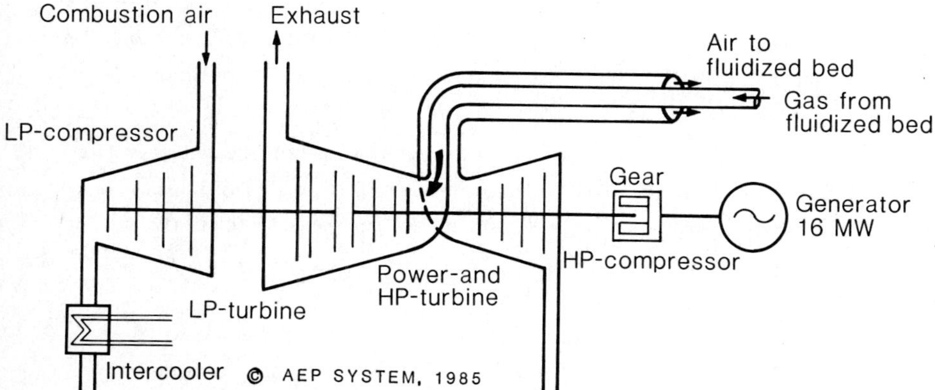

GT 35P GAS TURBINE

Figure 4

The clean gas then travels through the inner portion of the coaxial pipe and into the gas turbine. The ASEA STAL gas turbine, shown in Figure 4, is a two shaft design utilizing a free spinning low pressure turbine to drive the low pressure compressor and a fixed speed higher pressure turbine to drive the high pressure compressor and a 15 MW generator through a reduction gear. The generator acts as a synchronous motor to drive the high pressure compressor during start-up.

The exhaust gas from the gas turbine travels through an economizer which transfers the waste heat to the boiler feedwater. The gases then flow through an electrostatic precipitator and to the stack.

Steam produced within the in-bed tubes is conveyed to the existing Tidd steam turbine at 1300 psi and 925°F to generate 56 MW. The modest steam conditions are required to match the design of the existing steam turbine. Condensate is returned to the in-bed boiler tubes through the existing condensate and feedwater system, and the new economizer.

The output of 70 MW is achieved at a net efficiency of 34.2%; the original Tidd net efficiency was 31.0%.

COMMERCIAL PLANT

Successful demonstration of PFBC combined cycle power generation will provide utilities with an economically and environmentally superior technology compared to a Pulverized Coal Fired (PCF) plant with a Flue Gas Desulfurization (FGD) system.

Figure 5 provides a cycle schematic for a 640 MW commercial PFBC plant using two PFBC combustor/gas turbine modules coupled to one steam turbine cycle. Each gas turbine will produce 75 MW and the steam turbine will produce 490 MW.

The projected net efficiency of this supercritical single reheat cycle (3700 psia/1000°F/1000°F) is 39.4%. Further optimizations are expected to bring this cycle efficiency to above 40%. By contrast, the net efficiency of a well designed PCF plant with a wet FGD system is 36.6%.

Table 2 provides a comparison of parameters between a 640 MW PFBC plant and a 640 MW PCF plant with FGD. These parameters are the basis of the comparison that follows.

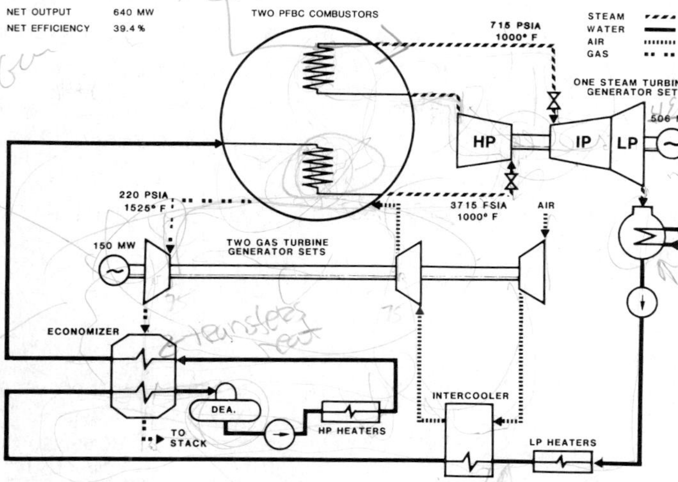

COMMERCIAL PFBC PLANT CYCLE (TWO MODULE)

Figure 5

	CONVENTIONAL PCF PLANT WITH AN FGD SYSTEM	PFBC COMBINED CYCLE PLANT
NET HEAT RATE (BTU/KWH)	9320	8660
STEAM CONDITIONS (PSIA/OF/OF)	3700/1000/1000	3700/1000/1000
SULFUR (%)	4	4
SULFUR REMOVAL (MINIMUM) (%)	90	90
SORBENT TYPE	LIME	DOLOMITE
CA/S MOLAR RATIO	1.05	1.6
COAL FEED (TPH) (FULL LOAD)	260	240
SORBENT CONSUMPTION (TPH) (FULL LOAD)	20	82
TOTAL SLUDGE & ASH PRODUCTION (TPH) (FULL LOAD)	149	130
COAL PRICE ($/MBTU)	2.00	2.00
LIME/DOLOMITE COST ($/TON)	50	10
WASTE DISPOSAL COST ($/TON)	6.25	4.0

COMPARISON OF PARAMETERS

640 MW PLANT DESIGN

Table 2

In addition to the higher efficiency, a PFBC plant is projected to have lower capital and operating costs compared to a PCF plant with a FGD system. Table 3 provides a comparison of the capital costs for a 640 MW conventional plant with FGD and a 640 MW PFBC plant. The PFBC plant costs include greater contingencies to allow for the new technology. As can be seen, the projected cost for a PFBC plant is 11% lower (on a $/KW basis) than for a conventional plant.

	$/Kw 1984 (DECEMBER) DOLLARS	
	PCF PLANT WITH FGD SYSTEM	PFBC COMBINED CYCLE PLANT
CAPITAL EXPENDITURES	1012	862
ENGINEERING & OVERHEAD	126	98
CONTINGENCIES	102	215
AFUDC	340	235
TOTAL PLANT INVESTMENT	1580	1410
PRE-PRODUCTION AND INVENTORY	65	60
TOTAL CAPITAL REQUIREMENT	1645	1470

NOTE 1: BASED ON 640 MW PLANT DESIGN.

CAPITAL COST COMPARISON[1]

Table 3

The capital reduction is due to the compact modular design of the combustor and a 25% reduction in erection time.

Table 4 provides a comparison of the bus bar power costs for the two plants burning the same coal. The high efficiency and reduced waste disposal costs (due to all dry waste products and reduced waste products) of PFBC results in a 9% advantage for the PFBC plant.

MILLS/Kwh 1984 (DECEMBER) DOLLARS

	PCF PLANT WITH FGD SYSTEM	PFBC COMBINED CYCLE PLANT
FIXED O & M COSTS	3.8	4.2
VARIABLE O & M COSTS	3.9	2.9
FUEL COSTS	18.7	17.3
PRODUCTION COSTS	26.4	24.4
LEVELIZED FIXED CHARGES (CAPITAL)	60.1	53.7
BUS BAR POWER COST	86.5	78.1
LEVELIZED BUS BAR POWER COST	115.6	104.9

NOTE 1: BASED ON 640 MW PLANT DESIGN WITH .65 CAPACITY FACTOR.

BUS BAR POWER COST COMPARISON[1]

Table 4

A comparison of the emissions of the two plants is indicated in Figure 6. The environmental advantages for PFBC include:

- 8% lower SO_2 emissions, with both plants operating at 90% SO_2 removal. This reduction is due to the efficiency advantage for PFBC: less coal is consumed, hence less SO_2 is produced.
- 54% lower NO_x emissions due to the low combustion temperature, hence thermal NO_x is virtually eliminated.
- 13% reduction in solid wastes. Further, all waste products are dry compared to the sludge produced by a wet scrubber.

PFBC technology also offers a modular concept which minimizes diseconomy of scale for smaller plant sizes. Figure 7 shows the economy of scale comparison between PFBC and a PCF plant with FGD for various plant sizes. This feature will allow utilities to build new plants in smaller sized increments which will more closely match load growth, hence reducing capital financing requirements, yet still maintain the $/KW cost for new plants at a reasonable level.

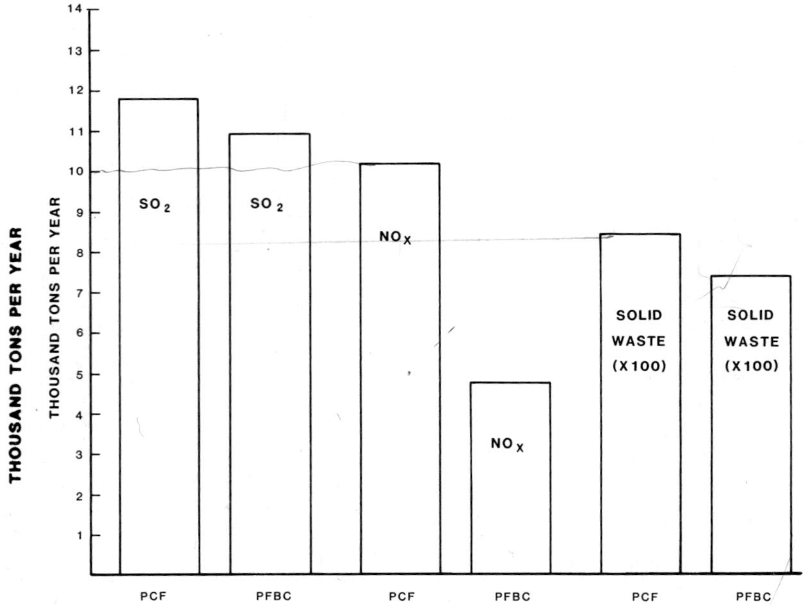

© AEP SYSTEM, 1985

ANNUAL EMISSION COMPARISON 640MW PLANTS: PFBC VS PULVERIZED COAL FIRED (PCF) WITH SCRUBBER

Figure 6

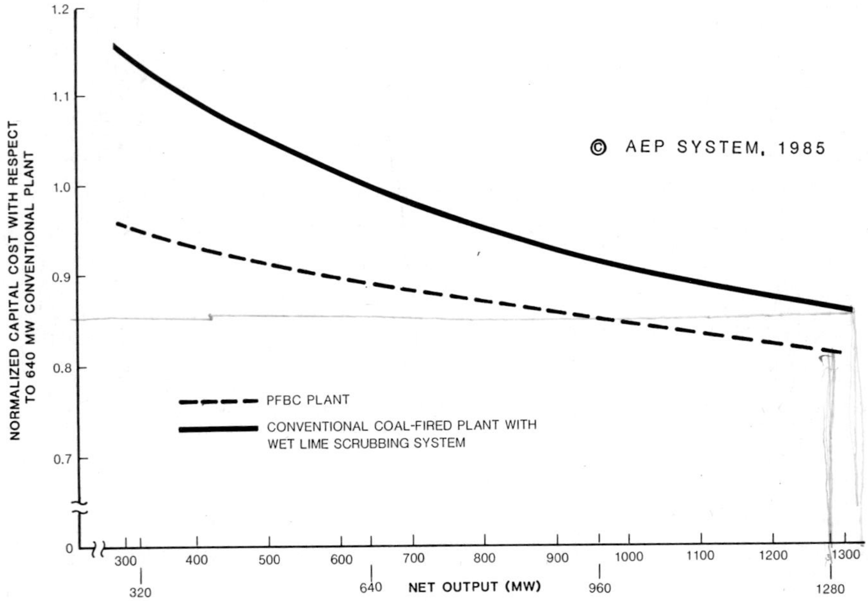

COMPARISON OF ECONOMY OF SCALE FOR VARIOUS SIZED PLANTS

Figure 7

CONCLUSION

Figure 8 shows the commercialization schedule for AEP's PFBC development program. To date, AEP and its partners have expended in excess of $32-million in developing PFBC technology.

The commercialization of PFBC combined cycle technology will ensure the continued use of high sulfur coal for electric power generation in an efficient, cost effective, and environmentally compatible manner.

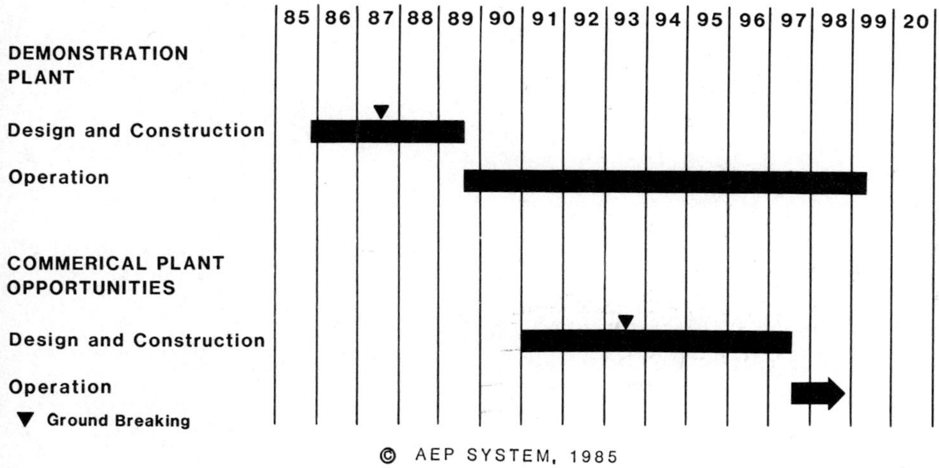

COMMERICALIZATION SCHEDULE

Figure 8

REFERENCES

1 J. J. Markowsky, J. R. Friedman; "Status of the AEP and ASEA-PFBC PFBC Development Program," Proceedings of the EPRI-PFBC Power Plants Utility Conference, 1984.
2 C. Olesen; "PFBC for Power Generation the ASEA PFBC Program," Proceedings of the EPRI-PFBC Power Plants Utility Conference, 1984.
3 S.A. Jansson; "PFBC Test Experience," Proceedings of the EPRI-PFBC Power Plants Utility Conference, 1984.
4 M. K. Guha, M. J. Mudd; "AEP's View of PFBC Technology for Utility Application," Proceedings of the Eighth International Conference on Fluidized Bed Combustion, 1984.
5 Electric Power Research Institute; "Guidelines for EPRI R&D Evaluation," 1982.

PART 8

CONTROL OF SULFUR DIOXIDE EMISSION

RECENT DEVELOPMENT IN FUNDAMENTAL STUDIES ON GAS ABSORPTION INTO SLURRIES

S. UCHIDA and H. AKAHORI

Department of Chemical Engineering, Shizuoka University

3-5-1 Johoku, Hamamatsu 432 (Japan)

ABSTRACT

In the development of more efficient and reliable slurry reactors in the processes such as lime or limestone slurry wet scrubbing processes and coal liquefaction processes, an understanding of the mechanism of mass transfer into slurries has become important for optimal design and operation of the processes. Therefore, recently, many theoretical and experimental studies on slurry reactors where solids are used either as reactants or as catalysts have been reported. Limited to the former case, fundamental works on gas absorption into slurries are reviewed and a study on the effect of solid particles present in the liquid phase on volumetric mass transfer coefficient for oxygen absorption into slurries of various kinds of particles in a bubbling stirred tank absorber is presented. The mechanism of the variation of the mass transfer coefficient is discussed and a correlation is given. Some comments are also given for the present and the future studies in this field.

INTRODUCTION

The problem of absorption of gases into nonhomogeneous solvents has drawn more attention than ever as it found in various fields of industries such as flue gas scrubbing, hydroprocessing, coal conversion and medicine. Therefore, there are many theoretical and experimental studies on slurry reactors where solids are used as reactants or as catalysts. The mechanism of mass transfer in these slurry reactors, especially catalytic slurry reactors, has been discussed throughly in the literature(refs.1,2). However, until recently, little attention has been paid to slurry reactors where solids are used as reactive materials. Limited to such slurry reactors as absorbers in lime or limestone slurry wet scrubbing processes, fundamental works on gas absorption into slurries are reviewed and a recent study related to slurry reactors is presented.

PREVIOUS STUDIES

Theoretical

In the case of gas absorption into a slurry containing fine particles, one of the following steps is considered to be rate controlling:

(1) Diffusion of gas through gas film

(2) Diffusion of absorbed(dissolved) gas through liquid film

(3) Reaction between absorbed gas and dissolved solid component

(4) Dissolution of solid

(5) Diffusion of dissolved solid through liquid film around solid particle

Steps(2) to (5) will be a series or a parallel process depending on several factors such as the partial pressure of the gas in the gas phase, the solubilities of the gas and the solid, the reaction rate, the size of particles etc. The simplest case is the absorption accompanied by the following irreversible instantaneous reactions:

A(gas) → A(aq) (1)

B(solid) → B(aq) (2)

A(aq) + B(aq) → Product (3)

This problem was first treated by Ramachandran and Sharma(ref.3) and a model based on the film concept was proposed. The absorption process is divided into two categories as follows.

Category I: Solid dissolution in the liquid film near the gas-liquid interface is not important. In this case, the particle size is large compared with the liquid film thickness and the solid concentration is low. The gas absorption process becomes series to the solid dissolution process.

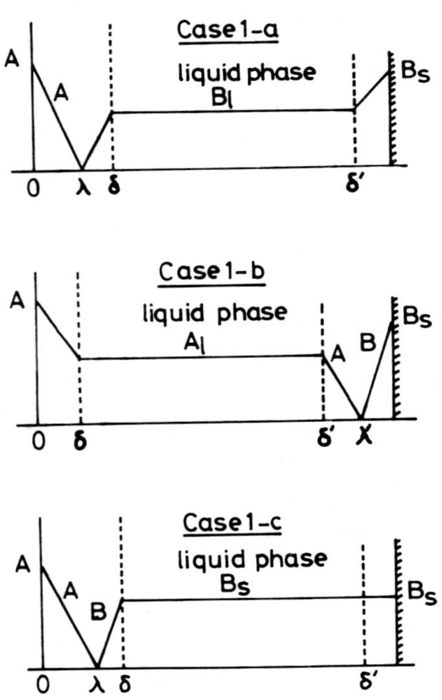

Fig.1 Concentration profiles . Solid dissolution in liquid film is not important.

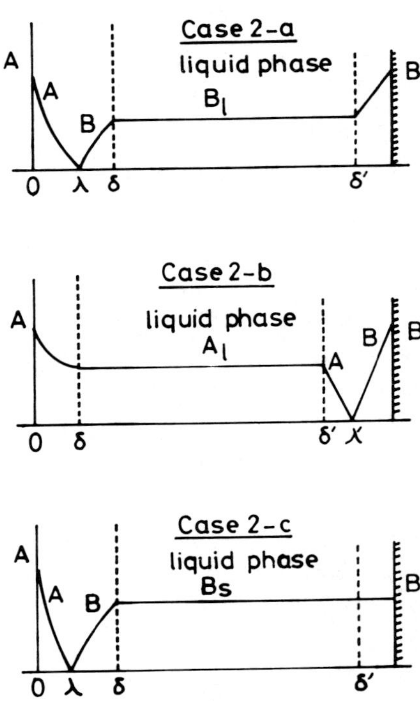

Fig.2 Concentration profiles. Solid dissolution in liquid film is important.

TABLE 1

Equations for gas absorption rates into slurries(ref.6).

Case 1-a;
$$R_A = \frac{A^* + \frac{D_B}{zD_A} B_s}{\frac{1}{k_L} + \frac{D_B a}{D_A k_s A_p}} \quad \text{for} \quad k_L a A^* < k_s A_p B_s$$

Case 1-b;
$$R_A = \frac{A^* + \frac{D_B}{zD_A} B_s}{\frac{1}{k_L} + \frac{D_B a}{D_A k_s A_p}} \quad \text{for} \quad k_s A_p B_s < k_L a A^*$$

Case 1-c;
$$R_A = k_L \left(1 + \frac{D_B}{zD_A} \frac{B_s}{A^*}\right) \cdot A^*$$

Case 2-a;
$$R_A = mD_A A^* \coth(m\lambda) + \frac{mD_B B_\ell}{z}\left(\coth(m\lambda) - \frac{1}{\sinh(m\lambda)}\right)$$
$$= \frac{mD_B B}{z} \coth m(\delta-\lambda) + m\left(D_A A^* + \frac{2D_B B_\ell}{z}\right)$$
$$x\left(\coth(m\lambda) - \frac{1}{\sinh(m\lambda)}\right)$$

where
$$B_\ell = \frac{\frac{m}{a} B_s}{\frac{1}{\sinh m(\delta-\lambda)} + \frac{m}{a}}$$

Case 2-b;
$$R_A = mD_A\left(A^* \coth(m\delta) - A_\ell\right) + \frac{mD_B B_s}{z}\left(\coth(m\delta) - 1\right)$$

where
$$A_\ell = \frac{A^* + \frac{D_B}{zD_A} B_s}{\cosh(m\delta) + \frac{m}{a} \sinh(m\delta)} - \frac{D_B}{zD_A} B_s$$

Case 2-c;
$$R_A = mD_A A^* \coth(m\lambda) + \frac{mD_B B_s}{z}\left(\coth(m\lambda) - \frac{1}{\sinh(m\lambda)}\right)$$
$$= \frac{mD_B B_s}{z} \coth m(\delta-\lambda) + m\left(D_A A^* + \frac{2D_B B_s}{z}\right)$$
$$x\left(\coth(m\lambda) - \frac{1}{\sinh(m\lambda)}\right)$$

Category II: Solid dissolution in the liquid film near the gas-liquid interface is important. In this case, the particle size is much smaller than the liquid film thickness and the concentration of particles in the film is high.

In their model, the physical dissolution of solid was assumed. However, Uchida et al.(ref.4) showed that the dissolution rate of limestone into acid solutions is enhanced by hydrogen ion, and proposed a modified model taking chemical dissolution into account(ref.5).

Uchida and Wen(ref.6) later classified the above two categories into 6 cases depending on the relative rate of gas absorption to the rate of solid dissolution and obtained a rate equation for each case as shown in Figs.1,2 and Table 1.

A penetration model for gas absorption into a slurry accompanied by an irreversible instantaneous reaction(Case 2-a in Table 1) has also been proposed by Uchida et al.(ref.7). The basic equations and the equations for absorption rate and enhancement factor are given as follows:

$$\frac{\partial C_A}{\partial t} = D_A\frac{\partial^2 C_A}{\partial x^2} - \frac{1}{z}k_sA_p\left(\sqrt{\frac{D_B}{D_A}} + z\sqrt{\frac{D_A}{D_B}}\cdot\frac{C_A}{C_{Bs}}\right)C_{Bs} \qquad (0 \leq x \leq x_r) \tag{4}$$

$$\frac{\partial C_B}{\partial t} = D_B\frac{\partial^2 C_B}{\partial x^2} + k_sA_p(C_{Bs}-C_B), \quad (x_r < x) \tag{5}$$

$$\text{at } t = 0,\ x \geq 0;\ C_A = 0,\ C_B = C_{Bs} \tag{6}$$

$$\text{at } x = 0,\ t > 0;\ C_A = C_A^* \tag{7}$$

$$\text{at } x = x_r(t),\ t > 0;\ C_A = C_B = 0, \quad zD_A\left(\frac{\partial C_A}{\partial x}\right) + D_B\left(\frac{\partial C_B}{\partial x}\right) = 0 \tag{8}$$

$$\text{at } x \to \infty,\ t \quad 0;\ C_B = C_{Bs} \tag{9}$$

$$R_A = 2(1+q_B)\left\{\sqrt{\frac{D_A}{\pi t}}\left(\frac{\sqrt{k_sA_pt}}{2} + \frac{1}{4}\sqrt{\frac{\pi}{k_sA_pt}}\right)\text{erf}\sqrt{k_sA_pt} + \frac{1}{2}\exp(-k_sA_pt)\right\}C_A^* \tag{10}$$

$$\beta = (1+q_B)\left\{\left(\frac{\sqrt{k_sA_pt}}{2} + \frac{1}{4}\sqrt{\frac{\pi}{k_sA_pt}}\right)\text{erf}\sqrt{k_sA_pt} + \frac{1}{2}\exp(-k_sA_pt)\right\} \tag{11}$$

Fig.3 Comparison of penetration model with film model.

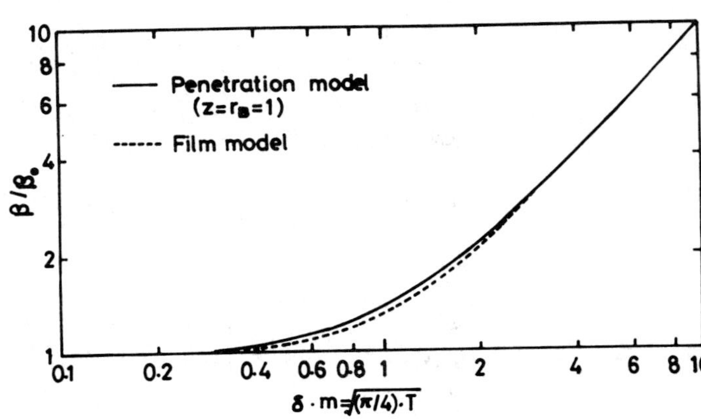

The numerical difference between the film model and the penetration model is, however, negligibly small as shown in Fig.3.

Generally, the conditions and the mechanism in practical cases are not so simple that these models can apply. Sada et al.(ref.8) reported a model for a finite reaction which is the extension of Ramachandran-Sharma's or Uchida's model. In this case, basic equations become complex and only an approximate or a numerical solution is obtained as the absorption rate. In the process of CO_2 absorption into alkali slurry, a second-order reaction, first-order with respect to CO_2 and OH^- respectively, takes place and this model may be applied.

Flue gases usually contains more than one acid gas such as SO_2 and NO_2. Sada et al.(ref.9) proposed a film model for the gas absorption accompanied by:

$$A_1 + \nu_1 B \rightarrow \text{Product(I)} \tag{12}$$

$$A_2 + \nu_2 B \rightarrow \text{Product(II)} \tag{13}$$

In SO_2-$Mg(OH)_2$ and SO_2-$MgSO_4$ added $CaCO_3$ slurry systems, the reaction between OH^- ion and HSO_3^- ion which is the reaction product is considered to be instantaneous as well as the reaction between SO_2 and OH^- and therefore there might be two reaction planes in the liquid film. Sada et al.(ref.10) analyzed these systems by the two reaction plane model and presented equations for the rates of gas absorption. The concentration profiles in the liquid film are shown in Fig. 4.

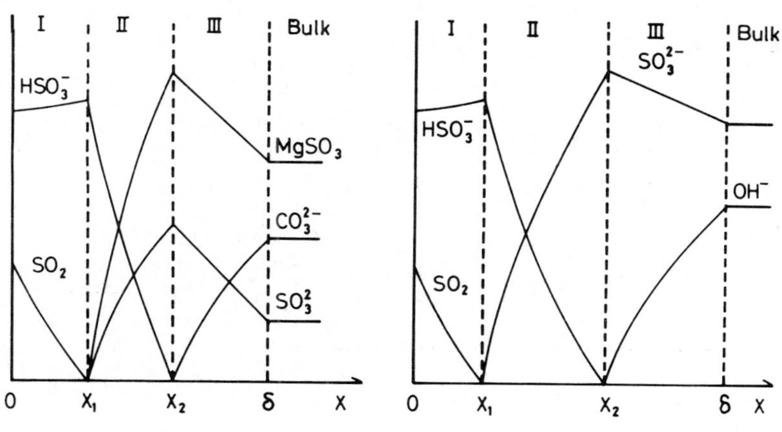

(a) SO_2-$Mg(OH)_2$ slurry system (b) SO_2-$CaCO_3$ with $MgSO_4$ slurry system

Fig.4 Concentration profiles in two reaction plane model.

Experimental

The first experimental work on gas absorption into a slurry was performed by Tadaki and Maeda(ref.11) using a wetted wall column for CO_2-lime slurry system. They reported that the solid dissolution affect the absorption rate as well as the diffusion resistances in the gas and the liquid films. The absorption mechanism was, however, not analyzed.

Bjerle et al.(ref.12) performed an experiment of SO_2 absorption into a limestone slurry using a liquid jet absorber. The absorption rate increased by 10% than that for SO_2-H_2O systm but they concluded that it was due to the reaction between SO_2 and dissolved HCO_3^- ion and that the contact time was so short that the dissolution of solid did not affect the absorption.

Uchida et al.(ref.13) used a stirred tank with a plane interface and showed that their model could approximately explain the experimental results. Sada et al.(ref. 8) also used a stirred tank to perform the experiment of the absorption of SO_2 and CO_2 into the slurry of lime particles. The extended model of Ramachandran-Sharma's and Uchida's models fit to the data.

Sada et al.(ref.10) performed experiments of SO_2 absorption into $Ca(OH)_2$ and $MgSO_4$ added $CaCO_3$ slurry and compared the results with two reaction plane model. They used a stirred tank with a plane interface and showed that the results fit to the model in which solid particles are not present between the gas-liquid interface and the first reaction plane. They also reported(ref.14) that the absorption rate of lean SO_2 gas into $CaSO_3$ aqueous slurry is enhanced by the surface reaction at the $CaSO_3$ particles floating in the liquid film.

As for the desulfurization processes for flue gases using lime or limestone slurry, the difficulty of finding the places for the disposal of the solid product such as $CaSO_4$ has arisen. To avoid this difficulty, the regenerable melamine slurry process has been recently reported by Gautney et al.(ref.15). In this system, the following reactions are considered to occur:

$$SO_2 + H_2O \rightarrow H_2SO_3 \tag{14}$$

$$C_3H_6N_6(s) + H_2O \rightarrow C_3H_6N_6(aq) \tag{15}$$

$$2C_3H_6N_6(aq) + H_2SO_3 + 4H_2O \rightarrow (C_3H_6N_6)_2 \cdot H_2SO_3 \cdot 4H_2O\downarrow \tag{16}$$

Reaction(16) proceeds quite readily and nearly irreversibly. If SO_2 is oxidyzed or SO_3 is present in the gas, then hydrated melamine sulfate is precipitated as follows:

$$SO_2 + 0.5O_2 \rightarrow SO_3 \tag{17}$$

$$SO_3 + H_2O \rightarrow H_2SO_4 \tag{18}$$

$$2C_3H_6N_6 + H_2SO_4 + 2H_2O \rightarrow (C_3H_6N_6)_2 \cdot H_2SO_4 \cdot 2H_2O\downarrow \tag{19}$$

In the regeneration step, the spent slurry is filtered and the melamine sulfite is decomposed thermally at 100~200°C to regenerate the melamine completely and yield SO_2 and water vapor as:

$$(C_3H_6N_6)_2 \cdot H_2SO_3 \cdot 4H_2O \rightarrow 2C_3H_6N_6 + SO_2\downarrow + 5H_2O\uparrow \tag{20}$$

Uchida et al.(ref.16) performed an experiment of SO_2 into a melamine slurry in a stirred tank absorber and showed that the absorption rate is correlated by the following equation:

$$Ra = \frac{Hp_G + (D_B/zD_A)C_{Bs}}{(H/k_G a) + (1/k_L a) + (D_B/D_A k_s A_p)} \tag{21}$$

The dissolution rate of melamine is fast compared with the absorption rate and the dissolution in the liquid film is not important. The variation of Ra with melamine concentration and comparison of the data with the values calculated by equation(21) are shown in Figs 5 and 6.

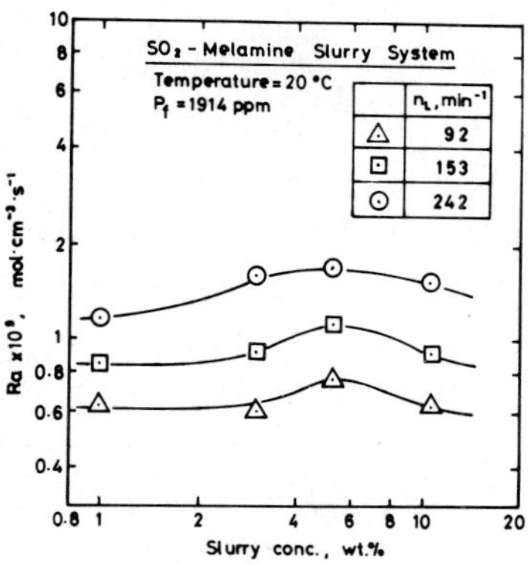

Fig.5 Effect of melamine concentration on SO_2 absorption rate.

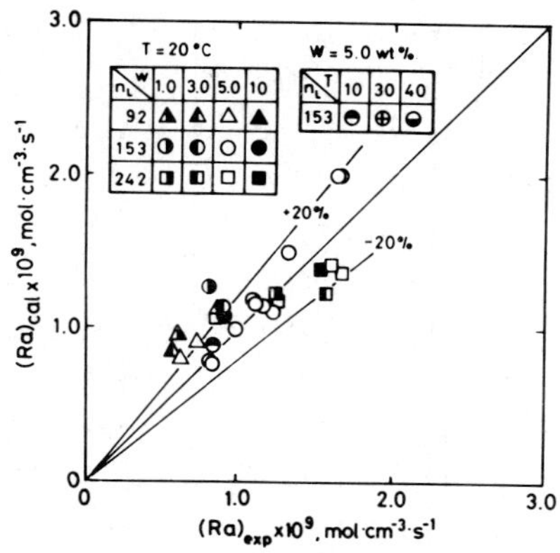

Fig.6 Comparison of experimental data with calculated values.

In addition to the fundamental studies reviewed above, many investigations related to flue gas wet scrubbing processes have been conducted from fundamental and applicable points of view(refs.17-28).

RECENT DEVELOPMENT

SO_2 absorption into limestone slurry

As mentioned above, there are many models proposed for SO_2 absorption into limestone slurry such as the two reaction plane model of Sada et al.(ref.10) taking into account the various reactions. Although the two reaction plane model can explain the experimental data for $Mg(OH)_2$ slurry, it can not apply to the limestone slurry system. Uchida and Ariga(ref.29) recently proposed a new model for the system.

When SO_2 is absorbed into a limestone slurry, it dissociates into ions as reactions with dissolved limestone components proceed in the liquid film near the gas-liquid interface. With the assumption that all reactions in the film are instantaneous and irreversible, the film is divided into three regions and the concentration profiles of all the components in each region are depicted as in Fig.7.

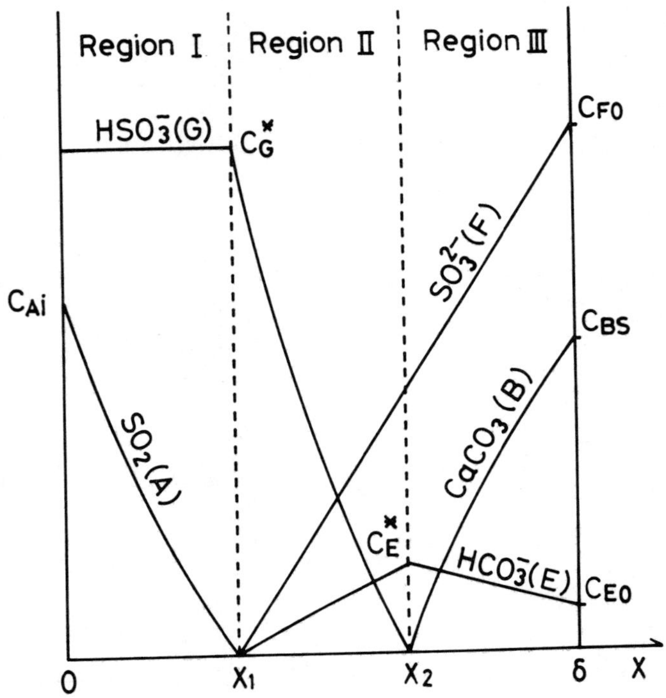

Fig.7 Concentration profiles in liquid film near gas-liquid interface for SO_2-$CaCO_3$ slurry system.

Taking material balances with respect to ions and components in each region yields the following set of equations:

Region I;

$$D_A\left(\frac{d^2C_A}{dx^2}\right) - k_sA_p\left\{1 + \left(\frac{D_AC_A}{D_BC_{Bs}}\right) + \left(\frac{D_GC_G}{D_BC_{Bs}}\right)\right\}C_{Bs} = 0 \tag{22}$$

$$D_G\left(\frac{d^2C_G}{dx^2}\right) = 0 \tag{23}$$

Region II ;

$$D_E\left(\frac{d^2C_E}{dx^2}\right) + k_sA_p\left\{1 + \left(\frac{D_EC_E}{D_BC_{Bs}}\right)\right\}C_{Bs} = 0 \tag{24}$$

$$D_F\left(\frac{d^2C_F}{dx^2}\right) = 0 \tag{25}$$

$$D_G(\frac{d^2C_G}{dx^2}) - k_sA_p\{1 + (\frac{D_GC_G}{D_BC_{Bs}})\}C_{Bs} = 0 \tag{26}$$

Region III ;

$$D_B(\frac{d^2C_B}{dx^2}) + k_sA_p(C_{Bs} - C_B) = 0 \tag{27}$$

$$D_E(\frac{d^2C_E}{dx^2}) = 0 \tag{28}$$

$$D_F(\frac{d^2C_F}{dx^2}) = 0 \tag{29}$$

$$\text{at } x = 0; \quad C_A = C_{Ai}, \quad \frac{dC_G}{dx} = 0 \tag{30}$$

$$\text{at } x = x_1; \quad C_A = C_E = C_F = 0, \quad C_G = C_G^{\star} \tag{31}$$

$$-D_A(\frac{dC_A}{dx})_{x_1^-} = D_E(\frac{dC_E}{dx})_{x_1^+} + D_F(\frac{dC_F}{dx})_{x_1^+} \tag{32}$$

$$D_G(\frac{dC_G}{dx})_{x_1^-} - D_G(\frac{dC_G}{dx})_{x_1^+} = 2D_F(\frac{dC_F}{dx})_{x_1^+} + D_E(\frac{dC_E}{})_{x_1^+} \tag{33}$$

$$\text{at } x = x_2; \quad C_B = C_G = 0, \; C_E = C_E^{\star}$$

$$D_B(\frac{dC_B}{dx})_{x_2^+} = -D_G(\frac{dC_G}{dx})_{x_2^-} = D_E(\frac{dC_E}{dx})_{x_2^-} - D_E(\frac{dC_E}{dx})_{x_2^+} \tag{34}$$

$$\text{at } x = \delta; \quad C_B = C_{Bs}, \quad C_E = C_{Eo}, \quad C_F = C_{Fo} \tag{35}$$

Solving the above set of equations, the enhancement factor β is obtained as follows:

$$\beta = (-\frac{m\delta}{r_Aq_A})\{(\frac{r_GY_G^{\star}+1}{\sinh(m\delta x_1)} - (\frac{r_Aq_A+r_GY_G^{\star}+1}{\tanh(m\delta x_1)})\} \tag{36}$$

where the dimensionless concentrations of HSO_3^- and HCO_3^-, $Y_G^{\star}$ and $Y_E^{\star}$, and the positions of reaction planes, x_1 and x_2, are obtained by the following relations.

$$r_GY_G^{\star} = \sinh\{m\delta(x_1-x_2)\}\cdot[\frac{2}{\tanh\{m\delta(x_1-x_2)\}} - \frac{1}{\tanh\{m\delta(1-x_2)\}}] - 1 \tag{37}$$

$$r_EY_E^{\star} = \frac{1}{(1-x_1)}\{(x_2-x_1)r_EY_{Eo} + (1-x_2)r_GY_G^{\star}\}$$

$$\frac{(r_EY_E^{\star}-r_GY_G^{\star})}{(x_2-x_1)} + \frac{r_FY_{Fo}}{(1-x_1)} - m\delta\,[\frac{(r_GY_G^{\star}+1)}{\tanh\{m\delta(x_1-x_2)\}} - \frac{1}{\tanh(m\delta x_1)} + \frac{1}{\sinh(m\delta x_1)} - \frac{2}{\sinh\{(m\delta(x_1-x_2)\}} + \frac{r_Aq_A}{\sinh(m\delta x_1)}] = 0 \tag{38}$$

$$\frac{(r_EY_E^{\star}-r_GY_G^{\star})}{(x_2-x_1)} + \frac{r_FY_{Fo}}{(1-x_1)} = 0 \tag{39}$$

In Fig.8, the enhancement factor β given by Equation(36) is plotted against the interfacial concentration of SO_2 divided by the saturated solubility of limestone $q_A(=C_{Ai}/C_{Bs})$, with the slurry concentration as a parameter. Comparison of the experimental data with the theoretical values calculated using present model is also shown in Fig.8 as well as those by the model of Sada et al.(ref.10).

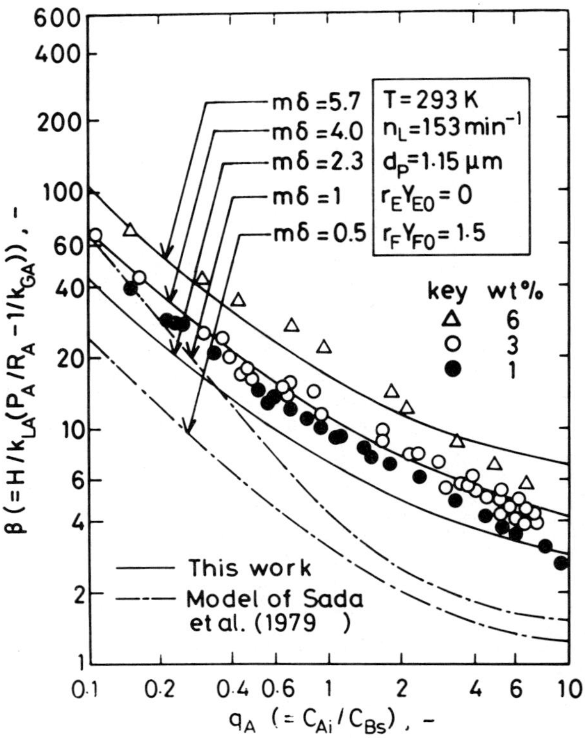

Fig.8 Comparison of experimental data with theoretical values calculated by equation(36).

Effect of solid particles on mass transfer coefficient

In the investigation of the absorption of gas into a slurry containing suspended fine solid particles, it usually was assumed that the liquid phase mass transfer coefficient was unaffected by the presence of the solids. The liquid phase mass transfer coefficient derived from applying the film concept is a function of the film thickness as given by $k_L = D_A/\delta$. The value of the film thickness will vary with the hydrodynamic conditions. It is reasonable to consider that the flow conditions of the liquid containing solid particles are different from those of liquid with no solids.

Joosten et al.(ref. 30) used nitrogen to strip helium from slurries containing several kinds of materials with different concentrations and sizes. In their experiment, nitrogen gas was introduced through three small orifices at the bottom of the vessel. They found that the volumetric mass transfer coefficient, k_La, was not influenced by the presence of solid particles(i.e. glass beads, polypropylene) if their concentrations were lower than about 20%. However, k_La decreases sharply with the concentrations of solid materials when they were higher than 20%. For a specific solid material, the values of k_La at a specified viscosity coincided for different particle sizes. Values of k_La for different kinds of solid materials did not so coincide. The least-dense solid caused the greatest decrease in k_La with increasing viscosity. Although it is not possible to give a full explanation of the variation of k_La with the volume fraction of solids, the sharp reduction in k_La is probably due in large measure to a reduction in the specific interfacial area, a.

One of the other earlier studies on the influence of suspended solid materials on the mass transfer coefficient is that of Alper et al.(ref.31). They used two different stirred cells to carry out absorptions of oxygen and carbon dioxide into solutions containing fine particles(i.e. powdered activated carbons, powdered quartz, oxirane-acrylic beads). Their data showed no effect of inert particles on k_L values. It was in complete agreement with the previously published data(ref.30). On the other hand, activated carbon increased k_L considerably due to its highly porous structure and high adsobing property. Kars et al.(ref.32) also obtained the same results.

Sada et al.(ref.33) performed an experiment of the absorption of N_2O into $Ca(OH)_2$ slurry using a stirred tank absorber with a flat interface and showed that k_L decreases sharply when the solid concentration exceed 3% which is much lower than the value for Joosten et al.(ref.30) In the same report, k_La for CO_2 absorption into $Mg(OH)_2$ slurry in a bubble tower was also shown to have the same tendency.

Although the results reported by some other investigators(refs. 34, 35, 36) are not necessarily consistent, it is obvious that the presence of solid materials does affect the mass transfer in slurries.

Recently, Uchida et al.(ref.37) conducted a physical absorption of O_2 into slurries containing inert particles of different densities and sizes using a bubbling stirred tank as shown in Fig.9. Pure O_2 was bubbled into the slurry

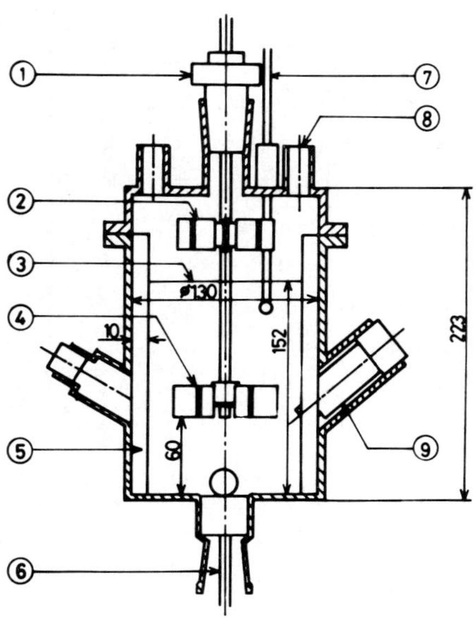

Fig.9 Details of stirred tank absorber.

1. Vacuum seal
2. Gas stirrer
3. Gas-liquid interface
4. liquid stirrer
5. Baffle
6. Gas inlet
7. Thermometer
8. Gas outlet
9. Oxygen probe

through a ball-filter and physically absorbed. The experimental conditions used in the study are summarized in Table 2.

TABLE 2

Experimental conditions

Particle	Glass Beads	Polystylene and Acryl Beads
Mean dia, m	25.6, 54.2, 104, 285	332, 771, 1001
Slurry conc.,vol%	0, 1, 3, 5, 10, 15, 20	0, 1, 5, 10, 11, 12, 15, 19, 20
Stirrer speed,1/s	5.0, 5.8, 6.7, 7.5	5.0, 5.8, 6.7, 7.5
Temperature,K	293	293
Gas flow rate,cm^3/s	20, 30, 40	20, 30, 40

Some typical results for the effects of the concentrations, the diameters and the densities of solid particles on the volumetric mass transfer coefficient, k_La, are shown in Figs. 10, 11 and 12. For glass beads and polystyrene particles, k_La decreases as the particle concentration increases. The coalescence of bubbles is enhanced by the presence of solid particles and it therefore results in the reduction of the gas-liquid contact area, while the solid particles increase the apparent density of liquid. For acryl particles whose density lies between the density of glass beads and that of polystyrene, the turbulence in the liquid around the particle may be significant and it cancels the effect of the reduction of contact area. The similar results are obtained for different gas flow rates, particle diameters, and stirrer speed. When the particle concentration is low, the effects of sizes and densities of particles on k_La are not observed. k_La increases with the particle size at the higher concentration.

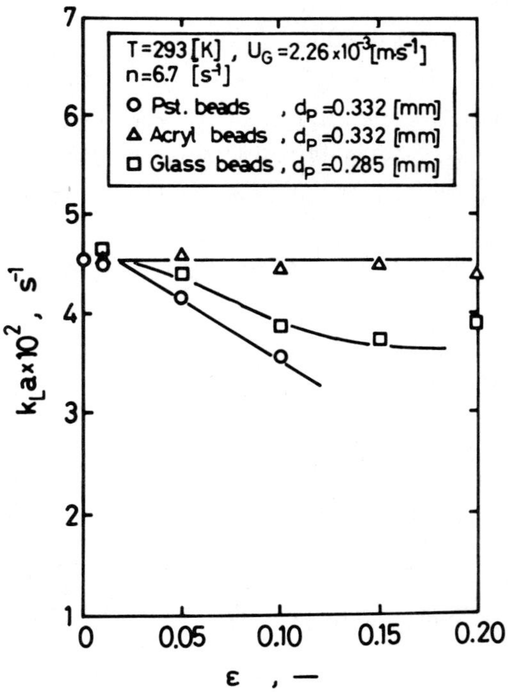

Fig.10 Effect of concentration of particles on volumetric mass transfer coefficient.

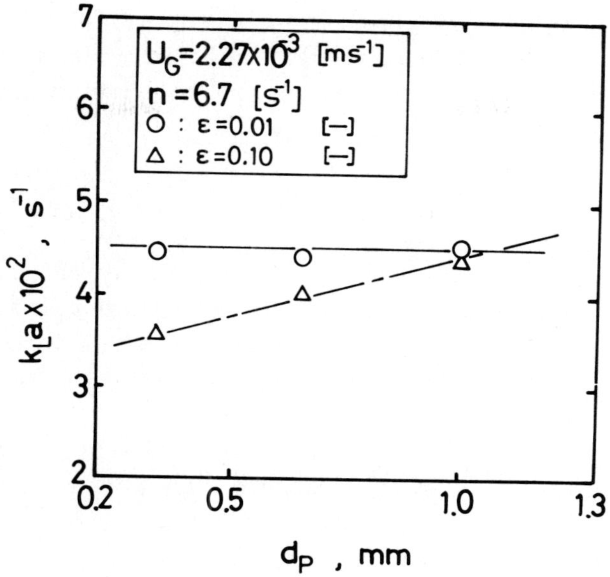

Fig.11 Effect of particle size on volumetric mass transfer coefficient.

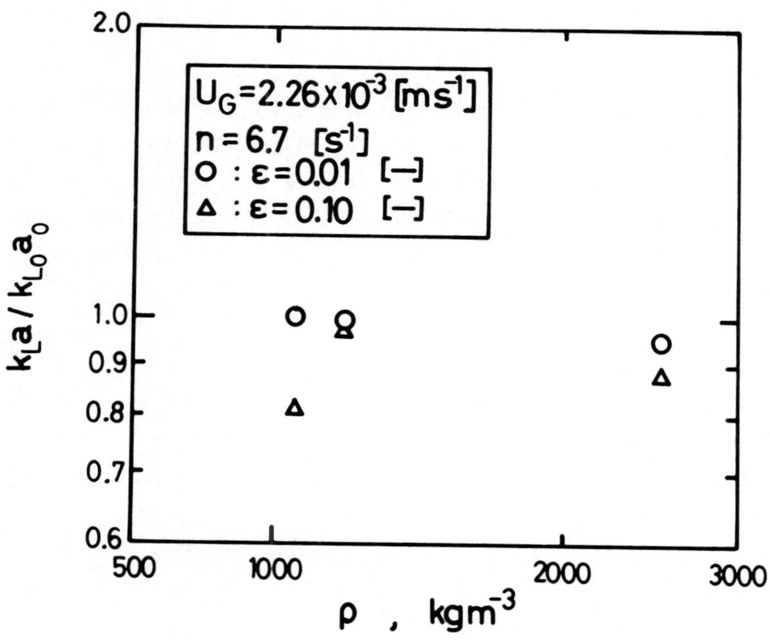

Fig.12 Effect of particle density on volumetric mass transfer coefficient.

Correlation for mass transfer coefficient

For a Newtonian liquid in a bubbling stirred tank absorber, the following relation is obtained by the dimensional analysis for the volumetric mass transfer coefficient(ref.38).

$$\frac{k_{Lo}a_o D_i^2}{D_L} = f\left(\frac{D_i^2 n\rho_L}{\mu_L}, \frac{D_i n^2}{g}, \frac{\mu_L U_G}{\sigma}, \frac{nD_i}{U_G}, \frac{\mu_L}{\rho_L D_L}\right) \tag{40}$$

Although the surface tension is considered to be important in bubble forming process, the effect of particles on it is difficult to estimate quantitatively and therefore the term including the surface tension is, in this study, removed. The experimental data for O_2 absorption into a distilled water can be correlated by the following equation with an average error of 2.76%:

$$\frac{k_{Lo}a_o D_i^2}{D_L} = 1.52\text{x}10^2\left(\frac{D_i^2 n\rho_L}{\mu_L}\right)^{0.73}\left(\frac{D_i n^2}{g}\right)^{0.80}\left(\frac{\mu_L}{\rho_L D_L}\right)^{0.50}\left(\frac{D_i n}{U_G}\right)^{-0.62} \tag{41}$$

For a slurry containing fine particles, the effect of particles on the volumetric mass transfer coefficient is assumed to be represented as:

$$Sh = Sh_o(1 + \delta) \tag{42}$$

where δ accounts for the effect of particles and is assumed to be given by:

$$\delta = -C_1\varepsilon Re^{-\alpha} + C_2\varepsilon Re_p^{\beta}\cdot Re^{\gamma} \tag{43}$$

The first term is the negative contribution to $k_L a$ due to the reduction of the interfacial area by the occupation of particles in the liquid phase, the coalescence of bubbles enhanced by the particles, and the increase in the apparent viscosity. The second term is the positive contribution due to the increase in the interfacial area by the motion of particles. The Reynolds numbers for a solution and a particle are given by:

$$Re = \frac{\rho_s D_i^2 n}{\mu_s} \tag{44}$$

$$Re_p = k\left(\frac{d_p\,_L g}{\mu_L^2}\right)^{\ell}\left(\frac{\Delta\rho}{\rho_L}\right)^m \tag{45}$$

Modifying equation(45) and introducing into equation(43) yield:

$$\delta = -C_1\varepsilon Re^{\alpha} + C_2\varepsilon\left(\frac{d_p^3\rho_L\Delta\rho g}{\mu_L^2}\right)\left(\frac{\Delta\rho}{\rho_L} - C_3\right)\cdot Re^{\alpha} \tag{46}$$

From the experimental data, constants in this equation are determined and thus:

$$\delta = -2.22\varepsilon Re^{-1.33} + 1.47\times10^{3}\varepsilon\left(\frac{d_p^3\rho_L\Delta\rho g}{\mu_L^2}\right)\left(\frac{\Delta\rho}{\rho_L} - 3.57\times10^{-3}\right)Re^{-1.33} \tag{47}$$

where $4.24\times10^4<(k_L a D_i^2/D_L)<1.57\times10^5$, $1.56\times10^4<(D_i^2 n\rho_S/\mu_S)<4.07\times10^4$,

$0.14<(D_i n^2/g)<0.43$, $4.49\times10^2<(\mu_S/\rho_S D_L)<7.93\times10^{-2}$ and

$0.09<(D_i^2 n/U_G)<0.28$.

The comparison of the experimental data with the calculated values by equation (47) is shown in Fig.13. The average error is 7.7%.

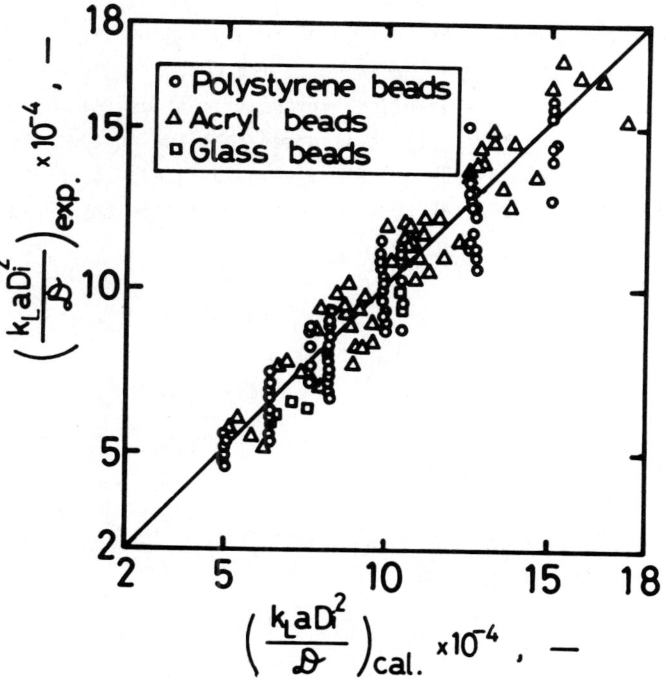

Fig.13 Comparison of experimental data with values calculated by equation(47).

CONCLUSIONS AND RECOMMENDATIONS

Slurry reactors are now very popular in various fields of industries and therefore, many aspects of them have been discussed theoretically and experimentally as seen above. Experimental data for gas absorption into slurries have also been obtained and analyzed by proposed mathematical models.

However, most of the experimental conditions used in these studies are too much different from those in practical cases. To promote the usefulness of the results obtained in academic studies, future study should include the following topics.

1) For practical purposes, data should be obtained under severer conditions as high pressure, high temperature and high concentration of solids which are encountered in a coal liquefaction process.

2) Regenerable processes such as the melamine slurry process seems to become promising in the trend toward the use of high sulfur coal.

3) The effect of particles on mass transfer should be investigated more throughly under various conditions using different types of reactors. The effects of particles on mass transfer coefficient, k_L and an interfacial area, a should be separately examined. Observation of the motion of the particles in the liquid film or near the bubbles in some way may clear the mechanism.

4) To say from the point of theoretical interest, most of the film models so far proposed have a contradiction. It is assummed that the another film is present around the particles which are in the main liquid film. It has physically nonsence. Penetration or surface renewal models may give an answer to this problem.

NOTATION

A	gas component or concentration of gas A
a	interfacial area
A_p	solid-liquid contact area
B	solid component or concentration of solid B
B_s	solubility of solid B
C_i	concentration of component i (i=A,B,E,F,G)
C_j	constant (j=1,2,3)
D_m	diffusivity of component m (m=A,B,E,F,G)
D_i	impeller diameter
D_L	diffusivity of gas in liquid phase
d_p	particle diameter
g	gravitational acceleration
H	Henry's constant
k_G	gas film mass transfer coefficient
k_L	Liquid film mass transfer coefficient
k_s	dissolution rate constant
m	$=\sqrt{k_s A_p/D_B}$
n	stirrer speed
p_G	partial pressure of gas

q_A	$=C_A/C_{Bs}$, dimensionless concentration of A
q_B	$=C_{Bs}/C_A^*$, dimensionless solubility of solid B
R, R_A	absorption rate
r_i	$=D_i/D_B$, (i=A,E,F,G), dimensionless diffusivity
Re, Re_p	Reynolds number
Sh	Sherwood number
t	time
U_G	superficial gas velocity
x	distance from interface
x_r, x_1, x_2	distance from interface to reaction plane
Y_i	$=C_i/C_{Bs}$, dimensionless concentration of i (i=E,F,G)
z	stoichiometric number

Greek letter

α	constant
β	enhancement factor or constant
γ	constant
δ	film thickness
ε	volume fraction of particles in slurry
λ	reaction plane given in Figs.1 and 2
μ	viscosity
ρ	density
σ	surface tention

Subscript

A	SO_2
B	solid component, $CaCO_3$
E	HCO_3^-
F	SO_3^{2-}
G	HSO_3^-
o	clear solution or solution containing no solid

Superscript

*	interface or dimensionless

REFERENCES

1 Y.T. Shah, Gas-Liquid-Solid Reactor Design, McGraw-Hill, New York, 1979, 373pp.
2 E. Alper(ed.), Mass Transfer with Chemical Reaction in Multiphase Systems, Vol.II, Three-Phase Systems, Martinus Nijhoff, The Hague, 1983, 399pp.
3 P.A. Ramachandran and M.M. Sharma, Chem.Eng.Sci., 24 (1969) 1681-1686.
4 S. Uchida, C.Y. Wen and W.J McMichael., Ch.I.Ch.E.J., 5 (1974) 111-114.
5 S. Uchida, K. Koide and M. Shindo, Chem.Eng.Sci., 30 (1975) 644-646.

6 S. Uchida and C.Y. Wen, Chem.Eng.Sci., 32 (1977) 1277-1281.
7 S. Uchida, M. Miyachi and O. Ariga, Can. J.Chem.Eng., 59 (1981) 560-562.
8 E. Sada, H. Kumazawa and M.A. Butt, Chem.Eng.Sci., 32 (1977) 1165-1170.
9 E. Sada, H. Kumazawa and M.A. Butt, Chem.Eng.Sci., 32 (1977) 1499-1503.
10 E. Sada, H. Kumazawa and I. Hashizume, Chem.Eng.Sci., 36 (1981) 639-642.
11 T. Tadaki and S. Maeda, Kagaku Kogaku, 27 (1963) 839-842.
12 I. Bjerle, S. Bengtsson and K. Tarnkvist, Chem.Eng.Sci., 27 (1972) 1853-1861.
13 S. Uchida, H. Moriguchi, H. Maejima, K. Koide and S. Kageyama, Can.J.Chem. Eng., 56 (1978) 690-697.
14 E. Sada, H. Kumazawa, I. Hashizume and H. Nishimura, Chem.Eng.Sci.,37 (1982) 1432-1435.
15 J. Gautney, Y.K. Kim and J.D. Hatfield, JAPCA 32 (1982) 260-265.
16 S. Uchida, M. Miyazaki and S. Masumoto, Chem.Eng.Sci., 39 (1984) 1527-1528.
17 P. Barton and V. Terdthai, Environ.Sci.&Tech., 10 (1976) 261-266.
18 C.S. Chang and G.T. Rochelle, AIChEJ., 28 (1982) 261-266.
19 C. Laohavichitra, K. Murayama, S.H. Weng and L.S. Fan, Chem.Eng.Sci., 37 (1982) 1572-1575.
20 G.T. Rochelle and C.J. King, I&EC Fundam., 16 (1977) 67-75.
21 P.A. Ramachandran and J.M. Smith, I&EC Fundam., 17 (1978) 17-23.
22 E. Sada, H. Kumazawa, M.A. Butt and T. Sumi, Chem.Eng.Sci., 32 (1977) 972-974.
23 E. Sada, H. Kumazawa and M.A. Butt, J.Chem.Eng.Japan, 12 (1979) 111-117.
24 E. Sada, H. Kumazawa and M.A. Butt, Chem.Eng.Sci., 34 (1979) 715-718.
25 E. Sada, H. Kumazawa and M.A. Butt, Chem.Eng.Sci., 35 (1980) 771-774.
26 E. Sada, H. Kumazawa, Y. Sawada and I. Hashizume, Chem.Eng.Sci., 36 (1981) 149-155.
27 E. Sada, H. Kumazawa, I. Hashizume and M. Kamishima, Chem.Eng.J., 22 (1981) 133-141.
28 W. A. Cronkright and W.J. Leddy, Environ.Sci.&Tech., 10 (1976) 569.
29 S. Uchida and O. Ariga, Can.J.Chem.Eng., to be published.
30 G.E.H. Joosten, J.G.M. Schilder and J.J. Janssen, Chem.Eng.Sci., 32 (1977) 563-566.
31 E. Alper, B. Wichtendahl and W-D. Deckwer, Chem.Eng.Sci., 32 (1977) 563-566.
32 R.L. Kars, R.J. Best and A.A.H. Drinkenburg, Chem.Eng.J., 17 (1979) 201-210.
33 E. Sada, H. Kumazawa and C.H. Lee, MASS TRANSFER OPERATION(1), Proceedings of PACHEC, Vol.1, Seoul, Korea (1983) 213-218.
34 K. Miyanami, K. Tojo and H. Iwanaka, FLUID MOTION,ibid. (1983) 169-173.
35 K. Ganguli and H. Vanden Berg, Chem.Eng.J., 16(1978) 193.
36 M. Miyachi, A. Iguchi, S. Uchida and K. Koide, Can.J.Chem.Eng., 59 (1981) 640-641.
37 S. Uchida, T. Seno and M. Nakane, to be submitted for publication.
38 H. Yagi and F. Yoshida, I&EC Process Des. Dev., 14 (1975) 488.

USE OF POLAR ORGANIC SOLVENTS TO CONTROL EMISSIONS FROM A SULFURIC ACID PLANT

Robert J. Demyanovich[1] and Scott Lynn[2]

[1]TERA Corporation, 2150 Shattuck Avenue, Berkeley, CA. 94704
[2]Department of Chemical Engineering, University of California, Berkeley, CA. 94720

ABSTRACT

Polyglycol ethers have a high absorption capacity for SO_2 and can be used in an absorber-stripper operation to remove and recover SO_2 from a gas stream. In a sulfuric acid plant this organic-absorption/SO_2-recovery process could recover SO_2 from the tail gas and transfer it to the air stream entering the plant, thereby recycling the SO_2 to the catalytic converters. The costs of a sulfuric acid plant so equipped are compared here to those for the standard double-contact/double-absorption (DC/DA) process.

Since the new process requires less catalyst, less heat-exchange area and a smaller inlet air blower, capital costs are about 8% lower. Furthermore, because the heat of reaction accompanying SO_2 conversion to SO_3 is more efficiently utilized, the credit for by-product electricity is larger resulting in a 4% reduction in operating costs.

INTRODUCTION

The purpose of this process design study is to evaluate the feasibility of controlling SO_2 emissions by absorption with organic solvents. Vapor-liquid equilibrium data for SO_2 in various organic solvents are presented in Demyanovich (ref. 1). This paper evaluates a process design using an organic solvent for SO_2 emissions control from sulfuric acid plants. An organic-absorption/SO_2-recovery process is compared with the current state of the art in H_2SO_4 manufacture, the double-contact/double-absorption (DC/DA) process. The new process incorporates an absorber-stripper for recycling SO_2 back to the sulfuric acid plant.

Choice of Organic Solvent

In this design evaluation, polyglycol dimethyl ethers, glymes, were chosen as the absorbents for SO_2 because of their low volatility, low toxicity, and relatively high absorption capacity. The pure glymes, such as the dimethyl ether of tetraethylene glycol (tetraglyme) and the dimethyl ether of triethylene glycol (triglyme), cost about \$5.50/kg (1984). However, a mixture of glymes trade-named Selexol* is commercially available and costs only about

*Selexol is a tradename of the Norton Chemical Company of Akron, Ohio.

\$2.90/kg (1984). Selexol consists primarily of tetra-, penta-, and hexaglyme and is manufactured by the Norton Chemical Company in Akron, Ohio. Since it is a mixture of high molecular weight glymes, Selexol has a lower vapor pressure than tetraglyme. The lower cost and vapor pressure of Selexol led to its choice as an absorbent for SO_2. Table 1 lists some available physical properties of Selexol.

Solubility data for SO_2 in Selexol are not available in the literature. Even though the composition of Selexol is proprietary, we may assume Selexol to be roughly equivalent to pentaglyme in solubility characteristics. Figure 1 illustrates how the solution capacity of SO_2 varies with increasing glyme monomer units (CH_2OCH_2). Extrapolating to beyond five monomer units (tetraglyme), indicates that the solution capacity would decrease by only 5% for pentaglyme. Thus the difference in solution capacity between tetraglyme and pentaglyme (and therefore Selexol) is probably less than 5%. As a first approximation, then, experimental vapor-liquid equilibrium (VLE) data for SO_2 in tetraglyme were used for SO_2 in Selexol. Vapor-liquid equilibrium data for SO_2 in tetraglyme are available in the literature (ref. 1, ref. 2).

Low cost is important because even though tetraglyme and Selexol have very low vapor pressures, the high volumetric flow rate of gas in the absorber and stripper in the proposed sulfuric acid plant design can lead to solvent losses that are economically unacceptable. The high molecular weight of these glymes (222 for tetraglyme and about 270 for Selexol) contributes heavily to such losses. As a result, the process flowsheet presented requires scrubbing the exiting gases from the absorber and stripper with water to recover Selexol. However, the flow of scrubbing liquid required, is small (<0.5 wt.%) relative to the total solvent flowrate.

SULFURIC ACID PLANT DESIGN

DC/DA Process

Sulfuric acid is produced by burning sulfur in air to produce SO_2:

$$S + O_2 \longrightarrow SO_2 \qquad (1)$$

The SO_2 formed is catalytically reacted with excess air to form sulfur trioxide:

$$SO_2 + 1/2\ O_2 \longrightarrow SO_3 \qquad (2)$$

Finally, sulfur trioxide is absorbed in a sulfuric acid stream containing water to yield a sulfuric acid product of 100% or less.

$$SO_3 + H_2O \longrightarrow H_2SO_4 \qquad (3)$$

TABLE 1 Physical properties of Selexol

	Metric Units	English Units
Molecular Weight	250 to 270	250 to 270
Freezing Point	-22.2 to -29 °C	-8.0 to -20.0 °F
Flash Point	151 °C	304 °F
Vapor Pressure		
25 °C (77 °F)	1.33 Pa	0.01 mm Hg*
50 °C (122 °F)	1.6 Pa	0.012 mm Hg
100 °C (212 °F)	150 Pa	1.1 mm Hg
Specific Heat (25 °C)	0.49 cal/g °C	0.49 Btu/lb °F
Density (25 °C)	1033 kg/m^3	8.6 lb/gal
Viscosity		
(25 °C)	5.8 cP	14.0 lb/ft-hr
(50 °C)	3.3 cP	8.0 lb/ft-hr
Thermal Conductivity	0.19 W/m°C	0.11 Btu/hr ft°F

*At low pressures mm Hg is the common "English" unit.

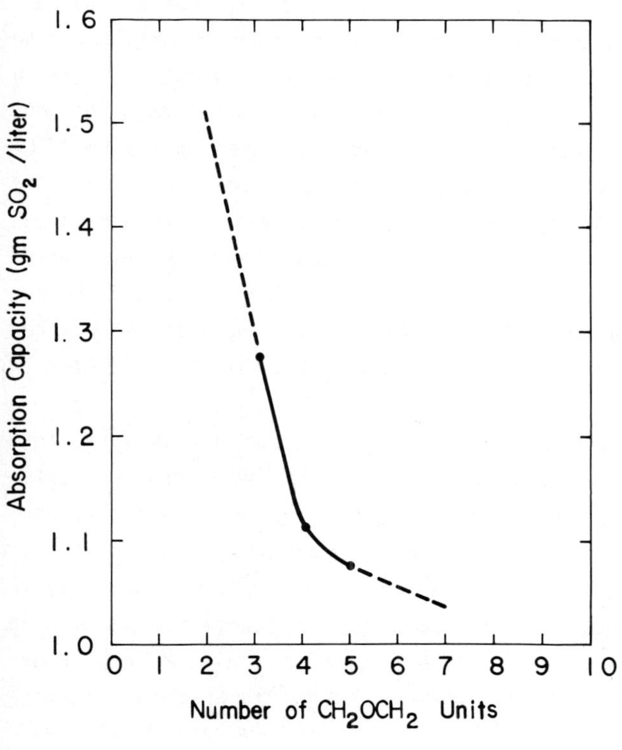

Fig. 1. SO_2 absorption capacity of glymes at P_{SO2} = 0.003 bar and at 55°C.

The process design presented produces 98% sulfuric acid by weight. (The potential production of oleum is not considered here.)

The double-contact/double-absorption (DC/DA) sulfuric acid plant is illustrated in Figure 2 (ref. 3). Stream flows for the production of 907 metric tons/day (1000 tons/day) of H_2SO_4 (100% equivalent) are shown in Table 2.

Incoming air is compressed to approximately 1.35 bars (1.33 atmospheres) and dried to prevent the formation of sulfuric acid mist further downstream. Sulfur is burned with this dry air stream to produce SO_2. The hot effluent gas is passed through a boiler that produces steam to be used for electrical generation. (In plants burning H_2S or many other sulfurous materials the drying step would follow the boiler.) The SO_2-rich gas (approximately 10% by volume) is then cooled to 435°C in Heat Exchanger #1 before entering the catalytic converter. Oxidation of SO_2 occurs over a vanadium pentoxide catalyst. Conversion of SO_2 is equilibrium-limited, however, since the reaction is highly exothermic and since the catalyst is not sufficiently active at less than 400°C. As a result, four catalytic stages are typically used to provide greater than 99.7% conversion of SO_2. To remove the heat of reaction, the gas leaving each of the first two stages is cooled, first in Heat Exchanger #2 and then in the superheater. The latter unit superheats steam from the boiler, which is then passed through a turbine for the generation of electricity. The exiting reactor gas from Catalytic Stage #3 is cooled in Economizer #1, which preheats boiler feed water, before entering SO_3 Absorber #1 for removal of product. Since the SO_3-lean gas exiting the absorber has been cooled to 45°C, much too cool for further conversion of SO_2 to SO_3 to proceed in a catalytic reactor, sensible heat is added to the gas in Heat Exchangers #1 and #2 to increase its temperature to 435°C prior to entering Catalytic Stage #4. There, residual SO_2 is oxidized; the gas is then cooled in Economizer #2 and the remaining SO_3 is absorbed in Absorber #2 where 98.0 weight percent H_2SO_4 is removed as product. The SO_2 content of the gas leaving Absorber #2 is 100 ppm or less.

The DC/DA process provides effective SO_2-emissions control. However, this control is effected at the expense of increased capital and operating costs. Vanadium pentoxide catalyst is very expensive, and Stage #4 essentially converts residual SO_2 to SO_3 only for emission control purposes. As a result, SO_3 Absorber #2 must absorb the small amount of SO_3 formed in the last catalytic stage since SO_3 cannot be released to the environment. Because the gas exiting SO_3 Absorber #1 is cooled to 45°C and must be reheated to 435°C, a large amount of heat-exchange area is required in the process. Furthermore, there is a significant degradation of the temperature level of the heat released in the reaction, since Economizer #2 does not recover all of this reheat.

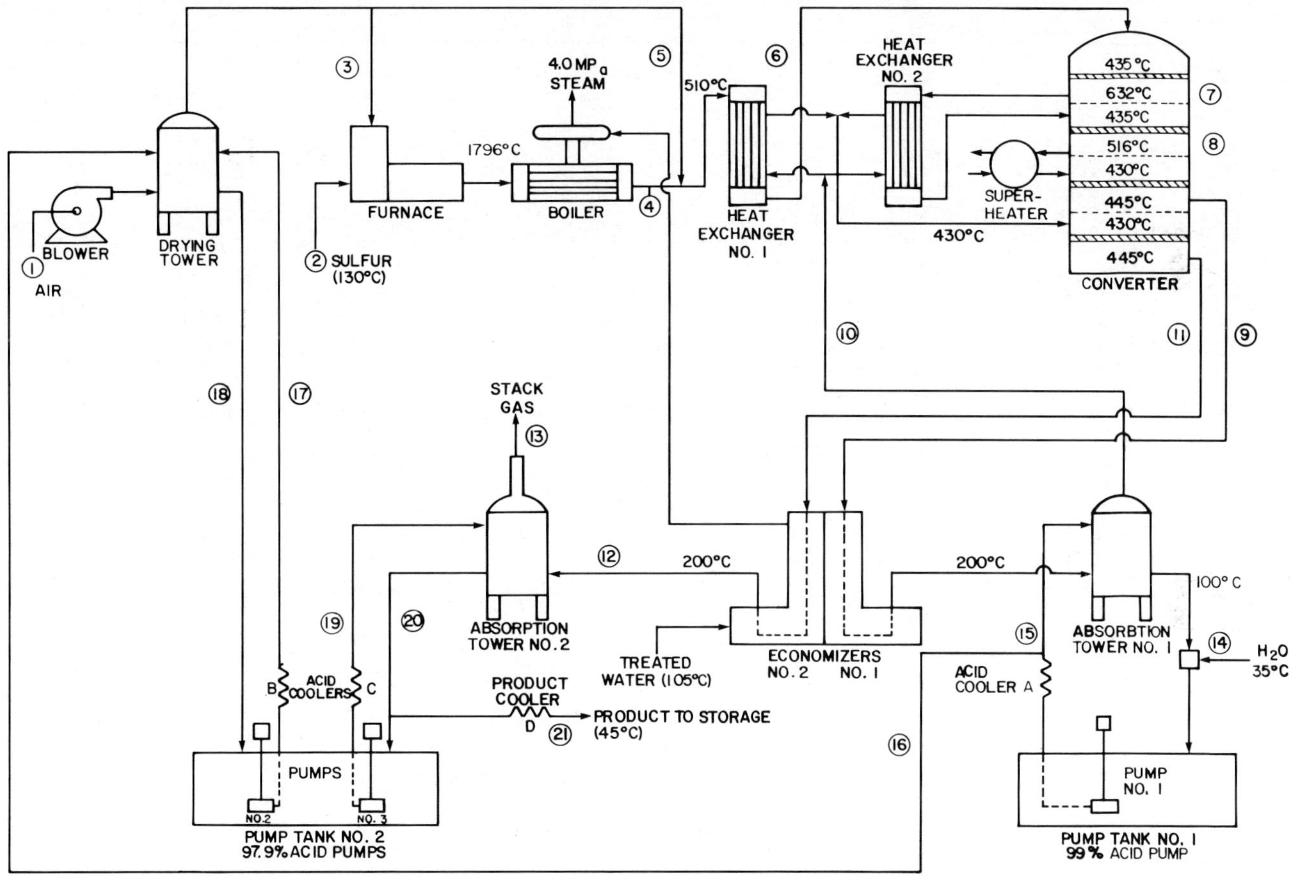

Fig. 2. Schematic diagram of an H_2SO_4 plant using the DC/DA process (ref. 3).

TABLE 2

Flow rates for DC-DA process

Stream	Component Flow (kg mol/s)							Temperature
	N_2	O_2	S	SO_2	SO_3	H_2SO_4	H_2O	^oC
1	0.908	0.240	--	--	--	--	0.012	35
2	--	--	0.120	--	--	--	--	130
3	0.454	0.120	--	--	--	--	--	45
4	0.454	--	--	0.120	---	--	--	907
5	0.454	0.120	--	--	--	--	--	45
6	0.908	0.120	--	0.120	--	--	--	435
7	0.908	0.0816	--	0.0432	0.0768	--	--	632
8	0.908	0.0654	--	0.0108	0.1092	--	--	516
9	0.908	0.0624	--	0.0048	0.1152	--	--	445
10	0.908	0.0624	--	0.0048	--	--	--	45
11	0.908	0.0601	--	9.6×10^{-5}	0.0047	--	--	445
12	0.908	0.0601	--	9.6×10^{-5}	0.0047	--	--	200
13	0.908	0.0601	--	9.6×10^{-5}	--	--	--	45
14	--	--	--	--	--	--	0.1212	35
15	--	--	--	--	--	2.089	0.1152	45
16	--	--	--	--	--	0.1152	0.0060	45
17	--	--	--	--	--	0.1979	0.0226	45
18	--	--	--	--	--	0.3131	0,0406	70
19	--	--	--	--	--	1.246	0.1437	45
20	--	--	--	--	--	1.251	0.1390	70
21	--	--	--	--	--	0.1199	0.0133	45

This process design comparison seeks to answer the following question: Could a separate absorber-stripper for SO_2 emissions control reduce capital investment and production costs by replacing one or more catalytic stages, SO_3 Absorber #2, Economizer #2, and other process units. Such a scheme has several potential advantages. Reducing the amount of catalyst would lower not only initial capital investment but also replacement costs due to catalyst poisoning. Furthermore, a separate pollution-control section could provide increased decoupling of H_2SO_4 production from control of SO_2 emissions.

Organic-solvent/SO_2-recovery process

Figure 3 shows the organic-solvent/SO_2-recovery process. Table 3 lists the stream flows for production of 907 metric tons/day (1000 tons/day) H_2SO_4. This flowsheet, while similar to the DC/DA process, differs in several important respects. First, since only two catalytic stages are used, the conversion of SO_2 to SO_3, while equally high overall, is only 87% per pass compared with 99.7% in the DC/DA process; thus reactor capital cost, initial catalyst cost, and catalyst replacement cost savings are realized. Second, since only one SO_3 absorption column is needed for the production of H_2SO_4 product, Economizer #2, SO_3 Absorber #2, Acid Cooler C, and an acid pump are eliminated. Furthermore, since the cool gas exiting the SO_3 absorber is sent to the SO_2 emission control section and not back to the catalytic reactor, the gas-gas heat exchange of Heat Exchangers #1 and #2 of the DC/DA process can be avoided.

The effluent stream from the SO_3 absorber, containing approximately 1-2% SO_2 by volume, is sent to the SO_2 absorber-stripper section. Sulfur dioxide is absorbed by a suitable organic solvent (Selexol for this analysis) and then stripped out for recycle back to the sulfuric acid plant. Stripping gas is provided by diverting the incoming compressed air required for the production of sulfuric acid to the stripper. To minimize column pressure drops, the SO_2 absorber and stripper are packed with 3.8 cm (1.5-inch) Intalox saddles. The exiting gas from the SO_2 absorber contains less than 100 ppm by volume of SO_2.

The choice of solvent is not necessarily the solvent with the highest absorption capacity. In this case, a solvent with a very low vapor pressure, low toxicity yet reasonably high absorption capacity is adequate. The reason is that since the stripping gas is the air stream entering the plant, it is essentially free. The amount of stripping gas (V_s) required is a function of the incoming gas flowrate (V_a) to the absorber and the ratio of equilibrium constants (Henry's constants if the tower pressures are equal) at the absorber temperature and the stripper temperature. Generally, high-capacity solvents are chosen because the slope of the solute vapor pressure curve with 1/T is larger, leading to a smaller ratio of equilibrium constants between the absorber and stripper. For example, at 40° and 70^oC and at equal absorber and

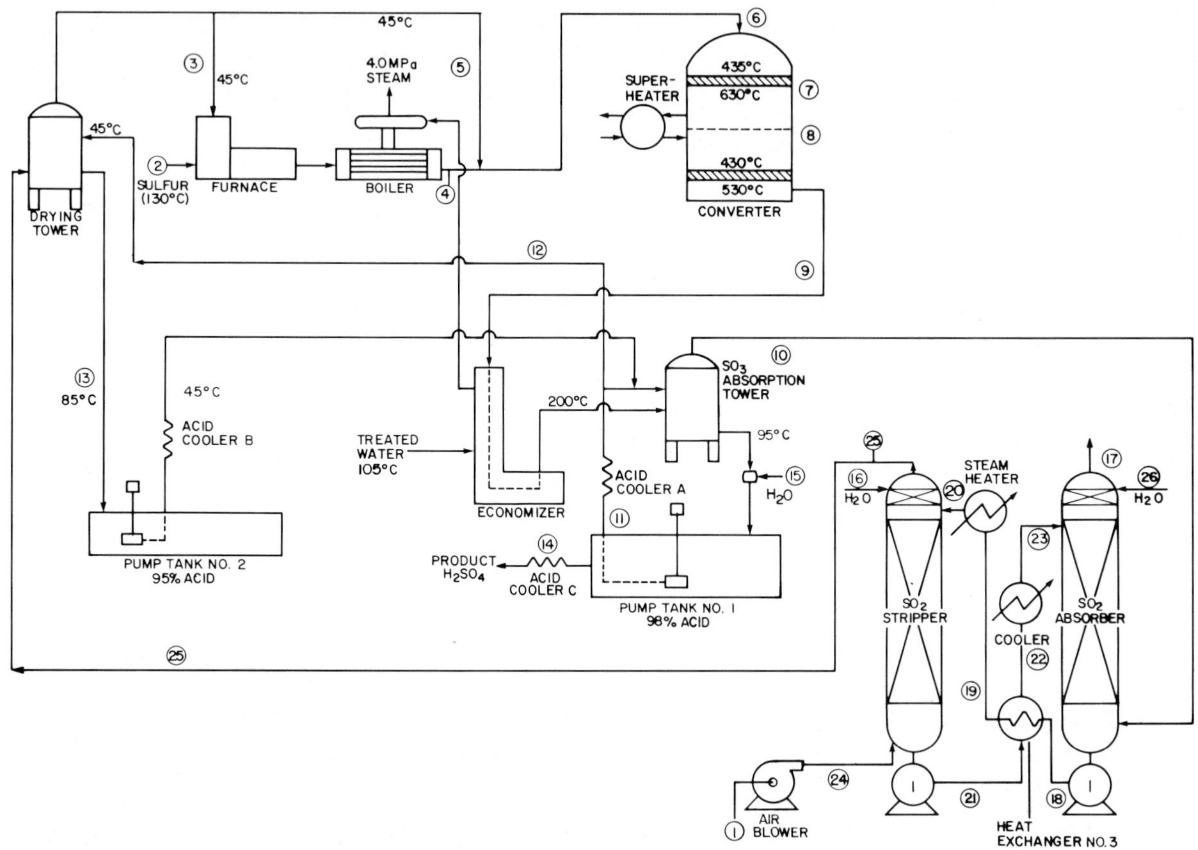

Fig. 3. Schematic diagram of an H_2SO_4 plant using the organic-absorption/ SO_2-recovery process.

TABLE 3

Flow rates for organic-absorption/SO_2-recovery process

Stream	Component Flow (kg mol/s)								Temperature °C
	N_2	O_2	Sulfur	Selexol	SO_2	SO_3	H_2SO_4	H_2O	
1	0.908	0.240	--	--	--	--	--	0.012	35
2	--	--	0.120	--	--	--	--	--	130
3	0.454	0.120	--	--	0.0089	--	--	--	45
4	0.454	--	--	--	0.1289	--	--	--	740
5	0.454	0.120	--	--	0.0089	--	--	--	45
6	0.908	0.120	--	--	0.1378	--	--	--	435
7	0.908	0.080	--	--	0.0578	0.080	--	--	630
8	0.908	0.080	--	--	0.0578	0.080	--	--	430
9	0.908	0.0601	--	--	0.0179	0.1199	--	--	530
10	0.908	0.0601	--	--	0.0179	--	--	--	45
11	--	--	--	--	--	--	2.293	0.255	95
12	--	--	--	--	--	--	0.182	0.0200	45
13	--	--	--	--	--	--	0.182	0.0520	85
14	--	--	--	--	--	--	0.1199	0.0133	45
15	--	--	--	--	--	--	--	0.1012	35
16	--	--	--	--	--	--	--	0.0200	35
17	0.908	0.0601	--	--	9.6×10^{-5}	--	--	0.017	35
18	--	--	--	0.562	0.0178	--	--	--	37
19	--	--	--	0.562	0.0178	--	--	--	63
20	--	--	--	0.562	0.0178	--	--	--	70
21	--	--	--	0.562	0.0000	--	--	--	68
22	--	--	--	0.562	0.0000	--	--	--	42
23	--	--	--	0.562	0.0000	--	--	--	35
24	0.908	0.240	--	--	--	--	--	0.012	50
25	0.908	0.240	--	negl.	0.0178	--	--	0.032	25
26	--	--	--	--	--	--	--	0.017	35

stripper pressures, the ratio of Henry's constants for N,N-Dimethyaniline (DMA), a high capacity solvent for SO_2, is 0.17 whereas the ratio for Selexol is estimated as 0.25 (ref. 1). Therefore, at equal absorption factor (m_aV_a/L) and stripping factor (m_sV_s/L) where m is the equilibrium constant and L is the solvent flowrate, the amount of stripping gas required for Selexol is almost 50% more than that required by DMA. Furthermore, solvents with a high solute capacity such as DMA have a smaller liquid flowrate, which decreases the size of the intercolumn heat exchanger.

In this case, however, since the volume of available stripping gas is large and the costs associated with its use are small, the solvent choice is perhaps better dictated by very low vapor pressure and toxicity. DMA is a very toxic substance and its vapor pressure is over 100 times greater than that of Selexol at 70°C. Since more stringent environmental regulations are foreseen in the future, low toxicity is expected to be a major advantage. Furthermore since DMA and water are not miscible, use of DMA would require a complicated solvent recovery section based on sulfuric acid. Glyme solvents are miscible with water and can be easily recovered in the top section of a tower using a water wash. However, these advantages are offset somewhat by the increased capital cost of the intercolumn heat exchanger caused by the higher solvent flowrate required when glymes are used.

Based on the foregoing considerations, Selexol was chosen as the solvent for the organic-absorption/SO_2-recovery process. Unreacted SO_2 is absorbed in Selexol, stripped by compressed air and recycled back to the furnace and catalytic converter.

Since a large amount of stripping gas is available in the form of incoming air to the sulfuric acid plant, sizing of equipment for the absorber-stripper section was based on total use of this incoming air. The absorber operates at 35°C with an absorption factor (L/m_aV_a) of 1.65. The stripper operates at a temperature of 70°C with a stripping factor (m_sV_s/L) of 2.23. Two major advantages are evident by total use of this air stream. First the temperature difference between the absorber and stripper is small relative to typical absorber-stripper operations. Second, a larger absorber solvent flow can be used to reduce the height of the absorber, because the cost of stripping gas is not a major factor.

Since the SO_2 stripper operates at about 70°C losses of Selexol solvent could be significant. At 70°C the vapor pressure of Selexol is about 100 ppm (by volume). Saturating the stripping air with that much Selexol would cause an intolerable economic burden of almost 2.5 million dollars per year (based on a Selexol molecular weight of 270). Furthermore, if Selexol were allowed to enter the dryer, product sulfuric acid would be contaminated since sulfuric acid is a very efficient absorbent for polar organic solvents. A Selexol loss of 100 ppm by volume in the stripper gas would result in about 2500 ppm (by weight) Selexol in the product sulfuric acid. Even a 10 ppm Selexol loss results in about 250 ppm of contaminant in the product H_2SO_4.

A water wash or "knock-down" section in the top of the stripper should reduce Selexol losses from the stripper to less than 1 ppm by volume. Water used for scrubbing would displace process water (Stream 15) required for H_2SO_4 production. The gas entering the water wash would exit saturated with water at approximately 25°C. Since this water is removed in the dryer, it replaces entering process water. Calculations indicate that only 25% of the process water would be displaced by the stripper wash water.

At 25°C the partial pressure of pure Selexol is estimated (by extrapolation) as less than 1 ppm. Since Selexol and water are miscible, further reduction in Selexol vapor pressure is certain. As a conservative estimate then, the loss of Selexol from the stripper was taken as the loss over pure Selexol and not the dilute solution. This estimate is most likely high, but is made because VLE data of Selexol and water are not readily available. As shown later, this loss does not have a significant impact on the operating costs of the organic process relative to the DC/DA process.

A water wash was also included for the absorber since at 35°C the loss of Selexol would be about 3 ppm ($66,000/yr). Addition of the water scrubbing section cools the gas to about 16°C and yields an equivalent reduction in pure Selexol vapor pressure of about 70%. The payback period for adding a single stage to reduce this Selexol loss by 70% is less than 1 year.

Each Selexol scrubbing section is assumed to produce a stream containing 10 mole % Selexol in water that is recycled to the main packed section. This is equivalent to a 62.5% Selexol stream on a weight basis. The water added to the main solvent flow by the combined water feeds to the Selexol scrubbing sections in both columns is less than 5,000 ppm (by weight). All of this water is evaporated into the gas streams leaving the two columns.

Because water in the knock-down section of the stripper contacts a gas stream containing SO_2, aqueous SO_2 is formed. Formation of aqueous SO_2 requires that the Selexol scrubbing section be rubber lined, which typically adds 25% to the scrubbing section cost. The stripper is constructed of carbon steel since the total solvent flow will contain less than 0.5% water (by weight) and is therefore noncorrosive.

Plant operating data required for equipment sizing for the DC/DA process were provided by Mr. Henry Borghi, plant manager of the Allied Chemical DC/DA process for H_2SO_4 manufacture in Richmond, California (ref. 4). These data are listed in Table 4. For the organic-solvent/SO_2-recovery process sections

TABLE 4

Typical operating parameters for the DC/DA process

Unit		Operating Parameters
Compressor	Outlet Pressure	1.34 bar (130 in. H_2O gauge)
Furnace-Boiler	Pressure drop	1 kPa (4-5 in. H_2O)
Reactor	Gas Velocity	1.2 m/s (4 ft./s)
	Ave. pressure drop/bed	2-3 kPa (8-12 in. H_2O)
	Catalyst Cost	\$2,600/m^3
SO_3 Absorbers	Packing	3.8 cm (1.5 in.) Intalox saddles
	Packing Cost	\$550/m^3 (\$15/ft^3)
	Packed Height	4.6 m (15 ft.)
	Pressure Drop	1.5-2.5 kPa (6-10 in. H_2O)

TABLE 5

Some equipment specifications for the organic-absorption/ SO_2 - recovery process

Unit		Operating Parameters or Equipment Specifications		
Inlet Air Blower	Outlet Pressure	1.20 bar	or	80 in. H_2O gauge
SO_3 Absorber	Diameter	6.5 m	or	21 ft.
	Packed Height	4.6 m	or	15 ft.
	Pressure Drop	0.015 bar	or	6 in. H_2O
SO_2 Absorber	Diameter	5.1 m	or	16.8 ft.
	Packed Height	11.0 m	or	36 ft.
	Pressure Drop	0.03 bar	or	12 in. H_2O
SO_2 Stripper	Diameter	5.3 m	or	17.5 ft.
	Packed Height	7.3 m	or	24 ft.
	Pressure Drop	0.02 bar	or	8 in. H_2O
Heat Exchanger #3	Area	785 m^2	or	8,450 $ft.^2$
Steam Heater	Area	13.7 m^2	or	150 $ft.^2$
Water Cooler	Area	467 m^2	or	5,000 $ft.^2$

similar to the DC/DA process were sized based on the data provided in Table 4. Specifications for the pollution abatement section are listed in Table 5.

ECONOMIC EVALUATION

Capital cost and production cost estimates for the production of 907 metric tons/day (1000 tons/day) of H_2SO_4 for both processes are shown in Tables 6 and 7 respectively. The costs presented are for November of 1984. The cost data of Peters and Timmerhaus (ref. 5) were used to make these estimates. Working capital was estimated as 15% of purchased equipment capital for both processes. Additional working capital for the organic-solvent/SO_2 - recovery process was included to cover the cost of Selexol inventory. The DC/DA working capital includes a present-worth capital charge for future catalyst replacement over a 30 year lifetime.

Tables 6 and 7 are intended for comparison purposes only. The first table indicates that the separate pollution control process reduces the total capital requirement by about 8% over the DC/DA process. However, it must be noted that the capital cost saving is based on the flowsheet presented by Browder in 1971 (ref. 3). Process improvements in the DC/DA process since that time have not been accounted for. Overall production costs appear to be about 4% lower for the SO_2 recovery process. This savings is mainly in the form of increased electricity credit as well as lower electricity use by the new process.

There are two major reasons for the difference in electrical and steam generation. First, SO_3 removal in SO_3 Absorber #1 in the DC/DA process cools the gas from 200° to about 45°C. Heat from the combustion of sulfur and from stages of SO_2 conversion to SO_3 are used to heat this cool gas to 435°C for adequately fast kinetics in stage #4. This heat, which is added in Heat Exchangers #1 and #2, is only recovered to 200°C in Economizer #2. The additional 155°C of heat is lost to the cooling water of SO_3 Absorber #2. The SO_2 recovery process, however, uses this heat to boil up extra steam thus increasing the amount of steam and electricity generated. Second, the DC-DA process requires a higher exiting pressure from the inlet blower to cover the system pressure drops. Approximately 40% more electricity is required, which for the DC/DA process amounts to over 40% of the electricity generated by the turbine.

Despite elimination of acid cooler C in the DC/DA process, cooling water requirements are only 5% lower because of the addition of a Selexol cooler in the pollution control section of the new process flowsheet. Furthermore, even though the new process produces more steam because of more efficient use of heat from the waste heat boiler and catalytic converter, additional steam is required by the steam heater in the pollution abatement section. The result is

TABLE 6

Comparison of capital costs for the DC/DA and organic-absorption/SO_2-recovery (OA/SO_2) processes

Component	DC/DA ($)	OA/SO_2 ($)
Inlet Air Blower	458,600	201,100
Air Dryer	122,100	122,100
Furnace and Boiler	585,200	560,000
Heat Exchanger #1	45,000	---
Heat Exchanger #2	70,100	---
Reactor	223,700	184,800
Superheater	109,400	231,900
Economizer #1	115,300	137,300
Economizer #2	96,800	---
SO_3 Absorber #1	185,500	185,500
SO_3 Absorber #2	185,500	---
Acid Tank #1	8,600	8,600
Acid Tank #2	8,600	8,600
Acid Pumps	47,400	31,200
Acid Coolers	608,500	464,200
SO_2 Absorber	---	190,000
SO_2 Stripper	---	183,500
Absorber Pump	---	11,200
Stripper Pump	---	11,200
Heat Exchanger #3	---	87,200
Water Cooler	---	71,600
Steam Heater	---	8,600
Turbine	356,300	400,700
Purchased Equipment		
Cost (PC) (1984 $)	3,226,600	3,099,300
Direct Fixed Capital 4(PC)	12,906,400	12,397,200
Working Capital (see text)	3,764,100	2,936,400
Total Fixed Capital (in Nov. 1984 $)	16,670,500	15,333,600

TABLE 7

Comparison of the cost of manufacture for the DC/DA and organic-absorption/SO_2 recovery (OA/SO_2) processes

Expense	DC/DA ($)	OA/SO_2 ($)
Raw Materials		
Sulfur ($138.8/m. ton)	15,315,000	15,315,000
Water ($0.25/m^3)	15,600	15,600
Selexol ($2.90/kg)	---	24,700
Total Raw Material	15,330,600	15,355,300
Utilities (credit)		
Steam (10 bar) ($8.6/1000 kg)	(3,985,400)	(4,032,700)
Electricity ($0.05/kW-hr)	(688,000)	(1,133,600)
Cooling Water ($0.05/m^3)	889,200	846,600
Total Utility	(3,784,200)	(4,319,700)
Other Expenses (Direct and Indirect)		
Labor (L)	300,000	300,000
Fringe Benefits (.22L)	66,000	66,000
Supervision and Clerical (.18L)	54,000	54,000
Operating Supplies (.10L)	30,000	30,000
Laboratory (.15L)	45,000	45,000
	495,000	495,000
Maintenance (0.06 DFC*)	774,400	743,800
Depreciation (0.1 DFC)	1,290,600	1,239,700
Insurance (0.01 DFC)	129,100	124,000
Local Taxes (0.02 DFC)	258,100	247,900
Factory Expense (0.05 DFC)	645,300	619,900
Total Other Direct and Indirect	3,097,500	2,975,300
Total Manufacturing Cost	15,138,900	14,505,900

*DFC - Direct Fixed Capital in Table 6.

no net increase in steam by-product credit.

CONCLUSIONS

A process flowsheet was evaluated that uses polyglycol ethers as solvents to control SO_2 emissions from sulfuric acid plants. The capital requirement for a new plant is about 8% less than for the conventional DC/DA process. Equipment savings in the form of reduced catalyst inventory, a 5% overall reduction in heat exchange area, and a 40% reduction in inlet blower size largely account for this difference. Production costs are estimated as 4% lower. Because extra heat is recovered in the new process in the form of steam superheat, more electricity is generated for sale to local utilities. Furthermore, the new process uses 40% less electricity since the air blower is significantly smaller. The process evaluation presented is for the construction of a new plant only. Retrofits to existing sulfuric acid plants using the DC-DA process would probably not be cost-effective since the extra capital equipment would provide no increased SO_2 removal, and since retrofit costs are generally higher than the costs of equivalent equipment for new plants.

ACKNOWLEDGEMENTS

This work was supported in part by the Morgantown Energy Technology Center, Assistant Secretary for Fossil Energy, Office of Coal Utilization, Advanced Research and Technology Development, Division of Surface Coal Gasification, through the U.S. Department of Energy under Contract No. DE-AC03-76SF00098. The authors gratefully acknowledge Dow Chemical for providing quantities of diethylene glycol, methyl ether used in this research.

NOMENCLATURE

L = Liquid flow rate (moles/time)
T = Temperature (^{o}K)
Va = Absorber gas flow rate (moles/time)
Vs = Stripper gas flow rate (moles/time)
m = equilibrium constant

REFERENCES

1 Demyanovich, R. J., Ph.D. Dissertation, University of California, Berkeley, CA., 1984.
2 Albright, L. F., et. al., Chem. Eng. Prog. Symp. Series, 1964, 59, 66.
3 Browder, T.J., Chem. Eng. Prog. Tech. Man., "Sulfur and SO_2 Development"; 1971, 91.
4 Borghi, H., Allied Chemical, P.O. Box 1712, Richmond,CA. 94802.
5 Peters, M.S., and K.D. Timmerhaus, Plant Design and Economics for Chemical Engineers, McGraw-Hill, New York, N.Y. (1980).

METHODS USING ACTIVATED COKE FOR REMOVAL OF SO_2 AND NO_x FROM THE FLUE GASES OF FIRING SYSTEMS

E. RICHTER, K. KNOBLAUCH, and H. JÜNTGEN
Bergbau-Forschung GmbH, Dept. Coal Chem., Franz-Fischer-Weg 61, 4300 Essen (Fed. Rep. of Germany)

ABSTRACT

Bergbau-Forschung GmbH developed several processes for the removal of SO_2 and NO_x from the flue gases of firing systems. Descriptions of the processes are given and their applications discussed.

INTRODUCTION

The operation of big firing plants is no longer conceivable without measures for environmental protection and relevant disposal techniques. To begin with, there are the primary measures intervening in the geometry of the combustion chamber but reducing the formation of nitric oxides only to a limited extent. The secondary measures for SO_2 and NO_x reduction then intervene in the flue gas transfer either between economizer and air preheater or between electric precipitator and stack. At the first of these points the flue gases are at temperatures above 300 °C and (for coal-based firings) at dust contents as high as 20 g/m^3. Behind air preheater and precipitator the flue gases are down at 150 °C.

To remove NO_x by means of a selective catalytic reaction (SCR process) a catalyst is arranged before the air preheater, and ammonia used as a reducing agent. Inconvenients are frequent as far as retrofitted plants are concerned as well as from the effects of ammonia breakthrough on subsequent plant components. The technique has so far been used in Japan only and still waits for adaptation to the operational conditions in other countries (ref. 1,2).

For flue gas desulphurization have been used so far mostly scrubbing and spray-drying methods. Scrubbing methods function on solutions of lime, limestone, ammonia or sodium sulphite, whereas for spray sorption calcium hydroxide solutions are used. For scrubbing, a current technique in the Federal Republic of Germany is to cool the flue gases - behind the precipitator - in a heat exchanger (functioning mostly according to the Ljungström principle) down to a temperature between 45 and 65 °C which is close to that one of the scrubber. Before being released into the stack the flue gases are reheated in the aforementioned heat exchanger.

Using activated coke, the sulphur oxides can be removed by adsorption at temperatures below 120 °C. This flue gas desulphurization according to the Bergbau-Forschung GmbH invention intervenes behind the precipitator and before the stack. Since the flue gases need not be cooled down, reheating can be done with as well (ref. 4).

Besides, the activated coke may be used as a catalyst with NO_x reduction by means of ammonia at temperatures between 80 and 150 °C. There is a wide range of applications for the process, e.g.:

- simultaneous removal of SO_2 and NO_x from the flue gases of coal- and oil-based firing plants;
- NO_x and SO_2 reduction in conjunction with other methods for flue gas desulphurization.

The methods using activated coal for SO_2 removal rely on regeneration. From the SO_2 rich gas yielded during regeneration of the activated coke may be recovered chemical feedstock as elementary sulphur or sulphuric acid.

If the activated coke method intervenes behind a flue gas desulphurization, the SO_2 rich gas may as well be recycled to upstream of the flue gas desulphurization. In the following chapters will be explained the fundamentals of the process as well as different methods developed by Bergbau-Forschung.

ACTIVATED COKE

Activated coke is produced from hard coal and serves for flue gas cleaning. The appertaining production steps are (ref. 3):

- grinding,
- air oxidation at temperatures below the ignition point,
- blending of oxidized coal with binders,
- extruding,
- low-temperature cokefaction in inert atmosphere.

The activated coke is of a cylindric shape with diameters up to 9 mm. Hereafter the characteristics of the activated coke as produced according to the process of Bergwerksverband, Essen:

- high adsorptive properties for SO_2,
- rapid adsorption kinetics,
- high catalytic activity for ammonia based NO_x reduction, even at low temperatures,
- low reactivity with oxygen,
- no activation losses due to regeneration,
- high mechanical strength.

The activated coke used for SO_2 removal has a pore volume which grows with the progression of adsorption/regeneration cycles. This still improves the adsorptive properties in the course of the process.

PROCESS BASES

The following paragraphs contain some basic remarks on adsorptive SO_2 removal and catalytic nitric oxide reduction on activated coke as well as on the reactions taking place during thermal regeneration of the loaded activated coke.

SO_2 Removal and NO_x Reduction

When developing the process we had to deal first with SO_2 removal on activated coke. SO_2 will be adsorbed on activated coke over a wide thermal range, starting already at ambient temperature and extending through 150 °C, where it reacts with oxygen and steam to yield sulphuric acid, as by the following expression:

$$SO_2 + \frac{1}{2}O_2 + H_2O \longrightarrow (H_2SO_4)_{ads} \qquad (1)$$

Sulphuric acid is retained by the pore system of activated coke. Peak loads between 10 and 20 % by wt. are attained even at temperatures of no higher than 120 °C as sulphuric acid has an extremely low vapour pressure (ref. 5).

In the flue gases from firing plants, nitric oxides (NO_x) are composed mainly by nitrogen monoxide (NO) and only by 5 to 10 % of nitrogen dioxide (NO_2). The latter is reduced quite fast by reacting with the carbon, at temperatures about 80 °C.

When adding ammonia to the flue gases, the NO reduction is subjected to a catalytic acceleration by activated coke, as per following expression:

$$6\ NO + 4\ NH_3 \longrightarrow 5\ N_2 + 6\ H_2O \qquad (2)$$

Above 80 °C any reaction between NO and NH_3 to give ammonium nitrite will be excluded as the latter substance is unstable within that temperature range.

Except for gas firing systems, the flue gases will contain besides nitric oxides also sulphur oxides which can never be gotten 100 % rid of, not even by preceding SO_2 removal steps. Furthermore ammonium will react additionally with the adsorbed sulphuric acid formed during SO_2 removal, to give ammonium bisulphate and ammonium sulphate:

$$NH_3 + H_2SO_4 \longrightarrow NH_4HSO_4 \qquad (3)$$

$$NH_3 + NH_4HSO_4 \longrightarrow (NH_4)_2SO_4 \qquad (4)$$

Thermal Regeneration

The activated coke, loaded by sulphuric acid and relevant ammonium salts, needs thermal regeneration after having served during more than 60 hours, in order to recover its full separating capacity for the next cycle of flue gas cleaning. From roughly 300 °C onward sulphuric acid will react with the carbon of the activated coke and give rise to the formation of sulphur dioxide and carbonaceous surface oxides C...O as an intermediate product, as per following expressions:

$$2\ H_2SO_4 + 2\ C \longrightarrow 2\ SO_2 + 2\ H_2O + 2\ C...O \qquad (5)$$

$$2\ C...O \longrightarrow C + CO_2 \qquad (6)$$

Ammonium salts decompose (opposed to the reaction paths of (3) and (4)) into sulphuric acid and ammonia. The latter will spontaneously reduce SO_3 (intermediate product of the degradation of sulphuric acid) or surface oxides as per:

$$2\ NH_3 + 3\ C...O \longrightarrow N_2 + 3\ H_2O + 3\ C \qquad (7)$$

This is why carbon losses are minimized during regeneration. On the other hand the CO_2 formation is beneficial to the development of the pore system which otherwise could be attained exclusively by an expensive activation process (e.g. by partial steam gasification) during the production process of the carbonaceous adsorbent.

APPLICATIONS OF METHODS BASED ON ACTIVATED COKE

The development work deployed by Bergbau-Forschung GmbH, Essen, has confirmed that methods based on activated coke can be used successfully for various applications with different targets. In the following will be highlighted the state of the art of some of these applications.

Fig. 1 shows, by way of example, four process approaches pursued and tested on various scales. All of them are based on catalytic NO_x reduction on activated coke and using ammonia. The alternatives differ from each other by their type of SO_2 removal. Selection of the most suitable approach depends on the purpose to be served (SO_2 content of flue gases and utilization of the sulphur-containing final products; existing plant/new plant; flue gas desulphurization existing or not; etc). It can be said, in general, that the method relying on activated coke will be most advantageous with SO_2 contents below 4000 mg SO_2/m^3 (operational conditions). With higher SO_2 concentrations the method should be linked to preceding flue gas desulphurization units whose SO_2 removing efficiency can be relatively low.

The activated coke loaded by sulphur compounds has to be regenerated at certain intervals which depend on operational conditions (essentially on the SO_2 contents before the desulphurization step). The SO_2 rich gas yielded during thermal regeneration will either be subject to secondary treatment or recycled to upstream of the SO_2 removal step of the firing system. The figure shows the following operations:

Simultaneous Removal of SO_2 and NO_x

The flue gases of a coal- or oil-fired boiler pass through the air preheater (Luvo) and electric precipitator (EF) and are fed at 120 °C to the flue gas cleaning plant (fig. 1). The latter consists of two steps operating on activated coke, of the regeneration step for activated coke and of the secondary treatment

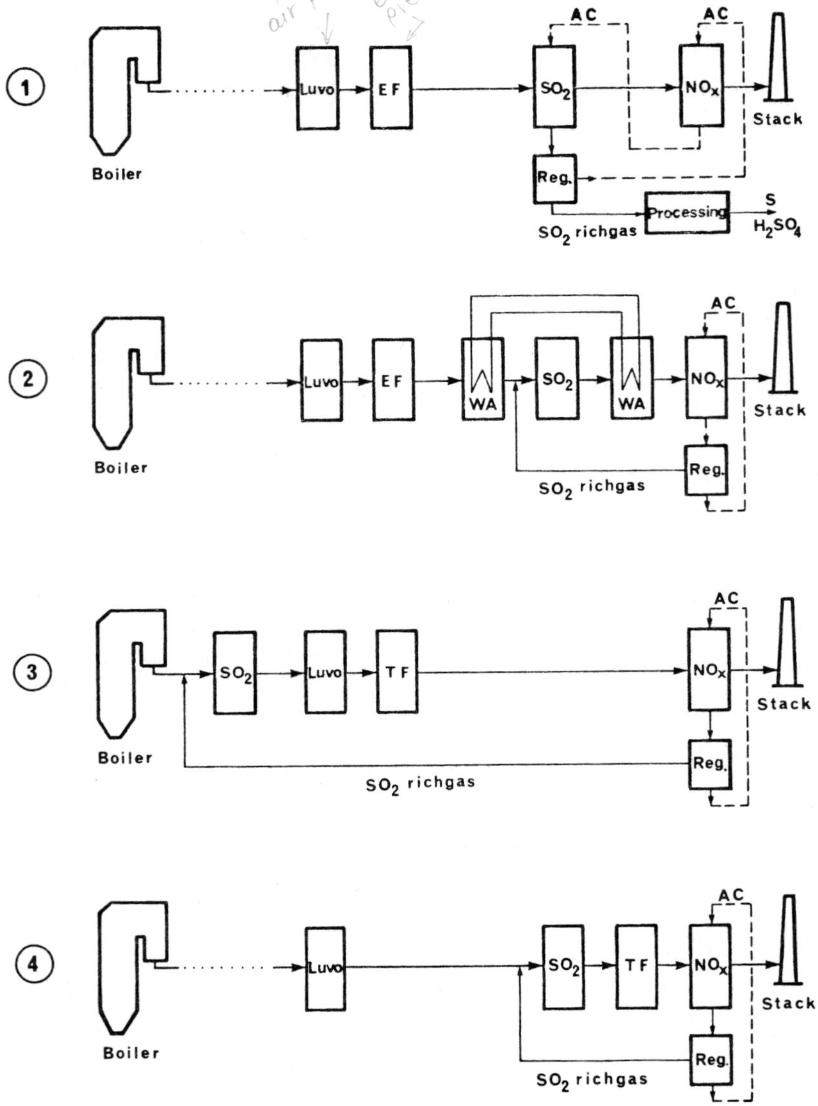

Fig. 1 Activated Coke Processes for SO_2/NO_x-Removal

unit transforming the SO_2 rich gas either into elementary sulphur, sulphuric acid or liquid SO_2. The configuration of reactors and activated coke cycles is shown more in detail on Fig. 2. The moving bed reactor for flue gas cleaning consists of two beds which may be superimposed. The first (lower) bed serves for separation of a major part of the sulphur dioxide (according to equation (1)) and nitrogen dioxide. No ammonia will be added before this bed since the separation capacity of the activated coke alone is sufficient. Prior to the second (upper) bed some ammonia will be dose-fed for reduction of the nitrogen monoxide (according to equation (2)). Furthermore the residual sulphur dioxide will be removed (according to equations (1), (3) and (4)). The formation of ammonium nitrate from nitrogen dioxide and ammonium is precluded since sulphur dioxide is present and the nitrogen dioxide had been almost completely removed already during the first step.

As in the upper moving bed a feeble amount of sulphur compound load only gets into the activated coke the latter can, upon leaving the upper bed, still be used for SO_2 removal in the lower bed. The activated coke can thus take more SO_2 than in a single-step process.

The spent material is thermally regenerated for reuse. Regeneration can happen e.g. in a column-type desorber a simplified representation of which is shown on fig. 2. The activated coke is heated in an upper section of tubes through which it passes, in order to give off its gases in an intermediate section when being at 400 or 450 °C. The SO_2 rich gas released hereby contains between 25 and 30 % by vol. SO_2 which will be cleaned in a separate, subsequent unit and transformed either into S, H_2SO_4 or liquid SO_2. Prior to being reused in the moving beds the activated coke is cooled down to approx. 100 °C in a tube-type cooler. The process lends itself not only to newly to be erected power stations but can also be retrofitted without any problems to existing ones since the components easily fit into the flue gas transfer, midway between electric precipitator and stack.

SO_2/NO_x Reduction Behind Flue Gas Desulphurization Units

SO_2/NO_x Reduction Behind Flue Gas Scrubbers. Flue gas scrubbers are worldwide the most common facilities for SO_2 separation. As far as the Federal Republic of Germany is concerned,

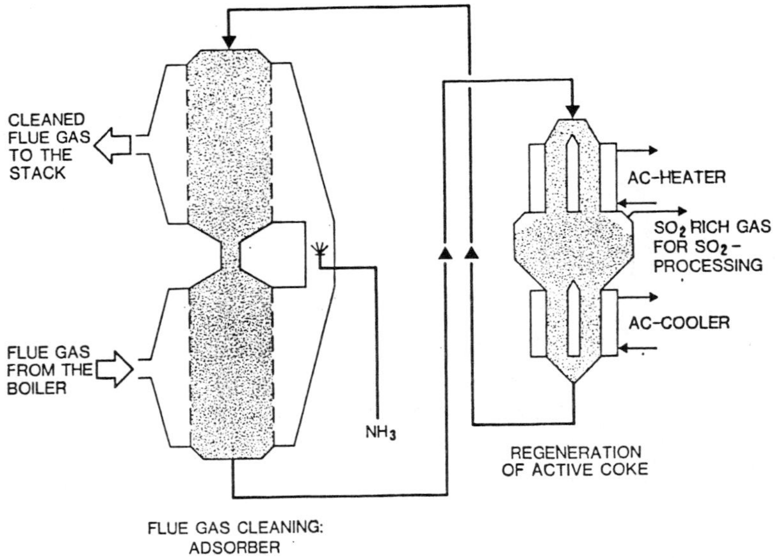

Fig. 2 Flowsheet of the BF-Process for SO_2- and NO_x-Removal from Flue Gases

Fig. 3 Photograph showing the Flue Gas Purification Plant at Omuta/Japan (BF-Mitsui-Process)

they are equipped with heat exchangers (WA) for reheating the flue gases cooled down by scrubbing (see alternative 2 on fig. 1,(ref. 3). Upon reheating, the flue gas is again between 85 and 95 °C which is sufficient for reducing the nitric oxides by means of ammonia in a subsequent reactor operating on activated coke.

Flue gas desulphurization equipment of West-German high-capacity firing systems of more than 300 MW have to be operated in a way as to bring the emissions down to below 400 mg SO_2/m^3 (operational conditions). The presently operative units usually stay below such level, some of them even far below. The SO_2 load on the activated coke method is, therefore, quite moderate so that only the reactor for NO_x removal will be required. This reactor will normally be designed as a moving bed reactor. The withdrawing rate of the activated coke can be kept very low, however, so that the regeneration unit may be laid out in a size which is considerably reduced against that required for the activated coke method for simultaneous removal of SO_2 and NO_x (alternative no. 1). The rich gas released during regeneration is recycled to upstream of the flue gas cleaning in order to be liberated of its SO_2. Any separate further treatment of the SO_2 can therefore be done with since one relies on the flue gas scrubbing technique. In this case can be recurred to simpler means of thermal regeneration since here the SO_2 concentration in the rich gas plays a negligible part so that separate rich gas cleaning is unnecessary.

If flue gases of quite high SO_2 contents are involved it can be useful to design the flue gas scrubbing plant for a relatively low efficiency and run it as a preliminary step to the two-step activated coke method.

As far as NO_x removal is concerned, the procedure is advantageous in so far as it is considerably easier to integrate in a power station than the selective catalyst-based reduction (SCR) working on temperatures above 250 °C. Moreover expenditures for the catalyst are lower notwithstanding the losses in activated coke in the course of recycling (guaranteed service life of SCR catalysts is between two and three years, at catalyst costs of about 40 million DM for a block of 750 MW_{el}).

NO_x Removal Behind Additive/Spray Sorption Methods for SO_2 Separation. Particularly for smaller boiler plants and those boilers burning a high-sulphur coal, additive methods are an appropriate alternative for SO_2 removal since the methods simply consist in the injection of a (usually) lime compound binding the sulphur oxides. The sulphur-containing final product is then separated by subsequent filters. They are mostly fabric filters (TF) where the building-up filter cake helps towards desulphurization. One of the advantages is that the product does not have to be treated at the site and cooling of the flue gases is negligible so that reheating, as it is usually required with scrubbing, can be done with.

Fig. 1 shows two options of installing and combining the procedure with activated coke methods for NO_x removal.

As to option 3 (medium temperature additive method) the additives (lime compounds and fly-ash) are introduced in the flue gas in a medium thermal range (300 to 500 °C). The products are separated by fabric filters. The activated coke method for catalytic NO_x reduction functions at temperatures between 100 and 150 °C. The SO_2-containing gas yielded during regeneration of the activated coke will be recycled to upstream of the additive injection point.

In case of alternative no. 4, the additives are injected as a suspension behind the air preheater (spray sorption method). The sulphurous final products are separated as a solid material in subsequent fabric filters. NO_x removal by means of activated coke follows as well.

In case of rather high SO_2 contents in the flue gases it may be useful, again, to run the additive and spray sorption methods at low efficiencies and arrange behind them the two-step activated coke method for SO_2/NO_x removal.

STATE OF THE ART

The BF process for flue gas desulphurization was at the beginning of these developments. The process was operated successfully during more than 16 000 hours at the demonstration plant of the STEAG power station "Kellermann", Lünen, for a throughput of 150 000 m^3/h (operational conditions) of flue gas, at intermittent load. The plant provided, in particular, any and all process data required for plant up-scaling, the know-how on equipment sets required, resistance to wear, tear, and corrosion as well as on the availability of the plant components and transport systems. In a separate test phase was tried to simultaneously remove SO_2

and NO_x by two successive individual adsorber elements. A SO_2 separation efficiency of more than 98 % was attained (same efficiency as when combining the activated coke method for NO_x removal to other SO_2 removing processes). The NO_x separation efficiency is adaptable to individual requirements via the reactor layout and addition of ammonia (e.g. smaller than 200 mg NO_2/m^3 in operational conditions).

The Japanese company Mitsui Mining Corporation Ltd. holds a licence of Bergbau-Forschung GmbH for the above process and has been operating, from October 1984, the commercial plant represented on Fig. 3 for a throughput of 30 000 m^3/h flue gas. The first commercial plant of this type in the Federal Republic of Germany is under construction at the Arzberg power station run by Energieversorgung Oberfranken. Here, two lignite-fired generator blocks yielding together 1.1 million m^3/h flue gas are to be served, with a scheduled reduction of nitric oxides to below 200 mg NO_2/m^3 (operational conditions). The order is executed by Uhde GmbH, Dortmund.

The alternatives of fig. 1 for NO_x removal from gas-fired plants or behind other upstream flue gas desulphurization units are under trial at several pilot plants in order to find out the specific layout data. In this case can be recurred to the technology having proved its usefulness in the second step of the simultaneous SO_2/NO_x removal process. One particular point to be investigated at the pilot plant is the influence of aerosols or dust constituents originating from upstream desulphurization units, on the catalytic properties of the activated coke. Contamination of the activated coke (contrary to SCR catalysts) is not apprehended since during thermal regeneration for getting rid of the sulphate load, the activated coke is liberated from any other contaminants whatsoever as well.

The methods described above can be integrated both in new and in existing plants. Retrofitting an activated coke unit to an existing plant will be easy in any case as intervention in the flue gas transfer takes place only just before the stack. This is of special importance if an existing plant has already a flue gas desulphurization unit under construction or operative which is to be completed by the activated coke method for NO_x removal.

As appears from calculations, the cost of the simultaneous SO_2/NO_x removal process is below that of any of the scrubbing processes for SO_2 removal linked to the Japanese SCR technology for NO_x reduction. The same cost relation appears when comparing the NO_x removal on activated coke basis with the SCR technology.

The methods are being improved by Bergbau-Forschung GmbH and their licensees.

REFERENCES

1 H. Jüntgen: Measures for reducing NO_x in the flue gas, - state of technology. VDI-Berichte no 495, pp 127/132; VDI-Verlag, Düsseldorf 1984.

2 H. Eipper: Influence of NO_x reduction on subsequent plant components. Techn. Mitteilungen (Essen) 78 (1985), issue 1/2, pp 28/32.

3 E. Richter: State of technology of flue gas desulphurization in the Federal Republic of Germany. Umwelt (VDI-Verlag, Düsseldorf), 1984, issue no. 3, pp 191/205.

4 E. Richter, K. Knoblauch: "Unloading of flue gases": methods of flue gas desulphurization and NO_x removal. Umweltmagazin (Würzburg) 1984, issue 6, pp 40/42.

5 H. Dratwa, H. Jüntgen: Flue gas desulphurization using adsorptive coke of varying properties. Staub - Reinh.d.Luft 27 (1967), issue no. 7, pp 301/307.

INVESTIGATION OF THIOSULFATE AS AN OXIDATION/SCALING INHIBITOR IN LIME/LIMESTONE SCRUBBERS FOR FLUE GAS DESULFURIZATION

J. F. Miller, J. H. Oxley, H. S. Rosenberg, and H. K. Nuzum; Battelle-Columbus Laboratories, 505 King Ave., Columbus, Ohio 43201 (U.S.A.)

ABSTRACT

In view of the conflicting reports in the late 1970's regarding the effects of thiosulfate in FGD scrubber-type systems, this investigation was undertaken with the objectives of determining whether or not thiosulfate additive inhibits oxidation and scaling in lime/limestone FGD scrubber systems and, if so, the range of conditions under which it can be utilized.

The results of the experimental laboratory batch scrubber study, in which the effects of magnesium, chloride, fly ash, and oxygen concentration in the flue gas also were investigated, indicated sodium thiosulfate, added in concentrations corresponding to about 20 pounds or more per ton of lime, can be an effective sulfite oxidation inhibitor. Including all conditions investigated, average oxidations with thiosulfate additions were down to 40 percent of the uninhibited value when magnesium was present in the lime. In individual experiments, much stronger inhibition was observed--to as little as 1.4 percent of the uninhibited level. Lowering the oxidation level reduces the rate of sulfate formation and thus the potential for gypsum scaling.

Experimental results and the results of analysis of variance of the data also indicate that

- thiosulfate additions strongly inhibit gypsum saturation,
- thiosulfate additions can slightly enhance SO_2 absorption efficiencies,
- among the variables investigated magnesium (promoter) and thiosulfate (inhibitor) exert major effects on oxidation, while only minor effects are associated with chloride and fly ash in the ranges investigated.

The general conclusions are supported by reports of experience with thiosulfate additions in the operation of scrubbers at the Phillips and Elrama Stations of Duquesne Light and the Paddy's Run Station of Louisville Gas and Electric and by the results of research at the University of Texas using synthetic solutions. The results of limited experiments at Battelle also indicated that thiosulfate inhibitor functions similarly in a limestone scrubbing system, as did pilot plant work at the Shawnee Station at the Tennessee Valley Authority, which was carried out by Bechtel concurrently with the Battelle study.

In recent years, the problem of scaling appears to have decreased somewhat as more has been learned about design and operation of lime/limestone scrubber systems. Scaling is a complex phenomenon, however, and can occur even where good design practice has been used. As further efforts are made to tighten the water loops in the scrubber operations, scaling difficulties may reappear. Thus, thiosulfate may be employed in the future as insurance to correct scaling conditions when they are encountered unexpectedly. There are also a number of presently operating plants of older design and construction where thiosulfate additions might prove helpful now.

INTRODUCTION

In view of the conflicting reports in the literature regarding the effects of thiosulfate in lime/limestone wet scrubbers for flue gas desulfurization (FGD), the U.S. Department of Energy through the Morgantown Energy Technology Center (METC) requested Battelle-Columbus to undertake a laboratory study to attempt to clarify the situation. The objectives of this study were to determine whether or not thiosulfate additive inhibits oxidation and scaling in lime/limestone FGD systems, and if so, the range of conditions under which it can be utilized. This study was carried out during 1980 and 1981 at a time when many large-scale FGD systems were having serious operational problems.

PROBLEM DEFINITION

Due to its relative simplicity, lime/limestone scrubbing is widely used to meet the SO_2 emissions standards in the U.S. and all over the world today. Over 100 coal-fired power plant boilers will have applied this method in the U.S. by the end of 1985; more than half the installations use limestone as the reactant. In Japan, the situation is similar regarding the extent of utilization of the systems, and in Western Europe a significant number of plants are starting operation. Although these statistics are factual, it is an exaggeration to state that all of these systems accomplish their purpose without many difficulties. Various technical problems remain to be solved.

Before considering lime/limestone FGD systems to be fully commercially satisfactory, the following conditions must be met: (ref. 1)

- High removal efficiency to meet emission standards even when burning high sulfur and high ash coal,
- High reliability on a commercial scale, to keep the expense for maintenance low and to meet the emission standards over the long-term,
- Operation in a closed-loop mode, preventing discharge of process water to natural water systems.

In general, especially for the very restrictive SO_2 emission standards applicable in western Pennsylvania, meeting these conditions has proven to be a formidable goal.

Process Chemistry

The SO_2 removed in a lime-limestone-based FGD system is converted to sulfite, generally represented by the overall reactions

$$SO_2 + CaCO_3 \longrightarrow CaSO_3 + CO_2$$

$$SO_2 + Ca(OH)_2 \longrightarrow CaSO_3 + H_2O$$

when using limestone and lime, respectively. Part of the sulfite is oxidized with the oxygen content in the flue gas to form sulfate:

$$CaSO_3 + 1/2O_2 \rightarrow CaSO_4 \ .$$

Although the overall reactions are simple, the chemistry is quite complex and not well defined. In a comprehensive theoretical and experimental study by Radian (ref. 2), it was indicated that 41 neutral and ionic species in solution and 7 solid components are of importance. Therefore, it can be inferred that the reaction rate and the influence of design and process variables on composition in this complicated three-phase system are almost impossible to determine. The complex chemistry may also explain why various studies have led to different conclusions.

When the slurry flows down through the absorber and contacts the SO_2-laden flue gas, SO_2 may react with at least 15 different basic species that may be present in a hundredfold higher total concentration than the inlet concentration of dissolved lime or limestone. The choice between lime and limestone, the type of limestone, and method of calcining and slaking can influence the gas-liquid-solid reactions taking place in the absorber.

In order to achieve optimum absorption of SO_2, it is necessary to understand the process chemistry. However, much of the present knowledge has been developed mainly on an empirical basis. Some important operational aspects that can be controlled by understanding the process chemistry are:

- Removal efficiency
- Plugging and scaling
- Reagent utilization.

High removal efficiency is favored by a high pH; a low pH decreases the reaction rate and increases the SO_2 back-pressure from the scrubber solution. On an empirical basis, it has been established that for lime systems the reactions are controlled when the pH is kept in the range of 6.9 to 8.9. For limestone, it is not possible to achieve pH values above 7.

Interest has also been focused on the effect of magnesium on the absorption rate. Because magnesium sulfite is about 300 times as soluble as calcium sulfite, the concentration of the former can build up in the scrubber loop. This considerably increases the amount of available basic species which, in turn, increases the removal efficiency. The source of magnesium can be as an additive, for instance the oxide or sulfate, but alternatively a lime type with a high magnesium content can be used (calcined dolomite). Small amounts of magnesium, on the order of 5 percent MgO in CaO, are enough to build up a magnesium to calcium ratio as high as 100 in the scrubbing liquor. The actual value is determined by the bleed-off from the absorber loop. If uncalcined dolomite is utilized, the absorption efficiency instead will be depressed due to the extremely slow dissolution rate of associated magnesium and calcium carbonates. The magnesium content in the lime is considered in the choice of raw materials and a specially selected lime with 3 to 9 percent MgO is usually used. (ref. 3)

The concentration of undissolved reactant in the absorber loop is of importance for the removal efficiency in the cases where the dissolution rate is the rate controlling step. Thus, an increased concentration, as well as a decreased particle size, will increase the absorption efficiency in these cases.

Lime Versus Limestone

In most cases, 75 to 90 percent of the lime/limestone is utilized, but a value as high as 98 percent has been reported for a limestone system installed on a utility boiler in Japan. In general, reagent utilization is higher for lime than for limestone. The choice between lime and limestone is usually based on economic considerations. When utilizing lime, the expenses for calcination, transportation, and slaking must be considered. With regard to limestone, transportation and grinding are of importance. The cost for lime is much higher due to the calcination. The cost ratio of lime to limestone on a molar basis may vary from 2 to 4 depending on the transportation distance. Although the carbon dioxide in limestone contributes to an increase in the molecular weight, the costs for transportation and handling tends to be lower than for lime since it can be transported in open trucks. (ref. 4)

In the past, when few full-scale units were in operation, it was generally believed that less scaling and higher removal efficiency could be attained with lime rather than limestone. This, however, has never been definitely proven to be the case. The trend today is slowly turning towards utilization of limestone instead of the more expensive lime. Of the plants under construction or contracted in the U.S., twice as many are based on limestone as on lime. Very few, if any, lime systems are under consideration for projected plants.

Nevertheless, some differences between lime and limestone have been observed. The dissolution rate of limestone is much slower than for lime. Mass transfer data also indicate a greater liquid-side resistance for the former. Although it has not been proven, the faster dissolution rate of lime probably leads to better load-following capability. An unsaturated mode is somewhat easier to attain with lime, as well. More important than the previously mentioned factors is the fact that in some cases, instability of pH has occurred with limestone due to the slow dissolution rate.

Scaling

Scaling is a very complex phenomenon that depends on design parameters and process variables as well as process chemistry. Hard scale buildup causes increased pressure drop and plugging in the equipment and, therefore, it is one of the main causes for shutdowns. Three different types of scaling are commonly referred to:

- Carbonate scaling
- Sulfite scaling
- Sulfate scaling.

It has been found by experience that the former two types can be controlled to a large extent by keeping the pH well below 9. In practice, sulfite scaling seldom occurs in limestone scrubbing because the pH is lower and because sulfite supersaturates to such a degree that crystallization does not occur until the slurry reaches the reaction tank. However, sulfate scaling is much more difficult to control. Unlike sulfite, the decrease in pH down through the scrubber does not help hold sulfate in solution. Thus, if there is a considerable amount of sulfite oxidation in the scrubber, sulfate crystallization can occur. Since the oxidation rate of sulfite to sulfate increases with decreasing pH, it is generally assumed that sulfate scaling can also be prevented by adjusting the pH. However, this results in higher reagent costs and often has caused mist eliminator scaling.

Sulfate is continuously formed in the absorber loop and most of it precipitates in the holding tank. It has been found that sulfate can supersaturate to a factor of 1.4 before massive nucleation starts (ref. 5), but deposition can still occur below 1.4. Therefore, the sulfate concentration should be kept below the saturation level to insure against spurious deposition. However, the saturation level may also be dependent on the solids composition. If the degree of oxidation is below 20 percent, only mixed crystals of sulfite and sulfate will form so that the slurry remains in an unsaturated state with respect to sulfate. This is, however, difficult to control, especially for low sulfur coals or old boiler units with high excess air since both cases give a relatively high oxygen to SO_2 ratio in the flue gas.

The ratio of sulfate to sulfite in an FGD system can be predicted to a certain extent from the method of fly ash removal, the pH of the scrubbing liquor, the ratio of oxygen to SO_2 and the content of nitrogen oxides in the flue gas. Other factors also are of importance, such as the presence of magnesium and chloride ions, and it is difficult to prevent sulfate scaling by control of the sulfate concentration. In some cases, carbide lime, a byproduct of acetylene generation from caldium carbide, is utilized as the scrubbing reagent. These systems are noteworthy for a lack of oxidation which permits them to operate in an unsaturated mode with respect to sulfate ion.

An alternative method for sulfate scale prevention is to allow large amounts of gypsum to recirculate so that deposition starts on the crystals present instead of on scrubber surfaces. A solid gypsum concentration of 5 percent has been found to be sufficient for this purpose. One method utilized to achieve the desired 5 percent concentration of gypsum is forced oxidation. Fly ash also helps to prevent scaling by providing sites for deposition; according to a TVA study (ref. 6), a total solids content exceeding 10 percent causes crystallization of the sulfate on the solids. However, an increased solids content in the slurry loop makes it difficult to pump the high volumes of slurry required for high SO_2 removal, so that an upper limit of 15 percent is generally accepted. In most of the FGD systems in the U.S., a 7 to 15 percent solids content exists in the recirculating slurry, but in some cases, values as low as 3 percent occur. In Japan, and sometimes elsewhere, gypsum seed-crystals are often added to the process to provide sites for sulfate nucleation.

Sodium thiosulfate was previously claimed to be the impurity in carbide lime that prevents scaling at the Paddy's Run Power Station (Louisville Gas and Electric) (ref. 7). Later, the Duquesne Light Company added thiosulfate as an oxidation/scaling inhibitor at the Phillips Station for more than a year apparently with considerable success (ref. 8). At the time of this Battelle study, the situation with regard to thiosulfate was very confused, as will be discussed below.

PREVIOUS WORK WITH THIOSULFATE

Because oxygen is present in flue gas, there is some degree of oxidation of sulfite to sulfate in any flue gas desulfurization system. The degree of oxidation, which initially was accepted at whatever steady-state level it reached, varies with a number of factors. However, designers and managers now are beginning to take steps to either promote or inhibit oxidation. To date, more has been done with promotion of oxidation because of beneficial effects with respect to solids dewatering and disposal. However, there has been some work on the development of oxidation inhibition, and in 1980, two utilities were using inhibition in commercial systems for its apparent reduction of scaling and improvement of SO_2 absorption efficiency. A third utility began using oxidation inhibition in 1984.

Louisville Gas & Electric

As an alternative to lime from limestone, calcium hydroxide obtained as a byproduct of calcium carbide decomposition has been, and is being, used in scrubbers where it is available. Calcium carbide is produced from coke and lime in an electric furnace and its decomposition with water produces acetylene and slaked lime.

$$CaC_2 + 2H_2O \longrightarrow CHCH + Ca(OH)_2.$$

Sulfur in the coke contaminates the carbide, and sulfur compounds are present in the raw acetylene and byproduct lime. When this lime was used instead of limestone derived lime in the scrubber at the Paddy's Run Station of Louisville Gas and Electric, virtually scale-free operation was obtained.

Calcium sulfate (gypsum) scaling is commonly experienced since part of the sulfite formed by absorption of the SO_2 in the flue gas is oxidized by oxygen also present in the flue gas. The source of oxygen, present in amounts of 4 to 15 volume percent, is the excess air used in the combustion system augmented by the leaking of air into the system. The amount of oxygen is more than adequate to oxidize all SO_2 present. Thus, the amount of sulfate formed is dependent on the kinetics of sulfite oxidation. When lime from carbide decomposition used at Paddy's Run was observed to reduce the scaling potential and by inference the sulfate concentration in the scrubber liquor, the most reasonable hypothesis was that this lime contained an oxidation inhibitor not present in limestone derived lime. Based on this hypothesis, an EPA supported study was initiated by Radian Corporation to determine possible inhibitors in carbide derived lime. (ref. 7) Organic compounds were found in the carbide lime, but all were present in concentrations considered too low (parts per billion range) to affect oxidation significantly. Analyses for reduced sulfur compounds which also were made revealed no sulfide. But of the 2.14 millimoles per liter of total sulfur in the water extract of the carbide lime, 1.52 millimoles per liter of total was present as thiosulfate. In bench-scale oxidation experiments, calcium sulfite derived from commercial lime was oxidized fully in about 20 minutes, whereas with 0.82 millimoles per liter of thiosulfate present, only about 30 percent was oxidized in 2 hours. Carbide lime gave results between these two extremes, presumbably because it supplied less than the 0.82 millimoles per liter of thiosulfate. Also, in other laboratory tests in which air was bubbled through sulfite solutions, with and without thiosulfate present, the inhibiting effect of thiosulfate was shown. (ref. 7)

Duquesne Light Company

The Frank R. Phillips Station of the Duquesne Light Company uses a lime-based scrubber system. Each scrubber module has only a single venturi and no reaction tank. With this system, using lime, SO_2 removal efficiency at startup was in the 60 percent range and scaling was severe, forcing frequent shutdown to clear plugged modules. In 1977, Duquesne Light shifted to magnesium lime containing about 5 percent MgO with dolomitic lime added to bring the MgO level up to about 9 percent for the purpose of increasing SO_2 removal efficiency. The high MgO level did increase the removal efficiency to about 83 percent and also reduced scaling in the scrubbers. However, scaling continued to be a significant problem in the mist eliminators and downstream ducting. The tendency for scaling is greater than normal at Phillips; oxygen concentration in the flue gas is very high (about 12 percent, compared with the usual 4 to 6 percent), with low L/G, low slurry density, and no holding tank.

In view of the scale-free operation at Paddy's Run using carbide lime and the results of the EPA study which indicated thiosulfate to be the oxidation inhibitor in carbide lime, Duquesne Light decided upon a full-scale test of thiosulfate addition to the scrubber system at Phillips. Results of the full-scale testing, which was carried out in August and September, 1979, indicated that scaling was reduced; there was very little scale deposition in the mist eliminators, downstream ducts, and fan housings. Although the degree of oxidation apparently was not determined, gypsum saturation in the slurry, which would be reduced by inhibition of oxidation, was found to be down to about 24 percent. Calcium ion concentration was less than 200 ppm, and sulfate ion concentration was about 10,000 ppm during the thiosulfate test additions. When thiosulfate addition was suspended, gypsum saturation climbed to almost 100 percent, calcium ion concentration rose to the 500 to 700 ppm range, while sulfate ion concentration rose to 20,000 ppm. In addition, when thiosulfate was added in these tests, alkalinity increased from the 500 ppm range to the 1800 ppm range, improving SO_2 absorption conditions and SO_2 removal efficiency by 4 to 6 percentage points. Seeing these improvements, Duquesne Light has continued thiosulfate additions at the Phillips Station and has begun to add thiosulfate during low load (high oxygen) periods at the Elrama Station.

Environmental Protection Agency

Contrary to the findings at Phillips, results of a pilot-plant study at EPA/RTP conducted in 1978 showed thiosulfate additions to be ineffective for inhibiting oxidation. (ref. 9) Flue gas oxygen content was lower (5 volume percent) than in the Phillips Power Station tests, the chloride content of the liquor varied widely, and a turbulent contact absorber was used rather than the venturi scrubber employed at Phillips. Other operating conditions also differed. The conflicting results of these two studies suggested that oxidation inhibition by thiosulfate might be a function of the type of equipment used, flue gas composition, operating conditions, and any number of other factors.

Laboratory Studies

Reported results of chemical research conflict with one another and with the just-discussed full-scale plant and pilot plant results. For example, thiosulfate has been reported to catalyze sulfite oxidation, (ref. 10) while it also has been reported to inhibit oxidation. (ref. 11) For example, Schroeter states that "thiosulfate serves as a powerful catalyst in the air oxidation of aqueous solutions of sulfurous acid salts". The mechanism presented for the catalysis involves trithionate ion ($S_3O_6^=$), produced from interaction of thiosulfate with an OH, O_2^-, or HSO_3 radical. The trithionate ion is said to be readily hydrolyzed to thiosulfate and sulfate, thereby resulting in catalysis of the oxidation.

On the other hand, J. L. Hudson (ref. 11) at the University of Virginia did some work with sodium thiosulfate in calcium sulfite solutions. The conditions of the tests are not described very well other than that the temperature was 50 C and the pH 4.0. It was reported that at $S_2O_3^=$ concentrations of 100 to 250 ppm, no oxidation was observed. At 500 ppm, addition of 10 ppm Mn did not cause oxidation. When the thiosulfate was reduced to 25 ppm about 10 percent oxidation occurred after an hour (bench-scale batch tests). Increasing the pH to 5.0 (at 25 ppm thiosulfate) gave about 25 percent oxidation in an hour.

The direct applicability of both of these results to the forming of conclusions regarding scrubber reactions is questionable. Schroeter does not cite any data to support the catalytic performance of thiosulfate. Hudston's tests were run in dilute solutions rather than slurries containing calcium sulfite solids. The pH was 4.0, a level at which oxidation normally is inhibited in sulfite solutions, and also, no baseline tests were made without thiosulfate for comparison.

EXPERIMENTAL BATTELLE PROGRAM

Through consideration of the background technology and the processes involved, a number of variables can be identified that can have significant effects on SO_2 absorption, sulfite oxidation, and scaling in the lime/limestone scrubbing process. These include the type of slurry (lime or limestone); the Mg^{++}, Cl^-, $S_2O_3^=$, and fly ash concentrations in the slurry; the pH of the scrubber slurry; the SO_2, O_2, NO, and NO_2 concentrations in the flue gas; and the source of fly ash. In the initial matrix of experiments, conditions were essentially similar to those at the Phillips Station of the Duquesne Light Company. A nominal 5 wt. percent slaked lime slurry was used along with a typical flue gas composition (nominal 12 vol. percent CO_2) except for the variation of oxygen concentration. The SO_2 and NO_x concentrations in the flue gas were held relatively constant at 2500 ppm for SO_2, 580 ppm for NO, and 70 ppm for NO_2. A matrix of experiments was designed to investigate the effects of concentration of thiosulfate (0, 1, and 10 wt. percent $NaS_2O_3 \cdot 5H_2O$ in CaO), presence and absence of magnesium (0 and 0.35 wt. percent $Mg(OH)_2$ in the slurry), presence and absence of chloride (0 and 2,500 ppm $CaCl_2$ in the slurry), presence and absence of fly ash (10 and 1,000 ppm fly ash in the slurry), and concentration of oxygen (5 and 10 vol. percent in the flue gas).

The apparatus used in this study is shown in Figure 1 and consisted of a manifold for mixing gases from cylinders to obtain a simulated flue gas, a humidifier to add water vapor to the gas stream, a 0.1 liter Pyrex reaction vessel (experimental scrubber), appropriate traps and valves for cleaning and controlling the gas stream, and appropriate instruments for temperature control and measurement. In addition, the system contained instrumentation for continuous monitoring of slurry pH and SO_2 and NO_x concentrations in the exhaust gas. The reaction vessel was held at a constant temperature of 125 F (52 C) and stirred at a constant speed of 800 rpm (984 rad/sec). The flue gas flow rate was held at a relatively constant 110 cm^3/sec. Samples for analysis were withdrawn from the reactor vessel during the steady state portion of a run at constant pH and SO_2 absorption. The slurry samples were analyzed for sulfite and sulfate to determine the degree of oxidation in terms of mole percent of sulfite converted to sulfate.

The experimental results are shown in Figure 2 which is a bar graph of the degree of oxidation for the lime scrubber conditions investigated. In this figure, the data for experiments run under like conditions, except for the thiosulfate concentrations, are placed together in groups of three. It can be readily seen that in 14 of the 16 groups the degree of oxidation is higher in the absence of thiosulfate, and that in all 16 sets of conditions, oxidation is inhibited by the addition of thiosulfate. The figure also clearly shows (1) the oxidation-enhancing effect of Mg^{++}, and (2) the lack of effect of the increase of concentration of oxygen in the flue gas. The latter is believed to reflect the surface-area or mass-transport limitations of the experimental scrubber. Interestingly, at 10 vol. percent oxygen, magnesium always reduced the oxidation level as compared to 5 vol. percent oxygen in the flue gas.

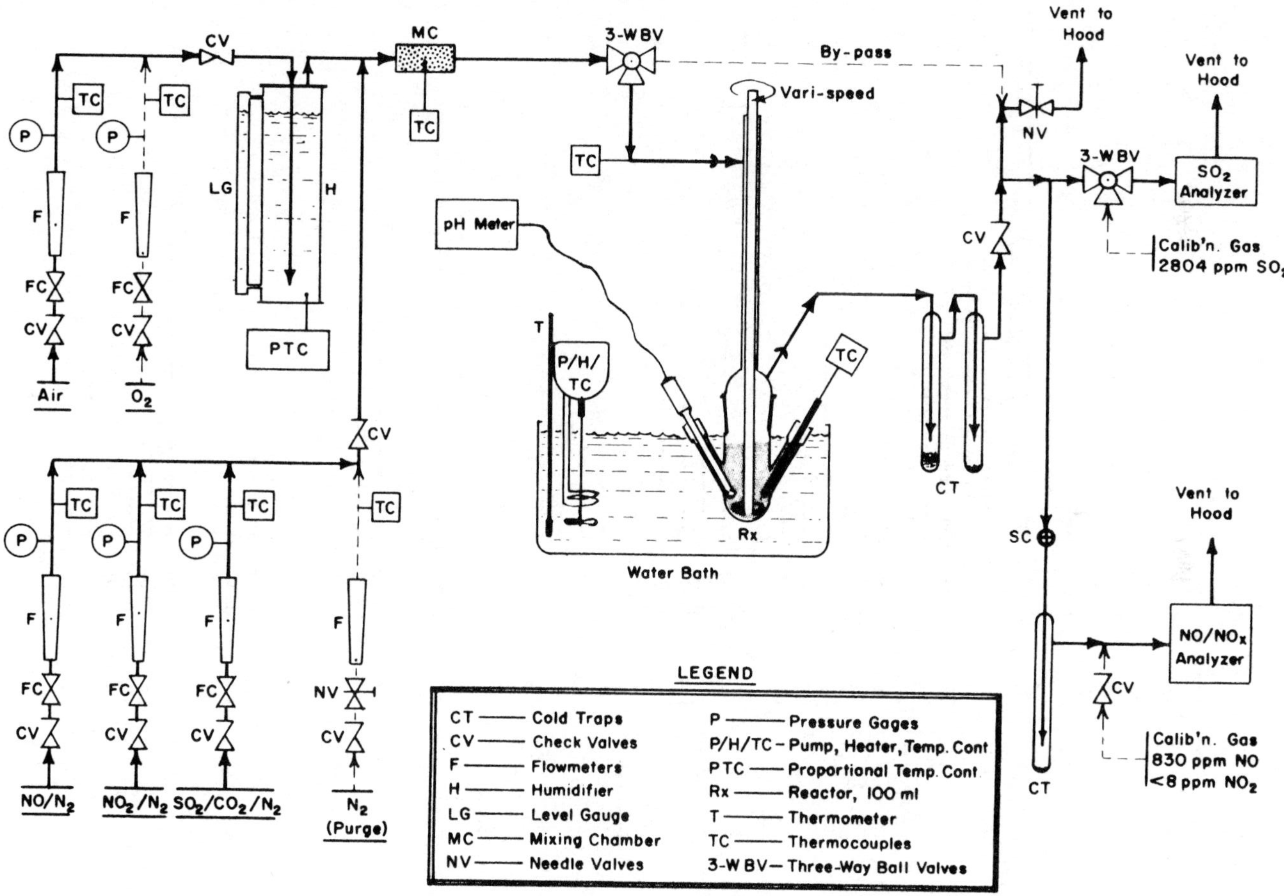

Fig. 1. Schematic diagram of the laboratory apparatus used for METC thiosulfate study at Battelle

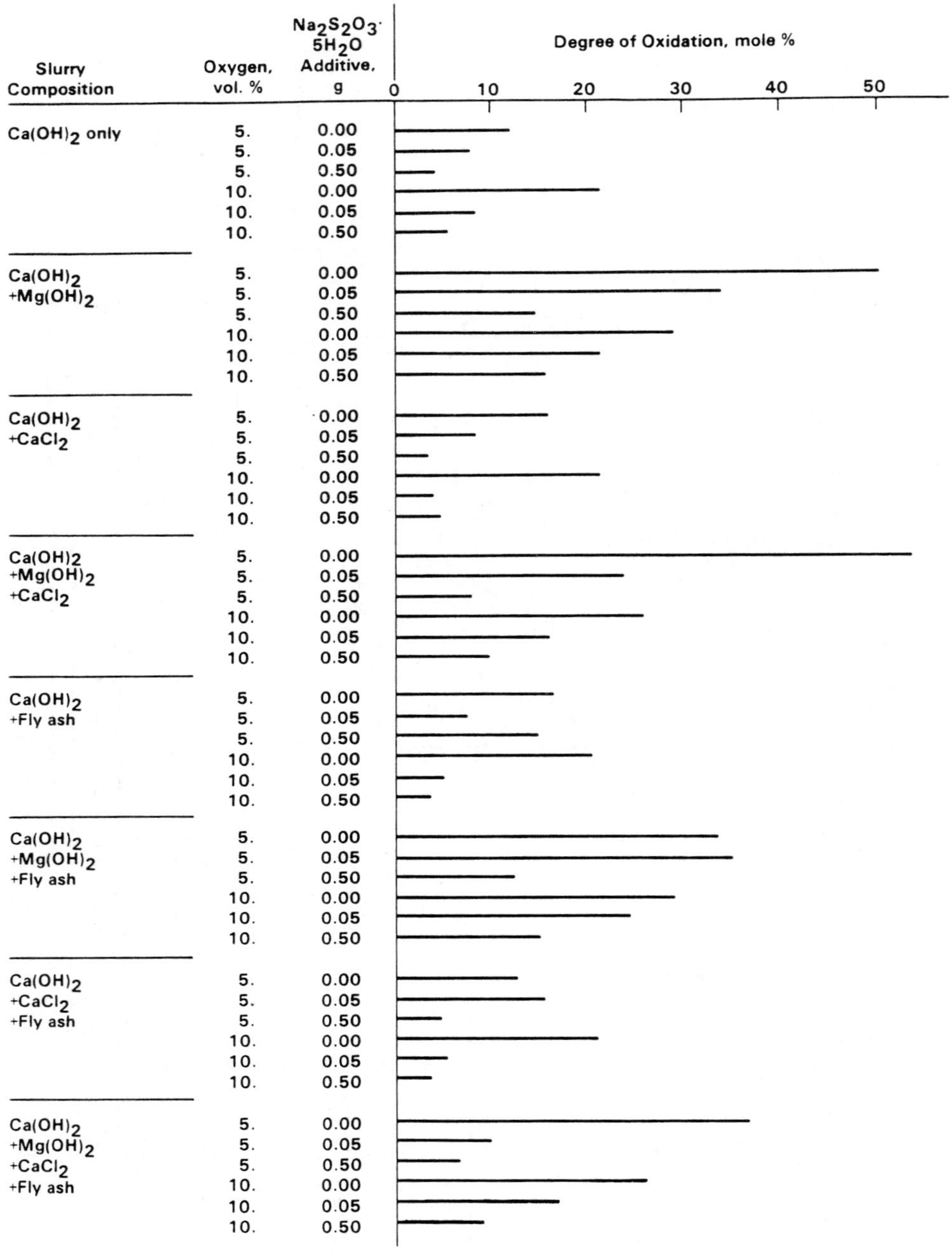

Fig. 2. Degree of oxidation from solids and liquids analysis for lime batch scrubber experiments

CONCLUSIONS

Including all conditions investigated, average oxidation with thiosulfate additions were down to 40 percent of the uninhibited value for pure lime scrubbing and to 50 percent of the uninhibited value when magnesium was present in the lime. In individual experiments, much stronger inhibition was observed--to as little as 1.4 percent of the uninhibited level. Lowering the oxidation level reduces the rate of sulfate formation and, thus, the potential for gypsum scaling. Experimental results and the results of analysis of variance of the data also indicate that, among the variables investigated, magnesium (promotor) and thiosulfate (inhibitor) exert major effects on oxidation, while only minor effects are associated with chloride and fly ash in the ranges investigated. The general conclusions are supported by reports of experience with thiosulfate additions in the operation of the magnesium-promoted lime scrubbers at the Phillips and Elrama Stations of the Duquesne Light Company. The results of limited experiments at Battelle indicate that thiosulfate also functions as an inhibitor in limestone scrubbing systems.

ACKNOWLEDGEMENTS

Among the various people who helped in providing information for this report, Battelle would like to thank Messrs. Leon Steed, Tom Zoghann, Egon Klatt, and Leroy Johnson of Duquesne Light Company, Bob Van Ness of Louisville Gas and Electric Company, Bill Elder and Gerald McGlamery of the Tennessee Valley Authority, Bob Borgwardt and Frank Princiotta of the Environmental Protection Agency, and Chuck Dene, now with the Electric Power Research Institute.

The early pioneering work by L. J. Holcombe and K. W. Luke of the Radian Corporation, which was funded by Julian Jones of the Environmental Protection Agency, should also be mentioned for special acknowledgement. The input of the various consultants, including Archie Slack, Jim Jonakin, and Gary Rochelle, helped immeasurably in providing broad perspective as to the ramifications of this work. The information provided by Dewey Burbank and his associates at Bechtel concerning related work at Shawnee was also most helpful in support of the Battelle efforts. Finally, special credit is due Bob O'Hara and Steve Pernick who recognized the potential of the Radian results and proceeded to test them out in commercial operation at the Phillips and Elrama Station of the Duquesne Light Company.

The Battelle study to further investigate the possible general applicability of thiosulfate addition was first suggested for DOE support by Gary Friggens and Bob Stewart of Morgantown Energy Technology Center (METC). Gary also initially managed the Battelle project. Later, John Williams became the project manager after his transfer from EPA to DOE. All three personnel reported to Jack Halow, new Chief of the Gas Stream Clean-up Projects Branch of

METC, during their supervision of the Battelle investigation. The support of all four individuals during the course of the study was a significant factor in meeting the goals of the project at costs significantly under budget allotment.

REFERENCES

1 H. S. Rosenberg, "How Good is Flue Gas Desulfurization?", Hydrocarbon Processing 57 (5): 132 (1978)

2 P. S. Lowell, "A Theoretical Description of the Limestone Injection-Wet Scrubbing Process", Vol. 1, PB-193029, National Technical Information Service, Springfield, VA. 1970.

3. J. C. Yarze and H. Hurwitz, "Scrubbing Gas from High-Sulfur Coal", Electrical World, 190 (8): 76 (1978).

4. A. V. Slack and G. A. Hollinden, Sulfur Dioxide Removal from Waste Gases, Second Edition, Noyes Data Corporation, Park Ridge, NJ 1975.

5 J. L. Phillips, "Precipitation Kinetics of $CaSO_4 \cdot 2H_2O$", presented at Second International Lime/Limestone Wet-Scrubbing Symposium, New Orleans, LA Nov. 1971.

6 A. V. Slack and J. D. Hatfield, "Removal of Sulfur Dioxide from Stack Gases by Scrubbing with Limestone Slurry: Operational Aspects of the Scaling Problem", presented at Second International Lime/Limestone Wet-Scrubbing Symposium, New Orleans, LA Nov. 1971.

7 L. V. Holcombe and K. W. Luke, "Characterization of Carbide Lime to Identify Sulfite Oxidation Inhibitors", EPA-600/7-78-176.

8 R. D. O'Hara, private communication.

9 R. H. Borgwardt, et al, "Limestone Scrubbing of SO_2 at EPA/RTP Pilot Plant", EPA Progress Report 36, october, 1978.

10 L. C. Schroeter, Sulfur Dioxide, Pergammon Press, New York, 1966, p 54.

11 J. L. Hudson, "Sulfur Dioxide Oxidation in Scrubber Systems", EPA-600/7-80-083, April, 1980.

OXIDATION OF SULFITE TO SULFATE IN FLUE GAS DESULFURIZATION SYSTEMS

H.T. KARLSSON[1], S. BENGTSSON[2], I. BJERLE[1], J. KLINGSPOR[1],
L-I. NILSSON[3] and A-M. STRÖMBERG[1]

[1]Dept. of Chemical Engineering II, Lund Institute of Technology, P.O.Box 124, S-221 00 Lund (Sweden)

[2]Fläkt Industri AB, S-351 87 Växjö (Sweden)

[3]ASEA-PFBC, S-612 20 Finspong (Sweden)

ABSTRACT

The oxidation of sulfite to sulfate is an important link in flue gas desulfurization systems. Depending on type of systems employed, oxidation of sulfite may be advantageous in order to improve byproduct properties. On the other hand, substantial oxidation may complicate the operation of some types of systems. In fact, both oxidation inhibition and forced oxidation are utilized on commersial flue gas desulfurization systems.

There are a great number of mechanisms for oxidation of sulfite to sulfate depending on type of flue gas desulfurization system. This paper describes experimental results on catalytic as well as noncatalytic oxidation of sulfite in all states, i.e. gas phase, liquid phase, liquid/solid phase and solid phase. The studies address a number of flue gas desulfurization processes in the area of wet scrubbing and spray dry absorption.

INTRODUCTION

For most types of flue gas desulfurization (FGD) systems it is of great concern if a substantial amount of the sulfur dioxide captured from the flue gas is oxidized to sulfate somewhere in the system. Obviously, the extent to which oxidation to sulfate occurs, will determine the major properties of the end product from the FGD system. However, the oxidation rate in a scrubber may also impact the cost and the ease of operating the scrubber.

The fate of sulfur in an FGD system depends on parameters inherited from the nature of the system design as well as its mode of operation. The list of parameters of major importance for the oxidation rate in a wet system includes the scrubbing liquor pH-value, oxidation catalysts (i.e. trace elements) present in the liquid phase, the NO/NO_2 ratio in the flue gas and the magnesium to chloride ion ratio in the liquor. On the other hand, various techniques are applied on a commercial scale to enhance or inhibit oxidation as shown in Table 1.

The problem area is very complex, especially since various objectives according to the table may lead to conflicting modes of operating an FGD system. This is especially true for lime/limestone based wet scrubbers. For instance

TABLE 1

Objectives of applying oxidation inhibition (A) or forced oxidation (B) in FGD systems

Mode	FGD system	Principle	Objective
A	Limestone scrubbing Regenerable scrubbing	Thiosulfate addition Makeup water control	Scale suppression Recovery of SO_2
B	Lime scrubbing Limestone scrubbing Spray dry absorption	Air sparging In situ Thermal oxidation	{Sludge dewatering {Produce gypsum byproduct Produce anhydrous sulfate

it is now generally accepted that increased scrubber reliability can be obtained by adding an inhibitor such as thiosulfate to the scrubbing liquor. The inhibitor maintains the sulfate level below the supersaturation point which suppresses problems with scaling. On the other hand, forced oxidation in situ reduces sludge volume, simplifies sludge dewatering and enables the recovery of gypsum as a useful byproduct.

In some regenerable wet scrubbing processes it is of utmost importance to keep the sulfate concentration on a low level. Once the sulfite is oxidized to sulfate, it is more difficult to recover it as sulfur dioxide. High sulfate levels require a large bleed stream from the scrubber to remove the sulfate. A parameter of major importance in this context is the concentration of trace elements present in the solution. High concentration of a number of metal ions catalyze the oxidation. This undesirable oxidation can be supressed to some extent by proper control of makeup water for the FGD system.

One emerging technology is the thermal oxidation of the end product from spray dry absorption. Special considerations concerning the disposability of FGD end product in some countries in Europe require the sulfite to be oxidized to sulfate. This enables the use of the sulfate in cement products. In these cases dumping of the end product is not allowed and sulfite does not meet the standards of the cement.

Sulfite oxidation has been studied intermittently over the past 10 years at the Lund Institute. Only work on liquid phase oxidation has been published previously (ref. 1). The present paper describes a number of experimental programs covering work on:

- catalytic gas phase oxidation of SO_2 on shale ash
- oxidation of sulfite in citrate solutions
- catalytic oxidation of sulfite in a spray dry absorber
- oxidation of sulfite in a slurry
- thermal oxidation of spray drying end product

Some of the studies are of engineering type, but some focus on fundamentals such as kinetics and mechanisms. However, they all address a number of FGD processes in the area of wet scrubbing and spray dry absorption.

CATALYTIC GAS PHASE OXIDATION OF SO_2 ON SHALE ASH

Background

Gas phase oxidation of SO_2 was studied in connection with Swedish plans to utilize domestic deposits of uranium bearing black shale. An integrated recovery program included the uranium content, as well as minerals and the fossil content of the shale. One route to utilize the fossil content was to combust the shale for energy production, and produce sulfuric acid from the SO_2 generated in the combustion process. The integrated recovery program called upon large amounts of sulfuric acid for leaching purpose of the shale.

The flue gases from shale combustion will contain several percent of SO_2. The shale can thus be classified as a real high sulfur coal since the ratio of sulfur content to energy content is 20 times higher than for a 1 percent sulfur coal. One feasible option is to oxidize the SO_2 to SO_3 in the gas phase catalytically and then to absorb the SO_3 to produce sulfuric acid.

The objective of our experimental program was to outline if the shale ash can be utilized as a cheap throwaway catalyst for the gas phase oxidation.

Experimental

The rate of the forward oxidation reaction, i.e.:

$$SO_2(g) + \tfrac{1}{2}O_2(g) \rightarrow SO_3(g) \qquad (I)$$

was studied by passing simulated flue gas through a fixed bed of shale ash. The experiments were performed at a velocity of 0.8 m/s with 2 percent SO_2 and 5 percent O_2 in the gas.

The catalyst was obtained by burning raw shale at 600-950 ^{o}C, and by sieving the ash to obtain three fractions, i.e.:

size (mm)	mesh	mean (mm)
0.21-0.30	65/48	0.25
0.84-1.19	20/14	1.0
1.68-2.00	10/9	1.8

Some test runs were also performed with shale ash impregnated with vanadium pentoxide, at a KOH/V_2O_5 ratio of 2; the potassium hydroxide acts as a promoter.

Theory

There is no generally accepted rate expression for the catalytic gas phase oxidation on a specific catalyst. In order to enable a comparison with a commercial vanadium pentoxide catalyst, the evaluation was based on the following rate expression (ref. 2):

$$r = -k\sqrt{(SO_2)/(SO_3)}\left\{(O_2) - [(SO_3)/(SO_2)K]^2\right\} \tag{1}$$

where K is the equilibrium constant for reaction (I), and k is the intrinsic rate constant, including the effect of porous diffusion within the shale ash particles.

Under conditions pertinent to shale ash combustion, only a small fraction of the generated SO_2 is oxidized to SO_3 in situ, because the reverse reaction dominates at typical combustion temperatures. In order to oxidize at least 90 percent of the SO_2, the temperature has to be lowered down to 400-500 ^{o}C from a thermodynamic point of view.

Experimental results

General. The shale ash, which contains most of the elements on the periodic table, was shown to be a good catalyst for the gas phase oxidation of SO_2 to SO_3. Under most of the experimental conditions, the reaction was shown to be controlled by porous diffusion within the shale ash. This was concluded from three observations. Firstly, the intrinsic rate constant increased with decreasing particle size. Secondly, the activation energy for the unimpregnated shale ash was roughly 30 kJ/mol; commercial catalysts all have activation energies from 40 to 100 kJ/mol. From simplified diffusion theory it is known that the activation energy for diffusion controlled reactions are about 50 percent of the activation energy for reaction without diffusion controll. Lastly, at low temperatures and small particles, the slope in a ln(k) vs. 1/T plot was larger by a factor of roughly two, as compared to high temperatures and large particles.

Combustion temperature. The temperature at which the shale ash was combusted had only a small impact on the catalytic activity from an engineering point of view. Hence, the porous structure probably undergoes only minor changes during combustion.

Particle size. At higher temperatures, the two larger particle fractions yielded a ratio of the intrinsic rate constants of 0.23 to 0.37. The highest value corresponds to 600 ^{o}C combustion temperature. However, the ratio was constant within the reaction temperatur range from 500 to 700 ^{o}C, provided the two fractions compared were combusted at the same temperature. From simplified diffusion theory, the ratio of the intrinsic rate constants should be equal to the size ratio, i.e. 1.0/1.8 = 0.56. However, the larger particle fraction was shown to contain only flat needle shaped particles while the 1.0 mm fraction was represented by irregular particles of spherical shape. Assuming a geometrical factor for a flat plate (1.8/2 mm) and similar for a sphere (1.0/6 mm), diffusion theory predicts a ratio of (1.0/6)/(1.8/2) = 0.19. The observed ratio of 0.23 to 0.37 falls within the range 0.19 to 0.56.

Reactivity. From Arrhenius plots, i.e. ln(k) as a function of 1/T, the slope was constant for high temperatures. For small particles the slope gradually increased by a factor of two below 500 ^{o}C. From this it was possible to estimate the true activation energy to E_a=60 kJ/mol for unimpregnated shale ash. By treating the shale ash with vanadium pentoxide, the reaction rate increased considerably. However, the activation energy decreased with increasing amount of promoter. 0.1 and 1.0 percent V_2O_5 yielded an activation energy of 52 and 44 kJ/mol, respectively.

Figure 1 shows a comparison of the rate constant for shale ash with the rate constant for a commercial vanadium pentoxide catalyst (ref. 2). The data are valid for the kinetic regime. The commercial catalyst has an activation energy of 99 kJ/mol. As can be seen, the activity for the shale ash is more than one order of magnitude lower at temperatures above 500 ^{o}C. However, at low temperatures where high conversion of SO_2 can be obtained, the situation is not so unfavourable. In fact, impregnated shale ash has a reactivity exceeding that of the commercial catalyst at 400 ^{o}C.

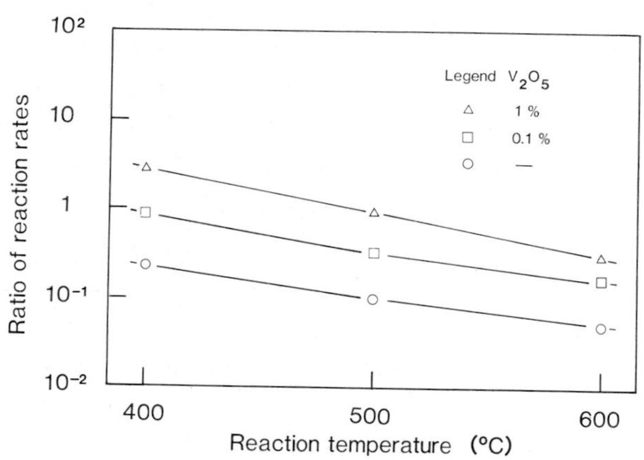

Fig. 1. Catalytic gas phase oxidation of sulfur dioxide; the ratio of the reaction rate on shale ash to the reaction rate on a commercial vanadium pentoxide catalyst (ref. 2) as a function of the temperature; comparison of shale ash with shale ash promoted with vanadium pentoxide.

Significance and conclusions

Obviously the shale ash is not sufficient to oxidize any substatial amount of SO_2 to SO_3 in situ when burning shale. Thermodynamic considerations set a limit, because the reverse reaction dominates at typical combustion temperatures.

It is conceivable to assume that oxidation starts if the flue gas and the shale ash are cooled down to say 400 to 500 ^{o}C in some point of the combustor. However, from a kinetic point it can be shown that a tremendous contact time is required. It can be calculated from the experimentally determined kinetic data, that the flue gas has to pass through a bed of 1-2 days of accumulated shale ash from the combustor, in order to oxidize 90 percent of the SO_2 at 400 to 500 ^{o}C.

On the other hand, shale ash with vanadium pentoxide promoter is suitable as a catalyst for SO_2 oxidation in a conventional process producing sulfuric acid.

OXIDATION OF SULFITE IN CITRATE SOLUTIONS

Background

One option for sulfur control when burning high sulfur coal is absorption of SO_2 in a lean solution of sodium citrate. The SO_2 is then stripped off from the solution, in order to enable the recovery of SO_2 as a useful byproduct.

It is highly undesirable that the sulfite present in a citrate solution is oxidized to sulfate. When it once has been oxidized, it can not be stripped off as SO_2. This leads to less recovery of SO_2. However, the most severe impact of oxidation is that a large bleed stream has to be withdrawn from the scrubber loop. From this bleed stream, sodium sulfate has to be separated and the citrate has to be returned to the scrubber to keep the makeup requirements of citrate on an acceptable level. The higher the oxidation rate, the larger the bleed stream has to be in order to prevent the sulfate concentration from building up in the scrubber loop. A large bleed stream is costly to handle.

The rate of sulfite oxidation in citrate solutions may depend on the concentration of various transition metals present in the solutions; metals such as cobalt, in general, act as catalysts for sulfite oxidation. The concentration of transition metals can be controlled to some extent by controlling the composition of the makeup water.

Typical levels of citrate in commercial processes under consideration lie around 1 M (mol/l). The presence of such large amounts of dissolved species will affect the amount of oxygen dissolved in the solution. Citrate suppresses the equilibrium concentration of oxygen, and hence, may lower the rate of sulfite oxidation.

Another aspect of importance is the high concentration of dissolved sulfite (and bisulfite). This will impact the rate of oxidation considerably.

The present study was undertaken to investigate the oxidation of sulfite to sulfate in a citrate solution under typical scrubber conditions. A large experimental data base was obtained on the reaction rate as a function of a number of parameters, i.e. the temperatur and the concentration of sulfite, citrate and oxygen. Furthermore, the impact of the concentration of the following metal ions was studied: Zn^{2+}, Ni^{2+}, Cu^{2+}, Co^{2+}, Fe^{3+} and Mn^{2+} in the 0.33 to 3.3 mM range.

Experimental

The experiments were performed in a well stirred batch at a controlled temperature. Gas with a known oxygen concentration was bubbled through the batch reactor with citrate solution. The oxygen concentration and the pH were followed by electrodes (oxygen: YSI type 57). A typical experiment was performed by adding chemicals to the citrate batch (gas bubbling turned off) and recording the depletion of the oxygen concentration by the electrode. The sulfite concentration was known from the amount added and by analyzing the final concentration by titrating with thiosulfate.

The chemicals used were of pure grade; i.e. sodium bisulfate from Fischer Scientific Company and sodium citrate from Merck's. However, since large concentrations of citrate were used, it was impossible to avoid traces of impurities; it is difficult to purify citrate because of its tendency to complex bind ions. According to the manufacturer, the citrate contains about 5 ppm lead which corresponds to $7 \cdot 10^{-6}$ M in the solution.

Experimental findings

Impact of citrate. Most of the experiments were performed at 42 ^{o}C in a 1 M citrate solution, as this corresponds to typical scrubber conditions. At these conditions, oxygen dissolves to a concentration of 55 percent of the concentration in pure water.

Base line tests were performed by varying the citrate concentration from zero to 1 M. These experiments were performed at different concentrations of sulfite and oxygen.

Figure 2 displays the influence of the citrate concentration on the oxidation rate. As shown by the curves, the behavior is extremely complex. In the absence of citrate, the rate is independent of the oxygen concentration. By continuously increasing the citrate concentration, the rate becomes increasingly inhibited at low oxygen concentration. At higher concentrations of oxygen, citrate enhances the oxidation reaction.

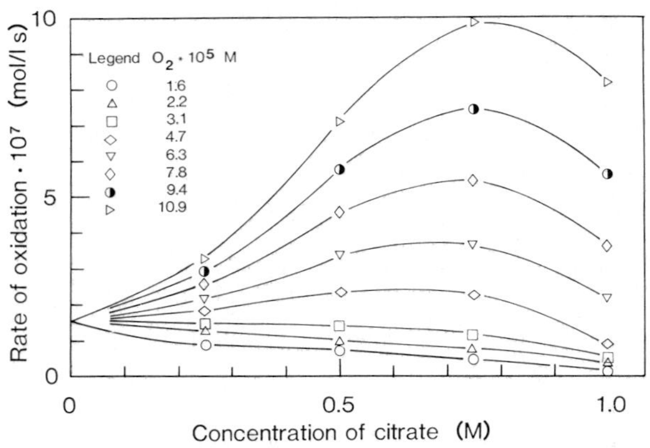

Fig. 2. The rate of sulfite oxidation as a function of the citrate concentration at 42 ^{o}C, pH=3.5 and 31 mM of sulfite.

The exact cause of the influence of citrate on the oxidation rate can only be speculated on. Obviously, there are both catalytic and inhibiting effects. It is known that organic species may inhibit the oxidation reaction. On the other hand, impurities of trace amounts in the citrate most likely catalyze the reaction. However, the highest concentration of metals originating from the citrate was estimated to be $7 \cdot 10^{-6}$ M. This may not explain a 100 fold enhancement of the oxidation rate.

Noncatalytic oxidation. The sulfite oxidation reaction was studied for a wide range of operating conditions, without any addition of transition metals. The citrate concentration was set at 1 M. The other parameters were varied over the following ranges:

pH: 3.5 - 4.5
sulfite: 0.0015 - 0.060 M
oxygen: $1.5 \cdot 10^{-5}$ - $1.1 \cdot 10^{-4}$ M
temperature: 25 - 42 ^{o}C

Here, sulfite refers to total sulfite including bisulfite.

A complete matrix of roughly 150 data points was extracted from the experiments by successively varying the concentration of sulfite, hydrogen ion and oxygen, respectively. The resulting reaction rate data covered a range of four orders of magnitude.

By varying only the concentration of one of the species at the time, it was possible to fit the data points to a rate expression of the irreversible nth order type. In the case of the oxygen dependency, there was a clear change in the reaction order in all data series at a concentration of about $3.5 \cdot 10^{-5}$ M. This is shown in Fig. 3a for some of the data.

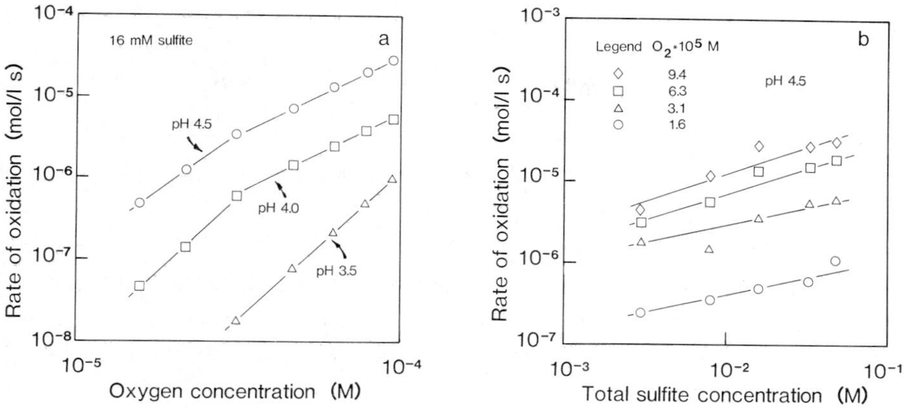

Fig.3. The rate of sulfite oxidation at 42 ^{0}C as a function of (a) the oxygen concentration with pH as a parameter and (b) the sulfite concentration with the oxygen concentration as a parameter.

However, as soon as a second variable was varied, the reaction order of the first variable was changed. So it was virtually impossible to fit the data to a rate expression, unless two of the three variables were kept constant.

For oxygen, the reaction order varied from 0.5 to 3.5, showing a maximum at low oxygen concentration and pH=4. The weakest dependency was at high pH and high oxygen concentration.

Some of the data for the sulfite dependency at pH=4.5 are shown in Fig. 3b. Here the reaction order increased only slightly with the oxygen concentration. The highest reaction order with respect to sulfite (2.7) occurred at at low pH and low oxygen. The lowest value (0.5) was obtained at high pH and low oxygen concentration.

The dependency of hydrogen ion showed less variation; high sulfite yielded a reaction order of -1. At low sulfite levels, the reaction order increased from -2 to -1 when increasing the oxygen concentration.

The activation energy for the oxidation reaction was determined to 76 kJ/mol in the temperature range 25 to 42 ^{0}C.

Catalytic oxidation. When the oxidation reaction was carried out in the presence of transition metals, the dependency of sulfite, oxygen and hydrogen ion was very similar to the dependency in the noncatalytic case.

The experiments revealed both strong catalytic effects as well as inhibiting effects by the metal ions. The strongest inhibiting effect was obtained with Fe^{3+}. In fact, at high oxygen concentration, the reaction rate decreased with increasing Fe^{3+} concentration. Under all other conditions, and for all other metal ions, the rate increased with increasing concentration in the investigated concentration range. However, some of the ions still acted as inhibitors when compared to the noncatalytic system.

The strongest oxidation catalysts were shown to be Co^{2+} and Mn^{2+}. The reaction order with repect to the catalysts varied slightly with the experimental conditions, except for Fe^{3+} which showed strong variations. However, under typical flue gas conditions, i.e. 5-7 percent oxygen, the reaction order with respect to Fe^{3+} and Mn^{2+} was shown to be roughly 0.25. In the case of Cu^{2+} and Co^{2+}, the reaction order was found to be 0.5.

Table 2 displays the factor by which the oxidation reaction rate changes in the presence of various transition metals. As shown, the rate may vary over a range of 10^3, depending on conditions and types of metal ions present.

TABLE 2
The factor by which the rate of sulfite oxidation is changed by adding 0.67 mM of transition metals at S_t= 16 mM, pH=4.0 and 42 ^oC.

$O_2 \cdot 10^5$ (M)	1.6	2.2	3.1	4.7	6.3	7.8	9.4	11
Zn	1.5	1.5	1.5	1.6	1.6	1.6	1.6	1.7
Ni	1.7	1.8	1.8	1.8	2.3	2.9	3.4	4.0
Cu	2.9	1.8	1.0	0.55	0.36	0.26	0.20	0.16
Co	22	23	23	24	25	25	26	-
Fe	0.67	0.40	0.23	0.12	0.080	0.057	0.071	0.13
Mn	13	13	10	25	32	38	45	51

Discussion of results

Signigicant for this study on catalytic as well as on noncatalytic oxidation of sulfite in the presence of citrate is the very strong impact of the oxygen and the sulfite concentration on the reaction rate. Typical reaction orders with respect to sulfite reported in the literature (ref. 3,4) do not exceed 1.5. In this study the reaction order varied from 0.5 to 2.7. A reaction order with respect to oxygen of two, has been reported in the literature (ref. 5). Present values cover a range of up to 3.5. All this is due to the presence of citrate.

Activation energies reported in the literature cover a range from 57 to 147 kg/mol (ref. 1,4,6). The present value obtained was 76 kJ/mol.

Similarly, the dependency of the metal ions agree well with data from the literature. The value for Cu^{2+} and Co^{2+} has been reported to be 0.5 (ref. 7) and (ref. 1), respectively, which is the same value as in this study at typical conditions.

Conclusions

Although a very complex behavior of the sulfite oxidation reaction was observed in the presence of citrate, some clear observations of considerable significance can be pointed out. As compared to systems without citrate, the following is valid:

- In general, 1 M citrate has a very small impact on the oxidation rate at typical scrubber conditions
- Excess oxygen leads to a very fast oxidation reaction
- High sulfite concentrations at high pH contributes strongly to the oxidation reaction
- Mn^{2+} and Co^{2+} are very strong catalysts
- Fe^{3+} inhibits the reaction strongly, especially at high oxygen concentrations

SPRAY DRY ABSORPTION: CATALYTIC OXIDATION OF SULFITE IN SITU

Background

If a modified spray drying process would enable the use of limestone as an absorbent for SO_2 removal, one may anticipate large cost cuts. The shortcoming of limestone is its slow dissolution rate, except for very small particles or at very low pH-values. One potential way to go about this shortcoming is to oxidize all the sulfite catalytically in situ, to obtain a low pH-value within the droplets and thereby enable a high dissolution rate of the limestone.

Hence, an experimental program was undertaken to outline the feasibility of limestone utilization in a spray dryer by in situ oxidation of sulfite. Screening tests of candidate catalysts were performed. The most promising candidates were tested with respect to their reactivity on a fabric filter, i.e. the dry reaction state, as well as their performance in a spray dryer.

From an economical point of view, as well as from an environmental point of view, the oxidation catalyst may have to be recovered from the dry end product. The effort required to accomplish this recovery was also studied experimentally.

The dry reaction. Screening tests were performed with a device that simulates the dry reaction on a fabric filter; this equipment has been described elsewhere (ref. 9). These basic experiments revealed that the dry reaction of SO_2 with limestone can be enhanced in the presence of an oxidation catalyst. Typically, the reaction rate is increased more than one order of magnitude under optimal conditions. Furthermore, the reaction rate will be in the same order as that of lime if a proper limestone is selected.

There is a strong correlation between the rate of the SO_2 limestone reaction and the degree to which sulfite is oxidized to sulfate. Among the most efficient oxidation catalysts, cobalt in the form of nitrate is the best option.

The catalytic effect was found to reach a limiting level at 5 percent $Co(NO_3)_2 \cdot 6H_2O$ by weight. Further addition of catalyst does not increase the rate of the reaction between SO_2 and limestone.

The catalytic effect is more pronounced at higher relative humidities of the gas. Furthermore, the catalytic effect depletes with increasing limestone particle size. In fact, the effect of adding catalyst is very small for particles larger than 70 microns, but below this size the catalytic effect increases drastically.

Spray drying. The concept with sulfite oxidation catalyst was tested with limestone slurry in a spray dryer pilot plant treating a flue gas amount corresponding to roughly 0.5 MW thermal. The spray dryer is equipped with a single nozzle with five holes and atomization of the slurry is accomplished by pressurized air. The spray cooled flue gas is passed through 10 bags. The pilot plant is described in greater detail elsewhere (ref. 10).

Fig. 8a shows the removal efficiency of SO_2 over the spray dryer as function of the relative humidity for limestone slurry with 5 percent $Co(NO_3)_2 \cdot 6H_2O$ on a dry basis. For comparison, data for slurries of hydrated lime and limestone have been included in the diagram. Similar to the dry reaction, the impact of the catalyst is more pronounced at higher relative humidities. Fig. 9b displays the influence of the stoichiometric ratio on the removal efficiency.

Data for the fabric filter have not been included in the diagrams, because the efficiency for the fabric filter depends on a number of parameters such as the air to cloth ratio and the pressure drop allowed over the filter before cleaning.

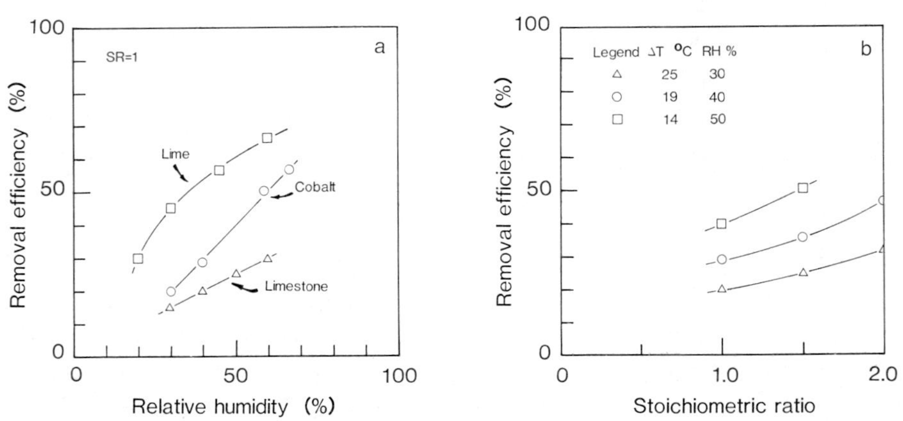

Fig. 4. The spray dryer removal efficiency of sulfur dioxide at 150 °C and 1000 ppm sulfur dioxide. (a) shows the dependency of the relative humidity; comparison of lime and limestone with cobalt promoted limestone. (b) shows the impact of the limestone stoichiometry.

Catalyst recovery. In order to separate the catalyst from the dry product, leaching tests were performed with a number of different leaching systems. These screening tests revealed that sulfuric acid is the most efficient option for separating the cobalt from the dry product. A leaching test program was thus performed with sulfuric acid in a batch system. Samples of the dry product were reslurried in a beaker and the pH-value was kept at a preset value by adding sulfuric acid by pH-stat equipment. After stirring for one hour, the slurry was filtered and washed with water.

The percentage cobalt recovered from the dry end product by sulfuric acid leaching is shown in Fig. 5. The temperature at which the leaching is carried out is of vital importance. The same fact holds for the pH-value. When less than 1 percent of the cobalt remains in the solid phase, the concentration level is on the same order as the naturally occurring cobalt in the limestone and/or in the fly ash. As comparison it can be mentioned that leaching with pure water enables a recovery of 60 to 70 percent of the cobalt.

The leachant phase will contain undesirable ions, such as chloride and various metals. If the leachant phase is returned to the spray dryer as makeup water, the concentration of the species mentioned will build up in the system. Hence, further tests were made to separate the cobalt from the leachant by means of liquid/liquid extraction using DEHPA (di-ethyl-hexyl-phosphoric-acid) as a reagent. The experiments revealed that more than 85 percent of the cobalt can be extracted in a single stage. However, the method is not without problems. Firstly, the selectivity is not as good as desired. Secondly, the settling rate of the dense phase is very slow. Further work with this problem will be undertaken, using ion exchange resins instead.

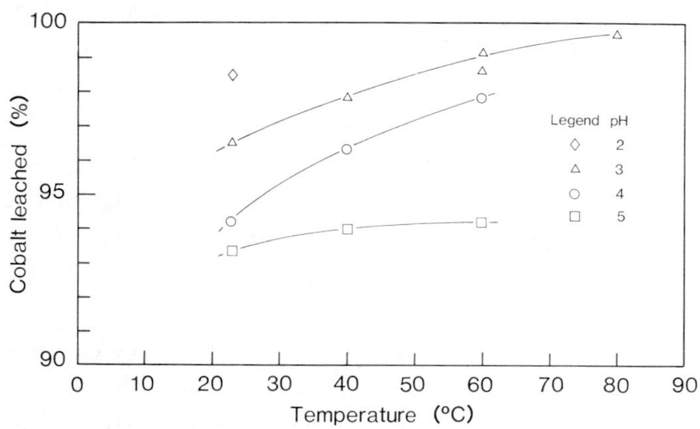

Fig. 5. Percentage cobalt leached from the dry end product as a function of the temperature with pH as a parameter.

Conclusions

The experimental program with sulfite oxidation catalyst has shown that SO_2 can be removed in a spray dryer by a limestone slurry, provided the type of spray dryer employed can be operated comparatively close to the adiabatic saturation point. This latter constraint stems from the fact that the catalytic effect is most pronounced at high relative humidities.

The main features of the process concept is that limestone can replace lime in a spray dry absorption process and that the final product is gypsum. The quality of the gypsum will depend on to what extent fly ash is collected in the flue gas, upstream from the spray dryer. However, it may be anticipated that the concept is not applicable for high sulur coals.

Some steps of the process concept have to be tested further. It has been shown that the cobalt can be recovered from the dry product by sulfuric acid leaching. In order to bleed undesirable species that will complicate the water chemistry, the cobalt has to be separated from the cobalt enriched leachant.

Further investigations are required, especially concerning the recovery of cobalt from the leachant phase and the behavior of the water chemistry on a continuous scale.

SLURRY OXIDATION OF SPRAY DRYING SULFITE

Oxidation of sulfite slurry has been in focus for many experimental studies. However, all studies have focused on sulfite obtained from wet scrubbers.

Results on slurry oxidation presented in the literature may in some cases point at conflicting conclusions. Hence, an experimental study was undertaken with the objective to outline the underlying mechanism of sulfite slurry oxidation. Furthermore, the study was based on calcium sulfite from spray dry absorption of SO_2 in order to highlight any parameter inherited from the spray dryer that might influence the oxidation rate.

Experimental

The sulfite product was obtained from the spray dryer described in the previous section. The oxidation experiments were performed in a sparged stirred tank absorber, keeping the pH-value constant by means of pH-stat equipment. Sulfite samples were reslurried in nitrogen sparged water, maintained at a constant temperature and at a constant pH-value. The slurry was then sparged with gas mixtures of various composition. Small slurry samples were continuously withdrawn from the absorber and the sulfite oxidation in the samples was freezed with potassium iodine. Titration with sodium thiosulfate was made subsequently in order to determine the amount of unoxidized sulfite.

Experimental findings

The experiments were performed at a sparging rate in which maximal mass transfer rate was obtained, i.e. where the oxidation rate is independent of the sparging rate.

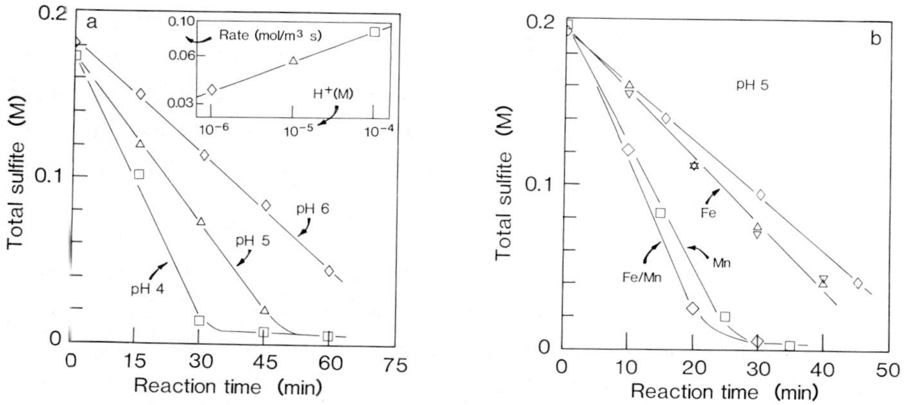

Fig. 6. Total slurry sulfite concentration as a function of elapsed time for oxidation at 50 ^{o}C. (a) shows the pH dependency. (b) shows the impact of metal catalysts; legends refer to Table 3.

Some of the experimental results are shown in Fig. 6. From Fig. 6a, a strong influence of the pH is observed, as expected. Inserted in the diagram is a cross plot of the slopes, yielding a 0.185 power dependency of the hydrogen concentration on the intrinsic oxidation rate.

The overall oxidation reaction was found to be rate determined by simultaneous mass transfer and chemical reaction in the liquid film of the gas bubbles. This conclusion was deduced from observations of the following parameters:

- sulfite conversion
- slurry concentration
- oxygen concentration
- catalytic effects

As shown by Fig. 6a, the rate is independent of the sulfite conversion, since the slopes do not change with time. Hence, mass transfer to or from the solid particles which contain sulfite can not be rate determining, since such constraints lead to a depletion of the oxidation rate as the sulfite is consumed.

Experiments with a slurry concentration ranging from 8 to 15 percent by weight, yielded a weak depletion in the intrinsic oxidation rate with increasing slurry concentration. Processes rate determined by liquid/solid mass transfer predicts a strong increase in the rate when increasing the solids content of the slurry. Furthermore, a weak depletion in the gas/liquid mass transfer area with increasing slurry concentrations is a well documented phenomena in chemical engineering. The experimental result is thus a strong indicator for mass transfer with chemical reaction in the liquid film at the gas interface.

Sulfite oxidation has been shown by many investigations to be a zero order reaction with respect to oxygen. Mass transfer theory predicts the absorption rate to vary with the square root of the oxygen concentration at the gas/liquid interface for such a reaction. By sparging with pure oxygen, the intrinsic rate was found to increase by a factor of 2.3 as compared to air sparging. This is consistent with the square root of the ratio of the oxygen concentrations, i.e. $(1/0.208)^{\frac{1}{2}} = 2.2$.

Further experiments were made in the presence of transition metals as sulfite oxidation catalysts. The result is shown in Fig. 6b. As can be seen, the intrinsic rate is enhanced by the metal ions, again indicating the oxygen mass transfer with chemical reactions to be the rate determining step.

The enhancement caused by the catalysts is summarized in Table 3; data from literature (ref. 8) have also been included for lean sulfite solutions. There is an agreement to some extent. The enhancement for Fe is in both cases independent of the catalyst concentration, but the level is lower in the slurry case. For the Mn case, where the enhancement depends on the catalyst concentration, the two investigations agree.

Discussion

Oxidation experiments with dry product from spray drying removal of SO_2 with hydrated lime have revealed that the intrinsic oxidation rate is rate determined by mass transfer of oxygen with simultaneous oxidation of sulfite in the liquid film at the gas/liquid interface. The conclusion holds for oxidation rates as high as 0.25 mol/m^3 s and at extremely low slurry concentrations. In fact, the rate was shown to be constant from zero to greater than 90 percent conversion of the solids. This latter value corresponds to a slurry concentration of less than 1 percent solids.

Other studies on oxidation of sulfite slurry from wet scrubbers have led to different results. In one case (ref. 11) the mechanism was found to be the same as in the present case. Another experimetal study (ref.12) revealed an influence of the dissolution rate of sulfite on the intrinsic oxidation rate at overall rates comparable to the present data.

Presumably, various types of calcium sulfite products behave differently with respect to oxidation. Furthermore, large differences between sulfite from spray dryers and sulfite from wet scrubbers may be anticipated. This can be understood from properties inherited from the mechanism for sulfur capture.

TABLE 3

Catalytic enhancement in slurry (present study) compared to lean solution (ref. 15) at 50 ^{0}C and pH=5. The catalysts were added as $FeSO_4$ and $MnSO_4$, respecteively.

Legend	Catalyst	Conc.(mM)	Enhancement	
			Slurry	Solution
◇	-	-	1	1
△	Fe	1.6	1.2	2.9
▽	Fe	3.2	1.2	2.9
□	Mn	1.6	2.1	2.1
◇	Fe/Mn	1.6/1.6	2.5	-

In wet scrubbers calcium sulfite occurs in the form of individual particles, because calcium sulfite precipitates in the scrubbing liquor. In a spray dryer, the crystallization mechanism is different, and part of the sulfite is produced in a semidry phase, leading to very small sulfite crystals. During reslurrying and adjustment of pH, the lime is dissolved very rapidly and subsequently extremely small calcium sulfite crystals are released. The dissolution rate for these crystals is very fast, as compared to the larger particles from a wet scrubber, see Fig. 7.

THERMAL OXIDATION OF SULFITE

Background

In a number of countries in Europe, spray drying absorption is considered for SO_2 control at smaller power stations. In some areas dumping or disposal of the calcium sulfite containing dry product has been precluded from the list of options. One emerging technology is to oxidize the spray drying end product thermally, in order to obtain anhydrous calcium sulfate. This enables the use of the dry end product as an additive in cement.

An experimental program was initiated to study the kinetics of the thermal oxidation of calcium sulfite to calcium sulfate. The objective was to determine the influence of a number of parameters on the residence time required to complete the oxidation.

Chemistry and Thermodynamics

When the sulfite containing product from spray drying is heated up in the presence of oxygen, it starts to react. The reactions of major importance are:

$$CaSO_3(s) + 0.5\ O_2(g) \rightarrow CaSO_4(s) \qquad (II)$$

$$Ca(OH)_2(s) \rightarrow CaO(s) + H_2O(g) \qquad (III)$$

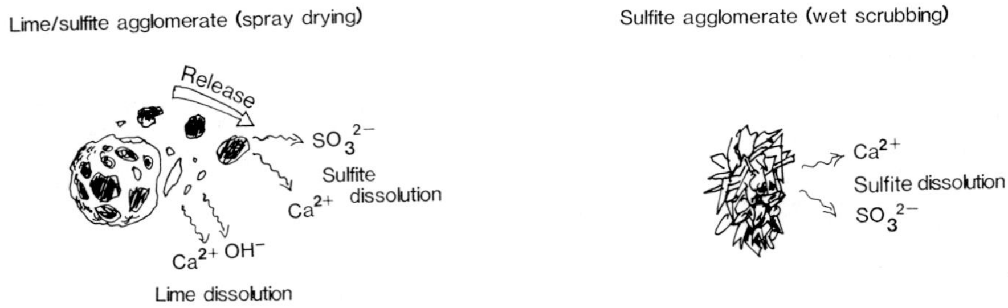

Fig. 7. When agglomerates originating from spray drying are exposed to hydrogen ion and oxygen, the limestone dissolves and small sulfite crystals are released from the agglomerates; these crystals dissolve rapidly when the sulfite is oxidized. Wet scrubbing sulfite agglomerates dissolve more slowly due to the larger size inherited from the crystallization mechanism in the scrubbing liquor.

The first reaction is strongly exothermic, with a heat of reaction of 255 to 260 kJ/mol in the range from 500 to 800 ^{0}C (ref.13). The second reaction consumes heat to an extent of 112 kJ/mol in the same temperature range. Hence, a considerable amount of the energy required to heat the dry product to accomplish the oxidation is obtained from the reactions.

Other reactions such as the liberation of water of hydration, are all weakly endothermic. Experimental results also indicate that there are no such species present in the complex mixture which constitutes the dry end product.

If oxygen is not in excess, several other reactions may also be involved:

$$4CaSO_3 \rightarrow 3CaSO_4 + CaS \qquad (IV)$$

$$CaS + 2O_2 \rightarrow CaSO_4 \qquad (V)$$

$$2CaS + 3O_2 \rightarrow 2CaO + 2SO_2 \qquad (VI)$$

$$CaSO_3 \rightarrow CaO + SO_2 \qquad (VII)$$

Experimental

The dry end product was obtained from the small spray dryer described in greater detail elsewhere (ref.10). The spray dryer pilot was operated at a stoichiometric ratio of one and a removal efficiency of about 75 percent. The product hence consisted of a molar ratio of calcium sulfite to calcium hydroxide of roughly three.

The oxidation experiments were performed in a vertical fixed bed reactor, heated by an electrical furnace. The samples were placed on a coarse glass frit and a gas mixture was passed through the bed from the top. The gas mixture was obtained by mixing O_2, N_2 and CO_2 from gas cylinders and subsequently evaporating water in the gas stream at a controlled rate.

Preliminary experiments were performed at various gas flow rates in order to elaborate with the elimination of the gas film resistance. After each experiment the bed was cooled in a gas stream of nitrogen and subsequently analyzed for sulfite by titration.

Experimental findings

When the sulfite was heated in the absence of oxygen, sulfur dioxide started to evolve at a substantial rate above 600 ^{o}C. At the same time sulfide was produced in the sample, giving a gray color to the product. In fact, already very small amounts of sulfide yielded a considerable shift in the color of the product, so this was considered as a good indicator for sulfide formation.

The oxidation was shown to increase considerably in the range 400 to 600 ^{o}C, and the most interesting result was that oxygen and water vapor showed similar types of impact on the oxidation rate. For instance, when the water vapor concentration in the gas was increased while the oxygen concentration was kept at a relatively low but constant level, the reaction rate increased considerably. At about 50 percent water vapor, a limiting level of the oxidation rate was reached. Similarly, by increasing the oxygen concentration, the oxidation rate increased very rapidly at high water vapor concentrations to reach a limiting level. The same level was reached at low water vapor concentrations as well, but required a much higher oxygen concentration.

Kinetic evaluation. About 96 percent of the sulfite was oxidized comparatively easily. However, the remaining 4 percent were more difficult to convert, even at very long exposure times. This may be due to the inherited nature of the type of dry product used for the experiments.

It was possible to model the sulfite conversion by a kinetic shrinking core equation of the following type:

$$1 - (1-X)^{1/3} = k \cdot t \qquad (2)$$

where X is the conversion of the 96 percent of easily convertible sulfite, t is the raction time and k is an overall rate constant.

A plot of experimental data using the parameters of Eq. (2) is shown in Fig. 8. The data were obtained from experiments at the limiting conditions where an increase in the concentration of oxygen and water vapor, respectively, has very little impact on the oxidation rate, i.e. 5 percent O_2 and 50 percent H_2O. The rate constant was obtained from the slopes of the curves. A cross plot in terms of ln(k) as a function of 1/T is shown in Fig. 9. As can be seen, the experimentally determined rate constant may well be described by an Arrhenius type of equation:

$$k = k_0 \cdot \exp(-E_a/RT) \tag{3}$$

where $k_0 = 5.4 \cdot 10^3\ min^{-1}$ and the activation energy is $E_a = 85$ kJ/mol.

The value of the inverse of the rate constant may be interpreted as the time required for complete oxidation in a batch or plug flow mode of operation. The values for 1/k are displayed in Fig. 9.

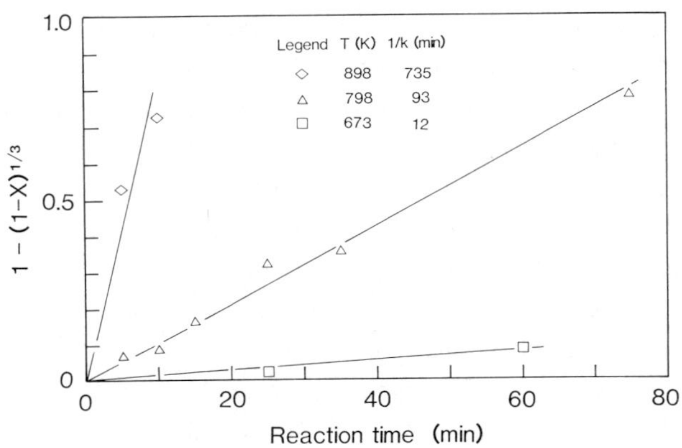

Fig. 8. Plot according to shrinking core model for three reaction temperatures.

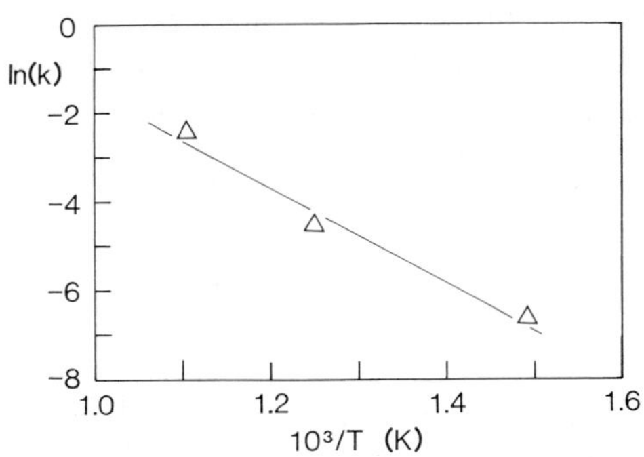

Fig. 9. Arrhenius plot of data for thermal oxidation of sulfite.

Engineering considerations

Flow pattern. The dry end product may be oxidized in a rotary furnace or in a fluidized bed. Depending on what type of equipment is utilized, the flow pattern or residence time distribution of the solids will influence the total conversion of sulfite to sulfite. Simple plug flow performance is given by Eq.(2). Complete back mixing as in the fluidized bed case can be described by the classical segregation model (ref. 14), since each particle is a segregated system. Taking the residence time distribution of a "well stirred tank" (complete macro mixing), the performance is given by:

$$X = 3n - 6n^2 + 6n^3(1-\exp(-1/n)) \tag{4}$$

where n is a multiple of the residence time required for complete conversion in a plug flow device, i.e. if t is the mean residence time in the back mixed system, then $n=k\cdot t$.

The back mixing has a severe impact on the conversion, which is shown by the following series of calculated data:

n =	1	2	3	4
X =	0.79	0.89	0.92	0.94

while X=1 at n=1 for a plug flow system.

Energy requirement. If the oxidation is carried out in a countercurrent flow pattern so that the incoming streams are heated up by the outgoing streams, no external heat supply would be needed from a theoretical point of view. Furthermore, an adiabatic temperature exceeding 1000 °C can be achieved if the excess air is not too large. The actual energy requirement will depend on the type of equipment utilized, and the operating conditions employed.

Operating conditions. The oxidation should preferably be carried out at 600 °C or higher. The oxygen concentration should be kept above 5 percent in all points of the oxidizer. Steam added to the gas to enhance the oxidation rate is payed back if oxygen is present in low concentration. The steam added will lower the energy consumption and/or decrease the reactor size required.

ACKNOWLEDGEMENT

This paper is base on work financially supported by Sydkraft AB, Fläkt Industri AB and AB Svensk Alunskifferutveckling.

NOMENCLATURE

E_a activation energy

H^+ concentration of hydrogen ion

k rate constant

k_0 frequency factor

n	residence time, times the rate constant
O_2	oxygen concentration in liquid phase; partial pressure of oxygen
R	gas constant
RH	relative humidity
r	reaction rate
S_t	total sulfite concentration, including bisulfite
SO_2	partial pressure of sulfur dioxide
SO_3	partial pressure of sulfur trioxide
T	temperature
ΔT	approach to adiabatic saturation point
t	time
X	sulfite conversion

REFERENCES

1 S. Bengtsson and I. Bjerle, Chem. Eng. Sci., (1975) 1429-1435.
2 W.W. Duecker and J.R. West, The Manufacture of Sulfuric Acid, Reinhold Publishing Corporation, New York, 1959, 515 pp.
3 R.D. Srivastava, A.F. McMillan and I.J. Harris, Can. J. Chem. Eng., 46 (1968) 181-184.
4 A. Matsuura, J. Hrada, T. Akehata and T. Shirai, J. Chem. Eng. Jpn., 2 (1969) 199-205.
5 S.H. Greenhalgh, W.J. McManamey and K.E. Porter, Chem. Eng. Sci., 30 (1975) 155-157.
6 T-I. Chen and C.H. Barron, Ind. Eng. Chem., Fund., 11 (1972) 466-470.
7 C.H. Barron and H.A. O'hern, Chem. Eng. Sci., 21 (1966) 397-404.
8 K.R. Ulrich, R.E. Prada and G.T. Rochelle, Oxygen Absorption into Bisulfite Solutions Containing Single and Synergistic Metal Catalysts, Presented at the AIChE National Meeting, Houston 1983.
9 J. Klingspor, A-M. Strömberg, H.T. Karlsson and I. Bjerle. Chem. Eng. Process., 18 (1984) 239-247.
10 H.T. Karlsson, J. Klingspor, M. Linné and I. Bjerle, APCA J., 33 (1983) 23-28.
11 R.H. Borgwardt, "IERL-RTP Scrubber Studies Related to Forced Oxidation", Presented at the EPA FGD Symposium, New Orleans, 1976.
12 J. Erwin, C.C. Wang and J.L. Hudson, In: Flue Gas Desulfurization (Hudson, J.L. and Rochelle, G.T., ed.), ACS Symp. Ser. 188, (1982) 191-220.
13 I. Barin, O. Knacke and O. Kubaschewski, Thermochemical Properties of Inorganic Substances, Springer-Verlag, Berlin, (1973) 921 pp.
14 O. Levenspiel, Chemical Reaction Engineering, John Wiley & Sons, New York, (1972) 578 pp.

SULFITE OXIDATION INHIBITED BY THIOSULFATE

G. T. Rochelle and D. R. Owens
The University of Texas at Austin
Department of Chemical Engineering
Austin, TX 78712

ABSTRACT

Thiosulfate was studied as an inhibitor of sulfite oxidation at conditions typical of limestone scrubbing for flue gas desulfurization. Rates of thiosulfate degradation and sulfite oxidation were measured in a highly agitated solution of 0.3 M Na_2SO_4 at 50°C with pH 4.5 to 6, 3 to 30 mM S(IV), 0 to 1 mM Mn, 0 to 0.05 mM Fe, 0.01 to 0.88 atm O_2, and 0.05 to 9.5 mM sodium thiosulfate. Sulfite oxidation occurred by an autocatalytic, branching mechanism when catalyzed by Fe^{++}. At high levels of thiosulfate the rate of sulfite oxidation (mM/hr) was given by the theoretical expression (with $[Fe^{++}]$ in mM):

$$r_{ox} = 4([SO_3^-]/[S_2O_3^-])/(1-50\ [Fe^{++}][SO_3^-]/[S_2O_3^-])$$

At lower levels of thiosulfate, the rate of sulfite oxidation was better represented by the empirical expression:

$$r_{ox} = 5000([SO_3^-]/[S_2O_3^-])^2[Fe^{++}]$$

The rate of sulfite oxidation does not appear to be an additional function of oxygen or manganese concentrations. The rate of thiosulfate degradation to trithionate was given by the expression:

$$r_{deg}/r_{ox} = (k_t/2k_{ox})[S_2O_3^-]/[SO_3^-]$$

where: $k_t/2k_{ox} = 1/(8 + 0.13 \times 10^{pK_a - pH})$ and pK_a=6.76 in 0.3 M Na_2SO_4.

INTRODUCTION

Sulfite oxidation is an important side-reaction in flue gas desulfurization processes using aqueous solutions. In slurry scrubbers using lime or limestone, gypsum ($CaSO_4 \cdot 2H_2O$) scaling on scrubber internals and mist eliminators has been a serious reliability problem. Thiosulfate ($S_2O_3^-$) has been identified and tested as an effective oxidation inhibitor in slurry scrubbing systems. It reduces gypsum saturation and thereby reduces scaling. It also enhances the dewatering properties of calcium sulfite solids, enhances SO_2 removal when used with MgO, and reduces oxidative degradation of organic acid additives (ref.1).

Previous work on inhibitors of sulfite oxidation has studied organic free radical scavengers such as mannitol (ref.2), hydroquinone (ref. 3), and ethanol (ref. 4) and metal chelating agents such as ethylenediaminetetraacetic acid (refs. 5 and 6). Work with thiosulfate, a free radical scavenger, has been limited. Hudson (ref. 7) found that 0.2 mM thiosulfate practically stopped sulfite oxidation at pH 4. Rochelle et al. (ref. 8) found that 1 mM thiosulfate stopped oxidation of calcium sulfite slurry and organic acid degradation. Ulrich et al. (ref. 6) showed that 0.1 mM thiosulfate significantly reduced enhanced oxygen absorption in sulfite solutions catalyzed by 30 mM manganese (Mn). Several investigators have tested thiosulfate in integrated slurry scrubber systems at the bench, pilot, and commercial scale (refs. 1,9-11).

The objective of this work was to measure and quantify the effects of pH, and conentrations of thiosulfate, Fe, Mn, S(IV),and oxygen on the rates of sulfite oxidation and thiosulfate degradation. This work is reported in greater deatail in ref. 12.

THEORY

Thiosulfate appears to function as a free radical scavenger rather than a complexer of S(4)(sulfite plus bisulfite) or metal ions.. Its effective concentration is much less than S(4), therefore it probably does not form an unoxidizable complex with S(4). Using literature values for the chelating constant ([Fe chelate]/([Fe^{+++}] [$S_2O_3^{=}$]) = 447 M^{-1} (ref. 13)) only 50% of the Fe^{+++} should be chelated, assuming 1 mM thiosulfate and less than 0.1 mM Fe^{+++}. Other metals are even less strongly chelated. Furthermore thiosulfate appears to degrade as it functions, consistent with a role as a free radical scavenger, but inconsistent with complexing.

A probable mechanism for S(IV) oxidation in the presence of thiosulfate may include limiting propagation reactions such as:

$$SO_3^{=} + HSO_5\cdot \xrightarrow[k_{SO3}]{} SO_3^{-}\cdot + HSO_5^{-} \qquad (1)$$

$$HSO_3^{-} + HSO_5\cdot \xrightarrow[k_{HSO3}]{} HSO_3\cdot + HSO_5^{-} \qquad (2)$$

The equilibrium of sulfite and bisulfite gives sulfite as a function of pH and S(IV):

$$[SO_3^{=}] = [S(4)]/(1 + [H^+]/K_a) \qquad (3)$$

pK_a= 6.76 at 50°C in 0.3 M Na_2SO_4 (ref. 12)

For convenience the rate of sulfite oxidation can be approximated by:

$$r_{ox} = k_{ox}\,[SO_3^{=}]\,[HSO_5\cdot] \qquad (4)$$

where: $k_{ox} = k_{SO3} + k_{HSO3}10^{6.76-pH}$

The steady-state concentration of the intermediate peroxomonosulfate should be proportional to the oxidation rate:

$$[HSO_5^{-}] = k_s r_{ox} \qquad (5)$$

Free radicals may be initially generated by some constant rate reaction ($r_{i,0}$) and by a reaction of Fe^{++} with HSO_5^{-}:

$$Fe^{++} + HSO_5^{-} \longrightarrow Fe^{+++} + SO_4^{-\cdot} + OH^{-} \qquad (6)$$
$$r_i = r_{i,0} + k_i[Fe^{++}][HSO_5^{-}] = r_{i,0} + k_i k_s [Fe^{++}] r_{ox} \qquad (7)$$

Free radicals are terminated by reaction with thiosulfate to give tetrathionate:

$$S_2O_3^{=} + HSO_5\cdot = S_2O_3^{-\cdot} + HSO_5^{-} \qquad (8)$$
$$2S_2O_3^{-\cdot} = S_4O_6^{=}$$
$$r_t = k_t [S_2O_3^{=}] [HSO_5\cdot] \qquad (9)$$

The thiosulfate radical has been studied by Hayon (ref. 14) and oxidation of thiosulfate by iodine or other oxidants typically gives tetrathionate(ref. 15). In the presence of sulfite, tetrathionate is rapidly converted to trithionate by the reversible reaction(ref. 1):

$$S_4O_6^{=} + SO_3^{=} \longleftrightarrow S_3O_6^{=} + S_2O_3^{=} \qquad (10)$$

The net effect of short term experiments is the degradation of thiosulfate with the stoichiometry (ref. 1):

$$S_2O_3^{=} + SO_3^{=} + 0.5O_2 + H_2O = S_3O_6^{=} + 2OH^{-} \qquad (11)$$
$$r_{deg} = r_t/2 \qquad (12)$$

Assuming that $r_i = r_t$, this oxidation mechanism gives a rate equation of the form:

$$r_{ox}[S_2O_3^{=}]/[SO_3^{=}] = (r_{i,0}\, k_{ox}/k_t) \,/\, (1 - (k_i k_s k_{ox}/k_t)[Fe^{++}][SO_3^{=}]/[S_2O_3^{=}])) \qquad (13)$$

At high levels of thiosulfate this reduces to:

$$r_{ox}[S_2O_3^{=}]/[SO_3^{=}] = (r_{i,0}\, k_{ox}/k_t) \qquad (14)$$

At low levels of thiosulfate the predicted rate of oxidation goes to infinity, because this mechanism does not include any other termination reactions. The degradation rate of thiosulfate can be modeled by ratioing Eqs. 9 and 4:

$$r_{deg} = (k_t/2k_{ox}) r_{ox}[S_2O_3^{=}]/[SO_3^{=}] \qquad (15)$$
$$= (r_{i,0}/2)/(1 - (k_i k_s k_{ox}/k_t)[Fe^{++}][SO_3^{=}]/[S_2O_3^{=}])) \qquad (16)$$

At long times in commercial systems, trithionate hydrolyzes to regenerate thiosulfate (ref. 1):

$$S_3O_6^= + H_2O \longrightarrow S_2O_3^= + SO_4^= \qquad (17)$$

$$r_h = k_h[S_3O_6^=]$$

$$k_h(hr^{-1}) = 1.5x10^{12} \exp(-20{,}200/RT)$$

The net effect of thiosulfate oxidation (Rxn. 12) and hydrolysis (Rxn. 17) is oxidation of S(IV) to sulfate, with no loss of thiosulfate. In a scrubber system where thiosulfate oxidation and hydrolysis both occur in the same well-mixed solution volume, for example the hold tank, the trithionate concentration should reach a steady-state concentration given by:

$$[S_3O_6^=] = r_{deg}t/(1 + k_h t) \qquad (18)$$

The system residence time, t, is defined as the ratio of system solution volume(hold tank + thickener) to the rate of solution loss from the system (filter cake + other losses). With large residence time (>100 hr) , typical of systems with good dewatering, the steady-state trithionate concentration is independent of residence time and is given by:

$$[S_3O_6^=] = r_{deg}/k_h \qquad (19)$$

EXPERIMENTAL METHODS

The pH-stat method (refs. 6 and 16) was used to measure the rate of S(IV) oxidation. The pH was controlled by titrating with a solution of 1.5 M Na_2SO_3 giving the overall stoichiometry:

$$SO_3^= + 0.5\ O_2 \longrightarrow SO_4^= \qquad (20)$$

In the absence of other reactions involving pH changes, the rate of S(IV) oxidation is equal to the titration rate of Na_2SO_3 and the pH and S(IV) concentration should be constant.

Oxidation was carried out in 1 liter of solution agitated at 1600 rpm and sparged with 1 liter/min of oxygen or oxygen/nitrogen mixtures at atmospheric pressure. The reactor temperature was controlled at 50±2°C and pH was controlled within 0.05 units by a Fisher tritrimeter. Both pH and cumulative titrant volume were recorded as a function of time. Thiosulfate and S(IV) concentrations were determined as a function of time by iodimetric titrations with and without formaldehyde. Additional description of the apparatus and procedure can be found in refs. 12 and 16.

All experiments were performed with an initial background electrolyte of 0.3 M Na_2SO_4. One analysis of this solution and the Na_2SO_3 titrant gave background levels of 0.0006 and 0.0014 mM Fe, respectively. Fe, Mn, and thiosulfate were added as solutions of $FeSO_4$, $MnSO_4$, and $Na_2S_2O_3$, respectively.

Initial S(IV) oxidation rates with air without addition of thiosulfate were 54 to 144 mM/hr. Because these rates were independent of Mn and Fe and did not decrease with small additions of thiosulfate, they probably correspond to the limet of oxygen mass transfer into the solution. Therefore no kinetic data could be obtained at fast rates. With the addition of significant quantities of thiosulfate, the apparaent oxidation rates initially decreased to values as low as 0.1 mM/hr, then increased to the mass transfer limit as the thiosulfate degraded over a period of 10 to 250 min.

Because thiosulfate degradation results in some consumption of S(IV), both The oxidation rate and S(IV) concentration must be corrected:

$$[S(IV)] = [S(IV)]_i - 2(1+f)([S_2O_3^-]_i - [S_2O_3^-]) \tag{21}$$

$$r_{ox} = r_{titrant} + r_{deg}(1+f) \tag{22}$$

where $f = K_a 10^{pH}$

GENERAL RESULTS

Results of typical experiments are given in Figures 1 and 2. Figure 1 shows how the cumulative apparent sulfite oxidation varies with with time. Initially, without thiosulfate (line A), the oxidation rate is fast and controlled by absorption of oxygen. With the addition of an increment of thiosulfate(point B), the oxidation rate decreases, then steadily increases as thiosulfate is degraded. At point C the oxidation increases very rapidly as thiosulfate is completely degraded, but a high oxygen remains in the solution. When the dissolved oxygen is used up, the oxidation rate recovers to its initial steady-state value controlled by oxygen absorption. The time at point D is defined as t_r.

Figure 2 shows thiosulfate concentration as a function of time in two typical experiments. Two distinct types of behavior were observed. In the absence of Fe or at high levels of thiosulfate, the rate of thiosulfate degradation was constant throughout the experiment. With the addition of Fe and/or at low thiosulfate levels, the degradation rate increased as thiosulfate concentration decreased and oxidation rate increased.

Tables 1 and 2 summarize the results obtained. Table 1 gives results in experiments where thiosulfate was not determined by chemical analysis. Each experiment includes several progressive additions of thiosulfate to the same solution, with accumulation of thiosulfate degradation products. The sulfite oxidation rate was determined immediately after each addition of thiosulfate. For most of these experiments, the thiosulfate degradation constant was calculated for each addition of thiosulfate by:

$$k_t/2k_{ox} = [S_2O_3^-]_{initial}/([SO_3^-]_{initial} t_r r_{ox,initial}) \tag{23}$$

This relationship is valid when the rate of sulfite oxidation is inversely proportional to thiosulfate concentration. When sulfite oxidation rate varied as $[S_2O_3^-]^{-2}$, the thiosulfate degradation rate constant was estimated by:

$$k_t/2k_{ox} = [S_2O_3^-]_{initial}/(2[SO_3^-]_{initial} t_r r_{ox,initial}) \tag{24}$$

Table 2 gives results of experiments in which thiosulfate was added once at a fairly high concentration, then determined by iodometric titration as degradation proceeded. The rate of sulfite oxidation was determined at each analysis of thiosulfate. The thiosulfate degradation rate for each experiment was calculated by the integral relationship:

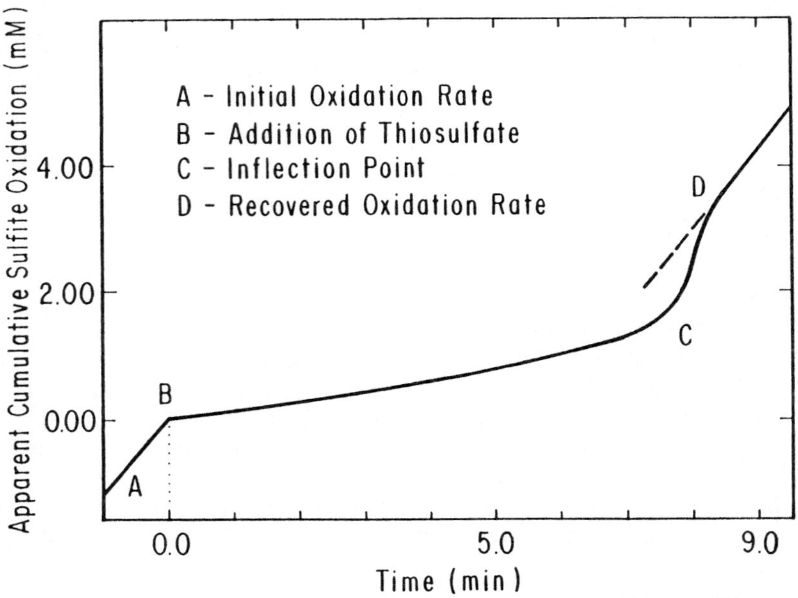

Fig. 1. Typical result for sulfite oxidation, Run S8, pH 5, 0.2 mM $S_2O_3^{=}$, 0.01 mM Fe, 10 mM S(IV).

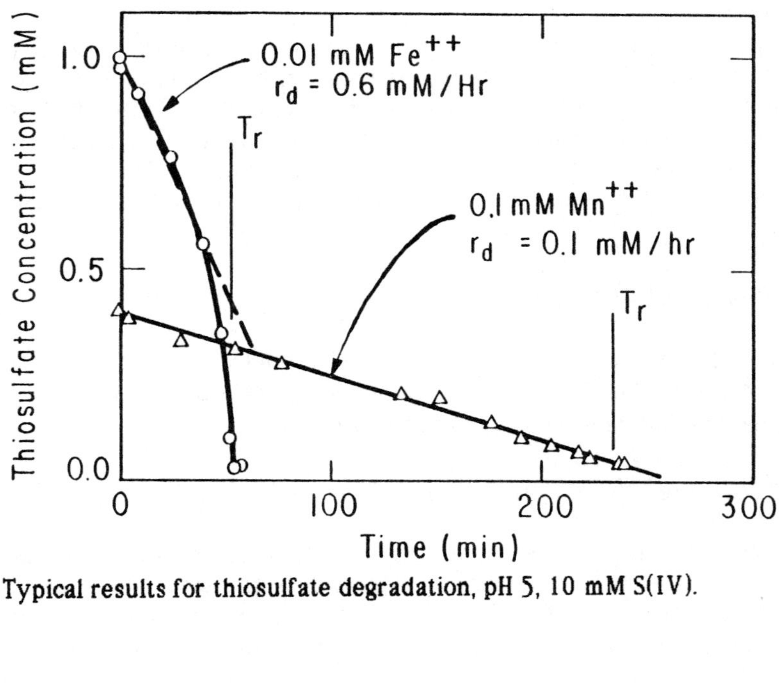

Fig. 2. Typical results for thiosulfate degradation, pH 5, 10 mM S(IV).

Table 1: Summary of experiments without analysis of thiosulfate, 10 mM S(IV)

Run number	S1	S2	S3	S4-1	S4-2	S5	S6-1	S6-2	S7
pH	5	5	5	5	5	5	5	5	5
O2 (%)	21	5	100	10	10	1	21	21	21
Mn(mM)	0.1	0.1	0.1	0.1	0.1	0.1	0.01	1	0.01
Fe added (mM)	0	0	0	0	0.01	0	0	0	0
Nominal Fe (mM)	0.0002	0.0002	0.0002	0.0002	0.01	0.0002	0.0002	0.0002	0.0002
Ave. kt/2kox	0.061	0.099	0.066	0.106	0.382	0.136	0.103	0.087	0.099
Ave. r1 (mM/hr)	13.2	2.3	5.3	5.2	110.8	1.2	1.8	6.0	2.1
Ave. r2 (1/hr)	19230	1401	3623	4014	561	1927	2065	5181	2269
Run number	S8-1	S8-2	S9	S10	S11	S12	S13	S14	S15
pH	5	5	5	5	5	5	5	5.5	5.5
O2 (%)	21	21	21	21	21	21	21	21	21
Mn(mM)	0	1	0	0	0	0	0	0	0
Fe added (mM)	0.01	0.01	0	0.0011	0.002	0.004	0.01	0.001	0.01
Nominal Fe (mM)	0.01	0.01	0.0002	0.0011	0.002	0.004	0.01	0.001	0.01
Ave. kt/2kox	0.184	0.077	0.109	0.104	0.129	0.129	0.119	0.301	0.183
Ave. r1 (mM/hr)	14.8	32.6	1.3	5.4	6.7	9.4	13.4	3.9	9.3
Ave. r2 (1/hr)	178	526	1090	522	417	334	314	711	264

TABLE 2: Summary of experiments performed with analysis of thiosulfate, 21% oxygen.

Run Number	D6	D4	D2	D10	D9	D11	D7-2	D12-2	D1	X1
Fe Added (mM)	0	0.1 Mn	0.1 Mn	0.002	0.002	0.002	0.01	0.01	0.01	1% O2
Nominal Fe (mM)	0.0005	0.0005	0.0005	0.002	0.002	0.002	0.01	0.01	0.01	0.01
pH	5	5	5	5	5	5	5	5	5	5
Initial S2O3= (mM)	0.25	0.1	0.4	0.31	0.3	1.3	0.32	0.52	1	0.5
Initial S(IV) (mM)	10	10	10	3	10	30	3	3	10	10
kt/2kox	0.046	0.038	0.062	0.045	0.074	0.087	0.045	0.071	0.074	0.077
Ave. r1 (mM/hr)	11.0	9.6	5.6	6.5	10.7	7.5	11.2	3.7	15.5	20.1
Ave. r2 (1/hr)	17310	6340	6920	8058	4595	5161	5876	1872	5237	3227
Run Number	X2	D13-1	D13-2	D8-1	D8-2	D14	F1	F1	F9	F7
Fe Added (mM)	10 % O2	0.01	0.01	0.01	0.01	0.01	0.0005	0.0005	0.001	0.004
Nominal Fe (mM)	0.01	0.01	0.01	0.01	0.01	0.01	0.0005	0.0005	0.001	0.001
pH	5	5	5	5	5	5	6	6	6	6
Initial S2O3= (mM)	0.5	1.1	1.23	1.05	1.03	3	1	1	1	0.45
Initial S(IV) (mM)	10	20	20	30	30	30	10	10	10	3
kt/2kox	0.063	0.066	0.082	0.072	0.081	0.072	0.118	0.118	0.149	0.135
Ave. r1 (mM/hr)	34.0	35.8	26.4	36.8	35.0	12.4	1.0	0.7	14.0	6.0
Ave. r2 (1/hr)	6207	6842	4613	5399	5195	4520	949	528	6493	2700
Run Number	F4	F8	F6	F2	F5	F3	L2	L1-1	L1-2	L4
Fe Added (mM)	0.004	0.004	0.01	0.01	0.01	0.05	0.002	0.01	0.01	0.01
Nominal Fe (mM)	0.001	0.001	0.001	0.001	0.001	0.001	0.002	0.01	0.01	0.01
pH	6	6	6	6	6	6	4.5	4.5	4.5	4.5
Initial S2O3= (mM)	1.7	9.5	0.5	2.5	8.5	2.5	0.5	0.21	0.45	1.2
Initial S(IV) (mM)	10	30	3	10	30	10	10	10	10	30
kt/2kox	0.142	0.049	0.146	0.078	0.109	0.099	0.028	0.061	0.060	0.056
Ave. r1 (mM/hr)	11.8	4.3	1.7	9.3	31.3	8.6	6.2	17.3	8.1	7.7
Ave. r2 (1/hr)	8340	7900	1010	10370	41570	8300	6616	2798	3323	1929

$$k_t/2k_{ox} = \frac{([S(IV)]_i - (1+f)[S_2O_3^=]_i)\ln([S_2O_3^=]_i/[S_2O_3^=]_i)}{\text{cum. app. oxid.} + [S_2O_3^=]_i - [S_2O_3^=]_f} \qquad (25)$$

Tables 1 and 2 also give calculated values of two oxidation rate parameters. The rate parameter, r_1, is based on the theory of Eq. 12 and should be constant at high levels of thiosulfate. It is defined by:

$$r_1 = r_{ox}[S_2O_3^=]/[SO_3^=] \qquad (26)$$

The rate parameter, r_2, is based partially on Eq. 12, but also on the empirical observation that oxidation rate varies as $[Fe^{++}][S_2O_3^=]^{-2}$. It is defined by:

$$r_2 = r_{ox}([S_2O_3^=]/[SO_3^=])^2/[Fe^{++}] \qquad (27)$$

Thiosulfate Degradation

Thiosulfate degradation is modeled very well by Eq. 15. The thiosulfate degradation rate constant, $k_t/2k_{ox}$, does not appear to be a strong function of S(IV), Fe, Mn, oxygen concentration. The data from Table 2 should be most reliable in determining $k_t/2k_{ox}$. Figure 3 gives these data as a function of pH. The average value of $k_t/2k_{ox}$ for a large number of experiments at pH 5 is 0.064. The weak dependence of $k_t/2k_{ox}$ on pH suggests that the primary propagation reaction is that with sulfite (Rxn. 1) rather than bisulfite (Rxn. 2). Including both reactions gives the following relationship for thiosulfate degradation:

$$r_{ox}/r_{deg} = (k_{SO3}[SO_3^=] + k_{HSO3}[HSO_3^-])/(k_t[S_2O_3^=]) \qquad (28)$$

Substitution of the sulfite/bisulfite equilibrium (Eq. 3) and rearrangement gives the relationship:

$$k_t/2k_{ox} = 1/\{(k_{SO3}/k_t) + (k_{HSO3}/k_t)10^{6.76-pH}\} \qquad (29)$$

The data in Figure 3 are correlated well with $k_{SO3}/k_t = 8$ and $k_{HSO3}/k_t = 0.13$.

Sulfite Oxidation

The most consistent data on sulfite oxidation are given in Figure 4. Empirically, the oxidation rate in the presence of Fe is correlated by the relationship:

$$r_2 = r_{ox}([S_2O_3^=]/[SO_3^=])^2/[Fe^{++}] = 5000\ hr^{-1} \qquad (30)$$

Correlation by the theoretical Eq. 13 gives:

$$r_1 = r_{ox}[S_2O_3^=]/[SO_3^=] = 4/(1-50\,[Fe^{++}][SO_3^=]/[S_2O_3^=]) \qquad (31)$$

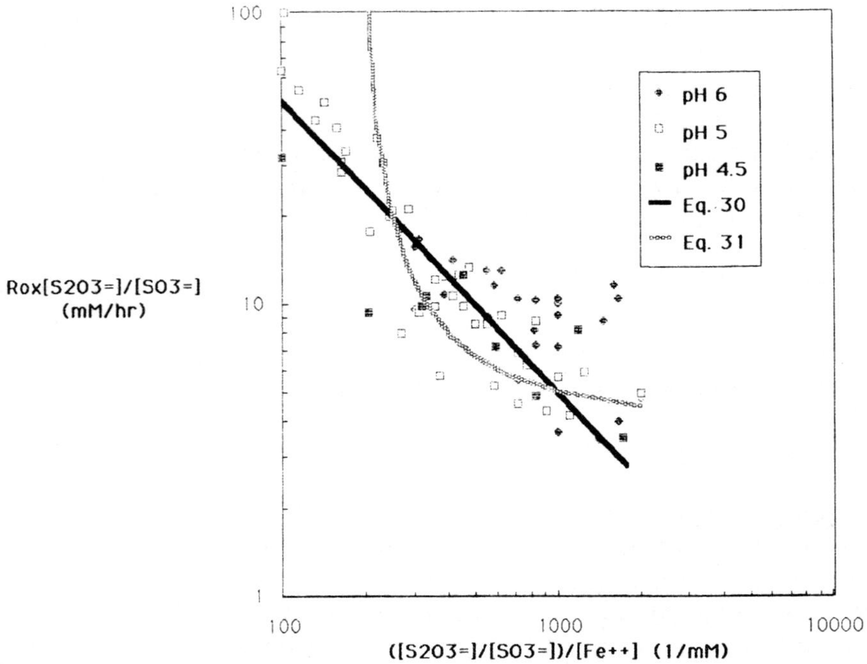

Fig. 3. Correlation of sulfite oxidation, selected data from Table 2.

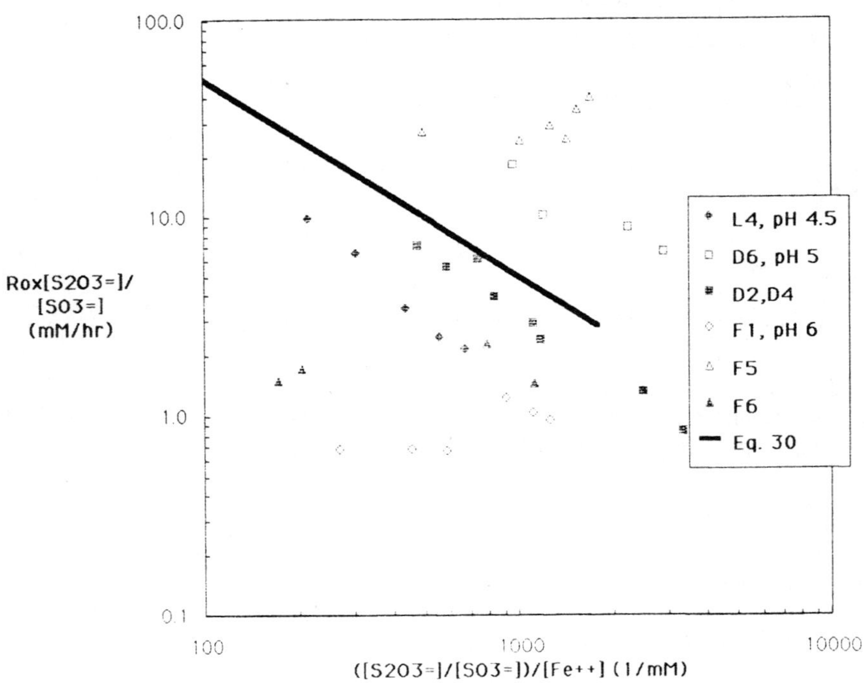

Fig. 4. Correlation of sulfite oxidation, other data from Table 2.

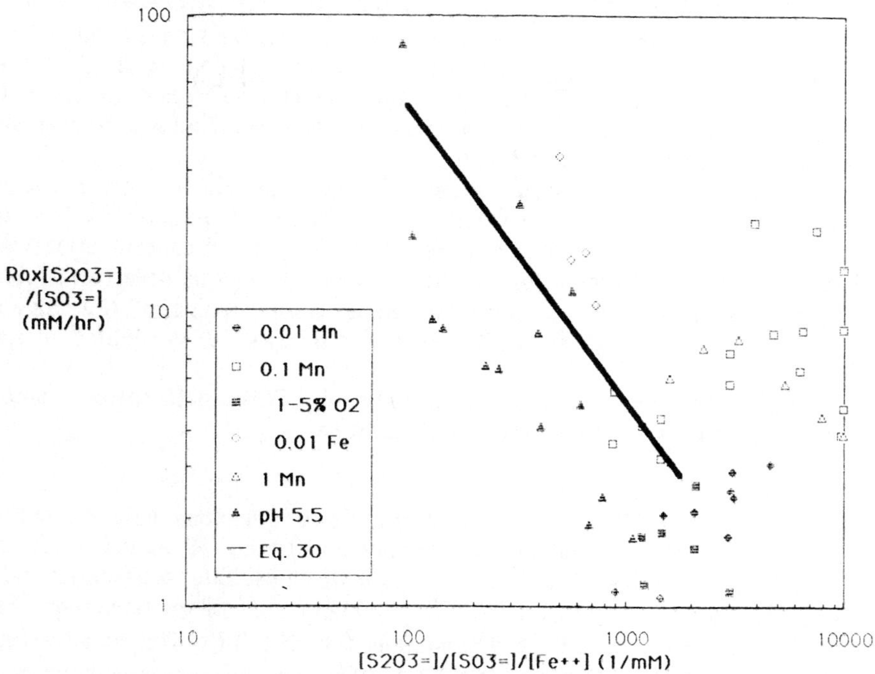

Fig. 5. Correlation of sulfite oxidation, data from Table 1 without thiosulfate analysis.

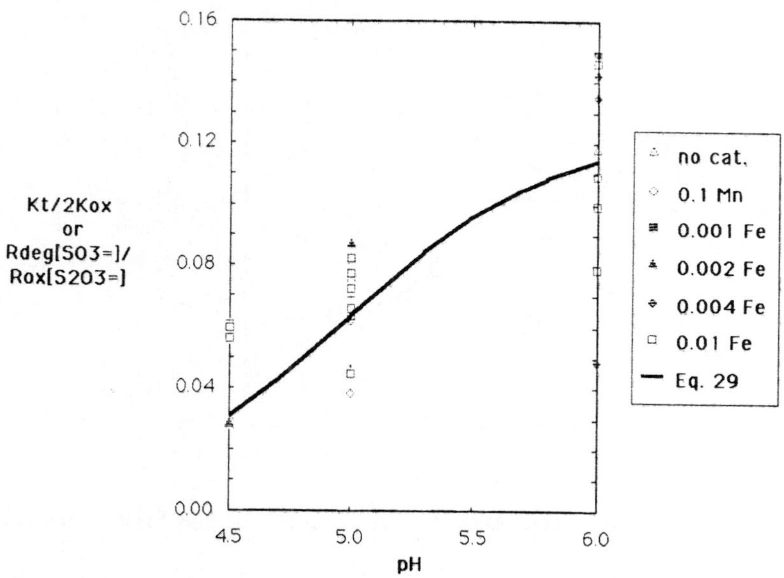

Fig. 6. Correlation of thiosulfate degradation, data from Table 2.

where $r_{i,o}\ k_{ox}/k_t$=4 mM/hr and $k_i k_s k_{ox}/k_t$=50 mM^{-1}. Eq. 31 does not work well at low values of thiosulfate because the theory does not account for termination of free radicals by mechanisms other than reaction with thiosulfate. The outstanding feature of these correlations is that they predict that the rate of free radical initiation is faster at higher rates of oxidation. This observation is consistent with a mechanism involving branching reactions such as Rxn. 6.

Figure 5 gives other detailed data from Table 2 that do not fit the correlations as well. Some scatter in the data is produced by uncertainties in the background level of dissolved catalysts, especially Fe. When no Fe was added, the Fe concentration was assumed to be .0005 mM. However higher Fe concentrations may be present in some experiments because of carryover from previous experiments. Data at pH 6 assume a maximum level of 0.001 mM dissolved Fe, even when more Fe was added to the initial solution.

Oxidation data from Table 1 are given in Figure 6. These data were modeled better by assuming a background level of 0.0002 mM Fe.

DESIGN IMPLICATIONS

Figure 7 illustrates the effect of thiosulfate in a typical limestone slurry scrubber system. The rate of sulfite oxidation was calculated using Eq. 30 at low levels of thiosulfate and Eq. 31 at higher levels. The rate of thiosulfate degradation was calculated by Eq. 15 with $k_t/2k_{ox}$=0.07. The total rate of sulfate formation was increased by the rate of thiosulfate degradation per Rxn. 17. The steady-state concentration of trithionate ($S_3O_6^=$) was calculated by Eq. 19. The molar makeup rate of thiosulfate will be equal to the loss of thiosulfate and trithionate in solution entrained with the filter cake. Therefere, the sum, $[S_3O_6^=]$ + $[S_2O_3^=]$, will be directly proportional to the makeup rate.

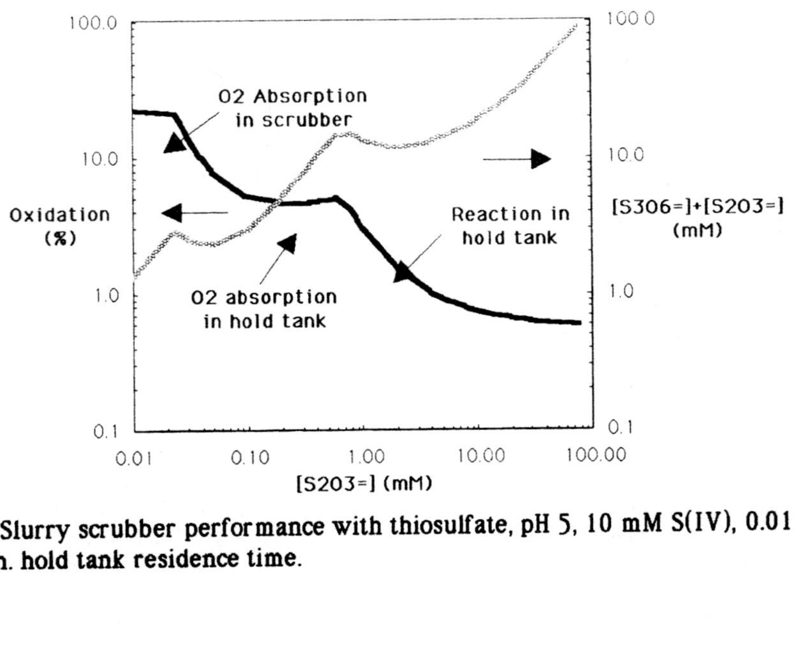

Fig. 7. Slurry scrubber performance with thiosulfate, pH 5, 10 mM S(IV), 0.01 mM Fe, 10 min. hold tank residence time.

At very low levels of thiosulfate, sulfite oxidation is controlled by oxygen absorption in the absorber. With moderate levels of thiosulfate, oxidation is completely inhibited in the absorber but is controlled in the hold tank by oxygen absorption and the carryover of dissolved oxygen from the absorber. With higher levels of thiosulfate, sulfite oxidation is controlled by reaction kinetics in the hold tank. The minimum level of oxidation at very high thiosulfate corresponds to the constant rate of free radical initiation and thiosulfate degradation. For the specific case illustrated by Figure 7, 0.1 mM thiosulfate gives 5% oxidation, 1 mM gives 3% and 10 mM gives 0.7%. Practically, sulfate levels in scrubbers will be somewhat higher because of SO_3 absorbed from the flue gas.

The effective makeup rate of thiosulfate does not increase monotonically with thiosulfate concentration. Because of the complicated interactions of sulfite oxidation and thiosulfate degradation, the sum of trithionate and thiosulfate concentrations decreases slightly as thiosulfate increases from 0.02 to 0.04 mM and from 0.8 to 2 mM. Furthermore, up to 1.2 mM thiosulfate, the thiosulfate concentration is less than 10% of the trithionate concentration, so that small changes in makeup rate will cause large, short-term changes in thiosulfate concentration. Therefore, it may difficult to operate the scrubber at steady-state thiosulfate concentration below 2 mM. In the case of Figure 7, a makeup rate equivalent to 12 mM thiosulfate will give a steady-state thiosulfate concentration of 2 mM and sulfite oxidation of1.5%.

CONCLUSIONS

1. Sulfite oxidation occurs by an autocatalytic, branching mechanism when catalyzed by Fe^{++}. At high levels of thiosulfate the rate of sulfite oxidation (mM/hr) is given by the theoretical expression (with $[Fe^{++}]$ in mM):

$$r_{ox} = 4([SO_3^{=}]/[S_2O_3^{=}])/(1-50\ [Fe^{++}][SO_3^{=}]/[S_2O_3^{=}])$$

At lower levels of thiosulfate, the rate of sulfite oxidation is better represented by the empirical expression:

$$r_{ox} = 5000([SO_3^{=}]/[S_2O_3^{=}])^2[Fe^{++}]$$

The rate of sulfite oxidation does not appear to be an additional function of oxygen or manganese concentrations.

2. The rate of thiosulfate degradation to trithionate is given by the expression:

$$r_{deg}/r_{ox} = (k_t/2k_{ox})[S_2O_3^{=}]/[SO_3^{=}]$$

where: $k_t/2k_{ox} = 1/(8 + 0.13x10^{pKa-pH})$ and pK_a=6.76 in 0.3 M Na_2SO_4.

3. Because of the complicated interactions of thiosulfate degradation and sulfite oxidation, there will be a minimum thiosulfate concentration that can be maintained at steady-state, typically 2 mM. At 2 mM thiosulfate, typical conditons in a slurry scrubber system should give 1.5% oxidation and would require thiosulfate makeup equivalent to solution losses containing 12 mM thiosulfate.

REFERENCES

1 G. T. Rochelle, D. R. Owens, J. C. S. Chang, and T. G. Brna, Thiosulfate as an Oxidation Inhibitor in Flue Gas Desulfurization Processes: A Review of R&D Results, presented at the Ninth Symposium on Flue Gas Desulfurization, Cincinnati, OH, June 4-7, 1985.

2 E. Fuller and R. Crist, J. Am. Chem. Soc., 63 (1941) 1644.

3 E. Altwicker, Trans. Inst. Chem. Eng., 55 (1977) 281.

4 T. Braga and R. Connick, ACS Symp. Ser. 188 (1982) 153.

5 A. Huss, P. Lim, and C. Eckert, J, Phys. Chem., 86 (1982) 4224.

6 R. K. Ulrich, G. T. Rochelle, and R. E. Prada, Submitted to Chem. Eng. Sci., 1985.

7 J. L. Hudson, Sulfur Dioxide Oxidation in Scrubber Systems. Monthly Progress Report, EPA Grant R805227-02, March 1979.

8 G.T. Rochelle, et al. Buffer Additives for Lime/Limestone Slurry Scrubbing. ACS Symp. Ser. 188:243-266, 1982.

9 J. B. Barkley et al., Evaluation of Sodium Thiosulfate Additive on the Performance of a Spray Tower Scrubber System Using Limestone Slurry. TVA/OP/EDT-84/7, November 1983.

10 J. F. Miller et al., Lime/Limestone Scrubber Modifications: Investigation of Thiosulfate as an Oxidation/Scaling Inhibitor. DOE/TIC-2005972, December 1981.

11 D. Laslo and E. Bakke, State-of-the-Art Design Application on a Closed Loop FGD System. Presented at the Ninth Symposium on Flue Gas Desulfurization, Cinncinnatti, Ohio, June 4-7, 1985.

12 D. R. Owens and G. T. Rochelle, Sulfite Oxidation Inhibited by Thiosulfate. M. S. Thesis, University of Texas at Austin. 1984. (Also Epa Draft Report on Purchase Order No. 3D3744NAEX, 1985)

13 L. G. Sillen and A. E. Martell, Stability Constants of Metal Ion Complexes. The Chemical Society, 1964.

14 L. Dogliotti and E. Hayon. Flash Photolysis Study of Sulfite, Thiocynate, and Thiosulfate Ions in Solution. J. Phys. Chem. 72: 1800-1807, 1968.

15 M. Goehring Die Chemie der Polythionsauren. Fortschr. Chem. Forsch. 2:444-483, 1952.

16 P. K. Chan and G. T. Rochelle, ACS Symp. Ser. 188 (1982) 75-98.

PART 9

BIOLOGICAL DESULFURIZATION OF COAL

MICROBIAL DESULFURIZATION OF COAL

J.D. ISBISTER and E. A. KOBYLINSKI
Atlantic Research Corporation, 5390 Cherokee Avenue, Alexandria, Virginia 22312

ABSTRACT

A unique microorganism (CB1) capable of removing a portion of the organic sulfur from coal has been developed at Atlantic Research Corporation. CB1 primarily attacks thiophenic sulfur in coal, removing the sulfur from the coal as a water soluble sulfate. The mechanism of action of CB1 on thiophenic sulfur has been defined via ^{14}C and ^{35}S studies on model compounds, sulfate analyses by ion chromatography and energy-dispersive x-ray microanalysis. In laboratory batch and bench-scale continuous systems, CB1 removed up to 47% of the organic sulfur from coal. The percent sulfur removal is highly dependent on the coal type, coal particle size and surface properties. A 2500 lb per day continuous pilot-scale desulfurization unit has been built and tested.

INTRODUCTION

Atlantic Research Corporation has developed a microorganism capable of removing a portion of the organic sulfur from coal. This organism, CB1, was developed from a mixed culture of naturally occurring soil microorganisms by mutagenic alteration of the microbial DNA. The microorganism created by these mutagen treatments is capable of oxidizing the sulfur in a model organic sulfur compound, dibenzothiophene (DBT), but does not break the carbon ring structure of the molecule. CB1 differs biochemically from other microorganisms of its species and is considered unique. A patent application for CB1 and its use in desulfurizing fossil fuels has been filed.

LABORATORY STUDIES

Growth Optimization

The original or parent CB1 culture was minimally capable of oxidizing dibenzothiophene (DBT) as shown in Table 1. Following mutagen treatment, CB1 was selected from a number of mutant colonies exhibiting significantly increased DBT oxidation. The first laboratory studies with CB1 were carried out to determine the effects of temperature on the oxidation of DBT. Results from these temperature studies are presented in Table 1. Although the greatest DBT oxidation efficiency occurred at 35°C, these data indicate that efficient microbial oxidation of DBT can be obtained over a temperature range of 25 to 40°C.

TABLE 1

Microbial Oxidation of DBT

Culture	Incubation Temperature, °C	Incubation Time, hrs.	Initial DBT Concentration (mg/L)	% Decrease in DBT Concentration
Orginial Culture	25	168	100	7
CB1	25	18	200	71
"	25	18	250	46
"	25	18	200	71
"	35	18	200	94
"	40	18	200	76
CB1 (After growth medium optimization)	25	18	400	80
"	25	18	500	81
"	25	18	800	70
"	25	18	1000	60

Other laboratory experiments were focused on optimization of the growth conditions for CB1 to increase the DBT oxidation efficiency. Modifications were made to the original salts medium for growth to obtain maximum organism production per molecule of carbon source, to increase the growth rate and to decrease the cost of the growth medium. The media modifications and subsequent microbial adaptation to the new medium resulted in a three-fold increase in the microbial growth rate and a decrease of 25% in the medium costs. As shown in Table 1, the adapted culture was capable of tolerating higher concentrations of DBT with good oxidation efficiency.

Mechanism of Action

The mechanism of action of CB1 on DBT was elucidated by employing ^{14}C-DBT and ^{35}S-DBT as a portion of the DBT treated by CB1. These studies showed that CB1 did not break the carbon ring structure, as no evolution of $^{14}CO_2$ or incorporation of ^{14}C into the biomass was detected. After microbial treatment, the ^{35}S from the ^{35}S-DBT had been released into the aqueous phase in the form of $^{35}SO_4^{-2}$. The increase in aqueous sulfate concentrations following treatment of DBT with CB1 was confirmed by ion chromatography. The organic product was determined to be 2,2'-dihydroxybiphenyl using gas chromatography-mass spectra analysis. These data show a mechanism of microbial activity which is very different from DBT oxidation pathways reported by other investigators. Other organisms break the carbon ring structure to yield 3-hydroxy-2-formylbenzothiophene (Hartdegen *et al*., 1984), but do not remove the sulfur from the molecule.

Microbial Treatment of Coal

Cleaned coals containing approximately 90% organic sulfur were used in the initial studies with CB1 to bypass problems associated with organic sulfur analysis. Total sulfur contents of these coals before and after treatment with CB1 were determined using the Fisher Total Sulfur Analyzer. Pyrite and sulfate analyses were performed according to ASTM method #D2492. As shown in Table 2, decreases in organic sulfur ranging from 18 to 28% were observed in these clean coals following treatment with CB1.

Several run of the mine (ROM) coals were also treated with CB1. As shown in Table 2, the percentage of organic sulfur removed from these ROM coals varied from 25% to 34%. The observed variation in the amount of organic sulfur removed from these coals is due to differing amounts of thiophenic sulfur in the coals and differences in particle sizes and surface characteristics. CB1 acts mainly on the thiophenic sulfur in coals although it may act to some extent on other organic sulfur forms. Thus, the maximum desulfurization that can be achieved for a coal will depend on the thiophenic sulfur content of the coal. The amount of desulfurization also depends on the ability of the microbe to contact the thiophenic sulfur in a manner that allows proper orientation of the enzyme(s).

TABLE 2

Laboratory-Scale Coal Desulfurization Studies with CB1

Coal Sample	Mesh Size	Before Treatment		After Treatment		% Reduction in Organic Sulfur
		Total Sulfur %	Organic Sulfur %	Total Sulfur %	Organic Sulfur %	
Wentz	60-200	0.67	0.6	0.56 0.52	0.49 0.45	18 25
Wentz	<200	0.67	0.6	0.53 0.50	0.46 0.43	24 28
Montcoal	< 60	0.80	0.72	0.62 0.63	0.54 0.55	25 24
Consolidation**	< 60	3.8	2.0	3.1*	1.5	25
Pittston**	< 60	4.0	1.9	3.3*	1.3	32
Sewickeley**	< 60	5.0	1.9	3.56*	1.26	34
Homer City Underflow†	<100	1.8	0.95	1.3*	0.5	47
Black Mesa	< 50	0.607	0.487	0.56	0.45	20
Peabody Illinois #6**	< 60	3.5	2.1	2.9*	1.6	24

* Total sulfur numbers reflect sulfate that was washed out of the original coal during treatment
** ROM coals
† Waste stream of fine coal following processing by heavy media cyclone

Therefore, microbial desulfurization is dependent on coal particle size (desulfurization increases with decreasing particle size) and surface properties (desulfurization is inhibited by surface oxidation).

Effects of Microbial Treatment on Coal Characteristics

The physical and chemical characteristics of coal before and after microbial treatment with CB1 were studied using a clean coal (Montcoal) and a ROM coal (Peabody Illinois #6). Electron micrographs of the treated coals indicated that microbial treatment seems to smooth the sharp edges prominent in coal particles before treatment. At least some of this "smoothing" may be the result of microbes collapsed on the coal particles during preparation of the treated coal for electron microscopy.

Particle size analyses of untreated coal and microbially treated coal showed an increase in particle size following treatment. The untreated coal was found to be 33.1% greater than 33.7 microns while the treated coal was 47% greater than 33.7 microns. Microbial hydration at the sulfur site during oxidation of the sulfur could account for the apparent increase in particle size observed in the treated coals. The apparent increase in particle size should have little or no effect on combustion of the treated coal.

Settling experiments compared an untreated coal sample with a treated coal sample. In both samples, the largest particles settled out immediately. However, as shown in Figure 1, the treated coal had a lower settling rate than the raw coal. This altered settling rate is probably due to the hydroxylation at the former sulfur site. If the microbially cleaned coal is to be used in coal water fuel, modification of the normal additive formulation will probably be necessary to compensate for the microbially mediated changes in the coal surface properties.

Combustion analysis of the coal before and after treatment indicated that the Btu contents of the untreated and untreated samples were comparable. The addition of the hydroxyl groups at the sulfur oxidation site would in theory reduce the Btu value of the treated coal a small amount. However, the organisms attached to the coal will contribute to the carbon content of the coal and may compensate for the small losses associated with desulfurization.

Direct Detemination of Organic Sulfur in Coal

In the standard method for determining organic sulfur in coal, the organic sulfur is calculated by subtraction of inorganic sulfur (pyrites and sulfates) from total sulfur. Pyritic sulfur is determined from the amount of pyritic iron in the coal. Interferences in pyrite determination, especially in treated coals, can lead to erroneous organic sulfur analyses. A method for direct determination

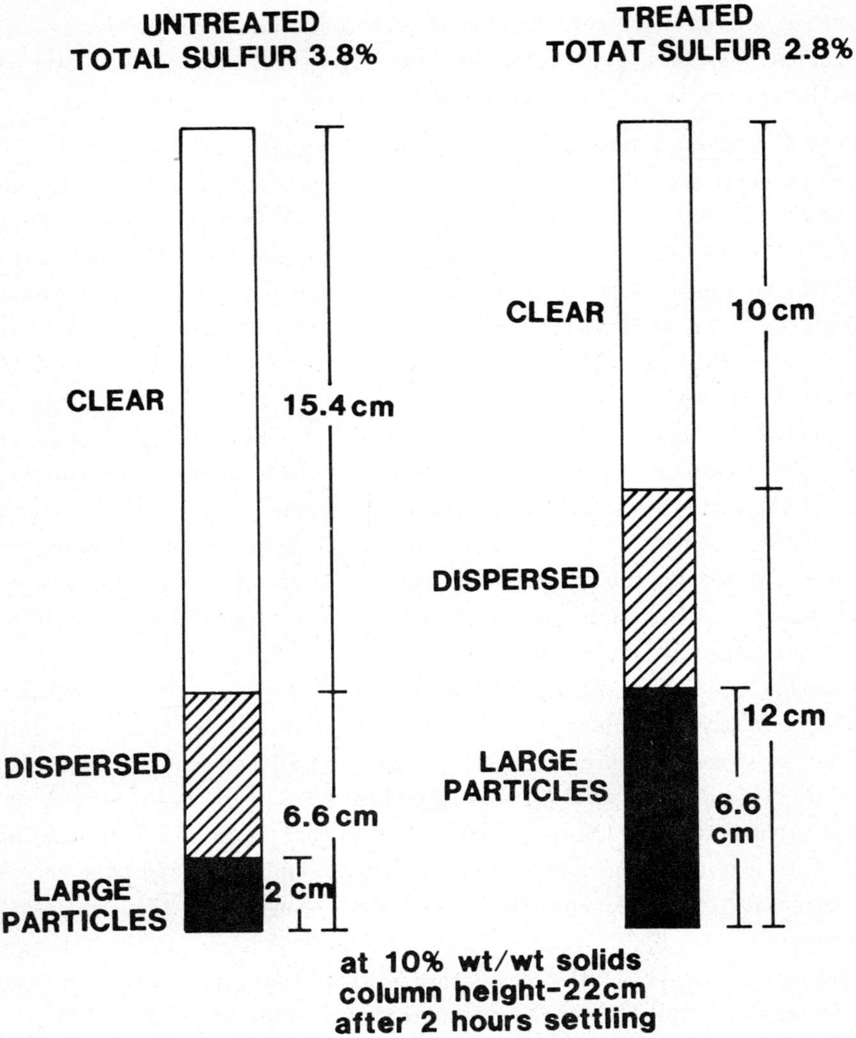

Fig. 1. Settling of Peabody Coal Before and After Treatment

of organic sulfur in coal has been developed (Straszheim _et al_., 1983) in which energy-dispersive x-ray microanalysis is combined with scanning electron microscopy. This method allows for differentiation between organic sulfur and pyrite in coals.

To confirm that CB1 removes organic sulfur and not pyritic sulfur from coals, a sample of ROM Homer City coal was ground in a micropulverizer for analysis by the EDS-SEM method. The pulverized coal was sieved and 32% was found to be -140 mesh (less than 106 um). This coal was treated with CB1, and total sulfur, pyrite and sulfate analyses were performed on both untreated coal and treated coal. Data from these analyses are presented below in Table 3.

TABLE 3

Analysis of ROM Coal Before and After Treatment with CB1.

Sample	% TS	% SO_4^{-2}	Pyrite	Organic S
ROM Coal Untreated	2.29%	0.2	1.3	0.79
CB1 Treated ROM Coal	1.83%	0.0	1.3	0.53

Based on these analyses, the organic sulfur had been reduced by 32.9% by treatment with CB1. These samples were analyzed by EDS-SEM for direct determination of organic sulfur by the method of Straszheim _et al_., (1983). Multiple samples of each coal were analyzed and a scan for each coal (treated and untreated) is presented in Figures 2 and 3. These scans are representative of the average numbers obtained from all analyses made on the coals. The data from these analyses demonstrate a 34% reduction in organic sulfur following treatment with CB1. Thus, the organic sulfur reduction as calculated from the EDS-SEM analysis is in agreement with ASTM analysis results and confirms that CB1 removes organic sulfur (and not pyritic sulfur) from coal.

CONTINUOUS MICROBIAL COAL DESULFURIZATION

In the laboratory studies, the coal was added directly to the medium in which CB1 was grown, for batch treatment. For a commercial microbial desulfurization process, a continuous system with maximum recycle of the salts contained in the growth medium is needed. A continuous microbial coal desulfurization system has been developed at Atlantic Research. The schematic of this process is presented in Figure 4.

In the bench-scale process, the initial step is dechlorination of the tap water prior to addition of the salts and the proprietary carbon source used to make the growth medium for CB1. UV light is used to keep microorganisms from contaminating the "feed". Fresh feed is continuously pumped into the fermenter

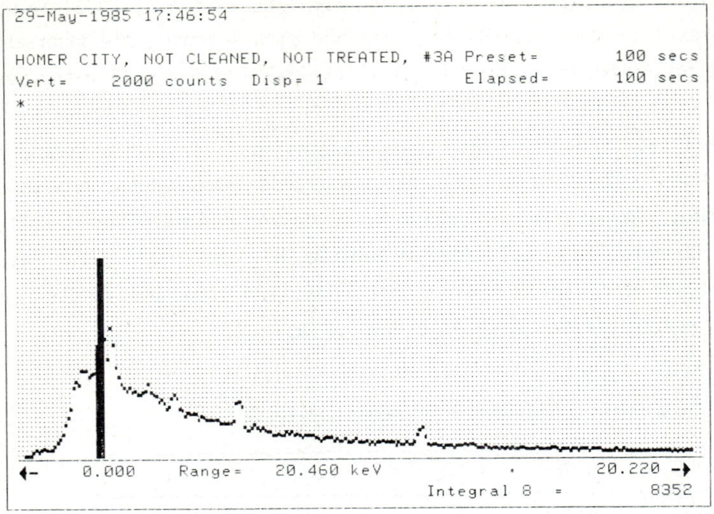

Fig. 2. EDS-SEM Analysis for Organic Sulfur in Untreated ROM Coal

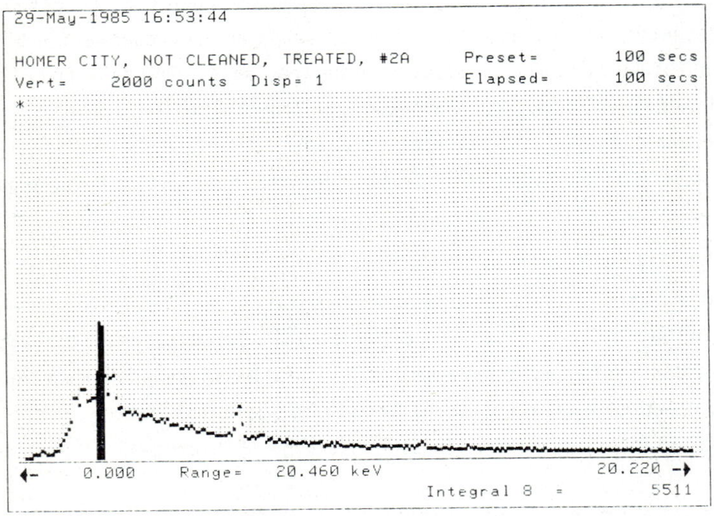

Fig 3. EDS-SEM Analysis for Organic Sulfur in Microbially Treated ROM Coal

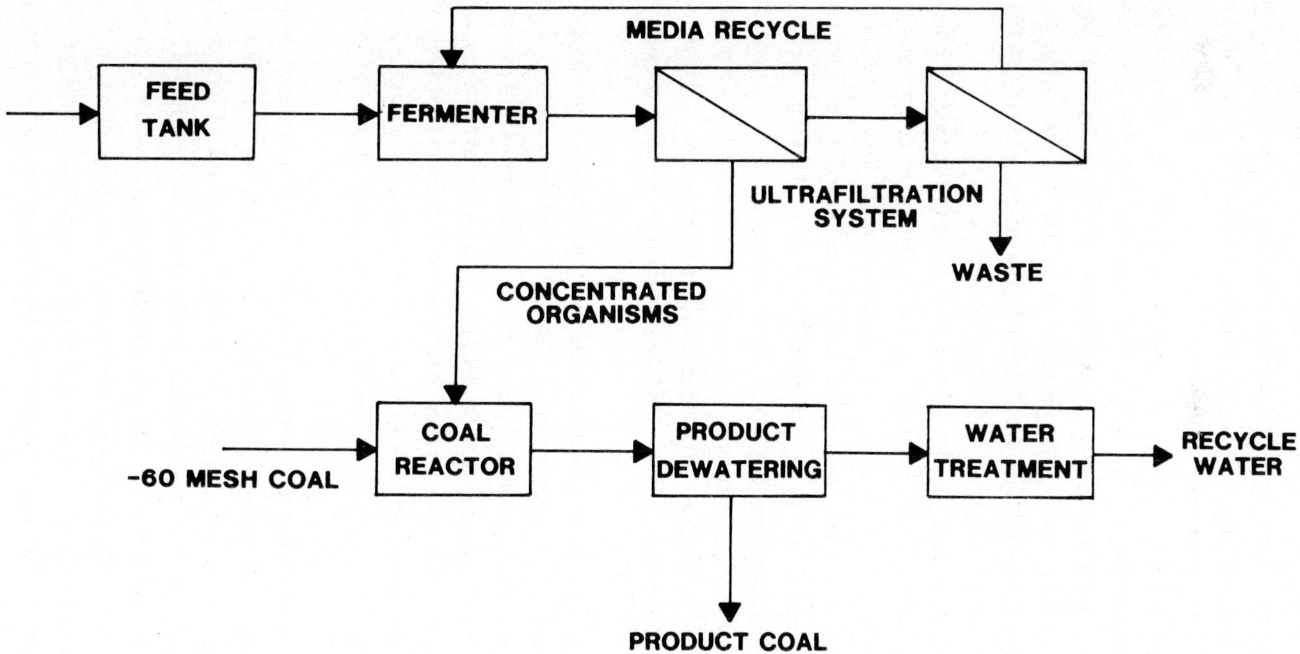

Fig. 4. Schematic of a Continuous Coal Desulfurization Unit

where CB1 is grown and the fermenter contents are maintained near pH 7.0 by a buffering system built into the growth medium. Fermenter temperature is maintained between 25 and 35°C with an average retention time for CB1 at 25°C in the continuous fermenter of 3.5 hours.

CB1 is then concentrated into a thick broth using an ultrafiltration system. The concentrated organisms are continuously metered into the coal reactor and the permeate containing the salts and buffers are recycled to the fermenter. Although tests have shown that greater than 90% of the salts medium can be recycled to the fermenter and provide good growth and activity by CB1, the bench-scale unit is only capable of 50% recycle.

In the coal feed tank, pulverized coal (-60 mesh) is added to water to form a slurry with a specified solids loading. This slurry is continuously pumped into the coal reactor where the coal is contacted with CB1. The coal reactor is continuously stirred and aerated, and the temperature is maintained between 25 and 35°C to provide optimal conditions for desulfurization.

The slurry leaving the coal reactor is dewatered on a vacuum filter and the soluble sulfate produced during desulfurization is washed from the coal. The water filtrate is treated with lime to precipitate the sulfate and any heavy metals leached from the coal. This treated water is then recycled to the coal feed tank.

Three coals have been treated in the continuous bench-scale unit. The data from these studies are presented in Table 4. Desulfurization achieved in the continuous unit was comparable to or better than desulfurization achieved in laboratory batch experiments. No decline in process efficiency was noted when the average retention time in the coal reactor was decreased from 48 hours to 24 hours, and then decreased from 24 hours to 12 hours. In addition, an increase in solids loading rates from 10 to 26% had no significant effects on process efficiency.

Data and operational experience from the continuous bench-scale unit were used to design a continuous pilot plant capable of treating 1000 to 2500 lb of coal per day. The process schematic of the pilot plant is basically the same as the schematic for the bench-scale unit shown in Figure 4. The primary differences in the two units are the ultrafiltration system, the heating system and the controls. A two-stage ultrafiltration system is used in the pilot plant to increase the flux rate and achieve greater than 90% recycle of the permeate to the fermenters. Computer control of the pilot plant acts on input signals to maintain proper process operations and limits the operator time required. Because this plant is located outdoors, heat to maintain the fermenters and coal reactor at the proper temperature is provided by recirculating hot water.

The pilot unit is shown in Figures 5 and 6. Initial shakedown of this unit was conducted in November and December, 1984.

TABLE 4

Bench Scale Pilot Plant Coal Desulfurization

Operation	Coal	Retention Time Hours	% Solids	% Total Sulfur Before	% Total Sulfur After	% Organic Sulfur Reduction
5 weeks	Illinois #6	48	10	3.5	2.9	24%
		24	10	3.5	2.8	29%
2 weeks	Consolidation	24	10	3.8	3.1	25%
		12	10	3.8	3.1	25%
1 week	Illinois #6	12	16	3.65	2.95	19%
1 week	Homer City Underflow	12	26	1.4	1.0 (PS .7%) (OS .3%)	57%

	% Total Sulfur	% Pyrite	% Sulfate	% Organic Sulfur
Illinois #6	3.5	1.3	.1	2.1
Consolidation	3.8	1.6	.2	2.0
Illinois #6	3.65	1.2	.3	2.15
Homer City Underflow	1.4 (Input)	0.7	-	0.7

Fig. 5. Photograph of Pilot Plant Fermenter System

Fig. 6. Photograph of Coal Desulfurization Unit

ECONOMICS

Initial cost estimates of $117 for medium to produce the microorganisms to treat a ton of coal indicated that significant cost reductions would be required if the process was to ever become commercially viable. A number of ways to reduce the costs were investigated. Optimization and modification of the medium resulted in a formulation of substantially lower cost. Recycle of the salts portion of the medium also seemed possible and was investigated. Laboratory tests showed that recycle of 90% of the medium was feasible with a concurrent medium cost reduction to $50/ton of coal.

Further laboratory studies focused on determining the minimum microorganism dose required to treat a ton of coal at a specific organic sulfur content. Data from these studies showed that a six-fold reduction in the microorganism dose did not adversely affect organic sulfur removal. As a result of the decreased microorganism dose, the medium cost was reduced to $8/ton. Additional studies investigated the use of a less expensive carbon source. Continuous growth with the alternate carbon source resulted in good microorganism production and activity, and further reduced the medium cost to $2.34/ton.

Following reduction of the microorganism medium cost to a more reasonable figure, a complete coal cleaning process to treat 100 TPH of fines from a coal cleaning plant was designed and costed. This plant uses a column froth flotation process to remove the ash and pyrite from the fines. The froth flotation portion of the plant was designed to operate 16 hours per day 300 days per year. Once ash and pyrite have been removed from the fines, the organic sulfur is removed from the coal by the microbes. The organic sulfur removal portion of the plant operates 24 hours per day 300 days per year. This plant would thus produce 323,376 ton per year of clean coal from a refuse which would normally be landfilled.

A general schematic of this plant is shown in Figure 7 along with the operating costs per ton (including capital amortization) for each unit process. Two options are presented in Figure 7 for the final product. One option is to dry the coal to approximately 10% moisture and sell it as a dry pulverized coal. This option costs $25.50 per ton of product coal. The second option is to dewater the fines to 30% moisture and formulate the coal into a coal water fuel. Due to the savings in the drying costs, this second option has a cost of $21.09 per ton of product coal associated with it. As shown in Figure 7, the organic sulfur removal accounts for 45 to 54% of the process costs (depending on the desired final product). ARC is currently investigating ways to reduce the microbial desulfurization process costs even further.

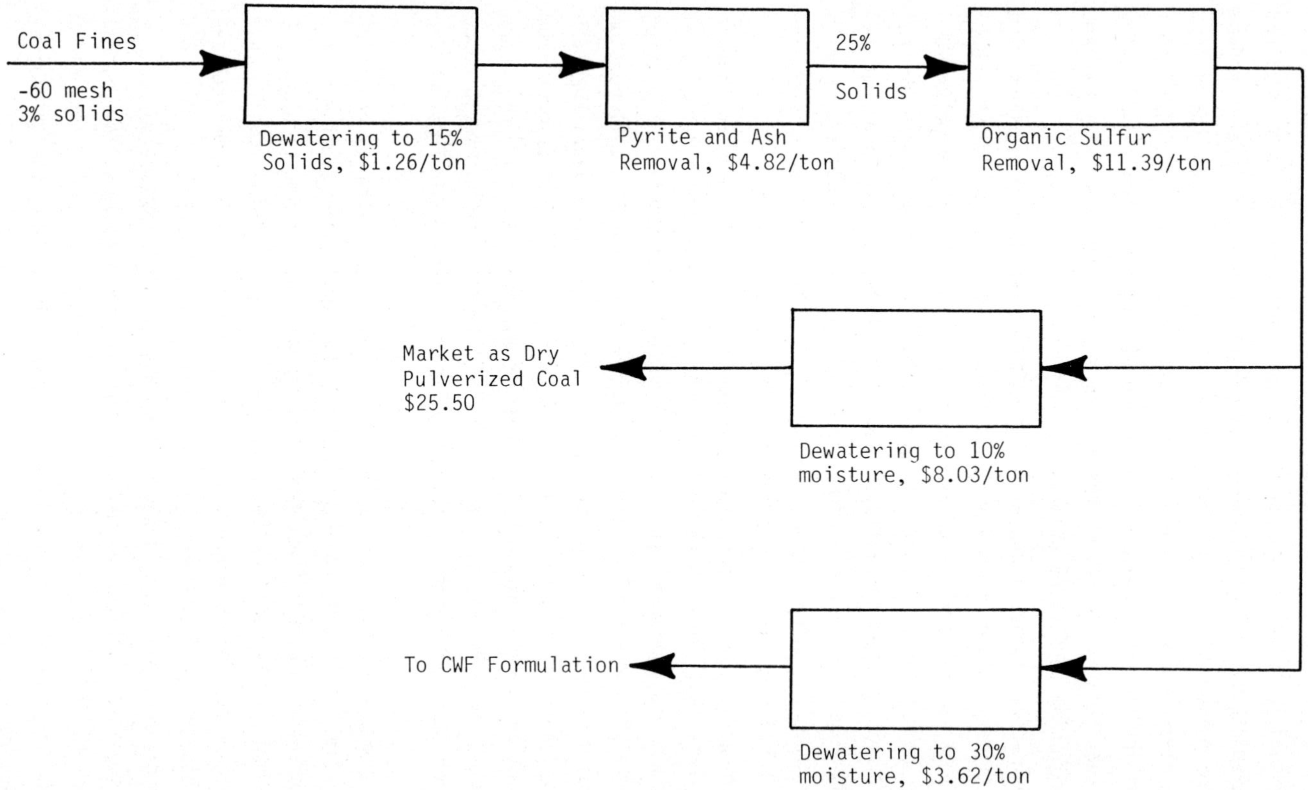

Fig. 7 . Process Costs for a Plant to Treat 100 TPH of Coal Fines to Produce 323,376 tons of Clean Coal per Year

FUTURE PLANS

Full-scale operation of the 2500 lb/day microbial desulfurization pilot plant is awaiting allocation of funds. Optimized growth conditions, reduced organism to coal treatment ratio, reduced retention time and increased solids loading in the coal reactor will be tested in the pilot plant.

Current efforts are focused on completing the optimization of CB1 for efficient and economical treatment of coal to remove thiophenic sulfur. Because of the variety of different organic sulfur forms found in coal, organisms capable of removing other organic sulfur forms are being evaluated using model sulfur compounds.

Reduction of organic sulfur alone in coal is not sufficient to produce a clean coal. Therefore, investigations aimed at combining a physical process for removal of ash and pyrite and microbial desulfurization have been undertaken. Preliminary data indicate that the two processes are compatible and result in a coal significantly lower in total sulfur and ash.

REFERENCES

1 Straszheim, W.E., R.T. Greer and R. Markuszewski (1983). Direct determination of organic sulphur in raw and chemically desulphurized coals. FUEL 62:1070-1075.

2 Hartdegen, F.J., J.M. Coburn, and R.L. Roberts (1984). Microbial desulfurization of petroleum. CEP, pp. 63-67, May, 1984.

COAL DESULFURIZATION BY MICROBIAL PROCESSING

JULIE MURPHY, EDWARD RIESTENBERG, ROBERT MOHLER, DEBORAH MAREK, BRUCE BECK, AND DUANE SKIDMORE

Chemical Engineering Dept., The Ohio State University, 140 W. 19th Ave., Columbus, Ohio 43210 (U.S.A.)

ABSTRACT

A process for microbial coal desulfurization based on the bacterium Sulfolobus acidocaldarius removes about 75% of total sulfur from finely-divided (approximately 73% minus 200 mesh) Appalachian high-sulfur coal in three to six days. The process might potentially operate on heat recovered from the coal combustion process.

The microbes acquire metabolic energy from oxidation of coal-sulfur forms with oxygen supplied in the air. Carbon dioxide from the air provides carbon for growth and reproduction. Stirring maintains the coal in stable slurry so that gas transfer to the microbes is facilitated.

The microbes are acid-loving and operate at pH near two. The low pH prevents premature precipitation of sulfate minerals. Water cleanup is effected by addition of calcium and aluminum ions to precipitate candle salt ($3CaO \cdot Al_2O_3 \cdot 3CaSO_4 \cdot 12\text{-}32\ H_2O$). Recycle water has a low concentration of sulfate ions.

The process appears to be economically feasible at $10-14/ton and organic sulfur removal in the 58% range is commonly observed.

INTRODUCTION

The concern about acid rain is well-documented in the public press. The relationship between consumption of high sulfur fuels and observed increases in precipitation acidity have motivated legislation and threatened more stringent controls aimed at reduction of sulfur oxide levels in effluent gases.

Several means of reducing sulfur oxide emissions are proposed. One solution is the increased use of low sulfur fuels, which in the case of coal in the eastern United States means importation. In addition, sulfur coals may be "deep cleaned" of sulfur, chemically leached or converted to gases or liquids. Finally, coal may be desulfurized microbiologically.

Prior work involved the microbes *Thiobacillus ferroxidans* (ref. 1), and *Sulfolobus acidocaldarius* (ref. 3,4). Each microbe has advantages and disadvantages. *The sulfolobus acidocaldarius* works fast, but at an elevated temperature. This work concentrates on a coal desulfurization process which solubilizes sulfur forms with S. acidocaldarius.

The laboratory work showed that the coal should be finely divided, slurried in a pre-acidified aqueous medium, inoculated with fast-acting, thermophilic bacteria, stirred and aerated. The liquor should be cleaned for recycle. Sulfate ions, metallic cations and H^+ ions should be removed for liquor recycle and the soluble trace elements precipitated. The air off the reactor should be cooled for condensation of acid vapors and scrubbed for mineral sulfate removal. The product coal should be washed clean of acid and sulfate precipitate and, if necessary, dried.

PROPOSED PROCESS FLOWSHEET

Figure 1 shows a proposed flow sheet in which microbes, inoculated into a semibatch reactor, react with sulfur from finely ground coal in an aerated system heated by steam to release sulfur forms as soluble sulfate ions. Associated with the sulfate ions are mineral ions exemplified by calcium (+2) and iron (+2 and +3). The products of the reaction are the desulfurized coal, dried and ready for market; and water vapor which escapes from the reactor or is released from the drying coal. Associated with the process proper is the water cleanup and recycle system which would utilize calcium and aluminum ions in a precipitation reaction to remove excess sulfate ions and trace elements. The reaction would produce a stable mineral sulfate which is readily separated from the recycled liquor and is projected for safe disposal in the environment.

The process has some added features. To avoid transferring the general acid rain problem to the immediate vicinity of the reactor, the off-gas, depleted in O_2 and CO_2 but rich in sulfuric acid vapor and mineral sulfates, passes through a condenser so that acidic vapors are condensed. A scrubber removes mineral sulfates and residual acids which are either added to the reaction liquor or sent to the water cleanup system. The coal is passed through a solids-liquid separator and then through a washer. It appears fairly certain that the process will require that the product coal be rinsed free of sulfate ions. Experiments showed that between 0.1% and 0.2% additional sulfur were removed in the final water wash. The dryer is an option if the coal is to be burned immediately. Otherwise, the same energy is consumed in the furnace as in the dryer which is a separate device. The separate dryer may be incorporated or not as economy dictates.

PROCESS DETAILS

The various steps in the process are now described in greater detail. Coal preparation generally involves the reduction of size, mineral matter, and moisture content. A finely ground coal is required as a feed to the process

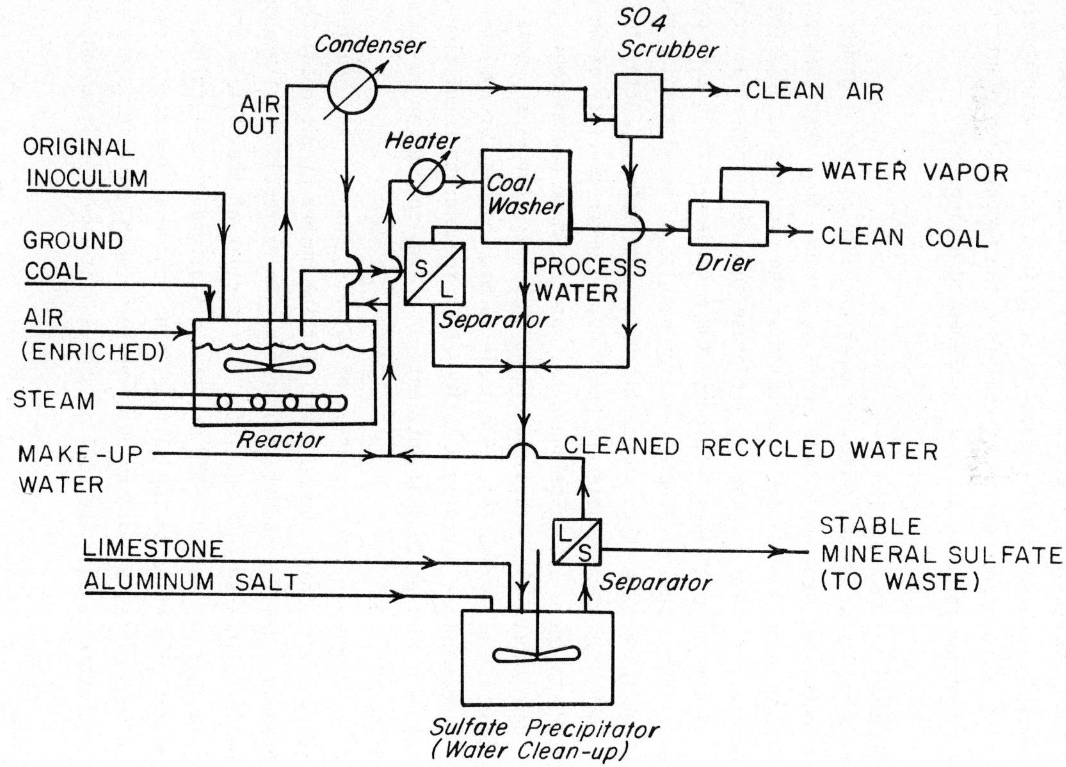

Fig. 1. Schematic of Continuous Process Flowsheet for Microbial Coal Desulfurization.

and some coal preparation is needed. The extent of the mineral matter and sulfur removal by conventional methods is a subject for economic consideration. The question of moisture content of coal from the preparation plant is a question of weight to be shipped, as compared to drying cost.

MICROBIAL GROWTH

Figure 2 shows the phases commonly experienced in microbial growth processes. Microbes inoculated into new growth media go through four describable phases. A lag phase occurs in which the metabolism of the microbes is adjusted to the immediate environment. Next is the phase in which exponential growth occurs. Each microbe divides to produce two microbes which in turn produce two more so that a branched chain reaction results. The plot which characterizes exponential growth shows constantly upturning numbers. After the exponential growth phase is a stationary phase during which the rate at which death occurs to the microbes balances the rate at which new microbes are produced. The microbes die because nutrients or energy sources essential for their production are exhausted, or a toxin is produced and accumulates. Finally, a death phase is reached during which the microbe population drops. Associated is a further loss of nutrients or energy sources or an accumulation of toxin.

The curves found in the production of sulfate ions from high sulfur coal closely parallel in shape curves which plot the number of microbial cells. Figure 3 shows a lag for sulfate production. An extremely rapid exponential growth phase follows so that most of the production of the sulfate occurs in a two to four day period. The plateau or stationary phase follows in which the net production of the sulfate is no longer an important consideration. In some plots, a death or decay phase is seen which is interpretable as a reprecipitation upon the coal of some sulfate previously solubilized.

SULFUR FORMS

This report deals with three different types of sulfur in coal. The percentage of toal sulfur is 4%. The sulfur types are organic, pyritic, and sulfate sulfur. Sulfate is 0.2 - 0.3%, pyrite sulfur is 1.6 - 1.8%, and organic sulfur is the remainder, or 1.9 - 2.2%. The results will demonstrate relative ease of sulfur removal by sulfur type.

Another variable of interest in the process is the distribution of particle sizes in the coal. Coal used in the process has been from Kentucky, Indiana, or Ohio. Normally, the coal was obtained from a power plant in the vicinity of Columbus, Ohio. The coal was specified at 70 - 75% minus 200 mesh. Another economically significant variable is slurry concentration. The runs have

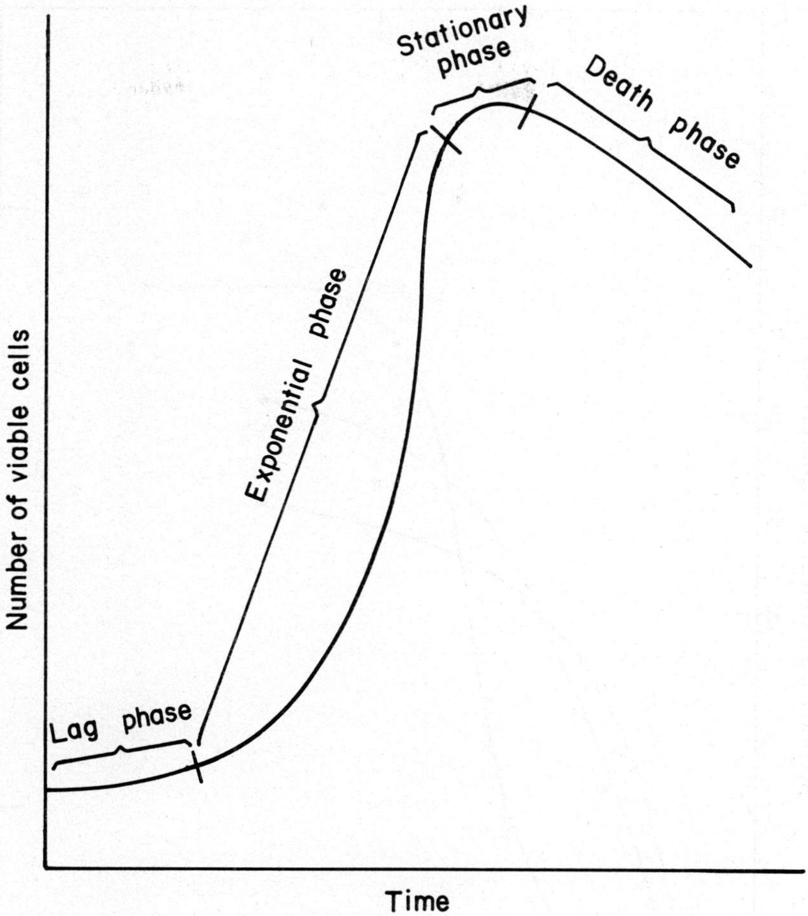

Fig. 2. Generalized Growth Curve of a Bacterial Culture

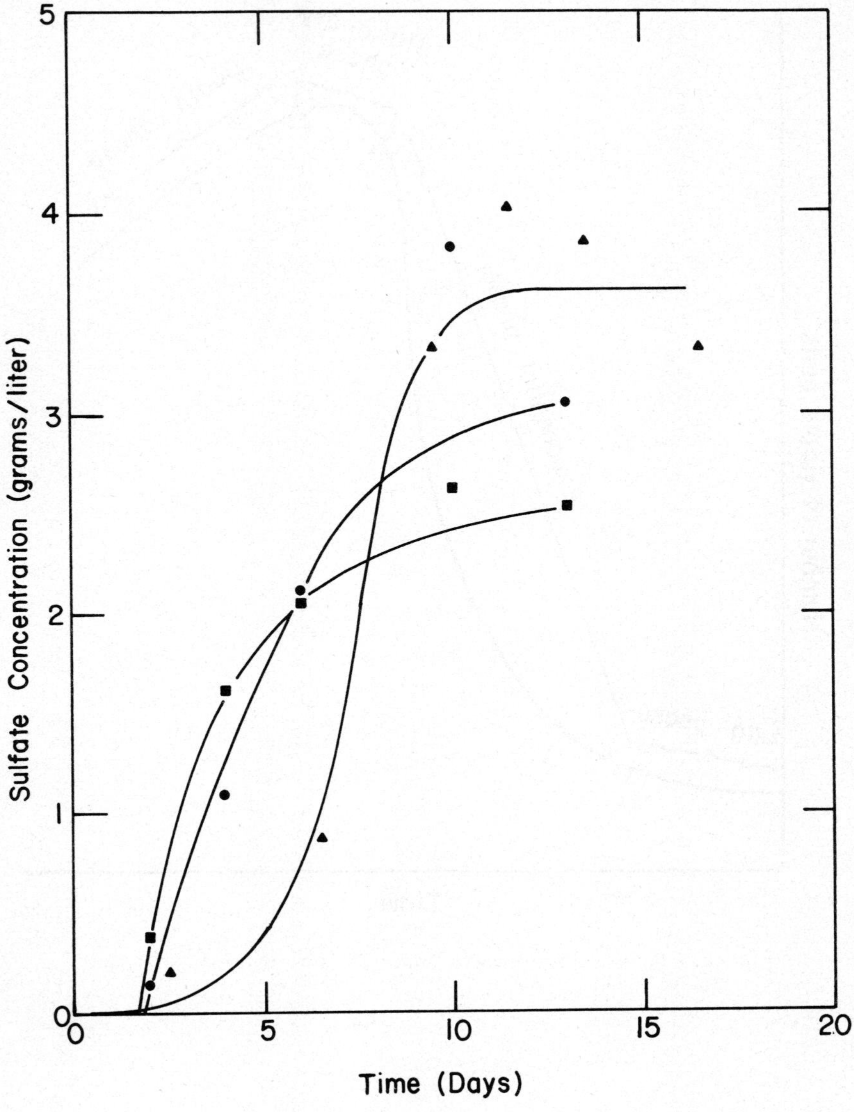

Figure 3. Sulfate concentration profiles for 5% slurries of: (▲) sub 200 mesh, (●) 150 mesh, and (■) 65 mesh Clarion 4-A coal.

included coal in water: 5 weight percent through 40 weight percent. The 5% is typically water-like in its viscosity and 40% is syrupy so that the viscosity change and gas diffusivity effects are readily visible. The viscosity increase seems to be a function not only of the weight percent according to the Einstein prediction, but is also dependent on bonding between the hydrophilic surface of the coal and the water.

INITIAL OPERATING CONDITIONS

The initial pH and inoculum count determine microbial growth lag time. A run without the optimum initial pH 2-3 is 9 to 15 days, but with the optimum initial pH and high inoculum counts, runs are typically in the range of two to five days. Fed-batch processing also reduces lag time. Surfactants may be added to improve the stability of the slurry, to promote the transfer of oxygen to the microbes, and to promote the adhesion of the microbes to the coal. Results to date show that addition of surfactants usually increases but occasionally diminishes the rate at which sulfur was oxidized, depending on surfactant type.

SELECTION OF MICROBE

Selection of the microbe is critical to successful operation of the process. Most of the published work describes results from *Thiobacillus ferroxidans*, a mesophilic microbe which has an optimum temperature (range) from 10 to 37 degrees C. *T. ferrooxidans* was isolated from acid mine drainage. This work, however, deals with *Sulfolobus acidocaldarius* obtained from Prof. James Brierley at the New Mexico Institute of Technology. The attached table compares and contrasts the two bacteria. Each is aerobic, requiring oxygen for its metabolism and also for the oxidation of the sulfur. The concentration of the microbes in the process is difficult to describe since about 90% of the microbes are estimated to adhere to the coal. The main difference in the microbes is the viable temperature range. The mesophile functions between 10 and 37 degrees C where the thermophile thrives between about 55 and 80 degrees C. The pH range is common to both in the order of 1 - 2, but the addition of nutrients in the process often do not increase, and sometimes decrease the rate at which the sulfur is released.

The air rate must be adequate to maintain equilibrium concentrations of carbon dioxide and oxygen. The oxygen concentration, for example, at room temperature and pressure, is about 8 parts per million. Equilibrium concentration at 70 degrees C is about 6 parts per million. An added measure of the high rate of sulfur release by the thermophile is its ability to operate at lower concentration of oxygen, indicating that sulfur is oxidized faster. The gas composition can be varied.

Stirring can occur either mechanically or by air movement. In a commercial process, propellors, turbines or air sparge reactors would be used. However, reactors in this work commonly used an excess of air and the air provided mixing. If the coal particles settle even to a small degree, sulfate production rate suffers. A prime requirement for the process is maintenance of a stable slurry.

The raw materials of the process include oxygen and CO_2. CO_2 is supplied by the air rather than from the carbon in the coal as might be surmised. A question remains whether the nitrogen either as ammonia or as nitrate is available from the coal by leaching or from the air. Pressure sensitivity is negligible up to four atmospheres for the Sulfolobus in that no diminution in rate resulted.

MODE OF OPERATION

Continuous runs have been tried, but not for times long enough to identify effects from accumulated toxins whether from the coal or from the materials of construction. Polypropylene, copper, 304 and 316 stainless steel, and glass were tested. Copper and 316 SS reacted excessively. In the laboratory at one atmosphere, reactor types included 3 necked flasks from 500 ml to 5,000 ml, glass carboys up to 20 liters, and polypropylene lined 55 gallon drums. The reaction seemed to proceed at similar rates according to the same mechanism in all three reactor types. The four atmosphere experiments were conducted with a glass insert in a stainless steel autoclave. Commercially the reactor could be a "swimming pool" mechanically mixed or air-sparged. At 70 degrees C and with a constantly circulating air stream, water evaporation is very high so that water levels have to be maintained to avoid concentration changes. Siphons have proven effective level controllers.

SULFATE REPRECIPITATION

Sulfate reprecipitated on the coal as jarosite or as calcium sulfate. Sulfate ions combined with iron or with calcium leached from the coal. Similarly, potassium, sodium, ammonium and OH^- ions were obtained at a pH from 1 to 2. Dilution provides an easy method for the prevention of accidental reprecipitation of the sulfate sulfur.

Wastewater must be recycled after sulfate and anion removal. A special system was devised to promote the recovery of the sulfate and to improve clarification of the recycled liquor. The system added calcium and aluminum ions to synthesize stoichiometrically a mineral called candle salt, $3CaO.\ Al_2O_3\ .\ 3CaSO_4\ .\ 12\text{-}32\ H_2O$. The sulfate ions were removed to the extent that the recycled water met the environmental standards for municipal drinking water.

FUTURE PLANS

Further work is required to extend the range of operating variables, to define effects of variable interactions, and to extend operations to about 6 months to demonstrate that the microbial population is stable; toxins don't accumulate, and the material is corrosion-resistant. Further work is needed on the prevention of sulfate precipitation on the coal.

KEY RESULTS

The most common question is the extent to which the process removed organic sulfur. We believe that the microbial process is capable of removing 50-60% of this "organic" sulfur, and about 90% of the sulfate and pyritic sulfur. Another question is the approach to so-called compliance goals. Compliance goals for these coals are in the order of .7 to .8% sulfur. Product coals are 1.1 to 1.2%. The compliance goal seems approachable with coal more finely ground: 90% through 200 mesh instead of 75%. Coal product is reduced in sulfur, moderately reduced in mineral matter, and moderately increased in heating value. Yields are above 98% by weight of the sulfur and ash free coal.

The microbes are necessary for the process. First, uninoculated runs produce negligible sulfate. Second, runs which are inoculated, but fed a toxin exemplified by sodium chloride above 2% or formaldehyde do not produce sulfate. The unmodified sytem with inoculum produces desulfurized coal in three to six days. The microbes are derived from the *Sulfolobus acidocaldarius* culture, adapted to grow on high sulfur coal without added nutrient.

Detz and Barvinchak published an estimated cost of $10 per ton in 1979. The $10 updates to $14 per ton. Incorporating all the savings and adding in all the additional features gives about $12 per ton in early 1985. The competition includes flue gas desulfurization from 20 to 40 dollars per ton, and chemical processes from 50 to $150 per ton. Chemical leaching is $45 per ton or more. The implication therefore follows that the process is competitive economically (ref. 2).

REFERENCES

1 P. R. Dugan and W. A. Apel, "Microbial Desulfurization of Coal", Metalurgical Applications of Bacterial Leaching and Related Microbiological Phenomena, (1978), pp. 223-246.

2 C. M. Detz and G. Barvinchak, "Microbial Desulfurization of Coal", Mining Congress Journal, 65 (7), (1979), pp. 75-86.

3 F. Kargi and J. M. Robinson, "Microbial Desulfurization of Coal by Thermophilic Microorganism Sulfolobus Acidocaldarius", Biotechnology and Bioengineering, 24, (1982), pp. 2115-2121.

4 J. Murphy, E. Riestenberg, D. Marek, R. Mohler, B. Beck, and D. Skidmore, "Microbial Desulfurization of Coal by Thermophilic Bacteria", paper presented at National Meeting of the American Institute of Mining, Metallurgical and Petroleum Engineers, New York, New York, February 26, 1985.

TABLE 1

Microbial Types, Variables, and Preferred Ranges of Variables for Microbial Leaching of Sulfur from Various Eastern Coals

Microbial Types	Thiobacillus ferrooxidans	Sulfolobus acidocaldarius
Aerobic	Yes	Yes
Initial Concentration (<)	5×10^8	10^8
Temperature Range	10 - 30° C	55° C - 80° C
pH	1 - 4 (2 - 3.5)	1 - 4 (1.5 - 2)
Nutrient Addition	Ammonia and Inorganic Salts of Phosphorous and Potassium	None
Air Rate	Adequate for Equilibrium	Adequate for Equil.
Air Composition (O_2)	20%	20%
Air Composition (CO_2)	.03%	.03%
Dissolved Oxygen	8 ppm (Equilibrium)	6 ppm (Equil.)
Composition Stirring	Stable Slurry	Stable Slurry
Synthesis Raw Materials	CO_2 and Leached Nitrogen Compounds	CO_2 and Leached Nitrogen Compounds
Inhibitors	Short Chain Carboxylic Acids	Non-Known
Reaction Time	9-16 Days	3-6 Days
Pressure Sensitivity	Unknown	Not Significant
Materials Leachant Sensitivity	Not Significant	Not Significant

REMOVAL OF PYRITE FROM COAL USING BACTERIA

E. BEIER
Advanced Technical College for Mining of the Westphalian Mining Fund, Herner Str. 45, D 4630 Bochum, Federal Republic of Germany

ABSTRACT

Investigations on the weathering of hard coals have proved that pyrite-oxidizing microorganisms play an important part in it. These microorganisms were used for a controlled oxidation of pyrite in hard coal using the submersing and percolating method. The influence of the pH-value and the nutrient matter in the microbe suspension (bio-suspension), of the kind of inoculation, of the carbon dioxide content in the air and in the carbon substance and of substances contained in the coal, of the coal particle size and of the coal content in the slurry were investigated. Furthermore, side effects of the depyritization were dealt with, some of which, e.g. the development of SO_2 occasionally noticed, will have to be investigated in detail. The bio-suspension, containing iron salts and sulphuric acids developed from the pyrite as well as substances dissolved from the coal by the sulphuric acid, can be used or disposed of in various ways. In the Federal Republic of Germany the pollution control regulations have become so strict that the depyritization of coal as a sole method of the reduction of SO_2-emissions of super power plants is not sufficient.

INTRODUCTION

Almost 30 years ago an investigation was started at the Bochum School of Mines with the aim of finding out how far the velocity of oxidation of hard coal at normal temperature depended on the air moisture and on other parameters (1). It was proven that the oxidation is fastest when the air is completely or almost completely saturated with vapour (2). An "interfering reaction" in connection with the determination of peroxides in coals had shown that iron(III)-salts oxidized hard coals (3). Since the iron(II)-salts developed in this reaction are oxidized again in the air the soluble iron salts can be taken as catalysts of the coal oxidation (4). These catalysts, soluble iron salts, are usually not present in coal coming straight from the mine, but pyritiferous iron ores, especially pyrite, are present. Pyrite does not oxidize in dry air, and in the presence of water no oxidation was detectable under sterile conditions within 13 days. Oxidation occurred under non-sterile conditions and if sour, microbe-containing mine water was used (5). In America acid mine water from mines causes problems to the water resources policy. Every year about 10 million tons of sulphuric acid from mines flow into rivers and lakes and thus affect about 10,500 miles of rivers and almost 22,000 acres of lakes (6).

The original purpose of the work, to contribute to the elucidation of the oxidation mechanisms, was achieved with the knowledge that in all coal-mines and also in coal and waste dumps obviously microbes are present which oxidize pyrite to soluble iron salts which again catalyze the oxidation of coal, and this increases the danger of spontaneous ignition.

If certain microbes spontaneously ignite pyrite, also a controlled oxidation must be possible. The controlled oxidation of pyrite with the intention to decrease the sulphur content of hard coal became the new task. The solution of this task is described here.

BIOLOGICAL OXIDATION OF IRON(II)-SULPHATE

Microbes capable of oxidizing pyrite were first described by A.R. Colmar and M.E. Hinkle (7) in 1947. These are Thiobacilli, especially *Thiobacillus ferrooxidans* (T. ferroox.) and *Thiobacillus thiooxidans* (T. thioox.). *T. ferroox.* oxidizes not only pyrite and other pyritiferous minerals, but also soluble iron(II)-salts. Since with available means chemical reactions could more easily be followed in solutions than in solid matter, the first investigations for the best living conditions for the T. ferroox. were carried out with iron(II)-salt solutions.

Test Conduct

The reactor presented in Fig. 1 was used (5,8). It was developed especially for the depyritization of hard coal. The main part of this reactor was a thermostatic cylinder not completely filled with solution with a pipe centrally arranged in it and open at top and bottom, whose top ended a little above liquid level. If air was pressed into this pipe from below, the liquid was transported by the airlift principle and at the same time saturated with oxygen dependent on the partial oxygen pressure. The nutrient solution, described by J.v. Beck (9), containing 50g/l $FeSO_4 \cdot 7\ H_2O$ was used. With sulphuric acid the pH-values were adjusted to 1.4; 1.6; 1.8; 2.0; 2.2 and 2.5. The mixtures were aerated for 90 hours.

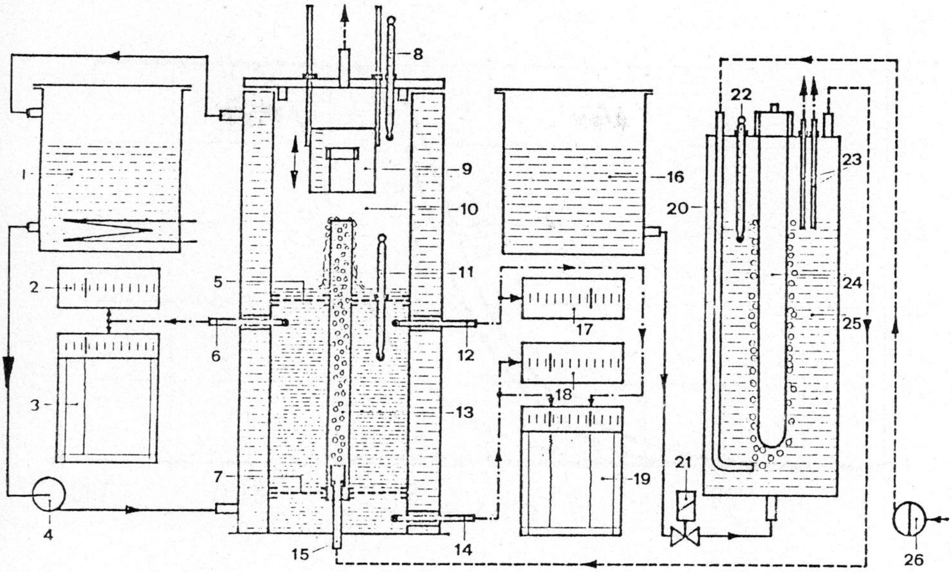

Fig. 1: Apparatus for microbial depyritization of coal

1 thermostat	10 percolator	19 oxygen recording instr.
2 pH measuring instrument	11 thermometer	20 inlet pipe
3 pH recording instrument	12 oxygen electrode	21 magnet valve
4 water pump (thermostat)	13 conveyor pipe	22 thermometer
5 screen	14 oxygen electrode	23 niveau electrode
6 pH electrode	15 air pipe	24 thermomastat —·—· signal
7 screen plate	16 store tank	25 evaporator —— water
8 thermometer	17 oxygen measuring instr.	26 diaphragm pump - - - - air
9 measuring glass	18 oxygen measuring instr.	

Results:

Fig. 2 shows that iron(II) oxidized fastest at pH 2.5, that the velocity of oxidation decreased slightly from pH 2.5 to 1.8, yet strongly from 1.8 to 1.6 and moved toward zero at 1.4.

It may be pointed out here already that the pH-value 2.5, at which the oxidation was fastest, is too high for the percolating method described below, because then precipitation occurs which plugs the pores in the coal particles.

In the same way as light is the source of energy for plants, the energy released at the oxidation of iron(II)-compounds is possibly the sole source of energy for T. ferroox. In order to find out whether it can live some hours without this source of energy, T. ferroox. was inoculated into a nutrient solution containing iron(II)-sulphate with the pH-value 1.6 and it was aerated. The oxygen content of the nutrient solution decreased at increasing

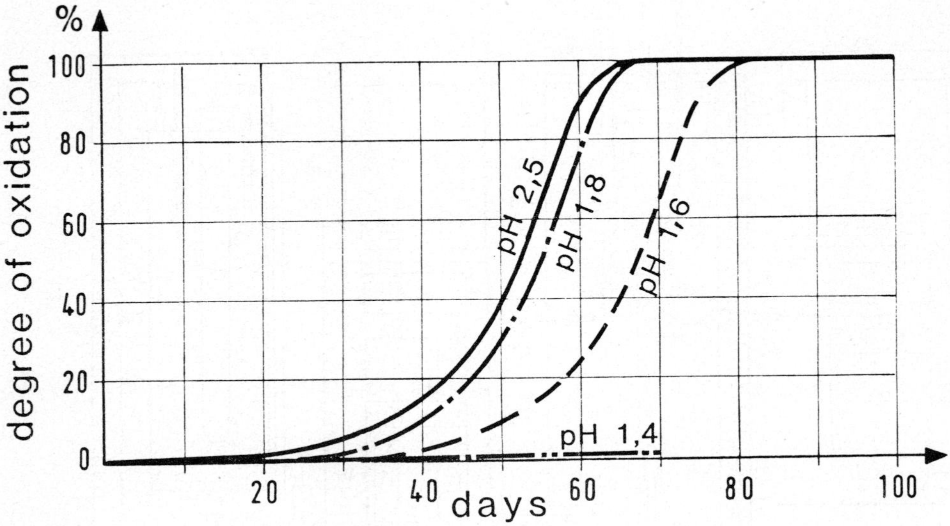

Fig. 2: Oxidation of iron(II)-sulphate on different pH-values

velocity and then again increased to the saturation value; this means all iron(II)-salt had oxidized and consequently T.ferroox. did not find any nutrition. Only 14 hours later iron(II)-sulphate was added to the bio-suspension. The oxygen content of the aerated bio-suspension decreased at the lower measuring point of the reactor within only a few minutes from the saturation concentration of 8 mg/l O_2 to below 1 mg/l, which means the bacteria were still active.

BIOLOGICAL OXIDATION OF PURE PYRITES

Test Conduct

Pyrite below 0,06 mm or below 0,04 mm particle size was moved by means of agitators or mechanical shakers by adding air at original pH-values 2 in different media.

Results:

After 13 days, oxidation in the media was as follows:

- without T.ferroox. and without Fe(III): no oxidation
- without T.ferroox., but with Fe(III): slight oxidation which stopped after reaching a certain Fe(III)/Fe(II) ratio
- with T.ferroox., but without Fe(III): slowly starting, but

continuing oxidation
- with T.ferroox. and with Fe(III): fairly quickly starting continuing oxidation

From these results, it can be concluded that at low pH-values oxidation by oxygen from the air occurs only in the presence of bacteria.

BIOLOGICAL OXIDATION OF PYRITE IN HARD COALS

Test Conduct

The tests were carried out in the following three ways:

1. Coals were mixed with bio-suspension in the mass ratio 2:1, put into a desiccator and left to themselves. Only a slight oxidation of pyrite occurred. Possibly this was due to the fact that the pH-value of the bio-suspension increased due to basic mineral matter in the coal which led to precipitation of iron(III)-hydroxide plugging the pores in the coal. These tests will not be dealt with in detail.

2. Whereas in the first test arrangement coal and bio-suspension remained unstirred, in the second test arrangement the bio-suspension was moved through the coal in various devices at 28 °C, and that:
- in the conventional (small) loop reactor with external airpipe for 60 g of coal
- in a newly developed (large) loop reactor (Fig. 1) with internally mounted airpipe for coal amounts up to 1 kg
- with coal heaps under a tent for coal amounts up to 100 kg

3. In these tests coals and bio-suspension were moved, and that
- in Erlenmeyer (conical) flasks in a mechanical shaker
- in round flasks with agitator
- in two different devices in which the coal was whirled up in the bio-suspension by air
- in an aerated pipe loop in which bio-suspension and coal were circulated by means of a centrifugal pump.

Influences

The following shows the effect of the physical and chemical influential values on the depyritization and the occurrences in connection with the depyritization. The description of the test conduct shows that different test arrangements were chosen. In the following text only results from the same test arrangement are compared unless no other statement is made.

Influence of the pH-Value. Tests with iron(II)-salts in the reactor as shown in Fig. 1 were described above. In the same reactor tests with 100 g hard coal each 0 0.2 mm with 2.42 % pyrite sulphur were carried out at five different pH-values which were kept constant. Fig. 3 shows the percent pyrite oxidation and compares with it the percent iron(II)-salt oxidation in dependence of the pH-value, and that after the same reaction time of two days. Whereas the rate of oxidation of the iron(II)-salt solution increases with increasing pH-value, with pyrite in the coal it increases only to a pH-value of 1.8 and then decreases again.

Since it appears that the decrease could be caused by precipitation of iron(III)-compound, nine iron(III)-sulphate solutions

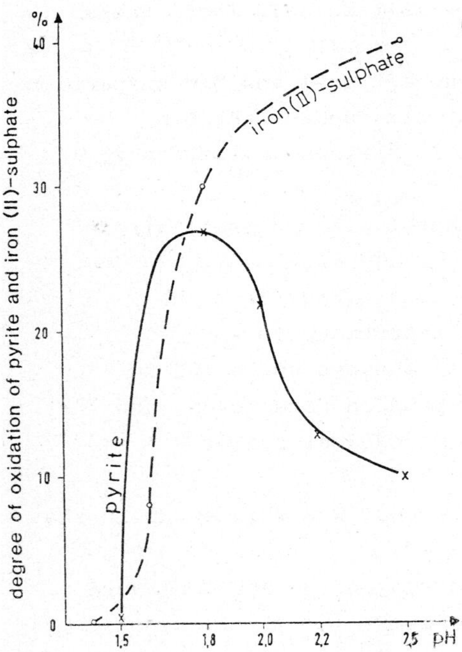

Fig. 3: Oxidation of pyrite in in coal and iron(II)-sulphate

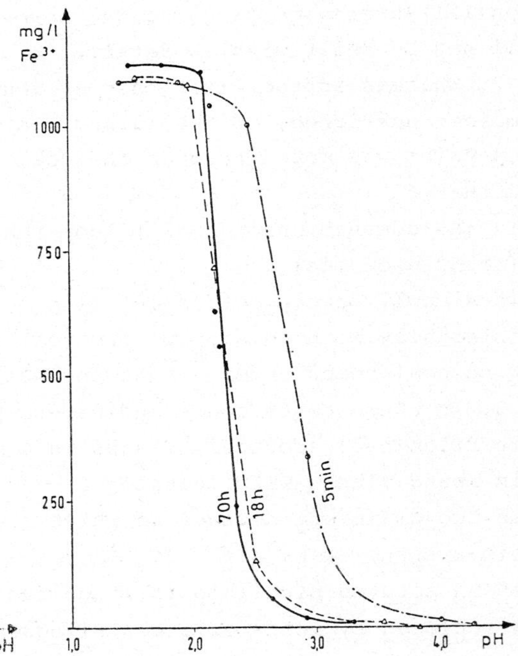

Fig. 4: Concentration of iron (III)-sulphate solution in dependence of the pH-value

with about 1,000 mg/l iron(III)-ions and pH-values between 1.5 and 4.5 each were mixed. The pure iron(III)-sulphate solution had the pH-value 2.3. The other pH-values had been adjusted by adding sulfuric acid or ammonia water. After adjusting the pH-value the solution was left to itself for five minutes, 18 hours or 70 hours at 22.5 °C and then filtered. In the filtrate the iron content was determined. Fig. 4 shows the iron salts are complete in solution, if the pH-value is 0 2,0. If the pH-value is a little higher the precipitation begins. If sufficient time is available for reaction, the precipitation is complete at pH-values of about 4.

Influence of the Nutrient Matter. The used coals obviously contained nutrient salts vital for the bacteria, since they could be depyritized without addition of nutrient matter to the bio-suspension. The velocity of decomposition, however, slightly increased when nutrient salts were added. Thus, a coal sample with 0,72% pyrite sulphur was depyritized in the 9-K-medium with high food value according to Silvermann and Lundgren (10) by 76 % in 14 days and in the medium with a low food value according to Leathen (11) by only 71%.

Influence of the Kind of Inoculation. The influence of the inoculated volume on the velocity of the oxidation of iron(II) was presented by M. Golomb and E. Beier (8). For the depyritization of coals the following bio-suspensions from cultures which had just reached the steady-state phase were used:

a) "Undiluted culture"
b) 1/10 culture + 9/10 nutrient solution
c) After "coarse filtration" (in Blaubandilter) bacterial material was filtered off in membrane filters (0,45 µm) and put into nutrient solution with a glass rod.

The coal used was bituminous coal with an extremely high pyrite content (8.5 % disulphide sulphur) from Plaßhofsbank Seam. The removal of pyrite was followed by measuring the mass of soluble iron salts (Fig. 5, curves) and by determination of the pyrite content of the coal after 10 and 20 days (Fig. 5, squares). With the undiluted culture depyritization starts much faster than with the diluted one, but after 20 days the results have become almost similar. The kind of inoculation in operational practice depends on how much time is available.

Influence of CO_2-Addition to Air. Laboratory air blown into the reactor always contained at least 0,05 % CO_2 by volume. By adding

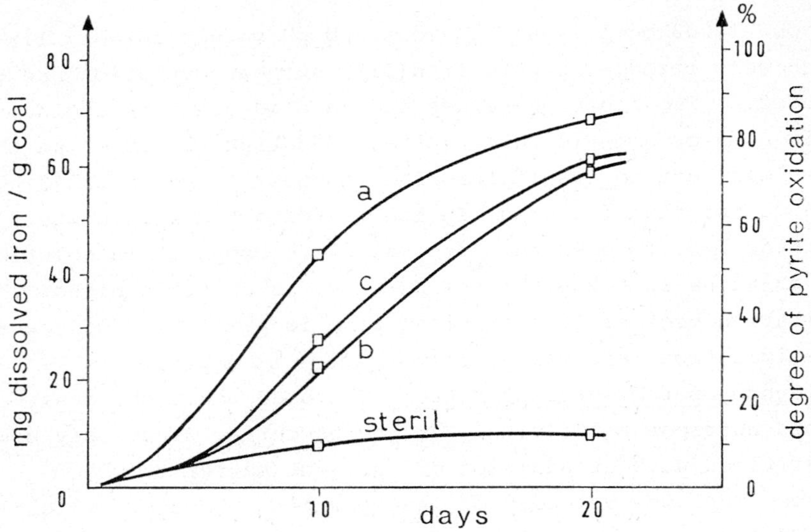

Fig. 5: Formation of soluble iron(III)-salts and oxidation of pyrite on coal

more CO_2 the velocity of depyritization could not be increased. Probably the bacteria cannot assimilate more CO_2 than offered through air and coal. (In all tests arrangements not only pyrite also coal is oxidized.)

Influence of the Coal Substance. When the coal in its natural state or after chemical processing contained no neutralizing matter, its addition to the bio-suspension did not decrease the activity of the bacteria.

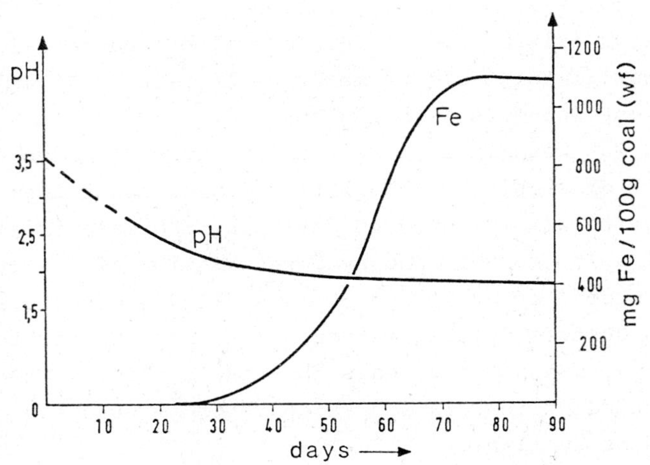

Fig. 6: Development of pH-value and formation of soluble ironsalt

Influence of Mineral Matter in Coal. Basic mineral matter led to an increase of the pH-value and thus, in agreement with. Fig. 3, at pH-values > 2 to a decrease of the velocity of pyrite oxidation. The resulting sulphuric acid, though developing slowly, usually led at last to a pH-value favourable to the depyritization. Fig. 6 shows for coal < 0.2 mm from Hagen Seam that the pH-value had decreased only after a four-week percolation that soluble iron salts developed. The coal contained 1.21 % pyrite sulphur when percolation began and 0.16 % when it ended.

Influence of Pre-treatment of Coals with Hydrochloric Acid. By pre-treatment with hydrochloric acid the velocity of depyritization could be increased with some coals at the beginning of the leaching process. But this process causes additional environmental load so that its operational application for a microbiological depyritization is not taken into consideration.

Influence of Pre-treatment of the Coal with Surface-active Substances. Already in 1964 Duncan et al. (12) reported that they could stimulate the liberation of copper from copper pyrite using the surface-active substance "Tween 20". Similar tests carried out with coal were not successful; the depyritization was even complicated by frothing and flotation effects.

It may be pointed out that T. Atkin and F. Pooley (13) wanted to ease a separation of pyrite by flotation, and that not by means of surface-active chemicals but by means of T.ferroox.

Influence of Particle Size of Coal. Bituminous coal with 3.4 m-% pyrite sulphur from Plaßhofsbank I Seam was crushed and divided into the following fractions: 2-1; 1-0.5; 0.5-0.2; 0.2-0.07; 0.07-0.04; < 0.04 mm. 10 g of all fractions were stirred for four weeks with 90 ml of nutrient solution and 10 ml of a culture developed on pyrite. From the pyrite sulphur content of the separate fractions before and after the treatment the depyritization coefficient was calculated.

The arithmetical mean from the upper and lower particle size is taken as the mean particle size. Plotting the depyritization coefficient over the mean particle size results in the thick curve in Fig. 7. This curve, read from left to right, does not reach the depyritization coefficient 100 %. This is possibly due to the fact that the determination of disulphide sulphur (according to the Laboratory Regulations for the Mining Industry LV 20/14/21) covers substances which are not acted upon by the microbes, or that this

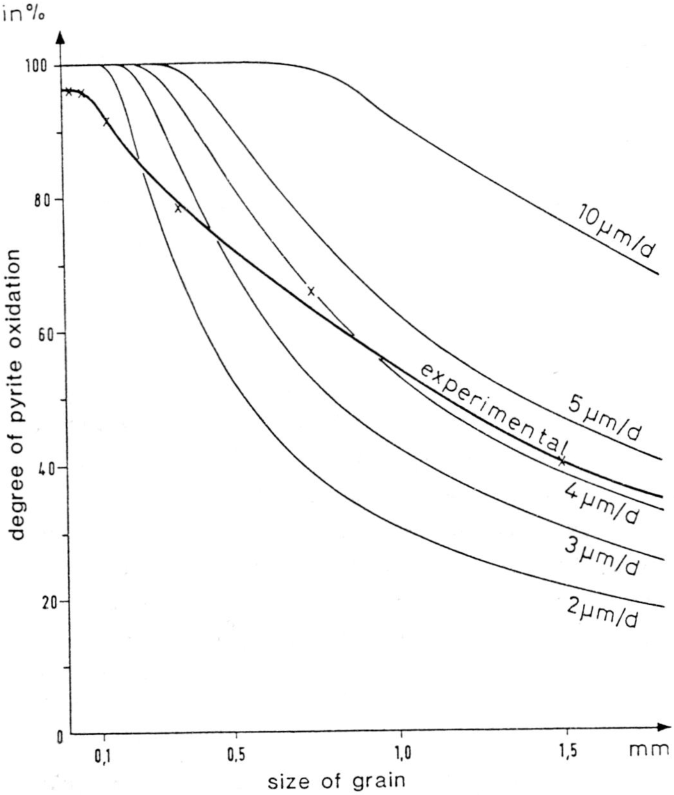

Fig. 7: Dependence of pyrite oxidation on the size of grains - calculated, - experimental reaction time: 28 days

is closely encapsulated pyrite.

In order to facilitate rough estimations the expression "rate of advance" used in mining has been introduced. In the same way as miners win a coal seam facing it, the bacteria act upon the pyrite intercalations. This does not mean that the bacteria swallow the pyrite and excrete the metabolic product. They secrete substances which oxidize the pyrite; the oxidation energy in the secretes leads to the development of matter rich in energy which again is absorbed by the bacteria. The details of the conversion are of no importance for the following considerations. We only have to accept that the pyrite is homogeneously distributed in the coal and accessible for the bacteria and that the bacteria can only attack a coal particle from outside, but cannot enter it through its pores and decompose the pyrite from inside.

This theory is used for the development of an equation on the degree of depyritization

r = radius of grains
t = time of depyritization
v = velocity of advance

$$\eta = \frac{m \text{ pyrite oxidized}}{m \text{ pyrite before oxidation}} = \frac{V \text{ pyrite oxidized}}{V \text{ pyrite before oxidation}} = \frac{r^3 - (r - vt)^3}{r^3} = \frac{3r^2vt - 3rv^2t^2 + v^3t^3}{r^3}$$

With this equation the thin curves in Fig. 7 are calculated. The thick experimental curve coincides with the calculated curve for the rate (velocity) of advance from 4µm/d for grains size over 1 mm. Possibly the experimental curve cuts the calculated curves for velocities < 4 µm/d, because the number of bacteria in the applied bio-suspension is not sufficent for the complete occupation of the surface of the small grains. If the number of bacteria is sufficent, the time of depyritization of coals should be independent of the pyrite content of the coals. This question will be examined in the near future.

Influence of the Coal Content of Slurries. Different amounts of coal < 0.2 mm from Plaßhofsbank I Seam with 3.4 % pyrite sulphur were stirred at an initial pH-value of 1.80 with bio-suspension for four weeks. Fig. 8 shows that the degree of depyritization at coal contents > 20 m-% deteriorates considerably. The reason for this is to be found probably not only in the less favourable distribution of the coal, but also in the more noticeable change of the chemical environment, e.g. greater increase of the pH-value due to mineral matter present in the coal.

In technical operation hydrous suspensions are used containing up to 70 m-% solids. If these suspensions exist for a longer time, as e.g. during transport in pipelines, a depyritization with bio-suspension could occur. In this case oxygen would have to be introduced into the pipeline at various places and that, due to the strongly corrosive medium, different steel would have to be used than at present. Furthermore, the bio-suspension must be separated before further use of the coal, because it contains the original pyrite sulphur of the coal now in the form of sulphuric acid and iron sulphate.

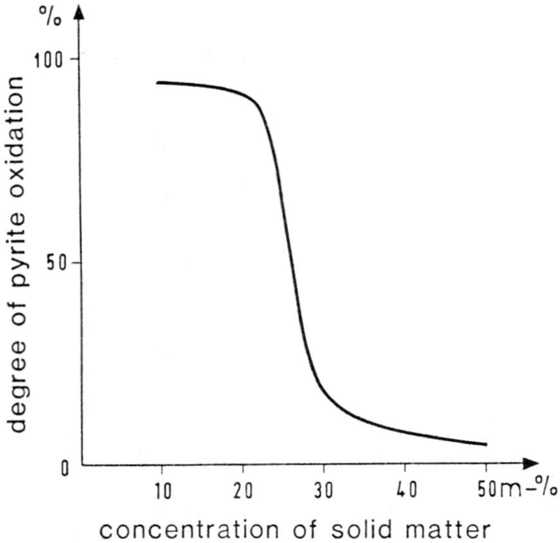

Fig. 8:
Dependence of pyrite oxidation on the concentration of solid matter

Side Effects

Particle-size Reduction (Increase of the Exterior Surface) During Depyritization. Two different test series with a 10% suspension were prepared with three different bituminous coals of the particle size fraction 0.5 to 0.2 mm, one stirred for four weeks in a bio-suspension and the other one sterile only in a nutrient solution. After that the coals are divided in three fractions: 0.5 - 0.2; 0.2 - 0.07 and < 0.07 mm. The pH-values adjusted at the beginning of the test to 1.9 for all test series changed differently depending on the coal admixtures and the presence or absence of bacteria. In the most extreme cases it climbed to 5.5 and then fell to 1.6. Fig. 9 shows that in all test series a particle-size reduction has occured, and that in every case down to the fraction < 0.07. In the test series with T. ferroox the amount of particles of the smaller fractions was on an average about twice as big as in the sterile test series. The real reduction is greater than these figures may show, because if e.g. a cube of 0.5 mm edge length is separated into 8 smaller, equally big cubes, then all eight cubes are still in the particle size fraction of 0.5 - 0.2 mm. The particle size reduction during depyritization can be explained by the falling apart of particles formerly glued together by pyrites.

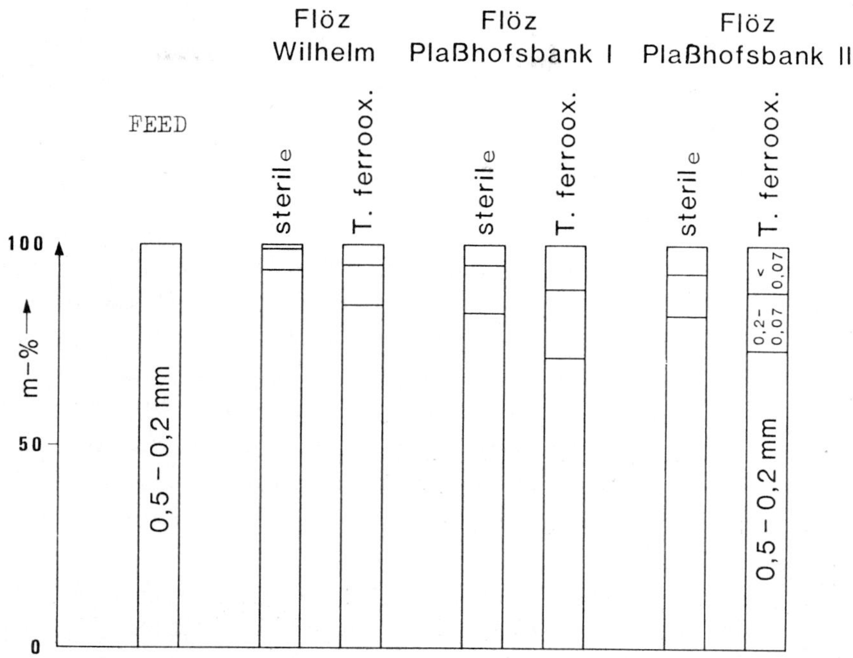

Fig. 9: Size reduction of coal grains by depyritization

A further test, in which 10 g coal from Wilhelm Seam (0.72% pyrite sulphur) each was stirred with 100 ml bio-suspension for 14 days, showed even greater particle-size reduction (Table). With bacteria a degree of depyritization of 85% was achieved here. This was probably only possible because of the considerable particle-size reduction.

Positive and negative aspects of particle-size reduction should be considered before the design of depyritization plants.

Table: Particle-size change by stirring coal suspensions for 14 days.

Size fraction mm	used coal %	stirred in a sterile medium %	stirred with bacteria %
0.5 - 0.2	100	89.0	20.3
0.2 - 0.08	0	8.8	36.9
0.08	0	2.2	42.8

Increase of Interior Surface. The exterior surface increases due to particle-size reduction, and the interior surface due to the development of pores when pyrite is dissolved. By slight decomposition of pyrite from 4.2 to 3.2% disulphide sulphur, the interior surface of one coal had increased by almost 50%. Further investigations could be carried out in order to test the suitability of depyritized coal for the production of activated coal.

Decrease of the Ash Content of Coal. By the withdrawal of pyrite and the dissolution of sulphuric-acid-soluble mineral matter the ash content of few coals has decreased by more than 50 % of the initial value. One has to bear in mind that the dissolved substances may present an additional load on sewage water. On the other hand, these substances obviously offer *T.ferroox* the necessary nutrient matter (like a fertilizer), because the coals were also depyritized when no nutrient matter was added to the bio-suspension.

Precipitation of Sulphurous Substances on the Coal. If at the percolating method the pH-value is > 2.0, basic iron sulphates can be precipitated which completely destroy the success of sulphur decomposition. In connection with this problem see Fig. 3.

Development of Sulphur Dioxide. While leaching a heap of hard coal SO_2 was smelt. After that a further test with a coking coal of the particle size < 10 mm was carried out under a plastic tent. Fig. 10 shows the increase of iron(III)-salt content in the bio-suspension by pyrite oxidation and the SO_2-content of the air, measured by Dräger-tubes (1/a) above the coal heap. The strongest SO_2-development was measured when the velocity of pyrite oxidation had already passed its climax. SO_2 obviously developed when the coal was not kept completely wet. In later percolating tests, carried out in an enclosed apparatus during which the coal was wet all the time, no SO_2 could be proved during the whole process of depyritization. This does not prove that no SO_2 developed, because if SO_2 develops it can be washed out and oxidized by the bio-suspension. Since the literature does not mention SO_2-development by *T.ferroox*, it is likely that it does not develop biologically but purely chemically.

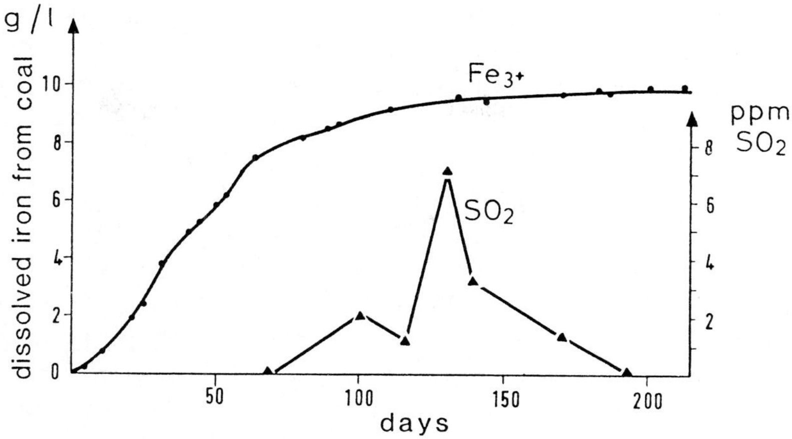

Fig. 10: SO_2-concentration in the atmosphere above coal heap and the generated dissolved iron(III)-salts

FURTHER PROCESSING OF THE BIO-SUSPENSION

Decanted bio-suspension must not be fed into the drainage ditch, but it can be applied in clarification plants for phosphate precipitation and as a flocculating agent. Iron hydroxides separated from the bio-suspension can be added to the coking coal. If the iron hydroxide has been precipitated with lime, special attention has to be paid to the precipitation process so that no calcium sulphate enters the product.

TECHNICAL DEPYRITIZATION PLANTS

Principally it is possible to depyritize coal floating in bio-suspension (submersing method) or by percolating bio-suspension (percolating method). Also, since for the faster submersing method a depyritization time of several hours is necessary for the smallest particle sizes involved, this method is ruled out because of the costs for the necessary huge installation and the energy costs. For the percolating method plants were designed which could work at the same costs as flue gas cleaning plants - provided no complications occur which are not known from laboratory tests. After the issue of the Regulation for Large Scale Furnaces for the Federal Republic of Germany on 22.06.1983 this method has been of little interest for the territory of the Federal Republic, because according to this regulation it is not sufficient any longer to depyritize the coal in order to fulfill the environmental requirements. With this method the organic sulphur remains in the coal, and therefore a flue gas cleaning plant is generally necessary when depyritized coal is burnt in power stations over 50 MW. And if a flue gas cleaning plant has to be installed it is cheaper to burn all the coal sulphur and wash it out in the cleaning plant than to depyritize the coal and wash out the SO_2 developing when the coal containing the organic sulphur is burnt. Furthermore it has to be considered that also the NO_x-content of the flue gases must be reduced and that this can be achieved together with the SO_2-cleaning.

For the use of coal outside and for small scale plants in the Federal Republic of Germany the microbiological desulphurization could be of interest. Therefore in one example the space required and the choice of the size fraction will be dealt with.

In the tests presented in Fig. 7 the "rate of advance" was about 4 µm/d. If we assume a rate of advance of 4 µm/d, this means depyritization times of much more than one year for the complete depyritization of a particle of 10 mm diameter, that is the largest particle size for coking coal. For particles of 1 mm diameter this time is reduced to about four months. For particles of 1 µm diameter we get a time of about three hours, but with such small particles calculations bear a greater risk than with larger fractions, because here coal particles and bacteria (length 1 µm, diameter 0.5 µm) are the same size. And if the bacteria are to attack the coal particle equally from all sides without omitting any areas, as presumed in the calculations, they must settle down on all sides of the particle. If we assume that eight bacteria settle down on the coal particle - like eight adjacent atoms on a central atom at the cubic compactest sphere packing - this means that the mass of bacteria is several times bigger than the mass of the coal. According to this consideration the cost for the mill of coal and the problems caused by agglomeration of finest coal grain need not be dealt with in order to show that the bacteria known at present will not be able to depyritize the coal within a few hours at costs which do not surpass the value of the coal.

Since it is too expensive to keep coal suspended in a bio-suspension for days for depyritization, only the percolating method should be pursued for technical application. Here it has to be considered that piles of grain under about 50 µm are too dense for percolation , which means that only a part of the coal can be of this fraction. What size the biggest fraction may be depends on the space available for the depyritization plant. Let us suppose that a colliery produces 5,000 t coal per day on an average and that 10,000 m² are available for the depyritization plant. Without discussing the technical expedience it may be assumed that the coal should be piled 3 m high and leached biologically. At an assumed settled apparent density of 1 t/m³ (this high value for ease of calculation) 30,000 t could be placed on this pile. If all the output were piled, leached and taken away again, this would mean a dwell time of only 6 days, not including the time needed for transport to and from the pile and for rinsing the leached coal. During this time at a "rate of advance" of 4 μm/d only grain < 0.05 mm could be completely depyritized. This, however, is only a small part of the coal; the main part would not be completely depyritized.

W.P. Bethe (14) reported, in 1980 the coal produced in the Federal Republic of Germany contained 10 m-% grain smaller than 0.5 mm. At this percentage in 60 days 30,000 t coal of this grain are produced. Without consideration of the time for "side works", the coal could be leached for 60 days. In this time the bacteria penetrate into the coal grains from the surface to the center by 60 x 4 μm. From this total way of 250 μm a distance of 240 μm is covered. Therefore a degree of depyritization of more than 99.99 % is calculated, then

$$\eta = \frac{(250-240)^3}{250^3}$$

The assumption and the calculated values are summed up:

Area of coal pile:	10,000	m²
Height of coal pile:	3	m
Volume of coal:	30,000	m³
Settled apparent density:	1	t/m³
Mass of piled coal:	30,000	t
Daily output:	5,000	t
Portion of grain < 0,5 mm:	10	m-%
Rate of advance:	4	μm/d
Time of depyritization:	60	days
Degree of depyritization:	100.0	%

Since in most cases not a complete depyritization is aimed at, one has to consider that at a linear advance of the bacteria seven eighth of a ball is depyritized when the microbes have made half the way to the center of the ball. Furthermore it has to be considered that the calculations have always been made for the biggest particle size of the fractions. If seven eighth of the biggest particles are depyritized, then theoretically particles of half the size will be completely depyritized. Furthermore, the particle-size reduction caused by the depyritization has a positive effect on the velocity of depyritization.

Pure pyrite particles and particles with a high pyrite content should be separated mechanically from the coal if possible before the bacterial desulphurization. Adequate methods have been described by R. von der Gathen (15).

Finally a circulation of bio-suspension (Fig. 11) and a scheme are presented in which has been taken into account that bio-suspension decants and lost water has to be replaced (Fig. 12).

Raw coal is dumped, depyritized, rinsed and taken away as clean coal. The bio-suspension circulates (Fig. 11). It percolates the coal, is aerated and percolates the coal again. Since it builds up iron salts, sulphuric acid and substances separated from the coal, part of it is continually decanted. Then decanted and evaporated water is replaced, e.g. by mine water.

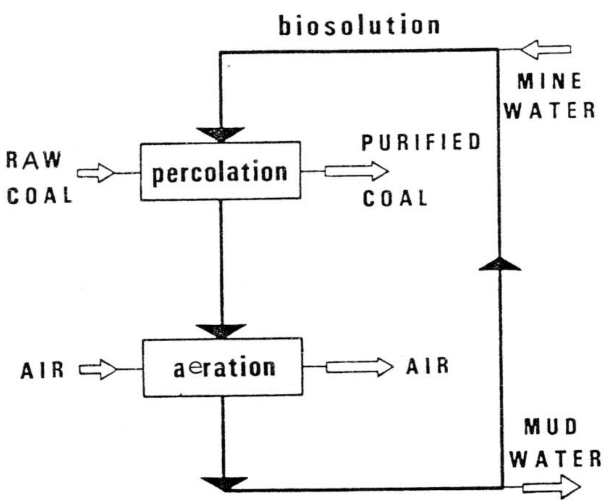

Fig. 11: Depyritization of coal during percolation

Fig. 12 shows that mine water containing barium and strontium salts is used. Barium and strontium are precipitated as sulphates and separated. The iron salts contained in the bio-suspension can be precipitated in different ways, here it is presumed, with soda slag from steel works, difficult to be removed. The soda slag contains sodium sulphide which also has to be oxidized. The process has to be carried out in a way so that on one hand the solution remains sour because of the activity of T. ferroox., on the other hand no H_2S is emitted. One advantage in contrast to precipitation with lime is the fact that iron hydroxide is not contaminated by calcium sulphate.

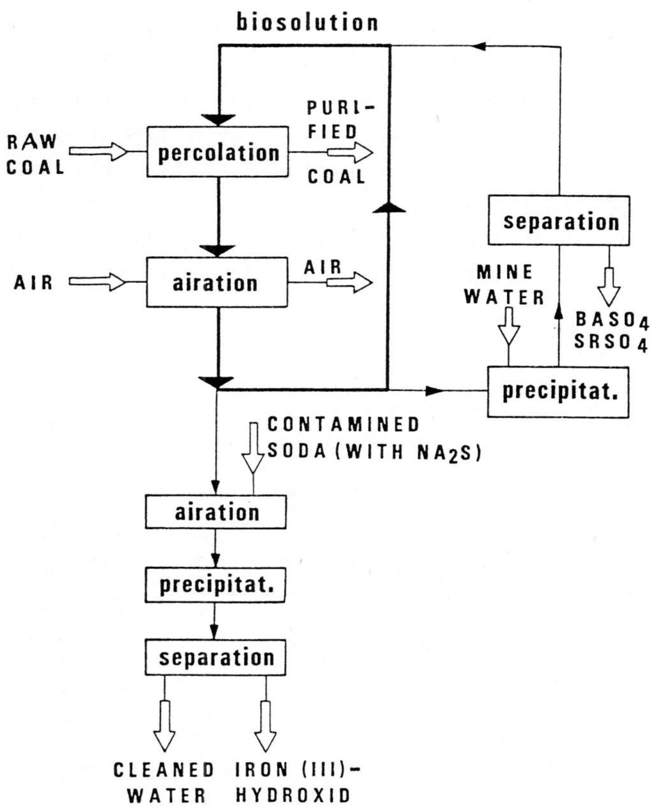

Fig. 12: Depyritization of coal, incl. cleaning of waste water

SUMMARY

The velocity of depyritization is influenced

- by the composition of the bio-suspension, especially by the pH-value and the kind of inoculation,
- by the particle size, interior surface and basic mineral matter of coal as well as by the coal content of the slurry in connection with the submersing method.

The following "side effects" were observed:

- particle disintegration,
- increase of the interior surface,
- decrease of the ash content,
- precipitation of sulphurous matter on the coal,
- production of sulphur dioxide.

Depyritization coefficients of up to 96 % were achieved. Only the percolating method seems to have a chance of technical application. In the case of some coals it seems sensible to separate pyrite mechanically before a bacterial depyritization.

The decanted bio-suspension must be clarified. In the sewage clarification plant it can be used for the precipitation of phosphates and as a flocculating agent.

LITERATURE

1 Beier, E.: Die Oxydation von Steinkohlen an Luft. Mitteilungen der Westfälischen Berggewerkschaftskasse Bochum 1962. Nr. 22, S. 1-105
2 Kröger, C. und Beier, E.; Brennstoff-Chemie 43 (1962), S. 328-335
3 Beier, E.: Bergbau 21 (1970) Nr. 4, S. 81-88
4 Beier, E.: Glückauf-Forschungshefte 34 (1973) Nr. 1, S. 24-32
5 Beier, E.: Glückauf-Forschungshefte 43 (1982) Nr. 4, S. 132-134
6 Lindbergh, C. und Provorse, B.: Coal. Scribe Publishing Corporation, Seattle 1980, S. 104
7 Colmar, A.R. und Hinkle, M.E.: Science 106 (1947), S. 22
8 Golomb, M. und Beier, E.: Bergbau 28 (1977) Nr. 10, S. 419-422
9 Beck, J.V.: J. Bacteriol. 79 (1960) S. 502-509
10 Silvermann, M.P. und Lundgren, D.G.: J. Bacteriol. 88 (1958), S. 642-647
11 Leathen, W.W. et al.: Science 114 (1951), S. 280-282
12 Duncan, D.W. et al.: Appl. Microbiol. 12 (1964) S. 122-126
13 Atkin, T., Pooley, Fr.: New Scientist 29. November 1984
14 Bethe, W.P.: Glückauf 116 (1980) S. 1117-1123
15 von der Gathen, R.: Glückauf 115 (1979) S. 112-118

BIOSURFACE MODIFICATION IN THE SEPARATION OF PYRITE FROM COAL BY FROTH FLOTATION

Y.A. ATTIA[1] AND M.A. ELZEKY[2]

[1]Mining Engineering Division, the Ohio State University, 148 Fontana Labs, 116 W. 19th Ave., Columbus, Ohio 43210 (U.S.A.)

[2]Suez Canal University, Metallurgical Engineering Department, Suez (Egypt)

ABSTRACT

Preliminary investigations on the feasibility of selective bio modification of the pyrite surface chemical behavior as an aid to its depression during the flotation of coal has been conducted using Thiobacillus ferrooxidans bacteria. The effects of variables such as bacterial conditioning time, concentration of bacterial cells, pH, temperature and nutrient solution on the flotation behavior of pyrite and coal suspensions have been studied. The feasibility of the concept was tested and has been confirmed. Preliminary results indicate that about 40% reduction in pyrite flotation was achieved with bacterial pre-conditioning in less than 15 minutes, while coal flotation was unaffected. The results will be discussed and the mechanism of the bacterial action will be reviewed.

INTRODUCTION

It is well known that the high sulfur content of many coal seams adversely impacts their utilization in an environmentally acceptable manner. The direct combustion of high sulfur coals liberates sulfur dioxide gases into the atmosphere, thus contributing to the objectionable environmental pollution known as acid rain (ref. 1). Removal of the sulfur components from high sulfur coal prior to their combustion would greatly alleviate this problem.

Advanced fine coal cleaning processes are needed, therefore, to treat high sulfur coals in an environmentally acceptable and cost effective manner. The need for developing the fine coal cleaning and desulfurization systems arises from the fact that in most coal deposits the "inherent ash" and a significant amount of the pyritic sulfur are intimately associated and very finely disseminated in the coal matrix. For separation of these impurities by physico-chemical means, their liberation from the coal matrix must be achieved. This is normally achieved by fine comminution of the coal matrix.

Standard commercial coal cleaning techniques such as gravity separation or froth flotation are not generally effective for the desulfurization (de-pyritization) of high sulfur coals due to the smallness of particle size necessary for pyrite liberation. In addition, the surface-chemical behaviors (e.g., flotation) of coal and pyrite are quite similar under the conditions employed in the commercial practice. Their separation by flotation is therefore inadequate. Of course, with the use of appropriate surface modifying reagents, selective flotation of the pyrite or the coal might be achieved, provided the pyrite is adequately liberated (ref. 2,4,5,6).

Sulfur removal from coals using biological techniques appear to offer an attractive possibility. The most widely applied biological technique at present is the bioleaching of the pyritic and some of the organic sulfur (ref. 1,3,7,8). The bioleaching techniques using mainly _Thiobacillus ferrooxidans_ and _Sulfolobus acidocaldarius_ are believed to have low capital and operating costs and are less energy-intensive than conventional chemical cleaning techniques.

BIOMODIFICATION OF PYRITE SURFACE BEHAVIORS

Vaaler and Attia (ref. 9) pointed out that if selective modification of the surface chemical and flotation behaviors of sulfur-containing components (chiefly pyrite) by biological conditioning using sulfur oxidizing bacteria such as _Thiobacillus ferrooxidans_, _Thiobacillus thiooxidans_, _Sulfolobus acidocaldarius_, etc., could be achieved, the separation of pyrite from coal by flotation might be possible. The use of biological techniques for selective surface modification of coal minerals appears to be even more promising than the bioleaching technique from the coal production rate (capacity) viewpoint. While bioleaching requires 4-12 days for dissolving the pyrite and other sulfur components, biomodification of surfaces should in principle require only a few minutes. Only oxidation of a few molecular layers on the mineral surface is required to change its surface characteristics, while bioleaching needs to oxidize the whole bulk of the mineral particles. Similarly, adsorption of the biomass on the mineral surface can lead to the desired modifications much more rapidly than bioleaching. It has been demonstrated by a group of Canadian researchers (ref. 7,8) that conditioning for only 15-30 minutes was sufficient to remove most of the pyrite from coal by oil agglomeration.

The objective of this work was to examine the conceptual feasibility of employing selective biological modification of the surface behavior of pyrite to achieve its separation from coal slurries by flotation. The ultimate goal of the research program is to achieve 90% or greater removal of pyritic sulfur and other sulfur containing components by flotation after conditioning the high sulfur coal slurry with the biological solution for only a few minutes.

EXPERIMENTAL METHODS

Materials

A clean coal sample from Pond Creek seam, Pikesville, Kentucky, containing 0.64% total sulfur and 7.2% ash was kindly supplied by Babcock and Wilcox Company, Alliance, Ohio, for this exploratory study. A typical analysis of this coal sample is shown in Table 1. Pure crystalline pyrite samples from Peru were purchased from Ward's Natural Science Establishment, Inc., Rochester, New York, for this study. Suspensions of each of pyrite and coal were obtained by wet grinding the mineral samples in a stainless steel ball mill for different time periods to minus 200 mesh (74 microns) particle size.

TABLE 1
The Pond Creek (Pikesville, Kentucky) Dry Analysis

Component	Percent	Component	Percent
Ash	7.2	Organic Sulfur	0.36
Sulfate	$<$0.01	Total Sulfur	0.64
Pyritic Sulfur	0.32		

Flotation reagents used in this study consisted of fuel oil and MIBC. Samples of these reagents were kindly supplied by Sherex Chemicals, Dublin, Ohio, and were blended together in a 1:2 ratio by volume of MIBC and fuel oil respectively. Sulfuric acid and sodium hydroxide solutions were used to adjust the suspension pH for both the flotation and the bacterial conditioning.

The bacterial culture, strain TFI-35 of *Thiobacillus ferrooxidans* which was originally grown on ferrous sulfate medium, was obtained from Dr. O.H. Tuovinen, Department of Microbiology at The Ohio State University. Tuovinen and Kelly's nutrient medium (ref. 10) was used for growth and adaptation of the microorganisms. This nutrient solution is prepared by mixing four parts of solution A with one part of solution B. Solution A is prepared by dissolving 0.5 grams of each potassium phosphate, magnesium sulfate, and ammonium sulfate in a liter of 0.01 N of sulfuric acid. Solution B is simply prepared by dissolving 167 grams of ferrous sulfate in a liter of 0.05 N, sulfuric acid.

Procedures

Flotation. A Galigher laboratory flotation machine, model LA-500, was used in this study with a cell capacity of 1.5 liters. During flotation, the impeller speed was kept at 1,000 rpm, and the air flow rate at 5 liter/minute. Approximately 1.5 liter slurries of coal and pyrite containing about 5% solids by weight were prepared; their pH values were adjusted and the flotation reagents were added while stirring was continued for one minute in the absence of air. Flotation was then allowed to take place for three minutes. Both the froth and residual slurry (tailings) fractions were collected, dried, and weighed.

Bacterial Culture Preparation and Adaptation. Two sets of stock culture of *Thiobacillus ferrooxidans*, strain TFI-35, were prepared in 250 ml erlenmeyer flasks, each containing 5 g pyrite (minus 200 mesh) substrate, 30 ml of bacterial solution, 5 ml Tuovinen and Kelly nutrient medium, and 65 ml double distilled water. The pH was continually adjusted to 2.0 and the bacteria were left to grow at 30° C in a shaking water bath at 100 strokes/ minute and aerated by air bubbling. One set was adapted for two weeks, and the other was adapted for six weeks to obtain two different microbial adaptations and cell growth (i.e., cell concentrations).

Bacterial Pre-Conditioning of Coal and Pyrite. The desired slurry was inoculated with 200 ml of the prepared bacterial solution and the pH was adjusted to 2.0. The mixture was aerated by air bubbling and vigorous stirring with a mechanical mixer at 30° C for the predetermined periods of time. The slurry was allowed to settle, and the solution over the solid was siphoned off. Tap water was added to the solids and the pH was adjusted to natural suspension pH before flotation (natural pH for pyrite was 7.5 and for coal it was 8.6).

Pre-Conditioning with Nutrient Medium. In order to better identify the bacterial action on flotation, the effect of the nutrient medium (without the bacterial cells present) on the flotation behavior of coal and pyrite was studied. The pre-conditioning and flotation procedures followed were the same as with the bacterial pre-conditioning above. The nutrient medium was prepared as mentioned earlier.

RESULTS AND DISCUSSIONS

Coal Flotation

The flotability of clean pond creek coal was studied using MIBC as a frother and fuel oil as a collector in the form of a blend with 1:2 ratio by volume of each respectively, following Key's procedure (ref. 11). The flotation was carried out at the natural coal suspension pH of 8.6 with flotation reagent consumptions varying from 0.74 to 9.12 lb/ton (0.33 to 4.14 kg/ton). The coal slurry was maintained at about 5% solids (using tap water) during the study of its flotation behavior. The results, shown in Figure 1-a, demonstrate clearly that coal recovery increased from about 28%

to 92% as the dosage of the flotation reagents was increased from 0.74 9.12 lb/ton. From Figure 1-a, it appeared that there was no significant increase in the coal recovery when the dosage of flotation reagents was increased to over 6.48 lb/ton; only about 4.5% more coal was recovered at the expense of 29% higher reagent consumption. Therefore, a 7.88 lb/ton reagent dosage in the flotation process gives nearly the maximum clean coal yield.

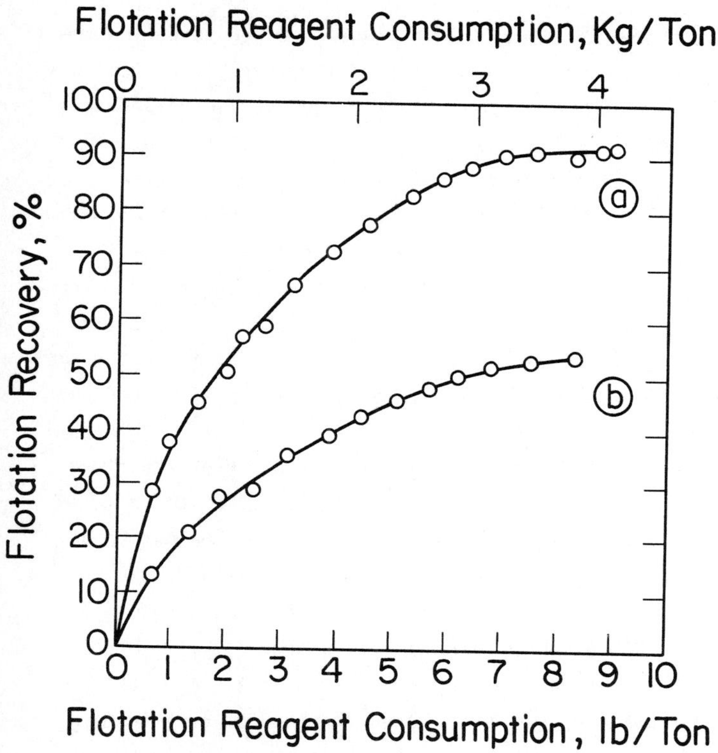

Figure 1. The Flotation of Coal and Pyrite at Natural pH with Different Flotation Reagent Dosages. a) coal at pH 8.6. b) pyrite at pH 7.5.

Pyrite Flotation

The flotability of pure, finely ground pyrite was also studied. Flotation reagent consumption was varied between 0.64 and 8.41 lb/ton (0.29 - 3.82 kg/ton). The pH was maintained at the natural value of 7.5, and tap water was added to maintain the desired level in the flotation cell. The results, shown in Figure 1-b, clearly show that pyrite recovery increased from 12% to 53% as consumption of flotation reagents increased from 0.64 to 8.41 lb/ton. From Figure 1, it is clear that clean coal is more flotable than pure pyrite at a constant flotation reagent consumption. For example, under similar flotation reagent consumptions (4.91 lb/ton or 2.33 kg/ton), coal recovery was about 80%, while pyrite recovery is only about 42%. Based on these results, selective flotation can be used in principle to remove a significant portion of the finely ground and liberated pyrite from coal. In practice, however, the pyrite flotation has to be depressed further to effect a high degree of sulfur removal.

Effect of pH on Coal and Pyrite Flotation

The effect of pH on coal and pyrite flotability was examined to define the optimum conditions for maximum coal recovery and pyrite removal. The coal was floated at two different reagent consumptions while pH was varied between 2 and 12 (Figure 2). The results on Figure 2(a,b) clearly indicate that recovery of the clean coal decreased with an increase in pH. That is with a flotation reagent consumption of 2.01 lb/ton (0.91 kg/ton), the recovery decreased from 87% to 34% as the pH was increased from 2 to 12 (Figure 2-a). Similar results were observed at a flotation reagent consumption of 3.9 lb/ton (1.77 kg/ton), where coal recovery decreased from approxmately 92% to 57% as the pH increased from 2 to 12 (Figure 2-b). Also, Figure 2(a,b) indicates that flotation reagent consumption influences coal recovery, but the effect of pH change still has the same general trend.

Similar experiments were performed to study the effect of pH on the flotability of pure pyrite. Two flotation reagent consumptions (2.03 lb/ton (0.91 kg/ton) and 4.9 lb/ton (2.23 kg/ton)) were used to float the pyrite at a range of 2 to 12. Figure 2-c, which shows pyrite recovery at 4.9 lb/ton (2.23 kg/ton) flotation reagent consumption, indicates that pyrite recovery decreased from about 93% to 35% as the pH was increased from 2 to 12. From Figure 2-d, it is also clear that the change of pH in the same range has the same tendency; i.e., the flotation recovery dropped from 85% at pH 2 to 20% at pH 12. These results clearly indicate that the pyrite recovery is highly affected by the suspension pH. It can be seen that lowering the pH from 7.5 (natural) to 2.0 increases the flotation recovery of pyrite from 45% to 93% at 4.9 lb/ton fuel oil/MIBC blend (Figure 2-c), and from 30% to 86% at 2.03 lb/ton fuel oil/MIBC blend (Figure 2-d). That is more than a 50% increase in the flotation recovery by just dropping the pH from 7.5 to 2.0.

Based on the results illustrated in Figure 2, it can be concluded that the change in pH has a similar effect on both coal and pyrite, but unfortunately has more influence on pyrite, especially at lower pH. These conclusions are demonstrated in Figure 2-b,d, where nearly the same flotation reagent consumptions were used (about 2.0 lb/ton). From these results, it is easy to see that the recoveries around pH 8 are about 56% and 28%, while at pH 2 they are about 88% and 86% (nearly equal) for coal and pyrite

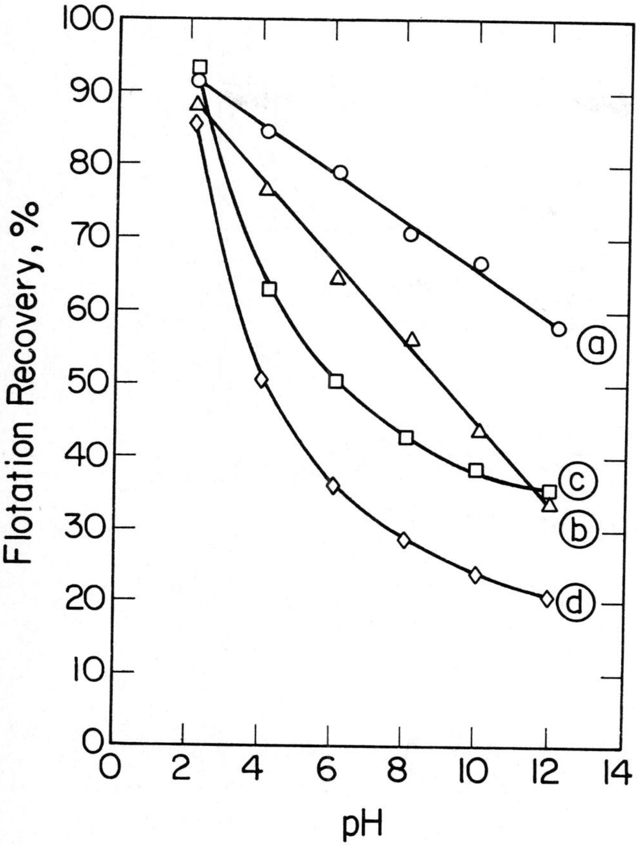

Figure 2. Flotation of Coal and Pyrite with Constant Flotation Reagent Dosage and at different pH. a) coal with 3.9 lb/ton. b) coal with 2.01 lb/ton. c) pyrite with 4.9 lb/ton. d) pyrite with 2.03 lb/ton.

respectively. In general, the obtained experimental flotation results are in good agreement with literature (ref. 6,10-12).

Effect of Bacterial Pre-Conditioning on Coal and Pyrite Flotation

The effects of pre-conditioning with aerated nutrient medium and with the bacterial solution of *Thiobacillus ferrooxidans* on the flotability of clean coal using flotation reagent dosage of 6.14 lb/ton (2.77 kg/ton) at pH 8.6 were studied to see if any depression effect on coal particles was apparent. Figure 3-a clearly shows that neither the nutrient medium (i.e., without the bacterial cells), nor the bacterial conditioning had any depression effect on the flotation recovery of clean coal.

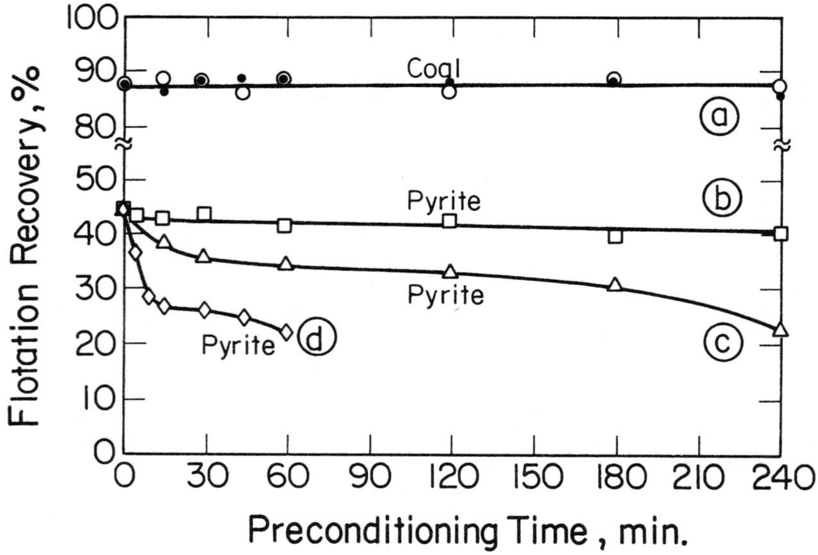

Figure 3. The Flotation of Coal and Pyrite at Natural pH after Pre-Conditioning with Nutrient Medium and T. Ferrooxidans. a) coal with aerated nutrient medium o and with T. Ferrooxidans •. b) pyrite with aerated nutrient medium. c) pyrite with 2-weeks old T. Ferrooxidans. d) pyrite with 6-weeks T. Ferrooxidans.

The effects of pre-conditioning of pyrite with the aerated nutrient medium only (i.e., without the bacterial cells) and with the bacterial solution on the flotability of pyrite was investigated. The results are shown on Figure 3(b,c,d). Figure 3-b shows the effect of pyrite pre-conditioning by aerated nutrient medium only at pH 2 and flotation recovery of pyrite at pH 7.5. The results of the flotation tests indicated only a slight decrease in the recovery of pyrite (about 3% less after 5 minutes of treatment to about 7% after 4 hours of treatment as shown in Figure 3-b). Some evolution of H_2S gas was noticed during the pre-conditioning with aerated nutrient medium as H_2SO_4 added to adjust the slurry to pH 2. This indicated that there was some chemical reaction between the sulfuric acid and the sulfide-sulfur in the pyrite sample.

The effect of bacterial pre-conditioning on pyrite flotation was studied using two different preparations of bacterial solutions over different periods of time (5 to 240 minutes). Figure 3-c shows the change in pyrite flotation recovery after pre-conditioning with bacteria previously grown for two weeks on pyrite substrate. As shown on the figure, the recovery of pure pyrite decreases by about 15% from that of untreated pyrite after just 15 minutes of treatment. After 4 hours of treatment, the pure pyrite decreased from about 44% to 22%; i.e., 50% reduction in pyrite flotation. Bacterial pre-conditioning with cells previously grown for six weeks on pyrite substrate decreased the recovery of pure pyrite by about 18% after only 5 minutes of treatment. Treatment with the 6-weeks old bacteria for 15 minutes decreased the flotation recovery by about 40% (from 44% recovery to 27% recovery) compared with 18% using the 2-weeks old bacteria. Extending treatment to 30 minutes decreased the recovery to 25%, and 60 minutes of conditioning decreased recovery to 22% (i.e., total of 50% reduction from that of untreated pyrite). Thus, the results indicate that pre-conditioning with *Thiobacillus ferrooxidans* has a good depression effect on pyrite flotation, but not on coal flotation. The results also indicate that previous adaptation and growth period of bacteria on pyrite substrate has an important effect on reducing the pre-conditioning time. However, for the same bacterial solution, doubling the pre-conditioning time has little effect on the flotation recovery; this is in agreement with Kempton et al. (ref. 8).

Mechanism of Bacterial Action on Pyrite Flotation

Although the exact mechanism for the bacterial action (i.e., depression) on the flotabilty of pyrite is not known at this time, some possible explanations can be offered. It is well known that *Thiobacillus ferrooxidans* are capable of oxidizing the pyrite and other sulfur components. Surface oxidation would change the adsorption and flotation behavior of pyrite. In this case, however, where fuel oil is used as the flotation collector, this surface oxidation is not likely to be the only operating mechanism.

The bacterial depression effect might also be explained by the fact that *T. ferrooxidans* are capable of producing surface active and polymeric substances (mainly polysaccharides and lipids) which could adsorb onto the surfaces of sulfide mineral substrates (ref. 8,14). Since the pyrite is a substrate, it should be possible to modify its surfaces to be more wettable or hydrophilic by microbial growth. Such growth is likely to cover the pyrite surface with the bacterial cells or with the cell-excreted compounds. The net effect of this bioadsorption will be the depression of pyrite flotation. The bioadsorption process is believed to be rapid enough to be done in a matter of a few minutes (ref. 8,9,14).

CONCLUSIONS

From this work, it can be concluded that: (a) the pre-conditioning of clean coal with bacteria does not affect its flotability, while pure pyrite is affected to a great extent under the same conditions; (b) the concentration and the previous adaptation period of the bacteria on pyrite has a significant role in the reduction of pyrite flotation recovery; (c) pre-conditioning for 15-30 minutes is probably adequate because doubling of treatment time does not significantly affect the pyrite flotation; (d) pH has the same effect on both coal and pure pyrite, but pyrite flotation is more favorable than coal at lower pH. It is recommended to work at higher pH; (e) the aerated nutrient medium has little effect on pyrite depression while bacteria perform the main role in the pyrite rejection from coal flotation; and (f) the suggested technique in this work (i.e., bacterial pre-conditioning followed by froth flotation) seems to be very promising. About 80% pyrite rejection is possible. It has economic and commercial potential because of the short time for bacterial treatment (approximately 15 minutes) and the effectiveness of froth flotation.

ACKNOWLEDGEMENTS

The authors greatfully acknowledge the financial support of Babcock and Wilcox Company, Alliance, Ohio, and the Egyptian Government's (through the Peace Fellowship Program) support of Dr. M. Elzeky's visit at The Ohio State University during this study.

REFERENCES

1 A. Saleem, Desulfurization of Coal, Ontario Hydro. Research Quarterly, Second Quarter, 1971, pp. 29-34.
2 T.D. Wheelock (Ed.), Coal Desulfurization, Chemical and Physical Methods, American Chemical Society, Symposium Series 64, Washington, D.C., 1977, pp. IX-XI, 101-120.
3 F. Kargi, Microbiological Coal Desulfurization, Enzyme Microb. Technol., 4 January (1982), pp. 13-19.
4 K.J. Miller and A.F. Baker, Flotation of Pyrite from Coal, Bureau of Mines, U.S. Dept. of the Interior, Tech. Report-51, 1972.
5 T.D. Wheelock and T.K. Ho, Modification of the Flotability of Coal Pyrites, AIME Annual Meeting, New Orleans, Louisiana, Feb., 1979, pp. 18-22.
6 F.F. Aplan, Use of the Flotation Process for Desulfurization of Coal, American Chem. Soc. Symposium Series No. 64, Coal Desulfurization, T.D. Wheelock (Ed.), 1977, pp. 70-82.
7 C.E. Capes, A.E. Mcilhinny, A.F. Sirianni, and I.E. Puddington, Bacterial Oxidation in Upgrading Pyritic Coals, Can. Inst. Min. Met. Bull., 66 (1973), pp. 88-91.
8 A.G. Kempton, N. Moneib, R.G.L. McCready, and C.E. Capes, Removal of Pyrite from Coal by Conditioning with Thiobacillus Ferrooxidans Followed by Oil Agglomeration, Hydrometallurgy, 5 (1980), pp. 117-125.
9 L.E. Vaaler and Y.A. Attia, Microbial Modification of Mineral Surfaces to Change Flotation Characteristics, Report of Investigation, Battelle, Columbus, Ohio, July 11, 1983.
10 O.H. Tuovinen and D.P. Kelly, Studies on the Growth of Thiobacillus Ferrooxidans, Arch. Mikrobiol., 88 (1973), pp. 285-298.
11 R.O. Keys, Improvements of In Coal Recovery Through Coarse Coal Flotation, Proceedings of the First Annual Coal Preparation Conference and Exhibition, March 27-29 (1984), pp. 11-29, Lexington, Kentucky.
12 H.F. Yancey and J.A. Taylor, Froth Flotation of Coal; Sulfur and Ash Reduction, U.S. Bureau of Mines Rep. Invest., 3262, 1935.
13 F.G. Miller, Reduction of Sulfur in -28 Mesh Bituminous Coal, Trans. AIME, 228 (1964), 338.
14 L.E. Murr and Y.K. Berry, Direct Observation of Selective Attachment of Bacteria to Low Grade Sulfide Ore and Other Mineral Surfaces, Hydrometallurgy, 2 (1976), pp. 11-24.

INHIBITION OF FERROUS ION OXIDATION BY *THIOBACILLUS FERROOXIDANS* IN THE PRESENCE OF OXYANIONS OF SULFUR AND PHOSPHORUS

MICHAEL R. HOFFMANN,[1] PAULA HILTUNEN,[2] and OLLI H. TUOVINEN[3]

[1]Environmental Engineering Science, W.M. Keck Laboratories, California Institute of Technology, Pasadena, California 91125 (U.S.A.)

[2]Department of Microbiology, University of Helsinki, SF-00710 Helsinki 71 (Finland)

[3]Department of Microbiology, The Ohio State University, Columbus, Ohio 43210 (U.S.A.)

ABSTRACT

Ferrous iron oxidation by growing cultures of *Thiobacillus ferrooxidans* was sensitive to negative catalysis (inhibition) by selected oxyanions of sulfur. The relative order of negative catalytic activity for the sulfoxyanions was $S_2O_5^{2-} > S_2O_3^{2-} > S_2O_4^{2-} > S_4O_6^{2-}$. Inhibition by the sulfoxyanions was greater at pH 1.5 than at pH 2.0. A mechanism involving the formation of radical intermediates, $HO_2^{\cdot}$, $\cdot OH$, $HSO_3^{\cdot}$, $HSO_5^{\cdot}$, $HSO_4^{\cdot}$, and $HS_3O_3^{\cdot}$, induced by Fe(II) and Fe(III) complexes is proposed to account for the observed inhibitory effects. Orthophosphate, pyrophosphate, and tripolyphosphate compounds were also inhibitory to ferrous iron oxidation by *T. ferrooxidans*. Orthophosphates were increasingly inhibitory at high concentrations (200-300 mM) whereas the apparent toxicity of pyrophosphate and tripolyphosphate was inversely related to concentration over the range of 50-300 mM with no apparent pH dependency at pH $\leq$ 2. Equilibrium models are presented to explain the relative effect of the phosphates in terms of phosphate precipitates, which were actually observed, and soluble complexes, which have an effect on the thermodynamics of electron transfer.

INTRODUCTION

Iron-oxidizing thiobacilli (*Thiobacillus ferrooxidans*) are inhibited by particulate materials such as pyrite (1,2). Soluble constituents leached with acidic solutions from pyrite are involved in causing the inhibition (2,3). In addition to the presence of toxic metals, other factors may also contribute to the inhibition, such as sulfoanions that are formed, sometimes transiently, during oxidation of iron sulfides. Nelson (4) and Goldhaber (5) reported the time dependent formation of SO_3^{2-}, S_8, and $S_4O_6^{2-}$ in their studies of the autoxidation of mackinawite and pyrite, respectively. Thiosulfate and $NH_2SO_3^-$ have been previously shown to inhibit the iron oxidation and oxygen uptake activity of *T. ferrooxidans* (6,7). The mineral nutrition of *T. ferrooxidans* has been characterized in several studies. In addition to its role as an essential nutrient required for phosphorylated cell constituents, phosphate has been suggested (8,9) to act as an iron chelating agent in ferrous iron oxidation. Several investigators have shown that phosphate interferes with the bacterial oxidation of pyrite (10,11) but some data are difficult to interpret since phosphate precipitates may form on the pyrite surface.

In the present work, chemical factors that result in an inhibition are emphasized. In coal desulfurization systems, the primary substrates for *T. ferrooxidans* are iron sulfides, Fe^{2+} and soluble S-intermediates. In order to study the factors that affect the lag times and rates in these systems, the primary process of Fe(II) oxidation and the effects of other reaction intermediates on this process need to be investigated. The present paper describes data that clearly show the inhibitory effect of oxyanions of sulfur and phosphorus on growing cultures of *T. ferrooxidans.* Sulfoanion inhibition is discussed in terms of the catalytic production of radical intermediates by Fe(III) or trace metals. Phosphoanion inhibition is discussed in terms of the formation of Fe(II) and Fe(III) phosphate complexes, precipitates and adsorbates that were predicted by computer-generated equilibrium models.

MATERIALS AND METHODS

T. ferrooxidans strain TFI-35 was cultured on 120 mM ferrous sulfate in a mineral salts solution adjusted to pH 2.0 and 1.5 with sulfuric acid (1). Cultures (100 ml) were maintained in shake flasks (180 rev/min) at 28°C in the dark; 5% inocula were used for subcultures.

Stock solutions of the test compounds were freshly prepared in distilled water before addition to cultures. Ferrous iron oxidation was monitored by titration of the residual ferrous iron with potassium permanganate (1). The ferrous iron oxidation is an established measure of growth of *T. ferrooxidans* (9). Appropriate controls were included to estimate the permanganate consumption by the test sulfoanion compounds at their final concentrations. Phosphate compounds were investigated over the concentration range of 50-300 mM.

All control cultures, which did not receive test compounds, completely oxidized 120 mM Fe^{2+} in two days at pH 1.5 and 2.0. The chemical oxidation of Fe^{2+} at this pH range was about 5% of that achieved in the presence of *T. ferrooxidans* in two days, and increased to about 10% upon prolonged incubation of the chemical controls for several weeks.

RESULTS AND DISCUSSION

Effect of sulfoanions on the rate of Fe^{2+} oxidation

Dithionate ($S_2O_6^{2-}$), persulfate ($S_2O_8^{2-}$) and pyrosulfate ($S_2O_7^{2-}$) did not influence the rate or lag time of bacterial iron oxidation at either pH 1.5 or 2.0 over the concentration range of 1 μM to 1 mM sulfoanion. A varying degree of inhibition was observed with other sulfur compounds tested as illustrated in Fig. 1A-E. Characteristically, the sulfoanions at concentrations > 10 μM prolonged the lag time as compared to control reactions but once growth and substrate utilization commenced, the oxidation rates with respect to each sulfur species were relatively unaffected (Table 1), although these rates were slower than the control. The apparent inhibition was reduced significantly by increasing the pH from 1.5 to 2.0.

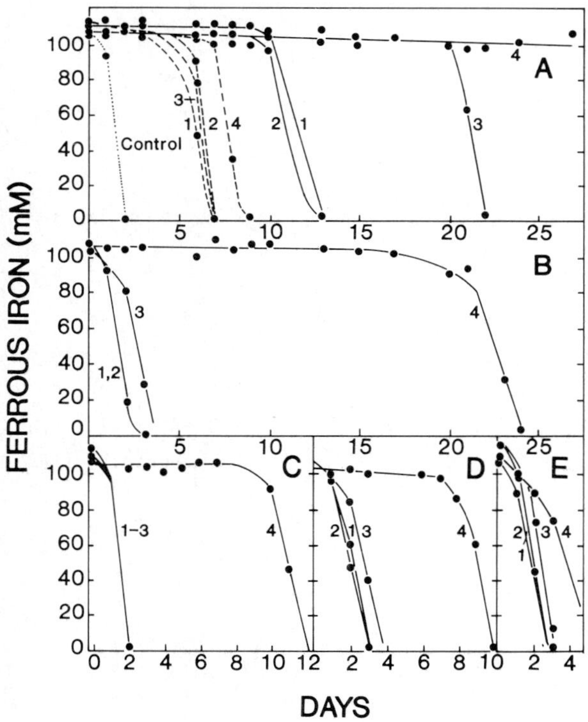

Figure 1. Ferrous iron oxidation by *T. ferrooxidans* in the presence of sulfoanions. Cultures growing at pH 1.5 (——) and 2.0 (----) received sodium metabisulfite (A); sodium thiosulfate (B); sodium dithionite (C); sodium sulfite (D) and potassium tetrathionate (E) at the following concentrations: 1, 10^{-6} M, 2, 10^{-5} M; 3, 10^{-4} M; 4, 10^{-3} M. Control cultures (····) oxidized the ferrous iron completely in two days at pH 1.5 and 2.0. The iron oxidation was not inhibited at < 10^{-3} M concentration of the test compound in B-E (data not shown).

TABLE 1

Observed lag times and rates of Fe(II) oxidation in the presence of selected sulfoanions.

Sulfoanion[a]	t_{lag} (days)		(mM/day)	
S_T	1.0 mM	0.1 mM	1.0 mM	0.1 mM
$S_2O_3^{2-}$	21	2	33	51
$S_2O_5^{2-}$	22	20	--	62
$S_2O_4^{2-}$	10	0	46	56
SO_3^{2-}	8	2	45	47
$S_4O_6^{2-}$	2	1	45	46
Control			90	

[a] Fe(II) = 0.12 M, pH = 1.5, S_T = 0.1 or 1.0 mM.

The rate of oxidation was established by taking the slope of the conc. vs. time curve at the beginning of the exponential phase, which corresponded to t_{lag}.

The relative order of negative catalysis or inhibition was as follows: $S_2O_5^{2-} > S_2O_3^{2-} > S_2O_4^{2-} > SO_3^{2-} > S_4O_6^{2-}$ based on their respective observed lag times. However, in oxygenated acidic solution at pH 2 none of the above sulfoanions are thermodynamically stable. Metabisulfite or disulfite will dissociate according to equations 1 and 2 to give bisulfite and aquated sulfur dioxide.

$$H_2O + S_2O_5^{2-} \underset{}{\overset{k_f}{\rightleftarrows}} 2HSO_3^- \tag{1}$$

$$H^+ + HSO_3^- \overset{K_{a1}^{-1}}{\rightleftarrows} SO_2 \cdot H_2O \tag{2}$$

where K_f = 14.3 (ref. 12) and $K_{a1}^{-1} = 8.13 \times 10^1$ (ref. 13) at μ = 0 and 25.0°C. Furthermore, aquated sulfur dioxide will rapidly equilibrate with the atmosphere in an open system:

$$SO_2 \cdot H_2O \rightleftarrows SO_2 + H_2O \overset{K_H^{-1}}{\rightleftarrows} SO_{2(g)} \tag{3}$$

where K_H^{-1} is the reciprocal of the Henry's Law constant (K_H^{-1} = 0.81).

Using the above constants, the fraction of disulfite, $\alpha_{1,2}$, can be determined readily according to the distribution equations given in Table 2. As determined from the equations in Table 2, concentrations of $S_2O_5^{2-}$ in solution for 1 mM HSO_3^- at pH 2.0 and 1.5 are 4.0×10^{-5} and 1.96×10^{-5}, respectively; therefore, the addition of metabisulfite results in the net addition of $H_2O \cdot SO_2$ and HSO_3^- with only a small fraction of $S_2O_5^{2-}$ remaining after equilibrium has been achieved (within a few minutes).

TABLE 2

Equations used to determine the fractional distribution of S(IV) species in a closed system.

---

1. $[S(IV]_t = [SO_2 \cdot H_2O] + [HSO_3^-] + [S_2O_5^{2-}] + [SO_3^{2-}]$

2a. $\alpha_1 = [SO_2 \cdot H_2O]/[S(IV]_T$

2b. $\alpha_1 = [HSO_3^-]/[S(IV]_T$

2c. $\alpha_{1,2} = [S_2O_5^{2-}]/[S(IV)]_T$

2d. $\alpha_2 = [SO_3^{2-}]/[S(IV]_T$

3. $\alpha_0 + \alpha_1 + \alpha_{1,2} + \alpha_2 = 1$

4a. $K_{a1} = [H^+]\,[HSO_3^-]/[H_2O \cdot SO_2] = 10^{-1.91}$

4b. $K_{a2} = [H^+]\,[SO_3^{2-}]/[HSO_3^-] = 10^{-7.18}$

4c. $K_d = [S_2O_5^{2-}]\,[HSO_3^-]^2 = 10^{-1.15}$

5a. $1/\alpha_0 = [S(IV)]_T/[SO_2 \cdot H_2O]$

5b. $1/\alpha_0 = 1 + [HSO_3^-]/[SO_2 \cdot H_2O] + [SO_3^{2-}]/[SO_2 \cdot H_2O] + [S_2O_5^{2-}]/[H_2O \cdot SO_2]$

Substitution of equations 4a, 4b, and 4c into equation 5b yields

6a. $\alpha_0 = (1 + K_{a1}/[H^+] + K_{a1}K_{a2}/[H^+]^2 + K_dK_{a1}[HSO_3^-]/[H^+])^{-1}$

6b. $\alpha_1 = \alpha_0 K_{a1}/[H^+]$

6c. $\alpha_2 = \alpha_0 K_{a1}K_{a2}/[H^+]^2$

6d. $\alpha_{1,2} = \alpha_0 K_d K_{a1}\,[HSO_3^-]/[H^+]$

In acidic solutions, thiosulfate decomposes to give aquated sulfur dioxide, colloidal sulfur (S_8, S_6), polythionic acids, and hydrogen sulfide (14). The rate of decomposition or displacement by water (eqn. 4) is acid catalyzed.

$$- \frac{d[SO_2 \cdot H_2O]}{dt} = k \, [S_2O_3^{2-}]^{1.5} \, [H^+]^{0.5} \qquad (4)$$

Davis (14) proposed that the decomposition proceeds through a series of nucleophilic displacements of sulfite by protonated thiosulfate and higher order polythionates on thiosulfate as follows:

$$S_2O_3^{2-} + H^+ \underset{}{\overset{K_{2a}^{-1}}{\rightleftarrows}} HS_2O_3^- \qquad (5)$$

$$HS_2O_3^- + S_2O_3^{2-} \overset{k_1}{\rightarrow} HS_3O_3^- + SO_3^{2-} \qquad (6)$$

$$HS_3O_3^- + S_2O_3^{2-} \overset{k_2}{\rightarrow} HS_4O_3^- + SO_3^{2-} \qquad (7)$$

$$HS_{n-1}O_3^- + S_2O_3^{2-} \overset{k_n}{\rightarrow} HS_nO_3^- + SO_3^{2-} \qquad (8)$$

$$HS_9O_3^- \rightarrow S_8 + HSO_3^- \qquad (9)$$

Using data obtained by LaMer and co-workers (15, 16) the half-life for the decomposition of $S_2O_3^{2-}$ under the conditions of the current experiments can be predicted at constant pH (1.5 or 2.0) as follows:

$$- \frac{d[S_2O_3^{2-}]}{dt} = \frac{d[SO_2.H_2O]}{dt} = (k \, [H^+]^{0.5}) \, [S_2O_3^{2-}]^{1.5} \qquad (10)$$

$$k' = 3.1 \times 10^{-2} = k \, [H^+]^{0.5} = 31 \, min^{-1} \, (10^{-2})^{0.5}$$

$$\int_{C_0}^{C_t} d\,[S_2O_3^{2-}] \; [S_2O_3^{2-}]^{-3/2} = k' \int_0^t dt \qquad (11)$$

$$\frac{2}{[S_2O^{2-}]_0^{\,0.5}} - \frac{2}{[S_2O_3^{2-}]^{0.5}} = k't \qquad (12)$$

where

$$t_{1/2} = \frac{\sqrt{2} - 1}{0.5k' \; [S_2O_3^{2-}]^{0.5}} \qquad (13)$$

For pH 2.0 and 1.5 the half-lives for 1 mM solutions of thiosulfate are 8.45×10^3 min (5.86 days) and 2.67×10^3 min (1.85 days), respectively, given the simple stoichiometry:

$$S_2O_3^{2-} \rightarrow \tfrac{1}{8}S_8 + SO_3^{2-} \qquad (14)$$

Dithionite, $S_2O_4^{2-}$, is not stable in aqueous solution (12, 17); it decomposes according to the following stoichiometry:

$$2S_2O_4^{2-} + H_2O \rightarrow S_2O_3^{2-} + 2HSO_3^- \quad (15)$$

and a small contribution from a secondary pathway involving the formation of the radical ion $SO_2^{\dot{-}}$ in alkaline, oxygen-free systems:

$$S_2O_4^{2-} \rightleftarrows 2SO_2^{\dot{-}} \quad (16)$$

The decomposition is rapid in acidic solution in the presence of air.

In strongly acidic solution polythionates decompose rapidly to give some elemental sulfur, $S_2O_3^{2-}$ and $SO_2{\cdot}H_2O$ in the absence of oxygen and some sulfate in the presence of oxygen.

$$3H_2O + 2S_4O_6^{2-} \rightarrow 2SO_3^{2-} + 3S_2O_3^{2-} + 6H^+ \quad (17)$$

The common feature of sulfoanions described above is that they are all sources of sulfite in solution at pH 2. A likely hypothesis for the shared inhibitory features of these sulfoanions is that the sulfite generated during decomposition undergoes a metal-catalyzed autoxidation with the formation of radical and peroxide intermediates at sufficient steady-state concentrations to either mask the net oxidation of Fe(II) [i.e., by reduction of Fe(III)] by bacteria or in turn for the radicals to interfere directly with microbial metabolism.

The oxidation of dissolved sulfur dioxide by oxygen to sulfate proceeds according to the stoichiometry of eqn. 18.

$$SO_2{\cdot}H_2O + 0.5O_2 \rightarrow H_2SO_4 \quad (18)$$

This reaction is sensitive to the influence of trace metal catalysis (18-23). In general, reactions of triplet ground-state oxygen with singlet spin-state reductants are slow in the absence of catalytic influences because they involve a change in spin multiplicity and because a significant amount of bond deformation or alteration of a permanent nature may occur. Autoxidations are accelerated frequently by transition metals such as Cu(II), Co(II), Fe(II), Fe(III), Mn(II), and Ni(II).

The commonly suggested mechanism for the catalytic autoxidation of $SO_2.H_2O$ in aqueous systems involves a free radical chain initiated by a metal ion (24).

Initiation: $$Fe^{3+} + SO_3^{2-} \xrightarrow{k_1} SO_3^{\dot{-}} + Fe^{2+} \quad (19)$$

Radical propagation: $SO_3^{\cdot-} + O_2 \xrightarrow{k_2} SO_5^{\cdot-}$ (20)

$SO_5^{\cdot-} + SO_3^{2-} \xrightarrow{k_3} SO_5^{2-} + SO_3^{\cdot-}$ (21)

$SO_5^{\cdot-} + HSO_3^- \xrightarrow{k_4} HSO_5^- + SO_3^{\cdot-}$ (22)

Oxidation: $SO_5^{2-} + SO_3^{2-} \xrightarrow{k_5} 2SO_4^{2-}$ (23)

Chain termination: $SO_3^{\cdot-} + SO_3^{\cdot-} \xrightarrow{k_6} S_2O_6^{2-}$ (24)

Modifications to this basic free radical mechanism have been proposed by Hayon et al. (25), Freiberg (26), Larson et al. (21), and Schmittkunz (27).

Hayon et al. (25) reported ESR evidence for the existence of $SO_4^{\cdot-}$ during the reaction whereas McCord and Fridovich (28) reported evidence for $O_2^{\cdot-}$ and $HO_2^{\cdot}$ during the course of the reaction.

Formation of superoxide and hydroperoxyl radicals may occur as follows:

$$ML^{n+} + O_2 \rightarrow ML^{(n+1)+} + O_2^{\cdot-} \quad (25)$$

$$O_2^{\cdot-} + H^+ \underset{}{\overset{pk_a = 4.7}{\rightleftarrows}} HO_2^{\cdot} \quad (26)$$

$$HO_2^{\cdot} + HSO_3^- \rightarrow HSO_4^- + \cdot OH \quad (27)$$

$$\cdot OH + HSO_3^- \rightarrow HSO_4^- + H\cdot \quad (28)$$

$$H\cdot + O_2 \rightarrow HO_2^{\cdot} \quad (29)$$

$$2HO_2^{\cdot} \rightarrow 2H_2O_2 \quad (30)$$

Inhibition of bacterial iron oxidation could be manifested in two ways. The first alternative would involve the cyclic catalytic oxidation and reduction of iron in the presence of the oxidizable oxyanions. For example, the first process in the sequence would be the bacterial oxidation of Fe^{2+} to Fe^{3+}.

$$2Fe^{2+} + 0.5\ O_2 + 2H^+ \rightarrow 2Fe^{3+} + H_2O \quad (31)$$

This reaction would be followed by the reduction of Fe^{3+} in the catalytic initiation of the sulfite radical chain processes as follows:

$$Fe^{3+} + HSO_3^- \rightarrow Fe^{2+} + HSO_3^{\cdot} \quad (32)$$

A non-enzymic reduction of Fe^{3+} in T. ferrooxidans cultures in the presence of sulfur compounds has been reported (29, 30). However, since the concentration of sulfite is at least two orders of magnitude less than the Fe^{2+} concentration, this explanation is not fully satisfactory in the present context. More likely, the free radical intermediates generated during the catalytic autoxidation of $SO_2 \cdot H_2O$ prove to be inhibitory to T. ferrooxidans. These intermediates will persist as long as sufficient sulfite remains to continue the radical-generating sequence of steps.

Hydrogen peroxide, hydroxyl, hydroperoxyl, superoxide, and singlet oxygen radicals are toxic to bacterial cells in general. The inhibition by H_2O_2 of the respiration and growth of T. ferrooxidans has been reported (31) in a system that actually generated radicals ($\cdot OH$, $HO_2^{\cdot}$) owing to the presence of ferrous iron. Although not comparable with the design of the present experiments, the reported data (31) confirm the toxicity of H_2O_2 and radicals even on a transient exposure of T. ferrooxidans. It is not known whether T. ferrooxidans possesses protective enzymes (e.g., catalase, peroxidase, superoxide dismutase) against the cytotoxic efects of H_2O_2 and highly reactive radicals.

It was not possible to demonstrate the alleviation of sulfoanion toxicity by radical-scavenging agents such as mannitol and methylene blue because these compounds were also toxic to the bacteria at the concentration range tested (0.1 mM). Moreover, catalase was immediately inactivated upon the addition to the culture solution.

To quantitatively estimate the time required to generate bisulfite radicals in microbial systems, the kinetic data of Martin et al. (32) on the Fe(III) catalyzed autoxidation of sulfur dioxide can be used. Martin and co-workers (32) found that the following differential equation (empirical rate law) could be used to describe the iron catalyzed autoxidation:

$$-\frac{d[SIV)]}{dt} = k\,[Fe(III)]\,[S(IV)] \tag{33}$$

where $k = 8.2 \times 10^{1}\ M^{-1}\ sec^{-1}$. When the initial concentrations of Fe^{3+} and S(IV) are not equal and the stoichiometry of eqn. 33 is assumed to be valid, the integrated solution to eqn. 34 is:

$$\frac{1}{[S(IV)]_0 - [Fe(III)]_0} \ln \frac{[Fe(III)]_0 \{[S(IV)]_0 - x\}}{S(IV)_0 \{[Fe(III)]_0 - x\}} = kt \tag{34}$$

where x is the amount of S(IV) that has disappeared per unit volue in time, t, and according to eqn. 32 the amount of $HSO_3^{\cdot}$ that has appeared in unit time due to radical initiation alone. If $[Fe(III)]_0 = 1\ \mu M$ and $[S(IV)]_0 = 10^{-3}$ M, then the time required for 95% of the Fe(III) to be utilized would be 36 sec. Thus generation of radical intermediates takes place rapidly relative to the oxidation of Fe^{2+} by microbial catalysis.

Low, steady-state concentration of radicals may be sufficient to inhibit bacterial growth and consequently the oxidation rate of Fe^{2+}. The sulfoanions tested could also have an inhibitory effect by virtue of their ability to reduce cytochromes, thereby slowing down the flow of electrons from ferrous ion to oxygen in the electron transport chain of _T. ferrooxidans_. Similarly, elemental sulfur in the growth medium results in a prolonged lag period and decelerates the rate of iron oxidation by _T. ferrooxidans_ (3, 33). This effect may be comparable with that of 10 mM ascorbic acid which has been mentioned to inhibit the activity of _T. ferrooxidans_ (34). In the present work 0.1 mM ascorbic acid had no effect on the bacterial oxidation of Fe^{2+} whereas at 1 mM concentration of ascorbic acid the oxidation rate declined. However, it should be emphasized that ascorbic acid may not only act as a reducing agent on the electron transfer components of _T. ferrooxidans_, but at concentrations above 1 mM it is likely to also impair the cell envelope or cell membrane in a fashion similar to that described for other organic acids (35, 36). In the present work, ascorbic acid was not investigated further.

Computer-generated model for the chemical speciation of Fe^{2+} and Fe^{3+} in the mineral salts

The computer code, SURFEQL (37-39), was used to generate an equilibrium model for the detailed speciation of Fe^{2+} and Fe^{3+} in the mineral salts medium over the pH range of 2.5 to 1.5. Results of these computations in terms of actual concentrations at pH 2.0 and 1.5 are given in Table 3. The discernible trends over this pH range are the following.

With the sulfite system ($[SO_3^{2-}]_T$ = 1 mM) the major ferric iron species are $Fe(H_2O)_6^{3+}$ (3.6-7.5%), $FeSO_4^+$ (61.9-73.4%), $Fe(SO_4)_2^-$ (24.7-17.5%), $Fe(OH)^{2+}$ (6.0-1.0%) and $Fe(HSO_3)^{2+}$ (1.4 to 0%). Similar trends are observed for the thiosulfate sytem with the principal exception being a negligible calculated concentration for the $FeHS_2O_3^{2+}$ species. The percentage contribution of free hexa-aquo ferric iron to the total Fe(III) concentration is small. It is also clear that complexation of ferrous iron by sulfite and thiosulfate cannot play a significant role in the apparent inhibition of Fe(II) oxidation. The dominant Fe(II) species over the entire range are $Fe(H_2O)_6^{2+}$ and $FeSO_4^0$.

If the principal mode of catalysis is through Fe^{3+}, the actual concentration must be ascertained in order to make quantitative estimates. For example, for a system at 1.0 M total iron the free Fe(III) concentration will be 4.67×10^{-8} M at pH 2.0 and 7.58×10^{-8} M at pH 1.5. With these concentrations the approximate life-time for SO_3^{2-} at 1.0 mM would be on the

TABLE 3

Dominant chemical species for 1 μM Fe(III) and 120 mM Fe(II) in the mineral salts medium.

Species[a]	Concentration ($-\log C_M$)[b]			
	at pH 2.0		at pH 1.5	
	A	B	C	D
Fe^{2+}	1.15	1.15	1.09	1.10
$FeSO_4^0$	1.30	1.31	1.30	1.40
$FeSO_3^0$	6.41	----	7.36	----
$FeS_2O_3^{2-}$	----	3.23	----	3.23
Fe^{3+}	7.33	7.32	7.12	7.13
$FeSO_4^+$	6.16	6.16	6.16	6.13
$Fe(SO_4)_2^-$	6.63	6.63	6.76	6.75
$Fe(OH)^+$	7.61	7.61	7.92	7.92
$FeHSO_3^{2+}$	8.24	----	8.54	----
$FeHS_2O_3^{2+}$	----	9.8	----	9.16

[a]Total concentrations of media components used as input values for the computer model were as follows (mM): K^+ = 4.59; Mg^{2+} = 1.62; NH_4^+ = 6.05; PO_4^{3-} = 2.35.
[b]$[SO_3^{2-}]_T$ = 1.0 mM for A and C; $[S_2O_3^{2-}]_T$ = 1.0 mM for B and D.

order of 80 sec. For comparison, the lower limit of concentration for effective metal catalysis of sulfite oxidation is 10^{-12} M in active metal (23). At this limit, assuming a radical chain length of 10000, the life-time for 1.0 mM SO_3^{2-} will be on the order of approximately 45 days whereas at 10^{-10} M this time would be reduced to 4.5 days. Therefore, a relatively small fraction of Fe(II) must be converted to Fe(III) to initiate a free radical chain process that may persist for several days. These estimates of time are consistent with lag times observed in this study.

Effect of phosphoanions on the rate of Fe(II) oxidation

The effects of orthophosphate, pyrophosphate and tripolyphosphate on *T. ferrooxidans* are illustrated in Fig. 2. Fig. 2A-C shows that the inhibition by 50 and 100 mM orthophosphate was slight but the inhibition increased as the added phosphate concentration was increased from 100 mM to 300 mM. At 200 and 300 mM total phosphate a white precipitate formed during the course of the reaction. A slightly greater inhibition was observed when the counter ion was K^+ as compared to Na^+.

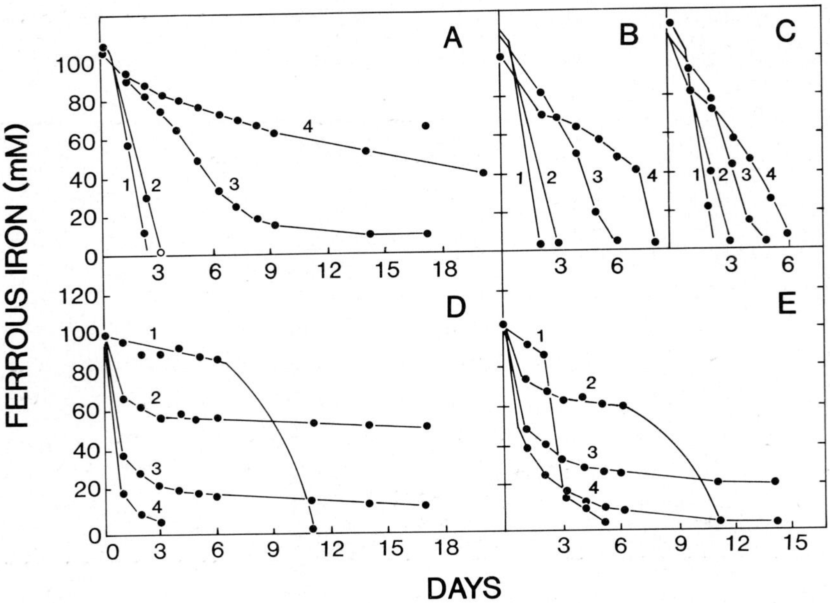

Figure 2. Ferrous iron oxidation by T. ferrooxidans in the presence of phosphate compounds. Cultures growing at pH 1.5 received potassium phosphate (A, K_2HPO_4), sodium phosphate (B, Na_2HPO_4), phosphoric acid (C, H_3PO_4), sodium pyrophosphate (D, $Na_4P_2O_7 \cdot H_2O$), and sodium tripolyphosphate (E, $Na_5P_3O_{10}$) at the following concentrations: 1, 50 mM, 2, 100 mM, 3, 200 mM, 4, 300 mM. The change of pH during growth was ± ΔpH 0.15. The inhibitory patterns changed only slightly when an initial pH of 2 was used and by substitution of monobasic phosphates for the dibasic phosphates (A, B). The control culture oxidized the ferrous iron completely in two days. Less than 10% of the ferrous iron was oxidized in the respective uninoculated chemical controls within the time course.

Pyrophosphate and tripolyphosphate inhibition was most pronounced at low concentrations (50 and 100 mM) compared to elevated concentrations (200 and 300 mM) as shown in Fig. 2 D-E. In the lower concentration ranges (50 - 100 mM) of $P_2O_7^{4-}$ and $P_3O_{10}^{5-}$, the concentration vs. time profile for Fe^{2+} shows a distinct plateau for a period 2 to 6 days. This plateau is followed by a rapid decrease of Fe^{2+} within the period of a few additional days.

Insight into the mechanism of inhibition can be gained by a computer analysis of the chemical speciation in the various reaction media with the aid of SURFEQL. Using the media compositions as input data, the dominant chemical forms of Fe(II) and Fe(III) were calculated for a system in which 5% of the initial Fe(II) has been converted to Fe(III) at pH 1.5. No further changes in redox chemistry were considered. Results of these analyses are given in Table 4. These results indicate in the case of orthophosphate addition that a precipitate of ferric phosphate, $FePO_4$, will form as the concentration of phosphate is increased from 2.3 mM (background concentration in the media) to 300 mM. At 300 mM of PO_4^{3-}, the solid precipitate will comprise 100% of the total Fe(III). Phosphate addition affects the Fe(II) speciation by forming a soluble complex, $FeH_2PO_4^0$.

TABLE 4

Chemical speciation vs. added orthophosphate at pH 1.5

$[PO_4^{3-}]_T$ added:	2.3 mM (μ=0.27)		300 mM (μ=0.27)		50 mM (μ=0.26)	
Species	% total metal	[C] mM	% total metal	[C] mM	% total metal	[C] mM
(III) $FePO_{4(s)}$	36.2	2.17	99.8	5.99	100	6.0
(III) $FeSO_4^+$	47.1	2.83	0	(7.6 μM)	0	(0.80 μM)
(III) $Fe(SO_4)_2^-$	11.0	0.66	0	(2.0 μM)	0	(0.26 μM)
(III) $Fe(H_2O)_6^{3+}$	4.8	0.29	-	(0.7 μM)	0	(0.08 μM)
	99.1	5.95	99.8	5.99	100	6.0
(II) $FeH_2PO_4^0$	0	0.082	22.7	25.9	75.8	86.3
(II) $FeSO_4^0$	32.4	26.9	27.4	31.2	9.6	10.9
(II) $Fe(H_2O)_6^{2+}$	67.6	77.0	49.9	56.9	14.6	16.7
	100	113.98	100	114	100	113.9

A likely explanation for the observed inhibition with increasing orthophosphate concentrations is as follows. First, the solid product begins to form in the finely divided state hexa-aquo Fe(II) and its complexes adsorb to that surface; second, these surface-adsorbed complexes of Fe(II) are initially unsuitable as electron donors for _T. ferrooxidans_. Consequently, the depletion rate of total Fe(II) is retarded. Third, _T. ferrooxidans_ may adapt to utilize the surface-adsorbed Fe(II) species by excretion of appropriate enzymes or chelating agents similar to the ferrichromes, thereby increasing the observed Fe(II) depletion rate. Experimental observation of a white precipitate lends support to this hypothesis.

A sequence of reactions which put the above hypothesis into a convenient notation are as follows:

$$Fe(II) + 0.25O_2 + H^+ \xrightarrow{\underline{T.\ ferrooxidans}} Fe(III) + 0.5H_2O \quad (35)$$

$$Fe(III) + H_2PO_4^- \rightarrow FePO_4 + 2H^+ \quad (36)$$

$$FePO_{4(s)} + Fe(II) \rightarrow Fe(III)PO_4Fe(II)^{2+} \quad (37)$$

$$Fe(III)PO_4 + FeH_3PO_4^0 \rightarrow Fe(III)PO_4Fe(II)H_2PO_4^0{}_{(s)} \quad (38)$$

The adsorbed complexes of eqns. 37 and 38 would be chemisorbed species rather than weakly bound ion pairs. Binding would take place most likely through phosphate bridges (Fig. 3).

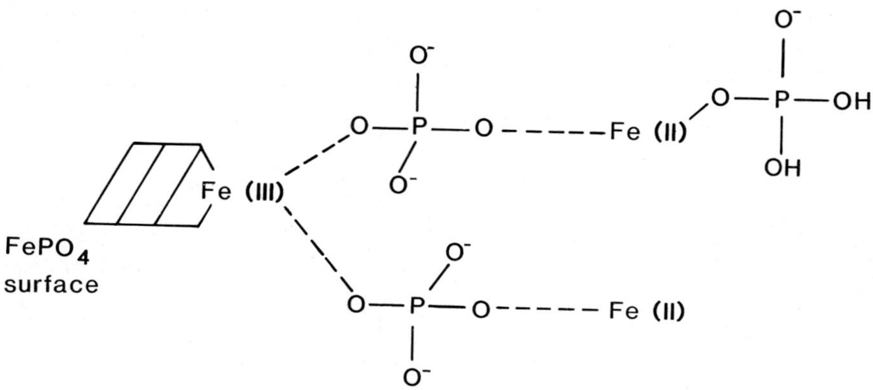

Figure 3. Postulated phosphate binding of iron.

In the case of $P_2O_7^{4-}$ additional SURFEQL calculations showed that no solids were formed and that the major Fe(III) species were (1:1) pyrophosphate complexes, $Fe(H_2P_2O_7)^{2+}$ and $Fe(H_3P_2O_7)^{2+}$. Ferrous iron is found primarily as the hexa-aquo species and secondarily as the sulfato complex, $FeSO_4^0$. However, in all flasks in which pyrophosphate and tripolyphosphate were added white precipitates were observed. In the case of $P_2O_7^{4-}$, a solid, $Fe_4(P_2O_7)_3$, with a log K_{s0} of -9.24 at μ = 0.45 was considered but its computed saturation index at pH 1.5 was 10^{-62}. Most likely the observed white precipitates were again due to $FePO_4$ formation in which the excess PO_4^{3-} came from the metal catalyzed hydrolysis of $P_3O_{10}^{5-}$ and $P_2O_7^{4-}$ as follows:

$$P_3O_{10}^{5-} + H_2O \overset{k_1}{\rightarrow} HP_2O_7^{3-} + HPO_4^{2-} \quad (39)$$

$$HP_2O_7^{3-} + H_2O \overset{k_2}{\rightarrow} HPO_4^{2-} + H_2PO_4^- \quad (40)$$

Clesciri and Lee (40) found that the k_1 was 1.4 x 10^{-4} min^{-1} at 25°C near neutral pH. This value would give a half-life for $P_3O_{10}^{5-}$ of roughly 82.5 hours or 3 days. Thilo (41) has shown that hydrolysis according to eqn. 39 is accelerated greatly by the addition of divalent metal ions such as Cu^{2+}, Zn^{2+} and Pb^{2+}. High concentrations of $P_2O_7^{4-}$ and $P_3O_{10}^{5-}$ may effectively bind catalytic sites for hydrolysis and prevent a rapid breakdown hydrolysis to give PO_4^{3-}. This latter effect needs to be examined in greater detail before a definitive hypothesis for the inverse concentration relationship can be tendered.

CONCLUSIONS

The data demonstrate the negative catalytic role of inorganic sulfur compounds in the microbiological oxidation of ferrous iron. With the current knowledge of the chemical pathways of sulfoanions, it has been possible to show that $HSO_3^{\cdot}$, $\cdot OH$ and $HO_2^{\cdot}$ are likely to be formed upon the decomposition of sulfur compounds in iron-containing culture solutions of *T. ferrooxidans*. The inhibition of *T. ferrooxidans* caused by these radicals was pronounced and concentration-dependent even though the radicals do not persist in the culture solution. These observations suggested a high degree of toxicity in accordance with their influence on other biological systems. The relative toxicities of the various oxyanions were consistent with the known pathways of radical formation and with the chemical speciation of Fe(II)- and Fe(III)-compounds. Similarly, phosphate compounds were examined and their inhibitory effects were due to the formation of iron surface complexes and precipitates.

Thus, data reported in the present work indicate that radicals and iron complexes may adversely influence the bacterial regeneration of ferric sulfate in acid leach liquors.

ACKNOWLEDGMENTS

The work was supported with funds from the Ministry of Trade and Industry, Finland (O.H.T. and P.H.), the Ohio Coal Research Laboratories Association, Contract OCRLA-6 (O.H.T.), and the Ohio Air Quality Development Authority (O.H.T.) We thank Mary Fuller for typing the manuscript.

REFERENCES

1. P. Soljanto, P. Rehtijärvi and O.H. Tuovinen, Geomicrobiol. J., 2 (1980), 1-12.
2. O.H. Tuovinen, M.R. Hoffmann, F.A. Panda and H.M. Tsuchiya, in P.A. Trudinger, M.R. Walter and B.J. Ralph (Eds.), Biogeochemistry of Ancient and Modern Environments, Springer-Verlag, New York, 1980, pp. 529-536.
3. A.A. DiSpirito, P.R. Dugan and O.H. Tuovinen, Biotechnol. Bioeng., 23 (1981), 1235-1243.
4. M. Nelson, Ph.D. Thesis, Stanford University, California (1977).

5. M.B. Goldhaber, Amer. J. Sci., 283 (1983), 193-217.
6. W.E. Razzell and P.C. Trussell, J. Bacteriol., 85 (1963), 595-603.
7. J.R. Lusty, Arab. J. Sci. Eng., 5 (1980), 101-105.
8. P.R. Dugan and D.G. Lundgren, J. Bacteriol, 89 (1965), 825-834.
9. O.H. Tuovinen and D.P. Kelly, Z. Allg. Mikrobiol., 12 (1972), 311-346.
10. J.V. Beck, in D. Perlman (Ed.), Fermentation Advances, Academic Press, New York, 1969, pp. 747-771.
11. M.R. Hoffmann, B.C. Faust, F.A. Panda, H.H. Koo and H.M. Tsuchiya, Appl. Environ. Microbiol., 42 (1981), 259-271.
12. F.A. Cotton and G. Wilkinson, Advanced Inorganic Chemistry, Wiley, New York (1980).
13. R.M. Smith and A.E. Martell, Critical Stability Constants, vol. 4, Plenum Press, New York (1976).
14. R.E. Davis, J. Amer. Chem. Soc., 80 (1958), 3565-3569.
15. E.M. Zaiser and V.K. LaMer, J. Colloid. Sci., 3 (1948), 571-598.
16. V.K. LaMer, R.H. Dinegan and R.H. Smellie, J. Amer. Chem. Soc., 13 (1951), 2050-2054.
17. W.J. Lem and M. Wayman, Can. J. Chem., 48 (1970), 2778-2785.
18. S. Beilke and G. Gravenhorst, Atmos. Environ., 12 (1978), 231-239.
19. D.A. Hegg and P.V. Hobbs, Atmos. Environ., 12 (1978), 241-253.
20. A. Huss, P.K. Lim and C.A. Eckert, J. Amer. Chem. Soc., 100 (1978), 6252-6253.
21. T.V. Larson, N.R. Horike and H. Harrison, Atmos. Environ., 12 (1978), 1597-1611.
22. M.R. Hoffmann and B.C. Lim, Environ. Sci. Technol., 13 (1979), 1406-1414.
23. M.R. Hoffmann, Environ. Sci. Technol, 14 (1980), 1061-1066.
24. H. Bäckström, Z. Phys. Chem., 25B (1934), 122-138.
25. E. Hayon, A. Treinin and J. Wilf, J. Amer. Chem. Soc., 94 (1972), 47-57.
26. J. Freiberg, Atmos. Environ., 9 (1975), 661-672.
27. H. Schmittkuntz, Ph.D. Thesis, University of Frankfurt/Main (1963).
28. J.M. McCord and I. Fridovich, J. Biol. Chem., 244 (1969), 6056-6063.
29. T.D. Brock and J. Gustafson, Appl. Environ. Microbiol., 32 (1976), 567-571.
30. K. Kino and S. Usami, Agr. Biol. Chem., 46 (1982), 803-805.
31. C.L. Brierley, Soc. Min. Eng. AIME, Preprint 77-B-333 (1977).
32. L.R. Martin, D.E. Damschen and H.S. Judeikis, U.S. EPA Report 68-02-2702 (1980).
33. K. Kino, N. Sugimoto and S. Usami, Hakkokogaku, 58 (1981), 415-420.
34. H. Tributsch and J.C. Bennett, J. Chem. Technol. Biotechnol., 31 (1981), 627-635.
35. J.H. Tuttle and P.R. Dugan, Can. J. Microbiol., 22 (1976), 719-730.
36. J.H. Tuttle, P.R. Dugan and W.A. Apel, Appl. Environ. Microbiol., 33 (1977), 459-469.
37. F.M. Morel and J.J. Morgan, Environ. Sci. Technol., 6 (1972), 58-67.
38. J. Westall, J. Zachary and F. Morel, Ralph M. Parsons Laboratory Technical Note No. 18, MIT, Cambridge, Massachusetts (1976).
39. J. Westall and H. Hohl, Adv. Colloid. Interf. Sci., 12 (1980), 265-294.
40. N.L. Clesciri and G.F. Lee, Int. J. Air Water Pollut., 9 (1965), 723-743.
41. E. Thilo, Adv. Inorg. Chem. Radiochem., 4 (1962), 1-75.

COMMERCIAL MICROBIAL DESULFURIZATION OF COAL

V.A. VASEEN

Technometrics, Inc., Wheat Ridge, Colorado 80033 (U.S.A.)

ABSTRACT

The microbes, Thiobacillus ferrooxidans, and related species, have long been known to have the ability to oxidize insoluble sulfides and organic sulfur compounds to soluble sulfates. Industrial laboratory research has in general now determined the design parameters required for the potential of commercializing the use of TBF in removing sulfur from finely powdered coal, as soluble sulfates. The bioreaction vessel within which the efficiency of these biohydrometallurgical reactions are controlled, and now offers probable commercial status to this process, is the horizontal rotating tube or drum, within which, continuous processing is possible. Controlled are: temperature, pressure, mixing microbial seeding, microbial nutrients, chemical adjustments, and oxygen (air) supply. Thus faster and more efficient biological/biochemical process reactions.

INTRODUCTION

Coal, a principal source of world energy from the late 1880's through the 1940's; then took second place through the 1950's to oil; then third place by the 1960's to oil and gas; is expected by the 1990's to again assume first place.[1] To do so, however, we must take a number of factors into consideration when considering coal as our principal energy source. This paper will expound on one of these, the desulfurization of high sulfur content of coals, by use of the microbe- Thiobacillus ferrooxidans.

Coal is found throughout most of the world; however, the United States, the Soviet Union, and China own nearly two-thirds of all the known coal reserves.[1]

The National Geographic Special Report on Energy for 1981 advises, "At present rates of consumption, these reserves would last the world more than 200 years, according to conservative estimates. Furthermore, geologists think the world probably has 15 times this much coal." The United States is estimated to own over 786 billion[1] tons of the world's total reserves of 3,000 billion tons.

In 1979 the U.S. produced 780 million tons of coal, of which 680 million tons or 77% was burned by electric utilities.[1]

Production of coal in the U.S. -- 810 million tons in 1981 -- is projected to double by 1995.[2] Historically, coal has been transported by rail. Some sixty (60%) per cent of all coal mined in the United States moves by rail. Coal mining and energy interests have been trying to find a competitive alternative to rail transportation for a number of years. The most prominently promoted system is the coal slurry pipeline, which makes use of a relatively simple technology to move a mixture of coal and water, or other liquid through a pipeline.

The following table illustrates the potential for coal transportation by pipelines now[2] being considered.

TABLE I

	Pipeline System	Length (miles)	Annual Capacity (tons)
1.	Black Mesa (operating)	273	4,800,000
2.	Allen-Warner Valley	183	11,600,000
3.	Pacific Bulk	650	10,000,000
4.	Nices	1,100	25,000,000
5.	San Marco	900	10,000,000
6.	Texas Eastern	1,260	22,000,000
7.	ETSI	1,387	25,000,000
8.	Coalstream	1,500	25-55,000,000
9.	VEPCO	350	5,000,000
10.	Ohio	108	1,300,000
11.	Powder River	1,300	25-36,000,000

The pipeline slurry is composed of 100% No. 14 mesh, 50% coal, and 50% liquid.

Desulfurization of coal proposed for pipeline coal/slurry transportation is applicable at both the slurry production, influent end of the pipeline, or at the slurry receiving, effluent end of the pipeline; by means of *Thiobacillus ferrooxidans*, fermentation.

About a fifth of the 100 million tons of coal normally mined each year in Britain is composed of fine particles less than 0.5 mm (0.02 in.) in size.[3] A method to reduce acid rain pollution by taking more ore sulfur out of coal before it is burned in power stations has been devised by British Scientists at North Staffs Polytechnic, stoke on Trent, England.[3] The technique uses bacteria to convert pyrites, solid sulfur compounds into sulfuric acid that can be washed away.[3]

The Netherlands, at Technische Hogeschool Delft; Afdeling der Scheikundige Technologie, Valgroep Bioprocestechnologies; as of June 1984; were considering the parameters for "scale up" of use of TBF in desulfurizing first a pilot plant of 100,000 metric tons per year of 2.3% sulfur coal; then a commercial plant for 1,000,000 metric tons per year.[4]

The illustrations of coal used quantities, and the world-wide demand for reduction in $(SO_3)^{2}$ in combustion discharge gases, offers a small picture of the needs to use alternative methods for desulfurization of high sulfur coals, along with the interest now evidenced as to the potential for use of microbial processing of powdered coal prior to combuston.

MICROBIAL DESULFURIZATION OF COAL

Desulfurization of coal herein is considered to be a biohydrometallurgical process, as the principal chemistry is with the pyritic and organic sulfur compounds, thus metallurgical. Because the process is accomplished with microbes or bacteria, which must, by nature, live in an aqueous environment; thus the term "Hydro." The biological oxidation reactions of sulfides, such as those taking place in the microbial leaching of arsenopyrites, are well documented, since the late 1970's, in both metallurgical[2,17] as well as biological literature.

Laboratory confirmation of the feasibility of the TBF use in commercial biohydrometallurgical has been principally by "shaker flasks," "mechanical mixing of laboratory-scale vessels," and/or via "air suspensions of coal/slurry." It is the purpose of this presentation to outline the probable commercial reactor vessels capacities required by "scale up" of the laboratory data. Laboratory examinations have been with, for the most part, <u>batch</u> operations, interested in the "kinetics" of the process, rather than the "engineering design factors."

The biochemistry of the process is rather simple, based on the insoluble sulfidic minerals, principally FeS_2, and organic sulfur compounds, being microbiologically oxidized to soluble sulfates. The two chemical equations are therefore:

$$4FeS_2 + 15O_2 + 6H_2O > 4F_3^{+3} + 8SO_4^{-2} + 4H_3O^{+}$$

$$CO_2 + H_2O + \text{Minerals (nutrients)} > \text{Biomass (microbes)*}$$

*Elemental biomass cell composition has been determined for aerobic metabolism as follows: Neglecting Phosphorus (which is always present) $C_5H_7O_2N$ (Busch, et al., WPCF, Aprin, 1962. Including Phosphorus: $C_{118}H_{170}O_{51}N_{17}P$ (rich unit process, Wiley, N.Y., 1963).

Three papers (15,16,17) provided by Dr. T. F. Huber, June 19, 1984, are the source of the "kinetics" and coal physicals used in the "scale-up" illustration.

Factors required for consideration in "scale-up" from laboratory to pilot to commercial processing are in general the following:

1. Direct contact between the microbes of TBF and the sulfide mineral or sulfur compounds are essential for the catalytic activity when the TBF are in aquatic environment with a mixed culture of the autotropic microorganisms of which TBF predominates. Pulverization of coal is therefore necessary so as to release the maximum surface area possible of the pyrite, et al., to the microorganism[(9,10)] Coal ground to particle sizes used in commercial "powder coal" burners of less than 100 microns generally suffices.[(7,8,11)]
2. TBF are thermophilic bacteria and react most favorably in the temperature range of 20°-35°C. It is believed a practical operating temperature for commercial processing is 30°C.[(7,8,11)]
3. The natural environment, due to the sulfate product of reaction from TBF, produces a pH of 1.2 to 2.5.[(17)] It now appears an operation pH of 1.7 is the most efficient.
4. Oxidation of the sulfide to sulfates, by TBF, requires oxygen availability. In "batch" processing air is used to retain the slurry/suspension of coal and thus also provides the aerobic environment. Commercial processing and use of air for coal/slurry suspension is expensive and wasteful of energy; therefore, mechanical mixing which provides for coal/slurry suspension with only sufficient air to satisfy the stoichiometric oxygen requirement is favorable. Since the microbes require "carbon" in their metabolism, carbon dioxide carried with the air, also satisfies the "carbon" need. Pyrite oxidation appears to follow first order kinetics with respect to the pyrite surface area.[(7)] With coal (powder) slurry densities, at least to 20% (w/w), the first order reaction rate constant is not influenced.(12)

   The bioreactor design must compromise between mixed flow and plug flow.[(13)] Backmixing is required to prevent "washout" of bacteria, but due to the first-order kinetics, plug flow is desirable. Laboratory findings indicates a cascade of mixed flow reactors are probably best for commercial capacity processing. Three serious considerations need be taken into account with any commercial sized reactor; that is:

1. Coal suspensions at a pH of 1.2-2.5 are corrosive; therefore, moving parts should be avoided.
2. In case of extended power failure, the suspension will settle. Since air (gas) agitation for resuspension is more efficient than mechanical agitation, some air should be considered.
3. The scale of any commercial installation will be very large.

BIOREACTOR DESIGN FACTORS

With these foregoing factors in mind, we propose a "plug flow" vessel, being a horizontal long, narrow tube or drum. The tube or drum, having the following mechanical factors:

1. Supported on external rollers, one or more of which are driven so as to produce slow rotation of the drum around its center line.
2. Each end of the drum connected to dual piping which connects to the drum via dual flow rotary fittings, external from the drum. The influent maniford carrying in one pipe heated air; and the other the coal slurry along with return recirculated primary reactor slurry and bacteria "seed." The effluent manifold carrying in one pipe the treated desulfurized coal slurry, and the other the spent and surplus air.
3. The interior of the drum walls containing Rifling Ridges designed to both lift the slurry via the drum rotation, thus assuring suspension, as well as moving any settleable solids through the drum to "lift scoops" on the drum effluent end, thus assuring discharge of all materials from the drum.
4. The drum wall penetrated via numerous ports for introduction of air (oxygen), at the invert of the drum and removal of surplus and spent air at the crown of the drum, with air flow quantity controlled along the length of the drum in concert with the requirements of the "log" growth of microorganisms, and air distributed to the individual drum wall manifolds via automatic distribution valves mounted on the drum heads exteriors.

5. The storage of air requirements for several hours, as surplus air, not only provides means of continuing coal slurry suspension during power outages, but by air motors can also drive the drum rotation mechanism for further assurance of suspension.

All microbial processes have four phases: a "lag phase," or acclimation period; a "log phase," or maximum growth period; a "stationary phase," or static biomass growth period; and a "log death" phase, or rapidly declining biomass growth rate period. "Batch" processes must contend with a homogenous mixture of all four of these phases, thus the major inefficiency of the batch process. Continuous fermentation bypasses the "log" phase by recycle of quantities of return "seed" bacteria already acclimated, in concert with the raw feed (nutrient) material, thus the process immediately enters the log growth phase. The return cycle seed bacteria are also removed at maximum quantity, thus during "stationary" phase back to beginning of processing, thus shortening time required for log death phase. The drawing herewith, Figure I, of purposes of simplification, illustrates a single bioreactor. In commercial practice this would be a primary bioreactor retaining the coal slurry through the log phase, with seed bacteria obtained from a return of 5%-20% of the primary effluent slurry, with the balance of the effluent slurry split to two or more secondary bioreactors for completion of the stationary and death log phases.

Commercial processing of high sulfur coal for desulfurization of the coal is best explained by illustration. The design of a modular plant with a capacity of 100,000 tons per year. The coal used for this illustration is a Netherlands coal having the following characteristics:

Analysis of Coal #501GB28 Based on Dry Weight[27]

Ash	18.8%
Volatile Matter	27.0%
Fixed Carbon	54.2%
Calorific Value	26.6 MJ/kg
Sulfate Sulfur	0.9%
Pyritic Sulfur	0.8%
Organic Sulfur	0.6%
Total Sulfur	2.3

The bioreactor is designed for variable drum rotation speed, preferably from 1 to 20 rpm to accomodate different particulate sizes, and change in coal slurry versus air injection quantities. Drum design with an 80% liquid depth (of diameter) provides for 85.7% of the cross-section area, and thus provides for maximum filling the bioreactor to 85.7% of its liquid capacity. The maximum capacity is reduced by controlling the drum overflow

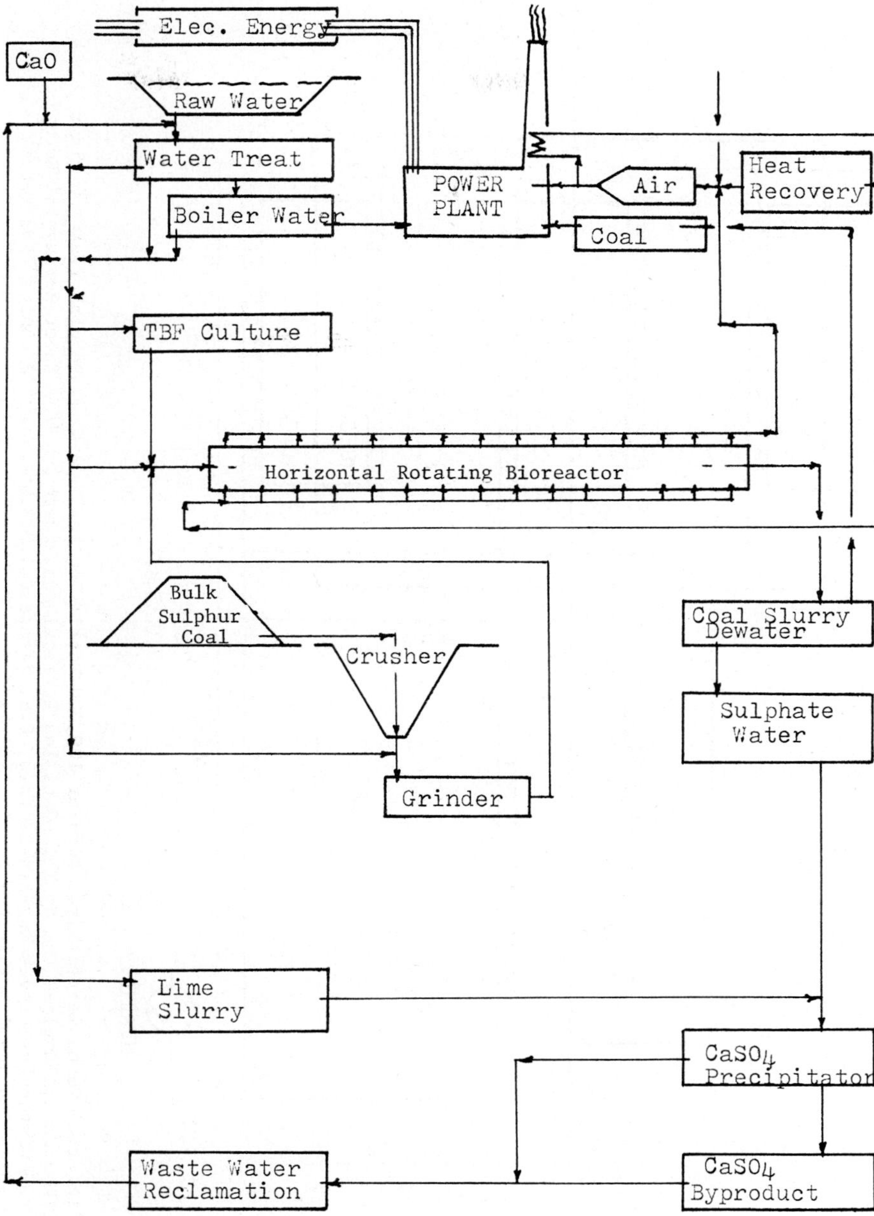

Fig. 1. BIOHYDROMETALLURGICAL DESULPHURIZATION OF HIGH SULPHUR COAL

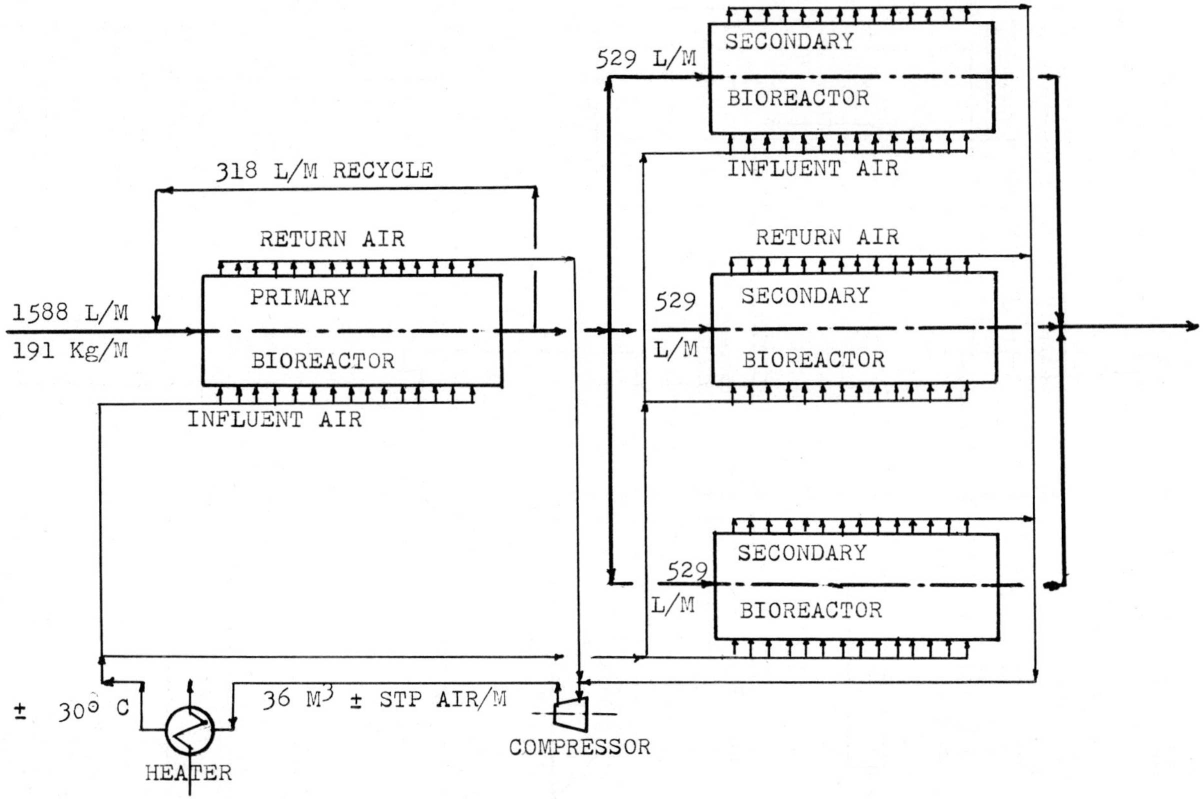

Fig. II. Technometrics Inc. Modular 100,000 Metric Ton Per Year-Coal Desulphurization Plant Bihydrometallurgical Use of Thio-Bacillus Ferrooxidans, Compound Flow Sheet-Horizontal Rotating Drum-Continuous Processing-Bioreactors, U.S. Patent # 4,223,094 Four (4) Units 7.32 Meters Diameter by 146.3 Meters Long.

liquid depth to the bioreactors minimum (without use of syphon or pressure/vacuum discharge means) quantity at approximately 50% of full-drum volume. Processes which require considerable reaction time, such as oxidation of sulfides by TBF requiring up to ten (10) days, are designed with greater length to diameter ratio, such as the 20:1 used in this illustration.

The flow sheet, Figure II, recommends use of a primary bioreactor, 7.32-meter (24-ft) diameter drum, 146.3 meters (480 ft) long, followed by three parallel operating secondary bioreactors; also (7.32) meters in diameter (24 ft), and 146.3 meters long. Design and use of single bioreactor are shown in Fig. III-V.

The following Table II illustrates the capacity range of this flow sheet. The proposed design possibility for the processing of 100,000 metric tons per year of 2.3% sulfur coal, with up to 20% return slurry, providing seed TBF, to the Raw Influent Flow calls for the following.

TABLE II

Primary Bioreactor

One Unit: 7.32 meters (24 ft) diam x 146.3 meters (480 ft) long

Coal Slurry Flow:

Raw Influent $\frac{3{,}176}{2}$ L/min = 1,588 liters/min
Return Flow @ 20% = 318 liters/min
Total Influent Flow 1,896 liters/min

Retention $\frac{5{,}274\ M^3}{2{,}730\ M^3}$ = 1.93 days (46.36 hr)

Coal: 274,368 kg/day = 90.52 kg/min

Three Units: 7.32 meters (24 ft) diam x 146.3 meters (480 ft) long.

Coal Slurry Flow: Primary Bioreactor Effluent

Influent Flow $\frac{1{,}588\ \text{liters/min}}{3}$ = 529 liters/min each

Retention $\frac{5{,}274\ M^3}{762\ M^3}$ = 6.92 days (166.1) hr each

Coal $\frac{190.52}{3}$ = 63.51 kg/min each

The following Tables (III-VII) illustrate the variation in design factors regarding capacity of bioreactors versus time required for satisfactory biochemical reaction efficiencies.

TABLE III

Horizontal Rotating Bioreactors -- Pilot and Commercial Capacities
Minimum (50% of Diam.) and Maximum (80% of Diam)
Metric Factors (Length to Diameter = 20 to 1)

Diameter		Length		Volume			Slurry Flow Rate - (liters/minute)									
				100% Diam	80% Diam	50% Diam	1 day 24 hr	Retention	4 day- 96 hr	Retention	6 day- 144 hr	Retention	8 day- 192 hr	Retention	10-day 240 hr	Retention
ft	meters	ft	meters	Cu meters	Cu meters	Cu meters	max	min	max	min	max	min	max	min	max	min
2	0.61	40	12.33	3.56	3.05	1.78	2.11	1.24	0.53	0.31	0.35	0.21	0.26	0.16	0.21	0.12
2½	0.76	50	15.24	6.95	5.96	3.48	4.14	2.42	1.04	0.61	0.69	0.40	0.52	0.30	0.41	0.24
3	0.91	60	18.29	12.01	1,030	6.01	7.15	4.17	1.79	1.04	1.19	0.70	0.89	0.52	0.72	0.42
3½	1.07	70	21.34	19.07	16.36	9.59	11.36	6.66	2.84	1.67	1.90	1.11	1.42	0.83	1.14	0.67
4	1.22	80	24.38	28.47	24.42	14.24	16.96	9.89	4.24	2.47	2.83	1.65	2.12	1.24	1.70	1.00
5	1.52	100	30.48	55.61	47.69	27.81	33.12	19.31	8.28	4.83	5.53	3.22	4.14	2.41	3.31	1.93
6	1.83	120	36.58	96.09	82.41	48.05	57.23	33.37	14.31	8.34	9.56	5.59	7.15	4.17	5.72	3.34
8	2.44	160	48.77	228	195	114	135	79.17	33.75	19.79	22.55	13.22	16.88	9.90	13.50	7.92
10	3.01	200	60.96	445	382	223	266	155	66.50	38.75	44.42	25.89	33.25	19.38	26.60	15.50
12	3.66	240	73.15	769	659	385	458	267	115	66.75	76.49	44.59	57.25	33.38	45.80	26.70
16	4.88	320	97.54	1822	1563	911	1085	633	271	158	182	106	136	79.13	108.50	63.30
20	6.10	400	122	3559	3052	1780	2119	1236	530	309	354	206	265	155	211.90	124
24	7.32	480	146	6150	5274	3075	3663	2135	916	534	612	357	458	267	366	214
30	9.14	600	183	12,011	10300	6006	7153	4171	1788	1043	1195	697	894	521	715	417

TABLE IV

Coal/Slurry Quantities
Horizontal Rotating Bioreactors-Pilot and Commercial Capacities
Aqueous and Coal Feed Rates for 20% Slurry
Metric Factors (Length to Diameter Ratio = 20 to 1)

Diameter		Length		Feed Rates - (at 80% Diam = 85.7% of full diameter depth)									
				1 day-24 hr Retention		4 day 96 hr Retention		6 day-144 hr Retention		8 day-192 hr Retention		10 day-240 hr Retention	
				Aqueous	Coal	Aqueous	Coal	Aqueous	Coal	Aqueous	Coal	Aqueous	Coal
ft	meters	ft	meters	l/min	kg/min	l/min	kg/min	l/min	kg/min	l/min	kg/min	l/min	kg/min
2	0.61	40	12.33	2.12	0.032	0.53	0.008	0.354	0.005	0.265	0.004	0.212	0.0032
2½	0.76	50	15.24	4.14	0.062	1.04	0.016	0.691	0.010	0.518	0.008	0.419	0.0062
3	0.91	60	18.29	7.15	0.107	1.79	0.027	1.19	0.018	0.894	0.013	0.715	0.0107
3½	1.07	70	21.34	11.36	0.170	2.84	0.043	1.90	0.028	1.42	0.021	1.136	0.0170
4	1.22	80	24.38	16.96	0.254	4.24	0.064	2.83	0.042	2.12	0.032	1.696	0.0254
5	1.52	100	30.48	33.12	0.497	8.28	0.124	5.53	0.083	4.14	0.062	3.312	0.0497
6	1.83	120	36.58	57.23	0.858	14.31	0.215	9.56	0.143	7.15	0.107	5.723	0.0858
8	2.44	160	48.77	135	2.035	33.75	0.51	22.55	0.340	16.88	0.254	13.50	0.2035
10	3.01	200	60.96	265	3.974	66.25	1.00	44.25	0.664	33.13	0.50	26.50	0.3970
12	3.66	240	73.15	458	6.867	115	1.72	76.49	1.147	57.25	0.86	45.80	0.6867
16	4.88	320	97.54	1085	16.28	271	4.07	181.00	2.719	136	2.04	109	1.6280
20	6.10	400	122	2126	31.79	532	7.95	355.00	5.309	266	3.97	213	3.1790
24	7.32	480	146	3663	54.94	916	13.74	612	9.175	458	6.87	366	5.494
30	9.14	600	183	7153	85.84	1788	21.46	1195	14.335	894	10.73	715	8.584

TABLE V

Minimum Probable Air Quantities Required
Horizontal Rotating Bioreactor-Pilot and Commercial Capabilities
Metric Factors - (length to diameter ratio 20 to 1)
Air Rates for Two Times Stoichiometric Requirements

Diameter		Length		Air Requirements - (2X stoichiometric) (@ S.T.P.)									
				1 day-24 hr Retention		4 day-96 hr Retention		6 day-144 hr Retention		8 day-192 hr Retention		10 day-240 hr Retention	
ft	meters	ft	meters	gm/min	liters/min	gm/min	liters/min	gm/min	liters/min	gm/min	liters/min	gm/min	liters/min
2	0.61	40	12.33	8.172	6.66	2.043	1.67	1.365	1.11	1.022	0.83	0.817	0.67
2½	0.76	50	15.24	11.804	9.63	2.951	2.41	1.971	1.61	1.48	1.20	1.180	0.96
3	0.91	60	18.29	19.98	16.26	5.00	4.07	3.337	2.72	2.50	2.03	2.00	1.63
3½	1.07	70	21.34	35.41	28.83	8.85	7.21	5.91	4.81	4.43	3.60	3.54	2.88
4	1.22	80	24.38	50.84	41.35	12.71	10.34	8.49	6.91	6.36	5.17	5.08	4.14
5	1.52	100	30.48	98.06	79.86	24.52	19.97	16.38	13.34	12.26	9.98	9.81	7.99
6	1.83	120	36.58	173	140	43.25	35	28.89	23.38	21.65	18	17.30	14
8	2.44	160	48.77	404	329	101	82	67.47	54.94	50.50	41	40.40	33
10	3.01	200	60.96	792	646	198	162	133	108	99	81	79.20	65
12	3.66	240	73.15	1364	1112	341	278	228	186	171	139	137	111
16	4.88	320	97.54	3240	2639	810	660	541	441	405	330	324	264
20	6.10	400	122	6322	5154	1581	1289	1056	861	790	644	632	515
24	7.32	480	146	11078	9006	2770	2252	1850	1504	1385	1126	1108	901
30	9.14	600	183	17052	13877	4263	3467	2848	2317	2132	1735	1705	1388

TABLE VI

Nutrient Reagents
(Based on Sulfur Content as Sulfur)

Diameter ft	Sulfur (gm/min) lb/min	H_2SO_4[2]	NH_3[1] (25% soln)	K_2HPO_4[1]	$FeSO_4$[1] 5.4 g/l (soln)
2	0.002		0.001	0.009	0.123
2½	0.003		0.002	0.013	0.185
3	0.005		0.003	0.022	0.31
3½	0.009		0.006	0.039	0.55
4	0.013		0.009	0.056	0.80
5	0.025		0.017	0.108	1.54
6	0.044		0.030	0.189	2.71
8	0.103		0.070	0.440	6.33
10	0.202		0.140	0.870	12.42
12	0.348		0.24	1.500	21.40
16	0.826		0.56	3.550	51.00
20	1.612		1.09	6.930	99.00
24	2.783		1.89	11.970	171.00
30	4.347		2.95	18.690	267.00
[1] Each lb of S.A. Gold Min Req's			0.6781	4.2948	61.4823

[2] H_2SO_4 to adjust pH as required to 1.7±.

TABLE VII

Horizontal Rotating Drum Bioreactor
Desulfurization of Coal via Biohydrometallurgy With TBF
Capacities - metric

Diameter		Length	Volume (@ 80% Depth = 85.7% Capacity)		Coal[1]		Slurry[2] (Aqueous)
ft	meters	meters	cu meters	kilograms	kg/day	metric tons/yr[3]	liters/min
2	0.61	12.33	3.05	3050	45.85	16.67	2.12
2½	0.76	15.24	5.96	5964	89.44	32.61	4.14
3	0.91	18.29	10.30	10302	154	56.34	7.15
3½	1.07	21.34	16.36	16364	245	89.47	11.36
4	1.22	24.38	24.42	24434	366	133.61	16.96
5	1.52	30.48	47.69	47711	716	260.89	33.12
6	1.83	36.58	82.41	82447	1236	450.83	57.23
8	2.44	48.77	195	195428	2930	1069	135
10	3.01	60.96	382	381699	5723	2087	265
12	3.66	73.15	659	659522	9889	3606	458
16	4.88	97.54	1563	1563339	23441	8548	1085
20	6.10	121.92	3052	3053378	45783	16696	2126
24	7.32	146.30	5274	5276308	79114	28851	3663
30	9.14	182.88	10300	10346477	123616	45071	7153

1 Based on 2.3% total sulfur reduced 90% in 10 days. Coal particle <100 μm with density of 1.5 g/cm^3.
2 Slurry density - 20% coal to 80% water (w/w)/24-hr flow.
3 1 ton (metric) = 2205 lb.

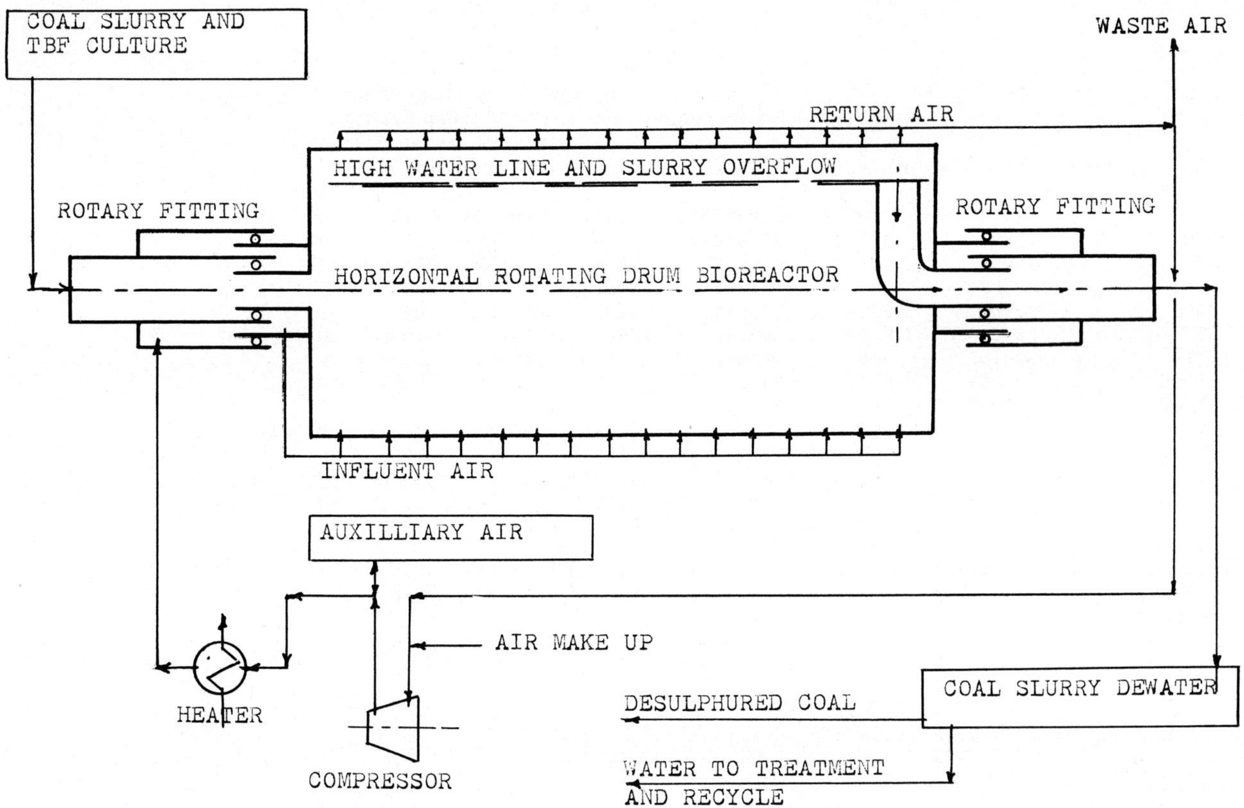

Fig. III. Technometrics Inc.-Simplex Flow Sheet-Biohydrometallurgical Desulphurization of Coal Via Use of Thiobacillus Ferrooxidans in Horizontal Rotating Bioreactor.

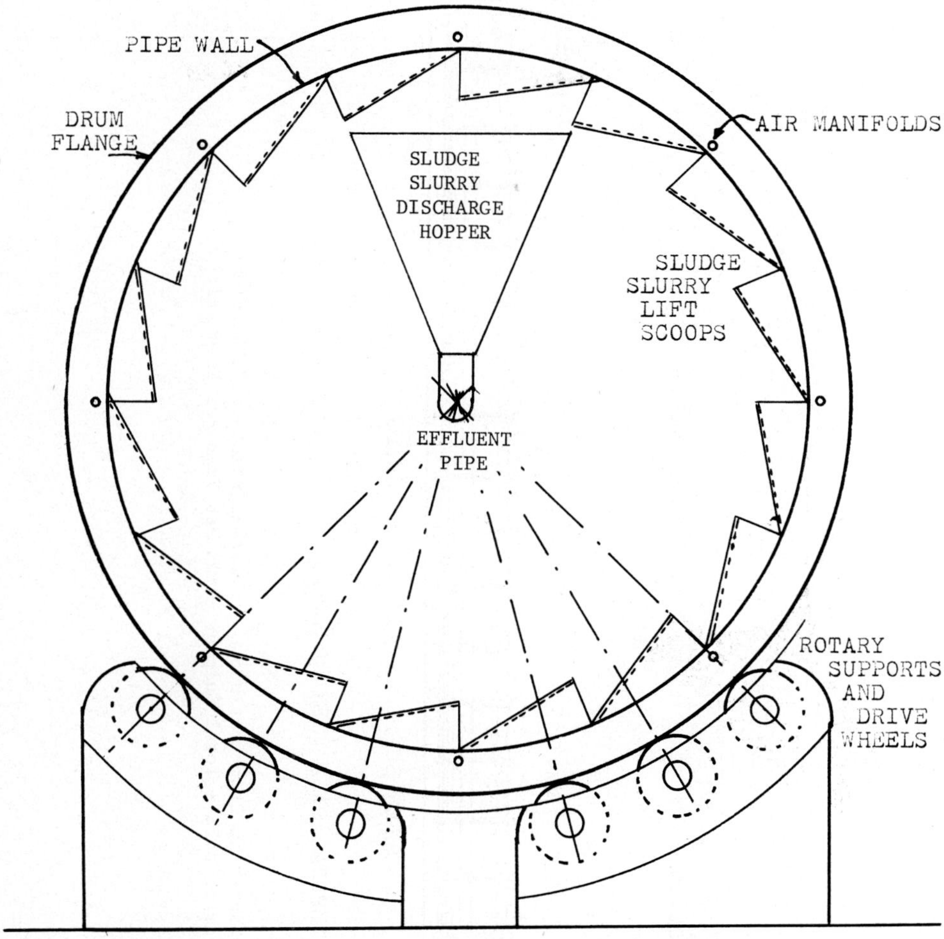

Fig. IV. Technometrics Inc. -Horizontal Rotating Drum Bioreactor Typical Section-At Effluent End-Sludge Slurry Lift Scoops, U.S. Patent # 4,233,094.

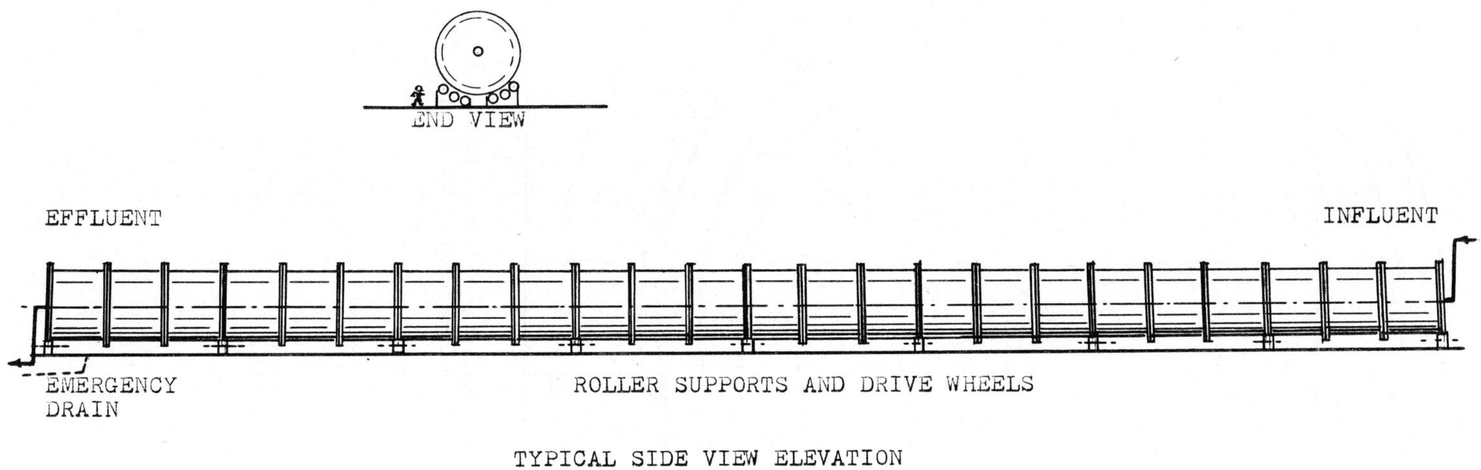

Fig. V. Technometrics Inc. Horizontal Rotating Drum-Continuous Process-Bioreactor (7.32 Meters dia. x 146.3 Meters), 25,000 Metric Tons of 3.2% + Total Sulphur -Coal Per Year Per Each Unit (U.S. Patent # 4,233,094).

REFERENCES

1 "A Special Report in the Public Interest-Energy-Facing Up to the Problem, Getting Down to Solutions." National Geographic, February 1981.

2 "The Coal Slurry-Railroad Controversy Remains Unsettled; Needed, Say Slurry Proponets, is Federal Eminent Domain." Compressed Air Magazine, March 1983.

3 "UK Method Removes More Sulfur from Coal." Research and Development, February 1985, p. 98.

4 T.F. Hubner, N.W.F. Kossen, P. Bos, and J.G. Kuenen; "Design and Scale-Up of a Reactor for Microbial Desulphurization of Coal: A Regime Analysis". Laboratory of Bioengineering, Dept. of Chemical Technology, Delft. University of Technology, Julianalaan 67,2628 B.C. Delft, Netherlands.

5 Kos, C.A., Poorter, R.P.E., Bos, P. and Kuenen, J.G., "Geochemistry of Sulphides in Coal and Microbial Leaching Experiments." Int. Conf. on Coal Science, Dusseldorf, 1981, 842-847.

6 Dugan, P.R. and Apel, W.A., 'Microbial Desulphurization of Coal" in: Metallurgical Applications of Bacterial Leaching and Related Phenomina. L.E. Murr, A.E. Torma and J.A. Brierly (BOS). Academic Press, New York, 1978, 223-250.

7 Hoffman, M.R., Faust, B.C., Panda, F.A., Koo, H.H., and Tsuchija, H.M., "Kinetics of the Removal of Iron Pyrite from Coal by Microbial Catalysis." Appl. Environ. Microbia., 42 (1981), 259-271.

8 Detz, C.M. and Barvinchak, G., "Microbial Desulphurization of Coal." Min. Congr. J., 65. (1979), 75-85.

9 Kargi, F., "Micobiological Coal Desulphurization." Enz. Micro. Technol., 4, (1982), 13-19.

10 Hiltunen, P., Vuorinen, A., Rethijarvi, P. and Tuovinen, O.H., "Bacterial Pyrite Oxidation: Release of Iron and Scanning Electron Microscopic Observations." Hydrometallurgy, 7, (1981), 147-151.

11 Rinder, G., and Bauer, E., "Micro Biologische Entpyritisierung von Kohlen in Suspension." Eroöl und Kohle-Erdgas-Petrochemie, 36,(1983), 170-174.

12 Kos, C.H., Bijleveld, W. Grotenhuris, T.,Bos P., Kuenen, J.G. and Poorter, R.P.E., "Composition of Mineral Salts Medium for Microbial Desulphurization of Coal." Proc. Int. Symp. Biohydrometallurgy, Cagliari, Italy 1-4, 1983.

13 Hubner, T.F., Russen, N.W.F., Bos, P. and Kuenen, J.G., "Modelling Design and Scale Up Reactor for Microbial Desulphurization of Coal." Proc. Int. Symp. Biohydrometallurgy Cagliari, Italy, v. 4, May 1983.

14 Chang, Y.C., and Myerson, A.S., "Growth Model of the Continuous Bacterial Leaching of Iron Pyrite by Thiobacillus Ferooxidans." Biotecknol. Bioeng., 24, (1982), 889-902.

15 T.F. Hubner, E. Ras, and N.W.F. Kossen, "Design and Scale Up of a Reactor for the Microbial Desulphurization of Coal: A Kinetic Model for Bacterial Growth and Pyrite Oxidation," Laboratory of Bioengineering Department, Dept. of Chemical Technology, Dept. Univ. of Technology, Julianalaan. 67, 2628 B.C. Delft, The Netherlands.

16 T.F. Hubner, N.W.F. Kossen, P. Bos and J.G. Kuenen, "Modelling Design and Scale Up of a Reactor for Microbial Desulphurization of Coal." First Symposium, on Biohydrometallurgy, Cagliari, Italy, May 1-4, 1983.

17 T.F. Hubner, N.W.F. Kossen, P. Bos, and J.G. Kuenen, "Design and Scale Up of a Reactor for Microbial Desulphurization of Coal: A Regime Analysis," Laboratory of Bioengineering, and Laboratory of Microbiology; Delft Univ. of Technology, Delft, The Netherlands.

18 E. Livesly-Goldblatt, P. Norman and D.R. Livesly Goldblatt, "Recent Progress in Biohydrometallurgy," 1983. Associazione Miner Aria Sarda, Iglasias (Italy) 983. (Table I, p. 636).

THE VALUE ADDED TO COAL BY MICROBIAL SULFUR REMOVAL

PATRICK R. DUGAN
Department of Microbiology, The Ohio State University, Columbus, Ohio 43210

ABSTRACT

The theoretical and experimental basis for microbiological coal desulfurization, both pyritic and organic, is well established. Analysis of the 1984 price of coal delivered to utilities in Ohio compared to total sulfur content indicates that a premium has been paid for low sulfur content. The premium paid is approximately $13/Ton/1% S reduction when total sulfur is in the range 1 to 3% and in excess of $26/Ton/1% sulfur reduction when the total sulfur was less than 1%.

INTRODUCTION

The theoretical and experimental basis for removal of both pyritic and organic sulfur from coal using the metabolic capabilities of microorganisms is well established (9, 10, 11, 16, 17, 22). Acidophilic iron and sulfur oxidizing autotrophic bacteria acting in concert with acid tolerant heterotrophic microbes are very active agents for removal of pyritic sulfur (1, 2, 7, 12, 14, 19, 21, 25, 26, 28). Both mesophilic and thermophilic microbes are effective with removal of pyrite, including the finely dispersed fraction which resists conventional cleaning (4, 10). Although newer physical and chemical "deep cleaning" techniques are capable of removing pyritic sulfur, they do so at considerable expense and with a significant loss of net revoverable energy (12) expressed as British Thermal Units (BTUs). Organic sulfur in coal consists primarily of dibenzothiophene, benzothiophene, disulfide groups linking both cyclic and alkyl carbon groups, and sulfide attached to various carbon groups (12, 24, 29). The relative amount of each organic sulfur form as well as the total organic sulfur content varies with the coal sample and progress has been made on the microbial removal of organic sulfur (3, 5, 13, 17, 18, 20, 21, 27). Microbes have also been isolated which have the ability to hydrolyze and/or oxidize the chemicals known to be organic sulfur forms in coal and oil and which may be capable of removing those substances from coal (24).

Mineral components in coal, including pyrite, are responsible for its ash content. A decreased ash content results in a higher BTU value per ton. The process of microbial pyrite removal converts insoluble pyritic sulfur to soluble sulfate in the form of both metal sulfate salts and sulfuric acid and the acid produced is capable of leaching some of the ash components out of the coal (8, 11). This, in addition to the growing concern with the contribution of high sulfur fuel combustion to the atmospheric sulfur load, has stimulated interest in microbial desulfurization.

It is postulated that both pyritic and organic sulfur can be removed microbiologically from coal via a two stage process.

Because the process requires either direct contact between the microbe (or its enzymes) and the sulfur form or in the case of pyrite, contact between ferric iron and reduced sulfur; the process requires pulverized coal in an aqueous slurry. The process is therefore compatable with transport by slurry pipeline and the emerging technology of coal-water fuels.

Major concerns which have delayed commercial development of the microbial processes for coal refining are: (A) the lack of understanding of the process within the coal mining and combustion industries and governmental agencies. The process does not directly adapt to either the equipment or the technology currently used for coal cleaning, i.e. it is more analogous to wastewater treatment technology. (B) A limited analysis of cost to benefit for the process in addition to a lack of incentive, both financial and environmental, to solve the high sulfur coal problem.

METHODS

It is assumed that coal which has been refined by removal of sulfur and ash becomes more valuable. One method of estimating the value added is to determine the premium which has been paid for low sulfur content and then to relate that premium to the value added via refining.

Evaluations were made of the 1984 price paid for coal deliveries to Ohio utilities in relation to the total sulfur content of the coal.

Many factors, in addition to sulfur content, contribute to coal cost, e.g. ash content, energy value, transportation, location and depth of the coal seam, spot purchase vs. long term contract and whether the coal is "captive" or "dedicated" to a specific user.

The relationship between sulfur content and price was established in a previous report (11). This report further defines the value added to coal by sulfur removal.

The premium paid for low sulfur coal can be estimated by charting the cost of coal versus the sulfur content with the following assumptions:
(A) coal ash content is related to its energy content and the influence of ash in reduced energy value is at least partially accounted for when price is reported as "cents per million British Thermal Units (BTU^s)."

(B) The influence of "captive" or "dedicated" coal on price can be largely avoided by eliminating from consideration the in-state deliveries to in-state utilities.

(C) Large price fluctuations were observed with small coal purchases of identical quality coal by a single utility, often from the same mine. This influence was reduced by considering only deliveries in excess of 5000 tons. Another arbitrary delivery tonnage could have been selected but higher tonnage units result in fewer data points and lower units re-introduce what appear to be "false" fluctuations.

(D) The transportation cost factor can be reduced by restricting consideration either to deliveries from a single state to Ohio or to deliveries from a single state to a single power company in Ohio.

RESULTS AND DISCUSSION

The influence of "captive" coal can be seen in Fig. 1 where coal deliveries of greater than 5000 tons from mines within Ohio to Ohio utilities have been charted as points representing cents per million BTU[s] vs. percent sulfur. Those points above the line represent deliveries identified as "captive coal" from two different mines. Some points below the line also represent captive coal. No further use of data from Ohio mine deliveries to Ohio utilities are made in this report.

Figure 2 represents data showing cost of coal vs. percent sulfur for deliveries greater than 5000 tons from mines in Kentucky, Pennsylvania and West Virginia to a single power company in Ohio. Each point represents the closest approximation to 10,000 tons delivered.

Greater than 5000 ton deliveries from a single state (Kentucky) to a different power company in Ohio are shown in Fig. 3 using the same criteria as for Fig. 2.

Similar data for deliveries from a different state to a third power company are presented in Fig. 4.

When cost data vs. percent sulfur for coal deliveries form individual states outside Ohio to power companies in Ohio are charted collectively, a series of linear relationships emerge (See Fig. 5) which indicates that there is a direct relationship between cost of coal and its sulfur content and the power companies have been paying a premium for low sulfur content, at least as far back as 1978 (11). The displacement of individual lines on the graph is not significant to the relationship of cost differential vs percent sulfur content. Rather, it is the slope of each line that defines cost differential vs percent sulfur and the slopes are quite similar for ranges of sulfur content. This indicates that cost differential vs sulfur content is surprisingly uniform.

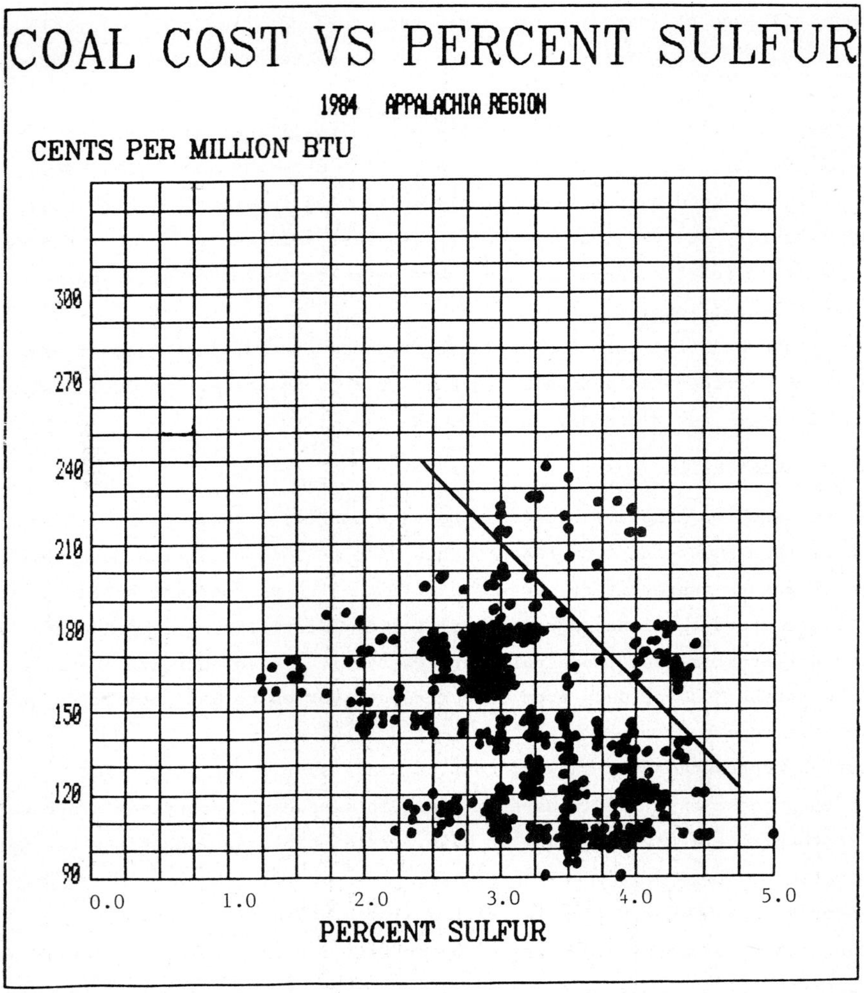

Fig. 1. Diagram showing lack of relationship of price vs. sulfur content of coal mined in Ohio and delivered to utilities in Ohio. Points above the line represent deliveries of "captive" coal from two mines. Each point represents approximation of 10,000 tons delivered and no deliveries of less than 5000 tons were used.

The amount of cost premium paid for reduced sulfur content is summarized in Fig. 6. This curve can be used to estimate the increased value of coal (in 1984 dollars) which results from coal desulfurization. For example, an estimate of the value increase per ton of coal which has been reduced from 4.5% sulfur to 1.00% sulfur via microbiological treatment is approximately $1.15 per million BTU[s]. If the coal was 13,000 BTU per pound, the increase in value

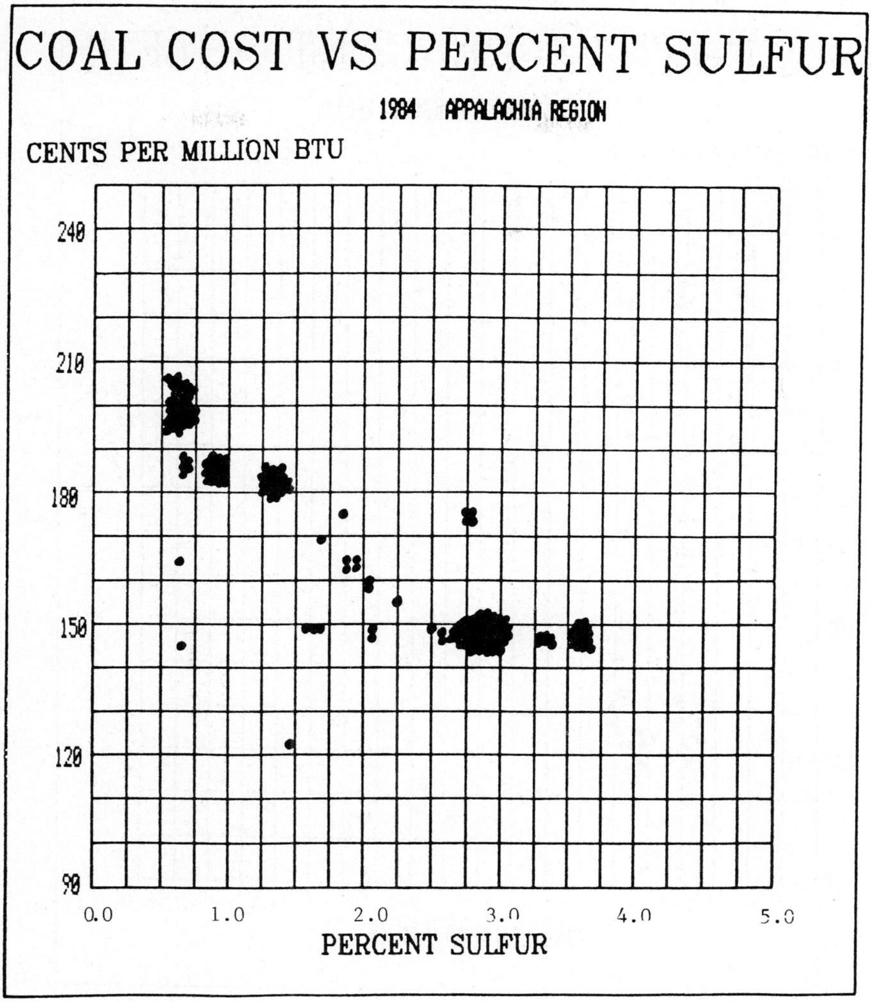

Fig. 2. Points represent greater than 5000 ton deliveries from Kentucky, Pennsylvania and West Virginia to a single power company in Ohio and show a relationship between cost vs. percent sulfur in the coal.

would be about \$30 per ton (\$32 per ton if the BTU value incresed to 14,000/lb as a result of treatment).

An early estimate of the cost of biological treatment, extrapolated from 1978 (\$12/ton) dollars to 1984 dollars, is approximately \$18 per ton (6,11) and provides continued incentive for process development.

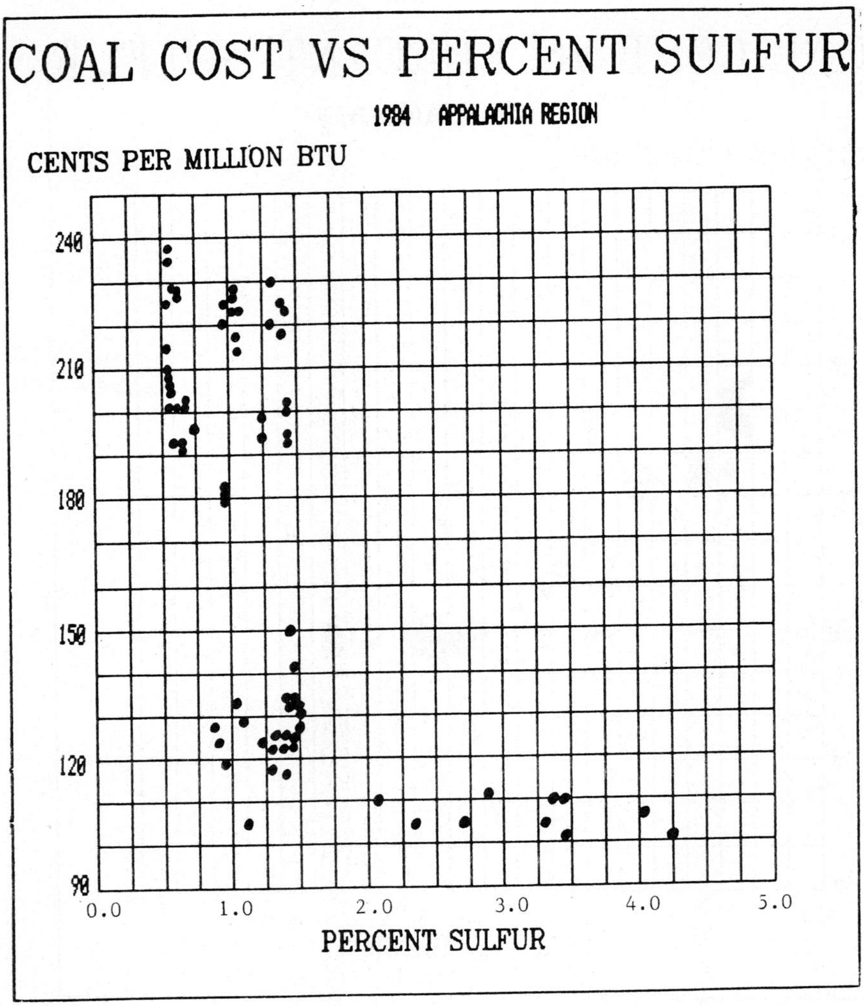

Fig. 3. Points represent greater than 5000 ton coal deliveries from Kentucky to a single power company in Ohio and illustrates a relationship between cost and percent sulfur.

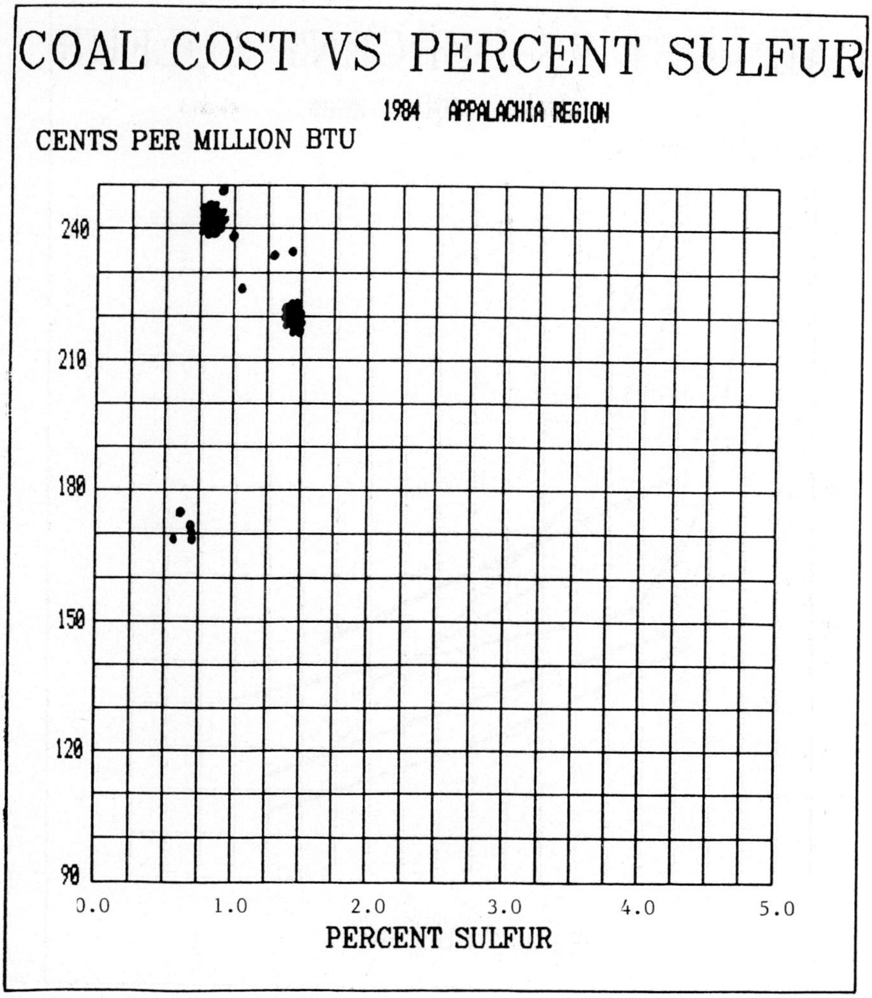

Fig. 4. Points illustrating greater than 5000 ton deliveries from Pennsylvania mines to a single power company in Ohio (different from company in Fig. 3).

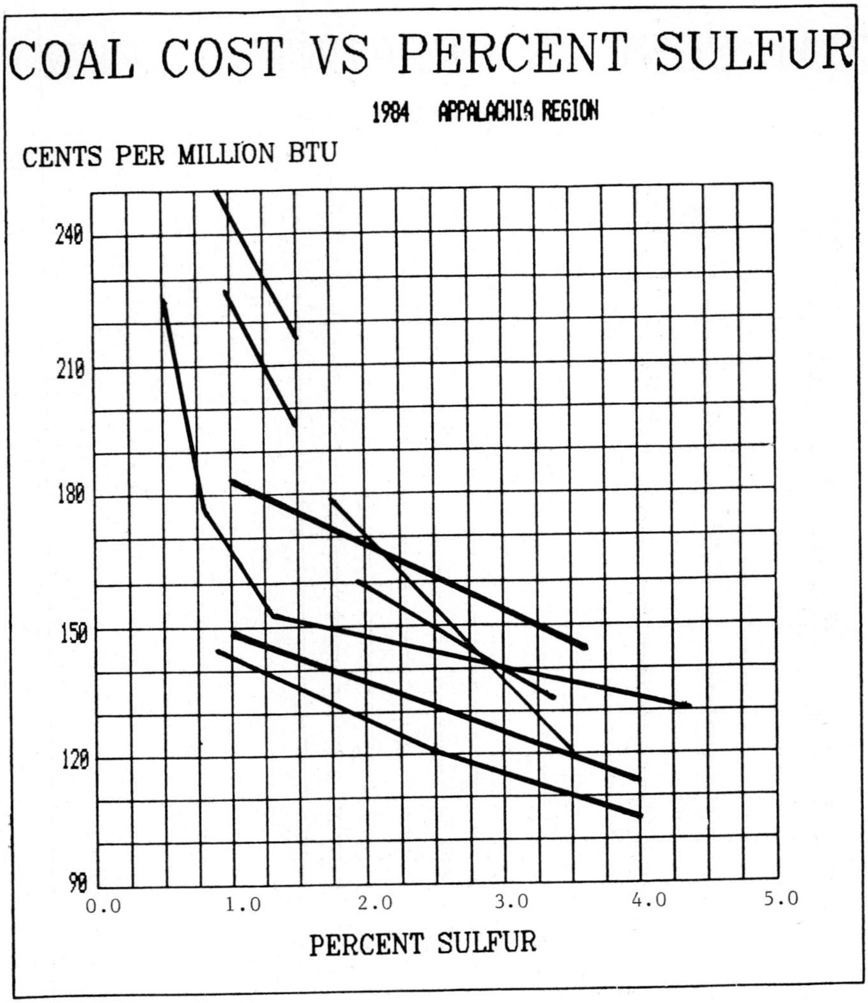

Fig. 5. Summary showing slopes of lines representing cost vs. percent sulfur relationships for greater than 5000 ton deliveries from various mines in Appalachia to individual power companies in Ohio.

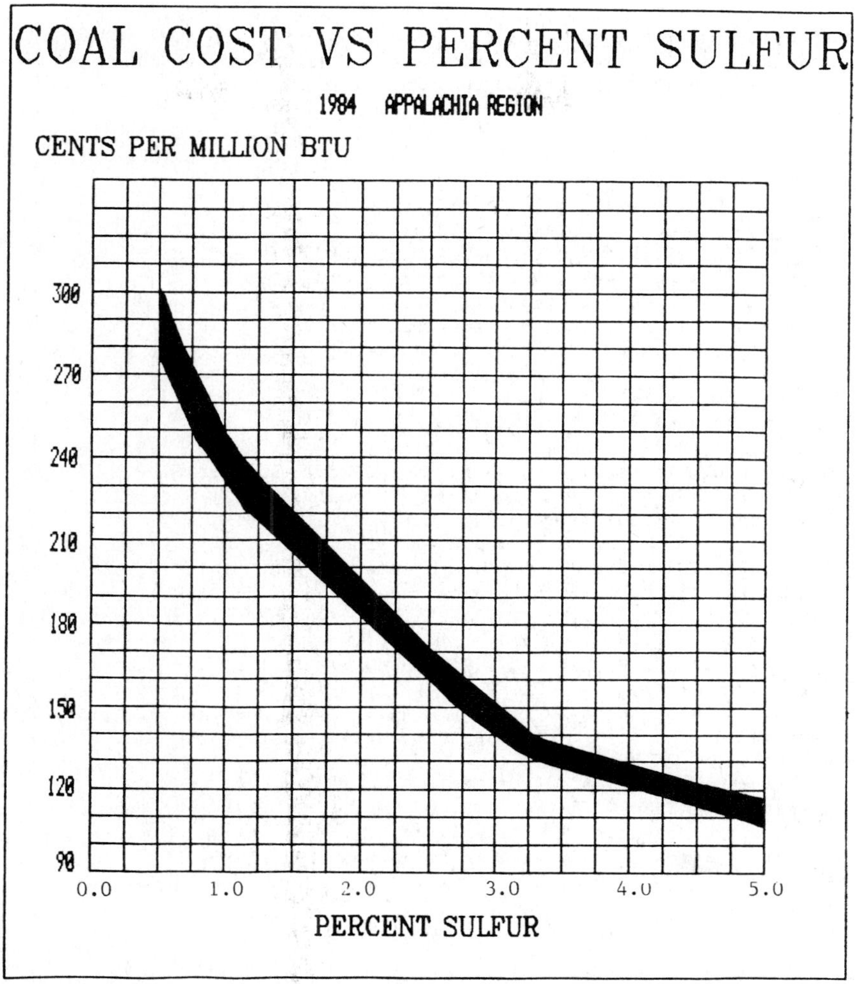

Fig. 6. Summary curve illustrating the relationship between cost of coal and percent sulfur content.

REFERENCES

1. G.F. Andrews and J. Maczuga, Biotechnol. Bioeng. Symp. No. 12, (John Wiley and Son's, N.Y., 1982), p. 337.
2. G.F. Andrews and J. Maczuga, Fuel, 63, 297 (1984).
3. J.A. Brierly, P.R. Norris, D.P. Kelly and N.W. LeRoux, European J. Appl. Microbiol. Biotechl, 5, 291 (1978).
4. J.A. Cavallaro, M.T. Johnston and A.W. Deurbrock, U.S. Dept. Interior, Bureau of Mines Rept. of Investigations, RI 8118 (1976), 323 pp.
5. D. Chandra, P. Roy, A.K. Mishra, J.H. Chakrabarti and B. Sengupta, Fuel, 58, 549 (1979).
6. C.M. Detz and G. Barvinchak, Mining Congr. Jour., 65, 75 (1979).

7. P.R. Dugan and W.A. Apel, in Proc. Third NCA/BCR Coal Conf. on Coal Preparation, National Coal Assoc., Washington, D.C., pp 10-21 (1977).
8. P.R. Dugan and W.A. Apel, In Metalurgical Applications of Bacteriol. Leaching and Related Microbiological Phenomena, L. Murr, A. Torma and J. Brierly, Eds. (Academic Press, N.Y., N.Y., 1977), pp. 223-246.
9. P.R. Dugan, in Microbial Chemoautotrophy, 8th Colloquium College of Biological Sciences, WR. STrohl and O. Tuovinen, Eds. (The Ohio State Univ. Press, Columbus, 1984) pp. 3-9.
10. P.R. Dugan, U.S. Patent No. 4,456,688 (1984).
11. P.R. Dugan, Biotechnol. Bioeng. (1985). In press.
12. U.S. Dept. Energy, Pittsburgh Energy Tech Ctr., Proceed. of Coal Water Fuel Tech. Workshop, Rept. No. BNL 51427, p. 96 (1981).
13. C.F. Gokcay and R.N. Yurteri, Fuel, 62, 1223 (1983).
14. M. Golomb and E. Beier, Bergbau 28, 419 (1977).
15. S.N. Groudev, F.N. Genchev and V.I. Grovdeva, XY Traditional Sci. Conf. on Mineral Processing, Inst. Mineral Processing, Cracow, Poland, (1981).
16. T.F. Huber, N.W.F. Kossen, P. Box and J.G. Kuenen, Int. Symp. Biohydromettalurgy, Italy, (1983).
17. J.D. Isbister, A.E. Desouza, E.A. Kobylinski, G.L. Anspach and J.F. Kitchens, Prog. 1st Ann. Pittsburgh Coal Conf. 463 (1984).
18. F. Kargi and J.M. Robinson, Biotechnol. Bioeng., 24, 2115 (1982).
19. F. Kargi and T.D. Cervoni, Biotech. Letters, 5, 33 (1983).
20. F. Kargi and J.G. Weissman, Biotechnol. Bioeng. 26, 604 (1984).
21. F. Kargi and J.M. Robinson, Appl. and Environ. Microbiol. 26, 687 (1984).
22. F. Kargi and J.M. Robinson, Biotech. and Bioengin. 27 (1985) 41-50.
23. C.H. Kos, R.P.E. Poorter, P. Box and J.G. Kuenen, Proc. Int. Conf. Coal Sci. Dusseldorf, p. 842, (1981).
24. D.J. Monticello, D. Bakker and W.R. Finnerty. Applied and Environ. Microbiol. 49 (1985) 756-760.
25. J. Murphy, E. Ristenberg, D. Marek, R. Mohler, B. Beck and D. Skidmore, Proc. Symp. on Microbiol. Process in Mineral Indust., AIME Feb. (1985) in press.
26. L.E. Murr and A.P. Mehta, Biotechnol. Bioeng., 24, 743 (1982).
27. K.A. Natarajan, Trans. Indian Inst. of Metals, 31, 224 (1978).
28. T.M. Olsen, P.R. Ashman, A.E. Torma and L.E. Murr, Biogeochem. of Ancient and Modern Environ., Australian Acad. Sci. p. 693 (1980).
29. C.L. Spriro, J. Wong, F.W. Lytle, R.B. Greegor, D.H. Maylotte, S.H. Lanson, Science, 226, 48 (1984).

PART 10

NEW APPLICATIONS FOR HIGH SULFUR COALS

THE USE OF COAL FOR CONVERSION OF SULFUR DIOXIDE TO ELEMENTAL SULFUR

D.S. MOULTON[1]

[1]Southwest Research Institute, 6220 Culebra Road, San Antonio, TX (USA)

ABSTRACT

Several processes are available, or under development, for removing sulfur dioxide from dilute sources such as flue gas. In many cases, the absorbing material can be economically regenerated, producing a stream of sulfur dioxide sufficiently concentrated that it can be used for elemental sulfur production. High-sulfur coal is an abundant and inexpensive reducing agent for the conversion reaction, but few processes are available because coal tars are co-produced which are difficult to separate from the product sulfur, and the agglomerating character of many coals complicates reactor design.

Previous work at Argonne National Laboratory indicates that high-sulfur bituminous coal can be used for the conversion reaction. Separation of the coal devolatilization reaction from char oxidation coupled with the use of fluidized-beds could provide a processing scheme to avoid coal agglomeration and reactor plugging. The catalytic properties of the coal ash could be used to reduce tar production to a small fraction of the sulfur production. For using sulfur in highway construction, small amounts of tar in the product may be permissable. Work pertaining to the use of coal for converting sulfur dioxide to elemental sulfur is reviewed, and an approach which uses a series of fluidized-bed reactors is suggested as a basis for initial process development.

INTRODUCTION

During the processing of materials, sulfur is frequently removed as either hydrogen sulfide (H_2S) or as sulfur dioxide (SO_2). Several well developed processes are in use for converting H_2S to elemental sulfur, but processes for converting SO_2 are used less frequently. A major reason for the dichotomy is the expense of reducing agents required for SO_2 conversion compared to the oxygen or air required for conversion of H_2S. Hydrogen is best suited for the production of high purity sulfur from SO_2, but it is also very expensive. High sulfur coal is the least expensive reducing agent available, but special processes need to be designed to avoid contamination of the product sulfur, and other problems incident to the use of coal.

SOURCES OF SULFUR DIOXIDE

Sulfur dioxide from most industrial processes is obtained in a stream which is too dilute for direct conversion. The first step in the conversion process is concentration of the SO_2 to about 10% or higher. This is most commonly done by an absorption-desorption sequence.

In fluidized-bed combustors, sulfur dioxide is absorbed by calcined limestone. Desorption can be accomplished by an absorbent regeneration process involving combustion under reducing conditions (sub-stoichiometric air) at a somewhat higher temperature than was used in the combustor. Process studies by Smyk and Snyder (1*) indicate that SO_2 can be concentrated to 8-25% depending on whether air or oxygen is used in the regenerator. The coal consumption required to produce the concentrated stream is only 2-3% as much as used in the fluidized-bed combustor.

In the Bergbau-Forschung process (2,3,4) sulfur dioxide is adsorbed on a char. Desorption is accomplished by steam stripping to provide an off-gas containing 20-30 percent SO_2 in steam. This process, now under development by Foster-Wheeler, is one of the few modern processes designed to use coal to convert SO_2 to sulfur.

The Sulf-X process, (5) under development by Pittsburgh Environmental and Energy Systems Inc. also uses coal to convert SO_2 to sulfur, but the chemistry is more complex. The SO_2 is absorbed in a solution of ferrous sulfide and oxidized to form ferrous sulfate and elemental sulfur. The ferrous sulfate is treated with sodium sulfide to regenerate ferrous sulfide and form sodium sulfate. The sodium sulfate solution is evaporated and the crystalline residue is calcined with coke to form the sodium sulfide required by the process. Makeup iron is supplied as iron pyrite, an inexpensive industrial by-product. The Sulf-X process has been tested in small-scale demonstration projects.

Other processes for absorbing and concentrating SO_2 are not tied to any particular sulfur conversion process. United Engineers and Constructors (6) have installed their Regenerative Magnesium Oxide FGD process on powerplants in the 350 mw range. The SO_2 product from the regenerator is suitable for conversion to sulfur or to sulfuric acid. The Davy McKee Wellman-Lord process uses a sulfite-bisulfite solution for absorption and a selective crystallization process for concentration of the sulfur dioxide. The Wellman-Lord process has been installed on several large utility powerplants.

Some potential processes could make a future contribution if successful development were to be carried out. Other materials which are being examined for their SO_2 absorption potential include copper oxide on an alumina support, and several liquid systems. The liquids generally involve bases such as aqueous solutions of sodium carbonate or some of the amines used in gas sweetening processes. It seems likely that expanded options may become available for acquiring SO_2 from dilute streams and concentrating it for further processing. This would provide opportunities to apply processes for converting SO_2 to elemental sulfur which are based on coal.

THE USE OF COAL FOR SO_2 CONVERSION

A major part of the cost of converting SO_2 to elemental sulfur is the cost of the reductant. To use low cost coal, two major technical problems must be overcome:

1. The swelling and caking properties of many coals cause plugged reactors.
2. The coal volatile matter contains tars which condense with the sulfur to produce an impure product.

---

* Numbers in parentheses refer to references at the end of this paper.

The reaction between SO_2 and dry coal is exothermic. The heat of reaction was estimated using equation (1). More detailed coal formulas could be used for specific coals in a similar equation, but as a general example the moisture and ash-free (MAF) coal was assumed to contain hydrogen and carbon in the ratio 0.85 on a mole basis.

$$1.21\,O_2 + CH_{0.85} \longrightarrow CO_2 + 0.43H_2O \qquad (1)$$

When SO_2 is substituted for oxygen to obtain reaction (2) for sulfur production, the difference in the heat of reaction can be calculated from the SO_2 heat of formation. For this example, it amounts to about 12040 Btu/lb MAF coal.

$$1.21\,SO_2 + CH_{0.85} \longrightarrow CO_2 + 0.43\,H_2O + 1.21\,S^o \qquad (2)$$

The heat of combustion, reaction (1), could be about 15,000 Btu/lb MAF coal, which implies a heat of reaction for reaction (2) as high as 2960 Btu/lb MAF coal under ideal conditions. If low rank or wet coal were to be used, the heat of reaction would be lower, but in many situations useful heat could be obtained from the process.

At Argonne National Laboratory, (7) the fundamentals of SO_2 reduction were investigated with the following objectives:

1. Find a processing scheme using coal or coal char as the reductant.
2. Make preliminary laboratory measurements of the reaction rates.
3. Determine the product purity to be expected from a simple processing scheme using coal.

If coal were to be used as the reductant, the coal could be used directly, or it could be gasified and the synthesis gas would be used. If the coal were to be used directly, the process would be simpler than if a separate gasification step should be required. Also, the SO_2 and coal volatiles would come in contact with the coal ash. It was therefore of interest to review the characteristics of SO_2 reduction catalysts to see if the coal ash could be useful. It was recognized that its low surface area would prevent a high specific activity, even if coal ash did have some of the characteristics of SO_2 reduction catalysts. However, the ash would be available at essentially no cost to the process, so having a high specific activity was less important for this application than for most catalyst applications.

Sulfur Dioxide Reduction Catalysts

The traditional catalysts for Claus processes are bauxites, naturally occurring aluminum oxides, which are also high in iron content. In the Claus process, the SO_2 is reduced with hydrogen sulfide. Bauxites with a high iron content are also effective for other SO_2 reduction reactions. A number of other metals besides iron are also active but have not had much commercial use. Under reaction conditions, the iron or other metal becomes sulfided, and the metal sulfide is generally thought to be the main catalytic agent.

Chowdhury and Datta (8) obtained evidence that aluminum oxide, the main bauxite component, also contributes to the catalytic activity. Commercially produced aluminum oxides are well known as attrition-resistant high-surface-area catalyst supports. Haas et al. (9) showed that the activity could not be due to its surface area alone and found that there was a sharply defined iron to aluminum oxide ratio for maximum activity. The optimum composition was especially effective for dilute reactant streams. This implies that aluminum oxide is also a catalyst for SO_2 reduction.

George (10) showed that the Claus reaction is catalyzed by bases in the activity order, Li Na K. It has been established by Peri (11,12) that the dehydrated gamma alumina, a form of aluminum oxide, has oxide ions exposed on its surface which are strongly basic. These interact with the SO_2, which is a Lewis acid. Chang (13) in an IR study of adsorbed SO_2 found strong evidence that on alumina, the SO_2 is adsorbed on oxide ions, forming SO_3^{2-} species.

In summary, to possess maximum activity, a SO_2 reduction catalyst should contain a metal, commonly iron, which becomes sulfided under reaction conditions, and material which is chemically a strong base, and these should be present in a optimum ratio.

Although ash compositions are extremely variable, the ash of many coals contains iron, aluminum oxide, and other bases. Hendrickson (14) gives the following average analysis of U.S. coal ash determined by the U.S. Bureau of Mines:

Compound	Wt.%
Silicon dioxide (SiO_2)	45.7
Aluminum oxide (Al_2O_3)	26.0
Iron oxide (Fe_2O_3)	18.1

Bases such as calcium, magnesium, sodium and potassium oxides are prominent ash constituents and have catalytic value. While the ash of many coals would undoubtedly lack the required properties, it is also true that many should have at least a low catalytic value because iron oxide, aluminum oxide and other bases are very common constituents of coal ash.

European experience with coke confirms the expectation that coal ash has a catalytic value. Coke made from coal has been used in Europe for SO_2 reduction for a long time. Macak and Pick (15) investigated cokes with different amounts of ash and found that coke with a high ash content was much more reactive than coke with a low ash content. Spectrally pure carbon was the least reactive. They also prepared synthetic cokes containing single specific ash components. Iron oxide, calcium oxide, and magnesium oxide caused the greatest increase in SO_2 reaction rates.

LABORATORY EXPERIMENTS

Rate Measurements

Some laboratory scale work was done at Argonne National Laboratory with coal, a devolatilized coal or char, coke, and ash to make preliminary measurements of the reaction rates, and to determine the purity of sulfur that could be produced. A small flow system was set up with a quartz U-tube reactor in an electric furnace. A fixed bed containing the reducing agent was placed in the reactor. Following the reactor, the gas flowed through a sulfur condenser and out to the vent. Conversion was measured with a gas chromatograph. Composition of the coal is given in Table 1:

TABLE 1. COAL PROPERTIES

Proximate Analysis		Ultimate Analysis*	
Moisture	1.11%	Carbon	81.3
Ash	19.06%	Hydrogen	5.5
Volatile Matter	36.56%	Nitrogen	1.3
Fixed Carbon	43.27%	Sulfur	6.8
		Oxygen	5.0

* Moisture and ash free basis

The ash was obtained by controlled oxidation of the coal at 750-850°C and contained 17% iron and 13% calcium oxide. The char was obtained from the cyclone separator of a fluidized-bed combustor, and had been devolatilized in the combustor at about 850°C. The coke was petroleum based. The gas stream was 9.6% SO_2 in N_2.

Reaction rates were measured at several temperatures. A small sample of the reducing agent was placed in the reactor and brought to temperature. The gas flow was adjusted to provide a 12 second apparent contact time. The extent of conversion was indicated by the SO_2 concentration remaining in the gas stream after passing through the reactor.

The first set of measurements were made to observe the catalytic value of the ash. Finely ground coke was mixed with ash and a control sample was made using sintered aluminum oxide (alpha alumina) which was believed to have very little, if any, catalytic activity, but its other properties resembled the ash. The results are shown in Figure 1. In the presence of the ash, lower curve, more sulfur dioxide was removed from the effluent than without the ash, upper curve. The temperature required to maintain a particular reaction rate was about 75°C (135°F) lower with the ash than with the inert diluent.

A similar set of measurements were made with 35% char in ash. The reaction rates were much higher than when the coke was used. The results, shown in Figure 2, indicate that the 12 second contact time was sufficient for reacting all the sulfur at 700°C (1290°F) or higher. With both coke and char, the reaction rates decreased rapidly while the reaction proceeded at constant temperature. The data shown in Figures 1 and 2 were obtained after the rates has become nearly constant, but some of the scatter can be attributed to the changing rates. Macak and Pick (15) attributed a similar rate change to a separate SO_2-

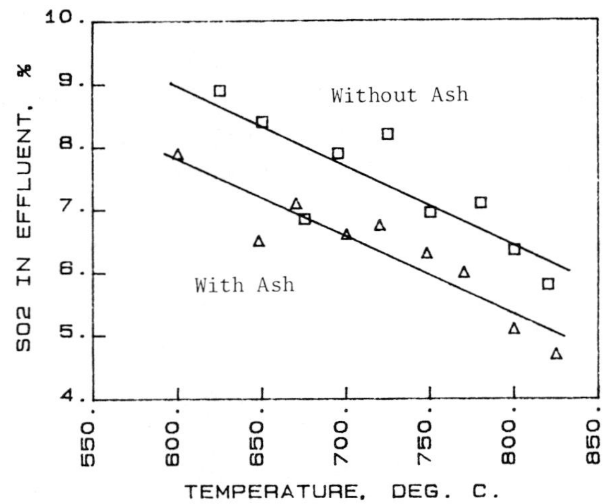

Figure 1. Increased Reactivity of Petroleum Coke for Reduction of SO_2 to Sulfur when Mixed with Coal Ash. Reactivity Indicated by SO_2 Consumption

carbon reaction which deactivated, or passivated, the carbon at a rate comparable to the sulfur reduction reactions.

A second set of measurements were made with the char-ash mixtures allowing a longer period of time to reach steady state. The relative rates measured at different temperatures are shown in Figure 3, an Arrhenius plot which indicates an activation energy of about 30 kal/mol. This implies that the reaction rate would double with a $48^{\circ}C$ ($87^{\circ}F$) increase in temperature. The rates are in units of mole of SO_2 per hour per mole of carbon in the bed. Reaction rate measurements were not attempted with coal because devolatilization to form char occurs rapidly.

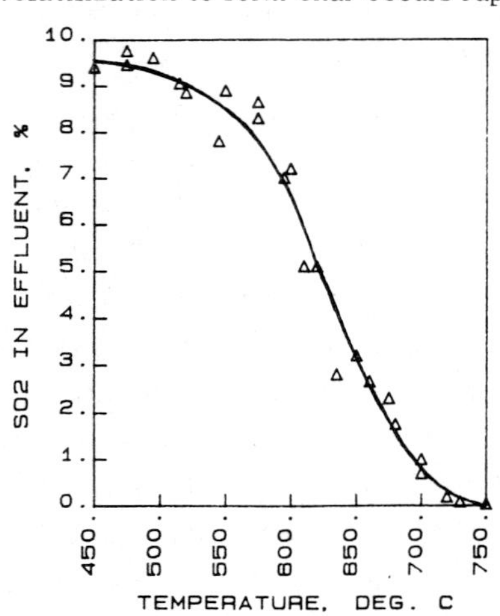

Figure 2. Reactivity of a Coal-Char Mixture for Reduction of SO_2 to Sulfur. Reactivity Indicated by SO_2 Consumption.

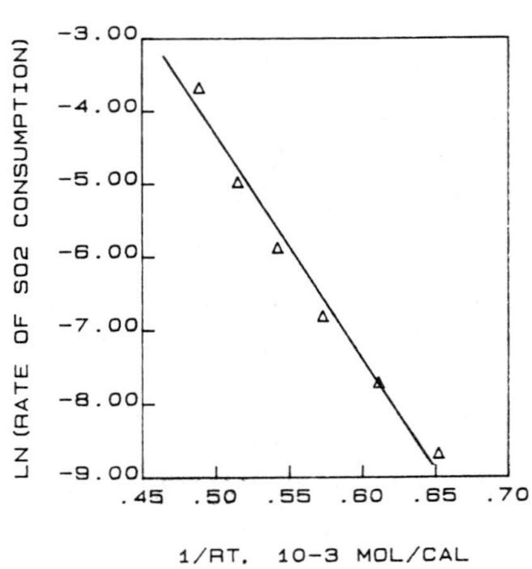

Figure 3. Temperature Dependence of the Rate of Reaction between SO_2 and Carbon in a Coal-Char and Coal-Ash Mixture.

Sulfur Purity

Attempts were made to produce a high purity sulfur using coal-char-ash mixtures. Each mixture was placed in the reactor and heated to reaction temperature at about 15-20°C per minute in nitrogen. The SO_2-nitrogen mixture was turned on and sulfur was collected in the condenser. Sulfur analyses are shown in Table 2.

TABLE 2. COMPOSITIONS OF REDUCTANTS AND SULFUR

Reaction Temp. °C	Reductant Composition, %			Sulfur Product Composition, %				
	Coal	Char	Ash	S	C	H	Alkali, ppm*	Other**
550	7.4	32.4	60.2	81.8	6.9	1.0	1340	10.2
600	0	35.0	65.0	97.3	1.6	0.3	95	0.8
600	2.9	34.0	63.1	-	2.1	0.2	190	-
600	17.4	28.9	53.7	70.3	23.8	2.5	825	3.3
650	0	35.0	65.0	97.9	0.6	0.0	40	1.5
650	0	35.0	65.0	99.6	0.2	0.1	40	0.1
650	1.0	34.7	64.3	98.0	0.6	0.1	125	1.3
650	7.4	32.4	60.2	97.5	0.9	0.1	10	1.5
700	0	35.0	65.0	98.7	0.4	0.1	400	0.8
700	2.9	34.0	63.1	98.6	0.6	0.1	35	0.7
700	0	35.0	65.0	98.9	0.9	0.1	30	0.1

* Includes sodium, potassium, and calcium
** Mostly unrecovered solvent

Results with the mixtures indicated that the coal volatile matter could be utilized in the reaction. Excessive amounts of coal caused the production of volatiles at a rate faster than SO_2 was available, and the product was contaminated. However, the results indicated that sulfur could be produced by reaction with coal in a mixed bed, such as a fluidized bed, with hydrocarbon contamination in the one percent range. The mixed bed would contain mostly char and ash. Fresh coal would be fed continuously at a rate slow enough for most of the volatiles to react with the incoming SO_2. Hopefully, this would also allow the fresh coal to be sufficiently dilute in the bed to prevent agglomeration of coking coals. However, this would require the reactor to be relatively large in size, and the solids overflow from the bed would contain a large fraction of unreacted char, so a significant fraction of the coal feed would not be used.

Another run was made with two separate beds to simulate separation of the coal devolatilization reaction from the reaction between SO_2 and the coal volatiles. The SO_2 first encountered a bed of fresh coal where it picked up coal volatiles, then the SO_2 and coal volatiles entered the second bed which contained only ash. For this run, the SO_2 stream composition was changed to make it more representative of what might come from an industrial process. The ash bed was maintained at 750°C (1382°F) but the coal bed was slowly lowered into the furnace during the run to simulate continuous coal addition in a reacting system. Compositions of the SO_2 stream and of the sulfur product are given in Table 3. The sulfur purity was better than with any run using a single bed. It was apparent that the use of at least two reactors would provide a better product.

TABLE 3. FEED AND PRODUCT COMPOSITION WITH SEPARATE ASH AND COAL BEDS

Feed, Vol%		Product, Wt%	
Nitrogen	66	Sulfur	98.5
Carbon Dioxide	16	Carbon	0.1
Steam	10	Hydrogen	0.1
Sulfur Dioxide	8	Alkali* (PPM)	115
		Other**	1.3

* Includes sodium, potassium, and calcium
** Mostly unrecovered solvent

PROCESSING SCHEME

Results from Coal Gasification

Some pilot scale work done on coal gasification at Bituminous Coal Research (BCR) is of interest for several reasons. (15,16)

1. BCR succeeded in preventing coal volatiles in the product gas by using separate reactors, in this case, a series of fluidized beds.
2. BCR demonstrated process feasibility using a eastern caking coal, without causing agglomeration in the bed.
3. Coal utilization was high.

The process, called the Tri-Gas low-BTU coal gasification process, used three fluidized beds. Dried, ground coal was devolatilized in the first bed at about 480°C (900°F). Agglomeration of coking coals was prevented by feeding fresh coal at a low rate so as to keep it a small part of the total bed. Devolatilized coal, or char, from the first bed was transferred to the second bed which was the main gasification reactor. It was operated between 870 and 980°C (1600-1800°F), and was the largest of the three beds. The remaining carbon and ash were transferred to the third bed which was operated at the same temperature as the second bed. Most of the remaining carbon was gasified in the third bed.

Air and steam were used to gasify the coal. The main air and steam entered the process at the third bed, then went to the first bed and left the process following the second bed. This arrangement provided an oxygen-rich gas to the third bed which helped complete the char consumption. Gas going to the first bed was already hot which helped to devolatilize the incoming coal and bring it to reaction temperature. Coal volatiles from the first bed were consumed in the second bed. Additional air and steam were supplied to the second bed. The product gas contained about 20% hydrogen and carbon monoxide, 12% carbon dioxide and the balance nitrogen. The only hydrocarbon reported was methane, generally less than two percent.

Research Suggested for Sulfur Dioxide Conversion Using Coal

The Tri-Gas processing concept appears to include solutions to the problems of sulfur reduction with caking coals. A series of three fluidized-beds could be used with functions similar to those in the Tri-Gas process. It may also be practical to use the same main gas routing. The SO_2 rich gas would be heated and partially reduced in the third bed, it would next pick up coal volatiles as it passed through the first bed, and finally the volatiles would be removed by reaction in the second bed. The product gas from the second bed would go through a cyclone for particulate removal, then to a condenser for recovery of the elemental sulfur. Particulates and combustible matter would be allowed in the incoming SO_2 rich stream.

This procedure may produce relatively pure elemental sulfur. If the impurities could be held to less than 1 percent it could probably be used for chemicals and fertilizer production. This would bring the highest price, but other potential uses may allow lower purity sulfur and some process simplifications. For example, the potential use as a paving material involves addition of asphalt and aggregate. These substances are somewhat similar to the coal volatiles and the ash particulates which could be expected as impurities with this processing scheme. It is likely, that higher levels of both coal volatiles and particulate matter could be allowed in sulfur destined for pavement than for fertilizer. It may be possible to make the same argument regarding sulfur to be used in other construction applications. This would reduce the cost of the product gas clean-up following stage two, and may allow greater latitude in the process controls. This would reduce the production costs, but economic studies would require a demonstration of process feasibility and pilot plant work. Allowable levels of both coal volatiles which contain mostly tars and high molecular weight phenols, and of particulates which contain coal ash, should be determined for the principal non-traditional sulfur uses. These limits should be established through laboratory testing.

Experimental work leading to a demonstration of process feasibility is suggested using small fluidized beds. The SO_2 necessary for two to four inch diameter beds could probably be handled in a laboratory setting. The demonstration should include steady-state operation with a series of fluidized beds, the cyclone or particle separator, and the sulfur condenser. Incoming feed gas should be blended to match the compositions reported for possible upstream sulfur concentration processes. The product sulfur should be tested for impurities and for potential uses in chemical and fertilizer manufacture, for paving material and for other uses.

CONCLUSIONS

1. Coal ash which contains iron and substances classed chemically as bases catalyze the reaction between SO_2 and coal to form elemental sulfur.
2. Sulfur with one percent or less hydrocarbon impurities can be produced from SO_2 and coal if the coal devolatilization reaction is separated from the char-SO_2 reaction. Coal ash also catalyzes the reaction between SO_2 and the coal volatiles.
3. A processing scheme similar to the Tri-Gas coal gasification process contains potential solutions to the problems of coal agglomeration, the production of coal volatile impurities, and low coal utilization.

4. Laboratory work leading to a demonstration of process feasibility is suggested using a series of fluidized beds. Successful development could provide an inexpensive process for converting SO_2 to elemental sulfur.
5. Laboratory testing to determine allowable levels of impurities in sulfur for non-traditional uses should be made concurrently with the process feasibility demonstration. Process development could include less expensive versions tailored to produce sulfur with lower purity requirements.

REFERENCES

1. E.B. Smyk and R.B. Snyder, Reductive Decomposition Process Studies, pp. 3-27 in Regeneration of Sulfated Limestone from FBC's and Corrosive Effects of Sulfation Accelerators in FBC's. Report No. ANL/CEN/FE-78-13, Argonne National Laboratory, Argonne, IL (1978) 65 pp.

2. Evaluation of Regenerable Flue Gas Desulfurization Process, Vol. II, Electric Power Research Institute, (Palo Alto, CA) Report No. EPRI-79-82 (1976) 79-82.

3. P. Steiner, H. Juntgen, and K. Knoblauch, in J.B. Pfeiffer (Ed.), Adv. Chem. Ser. 139, (1975) 180.

4. N.R. Raskin, Process for Preparing a Sulfur Dioxide Containing Gas Stream for Reaction with Coal, U.S. Patent 4, 477, 426. (1984).

5. W. Ellison, and E. Shapiro, 10th Energy Technology Conference, Proceedings, (1983) 925.

6. C. Murawczyk, and J.S. MacKenzie, Chem. Eng. Prog. 80, (Sept. 1984)62-67.

7. D.S. Moulton, and W.M. Swift, Alternative Regeneration Process Development, pp 22-46 in Regenerative Of Sulfated Limestone from FBC's, Report No. ANL/CEN/FE-79-13, Argonne National Laboratory, Argonne, IL (1979) 53 pp.

8. J.K. Chowdhury, and R.M. Datta, J. Ind. Chem. Soc. 20, (1943) 253.

9. L.A. Haas, et al., Removing Sulfur Dioxide by Carbon Monoxide Reduction, U.S. Bureau of Mines Report of Investigations 7483, Washington, D.C. (1971).

10. Z.M. George, in J.B. Pfeiffer (Ed.), Adv. Chem. Ser. 139, (1975) 75.

11. J.B. Peri, J. Phys. Chem. 69, (1965) 220.

12. J.B. Peri, J. Phys. Chem. 69, (1965) 211.

13. C.C. Chang, J. Catal. 53, (1978) 374.

14. T.A. Hendrickson, Synthetic Fuels Data Handbook, Cameron Engineers Inc., Denver, CO, (1975), pp 147-150.

15. Bituminous Coal Research, Inc. Test and Evaluate the Tri-Gas Low-BTU Coal-Gasification Process, Final Report, DOE Report No. DOE/ET/10254—88, Bituminous Coal Research Inc., Monroeville, PA (1980) 73 pp.

16. M.A. Colaluca, M.A. Paisley, and K. Mahajan, Chem. Eng. Prog. 75 (1979) pp 33-40.

THE USE OF HIGH-SULFUR COAL IN COPPER ELECTROWINNING

J.R. CRAYNON and G.R. SMITH

U.S. Department of the Interior, Bureau of Mines, Avondale Research Center, 4900 LaSalle Road, Avondale, MD 20782-3393

ABSTRACT

Research has been conducted to lower the energy requirement in copper electrowinning by replacing the anodic water oxidation reaction with a coal slurry oxidation system in a combined electrolysis cell. The cell, which employs a membrane to separate the two half-cells, was operated at 60° C at a cell voltage of approximately 0.5 V. This represents a 67 pct reduction in voltage from the 1.5 V required for conventional electrowinning. Using acetone washing and baking, the low-voltage electrolysis using the high-sulfur Pittsburgh #8 coal has been conducted for up to 102 h. Analytical results showed that the character of the coal remaining after electrolysis was only slightly lower in heat value while being significantly reduced in ash and sulfur content compared to the feed coal. Based upon the decrease in cell voltage, a decrease in the net cost of electrolysis energy could be achieved.

INTRODUCTION

Because the grade of exploited ore bodies has been steadily decreasing and their mineralogy has been becoming more complex, more efficient hydrometallurgical methods for recovering the metals in these ores are needed. Obvious benefits would occur from the use of sulfate electrowinning systems that include an alternative anode reaction with a lower half-cell potential than that of water oxidation (E_h^o= 1.23 V). In some cases (refs. 1-5), an electrochemical reaction producing a commercially useful product has been substituted. The advantage of substituting these electrochemical production reactions arises from the fact that the energy consumed in the electrolytic cell can be allocated to two products rather than one.

Considerable interest has been given to substituting anodic reactions that would lower the overall cell voltage required for electrowinning, such as the oxidation of SO_2 to H_2SO_4 (E_h^o + 0.17 V) (refs. 6-11), H_2 to H^+ (E_h^o = 0.0 V) (refs. 12-13), or C in coal slurries to CO_2 (E_h^o = 0.21 V) (refs. 14-15). The electrochemical characteristics of the latter reaction have been examined in depth.

Early work to characterize the electrochemical oxidation of carbon was done on graphite systems, since graphite electrodes have long been used in electrowinning (refs. 16-19). Initial electrolytic studies involving coal slurry oxidation pertained to use of the method as a means of producing low cost hydrogen (refs. 20-24). The technology was expanded to include the carbon oxidation reaction in metal electrowinning (refs. 14, 15, 25).

Many of the researchers found that the presence of a metallic ion, such as ferric ion (Fe^{3+}), was necessary to "catalyze" the oxidation of the coal slurry (refs. 22-24, 26-30). The reactions were characterized as the chemical oxidation of carbon in coal,

$$4Fe^{3+} + C + 2H_2O \rightarrow CO_2 + 4Fe^{2+} + 4H^+, \quad (1)$$

followed by the electrochemical oxidation of ferrous ions,

$$Fe^{2+} \rightarrow Fe^{3+} + e^- \ (E_h^o = 0.770 \text{ V}), \quad (2)$$

as the reaction at the anode which determines cell potential. It has been shown (ref. 28) that reaction (1) does not necessarily result in the production of CO_2, but it can result in the formation of other oxidized carbon species instead. The discovery that Fe^{3+} was responsible for the oxidation of carbon was later employed to improve the operation of an electrolytic cell combining coal slurry oxidation with copper electrowinning (ref. 31). The use of Fe^{3+} ions increased the oxidation rate of the coal slurry oxidation to such a degree that the anodic and cathodic current densities were compatible.

The latter work was conducted at electrolyte temperatures up to 300° C, significantly higher than those employed in commercial electrowinning. At these high temperatures, a pressurized system was required to keep the reactants in the aqueous state. In the present work, the temperature was maintained at 60° C, which permitted the use of a more conventional electrowinning cell. One common innovation in each of these experimental cells was the use of a cationic permselective membrane to separate the anolyte and catholyte.

A high-sulfur coal from the Pittsburgh No. 8 seam was employed as the oxidizable species in the research reported in this paper. Although widespread use of a high-sulfur coal for combustion is limited by environmental and chemical constraints, its suitability for use in electrowinning cells has been investigated. A study into the specific effect of a combined electrowinning-coal slurry oxidation process on the copper electrowinning efficiency has been conducted. In addition, changes in the characteristics of the coal after use in this electrolytic cell have been determined.

EXPERIMENTAL

Electrolytic cell

The rectangular experimental cell used in this study is shown schematically in Fig. 1. The 180-ml capacity cell was constructed from polyacrylic plastic.

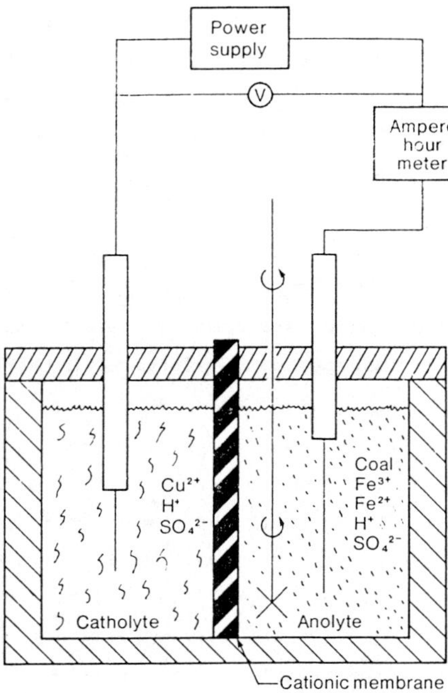

Fig. 1. Schematic diagram of combined electrowinning-coal slurry electrolysis cell.

An Ionac MC-3470[1] cationic membrane (0.38-mm thick) was used to separate the cell into equally sized anolyte and catholyte sections. This membrane was a heterogeneous type consisting of a sulfonated polystyrene-divinyl benzene copolymer exchange resin pressed in a polyolefin binder. Its resistivity was 6 ohm-cm^2 in 1.0 N NaCl solution. Approximately 0.27 dm^2 area of the membrane was exposed to the electrolytic current. This membrane was chosen because of its performance in previous research conducted by the Bureau of Mines (refs. 4-

[1]Reference to specific brands is made for identification only and does not imply endorsement by the Bureau of Mines.

5). The cathode material was a copper sheet with an area of 0.02 dm^2 for most experiments. The anode area was usually 0.4 dm^2, but was adjusted to a smaller area for some experiments in order to increase the anodic current density, while allowing the cathodic current density to remain unchanged. The anode was made from a smooth platinum sheet. The two electrodes were placed vertically in the center of each half-cell, spaced 50 mm apart, and positioned parallel to the membrane. The cathode was operated at 1.0 A/dm^2, while the anode was usually operated at 0.05 A/dm^2. A constant-current dc power supply controlled current to the cell. Cell voltages and total coulombs were recorded on a strip chart recorder and an ampere-hour meter. Temperature of the anolyte slurry and the catholyte was maintained at 60° C by partially immersing the cell in a constant-temperature water bath. The coal was suspended in the slurry using a polyacrylic paddle attached to a Teflon-coated stainless steel shaft and rotated by a constant speed ac motor.

Solutions

The catholyte was composed of 1$\underline{M}$ $CuSO_4$, 1$\underline{M}$ H_2SO_4, and 0.004 g/L thiourea $[(NH_2)_2C = S]$ and 0.001 g/L glue, which are commonly used as grain refining and levelling agents. The anolyte compositions consisted of either 1$\underline{M}$ H_2SO_4-2.0$\underline{M}$ $Fe_2(SO_4)_3$, or 0.6$\underline{M}$ H_2SO_4-0.5$\underline{M}$ $Fe_2(SO_4)_3$.

Coal

The Pittsburgh No. 8 coal was provided by the Consolidation Coal Company from its Ireland Mine near Wheeling, WV, and measured 1 1/4-in x 0 as prep plant product. The 3/8-in x 0 fraction was raw coal and was screened off prior to use. The remaining 1 1/4-in x 3/8 in fraction was crushed to minus 4-mesh (0.187 in) to provide a more uniform starting material. Immediately prior to use, the minus 28-mesh (0.023 in) fraction was removed. Preliminary experiments had shown that the stock coal had sufficient time in storage to be oxidized before use in the electrolytic cell. Therefore, the coal was prepared for electrolysis by washing with acetone and baking at 250° C (ref. 15). This procedure served to remove surface oils and oxidation products. An acetone wash of 30 min and a baking time of 2 h improved the duration of electrolysis by ~100 pct compared to the oxidized coal. Doubling the wash time and increasing the baking time to 26 h increased the period of low-voltage electrolysis by about 10 pct. The minus 4- plus 28-mesh fraction thus treated was then ground for two 5-min stages in a 4-in ceramic ball mill with the minus 150-mesh (0.004 in) fraction removed after each stage. This coal was then suspended in the anolyte for electrolysis.

Analysis and evaluation

Samples of the anolyte and catholyte were taken both before and after electrolysis to determine the effectiveness of the cationic membrane in separating the half-cells. Standard wet chemical and atomic absorption analytical techniques were used to determine the concentration of the ionic species in each of these solutions. Physical and chemical characteristics of the coal were determined using ASTM-standard methods and Fourier Transform Infrared Spectroscopy (FTIR). Copper electrowinning efficiency was calculated from the weight of the cathode deposit.

RESULTS AND DISCUSSION

A batch-type operation of the combined electrowinning-coal slurry oxidation cell was demonstrated for various periods of low-cell-voltage operation. As a part of these experiments, the efficiency of copper electrowinning and the duration of low-voltage electrolysis was determined for different anolyte concentrations, coal slurry densities, coal preparations, and anodic current densities. In addition, the degree of interdiffusion of the anolyte and catholyte solutions was determined. Data obtained from these experiments were used to compare the combined electrolysis method to conventional copper electrowinning.

Low-voltage electrolysis

In combined electrowinning-coal slurry oxidation experiments, cell voltages as low as 0.41 V were obtained. In general, the operating cell voltage was in the range of 0.4 to 0.6 V for some finite period of time in each of the electrolyses. After this period of low-voltage electrolysis, the cell voltage increased indicating a loss of coal reactivity. The duration of low-cell-voltage electrolysis will be referred to as the electrolysis time in this report. The operating cell voltage of the combined cell represents a reduction of 67 pct when compared to a cell operated in the standard commercial configuration, that is, with the oxidation of water as the anodic half-cell reaction. The extent to which the cell voltage was decreased was dependent upon the ratio of Fe^{2+}/Fe^{3+} in the anolyte as would be expected from the Nernst equation. The electrolysis time was dependent upon the total quantity of Fe^{2+} produced in the anolyte by reaction of Fe^{3+} with the coal. Oxidation of the carbon in the coal slurry by reaction (1) occurs rapidly at 60° C as indicated by the reaction rate curves shown in Fig. 2. Nearly 95 pct of the original Fe^{3+} was reduced within 2 h at this temperature. Significantly slower rates were observed at the lower temperatures.

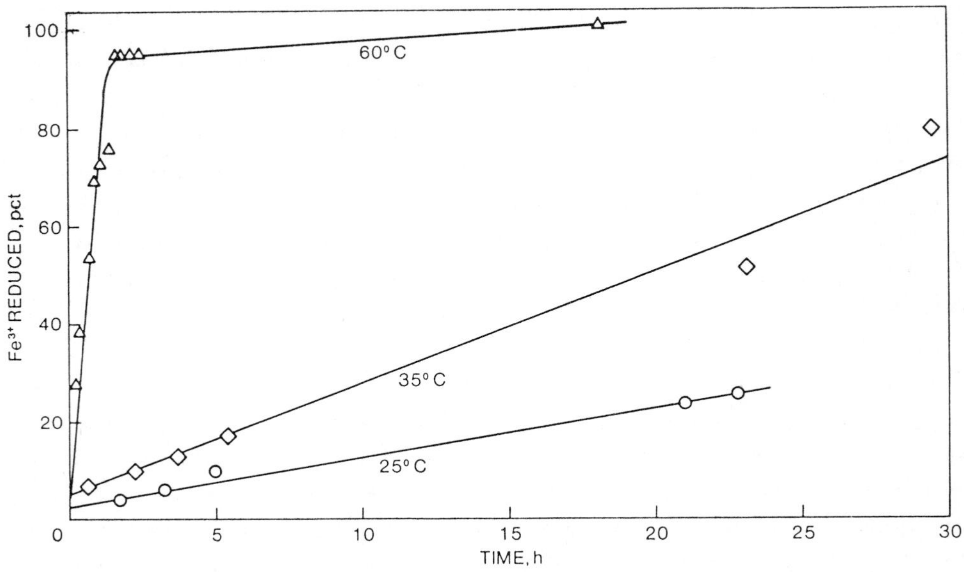

Fig. 2 Reaction rate at 3 temperatures for Pittsburgh No. 8 coal with 0.2M $Fe_2(SO_4)_3$-1.0M H_2SO_4 anolyte.

Table 1 shows the results of several combined electrowinning-coal slurry oxidation experiments comparing the electrolysis time to the anolyte composition, the coal slurry density, and the anode current density. Each test was conducted at a constant current of 0.02 A so that the electrolysis time was a direct indication of the total coulombs passed through the cell. Longer periods of electrolysis were achieved using an anolyte composition of 1.0M Fe^{3+} and 0.6M H_2SO_4. Based on equilibrium equations, reaction (1) was calculated to be thermodynamically more favorable at this anolyte composition than in the anolyte containing 0.4M Fe^{3+} and 1.0M H_2SO_4. Increasing the coal slurry density also increased the duration of low-voltage electrolysis.

TABLE 1

Duration of low-voltage electrolysis under various conditions.

Anolyte concentration		Coal slurry, g/L density	Anode current density, A/dm^2	Electrolysis time, h
Fe^{3+}, M	H_2SO_4, M			
0.4	1.0	80	0.05	25
.4	1.0	100	.05	40
1.0	.6	80	.05	62
1.0	.6	100	.05	80
1.0	.6	80	.05	102[1]
1.0	.6	100	0.1	36

[1]Two-stage electrolysis - 62 and 40 h.

Total coulombs passed through the cell during a low-voltage electrolysis period did not exceed the coulombic value equivalent to reoxidation of the Fe^{2+} produced in the anolyte during the initial reaction with coal. There apparently was little or no secondary reaction of the electrolytically produced Fe^{3+} to further oxidize the coal slurry, probably as a result of the buildup of oxidation products on the coal surface. However, a second stage electrolysis could be carried out by separating the spent coal from the anolyte, acetone-washing and baking the coal, and reusing it in a fresh anolyte. Approximately 102 h of electrolysis was achieved in two stages by this method.

When the current density on the anode was doubled to 0.1 A/dm^2, the electrolysis time was correspondingly halved and a charge equivalent to that of tests at 0.05 A/dm^2 was passed through the electrolytic cell. Anodic current densities of the same magnitude as those of the cathode are preferred in order to provide practical coupling with copper electrowinning. With similar current densities, minimal cell and electrode geometry changes would be required to incorporate the coal slurry oxidation system into present industrial electrowinning cells and to produce copper at a rate equivalent to that of present commercial cells. The results indicate that if the anolyte composition is properly controlled and the coal is maintained as a clean, reactive surface, extended low-voltage electrolysis at practical anode and cathode current densities is achievable.

Coal characteristics

Suitability of the post-electrolysis coal for further use as a fuel or for other purposes, e.g., as a source of organic oxidation products, was considered. Table 2 gives the proximate analysis for the Pittsburgh #8 coal before electrolysis and after two different coal preparations and electrolyses. Results showed that the energy content in the coal remained relatively high after 70 to 80 h of electrolysis. In addition, the sulfur and ash content of the coal were lower after electrolysis.

TABLE 2

Proximate analysis (dry basis) of Pittsburgh No. 8 coal before and after electrolysis with different pretreatment.

Coal preparation		Electrolysis, h	Ash, wt pct	Volatile matter, wt pct	Fixed carbon, wt pct	Sulfur, wt pct	Energy content, Btu/lb
Acetone wash, h	Bake, h						
0	0	0	7.90	42.08	50.02	4.48	13688
0.5	2	71	5.85	40.28	53.87	3.84	13496
0.5	26	80	6.29	39.46	54.25	3.82	12590

Results in Table 3 show that the reduction in the sulfur content of the coal was related to the total electrolysis time, as well as to the number of electrolysis periods and intermediate preparation stages to which the coal slurry was subjected. Nearly half of the original sulfur, equivalent to the pyritic sulfur content, was removed from the coal during 102 h of electrolysis that was conducted in two stages. In a practical operating system, acetone-washed and baked coal would be expected to be recycled into controlled electrolytes, subsequently providing additional hours of electrolysis and possibly removing more of the sulfur present in the coal.

TABLE 3
Reduction in the sulfur content of the Pittsburgh No. 8 coal as a result of electrolysis.

Electrolysis, h	Electrolysis stages	Sulfur, pct	Reduction in sulfur, pct	Equivalent reduction in pyrite, pct
0	0	4.48 (2.11 as pyrite)	NAp	NAp
80	1	3.82	14.73	31.28
24	3	2.91	35.04	74.41
102	2	2.29	48.88	103.79

NAp - Not applicable.

The oxidation of carbon compounds in the coal by ferric ion [reaction (1)] was verified by FTIR studies on the postelectrolysis coal. This analysis indicated notable increases in peaks at ~1,700, ~1,650, ~1,250, and 1,100 to 1,000 cm^{-1}. Similar results were obtained by Dhooge and Park (ref. 28). These peaks were attributed to the increased presence of -COOH, quinoids, phenols, and aliphatic ethers and alcohols, respectively.

Copper electrowinning

The efficiencies for copper electrowinning in the combined electrolysis cell were greater than 97 pct. In most cases, the copper was electrowon at nearly 100-pct efficiency. The copper recovered on the cathode was of high quality and was 99.96-pct pure.

Membrane performance

The membrane used in the combined electrowinning-coal oxidation cell allowed some interdiffusion of the anolyte and catholyte solutions. Table 4 shows that an average of 0.23 pct of the iron, initially present in the anolyte, migrated to the catholyte per hour of electrolysis. A similar migration of Cu^{2+} to the anolyte was observed. The quantity of ions diffused through the membrane is affected by the relative concentration of ions on each side of the membrane, the relative size and electrochemical charge on these ions, as well as the external electromotive force applied across the membrane in order to operate the electrolytic cell. Since hydrogen ions were present in the initial anolyte and were produced by reaction (1) during the electrolysis, these smaller, more mobile ions carried most of the current through the membrane. Other membranes or electrolyte combinations would exhibit different interdiffusion and current transport characteristics, however.

TABLE 4
Interdiffusion of metallic cations through the membrane during electrolysis.

Electrolysis time, h	Total Interdiffused, pct initial		Average diffusion, pct Cu^{2+}/h	Average diffusion, pct Fe^{2+},Fe^{3+}/h
	Cu^{2+}	Fe^{2+},Fe^{3+}		
66.5	17.3	15.0	0.26	0.23
43.2	15.2	9.5	.35	.22
95.0	13.8	23.7	.15	.25

Iron migrated to the catholyte predominantly as the Fe^{2+} ionic species. As a result, this migration did not detrimentally affect the copper electrowinning process. If a significant quantity of Fe^{3+} had migrated, the current efficiency for copper production would have decreased according to the chemical reaction:

$$2\ Fe^{3+} + Cu^{0} \rightarrow 2\ Fe^{2+} + Cu^{2+}. \qquad (3)$$

Essentially, a concentration equilibrium condition would be reached between the migrating ions. In an actual operating cell, it is, therefore, likely that Fe^{2+} and Cu^{2+} ions would be present in both the anolyte and catholyte as an integral part of the process.

Energy considerations

Although the scope of this research has not progressed to the extent where the specific engineering details of the process can be specified or a detailed economic evaluation of the process be conducted, several important considerations merit discussion at this stage. The reduction in the electrolysis energy consumption from 1.27 to 0.43 Kw•h/kg Cu, achieved by lowering the cell voltage in the combined cell, represents a signficant decrease in the cost of electricity required for the electrowinning of the copper. From experimental data, it was determined that the combined process uses approximately 7,750 Btu/kg Cu in the form of a reduction in the chemical heat energy of the coal. Since the electrolyzed coal has been lessened in ash content and sulfur content, with only a small reduction in heating value, it is assumed that this coal could be further used on site for heating purposes or to produce electricity or sold at a value equal to or higher than its initial price to consumers. Though operations, such as coal grinding and slurry pumping, would require energy, some of this energy could be derived from the post-electrolysis coal.

SUMMARY

Laboratory studies aimed at reducing the energy requirement in copper electrowinning by combination with coal slurry oxidation have led to the following conclusions:

(1) The cell voltage required for the combined cell can be reduced by 67 pct compared to conventional electrowinning. The operating cell voltage for the combined process can be maintained at about 0.5 V for up to 102 h of electrolysis conducted in two stages.

(2) High-sulfur coals, such as the Pittsburgh #8 coal used in this work, can be employed in this combined process. The amount of sulfur and ash contained in the coal can be significantly reduced in the course of being used in this process without a major decrease in the energy content of the coal.

(3) Copper of high purity and quality can be electrowon in the combined cell at current efficiencies greater than 97 pct.

REFERENCES

1 A.M. Pande, K.N. Gupta and V.A. Atelkar, Hydrometallurgy, 9 (1982) 57-68.
2 B. Verbaan and B. Mullinder, Hydrometallurgy, 7 (1981) 339-352.
3 W.A. Nattrass, Mintek Report, No. M62, May (1983), 38 pp.
4 G.R. Smith and W.R. Thompson, in D.J. Robinson and S.E. James (Eds.), Proceedings of the Electrolytic Processes Committee—Anodes for Electrowinning, AIME Annual Meeting, Los Angeles, CA, Feb. 28-Mar. 3, 1984, pp. 1-12.
5 G.R. Smith, in P.E. Richardson, S. Srinivasan and R. Woods (Eds.), Proceedings of the International Symposium on Electrochemistry in Mineral and Metal Processing, Proceedings v. 84-10, 165th Meeting of The Electrochemical Society, Inc., Cincinnati, OH, May 6-11, 1984, pp. 548-561.
6 A.J. Appleby and B. Pichon, J. Electroanal. Chem., 95 (1979), 59-71.
7 Gerald F. Pace and John C. Stauter, CIM Trans., LXXVII (1974) 51-57.
8 Zdenek Samec and Jan Weber, Electrochim. Acta, 20 (1975) 403-419.
9 Douglas J. Robinson, J. Met., January (1984) 43-47.
10 E.T. Seo and D.T. Sawyer, Electrochim. Acta, 10 (1965) 239-252.

11 K. Weisener, Electrochim. Acta, 18 (1973) 185-189.
12 R.J. Allen, W. Juda, R.W. Lindstrom and H.G. Petros, U.S. Dept. of Energy, Contract No. ET-78-C-02-4881, Report COO-4881-6 (1979), p. 72.
13 Amiram Bar-Ilan, Walter Juda, Robert J. Allen and Robert W. Lindstrom, U.S. Dept. of Energy, Contract No. DE-AC02-78ET25309, Report COO-4881-16 (1980), p. 37.
14 Mohammad Farooque and Robert W. Coughlin, Nature, 280 (5724)(1979) 666-668.
15 Mohammad Farooque and Robert W. Coughlin, Fuel, 58 (10)(1979) 705-712.
16 K.F. Blurton, Electrochim. Acta, 18 (1973) 869-875.
17 H. Binder, A. Kohling, K. Richter and G. Sandstede, Electrochim. Acta, 9 (1964) 255-274.
18 Gleb Mamantov, Donald B. Freeman, F.J. Miller and H.E. Zittel, J. Electroanal. Chem., 9 (1965) 305-310.
19 K. Kinoshita and J. Bett, Carbon, 11 (1973) 237-247.
20 Robert W. Coughlin and M. Farooque, Nature, 279 (1979) 301-303.
21 Robert W. Coughlin and Mohammad Farooque, Ind. and Eng. Chem., Process Des. and Dev., 19 (1980) 211-219.
22 K.E. Johnson, L. Verdet, F.W. Yerhoff, S. Barton and C. Dhem, Abstract 456, The Electrochemical Society Extended Abstracts, Montreal, Quebec, Canada, May 9-14, 1982, p. 746.
23 Richard P. Baldwin, Keith F. Jones, Joseph T. Joseph and John L. Wong, Fuel, 60 (1981) 739-743.
24 G. Okada, V. Gueuswamy and J. O'M. Bockris, J. Electrochem. Soc., 128 (1981) 2097-2102.
25 Robert W. Coughlin, U.S. Patent 4,268,363, May 19, 1981.
26 Patrick M. Dhooge, David E. Stilwell and Su-Moon Park, J. Electrochem. Soc., 129 (1982) 1719-1724.
27 Patrick M. Dhooge and Su-Moon Park, J. Electrochem. Soc., 130 (1983) 1029-1036.
28 Patrick M. Dhooge and Su-Moon Park, J. Electrochem. Soc., 130 (1983) 1539-1542.
29 Su-Moon Park, J. Electrochem. Soc., 131 (1984) 363C-373C.
30 K.E. Anthony and H.G. Linge, J. Electrochem. Soc., 130 (1983) 2217-2219.
31 Ronald J. Vaughan, U.S. Patent 4,405,420, September 20, 1983.

POTENTIAL PRODUCTS RESULTING FROM DESULFURIZATION OF HIGH SULFUR CONTENT COAL WITH SUPERCRITICAL ALCOHOLS

Charles B. Muchmore[1] , Albert C. Kent[1] and Ta-Chang Lin[1]

[1]Department of Mechanical Engineering and Energy Processes, Southern Illinois University, Carbondale, Illinois 62901, (618) 536-2396

ABSTRACT

During the supercritical desulfurization of high organic sulfur content Illinois coals with supercritical alcohols, three products result: a desulfurized solid product, a high BTU gas, and a liquid product that contains, in addition to unreacted alcohol, a variety of coal-derived liquids and solvent-derived sulfur compounds. The solid product appears suitable, based on volatile content and heating value, for use in coal fired power generating stations; the gas could also serve as a fuel source. The liquid product has been characterized by gas chromatography; there are many coal-derived compounds with potential uses as chemical feedstocks, and the solvent-derived sulfur compounds have market potential for uses such as sulfiding agents. The liquid product composition is dependent on the separation procedures employed following the supercritical reaction, as well as on the reaction conditions.

INTRODUCTION

Concern over environmental effects attributed to acid rain has resulted in a variety of legislative proposals to address the problem. Although differing in approach and scope, the net effect of any of the proposed requirements is a negative impact on the coal mining industry of the midwest. Thus, interest has grown in methods to effectively contend with the high sulfur content indigenous to many coals.

The existing technology that serves as a yardstick for economic comparison of any proposed approach to contend with the high sulfur content coal problem is the post combustion approach - flue gas desulfurization. Although the effectiveness of this approach has been demonstrated, several inherent problems exist with this technique: high capital and operating cost, reduced availability of the power generating plant, and production of significant volumes of waste sludge that, depending on plant location, can present a formidable disposal problem. Thus,interest has grown in recent years in development of pre-combustion desulfurization processes.

The pre-combustion process discussed here is one that utilizes the unique capability of alcohols,under supercritical conditions, to selectively remove organic sulfur from the coal matrix (ref. 1). Although other researchers have investigated supercritical solvent extraction of coals, the major objectives of those efforts have been to study conversion to liquid products, rather than desulfurization. Recent studies reported in the literature involving use of alcohols as supercritical extractants have also stressed extractability of coal liquids, and have generally not reported on the desulfurization aspect of the procedures employed (ref. 2,3).

The successes we have obtained in our laboratory with respect to desulfurization of coal thus appear to be unique in terms of application of supercritical extraction/reaction of coal with alcohols for the purpose of desulfurizing coal. The fact that total sulfur concentration reductions of about 50% have been attained on high organic sulfur content coals implies that organic sulfur is being removed under supercritical conditions. This suggests that application of the supercritical desulfurization process following a physical treatment process for removal of pyritic sulfur could produce a solid fuel meeting environmental requirements.

Process Characteristics

As indicated in Figure 1, the desulfurization of coal with supercritical alcohol results in three potentially useful products: 1) a solid product reduced in sulfur content, 2) a variety of coal derived liquid compounds and 3) a high BTU gaseous product. There are no major waste streams generated from the process, an inherently attractive feature from an environmental point of view.

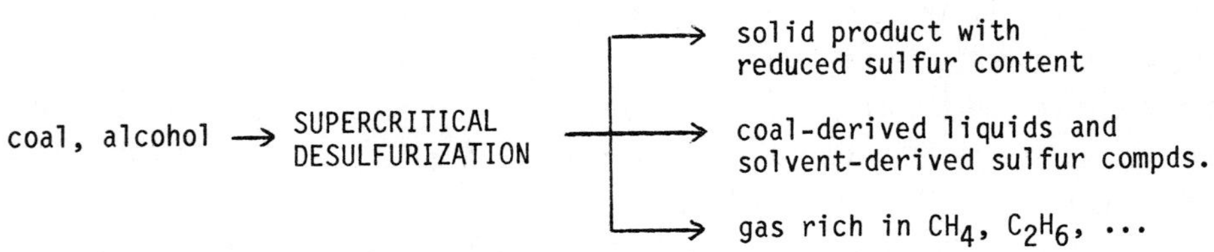

Fig. 1. Products resulting from supercritical desulfurization of coal with alcohol.

EXPERIMENTAL

The development of the desulfurization process in the laboratories of the Department of Mechanical Engineering and Energy Processes has utilized batch, semi-continuous, and continuous reactor systems, employing either ethanol or methanol as the supercritical extractant. The batch system consists of a 300cc stirred pressure reactor (Autoclave Engineers), volatile product condenser and collection system, and a series of three gas absorbers. Measured quantities of coal and alcohol were charged to the reactor, heated to reaction temperature, and held for the desired reaction time, generally one hour. After the reaction period, the fluid phase was vented through the condenser and liquid product collection system, after which the product gas was passed through the series of absorbers and vented through a wet test meter.

The semi-continuous reaction system consists of a Berty reactor that permits recirculation of the supercritical phase through a fixed bed of coal suspended in a basket within the reactor. The solvent being used for the extraction was continuously pumped through the reactor during a run. Details of the batch and semi-continuous reactor systems have been given previously (ref. 4,5).

The continuous reactor system that has recently been constructed is shown schematically in Figure 2. A coal-alcohol slurry is prepared in the mix tank, where agitation is provided by an impeller-baffle system and by recycle of the slurry. A high pressure pump withdraws slurry form the recycle line and pumps the mixture through a 2-liter stirred tank reactor to the product collection system, that operates at system pressure. This configuration avoids the inherent problems associated with pressure let-down of a three phase system on small equipment.

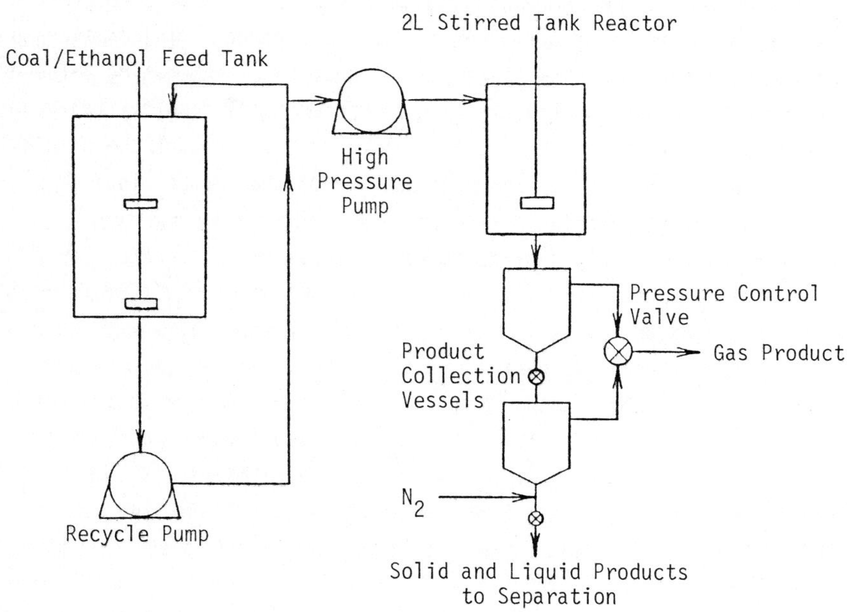

Fig. 2. Flow diagram of the continuous reactor system

The original system employed two collection cylinders in parallel that operated at system pressure; when one became nearly full of liquid and solid products a three way valve permitted switching to the parallel collection vessel while the first one was being emptied and repressurized. In this manner, continuous operation of the 2-liter stirred tank reactor was possible. However, plugging in the three way valve and horizontal lines leading to the collection vessels became an operational problem. As indicated in Figure 2, the current configuration involves a vertical drop of the solid and liquid products through two specially constructed collection vessels in series. During operation, the valve between the vessels remains open, and product condenses and falls to the lower chamber. When it is filled, the valve between the vessels is closed, permitting collection of product from the lower vessel, and repressurization of that vessel. The connecting valve is then opened.

RESULTS AND DISCUSSION

Batch and Semi-continuous Reactors

The characteristics of the solid products resulting from supercritical extraction of an Illinois No. 6 coal with ethanol or methanol over a temperature range of 240-400 C and pressures to 285 atm have been previously reported (ref. 5). The batch and semi-continuous reactors were used. Solid recoveries were generally in the range of 80 to 95% of the coal charged. In addition to the reduction in sulfur concentration of up to about 50%, the heating value of the solid product is enhanced slightly by the supercritical desulfurization process and a large portion of the volatile material (over 50% of the original value) is retained in the solid. Thus the quality of the product indicates its potential for use in existing coal fired power generating stations.

In the batch reactor, the gas produced during supercritical extraction of coal with alcohols results primarily from the decomposition of the solvent, particularly in the higher temperature ranges. When ethanol is used, ethane is a major component of the product gas; methane is the principal gas component when methanol is used as the supercritical fluid. Gas production in the batch reactor can be appreciable; from 10 to 30% of the alcohol may be converted during the reaction. For ethanol, major decomposition pathways include production of hydrogen and acetaldehyde, or ethylene and water. The semi-continuous reactor runs indicated only slight production of gaseous products.

Continuous Reactor

The preliminary results presented here demonstrate the operability of the continuous reactor system. Figures 3 and 4 present data on a 5 hr run at 300C, 83 atm, utilizing Illinois State Geological Survey standard coal no.3, that has

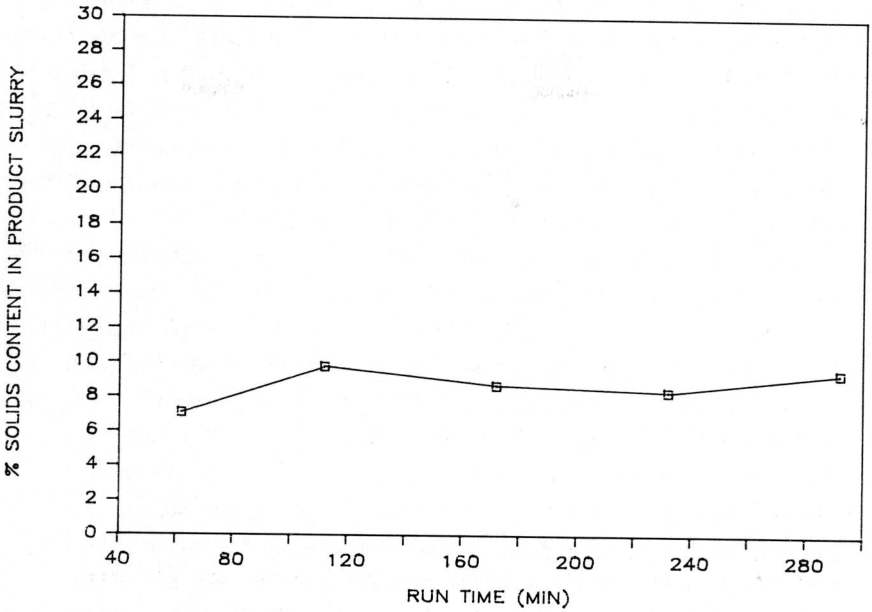

Fig. 3. Solids content of product samples collected from continuous reactor run at 300 C, 83 atm, ISGS standard coal no. 3, ethanol as supercritical phase.

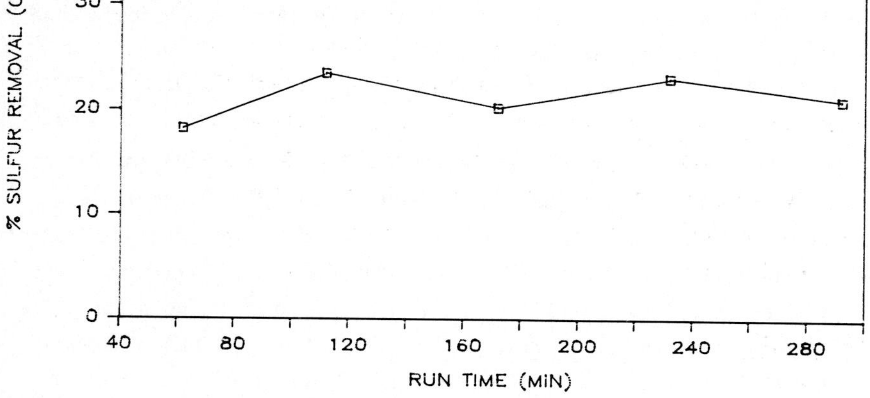

Fig. 4. Sulfur removal of product samples collected from continuous reactor run as described above.

a total sulfur content of 2.76% (organic S/pyritic S ratio =1.56). Figures 3 and 4 indicate that reasonably steady-state conditions were attained during the run. Product solids content stability is demonstrated in Figure 3, and desulfurizaiton data is given in Figure 4. The average value of 21 % sulfur reduction is comparable to results reported previously for this coal in the batch reactor with ethanol and at comparable temperature; the 17 min residence time in the continuous reactor compares to 1 hr in the batch unit.

As was noted in the semi-continuous reactor, little decomposition of solvent to gaseous products has been observed in the continuous reactor system. This is significant in terms of overall process economics, since a minimum amount of make-up alcohol would be required. It should be possible to recycle nearly all the ethanol used in the desulfurization reaction, after separation of the coal-derived liquids that are produced.

In the batch reactor, approximately 2 to 8% of the coal was converted to liquid product, depending on reaction conditions. A gas chromatogram of a typical liquid product resulting when ethanol was used as the supercritical extractant is given in Figure 5. The left trace gives the response of the flame photometric detector (FPD) that is specific for sulfur-containing compounds; the right trace gives the response of the flame ionization detector (FID) that is sensitive to most organic compounds. The FID data was used to determine the boiling range distribution of these liquids (ASTM procedure D2887-73). A comparison of boiling range distribution of liquid produced in the continuous reactor to that obtained from a typical batch reactor run is given in Figure 6. Several solvent-derived sulfur compounds have been identified in the liquid product. These include ethyl mercaptan, ethyl sulfide, ethyl disulfide, and ethyl thioacetal when ethanol is used as the supercritical solvent. When methanol is used, the analogous sulfur containing methyl derivatives are found (ref.5). The potential exists for separation of these compounds, to be marketed as sulfiding agents or for other uses.

CONCLUSIONS

In addition to a solid product with reduced sulfur content, the co-products resulting from the supercritical desulfurization of coal with alcohol are a high BTU gaseous product, that should have a market value approximating that of natural gas, and a variety of coal and solvent derived liquid products. These materials have potential marketability as chemical feedstocks. No major waste streams are produced by the supercritical desulfurization process, since all products from the reaction have significant market potential.

Thus, from an economic and environmental point of view, the supercritical desulfurization process has the potential to be competitive with the currently

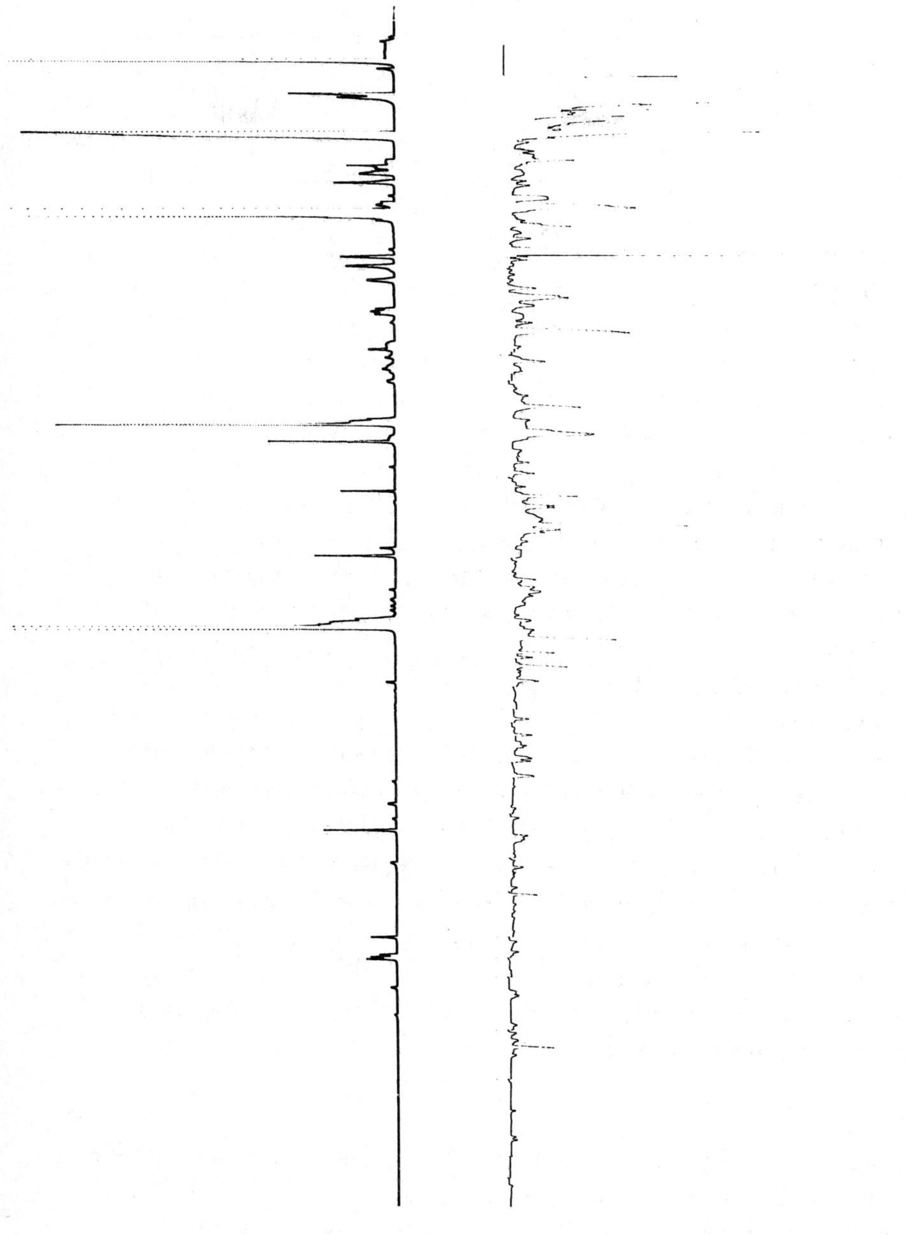

Fig. 5. Gas chromatograph of a liquid product from the batch reactor run at 350 C, 164 atm, Illinois no. 6 coal, ethanol as supercritical phase.

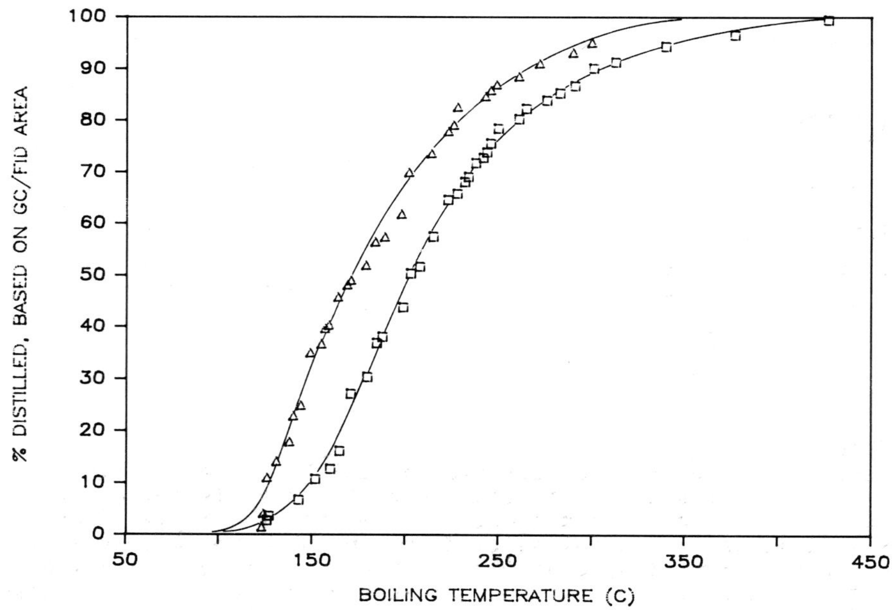

Fig. 6. Comparison of the boiling point distributions of liquid products from batch (△) and continuous (□) reactors.

available process option of flue gas desulfurization for combustion of high sulfur coals. Additional development work on the process is required to establish the material and energy balances under different operating conditions. This information, together with more complete characterization of the product streams, will permit a more accurate assessment of process economics.

ACKNOWLEDGEMENTS

Financial support from the Illinois Coal Research Board and the U.S. Department of Energy is gratefully acknowledged.

REFERENCES

1 C.B. Muchmore, J.W. Chen, and R.E. Templemeyer, Supercritical Extraction/ Reaction of Coal with Ethanol. Paper read at AIChE winter meeting, Orlando, FL (1982) 22 pp.

2 L. A. Amestica and E. E. Wolf, Supercritical Toluene and Ethanol Extraction of an Illinois No. 6 Coal, Fuel, 63 (1984) 227-230.

3 M. Makabe and K. Ouchi, Solubility Increase of Coals by Treatment with Ethanol; Fuel Processing Technology, 5 (1981) 129-139.

4 C.B. Muchmore, J.W. Chen, A.C. Kent, K.E. Templemeyer, Y.C. Chang, and V. Gupta, Coal Conversion and Deslulfurization Under Supercritical Conditions. Paper read at St. Louis Section, AIChE Symposium,St. Louis, MO (1984) 23 pp.

5 C.B. Muchmore,J.W. Chen, and K.E. Templemeyer, Removal of Organic Sulfur from Coal by Supercritical Extraction with Alcohols, Paper read at Division of Fuel Chemistry Synposium, American Chemical Society meeting, Miami Beach, FL (1985) 11 pp.

HIGH SULPHUR COAL: A REHABILITATED RESOURCE

W.J. WILSON[1] AND S.C. WILSON[2]

[1]Atlantic Analytical Services Ltd., 275 City Road, Saint John, N.B., E2L 3N3 (Canada)

[2]Atlantic Microbiology Ltd., 275 City Road, Saint John, N.B., E2L 3N3 (Canada)

ABSTRACT

Reasons for uneconomic operation of sulphur-recovery systems to date are discussed; it is concluded that this is primarily due to their reliance on production of either elemental sulphur or sulphuric acid-both commodities of low value per unit sulphur-as a means of recouping a portion of the cost of the scrubbing process. The Pritchard-Fraxon approach, based on further processing of recovered SO_2 to other commodities, is discussed, with particular reference to an integrated plant where it is proposed to manufacture K_2SO_4, $CaCl_2$, and liquid SO_2. This includes a discussion of the economics of the plant according to current forecasts. In such a plant, high levels of sulphur in coal are advantageous since more product can be produced per unit of flue-gas processed. It is stressed that any sulphur-recovery system must be examined in association with the particular character of the industrial base in the vicinity of the proposed plant. It is concluded that, given these conditions, high-sulphur coal is a valuable resource whose potential can now be realized.

INTRODUCTION

The necessity of coping with the sulphur content of coal is a problem which is an unavoidable obstacle in both the production and consumption of coal. The producer uses such physical methods as are available to reduce the sulphur content of this coal to acceptable levels. The consumer deals with the sulphur problem by the least expensive method at his disposal, for example, fire-side additives, lime scrubbing with landfill disposal of the product, or scrubbing followed by conversion to sulphuric acid or elemental sulphur. Both the producer and the consumer regard the sulphur present in the coal as a nuisance; neither one sees it as a potential resource. The purpose of this paper is to demonstrate how this sulphur can be transformed from a liability into an asset, and how a higher sulphur content in coal can increase its value; how, in fact, high-sulphur coals can be rehabilitated.

At present, all of the usual methods of SO_2 abatement have one feature in common: they do not produce a valuable end-product. (ref.1). The gypsum produced by lime-scrubbing is usually discarded, and both sulphuric acid and elemental sulphur, although produced in large quantities, are commodities which usually carry a relatively low value per tonne. (At present sulphur prices are abnormally high, at about $125. U.S. per ton (long ton).) They therefore offer a very small return on the investment involved in installing a scrubber; usually the system operates at a net loss and as a result, producers of SO_2 - that is, consumers of coal - have been reluctant to implement such technology, to the detriment of both their public image and that of the coal industry, and the environment. Obviously, if a means could be found whereby the SO_2 resulting from the combustion of coal could be put to profitable use with a reasonable return on the purchase price of the scrubbing system, emitters of SO_2 would be more willing to recover it, and producers of high-sulphur coal would be in a more favourable position in the marketplace, especially in those areas where local coal has been displaced by low-sulphur coal imported from some distance.

The authors' interest in this problem came about as a result of research done into the use of sulphur dioxide in the chemical production of such diverse materials as phosphates, sulphites, and sulphates, and as an alternative to mineral acids in ion-exchange systems. The interest in the potential of SO_2 arose due to work concerned with the recovery of sodium from spent sulphite liquor in the pulp and paper industry using solutions of SO_2 and water. It was discovered that the addition of a carbonyl compound such as acetone made the process very effective and economically viable. This process became the basis of the Pritchard-Fraxon process.

The Pritchard-Fraxon solvent, a solution composed of water, SO_2, and one of a number of carbonyl compounds in varying proportions, used in an industrial process, was found to have a number of interesting possibilities far beyond the purpose for which it had been designed. Its benefits include the fact that the resulting salts are sulphites and bisulphites, as opposed to sulphates. All bisulphites are soluble, all multi-valent sulphites are insoluble, and the acid is weak and volatile, leading to ease of handling and separation in the system. The Pritchard-Fraxon solvent also permits the splitting of salts of strong acids, such as NaCl, by either ion-exchange or direct chemical reaction, into salts of the weaker acid, $NaHSO_3$ and Na_2SO_3, The system therefore showed a new and valuable method for the manufacture of industrial chemicals in which the results achieved by the use of strong mineral acid can be accomplished without its use.

These results include the production of phosphoric acid and dicalcium phosphate from phosphate rock (ref. 2,3,4), recovery of sodium and potassium sulphites and bisulphites (ref. 5), recovery of base from spent sulphite liquor (ref.6), and the production of potassium sulphate (ref.7). The fact that SO_2 is an essential component of this solvent system thus led naturally to consideration of possible sources of SO_2; among the most visible is SO_2 produced as a by-product of the combustion of fossil fuels - in a word, coal.

The P.F. Scrubbing Process

The process proposed here can change the perception of SO_2 as a problem. It uses the SO_2 in stack-gases to produce, cheaply and efficiently, a high value-per-ton commodity for which there is a large and expanding market. The system is based on the Pritchard-Fraxon process outlined earlier, together with a modified Wellman-Lord Process; the product is potassium sulphate, K_2SO_4, widely used in agriculture, and selling for about $200. U.S. per ton at the Gulf Coast (ref.8).

The starting point for the process is a 4-molar solution of potassium chloride, KCl, which is converted to potassium acetone-sulphonate using the Pritchard-Fraxon solution (SO_2-acetone-water) as the eluent in an ion-exchanger. The IXR unit is loaded with KCl, giving a stream of HCl, then eluted with SO_2-acetone-water to give the stream of potassium acetone-sulphonate; the eluent also acts to regenerate the ion-exchange resin.

Heating the solution of potassium acetone-sulphonate strips out the acetone and SO_2, converting it to potassium sulphite, K_2SO_3. The solution is evaporated and crystallized, and the solid K_2SO_3 is fed to a stack-gas scrubber in which circulates a saturated solution of potassium bisulphite, $KHSO_3$. The K_2SO_3 dissolves in the solution and reacts with SO_2 in the incoming flue-gases to give $KHSO_3$, as follows:

$$K_2SO_3 + SO_2 + H_2O \longrightarrow 2KHSO_3$$

This is essentially a version of the Wellman-Lord process.

Since the solution is already saturated with $KHSO_3$, any newly-formed $KHSO_3$ promptly precipitates as potassium pyrosulphite, $K_2S_2O_5$, the solid form of $KHSO_3$, and is separated from the solution. Heat converts $K_2S_2O_5$ to K_2SO_3, with the recovery of SO_2 for recycle; then the K_2SO_3 is heated further in the presence of air in order to oxidize it to K_2SO_4.

A flow-chart of the plant is shown in Fig. 1. All of the units use standard technology; that is, no innovations have been made with regard to the actual equipment to be used. The evaporator, for example, is a multiple-effect system similar to those used in sugar refineries, in which the last effect acts as a crystallizer. The scrubber is related to the Wellman-Lord system, replacing $NaHSO_3/Na_2SO_3$ with $KHSO_3/K_2SO_3$, and using a non-scaling tower such as a turbulent-contact tower, or a Venturi scrubber. Eluent make-up is accomplished in a simple gas-absorption tower, absorbing SO_2 into a solution of acetone and water; the stripper amounts to a still (distillation column), and the converter/calciner is a rotary dryer. The ion-exchange unit is a continuous counter-current system, a Higgins Contacter, which is in commercial use at a number of installations.

The use of standard equipment makes this process far more attractive, both economically and technically, than processes which require the installation of specialized or custom-made units. The projected cost of this equipment is shown in Table 1; Tables 2 and 3 show the estimated annual operating cost for the process, and the estimated product value and return on investment.

TABLE 1: Fixed Capital Investment

Item	$U.S. (thousands of dollars)
Process System, Consisting of Scrubber Tower, Regenerator Tower, 2 Vessels, 4 Heat Exchangers, 8 Pumps, Stack, Ion Exchanger.	5600
Building Air System	310
Connection to Power Plant	300
Total Fixed Capital Investment	6210

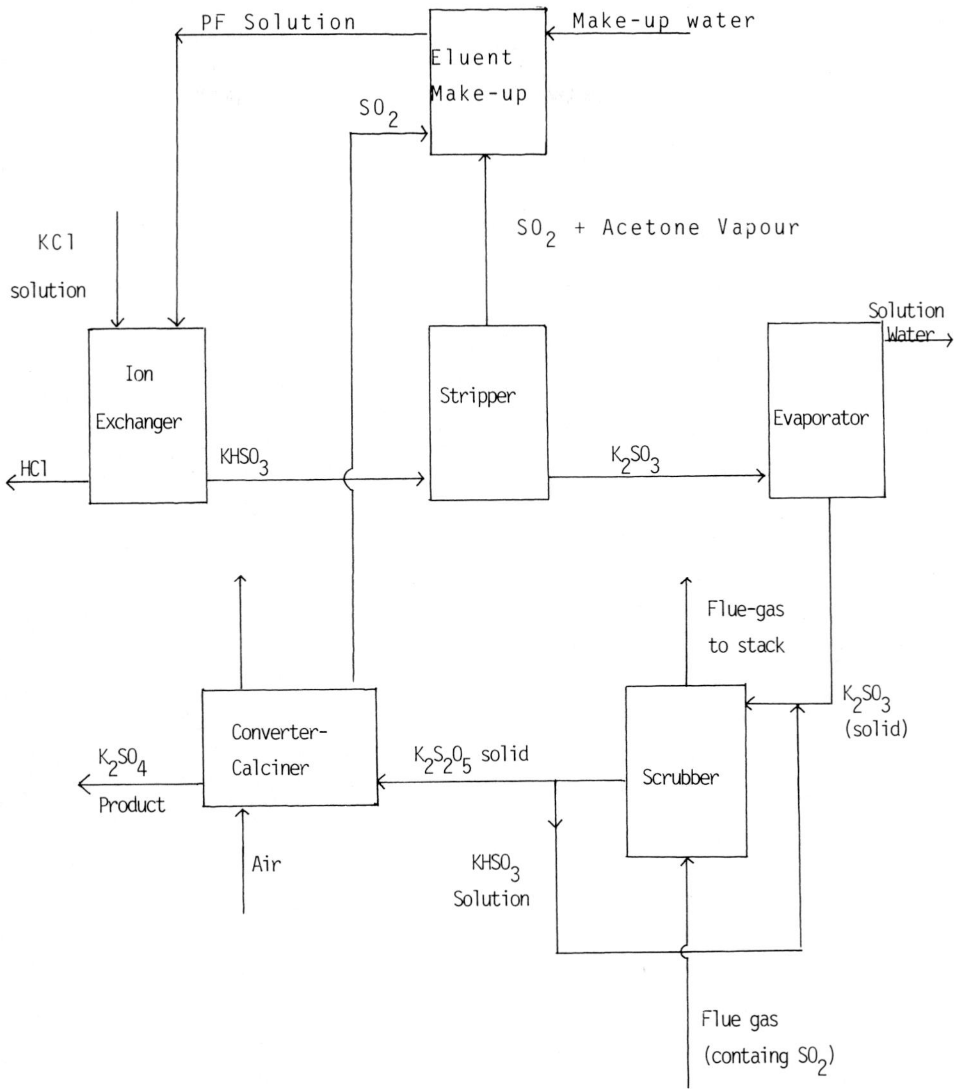

Figure 1: Flow Chart of Process

TABLE 2: Estimated Annual Operating Cost

Basis: 8400 hours/year of operation
K_2SO_4 production of 50,000 tonne/year
Fixed Capital Investment (FCI) = $6,210,000
SO_2 source: 5000 ppm SO_2 flue gases from 6-8% Sulphur coal

Item	Cost (thousands of $U.S.)
A) Direct Cost	
KCL: 43,000 tonne/year at $50/tonne	2150
Labour: 2 Foremen $25,000 ea.	50
9 Operators $20,000 ea.	180
Supervision: manager $35,000	35
Maintenance: 4% of FCI	250
Supplies: 15% of Maintenance	38
Utilities*	1890
Total Direct Cost	4593
B) Indirect Cost	
Payroll Burden: 15% of Labour & Supervision	40
Plant Overhead: 40% of Labour, Supervision, Maintenance and Supplies**	225
Total Indirect Cost	265
C) Fixed Costs:	
Depreciation: 10% of FCI	620
Taxes & Insurance: 5% of FCI	310
Total Fixed Cost	930
Total Operating Cost	5788

* Includes cost of 50,000 lb steam/hour at $4.50 per 1000 lb (congenerated steam)

** Includes clerical and sales staff

TABLE 3: Product Value and Return on Investment

Product	Amount	$U.S./tonne	Value (thousands of $)
K_2SO_4	50,000 tonne/ year	200	10,000
HCl	20,290 tonne/year	20	400
Total Income			10,400

Estimated Return on Investment:

Gross Earnings = Income - Operating Costs

= 4621

Payback Period = FCI / Gross Earings

$$= \frac{6210}{4613} = 1.35 \text{ years}$$

Income = 1.80 x operating cost

Process Economics and Product Markets

As can be seen from Table 3, the process is expected to pay back the initial investment in 1.35 years, with a groww income of 1.80 times the estimated operating cost. This expected profitability is in direct contrast to the commonly used flue-gas desulphurization systems, which operate at a net loss to the user. In addition, the process can easily be adopted to produce products other than K_2SO_4. For example, an increase in the size of the scrubber would permit the recovery of SO_2 from the flue-gas in an amount greater than that needed to provide make-up SO_2 to the process; this SO_2 could be recovered, condensed, purified, and sold as liquid SO_2, for which there is a moderate but steady demand and which commands a price per ton comparable to that of K_2SO_4. The intermediate products, K_2SO_3 and $K_2S_2O_5$, could be sold directly, without conversion to K_2SO_4; the sodium equivalents, Na_2SO_3 and $Na_2S_2O_5$ could also be manufactured, although they bring a lower price per ton.

The international market for K_2SO_4 is large and expanding fairly rapidly. It is used as a fertilizer in regions in which the use of KCl as a source of potassium is contra-indicated due to the sensitivity of certain crops - for example, citrus, potatoes, rice and vines - to the chloride ion. The need to provide sulphur to certain soils, notably tropical soils in the Third World (parts of Africa, Asia, and Latin America) and in various places in the "developed" world, including Western Canada, parts of Europe and of the U.S. West and Mid-West, also promotes the use of K_2SO_4, (ref. 9).

A study of the world K_2SO_4, done by the British Sulphur Corporation Ltd., in May 1984, for the Mineral Resources Branch of the New Brunswick Department of Natural Resources, gives a total world market for K_2O of 1.16 million tonnes in 1982/83. K_2SO_4 is approximately 50% K_2O, so this is equivalent to about 2.32 million tonnes of K_2SO_4. This demand is expected to rise to 1.55 million tonnes (as K_2O) in 1988/89, and 1.97 million tonnes (as K_2O) in 1995/96, (ref. 10). Although supply is approximately equal to demand at present (1982/83), production in the future is expected to lag behind consumption, creating favourable market conditions for new producers of K_2SO_4. The greatest growth in consumption is expected to occur in Asia, especially in the People's Republic of China, where the agricultural base is expanding and modernizing.

No mention has yet been made of the HCl effluent from the ion-exchanger. This may be evaporated and sold as commercial concentrated HCl, reacted with lime ($CaCO_3$) to form $CaCl_2$, which has a considerable market, especially for roads as a dust-settling agent, and as road-salt in the winter. In the worst case it can be neutralized with limestone and sewered, at a cost to the plant.

High Sulphur Coal Utilization

At a first glance the connection of this process to the use of high-sulphur coal may seem to be tenuous. However, its ability to profitably consume the SO_2 produced by the combustion of high-sulphur coals gives such coals an advantage over low-sulphur coals; in effect the consumer is buying two raw materials, both of which will contribute to the revenue of his plant, rather than a raw material containing an unwanted and difficult to remove impurity. The process also works more efficiently, and gives more productive use of gas-handling equipment when concentrations of SO_2 in the gases are higher, since the required amount of SO_2 make-up can be obtained from a smaller quantity of the gas. Even if the total SO_2 emissions of the plant were not consumed in the production of K_2SO_4, the profit from its production could offset the loss incurred through the operation of additional, non-profitable scrubbing systems. On the other hand, the amount of SO_2 removed from the shaved-off gas-stream might be enough to reduce the total SO_2 output from the installation to a level below the regulatory limit, as is frequently done with particulate emissions.

Naturally, the use of this or any other pollution-control system must be examined in terms of the particular case under consideration. The PF process is very versatile in terms of uses for the recovered SO_2; it is not a panacea for every SO_2 problem. Nevertheless, it is an improvement over previous systems due to its efficiency and profitability, and will, if adopted fairly widely, aid in high-sulphur coal industry and rehabilitate high-sulphur coal as an acceptable fuel supply.

Current Work

Currently further studies on plant costs and economics are being carried out specifically related to the construction of a demonstration plant of the size used in this paper. The sulphur dioxide source will be a New Brunswick generating plant burning 6-8% sulphur coal; the stack-gas emissions are at a level of 4,500-5,500 ppm SO_2.

Additionally, research has begun into an alternative system not using ion exchange, and recovering saleable calcium chloride, with a value of approximately 30% of that of potassium sulphate, as an additional product.

ACKNOWLEDGEMENTS

The authors would like to thank Mr. Gerry Clancy, of Pritchard Corporation, Kansas City, Kansas; Mr. Bob Neill, of Neill and Gunter Ltd., Fredericton, N.B., and Dr. Irwin Higgins, of C.S. Associates, Oakridge, Tennessee, for their assistance in the preparation of this paper and the research it describes.

REFERENCES

1 D.M. Ottmers Jr., et.al., Evaluation of Regenerable Flue-Gas Desulphurization Processes, Radian Corporation, Austin, Texas, 1976, 578 pp.
2 W.J. Wilson, Sulphur Dioxide, Acetone, Water in the Preparation of Fertilizers and Potassium Chemicals, Atlantic Analytical Services Ltd., Department of Regional Industrial Expansion (Canada) #3124, Saint John, New Brunswick, 1984, 106 pp.
3 McCullough, J.F., et.at., U.S. Patent #4,113,842, September 1978. Phillips, J.F., et.al., U.S. Patent #4,238,459, December 1980
4 A.H. Kislin (ed.), Chemical Week 134, 9 (1984), 16.
5 W.J. Wilson, Recovering Monovalent Cations from Spent Sulphite Liquor, U.S. Patent #3,248,278, April 1966.
6 B.S. Das and W.J. Wilson, Production of Alkali Metal Sulphites or Bisulphites, Canadian Patent #832,931, January 1970.
7 W.J. Wilson and S.C. Wilson, Plant Design and Cost-Benefit Analyses for the Manufacture of Sodium Sulphite or Potassium Sulfite using the Pritchard-Fraxon Process, Atlantic Analytical Services Ltd., Department of Regional Industrial Expansion #3141, 1984, 28 pp.
8 Chemical Marketing Reporter, 227 (20), 1985.
9 The British Sulphate Market Ltd., An outline of the World Potassium Sulphate Market, prepared for Mineral Resources Division, New Brunswick Department of Natural Resources, May 1984, 86 pp.

CHANGING SULPHUR PATTERNS

H. L. FIKE

President, The Sulphur Institute, 1725 K Street, N.W., Washington, D.C. 20006

ABSTRACT

Sulphur is used as a raw material in the manufacture of petroleum, rubber, paper, sugar, pigments, fibers and many chemicals. Until recently, its uses were so broadly based that economists used sulphur consumption to measure industrial activity. However, a combination of economic and environmental factors have changed sulphur use patterns. In 1984 fertilizer manufacture accounted for about two-thirds of the world's sulphur consumption and agricultural rather than industrial activities are the primary determinant of sulphur use.

During the same period, major changes also took place in sulphur supply. Sulphur produced as a by-product increased rapidly and now accounts for over half of sulphur production. The anticipated increase in the use of coal as an energy source is apt to increase the importance of involuntary sulphur production.

The above changes encouraged development of new uses for sulphur. One of these is already consuming substantial quantities of sulphur and others could have a significant impact on sulphur usage by 1990.

INTRODUCTION

Sulphur is a non-metallic element of atomic number 16 and in the same chemical family as oxygen. At ambient temperature it is an odorless, tasteless, non-toxic pale yellow solid with the following physical properties.

Melting Point	118°C	
Boiling Point	444°C	
Sp G	2.07	@20°C
	1.80	@120°C
Compressive Strength	3300 psi	
Tensile Strength	600 psi	

Sulphur is a raw material used widely in the manufacture of fertilizers, paper, rubber, metals, food, pharmaceuticals and a variety of other materials. Its uses are so broad that for many years it was considered an indicator of a nation's industrial productivity. However, some significant and often unnoticed changes have taken place during the past quarter century and these are the subject of this paper.

PRODUCTION OF SULPHUR

Elemental sulphur, commonly called brimstone, is mined from ores in many countries. The U.S.A., the U.S.S.R., Poland, Mexico and Iraq are the major producers.

Brimstone is also produced as a by-product from the processing of fossil fuels, mainly natural gas, petroleum and tar sands. Canada, the U.S.A., France, Japan, Saudi Arabia, the U.S.S.R. and other countries are important producers of by-product sulphur.

Sulphur compounds, particularly sulphide ores and sulphuric acid are also useful sources of sulphur. Iron sulphides (pyrites) are produced for their sulphur values, but other materials such as sulphuric acid produced at non-ferrous metal smelters and at a few coal-burning plants are often considered unwanted by-products.

Table 1 indicates that total sulphur production increased by only 160% during the past quarter century. However, elemental sulphur produced as a by-product from fossil fuels increased by 620% and is now the largest source of sulphur. Table 1 also shows that by-product sulphur accounts for over half the sulphur produced.

TABLE 1
World sulphur production, by type

	million metric tons			
	1960		1985 (e)	
	tons	%	tons	%
Elemental Sulphur				
Mined	8.2	36	14.1	25
By-product	3.1	14	21.3	39
Non-Elemental Sulphur				
Mined	8.2	36	11.0	20
By-product	3.2	14	8.9	16
	22.7	100	55.3	100

The geographical distribution of sulphur production has also changed. Together the U.S.S.R. and Eastern Europe account for 37% of world production compared with only 14% twenty-five years ago. North America and Western Europe have increased production but have declined in market share (Table 2).

TABLE 2

World sulphur production, by geographical area

	million metric tons			
	1960		1985 (e)	
	tons	%	tons	%
Area				
North America & Mexico	9.2	41	17.3	31
South America	0.2	1	1.0	2
Western Europe	6.4	28	8.3	15
Middle East	0.1	0	1.4	3
Far East	2.3	10	3.5	6
Africa	0.3	1	0.3	1
Oceania	0.2	1	0.2	0
Eastern Europe & USSR	3.1	14	20.5	37
China & North Korea	0.9	4	2.8	5
	22.7	100	55.3	100

SULPHUR CONSUMPTION

During the same quarter century, sulphur consumption increased from 22 million to about 57 million metric tons. Almost all of the growth resulted from increased usage of sulphur to manufacture phosphatic fertilizers (Table 3).

TABLE 3

World sulphur consumption

	million metric tons	
	1960	1985 (e)
End Use		
Phosphatic Fertilizer	5.5	36.0
Other Traditional Uses	16.5	20.0
New Uses	0.0	1.0
	22.0	57.0

Production of phosphatic fertilizer accounted for over 80% of the increase in sulphur usage and now accounts for about 60% of total demand. In the aggregate all other traditional uses increased by less than 4 million tons. Consumption of sulphur in many of the traditional end uses (rayon, pigments, steel) has and will continue to decline. Thus, changes in the agricultural area are of greater relevance in determining sulphur usage than changes in industrial activity.

Table 4 suggests that since 1960, growth in demand for sulphur used to manufacture phosphatic fertilizers is slowing. If this trend continues sulphur consumption will plateau and the impact on the sulphur industry will be great.

TABLE 4
Use of sulphur to manufacture phosphatic fertilizers

	Annual Growth Rate
1960 - 1970	9.8%
1970 - 1980	5.6%
1980 - 1984	2.5%

The apparent decline in the growth rate of phosphate may not be as serious as the table implies. Demand for sulphur to produce phosphate is exhibiting a greater cyclical movement (Table 5) than in the past. However, 1980 could have been an unusually good year and this could be responsible for at least part of the apparent reduction in growth rate in subsequent years.

TABLE 5
Sulphur used for manufacturing phosphatic fertilizer

	1980 = 100	
	North America	World (excluding North America)
1980	100	100
1981	91	93
1982	75	99
1983	84	106
1984	100	113

Recovery from the bottom of the cycle in North America and Western Europe was much slower than in India, China and other nations.

SULPHUR PRICES

Sulphur prices are quite responsive to change in the supply/demand balance and vary in the different geographical areas. However, the average U.S.A. price as reported by the U.S. Bureau of Mines (USBM), indicated in Table 6, are probably representative of world trends.

TABLE 6
Sulphur prices

Year	$U.S. per metric ton[1] current $	constant 1983 $
1960	22.76	70.00
1965	22.12	64.15
1970	22.77	53.70
1975	44.91	76.99
1980	89.06	107.65
1984	94.31	99.07

[1]Average price as reported by the USBM in Reference 1.

During the past 25 years the following trends have taken place in the sulphur industry.

1. Sulphur produced as a by-product from the processing of fossil fuels and metals has become the most important source of sulphur.
2. Coal-burning plants which could have become the major source of sulphur produced only insignificant amounts of sulphur in usable form.
3. The U.S.S.R. and Eastern European nations account for nearly 40% of world sulphur production and are very important in determining the sulphur/demand balance.
4. One single end use (phosphatic fertilizer manufacture) has become the most important determinant of sulphur consumption.
5. Growth in the consumption of phosphates has plateaued and became more cyclical in North America.

If these trends continue, the limited ability of the sulphur industry to enlarge production capacity in the short-term will make careful planning more vital to the health of both the phosphate and sulphur industries.

As the sulphur industry's future became more dependent upon the fortunes of the fertilizer industry, it began to seek new markets, particularly those which would:

- represent large potential markets for sulphur
- grow at higher rates than traditional uses
- exhibit different cyclical patterns

Three important potential new markets for sulphur are as a plant nutrient, in the manufacture of corrosion-resistant concrete and in sulphur paving materials.

NEW MARKETS

Plant-nutrient sulphur

Crops require large amounts of sulphur for growth and the soil serves as a sulphur bank. Withdrawals can only exceed deposits until the surplus is withdrawn and in many geographical areas so much sulphur has been removed that unless it is replaced, sulphur deficiencies will limit crop yields. The need for sulphur for crop growth around the world has increased substantially during the past quarter century. Crop removal of sulphur more than doubled between 1960 and 1985. It now totals 9.2 million tons and is expected to increase to 14.2 million tons by the year 2000.

We can divide the world into near- and long-term markets for plant-nutrient sulphur. Near-term markets are those in which sulphur deficiencies are becoming widespread and sulphur supplementation may be required by 1990 to achieve maximum crop yields. Long-term markets are those in which sulphur deficiencies are less widespread and it will probably be beyond 1990 before their potential is realized.

Development of near-term markets for sulphur requires educational and promotional efforts to create an awareness of the need for sulphur and for companies to aggressively market sulphur fertilizers. The educational efforts which provide information to the farmer, the fertilizer industry and the academic community are efficiently carried out through cooperation between agencies and The Sulphur Institute (TSI). TSI has been actively engaged in these activities since 1960.

One region of the world which is identified as a large, near-term potential market is North America. In 1960 sulphur deficiencies had been identified in only 13 states and 4 provinces of Canada.

Currently, they have been identified in 35 states and 5 provinces of Canada. The number of companies marketing plant-nutrient sulphur is one indication of the magnitude of the market. Companies marketing sulphur in

North America now total 35, a fivefold increase over the past 15 years.

A further indication of the market potential for plant-nutrient sulphur is a comparison with consumption of other nutrients. While in the past decade total phosphate consumption has changed little in the U.S.A. and total potash consumption has increased only slightly, deliberate application of sulphur as a plant nutrient has increased nearly fourfold. This is a rapidly growing market which has not yet achieved its potential. In the U.S.A., current deliberate consumption of sulphur as a plant nutrient is about 800,000 tons, but a near-term potential of an additional 500,000 tons by 1990 and a total potential of about 2.6 million tons is foreseen.

To more rapidly realize this potential, TSI cooperates with companies marketing plant-nutrient sulphur in North America by providing them with information for their advertising and promotional programs.

Companies marketing plant-nutrient sulphur co-sponsor numerous research projects to generate data demonstrating the geographical areas and crops in which application of sulphur will increase yields and profits.

Another rapidly growing area with a large potential market for plant-nutrient sulphur is Southeast Asia. In 1983 TSI co-sponsored a symposium in Bogor, Indonesia, which brought to the forefront the growing number of sulphur deficiencies in the region and the need for more research to better qualify and quantify its scope. The proceedings have been widely distributed throughout Asia, among international agencies, researchers, industry and in other tropical regions of the world, and have generated a tremendous amount of interest in sulphur.

India currently is the fourth largest consumer of fertilizers with a current consumption of about 8 million tons annually, a figure expected to double by 2000. Deficiencies of sulphur are currently widespread and will increase with the growing usage of sulphur-free fertilizers and increasing crop yields removing greater quantities of sulphur from the soil.

Current consumption of sulphur in India is about 350,000 tons and current need based on crop removal is estimated to be an additional 1 million tons. Since India has a fairly advanced agricultural system and network of fertilizer distribution, it is considered a near-term market. We are attempting to motivate the Indian agricultural leaders to include sulphur in their programs and develop promotional materials. A national symposium to be held in early 1987 will describe the need for sulphur in the different regions of India.

Another area in the tropics which has a near-term potential for increased plant-nutrient sulphur consumption is Brazil. Brazil's fertilizer consumption and agricultural output has increased substantially in the past decade. However, while current sulphur fertilizer consumption in all forms is about 340,000 tons, current need is estimated to be 870,000 tons, leaving a gap of about a half million tons.

In addition to the near-term markets, long-term markets with larger potential for future plant-nutrient consumption exist. Developing the long-term markets for plant-nutrient sulphur requires the motivating of research activities to identify sulphur needs; dissemination of information to governments, industry and researchers; and cooperating with international organizations to develop awareness and programs for plant-nutrient sulphur.

Long-term potential markets in Asia include Indonesia, the Philippines, Thailand and China.

TSI will be assisting other international agencies in supporting a symposium to be held in the latter part of 1986 to bring together scientists from China and other regions of the world to discuss plant-nutrient requirements in China.

Figure 1 lists world areas in which sulphur deficiency symptoms have been reported.

In summary, a very real opportunity to increase plant-nutrient sulphur consumption exists in those areas identified as near-term markets - North America, Western Europe, India, Brazil and China. Other areas of Asia and South America, and Africa are longer-term markets.

Corrosion-resistant concrete

A second potential new use for sulphur is in the manufacture of corrosion-resistant concretes.

These materials are produced by substituting sulphur cement for the portland cement normally used in concrete manufacture. Termed sulphur concretes (SC), they are as strong or stronger than conventional concrete and are almost impervious to acids and many other chemicals. Sulphur concrete is more expensive to produce than portland cement concrete but less expensive than polymer concrete, epoxy coatings, acid brick or other materials used in areas where conventional concrete fails prematurely.

As curiosities, SC has been known for many decades, but it was not until about 1975 that groups in Austria, Canada, Japan and the U.S.A. began to develop materials suitable for large-scale commerical use.

Although TSI cooperates with all of the current manufacturers of SC, our major involvement has been with the USBM. Cooperative efforts with the USBM led to the successful testing of SC in over 100 different industrial environments. Using information derived from these tests, TSI published a design manual and to date has distributed several thousand copies to groups in 27 different countries. Thus, interest in SC is already spread over a large geographical area and is continuing to expand.

SC is a generic term, and until specifications are established, the design manual and consultations with USBM or TSI staff assist potential users to determine whether SC will be a cost-effective material in their particular environment. TSI co-sponsors work with the USBM to provide more fundamental data needed to develop internationally-accepted standards and specifications. This work is directed toward:

1. Reducing costs.
2. Increasing durability.
3. Developing specifications.
4. Determining performance in new environments.
5. Analyzing premature failure.

The number of contractors engaged in the placement of SC has grown from two in 1982 to ten in 1985 and the number of commercial installations will reach about 150 this year. Commercialization of SC is still in an early growth stage. To insure that growth is sound as well as rapid we need to:

1. Develop more comprehensive environmental, safety and toxicity data on the material.
2. Cooperate in the promotion and publicity activities.

SC is a specialty material which will serve as a cost effective replacement for conventional concrete in areas in which the present materials fail prematurely. However, five billion tons of portland cement concrete are used annually worldwide and if SC eventually captures only 0.5% of the cement market, it would require five million tons of sulphur annually.

Sulphur paving materials

Asphalt pavements deteriorate from the effects of light, water, heat and air and the need to improve them has been recognized for a long time.

The release of a Sulphur Institute-funded study in 1971 spurred many agencies to evaluate sulphur as a paving material. These efforts subsequently led to the construction of about 100 sulphur-asphalt (SA) roads in the U.S.A. and 200 worldwide. Most of these roads are performing quite well.

One reason for the early interest in SA was concern over the future availability of asphalt particularly in North America. North America consumed about half the world's asphalt supply during the 1970's and if lack of road building funds, particularly in the U.S.A., had not reduced asphalt consumption, a shortage might have materialized. During the evaluation of SA materials, highway authorities found that including sulphur in the paving mix improves the durability of roads. Evidence indicates that properly designed SA roads are less susceptible to both cracking in the winter and rutting in the summer. As a result of this greater durability, SA roads are less expensive over the full life of the highway. In some areas, particularly Canada, France, Mexico and the U.S.A., sulphur permits use of sand to replace crushed stone. This improves the economics in geographical areas in which local sands can replace stone shipped from distant sources.

The Transportation Research Board recently published a detailed report emphasizing the need for more durable rather than minimum initial cost highways. This led the U.S. Government to establish a Strategic Highway Research Program (SHRP) which allocated $100 million to improving the properties and performance of asphalt. Sulphur has been more thoroughly evaluated than other asphalt enhancers. In addition, its ready availability and low cost suggest that it be given high priority. Of the five asphalt properties cited as requiring improvement, sulphur is known to increase stability, improve flexibility, and act as an aid in recycling. If SHRP findings substantiate the conclusions by highway officials reporting on the performance of SA roads, sulphur is apt to be routinely incorporated into paving materials.

IMPLICATIONS FOR THE COAL INDUSTRY

During the past quarter century sulphur produced as a by-product from the processing of petroleum and natural gas increased nearly 600% and is now the major source of sulphur. Although high sulphur coal represents an even larger potential source of sulphur, today its contribution to the sulphur supply is of no significance.

Environmental considerations will probably force users of high sulphur coal to further limit sulphur emissions. Whether this will result in sulphur being produced in usable form will depend upon the technology utilized. Gasification, liquification and other technologies produce sulphur as brimstone or sulphuric acid, which can be used by the sulphur-consuming industries. However, precombustion cleaning produces pyrites and most post combustion processes produce calcium sulphates. There is little chance that these materials will be used as a sulphur raw material.

If the coal-consuming industries produce sulphur in a usable form it may comfort them to recognize that the sulphur-consuming industries were able to utilize the large quantities of by-product sulphur produced in the past 25 years. This is indicated in Table 7 which shows sulphur production and inventory in Western Canada, the major producer of by-product sulphur. Although a significant build-up of inventory occured in Alberta it is now being sold to world markets. To avoid this situation the coal industry should initiate or cooperate in the development of new uses for sulphur before production of sulphur becomes sizable, as future sulphur consumption is somewhat dependent upon the development of new uses for sulphur. Table 8 summarizes my estimates on the short- and long-term status of new use programs. These programs suggest that they could accommodate sulphur produced by the coal-consuming industry.

TABLE 7
Western Canadian sulphur situtation

Year	million metric tons production	inventory-Year end
1960	0.4	0.3
1965	1.6	0.5
1970	4.3	3.6
1975	6.6	16.5
1980	6.2	19.1
1985 (e)	5.8	8.5

TABLE 8
Sulphur consumption

	million tons			
	1985	1990(e)	Potential	Status
Plant Nutrient	1.2	2.0	8	Large and growing
Sulphur Concrete	0.05	0.4	5	Small, but growing rapidly
Paving	0.05	0.4	10	Sleeping Giant
	1.3	3.1	26	

FIGURE 1

Areas reporting sulphur deficiencies in 1983

AFRICA

Cameroon
Central African Republic
Chad
Congo
Ghana
Guinea
Ivory Coast
Kenya
Mali
Malawi
Mozambique
Nigeria
Senegal
South Africa
Tanzania
Togo
Uganda
Upper Volta
Zaire
Zambia
Zimbabwe

AMERICAS

USA

Alabama
Alaska
Arkansas
California
Colorado
Florida
Georgia
Hawaii
Idaho
Illinois
Indiana
Iowa
Kansas
Louisiana
Maryland
Michigan
Minnesota
Mississippi
Missouri
Montana
Nebraska
North Carolina
North Dakota
Ohio
Oklahoma
Oregon
Pennsylvania
South Carolina
South Dakota
Tennessee
Texas
Virginia
Washington
Wisconsin
Wyoming

Canada

Alberta
British Columbia
Manitoba
Ontario
Saskatchewan

Latin America

Argentina
Brazil
Chile
Colombia
Costa Rica
El Salvador
Honduras
Puerto Rico
Venezuela
Windward Islands

ASIA

Burma
India
Indonesia
Lebanon
Malaysia
Sri Lanka
Taiwan
Thailand

AUSTRALAISA

Australia
Fiji
New Guinea
New Zealand
Solomon Islands

EUROPE

Belgium
Bulgaria
Czechoslovakia
Denmark
Finland
France
Germany
Iceland
Ireland
Italy
Netherlands
Norway
Poland
Spain
Sweden
UK
USSR
Yugoslavia

REFERENCE

1 D.E. Morse, Sulfur, preprinted from Bulletin 675, Mineral Facts and Problems, 1985 Edition, Bureau of Mines, U.S. Department of Interior.

MARKET PROSPECTS FOR COAL-DERIVED SULFUR

John M. Dale

Southwest Research Institute, PO Drawer 28510
6220 Culebra Road, San Antonio, Texas 78284 (U.S.A.)

ABSTRACT

A number of new use application technologies for sulfur have been developed over the last several years. Those that extend or replace asphalt as either sulfur-extended asphalt or chemically modified sulfur pavement binder (SULPHLEX®) have shown themselves to be technically feasible and potential markets for large quantities of sulfur. Today, the price of sulfur is so high that sulfur cannot enter the new markets. This paper looks at historical conditions and associated technology in the sulfur industry, and the price of sulfur, and attempts to relate these factors to sulfur that could become available from high-sulfur coal if suitable processes for its recovery through coal cleaning or SO_2 conversion are developed.

Identifying the market prospects for coal-derived sulfur can be approached in several ways. One can document the past and future supply and demand for sulfur in great detail, construct elaborate computer models and, from the projections, draw very specific conclusions. This has been done numerous times, and recent published [1-3] studies have had great difficulty in withstanding the test of time. It is not easy to pinpoint the exact reason for this, but it is clear that the sulfur industry in the United States is far from being a model for economic theory. In the last several years, with millions of tons of sulfur reserves in the ground and a slack market, six of ten Frasch sulfur mines have shut down, the price of sulfur has risen like never before, and there is talk of a sulfur shortage. The approach taken herein is to look at historical conditions and associated technology as they relate to changes in the sulfur industry and to consider from a broad perspective what conditions and technology exist today and what the market prospects might be for coal-derived sulfur.

Following the discovery of gunpowder in the 13th century, significant mining of sulfur began in Sicily. From that time until 1900, Sicily had a virtual monopoly on the world sulfur market. The Sicilian deposits were extensive, underground, and had a sulfur content of from 25 to 40 percent in a matrix of limestone, gypsum, and clay. The ore was mined and brought to the surface. The sulfur was freed from the gangue by the heat of its own combustion, and then transported to the coast for shipment to the world markets. At the end of the 19th century, the United States produced less than one-half of one percent of its own sulfur requirements. The remainder was imported from Sicily.

The Sicilian producers were so dominant in the world sulfur market that in 1904 they refused to believe the rumors and firsthand accounts affecting their industry. These rumors stated that the American Gulf Coast deposits of salt dome sulfur were very large, the newly developed Frasch process was practical and economical, and this new source of sulfur could be landed in Europe at 20 percent less than the cost of mining and extracting Sicilian sulfur at the mine. At that time, the price of Frasch-Gulf Coast sulfur was $18 a ton at the mine, regardless of the quality or destination. Within a few years, the Sicilian sulfur industry was devastated. It never recovered. Herman Frasch and the Gulf Coast sulfur producers became the dominant force in the world sulfur industry. It is said of Herman Frasch[(4)], "He believed that a reasonable price would extend the use of sulfur, and as early as 1898 he backed research aimed to broaden the market". From the beginning of the century until the mid 1960's, the Frasch sulfur industry was in the most enviable position possible. Its ability to make high returns on invested capital, year after year was phenomenal. The major producers were perennially in the "Fortune 500" and almost always in the top 50. It was a good business to be in.

In the early 1940's, the utilization of sour petroleum crudes and sour natural gases took on increasing importance. A number of satisfactory methods had been developed to remove hydrogen sulfide from the different streams. In most cases, hydrogen sulfide could not be utilized, and it was flared, resulting in sulfur dioxide pollution of the atmosphere. The first sour gas sweetening plant was placed in operation at McKamie, Arkansas, in 1944 by the Mathieson Corporation. This plant took hydrogen sulfide from natural gas and converted it by the Claus process to elemental sulfur. This technology not only solved a major problem by converting a hazardous and unwanted byproduct into a useful and salable item, but it also generated 5000 pounds of 300-psi steam for each ton of sulfur produced. Thus, began the proliferation of Claus process plants. Today there are 100 such plants in crude processing facilities and 53 such plants in natural gas processing facilities in the United States alone.[(5)] The owners of these plants are in the oil and gas business, and their byproduct sulfur is something to be disposed of at whatever price it will bring. They have, historically, built little or no inventories. In the late 1960's and early 1970's, gas producers in Western Canada became major producers of recovered sulfur. They were forced to hold large inventories which they tried to reduce by selling this sulfur in the United States. This severe threat to the United States Frasch producers was warded off in 1973. The Frasch producers convinced the United States Treasury Department and Tariff Commission to rule that eight of Canada's largest sulfur producers were dumping sulfur in the United States at less than fair market value. The nature and cost of producing recovered sulfur are such that it is difficult to see how it can ever be eased out of the market by some new technology. The recovered sulfur producers are limited in their output by the amount of raw high-sulfur petroleum and sour natural gas available for processing.

Consequently, the recovered sulfur producers have no real incentive for going after or developing any big new markets for sulfur. The impact of recovered sulfur production on the Frasch producers has not been as devastating as Frasch production was on Sicilian producers because recovered sulfur has appeared on the scene at a much slower rate. However, the ultimate outcome could well be the same. Figure 1 shows how recovered sulfur has moved into the market and displaced Frasch production.

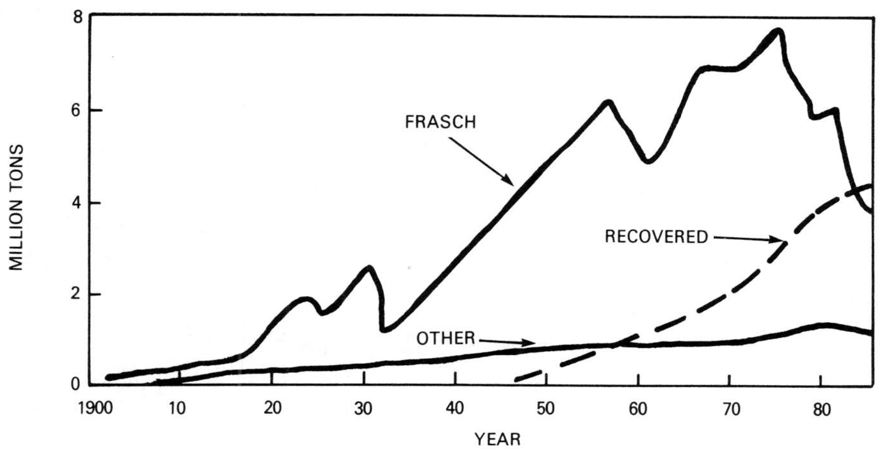

FIGURE 1. UNITED STATES SULFUR PRODUCTION

From time to time, announcements of new technologies have appeared that have each threatened to glut the sulfur market. These have included sulfur from gypsum (Elcor), bacterial sulfur processes, sulfur from coal cleaning, sulfur from stack gases, and sulfur from synfuels. For lack of technical, economic, and environmental imperative or a combination thereof, none of these technologies has yielded any significant amount of sulfur. The impact of recovered sulfur and a sluggish market has been sufficient to cause numerous Frasch mine closings. In 1982, ten Frasch mines were in operation--seven in Texas and three in Louisiana. Today, there are only four mines in production. Estimates vary, but underground Frasch reserves in the United States are on the order of 130 to 150 million tons. Approximately 40 million tons of these reserves are in West Texas; the remainder are along the Texas-Louisiana Gulf Coast. Certainly, there is no desperate situation, shortage, or tight supply of sulfur as one reads about[6]. All one has to do is to heat water, pump it into the ground with skill, and up comes sulfur. Discussions of Frasch aboveground inventories rising and falling are not very meaningful. Inventorying large quantities of Frasch sulfur above ground only increases its cost by virtue of the interest on the money it requires to move it from below to above ground, maintain it in storage, and rehandle it. While there are fluctuations in the demand for sulfur, brought about by fluctuations in the fertilizer industry, the long-term growth curve for sulfur was around 5 percent per year until the 1970's. Since that time, it has dropped to around 3 percent. It is hard to comprehend the trade journals' reference to "shortages". They may more reasonably say there is a "shortage of low-priced sulfur".

This brings up the "price" of sulfur, which was recently quoted at $162.50 per ton. This is the highest price sulfur has ever been, as can be seen in Figure 2.

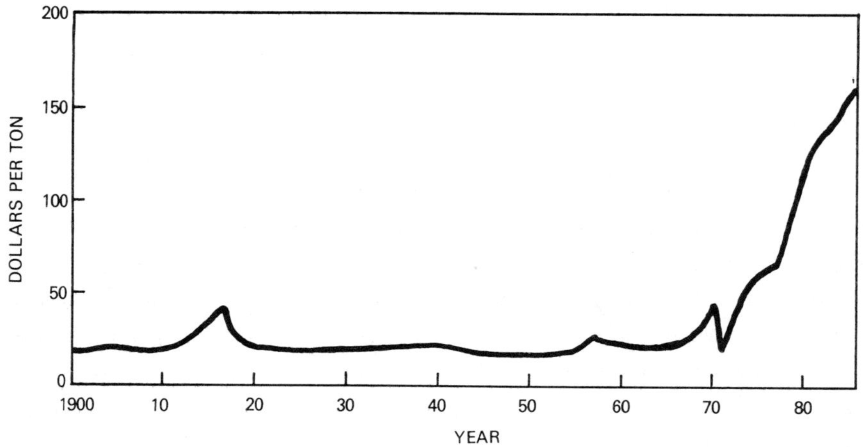

FIGURE 2. UNITED STATES SULFUR PRICE

Most often, prices are related to production costs. Given the involuntary byproduct nature of recovered sulfur, which is now the largest source of sulfur in the United States, it is not possible to relate the cost of production of recovered sulfur to the current sulfur price. The principal cost factors in mining sulfur by the Frasch process are the cost of energy to superheat water for mining, the cost of the water heating plant, the cost of drilling wells, and, in some instances, royalty fees. Detailed analysis[2] of the three major Frasch producers' production costs, showed that their average total cost of production in 1980, including royalties and transportation costs to the coast, was $48 per ton. At that time, the cost of energy was at a high. Today the energy business has changed drastically, and the buyer is in command. With energy prices down, it is difficult—if not impossible—to relate Frasch sulfur production costs with the current sulfur price. Even though the recovered sulfur producers are the major suppliers of sulfur, the Frasch producers set the price of sulfur. It appears that every time the consumption of Frasch sulfur declines, the Frasch producers raise the price of sulfur. The recovered sulfur producers continue to sell all the sulfur they produce.

The rise in the sulfur prices infuriated[7] the Florida fertilizer producers, who utilize sulfur as sulfuric acid to acidulate phosphate rock. The fertilizer producers responded by constructing sulfuric acid terminals[8] in Tampa so that they could import sulfuric acid from such places as Arizona and Japan. Under more normal circumstances, economics favor shipping sulfur to the point of consumption and converting it to acid there, because one ton of sulfur converts to three tons of acid. Considering that much of the sulfur in the Japanese sulfuric acid probably originated in Middle East crude oil, that sulfur has traveled three-fourths of the way around the world to get to its point of consumption. But, as was mentioned earlier in this paper, older methods of conducting business are no longer standard in the sulfur industry.

Since logical and rational explanations of the sulfur price are difficult to find, it is reasonable to look for some other explanation. Along these lines, it is to be recalled that after OPEC raised the price of oil, the oil companies, balance sheets began to run wild. The oil companies needed to invest their earnings somewhere quickly before the stockholders and the government became too interested and sought to share them. Accordingly, the three major Frasch producers were acquired by oil companies, who could have conceived that the sulfur industry in the United States was similar to the oil business in the world. The recovered sulfur producers were like the world oil producing companies; many in number with individually small reserves and limited individual production capacity. The Frasch producers were like OPEC, holding large reserves and unlimited production capacity. They had the ability to control the price because they could meet that last and vital segment of demand. The Frasch producers seemingly have more power than OPEC. This might be related to the fact that there is a spot market for oil whereas there is none for sulfur.

Whatever the explanation, the current system is working. It would seem that as long as Frasch producers can meet the last segment of demand, they will remain in control of the pricing system. When they lose the ability to meet the last segment of demand, they will very likely lose their control. Given time, and with the low projected growth rate for sulfur consumption, United States recovered sulfur and imported sulfur will probably totally displace Frasch production. It is impossible to suggest when this might happen. If and when it does happen, several possibilities come to mind. The Frasch reserves will very likely still be large and could become, in essence, a strategic reserve. Another possibility is that the Frasch producers might prefer to mine the sulfur and sell it at a lower price for some new use application.

In the first half of this century, the United States dominated the world supply of sulfur. However, it no longer does so. In 1984, the United States produced only 20 percent of the world's sulfur supply of 54 million tons. Competition for the export market is growing. There are some new faces, on the horizon, namely the Russians. While Canadian recovered sulfur production is declining, the Russians are hurriedly developing their Ostrakhan sour gas deposits. These deposits are reportedly enormous and are expected to yield far greater quantities of recovered sulfur than the Russians can use internally. Everyone is wondering how soon this sulfur will appear in the world market. Given the Russians' need for hard currency and the technical assistance they are receiving from the West, this sulfur should begin appearing on the world market by 1987. Resolution of the Iran-Iraq war, increased oil and gas production in the Middle East, and new mines in Poland and Mexico all have the potential for adding to the world supply. If these materialize, they could reduce or eliminate the United States' sulfur export market.

The conditions and technology in-place today in the sulfur industry would appear to be as follows:

1. Recovered sulfur is now the dominant source of sulfur in the United States and the world, and it will become even more dominant in the future. The producers of recovered sulfur are many. Technology and society give them no alternatives but to produce all the sulfur they physically can. Yet they are limited by their source as to how much they can produce. The recovered sulfur producers can fall back to any price, if forced to.
2. The Frasch industry in the United States is very much in control of the price of sulfur. The price of sulfur is the highest it has ever been and bears little, if any, relationship to the cost of production or availability.
3. Unless beset by some new internal source or outside force, the sulfur industry in the United States can proceed on its present course for years to come.

Given the above conditions, it is little wonder that the United States sulfur producers have shown little interest in investing their funds in developing new uses for sulfur. United States government agencies, in working to solve various problems, have, over the last 20 years, spent large sums of money on development programs that have resulted in a number of new uses for sulfur. These have included the development of sulfur foams, sulfur coatings, sulfur-aggregate concretes, sulfur-extended-asphalt, and chemically modified sulfur pavement binders (SULPHLEX).

Of the new use application technologies developed for sulfur, those that extend or replace asphalt as either sulfur-extended-asphalt(9) or SULPHLEX(10) have shown themselves to be technically feasible and to involve a large tonnage market. Asphalt consumption in the United States for paving applications is approximately 24 million tons per year. This is twice the size of the current United States consumption of sulfur. It is 2.7 times the amount of sulfur which might be extracted from emissions currently coming from coal-burning electric utilities and industrial boilers. Thus, if high-sulfur coal is burned, and the sulfur can be recovered there is a place for it to go. To enter this market, the price of sulfur needs to be less than two thirds that of asphalt, due to differences in specific gravity of sulfur and asphalt and other technical and economic reasons. Accordingly, with asphalt currently quoted at $160 per ton, sulfur should be less than $100 per ton to become viable in the paving market. The current United States price precludes this market from developing. The viability of this market has been demonstrated in Canada. In the last several years, the Canadians have used 40,000 tons of their sulfur in new use applications. Over 30,000 tons of this was used in sulfur-extended-asphalt. The Canadians were willing to accept a net back plant price on this sulfur of $65 per ton(11). The incentives to pursue the recovery of sulfur from coal are greater than what it might demand as an alternative to asphalt. The United States is still heavily dependent on foreign oil (33 percent); much of the asphalt being used today comes from foreign oil. More and more imported oil is being replaced by imported refined products, including asphalt. Using sulfur from coal as a paving binder could lessen our dependence on foreign oil, which costs our society far more than the apparent $26 or $28 per barrel.

Herman Frasch, in accepting the Perkin Medal in 1912 for his development of the Frasch process said, "...this severe criticism, while not agreeable, did not carry very much weight with me. I felt that I had given the subject more thought than my critics, and I went about my work as best I could, thoroughly convinced that he who laughs last, laughs best" (12). Today, 73 years later, Herman and his heirs in the Frasch industry are still laughing. They are laughing at you and me. The environmental imperative, the market, and the infrastructure to use sulfur from high-sulfur coal are all in place. All that is needed is a process that will economically recover sulfur from high-sulfur coal.

REFERENCES

1 Rangnow, D.G., Fasullo, P.A., "Rapid Growth Is Outlook for Recovered Sulfur", Oil and Gas Journal, September 28, 1981.

2 Manderson, M.C. "Sulfur Supply and Demand and Its Relationship to New Energy Sources", Presented at the 181st National Meeting of the Am. Chem. Soc., Atlanta, Georgia, April 2, 1981.

3 O'Brien, W.E., Anders, W.L., and Veitch, J.D., "Projection of 1985 Market Potential for FGD Byproduct Sulfur and Sulfuric Acid in the U.S.," EPA-TVA Interagency Energy/Environment R&D Program, EPA-60017-80-131, TVA EDT-115, July 1980.

4 Haynes, Williams, "The Stone That Burns", D. Van Nostrand Co., Inc., 250 Fourth Ave., New York, New York, 1942.

5 Cantrell, Ailleen, "World Petroleum-Derived Sulfur Production," Oil and Gas Journal, August 8, 1983.

6 "Tightening Supplies For the Long Term", Chemical Week, p. 26-32, March 27, 1985.

7 "IMC Roars Over TG's Sulfur Price Increase", Chemical Week, p. 12-13, September 16, 1981.

8 "Bringing Sulfuric to Florida" Chemical Week, p. 15-16, March 24, 1982.

9 Saylak, D. and Conger, W.E., "A Review of the State of the Art of Sulfur Asphalt Paving Technology", in ACS Symposium Series No. 183, "Sulfur: New Sources and Uses", p. 155, The American Chemical Society, Washington, D.C., 1982.

10 Harrigan, E.T. and Lentz, H.J. "Chemically Modified Sulfur Paving Binder", Public Roads, Vol. 46, p. 125, March 1983.

11 Annual Report-1984, Sulphur Development Institute of Canada, Calgary, Alberta T2P3E6, Canada.

12 Frasch, Herman, Address of Acceptance - Perkin Medal," Journal of Industrial and Engineering Chemistry, p. 99-104, Feb 1912.